FUSED PYRIMIDINES
Part II: PURINES

This is the twenty-fourth volume (Part II) in the series
THE CHEMISTRY OF HETEROCYCLIC COMPOUNDS

THE CHEMISTRY OF HETEROCYCLIC COMPOUNDS

A SERIES OF MONOGRAPHS

ARNOLD WEISSBERGER and EDWARD C. TAYLOR

Editors

FUSED PYRIMIDINES

Edited by

D. J. Brown

Part II

PURINES

J. H. Lister

Chester Beatty Research Institute,
Institute of Cancer Research,
London

With contributed essays on spectra by

R. L. JONES and P. D. LAWLEY
Chester Beatty Research Institute

and collaboration in part by

G. H. HITCHINGS and G. B. ELION
Wellcome Research Laboratories, Tuckahoe, N.Y.

WILEY—INTERSCIENCE
a division of John Wiley & Sons, Inc.
New York - London - Sydney - Toronto

No part of this book may be reproduced by any means, nor transmitted, nor translated into a machine language without the written permission of the publisher.

Copyright © 1971, by John Wiley & Sons, Inc.

Library of Congress Catalog Card Number: 68–4274

ISBN   0–471–38205–1

Printed in the United States of America

10 9 8 7 6 5 4 3 2 1

The Chemistry of Heterocyclic Compounds

The chemistry of heterocyclic compounds is one of the most complex branches of organic chemistry. It is equally interesting for its theoretical implications, for the diversity of its synthetic procedures, and for the physiological and industrial significance of heterocyclic compounds.

A field of such importance and intrinsic difficulty should be made as readily accessible as possible, and the lack of a modern detailed and comprehensive presentation of heterocyclic chemistry is therefore keenly felt. It is the intention of the present series to fill this gap by expert presentations of the various branches of heterocyclic chemistry. The subdivisions have been designed to cover the field in its entirety by monographs which reflect the importance and the interrelations of the various compounds, and accommodate the specific interests of the authors.

In order to continue to make heterocyclic chemistry "as readily accessible as possible", new editions are planned for those areas where the respective volumes in the first edition have become obsolete by overwhelming progress. If, however, the changes are not too great so that the first editions can be brought up-to-date by supplementary volumes, supplements to the respective volumes will be published in the first edition.

Research Laboratories   ARNOLD WEISSBERGER
Eastman Kodak Company
Rochester, New York

Princeton University   EDWARD C. TAYLOR
Princeton, New Jersey

Foreword to The Fused Pyrimidines

Originally it was intended to present all the fused pyrimidine systems in one volume of this series. Resurgence of interest in purines and quinazolines, the development of pteridine chemistry, and the wide exploration of a great many new fused systems embracing the pyrimidine ring, have made the task impossible.

The fused pyrimidines will now be covered in four parts, of which Dr. Armarego's *Quinazolines* was the first and Dr. Lister's *Purines* is the second. Two others, dealing with *Pteridines* and *Miscellaneous Fused Pyrimidines* respectively, are in active preparation. Eventually, this bracket of volumes will bring to the series the expertise of four senior authors and several coauthors with wide and diverse experience in the field.

It is a privilege to assist Dr. Weissberger, Dr. Taylor, and the authors in organizing this project and in maintaining a measure of uniformity and balance in its parts.

D. J. BROWN

The Australian National University
Canberra, Australia

EDWARD C. TAYLOR

Department of Chemistry
Princeton University
Princeton, New Jersey

Preface

This book, which forms part of the series devoted to Fused Pyrimidines, follows, insofar as subject matter permits, the format adopted in *The Pyrimidines*, the parent member of the series. As in previous volumes a critical approach to the subject is taken—treating theoretical aspects of the chemistry in outline and giving the major emphasis to practical aspects, which are supported by the appendix tables. The tables furnish melting points and selected source references for simple purine derivatives known up to the end of 1969. The literature has been surveyed in detail from Scheele's isolation of uric acid in 1776 to December 1969 with additional coverage of the leading journals to late 1970. As nucleoside forms are outside the terms of reference, these are only mentioned in the text where some interesting feature or reaction of the purine moiety is involved or to draw attention to some notable difference in behaviour between the latter and the free purine.

Purine chemistry as we know it today had its origins in Fischer's prodigous work, started nearly ninety years ago, in which the synthesis of all the naturally occurring purines then known was accomplished. This fruitful period, associated also with names such as Traube, Biltz, Gabriel, and Johns, was followed by a decline in interest in the chemistry of these nitrogenous bases which lasted for nearly four decades. In the early 1950's, however, following upon the determination of partial structures of the nucleic acids, a new period of expansion was initiated. This arose from the realisation that man-made purines and purine analogues could possibly interfere with nucleic acid biosynthesis and therefore act as growth inhibitors. Although many hundreds of purines have been prepared with this end in view, the results have been disappointing and only derivatives of 6-mercaptopurine have found any wide clinical application in this field of chemotherapy.

For help in the preparation of this work I am indebted to a number of colleagues and others who have contributed in various ways. Drs. G. H. Hitchings and G. B. Elion very kindly gave me access to an unpublished survey they had made of the early purine chemistry. My thanks are also due to my co-authors, Drs. R. L. Jones and P. D. Lawley for their informative essays on infrared and ultraviolet spectra, respectively. Colleagues who have sent me their results before publication and who, in so doing, have helped to reduce the "outdatedness" of the volume at the time of publication, include Professor Adrian Albert, Drs. G. B. Brown, A. Giner-Sorolla, R. Hull, Paul T'so, and T. L. V. Ulbricht.

I have also been greatly helped by the facilities afforded me at the Chester Beatty Research Institute and for these I wish to express my deep gratitude to the recently retired director, Sir Alexander Haddow, F.R.S. Among the people directly concerned have been Michael Docherty and other members of the Photographic Department, and the Library staff, who at all times have provided an efficient and willing service. It would be remiss of me, however, not to record a special word of praise for the Herculean efforts of Miss Margaret Foster and Mrs. Audrey Inglefield, who between them have carried out the arduous task of typing both the draft and final stages with great cheerfulness. My sincere thanks also are accorded to my editor and former colleague, Dr. D. J. Brown, for his guidance and encouragement, and last but by no means least to my wife for her constant support and for patiently enduring the trials of having an author in the family.

J. H. LISTER

Chester Beatty Research Institute,
Institute of Cancer Research: Royal Cancer Hospital,
London
October 1970

Contents

Contents   xix

Tables

FUSED PYRIMIDINES
Part II: PURINES

This is the twenty-fourth volume (Part II) in the series
THE CHEMISTRY OF HETEROCYCLIC COMPOUNDS

CHAPTER I

Introduction to the Purines

1. History

The origins of purine chemistry, because of their association with the names of Scheele, Liebig, and Wöhler, are an integral part of the beginnings of organic chemistry itself. Uric acid was the first purine to be obtained in a pure state, after extraction by Scheele, from gallstones, in 1776.[1] Elucidation of the structure was not to be achieved for a hundred years hence although in this interim period the work of Liebig[2] and Mitscherlich[3] established the correct empirical formula. Also, the results of intensive oxidative studies jointly by Liebig and Wöhler[4] and later by Bayer,[5] in which the products were characterised, enabled structural fragments to be ascertained and in turn led to various formulae being proposed, those of Fittig[6] (1) and Medicus[7] (2) being

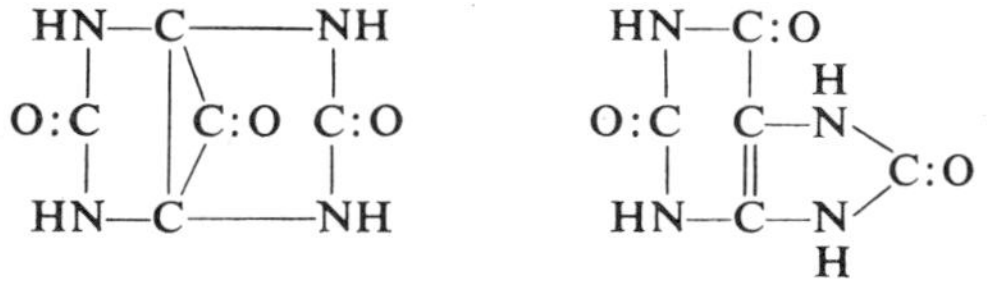

```
    HN—C———NH          HN—C:O
     |    \    |          |   |  H
    O:C   |C:O C:O       O:C  C—N
     |    /    |          |   ‖      C:O
    HN—C———NH          HN—C—N
                                  H
```

(1)   (2)

the most widely favoured. Although syntheses of uric acid by Horbaczewski,[8] in 1882, and Behrend and Roosen,[9] some six years later, were successfully accomplished, neither afforded complete structural proof. This proof was finally obtained from an unambiguous synthesis by Fischer[10] which showed that the formula proposed by Medicus was the correct one. Subsequently, the work of Fischer and his school[11] removed any remaining doubts by synthesising all of the fourteen N-methyl derivatives allowed by the Medicus structure thereby invalidating that of Fittig, which, by virtue of its symmetry, could only give rise to five N-methylated forms. Following this the same workers demonstrated the conversion of uric acid to a number of naturally occurring

1

purines[12] (Scheme A) which included xanthine (1,2,3,6-tetrahydro-2,6-

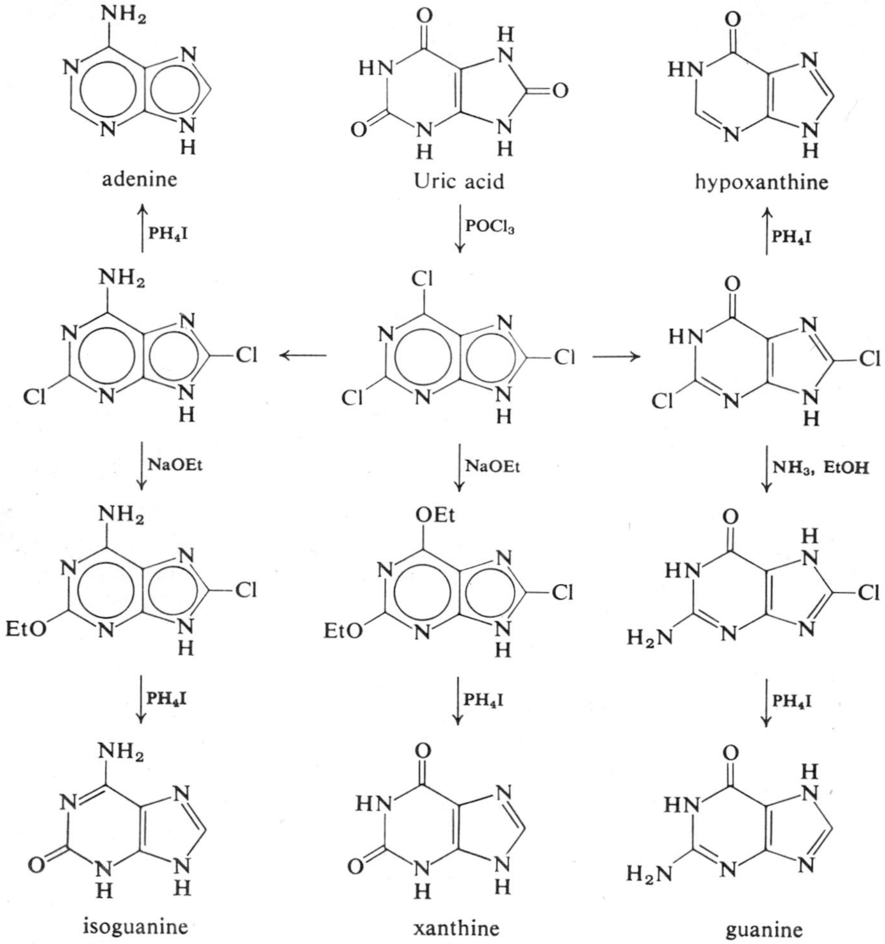

adenine   Uric acid   hypoxanthine

isoguanine   xanthine   guanine

SCHEME A

dioxopurine), isolated from gallstones in 1838,[4] guanine (2-amino-1,6-dihydro-6-oxopurine), a constituent of sea birds excreta used commercially as the fertiliser "guano,"[13] and the N-methylated xanthines, found in the beverage plants, theophylline (1,3-dimethylxanthine), theobromine (3,7-dimethylxanthine), and caffeine (1,3,7-trimethylxanthine).

Fischer's work laid purine chemistry on very secure foundations although it should be appreciated that his efforts were directed mainly to the interconversion of purines rather than to their outright syntheses from simpler precursors. Between the years 1882 and 1906 the results of this work are seen in the publication of nearly fifty papers[11] in which are described about 150 new purine derivatives. In addition to the many new reactions resulting from these studies, credit must also go to Fischer for discovering the first case of an alkali-induced isomerisation of a purine,[14] the mechanism of which still awaits resolution.

The synthetic routes devised by Fischer and Behrends, using pyrimidine precursors, were limited in scope and it was not until the appearance of the versatile synthesis due to Traube[15] that the preparation of a wide spectrum of purines was possible. Since then this route and its adaptations have been responsible for the majority of purines synthesised to date.

Following the groundwork laid by Fischer purine chemistry flourished and at the turn of the century the literature is prolific with papers predominantly from the German schools. Between the two world wars there was a period of decline with little new chemistry being produced, but a renaissance came about with the characterisation and elucidation of the basic structures of the nucleic acids in which the purines, adenine, and guanine, were found to be constituent bases. As the role played by nucleic acids in cell function became more understood, many programmes of purine syntheses were initiated to design antimetabolites capable of interfering with nucleic acid replication in malignant cells. Although much effort has been, and is still being, directed to this end few purines of promise have resulted, but mention must be made of "6-mercaptopurine" (1,6-dihydro-6-thiopurine) which has been extensively used clinically.

As nucleic acid is inherent to all living cells it is not surprising that purines have been detected in a diversity of forms of living matter. In some cases they appear as metabolic breakdown products, while in others they act as vitamin cofactors. Although *in vivo* they are usually present as the glycoside, isolation procedures often result in hydrolytic cleavage at $N_{(9)}$ and the free base is obtained.

In spite of the increase in knowledge in this branch of heterocyclic chemistry made over the past seventy years, a surprising feature is the paucity of comprehensive review articles. Of those dealing mainly with the chemical aspect the earliest is a detailed account of Fischer's own work[11] between the years 1882 and 1906. Twenty-five years later (1931) the appearance in a monograph on nucleic acids[16] of two chapters dealing with uric acid and other naturally occurring purines reveals what little progress had been made in this intervening period. A subsequent review,[17] also in a volume devoted to nucleic acids, records the state of knowledge a further twenty years hence. More recent reviews of the general chemistry include two in outline[18, 19] and two of a more

TABLE 1.   Trivial Names of Purines.

Trivial Name	Systematic or Descriptive Name
Adenine[a]	6-Aminopurine
Adenosine[a]	Adenine-9-β-D-ribofuranoside
Angustmycin A	Adenine-9-β-6'-deoxy-D-hex-5'-enofuran-2'-uloside
Angustmycin C	Adenine-9-β-D-psicofuranoside
Caffeine[a]	1,3,7-Trimethylxanthine
Chidlovine	v. Triacanthine
Coffeine	v. Caffeine
Cordycepin	Adenine-9-β-3'-deoxy-D-ribofuranoside
Crotonoside	Isoguanine-9-β-D-furanoside
Decoyinine	v. Angustmycin A
Deoxyadenosine	Adenine-9-β-2'-deoxy-D-ribofuranoside
Deoxyguanine	2-Amino-1,6-dihydropurine
Deoxyguanosine	Guanine-9-β-2'-deoxy-D-ribofuranoside
Deoxyinosine	Hypoxanthine-9-β-2'-deoxy-D-ribofuranoside
Deoxyxanthine	1,2,3,6-Tetrahydro-2-oxopurine
Epiguanine	7-Methylguanine
Eritadenine	6-Amino-9-(3-carboxy-D-erythro-2,3-dihydroxy-propyl)purine
Guanine[a]	2-Amino-1,6-dihydro-6-oxopurine
Guanopterin	v. Isoguanine
Guanosine[a]	Guanine-9-β-D-ribofuranoside
Guaranine	v. Caffeine
Herbipoline	7,9-Dimethylguaninyl betaine
Heteroxanthine	7-Methylxanthine
Hypoxanthine[a]	1,6-Dihydro-6-oxopurine
Inosine[a]	Hypoxanthine-9-β-D-ribofuranoside
Isoadenine	2-Aminopurine
Isocaffeine	1,3,9-Trimethylxanthine
Isoguanine[a]	6-Amino-2,3-dihydro-2-oxopurine
Isokinetin	2-Furfurylaminopurine
Isouric acid[a]	4(5)-Substituted uric acid
Kinetin	6-Furfurylaminopurine
Lentinacin	v. Eritadenine
Lentysine	v. Eritadenine
Leukerin	v. 6-Mercaptopurine
Mercaleukin	v. 6-Mercaptopurine
6-Mercaptopurine (6-MP)	1,6-Dihydro-6-thiopurine
Nebularine	Purine-9-β-D-ribofuranoside
Nebuline	v. Nebularine
Nucleocidin	Adenine-9-5'-O-aminosulphonyl-4'-fluoro-D-ribofuranoside
Paraxanthine	1,7-Dimethylxanthine
Psicofuranine	v. Angustmycin C
Purinethiol	v. 6-Mercaptopurine
Purinethol	v. 6-Mercaptopurine

TABLE 1. (Continued)

Puromycin	6-Dimethylaminopurine-9-3'-(p-methoxy-L-phenyl-alanylamino)-3'-deoxy-β-D-ribofuranoside
Purone	1,2,3,4,5,6,7,8-Octahydro-2,8-dioxopurine
Sarcine	v. Hypoxanthine
Sarkin	v. Hypoxanthine
Septacidin	6-Substituted-aminopurine
Spongopurine	1-Methyladenine
Spongosine	6-Amino-2-methoxypurine-9-β-D-ribofuranoside
Stylomycin	v. Puromycin
Theine	v. Caffeine
Theobromine	3,7-Dimethylxanthine
Theocyn	v. Theophylline
Theophylline[a]	1,3-Dimethylxanthine
Thioguanine[a]	2-Amino-1,6-dihydro-6-thiopurine
Togholamine	v. Triacanthine
Triacanthine	6-Amino-3-(3-methylbut-2-enyl)purine
Uric acid[a]	1,2,3,6,7,8-Hexahydro-2,6,8-trioxopurine
Vernine	v. Guanosine
Vitamin B$_4$	v. Adenine
Xanthine[a]	1,2,3,6-Tetrahydro-2,6-dioxopurine
Xanthosine[a]	Xanthine-9-β-D-ribofuranoside
Zeatin	6-(4-Hydroxy-3-methylbut-2-enyl)aminopurine

[a] Denotes names used in text.

comprehensive nature[20, 21] comprising 63 and 245 pages, respectively. Other works covering specialised aspects include treatment of the physico-chemical nature,[22, 23] syntheses by the two main routes from pyrimidines[24] and imidazoles[25] and a collection of papers covering both biological and chemical topics.[26]

2. Nomenclature and Notation

Before the work of Fischer most of the purines studied, being either simple derivatives of uric acid or xanthine, were named as such. The term "Purines" was coined by Fischer[27] to embrace all derivatives of this ring system.*

Although both the older Kekulé (3) and more recent structural (4) formula are currently in use, the outmoded "block" form (5), which has almost disappeared from English and American journals, is still

* "...und Währe für diese den Namen Purin welcher aus den Worten 'purum' and 'uricum' Kombiniert war."

extant in some European papers. The numbering sequence devised by
Fischer is retained even though it represents an exception to the general
rule for numbering of fused heterocycles. Under this system purine
itself becomes either 7H- or 9H-imidazo[4,5-d]pyrimidine (Ring Index
Nos. 1179 and 1180, respectively).[28] A controversial point among early
workers in the field was the assignment of the imidazole proton to either
the $N_{(7)}$- or $N_{(9)}$-atom. The fact that alkylation of some purines gives

(3)   (4)   (5)

7-alkyl derivatives while in others a 9-alkylated product is obtained gave
rise to the supposition that two reactive modifications of purine co-
existed. This led Biltz[29] to reserve the term purine for those derivatives
substituted at $N_{(7)}$ while for the $N_{(9)}$-analogues he introduced the term
"isopurine."* Subsequent authors have used the "iso" prefix for a
variety of purposes, this having led to some confusion in purine nomenclature. Even for indicating $N_{(9)}$-alkylated purines its application in one
paper[30] to a guanine derivative conflicts with accepted usage of the
name "isoguanine" for 6-amino-2,3-dihydro-2-oxopurine. It is even
more surprising to find in a well-known textbook[31] that this prefix is
used to denote the oxo tautomers of hydroxypurines. Some uric acid
derivatives, now of historic interest only, were named "isouric acids"
although isouric acid itself is unknown (see Ch. VI, Sects. 9a and b).
Present practice favours the term only for a few isomers of trivially
named purines, in addition to isoguanine, noted previously, "iso-
adenine" is applied to 2-aminopurine and other examples appear in
Table 1.

With purines possessing potentially tautomeric groups nomenclature
difficulties can arise. While some groups appear to exist predominantly
in one form, e.g., amino not imino, others such as hydroxy and thio,
show reactions characteristic of both tautomeric forms. Thus, while
xanthine may be described as either 2,6-dihydroxypurine (6) or 1,2,3,6-
tetrahydro-2,6-dioxopurine (7)† the latter notation will be employed

---

* Until fairly recently Chemical Abstracts perpetuated this misleading classification
by indexing 7- and 9-substituted purines under the heading of α and β-purines, respectively.

† In some current publications this is designated as 2,6-dioxopurine with the hydrogen
locations omitted. This practice, it should be noted, is not in accordance with current
I.U.P.A.C. convention.[32]

throughout this book as it represents the predominant tautomer in most reactions. Alternative, but by no means recommended, other namings of this purine in the literature include 2,6-dipurinol, purine-2,6-dione, and 2,6-dipurinone. For the same reason diverse nomenclatures are used with the sulphur-containing purines (Ch. VII) as, for example, 1,6-dihydro-6-thiopurine, which although usually referred to by the trivial name "6-mercaptopurine" is also found as purine-6-thiol or purine-6-thione. A more adequate exposition covering nomenclature and presentation of these compounds is given in the introduction to Chapter VI.

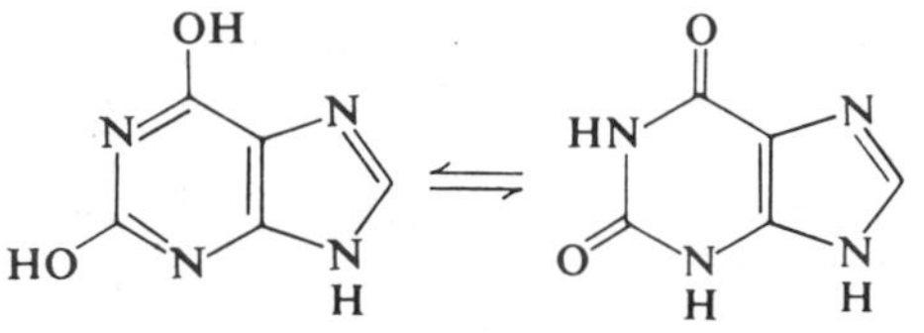

(6)   (7)

3. The Basis of Purine Chemistry

Many of the reactions undergone by purines can be explained in relatively simple terms if there is an understanding of the fundamental physico-chemical nature of this heterocyclic system. A consideration of the separate ring systems shows that whereas on the one hand the pyrimidine ring, because of the electron-localising effect of the two nitrogen atoms, is an example of a π-electron deficient system, the imidazole ring on the other hand, with both single and doubly bound nitrogen atoms, is an electron excessive system. The overall electron distribution in the unsubstituted purine molecule is produced by a sharing of imidazole ring π-electrons by the pyrimidine moiety; this balance can, however, be disturbed by insertion of appropriate strong electron-attracting or -withdrawing groups into either ring.

A. The Electrophilic Character of the 2-, 6-, and 8-Carbon Atoms

As a consequence of the localization of π-electrons around the ring nitrogen atoms, the adjacent carbon atoms show a pronounced degree of electrophilic character. Elements or groups attached to them will undergo nucleophilic displacement and the order in which this occurs is governed by the ionisation state of the molecule at the time of attack.

With the unionised molecule the 8-carbon atom appears to be the most electron-deficient and substituents at this position are displaced first, the complete order of reaction being $C_{(8)}$, $C_{(6)}$ then $C_{(2)}$. Proton loss from the imidazole nitrogen leads to anion formation and an increase in the electronegative character of the ring. A consequence of the induced negative charge is that nucleophilic attack is directed to the pyrimidine ring where $C_{(6)}$ is initially involved with subsequent displacements at the $C_{(2)}$- and then $C_{(8)}$-positions.

B. The Nucleophilic Character of the 8-Carbon Atom

Although, as stated above, the 8-carbon atom shows a considerable degree of electrophilic character, it can also undergo attack by electrophilic agents. A proviso to this is that electrophilic substitution is not a feature of simple purines, the presence of at least one strong electron-donating group being necessary to activate the 8-position. Thus, with adenine and hypoxanthine direct halogenation is only successful in forming the 8-bromo analogue whereas polysubstituted purines will both brominate and chlorinate. Direct nitration is possible with xanthine derivatives, and electrophilic substitutions of 8-thio groups by nitronium ions are known. Many purines undergo coupling with alkaline diazonium salts. The explanation for the ambivalent character of $C_{(8)}$ can be related with the fact that electron-donating groups must be present in the pyrimidine ring before electrophilic attack can take place. As the effect increases with the number of such groups present, this would suggest that their purpose is to restore the inherent π-electron deficiency of the pyrimidine ring and reduce the depletion of the imidazole ring π-electrons thereby restoring the imidazole ring to a state of near aromaticity.

C. Tautomeric Groups

Amino-, oxo-, and thiopurines are each capable of being depicted in two classical tautomeric modifications but, on physical evidence, appear to predominate as one or other tautomer. Thus, while oxo- and thiopurines exist mainly with the oxygen or sulphur atom in the double bound form, for example (8), the aminopurines on the contrary show little evidence for the presence of imino tautomers. Although the foregoing remarks about the oxo- and thiopurines suggest that aromaticity is lacking in these derivatives, it is pointed out[33] that resonance forms of the type **9** overcome this objection.

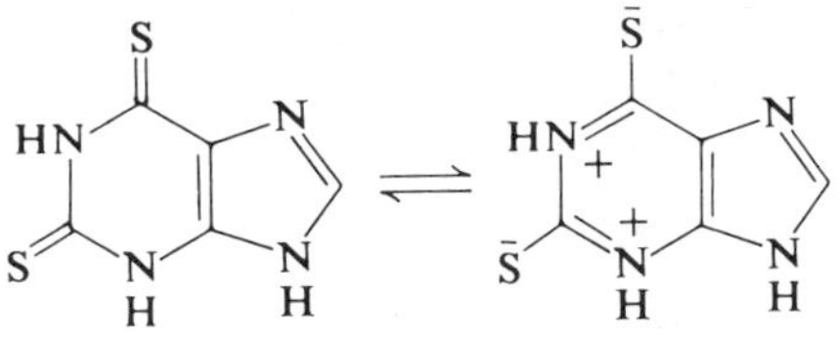

(8)   (9)

4. General Summary of Purine Chemistry

Two main synthetic routes to purines using either pyrimidine or imidazole precursors have been evolved, each with many variations and refinements. Of all procedures at hand the most extensively used is the Traube synthesis in which 4,5-diaminopyrimidines are condensed with simple derivatives, the purpose of which is to supply a one-carbon fragment to bridge the two pyrimidine amino nitrogen atoms. The other major synthetic approach, using imidazole derivatives, also utilises a one-carbon fragment to convert a side chain into a pyrimidine ring. The latter route is also of interest in that the mode of synthesis follows that of the biosynthesis of purine nucleotides.

The compensating effect of having a π-electron excessive and π-electron deficient system in juxtaposition produces a relatively stable bicyclic system. Further stability toward nucleophiles is produced by insertion of electron-releasing groups, such as oxo, thio, or amino, or by converting the molecule to the anion.

The majority of purine metatheses in the literature are of the nucleophilic displacement type. The limited number of examples of electrophilic substitution at a carbon atom involve attack at the 8-position only. It is convenient also to include the action of electrophilic reagents on ring nitrogen atoms, at the same time, as this forms an important section of purine chemistry.

A. Electrophilic Substitution

Under this heading electrophilic displacement at both carbon and nitrogen atoms will be included.

a. *Nitration* (Ch. X, Sect. 1Aa)

This necessitates the presence of a strongly nucleophilic carbon atom at the 8-position. So far only *N*-methylated derivatives of xanthine appear to satisfy this requirement although, theoretically, any purine with a number of strong electron-donating groups should do so. 8-Nitroxanthines can also arise from electrophilic substitution of an 8-thio group by nitronium ion. A spurious claim to the preparation of a series of 8-nitrosopurines has been made. (See Ch. X)

b. *Diazo Coupling* (Ch. X, Sect. 3Aa)

The weaker nucleophilic carbon required at $C_{(8)}$ for coupling with diazonium salts allows a wider range of purines to be used than is the case with nitration. Even the presence of one electron-releasing group may be sufficient.

c. *Halogenation* (Ch. V, Sect. 2)

Introduction of halogen atoms directly at the 8-position needs similar electronic requirements as for diazo-coupling reactions. With mono-substituted purines, e.g., adenine or hypoxanthine, bromine can be directly introduced but the less reactive chlorine requires disubstituted derivatives, e.g., xanthine. A side effect is that any *C*- or *N*-methyl groups present may be halogenated under these conditions.

d. *Alkylation*

It follows from the π-electron excessive character of the imidazole ring that either of the nitrogen atoms is potentially available for electrophilic substitution. With alkylating agents a 7- or 9-alkylpurine is produced. The presence of one alkylated site in the molecule facilitates further alkylation which usually takes place at a nitrogen atom in the conjugate ring. The foregoing, which assumes the purine to be in an anionic form at the time of attack by the electrophile, does not apply if nonionising conditions are employed. In this eventuality, substitution takes place at other sites and quaternisation of imidazole or pyrimidine nitrogens can occur.

B. Nucleophilic Substitution

Under this heading are found the majority of purine reactions, usually involving displacement of a halogen or thio group at the 2-, 6-, and 8-positions by a wide range of substituents. The mechanisms involved in metatheses of this type have been the subject of a number of investigations from which a satisfactory theory to explain the reaction sequence has resulted.

Chloropurines are the most versatile derivatives in which the order of replacement of the 2-, 6-, and 8-chlorine atoms by nucleophilic agents is governed by the ionic state existing at the time of reaction. With the anion of 2,6,8-trichloropurine attack by bases follows the order $C_{(6)}$, $C_{(2)}$ then $C_{(8)}$. In examples where such ionisation is precluded as with 7- or 9-methyl-2,6,8-trichloropurine, the sequence followed is $C_{(8)}$, $C_{(6)}$ then $C_{(2)}$. The latter order of substitution also holds where the cation is involved, as with reactions carried out in acid media. These effects can be explained in general terms by assuming that in the unionised molecule depletion by the pyrimidine ring of the imidazole π-electrons leads to a highly electrophilic 8-position. In the case of the anion the situation is reversed. Through deprotonation of the imino group of the imidazole ring the induced electronegativity renders the pyrimidine ring relatively more positive, thereby making $C_{(6)}$ the most electrophilic position. No clear-cut explanation to fully explain nucleophilic attack on cationic forms is available, as in many cases practical results conflict with those postulated for the protonated species assumed to be reacting. The presence of electron-releasing groups modifies the response to nucleophilic attack, the main effect being deactivation of the chlorine atom. Amino and oxo groups, situated in either ring, demonstrate this most effectively.

a. *Halogen Replacement by Amino Groups* (Ch. V, Sect. 5C)

Although halogen atoms in purines are relatively less reactive compared with those of nitropyrimidines, they can be replaced fairly easily with primary or secondary amines (and hydrazines) while ammonia shows its weaker basicity in requiring more forcing conditions to effect amination. Of the monochloropurines the 6-chloro is the most active, reaction with the alkyl- or aryl-amine taking place in boiling ethanolic solution. Whereas ammonia requires sealed tube conditions with temperatures above 100°, the 6-hydrazinopurine is formed at room temperature. In contrast, 2-chloropurines will not react with amines under these conditions and even preparations of the 2-hydrazino derivative require elevated temperatures. Both 2,6- and 6,8-dichloropurine give only the corresponding 6-aminochloropurine, 2,6,8-trichloropurine likewise only aminates at $C_{(6)}$, this effectively deactivates the remaining chlorine atoms. If chloropurines and the amines are heated together without solvent, complete amination to the di- and triaminopurines is achieved. An exception is 2,8-dichloropurine which, having no $C_{(6)}$-chlorine atom, might be expected to undergo substitution at $C_{(2)}$, but in practice gives 8-amino-2-chloropurine.

b. *Halogen Replacement by Methoxy and Other Alkoxy Groups* (Ch. V, Sect. 5E)

The usual conditions employ an alcoholic solution of the alkoxide either at room temperature or with mild heating. As found in reactions with amines the 2-chloro derivatives are remarkably inert and sodium methoxide in boiling toluene is needed to effect substitution. Less vigorous treatment usually suffices for preparing 6-methoxypurines, where at the most a water bath temperature is required. Both 2,6- and 6,8-dichloropurine give only the corresponding 6-methoxypurine under these conditions, while 2,6,8-trichloropurine initially undergoes a replacement at the 6-position but with an excess of the reagent forms the 8-chloro-2,6-dimethoxy derivative. The exception again is 2,8-dichloropurine which gives 2,8-dimethoxypurine, but if electron-releasing groups are present at the 6-position only the 2-alkoxy derivative is produced.

c. *Halogen Replacement by Oxo Group* (Ch. V, Sects. 5Fa and b)

Acid or alkaline conditions are equally suitable for hydrolytic procedures; normal hydrochloric acid or normal sodium hydroxide are employed. The alkaline hydrolysis of 2,6,8-trichloropurine was investigated in detail in order to ascertain the sequence of nucleophilic attack in purines. In some instances a better product results if conversion of the chloro purine to the corresponding methoxy derivative is effected before the hydrolysis.

d. *Halogen Replacement by Alkylthio Groups* (Ch. V, Sect. 5G)

Methylthiopurines are the most common alkylthio derivatives; these and related alkylthio analogues can be obtained by interacting the sodium derivative of the alkyl mercaptan in alcoholic solution. The alternative route through direct alkylation of thiopurines is generally the one of choice.

e. *Halogen Replacement by Thio Groups* (Ch. V, Sects. 5Ha and b)

An active halogen atom can be converted to a thio group by alcoholic or aqueous potassium hydrosulphide, usually freshly prepared by saturating potassium hydroxide with hydrogen sulphide. More conveniently thiourea and the chloropurine can often react and the thioureido adduct is broken down with alkali.

f. *Halogen Replacement by Thiocyanato and Cyano Groups* (Ch. V, Sect. 5I; Ch. IX, Sect. 4Aa)

In methanolic solution 6-chloropurine reacts with potassium thiocyanate producing 6-thiocyanatopurine. Unlike the thiocyanatopyrimidines thermal isomerisation to an isothiocyanate derivative has not been shown to occur in the purine series. The derivatives provide an indirect route to 6-thiopurines which are obtained on treatment with cold aqueous alkali. If cuprous cyanide in pyridine is heated with the more reactive 6-iodopurine, the product is 6-cyanopurine.

g. *Halogen Replacement by Sulpho Groups* (Ch. VII, Sect. 4A)

Aqueous sodium sulphite on warming with chloropurines gives rise to the corresponding purine sulphonic acid by direct replacement of the chlorine atom by the sulpho radical.

h. *Replacement of Methoxy, Methylthio, and Methylsulphonyl Groups*

Simple methoxy and methylthio groups are readily hydrolysed by acid to oxo groups, although methylthio derivatives are not usually used as precursors of oxopurines. Replacement by amino groups is, by contrast, a widely used feature of methylthio groups but is not usual with methoxy groups. In this reaction the methylthio group activity is usually inferior to that of a halogen atom at the same position. Aryloxy and arylthio groups have not been widely used in replacement reactions but benzyloxy and the thio analogue are used as protecting groups for the oxo and thio groups which they give on catalytic hydrogenolysis. Methylsulphonyl groups show a wide range of activity toward nucleophilic agents, which in some instances may parallel that of halogen atoms. Although conversion to oxo and amino groups is the most common synthetic procedure carried out with the above groups, the methylthio (and thio) group is capable of facile conversion to chlorine on treatment with hydrogen chloride gas in methanol. Procedural variations permit formation of the corresponding bromo- and iodopurines. The group can be oxidised to give sulphinic and sulphonic acid derivatives, the former being capable of removal by formic acid to give an unsubstituted position. Complete replacement of methylthio (or thio) group by hydrogen is, however, usually made by heating with Raney nickel in aqueous solution.

C. Group Interconversion

Under this heading are summarised the main group interchanges that may be brought about. No attempt has been made to place these reactions in systematic order other than that of degree of utility, and no general mechanistic system is followed.

a. *Interchange of Halogen Atoms* (Ch. V, Sects. 1Gb and 1F)

Such interchanges are sometimes necessary in order to provide a more reactive halogen than the one *in situ*. Only a restricted number of examples are known. Thus chloropurines treated with silver fluoride in high boiling solvents give the analogous fluoropurines. A specific route for 6-fluoropurines involves conversion of the 6-chloro analogue to the trimethylammonium quaternary chloride and reacting this with an alkali metal fluoride. Replacements of chlorine by iodine have been made using hydriodic acid.

b. *Oxo- to Aminopurines* (Ch. V, Sect. 1B)

Oxopurines may be converted to their dialkylamino analogues by heating with an excess of tertiary alkylamine in the presence of phosphoryl chloride. Similar replacements are known in related heterocyclic systems. The oxo groups of hypoxanthine and xanthine have been so converted as has the $C_{(6)}$-group in uric acid. A reaction of this type is best formulated as passing through two intermediate stages, comprising an initial normal chlorination of the oxygen-containing groups, this being followed by quaternary salt formation by means of the tertiary amine, with the quaternary salt then undergoing a Hofmann–Martius degradation to the dialkylaminopurine.

c. *Oxo- to Chloropurines* (Ch. V, Sect. 1B)

Chlorination of oxo groups is normally carried out by the general chlorination procedure for hydroxy heterocyclic compounds, that is, using phosphoryl chloride by itself or in combination with other reagents. These may be other chlorinating agents, e.g., phosphorus pentachloride, or substances whose presence is required in catalytic amounts such as *NN*-dialkylanilines or water. Aliphatic tertiary bases have been used for this purpose also, but suffer the disadvantage that amination of the oxo groups may occur (see preceding section). To a lesser extent thionyl chloride complexed with dimethylformamide (Bosshard reagent) is useful in that chlorinations in chloroform at lower temperatures can be effected. This reaction is of special value if acid-labile groups are present as, for example, with nucleosides (Ch. V, Sect. 1Ba). Phosphoryl bromide is used for preparing the analogous bromo-purines but attempted fluorinations with phosphoryl fluoride have not been successful.

d. *Oxo- to Thiopurines* (Ch. VI, Sect. 6B)

Heating oxopurines with phosphorus pentasulphide in a high boiling solvent at the reflux point is a general route. Although it is most effective for replacement of oxygen atoms on $C_{(6)}$ and $C_{(8)}$, it fails with simple 2-oxopurines. Nevertheless, with some *N*-methylated xanthines both $C_{(2)}$- and $C_{(6)}$-groups are replaced if elevated temperatures are employed. Among the solvents commonly employed are pyridine, toluene, xylene, tetrahydronaphthalene, and kerosene; the one of choice being that showing the best solubilising power for the oxopurine. A notable feature of the reaction is that oxo groups fixed in the doubly bound form, for example, theophylline, are also replaced under these conditions.

e. *Thio-* (*and Methylthio-*) *to Oxopurines* (Ch. VI, Sect. 1F)

As in related heterocyclic systems, such as pyrimidines, thio groups in purines show reluctance toward hydrolytic conversion to the oxo form. Some examples of the use of strong acids (e.g., nitric and nitrous acids) for this purpose are to be found but this approach has little practical application as more productive procedures are available. Of the methods outlined below the first two have general application and the others a more limited scope. (*a*) Mild oxidising agents convert a thio group to the sulphinic or higher oxidised sulphonic acid form. Either can be easily hydrolysed to oxo group with dilute acid. (*b*) *S*-Alkylation to the thioether followed by mild oxidation gives the methylsulphone which likewise is a readily hydrolysable group. (*c*) Direct acid hydrolysis of the alkylthiopurine often gives the oxo derivative. If alkylation with chloroacetic acid, rather than an alkyl halide, is made the carboxymethyl-thiopurine produced can be very acid labile. (*d*) Miscellaneous strong oxidising agents, which include hydrogen peroxide, manganese dioxide, and nitric acid have been used to replace the sulphur atom by oxygen, but the likelihood that the sulphydryl group may be replaced by hydrogen under these conditions is always present. Scheme B (R = purinyl) summarises these conversions.

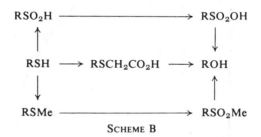

$$\text{S{\scriptsize CHEME}} \ B$$

f. *Thio-* (*and Methylthio-*) *to Halogenopurines* (Ch. V, Sects. 1C and 1Eb)

Both thio and the *S*-methylated-purines are directly transformed to the corresponding chloropurines by passing chlorine through their dry methanolic solutions. When thiopurines are used a prerequisite is the saturating of the methanol with hydrogen chloride. In this manner 2,6,8-trichloropurine results from the trithio analogue in good yield. Only oxidation of the thio group to the sulphone stage occurs if aqueous methanol is employed. Mixtures of bromine and hydrogen bromide under the same conditions afford the analogous bromopurines.

g. *Thio- to Aminopurines* (Ch. VII, Sect. 1Cf)

The case of the replacement of methylthio by amino has been dealt with already (Sect. 4Bh), this being the general means by which a sulphur atom undergoes nucleophilic displacement by an amine. The inertness shown by thio groups toward hydrolysis is reflected in a similar lack of activity toward amines. A number of cases of amination are known with *N*-methylated thiopurines, but the method is too limited to be of general applicability and is confined to replacement of 6-thio groups for which forcing conditions are required.

h. *Interchange of Amino Groups* (Ch. VIII, Sect. 1K)

Under vigorous conditions primary amino groups at the 6-position can be replaced by alkyl- or aryl-amino groups. A mixture of the alkylamine hydrochloride and alkylamine is employed, the latter, in excess, serving as solvent. Similar exchanges in pyrimidines and pteridines have been investigated and the amino group exchanges are presumed to take place through ring-opened intermediates. The fact that amino groups at $C_{(2)}$ do not appear vulnerable to this reaction supports a specific type of ring-opening mechanism obtaining.

i. *Amino- to Oxopurines* (Ch. VI, Sects. 1Da and b)

Primary amino groups undergo hydrolysis on prolonged heating with mineral acids, the strength of which is important as ring fission may succeed oxopurine formation. Many conversions are carried out in hot nitrous acid, guanine being converted to xanthine and adenine to hypoxanthine in this way. Substituted-aminopurines are more readily hydrolysed by means of aqueous mineral acids.

j. *Amino- to Halogenopurines* (Ch. VIII, Sect. 5B)

Such transformations are infrequent in the purine series. Sandmeyer type reactions have been attempted without success. The amino groups show weak aromatic character, only 8-aminopurines form isolatable diazonium salts. Evidence for the diazotisability of groups at other positions is seen in the fact that 2-aminopurines with nitrous acid and in the presence of fluoboric acid (Schiemann reaction) give 2-fluoropurines. With hydrochloric acid the 2-chloropurine has been obtained. When 2,6-diaminopurines are treated with hydrofluoric acid or fluoboric acid only the 6-amino-2-fluoropurine is formed.

D. Addition Reactions

As the purine ring system exhibits no pronounced unsaturated character, reactions of this type tend to be limited in character. The more important of these involve addition at a nitrogen atom.

a. *The Michael Reaction*

Alkenyl derivatives with simple 6-substituted purines, usually adenine or hypoxanthine, under alkaline conditions give the appropriate 9-alkylpurine. With theophylline or theobromine the respective 7- and 1-alkylpurines result. Although not examples of Michael additions, as they take place under acid conditions, the reaction of 2,3-dihydropyran with various purines giving the corresponding 9-(tetrahydropyran-2-yl)-purines is included here for convenience, the overall reaction being the same.

b. *Quaternisation*

Whereas alkylating agents in aqueous basic media give rise to N-alkylated purines, the same reagents in aprotic solvents, such as dimethylformamide or dimethylsulphoxide, and in the absence of base, will in many instances lead to formation of quaternary derivatives. The majority of examples arise through reactions at both nitrogen atoms of the imidazole moiety, the products being isolated either as the quaternary salt, such as **10** or as the betaine canonical forms (**11a** and **11b**) on treatment of the former with alkali. Hypoxanthine and xanthine derivatives and the thio analogues afford these derivatives readily while only a few examples are recorded for adenine derivatives. Cases of quaternisation of pyrimidine nitrogen atoms are also found in various purines.

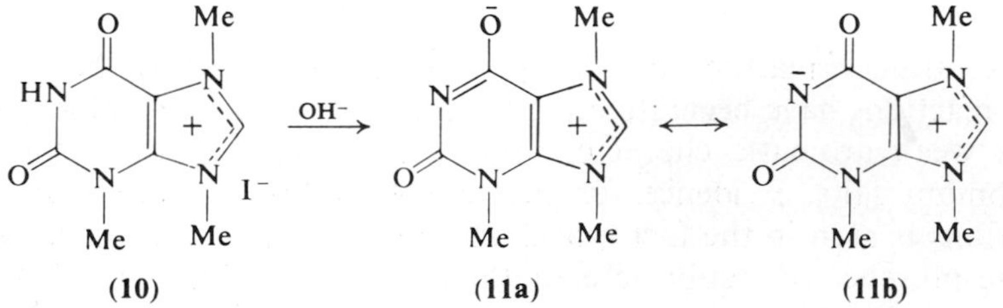

c. *Formation of N-Oxides*

Direct oxidation, the principal reagent for which is acidified hydrogen peroxide, gives the 1-oxides of purine, 6-methylpurine, adenine and other 6-aminopurines, 2,6-diaminopurine, and 8-oxopurines (Ch. XI, Sect. 1A). Guanine under similar conditions is converted to the 3-oxide (Ch. XI, Sect. 1B), as are also 6-chloro- and 6-alkoxypurines. The 7-oxides can be formed by various synthetic routes but not by direct oxidation (Ch. XI, Sect. 1C). A useful feature of the purine oxides is the increased electrophilic character induced in the ring carbon atom adjacent to the oxide nitrogen atom. This allows nucleophilic displacements at such carbon atoms to be made which would not be possible in the parent purines.

d. *Addition of Water and Alcohols*

In spite of intensive investigation no positive evidence for the addition of water across a double bond in a purine has been obtained.[34] This result is surprising on taking into account the fact that pyrimidines and various fused derivatives, for example, pteridines and quinazolines, show well defined hydration of this type. Furthermore, very rapid hydration is found with the cation of 8-azapurines at the $C_{(6)}$-$N_{(1)}$ bond. In line with these findings it is interesting to observe that isolatable adducts are obtained on ultraviolet irradiation of deoxygenated solutions of purine in methanol, ethanol, or propanol.[35] Addition of the alcohol α-carbon atom at $C_{(6)}$ of purine gives, for example, **12** in the case of ethanol. In the presence of even small amounts of water no reaction occurs.

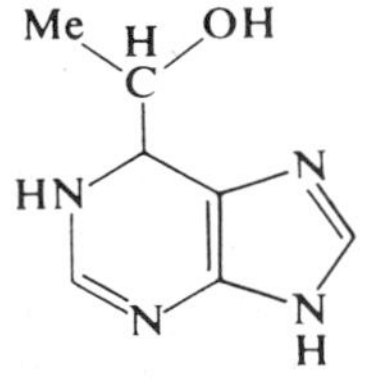

(12)

E. Modification of Substituents

Under this heading are grouped various reactions which modify a group without rearrangement or fission of the bond joining it to the nucleus.

a. *Amino Groups* (Ch. VIII)

Physical measurements show that in both the solid state and solution
amino groups are present as such ($\cdot$NH$_2$) rather than the imino (:NH)
form. Chemical behaviour is not, however, consistent with that of an
aromatic amine, the nearest approach to such character being shown by
8-aminopurines. The lack of reactivity of amino groups at the 2- and
6-positions is best explained by considering them as components of
cyclic guanidine and amidine systems, respectively.

Diazotizability is an established feature of 8-aminopurines, well
defined diazonium salts being given. Amino groups at the 2-position do
not show this facility although a 2-aminopurine has been converted to
the 2-chloro analogue *via* nitrous acid and hydrochloric acid. Similarly
when fluoboric acid is used a Schiemann reaction produces 2-fluoro-
purines. Primary amino groups at the 6-position do not appear to react
under these conditions but secondary amino groups with nitrous acid
can be modified to their *N*-nitroso form.

All exocyclic amino groups can be acylated and aroylated as can the
imino group in the imidazole ring. A feature of the 7- or 9-acylpurines
is their instability in aqueous conditions, hydrolysis to the unsubstituted imidazole being extremely facile. Alkylating agents, on the
other hand, usually react only with nuclear nitrogen atoms and examples of apparent direct alkylation of exocyclic amino groups are
almost always explicable through rearrangement of a ring nitrogen-
alkylated derivative. The formation of azomethine derivatives with
aldehydes is known to take place but it is not a general feature of
aminopurines.

b. *Oxo Groups* (Ch. VI)

The lack of reactivity shown by oxo groups toward alkylating and
acylating reagents reflects their predominant "oxo" rather than
"hydroxy" status. Alkylation gives normally only *N*-alkylpurines but
the nucleosides inosine and guanosine are *O*-methylated with diazomethane. Of the purine bases only derivatives of uric acid afford
examples of *O*-methylation which include formation of the 2- and
8-alkoxy analogues. Both of the above types of alkoxypurines can
undergo thermal rearrangement in which migration of a methyl or
other alkyl group to a ring nitrogen atom occurs. Isomerisations of this
kind transform the 2-methoxypurines to 3-methyluric acids and the
8-methoxy analogues to 7- or 9-methyluric acids.

c. *Thio Groups* (Ch. VII)

Compared with the corresponding oxo groups the more acidic character of the 2-, 6-, and 8-thio groups shows in their ready reaction with alkylating agents. Mild conditions generally suffice for *S*-alkylation while more vigorous treatment leads to concomitant *N*-alkylation which may be followed by *S*-dealkylation. Acylation and aroylation take place under similar conditions. Controlled oxidation of thiopurines can give rise to a variety of products including disulphides, *S*-oxides, and sulphinic and sulphonic acids. Alkylthiopurines similarly afford the corresponding alkylsulphinyl- or alkylsulphonylpurines.

d. *Methyl Groups* (Ch. IV)

Purines with 6-methyl groups have been most investigated. The fact that the activity of the hydrogen atoms is lower than might be expected for a methyl group located α and γ to ring nitrogen atoms, as for example, compared with that of 4- or 6-methylpyrimidines, may reflect the π-electron contribution from the imidazole to the pyrimidine ring. Stepwise replacement of the hydrogens by halogen is readily carried out with chlorine in chloroform. Chloral adds on to 6-methylpurine forming a well defined 3,3,3-trichloro-2-hydroxypropyl derivative (**13**). Direct oxidation to aldehyde or carboxyl groups is not usual but the more reactive methylene pyridinium iodide (**14**) is used for this purpose. Increased methyl group activity follows *N*-oxide formation at the 1-position. Acetic anhydride isomerises 6-methylpurine-1-*N*-oxide to the acetate of 6-hydroxymethylpurine (Ch. XI, Sect. 3C).

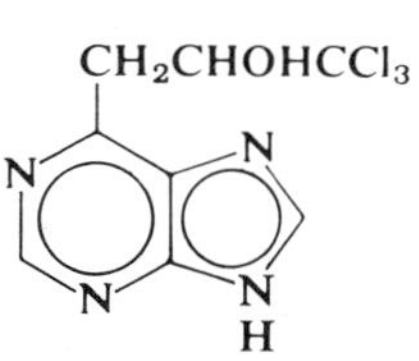

(**13**) 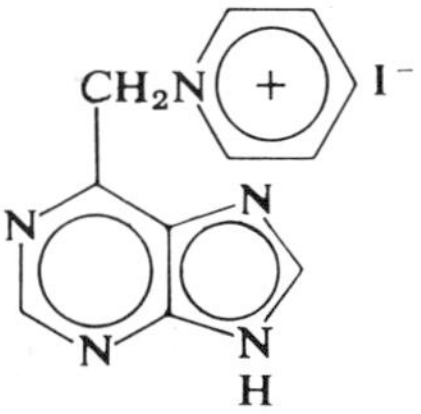   (**14**)

F. Reductive Reactions

A wide variety of reagents and methods is employed ranging in type from chemical to catalytic and electrochemical means. Practical applications are confined mainly to the reduction or reductive removal of extranuclear groups; complete reduction of the purine nucleus has produced unstable derivatives.

a. *Nuclear Reduction* (Ch. XII)

Catalytic hydrogenation in acid solution, carried out over a palladium catalyst, was successful in reducing purine and the 2-oxo derivative to the corresponding 1,6-dihydropurines. Under identical conditions 2,6,8-trichloropurine is dechlorinated and the pyrimidine ring reduced to the tetrahydro state giving rise to a highly unstable product. Additional evidence for the formation of 1,2,3,6-tetrahydropurines has come from polarographic studies (see later, Sect. 5E) on purine. Complete reduction to the perhydropurine is claimed with a platinum catalyst in acetic anhydride. Oxo groups at $C_{(6)}$ are not reduced by catalytic hydrogenation, but electrochemical reduction of these compounds is possible. In acid solution at a lead cathode reduction of the $C_{(6)}$-oxo function gives rise to 1,6-dihydropurines, termed in the older literature "6-desoxypurines," successful conversions being effected with guanine, xanthine, uric acid, theobromine, theophylline, and caffeine. No further reduction of the ring was observed except in the case of uric acid, which gives a tetrahydro derivative. Zinc dust and dilute acid will also give deoxypurines but the severity of the conditions is such as to lead to ring degradation in many cases.

b. *Removal of Groups*

Halogeno, thio, and alkylthio groups are removed under reducing conditions. Hydrogenolysis is the agent of choice for halogenopurines, a palladium catalyst being employed in the presence of a base, such as magnesium oxide, to remove hydrogen halides produced, the latter being conducive to nuclear reduction. Mono-, di-, and trichloropurines have been reduced to purine by stepwise removal of the chlorine atoms in this manner. Among the older methods zinc dust in aqueous solution is of limited use but the choice of halogenopurine must be made with care as adenine and hypoxanthine are degraded under these conditions. The use of hydriodic acid and phosphorus has now largely lapsed into historical significance. Other dehalogenating procedures include electrochemical reduction, at lead or mercury electrodes, sodium amalgams, activated nickel catalysts, and iodine in high boiling hydrocarbon solutions.

Both thio- and methylthio-purines are much favoured as precursors of the unsubstituted purine. Desulphurisation is effected by heating a solution or suspension of the purine in aqueous media with an excess of Raney nickel. Sulphur atoms at the 2-, 6-, and 8-positions are readily removed, illustrated by the conversion of 1,2,3,6-tetrahydro-2,6-dithiopurine to purine by this procedure. Hydrogenation in the presence of Raney nickel or Adam's catalyst is a means of reducing N-oxides to the parent purine.

c. *Reductive Modification of Groups*

Reductions of this kind fall into two classifications, the first being those which alter the function and type of the group, as in the reduction of a nitro to an amino group, whereas in the second removal of a protecting group leaves the parent group unchanged, as in debenzylation of benzylamines. Both categories use normal reducing methods, nitro groups being treated with amalgams, by catalytic means or with sodium hydrosulphide, the last also being used to reductively cleave 8-benzeneazopurines to their 8-amino analogues. Hydrogenation over nickel catalysts is used to convert 6-hydroxylaminopurines to the adenine analogues and likewise 6-cyanopurine to a 6-aminomethyl derivative.

Lithium aluminium hydride is the reagent of choice for reduction of acylated or aroylated amino groups to the secondary alkyl(aryl)amino form.

Examples of the second type usually involve removal of a benzyl group, this being the most common protecting group encountered for amino, oxo, and thio groups. Catalytic hydrogenation, in conjunction with a palladium or platinum catalyst, is recommended to expose the reactive group but the older procedure employing sodium dissolved in liquid ammonia is still widely favoured. Both endocyclic N-benzyl and exocyclic N-, O-, and S-benzyl groups are susceptible to removal by these methods.

G. Oxidative Reactions

Nuclear oxidation by chemical means predisposes toward ring disruption. Direct introduction of oxo groups is possible but this incurs free radical or enzymic oxidation reactions which are dealt with separately below.

a. *Chemical Oxidation*

When purine, or the 6-methyl or 2-amino derivative (**15**), are heated with sulphur, direct introduction of an 8-thio group is effected (**16**). Other cases of nuclear oxidation are the formation of N-oxides, using acidified peroxides as oxidants and the oxidation of dihydropurines to

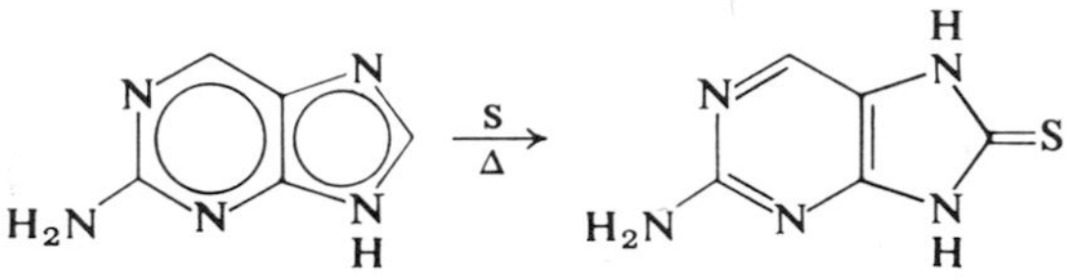

(**15**)   (**16**)

the parent compound. Oxidation of extranuclear groups parallels that found with the benzene analogues, thio groups being converted to sulphinic and sulphonic acid groups by hydrogen peroxide or permanganate solution, while if carried out in the presence of ammonia solution sulphamoyl groups result. Iodine has been used to convert thiopurines to their dipurinyl disulphide derivatives. Modified methyl groups are oxidised to their aldehyde and carboxyl forms. A $C_{(9)}$-propenyl group can be removed by oxidative hydrolysis using permanganate.

b. *Free Radical Attack*

Purines undergo C-oxidation in the presence of hydroxyl radicals generated either by X-irradiation of an aqueous solution or by treatment with Fentons reagent (ferrous sulphate-hydrogen peroxide). Oxidation occurs first at $C_{(6)}$ and subsequently at $C_{(8)}$ but further oxidation at $C_{(2)}$ does not take place. Longer exposure to these conditions results in degradation to pyrimidine derivatives.

c. *Enzymic Oxidation*

In the presence of the milk enzyme, xanthine oxidase, purine is converted stepwise to hypoxanthine, then xanthine and finally to uric acid. Oxidation is thus seen to follow the same order of reaction, that is at $C_{(6)}$, $C_{(2)}$ then $C_{(8)}$ as is found for nucleophilic substitution under basic conditions. It should be noted, however, that the oxidation sequence is not the same for all enzymic oxidation systems nor if different purines are used. Indeed, evidence exists to show that oxidation can follow two pathways simultaneously to give the same end product.

5. Physical Properties of Purines

As a result of the intense effort that has gone into the study of the structure of the nucleic acids, quite extensive physical data exist for adenine, guanine, and hypoxanthine and to a lesser extent on the methylated purines, theophylline, theobromine, and caffeine. Theoretical aspects, especially those concerned with the positional activity in terms of electronic structure, have received much attention from a number of schools, but complete agreement between observed experimental results and calculated theoretical activities has yet to come. As in pyrimidine chemistry, the main criteria have been ionisation constants and spectra, mainly in the ultraviolet, but crystal structure determinations have yielded confirmatory evidence of the predominant molecular configuration adopted in cases where tautomeric groups are present.

Apart from an outline review[23] of recent origin a current and commendable work is available[36] in which the determination and significance of these and other physico-chemical data is given detailed treatment.

A. Electronic Considerations

Calculations of the relative reactivities of the carbon and nitrogen atoms have been made by various schools of theoretical chemistry. The results, which are expressed in a variety of activity criteria including Hückel charge densities,[37-39] frontier electron densities,[40] indices[37] and localisation energies,[41, 42] show areas of disagreement. As all environmental factors influencing molecular states such as solvation effects or group influences at each position cannot be fully appreciated, these calculations have been based upon the isolated purine molecule[43, 44] and thus the validity of such results have been questioned. Attempts to overcome this objection have been made by use of a refinement of the linear combination atomic orbital–molecular orbital method (LCAO–MO). This method (SCMO—self-consistent molecular orbital) has given results[45] that show the most general agreement with those observed practically. However, until equations depicting the state of the purine molecule at the moment of reaction are available, some anomalies between observed and predicted behaviour must be expected.

B. Ionisation Constants

Useful structural data have been obtained from these determinations, which were carried out using either potentiometric titrimetry or ultra-violet spectrophotometry.[46] Constants of a large number of purines are contained in a review[47] embracing both purines and related heterocyclic systems. For a complete treatment of the relationship between structure, ionisation state, and ultraviolet spectrum the first essay by Lawley in Chapter XIII should be consulted. Purine (pK_a 8.9) is an acid comparable in strength with phenol, while the basic ionisation constant (pK_a 2.4) shows it to be weaker than aniline. The introduction of C-methyl groups produces little change in either acid or base strength. Stepwise insertion of oxo groups leads to a progressive increase in acidity, an effect even more pronounced if they are replaced by the more polar thio groups. An opposite effect obtains with aminopurines in which the basic strength increases with the number of amino groups present, 2,6,8-triaminopurine (pK_a 6.2) being one of the most basic purines known. Purine, also C-alkyl- and aminopurines form the anion through proton loss from the imidazole imino group, as evidenced by the failure of their 9-methylated derivatives to give an anion. The oxo- and thiopurines produce both mono- and dianions, the second ionisation involving a nitrogen atom in the conjugate ring. Some doubt exists with hypoxanthine as to the first ionisation site but other oxo-purines form the anion at $N_{(3)}$, the dianion following by loss of proton from the imidazole ring. If the $N_{(3)}$ is already substituted then the imidazole proton is lost first as in the alkyl- and aminopurines.

C. Crystal Structure

Fourier techniques, based on X-ray diffraction measurements, show that purines, including their N-alkylated variations, are planar molecules. This planarity is demonstrated by the fact that some purines with polycyclic hydrocarbons form crystalline 1:1 complexes,[48] the structures of which consist of molecular columns containing alternate hydrocarbon and purine units stacked in sandwich arrangement. These findings suggest an explanation for the fact that in aqueous solutions the solubility of polycyclic hydrocarbons is increased by the presence of purines.

Apart from demonstrating architectural features crystallographic studies provide configurational data and, where tautomeric modifications are possible, bond lengths may indicate which form is extant.

Amino groups in adenine and guanine are shown to be amino not imino, and the oxygen atom in guanine is predominantly double bonded in the crystal lattice. Sites of protonation can be inferred, although the techniques do not demonstrate the existence of a proton directly. The inter-unit distance in the lattice is a guide to probable hydrogen bond formation and this often allows assignment of the proton to a particular nitrogen atom. Purine, for example, in the solid state has the proton associated with $N_{(7)}$ rather than being the expected $N_{(9)}$ tautomer.[49]

D. Dipole Moments

Owing to the marked insolubility of most purines in nonpolar solvents, direct dipole moment measurements are seldom possible[50] and we must resort to theoretical values.[51, 52] Dipoles may be inherent in the molecule or may arise through external influences. The inherent dipole in purines is a consequence of the sharing of the π-electrons of the imidazole ring by the pyrimidine moiety. This produces a decrease of electron availability and induces a more positive state in the five-membered ring with a corresponding increase in the negative state in the pyrimidine ring. The greatest effect is exerted in the direction of the long axis of the molecule and can be modified by insertion of electron-attracting or -donating groups at the appropriate positions. The ultra-violet spectra of simple purines show a main broad absorption band in the 260 mμ region, the "x-band," which is attributed[53] to the dipole effects. As the band is composite in nature some contribution is presumed to be made by a minor transverse dipole acting at 90° to the longitudinal one. Induced dipoles arise from various effects such as the proximity of charged molecules, quaternisation with subsequent betaine formation, or ionisation. Dipole measurements or calculations provide indicative data concerning location of the imino group in the imidazole moiety. In purine itself association of the proton with $N_{(7)}$ rather than $N_{(9)}$ was originally based solely on crystallographic evidence[49] but the greater dipole moment of the former tautomer substantiates this assignment.[54] Studies with other purines[55] show that $N_{(9)}$, on the other hand, is the preferred site in the majority of cases investigated.

E. Polarography

Little evidence of structural behaviour has resulted from studies of this kind—the main emphasis of effort being applied to the development of these techniques for the rapid estimation of purines in small amounts.[56] In acid solution at a dropping mercury electrode purine shows a double wave which can be correlated with a two-stage reduction of the pyrimidine moiety, first to the 1,6-dihydro- then to 1,2,3,6-tetrahydropurine.[57] Under these conditions adenine exhibits a single wave but a tetrahydro derivative results, passing through the 1,6-dihydro stage during which the $C_{(6)}$-amino group is lost. Hypoxanthine shows only weak waves although reduction to dihydro derivative occurs, whereas guanine is not reduced under these conditions. Polarographic studies of 2,6-diaminopurine[58] and 6-thiopurine[59] have been made. A general review covering polarographic studies with purines is available.[60]

F. Solubility and Melting Point

Purine and its C-alkyl analogues show a high water solubility (purine 1 in 2 at 20°). The introduction of oxo groups provides greater facilities for interbase hydrogen bonding (i.e., between ring nitrogen atoms and oxygen atoms on adjacent bases). Formation of such bonding is manifested by a dramatic decrease in solubility. By contrast the aminopurines do not show so high a degree of insolubility, in this instance hydrogen bonding with water molecules rather than intermolecular bonding seems to be the predominant effect. A combination of both oxo and amino groups in the same purine, however, leads to a very high degree of insolubility indeed. Guanine, for example, exhibits a solubility in the cold of 1 in 200,000. A general background to the nature of and factors governing solubility in heterocyclic systems has been set out in a current review.[61] The simple purines have been investigated by Albert and co-workers, whose paper[46] should be consulted for specific physical data. Melting points follow the general rule, being low in monosubstituted purines, but increase sharply in compounds in which some degree of hydrogen bonding is possible. The majority of amino, oxo, and thio purines indeed show no well defined fusion points but undergo slow decomposition at elevated temperatures.

G. Spectra

This topic, embracing ultraviolet, infrared, mass, and nuclear magnetic forms is dealt with in Chapter XIII.

References

1. Scheele, "Examen Chemicum Calculi Urinarii," *Opuscula*, **2**, 73 (1776).
2. Liebig, *Annalen*, **10**, 47 (1834).

3. Mitscherlich, *Poggendorfs Annalen*, **33**, 335 (1834).
4. Wöhler and Liebig, *Annalen*, **26**, 241 (1838).
5. Baeyer, *Annalen*, **127**, 1 (1863).
6. Fittig, *Lehrbuch der Organischen Chemie*, 1877.
7. Medicus, *Annalen*, **175**, 230 (1875).
8. Horbaczewski, *Monatsh.*, **3**, 796 (1882); **6**, 356 (1885); **8**, 201 (1887).
9. Behrend and Roosen, *Annalen*, **251**, 235 (1888).
10. Fischer and Ach, *Ber.*, **28**, 2473 (1895).
11. Fischer, in *Untersuchungen in der Puringruppe*, Springer, Berlin, 1907.
12. Fischer and Ach, *Ber.*, **30**, 2208 (1897).
13. Unger, *Annalen*, **58**, 18; **59**, 58 (1846).
14. Fischer, *Ber.*, **31**, 542 (1898).
15. Traube, *Ber.*, **33**, 1371 (1900).
16. Levene and Bass, in *Nucleic Acids*, Chemical Catalog Co., New York, 1931, Ch. 4.
17. Bendich, in *The Nucleic Acids*, Eds., Chargaff and Davidson, Academic Press, New York, 1955, Vol. I, Ch. 3.
18. Baddiley, in *Thorpe's Dictionary of Applied Chemistry*, 4th ed., Longmans Green and Co., London, 1950, Vol. X, p, 274.
19. Ulbricht, *Purines, Pyrimidines and Nucleotides*, Ed., Sir Robert Robinson, Pergamon Press, London, 1964.
20. Howard, in *Chemistry of Carbon Compounds*, Ed., Rodd, Elsevier Publishing Co., London, 1960, Vol. IVc, Ch. 20.
21. Robins, in *Heterocyclic Compounds*, Ed., Elderfield, Wiley, New York, 1967, Vol. VIII, Ch. 3.
22. Albert, *Heterocyclic Chemistry*, 2nd ed., Athlone Press, London, 1968.
23. Lister, in *Advances in Heterocyclic Chemistry*, Ed., Katritzky, Academic Press, New York, 1966, Vol. VI, p. 1.
24. Lister, *Rev. Pure Appl. Chem. (Australia)*, **11**, 178 (1961).
25. Lister, *Rev. Pure Appl. Chem. (Australia)*, **13**, 30 (1963).
26. *Ciba Foundation Symposium on the Chemistry and Biology of Purines*, Eds., Wolstenholme and O'Connor, J and A Churchill Ltd., London, 1957.
27. Fischer, *Ber.*, **17**, 329 (1884).
28. Patterson, Capell, and Walker, *The Ring Index*, 2nd ed., American Chemical Society, New York, 1960.
29. Biltz, *J. prakt. Chem.*, **145**, 65 (1936).
30. Mann and Porter, *J. Chem. Soc.*, **1945**, 751.
31. Morton, in *Chemistry of Heterocyclic Compounds*, Ed., Hammett, McGraw-Hill, New York, 1946, p. 494.
32. *Handbook for Chemical Society Authors*, The Chemical Society, London, 1960.
33. Albert, in *Current Trends in Heterocyclic Chemistry*, Eds., Albert, Badger, and Shoppee, Butterworths, London, 1958, p. 24.
34. Albert and Armarego, in *Advances in Heterocyclic Chemistry*, Ed., Katritzky, Academic Press, New York, 1965, Vol. IV, p. 1.
35. Connolly and Linschitz, *Photochem. and Photobiol.*, **7**, 791 (1968).
36. Ts'o, in *Structure and Stability of Biological Molecules*, Eds., Fasman and Timascheff, Marcel Dekker Inc., New York, 1970, Vol. II, part 2.
37. Mason, in *Chemistry and Biology of Purines*, Eds., Wolstenholme and O'Connor, J and A Churchill Ltd., London, 1957, p. 72.
38. Pullman and Pullman, *Bull. Soc. chim. France*, **1958**, 766.
39. Pullman, Kochanski, Gilbert, and Denis, *Theor. Chim. Acta*, **10**, 231 (1968).
40. Fukui, Yanezana, and Shingu, *J. Chem. Phys.*, **20**, 722 (1952).

41. Fukui, Yanezana, Nagata, and Shingu, *J. Chem. Phys.*, **22**, 1433 (1954).
42. Veillard, *J. Chim. phys.*, **59**, 1056 (1962).
43. Lykos and Miller, *Tetrahedron Letters*, **1963**, 1743.
44. Pullman, *J. Org. Chem.*, **29**, 508 (1964).
45. Miller, Lykos, and Schmeising, *J. Amer. Chem. Soc.*, **84**, 4623 (1962).
46. Albert, *J. Chem. Soc.*, **1954**, 2060.
47. Albert, in *Physical Methods in Heterocyclic Chemistry*, Ed., Katritzky, Academic Press, New York, 1963, Vol. I, p. 1.
48. De Santio, Gigli, Liquori, and Ripamonte, *Nature*, **191**, 900 (1961).
49. Watson, Sweet, and Marsh, *Acta Cryst.*, **19**, 573 (1965).
50. Devoe and Tinoco, *J. Mol. Biol.*, **4**, 500 (1962).
51. Berthod and Pullman, *Compt. rend.*, **257**, 2738 (1963).
52. Berthod and Pullman, *Biopolymers*, **2**, 483 (1964).
53. Mason, *J. Chem. Soc.*, **1954**, 2071.
54. Pullman, Berthod, and Caillet, *Theor. Chim. Acta.*, **10**, 43 (1968).
55. Weiler-Feilchenfeld and Bergmann, *Israel J. Chem.*, **6**, 823 (1969).
56. Heath, *Nature*, **158**, 23 (1946).
57. Smith and Elving, *J. Amer. Chem. Soc.*, **84**, 1412 (1962).
58. Brown, in *Chemistry and Biology of Purines*, Eds., Wolstenholme and O'Connor, J and A Churchill Ltd., London, 1957, p. 311.
59. Horn and Zuman, *Coll. Czech. Chem. Comm.*, **25**, 3401 (1960).
60. Janik and Elving, *Chem. Rev.*, **68**, 295 (1968).
61. Pfleiderer, *Physical Methods in Heterocyclic Chemistry*, Ed., Katritzky, Academic Press, New York, 1963, Vol. I, p. 177.

CHAPTER II

Syntheses from Pyrimidines

Of the many purines synthesised to date the majority has been derived from pyrimidine precursors. These fall into three classes which in order of importance and utility are the derivatives of 4,5-diamino-, 5-amino-4-oxo-, and 4,5-dioxopyrimidines. Because of the extensive use made of Traube-type condensations pyrimidines of the first class have had, by far, the greatest application to purine syntheses.

1. Use of 4,5-Diaminopyrimidines (The Traube Synthesis)

A. History and General Application

As the name suggests the first general use of this route was due to Traube,[1] who heated 2,4,5-triamino-1,6-dihydro-6-oxopyrimidine with formic acid obtaining guanine. In the same year a succeeding paper[2] showed that the potentiality of the method had been appreciated, and in addition to formic acid, used to form 8-unsubstituted purines, ethyl chloroformate was utilised to produce 8-oxopurines. Within a short time, other reagents, notably urea and thiourea, were employed for the preparation of 8-oxo- and 8-thiopurines.[3]

This brief historical background outlines the usefulness of the procedure. The present-day availability of a wide range of cyclising reagents allows the Traube synthesis great versatility, and virtually any type of substituted purine can be made using it.

In outline it involves the introduction of a one-carbon fragment to bridge the nitrogen atoms of the amino groups at $C_{(4)}$ and $C_{(5)}$ of the pyrimidine ring* (Eq. 1). A considerable number of one-carbon reagents meeting these requirements exists, some of the more widely used being carboxylic acids (X = OH, Y = O), carboxamides (X = NH_2, Y = O),

---

* A parallel may be drawn with the Isay reaction for the synthesis of pteridines in which 1,2-dicarbonyl compounds are condensed with 4,5-diaminopyrimidines. It is

acid chlorides (X = Cl, Y = O), chloroformic esters (X = OEt, Y = Cl) and ureas, thioureas, and amidines (X = NH_2, Y = NH). As the by-products of the reaction are generally volatile (i.e., water, ammonia, hydrogen chloride, or mixtures of these) the purines are obtained relatively uncontaminated. By appropriate choice of reagents

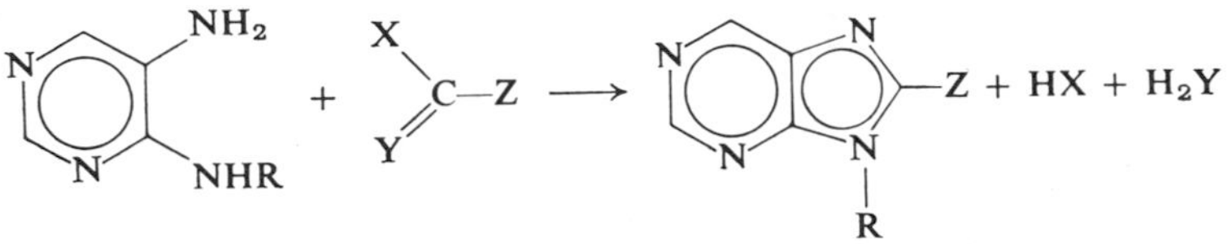

$C_{(8)}$ of the purine may be unsubstituted (Z = H) or carry alkyl, oxo, thio, or amino groups (Z = alkyl, O, S, or NH_2). Although special points of interest connected with the use of a particular reagent are detailed in the appropriate section some general remarks on the Traube synthesis, applicable to most types of reagent, are noted below.

Because of the fairly energetic conditions needed to effect ring closure, the likelihood of reaction between the reagent and other groups on the 4,5-diaminopyrimidine has to be considered. With halogenopyrimidines affording 6-halogenopurines on cyclisation concomitant hydrolysis of the halogen atom generally occurs giving rise to a hypoxanthine derivative. Some reagents, however, are available which permit direct cyclisation to 6-halogenopurines. Complications are also encountered in ring closures leading to 6-thiopurines—the products being wholly or partly the isomeric thiazolo[5,4-d]pyrimidines. In some cases rearrangement to the required purine follows treatment with alkali (Eq. 2). The

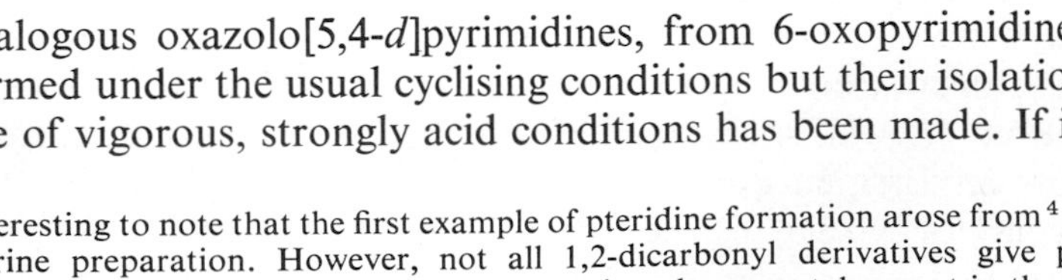

analogous oxazolo[5,4-d]pyrimidines, from 6-oxopyrimidines, are not formed under the usual cyclising conditions but their isolation after the use of vigorous, strongly acid conditions has been made. If in addition

interesting to note that the first example of pteridine formation arose from[4] an attempted purine preparation. However, not all 1,2-dicarbonyl derivatives give pteridines, as examples exist in which only one of the carbonyl groups takes part in the reaction, and a purine is formed.

to primary amino groups at $C_{(4)}$ and $C_{(5)}$ a secondary amino group is present at $C_{(6)}$ alternative cyclisation modes are possible leading to either 9-substituted adenine or a 6-substituted-aminopurine (Eq. 3).

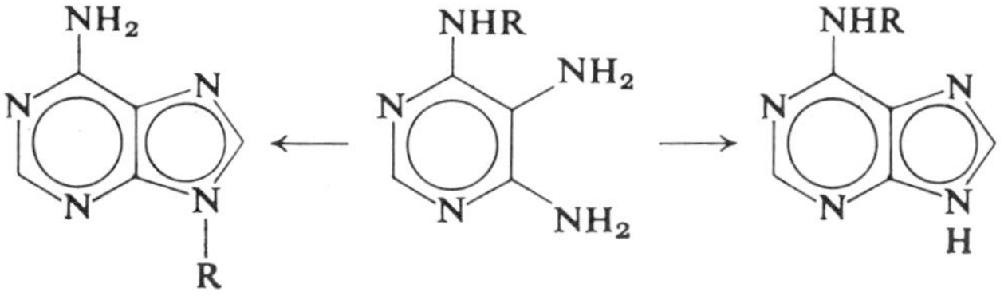

$$(3)$$

With most reagents the 9-substituted adenine is the major product, any isomeric material being removed by treatment of the mixture with alkali in which, by virtue of the acidic proton on the imidazole nitrogen atom, it is soluble. The trialkyl orthoesters are the exception, these reagents preferentially ring close across the two primary amino groups.

In the succeeding discussion of the Traube cyclisation Sections B to J and Section V are devoted to reagents used to form $C_{(8)}$-unsubstituted- or $C_{(8)}$-alkyl- or -arylpurines; the remainder is devoted to those giving 8-amino-, 8-oxo-, and 8-thiopurines.

B. Cyclisation with Formic Acid

This section is accorded fairly detailed treatment, many of the remarks made applying equally well to other cyclising agents described later. With this reagent a two-stage reaction follows in which the product of the first stage, a 4-amino-5-formamidopyrimidine (1), is isolated before final ring closure to the purine (2) is effected. That the site of acylation is the 5- rather than the 4-amino group has been proved by direct synthesis[5] of a known 5-acylaminopyrimidine and by the fact that a derivative of this type, for example 1, on reduction[6] affords a 5-alkylaminopyrimidine (3) capable of cyclisation to a known 7-alkyl-purine (4). Ease of formylation varies with the 4,5-diaminopyrimidine employed. Whereas 4,5,6-triaminopyrimidine is claimed to react in 5% aqueous formic acid solution,[7] 4,5-diamino-2-methoxypyrimidine is not affected and requires treatment in a mixture of formic acid and acetic anhydride,[8] the formylating agent in this case being formic acetic anhydride formed in situ. Generally formylation is carried out with formic acid under reflux conditions. In a minority of cases direct conversion of the pyrimidine to the purine takes place in the hot formylating mixture, in one example guanine is obtained in good yield on heating together 2,4,5-triamino-1,6-dihydro-6-oxopyrimidine, as the sulphate, and formic acid for some hours.[1, 9] Other single-stage conversions have been recorded for hypoxanthine[6, 10] and the 1-methyl,[11] 2-methyl,[12] and 9-methyl[13] derivatives. This means has given other purines[14] in-

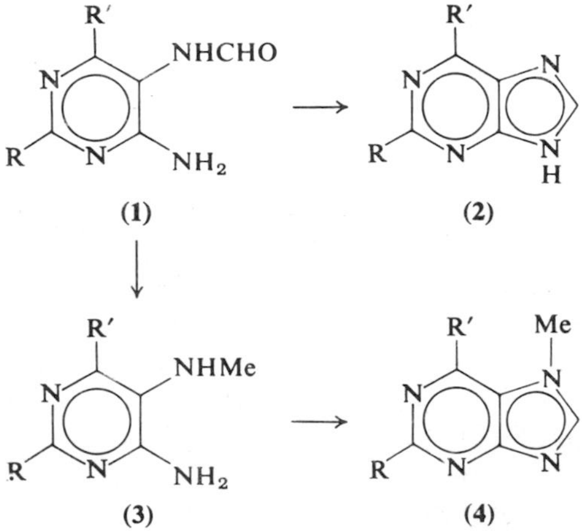

cluding 2-thio derivatives[15] but loss of a 2-sulphinyl group occurs during cyclisation of 4,5,6-triamino-2-sulphinylpyrimidine.[16] Ring closure is promoted by higher reflux temperatures for which purpose the addition of sodium formate or sodium acetate is sometimes made.[1, 2, 17] Alternatively, use of formic acid–acetic anhydride mixtures, an example of which is noted above, are employed in refractory cases such as pyrimidine carboxylic esters[18] and N-oxides.[19]

Direct closure to the purine is, however, more the exception than the rule and a two-step synthesis is generally expected. Conversion of the 4-amino-5-formamidopyrimidine to the final product can be achieved in a variety of ways, the main ones being by fusion, or by heating in aqueous alkali or in a high boiling inert organic solvent.

The first of these still finds occasional use but is mainly of historic importance. Due to the fairly high temperatures needed (over 200°) the procedure is limited to fairly robust pyrimidines. It has the further disadvantage in being suitable only for small-scale preparations owing to poor thermal conductivity across the melt leading to mixtures containing partly cyclised products if larger quantities are attempted.[20] Extensive use of fusion techniques is found in the early literature.[3, 4, 12] Among the more recent preparations, carried out in an atmosphere of carbon dioxide or nitrogen,[8] are adenine[21] and other aminopurines,[22]

alkylthio-,[23] and *N*-alkylatedpurines.[24] Purine itself is a further example.[25] Where the purine is a low melting one, as in the last example,[4, 8, 26] distillation or sublimation procedures can be incorporated into the fusion reaction. Other illustrations of this specialised route include the formation of 9-methyl-[27] and 2-methoxypurines.[8]

Oxo- and amino-oxo derivatives of 4,5-diaminopyrimidines formylate readily but the products are not amenable to cyclisation by fusion due to the strongly acid character they possess. Conversion to the sodium or potassium salt overcomes this difficulty as demonstrated in the synthesis of xanthine,[1] and also the 1-methyl,[28] 3-methyl[1] and 1,3-dimethyl[1] homologues. Many other oxopurines and *N*-alkyl oxopurines have resulted likewise[12, 29, 30-32] as well as oxo-thiopurines.[17] A convenient modification employs aqueous alkali, either sodium hydroxide or sodium carbonate, under reflux conditions, in place of the dry fusion, with excellent results. Derivatives of theophylline[33-36] and 6-amino-2-alkylthiopurines[37] provide illustrations of this route. In the case of 1,6-dihydro-6-thiopurine itself, a temperature over 100° was found necessary to effect cyclisation.[38]

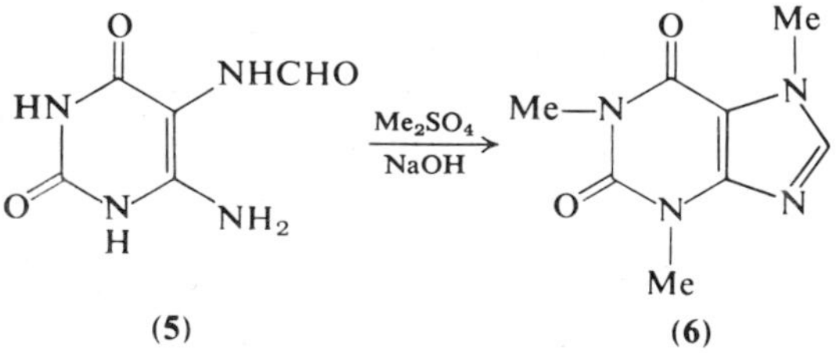

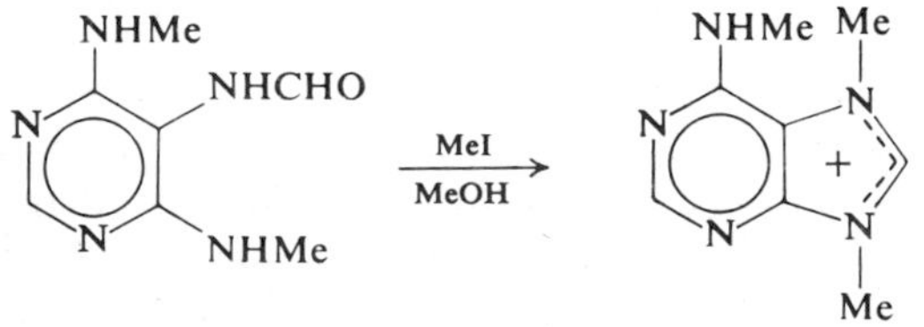

The addition of methylating agents before cyclisations in alkali extends the reaction to the formation of *N*-alkylated purines from nonalkylated pyrimidines. A good example is the formation of caffeine (6) from the pyrimidine (5).[33] Theobromine arises similarly from 4-amino-5-formamido-1,2,3,6-tetrahydro-3-methyl-2,6-dioxopyrimidine.[39] As *N*-methylation of the formamido group precedes cyclisations of this type, the diminished acid character thereby induced leads to the more facile ring closure of the formamidopyrimidine. That the alkaline conditions can be dispensed with altogether with more basic purines is seen in the isolation of 7,9-dimethyl-6-methylaminopurinium iodide (8) on heating (100°, 2h) 5-formamido-4,6-dimethylaminopyrimidine (7) with methanolic methyl iodide.[40] A further example is 7-benzyl-2,6-dichloro-purine from reaction between 4-amino-2,6-dichloro-5-formamido-pyrimidine and benzyl chloride in dimethylformamide at room temperature.[41] This product is somewhat noteworthy because hydrolysis of the halogen atoms does not occur during ring closure.

Cyclodehydration in high-boiling organic solvents is the third widely used procedure. Like the alkaline solution method it has the advantages over the fusion procedure because it is suitable for large-scale preparation and can be carried out at much lower temperatures. Solvents used are usually aliphatic amides or aromatic tertiary amines, but other stable, inert compounds have been found suitable. Although formamide was initially employed solely as a cyclising medium[42] the demonstration that it suffers decomposition as it approaches the boiling point[43] led to the realisation that it is an effective formylating agent in its own right (see Sect. H). Studies with 4,6-diamino-5-formamidopyrimidine, labelled with ^{13}C in the formamido group, showed adenine having only one-quarter of the original activity to be obtained after cyclisation in formamide.[44] With other amides, such as dimethyl formamide[7] or N-formylmorpholine[7, 45, 46] variable results are claimed but in no case is the total activity retained. However, this is achieved if nonamidic solvents, for example, diethanolamine or nitrobenzene, are used instead.[46] Although attempts[7] to explain these results by a "formyl group interchange" theory have been challenged,[46] such a mechanism is plausible in the light of experimental verification that has shown that interchange of 5-acylamino groups is possible. Definitive cases are the formation of 8-methylxanthine (10) from heating the 5-formamido-pyrimidine (9) with acetamide[47] and xanthine when the analogous 5-acetamidopyrimidine is treated with formamide.[47]

In addition to the use of formamide, N-formylmorpholine,[7, 8] diethanolamine,[46, 48] and nitrobenzene,[46] instanced above, successful closures of formamidopyrimidines are recorded in glycol, as in the formation of guanine,[49] and 2-amino-6-methylpurine was obtained in high-boiling paraffin.[50] Quinoline was a useful medium for forming 6-methylaminopurine and the 2-thio analogue.[20] Pyridine, with a lower boiling point, is not employed in simple cyclisations but when combined with phosphorus pentasulphide gives a medium in which oxo-5-form-amidopyrimidines are converted directly to the corresponding thio-purines. The preparation of both thioguanine (12, R = H)[51] and the 1-methyl homologue (12, R = Me)[52] from the respective pyrimidine (11, R = H and Me) illustrate the method, for which extensive applications are to be found in the literature. As 2-oxopurines are not readily transformed into the 2-thio analogues, the procedure is generally applicable only to the formation of 6-thiopurine derivatives. One advantage is that only 6-thiopurines appear to be obtained and thiazolo-

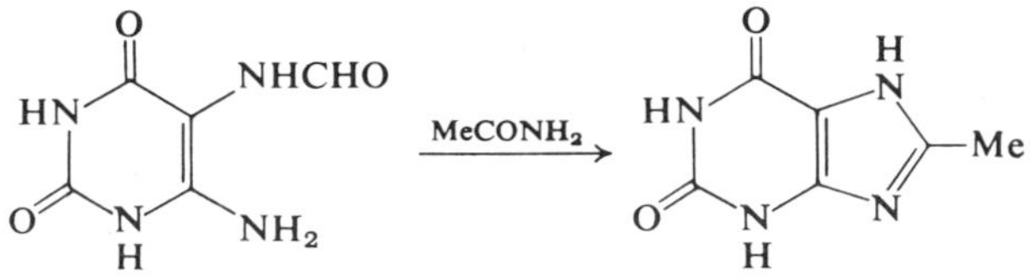

[5,4-*d*]pyrimidine formation appears limited or absent. This is not the case when 4,5-diamino-1,6-dihydro-6-thiopyrimidine in formic acid is heated for some hours, the product in good yield being 7-aminothiazolo-[5,4-*d*]pyrimidine (13).[53] The alternative ring closure to 1,6-dihydro-6-thiopurine takes place if the diamino pyrimidine is formylated at room temperature and heated with dilute sodium hydroxide.[38] An investigation of the factors controlling the cyclisation pathway has shown that a low pH and high temperature favours thiazolopyrimidine formation. Isomerisation of some thiazolopyrimidines to 6-thiopurines is possible and is discussed in Section 8A. Further examples of this type of ring closure are available.[13, 53–56]

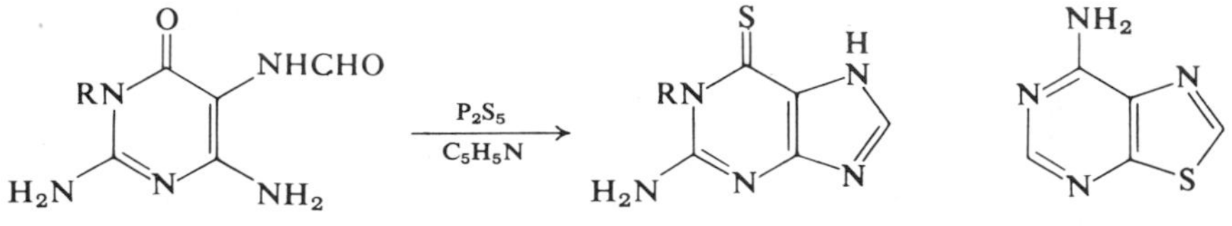

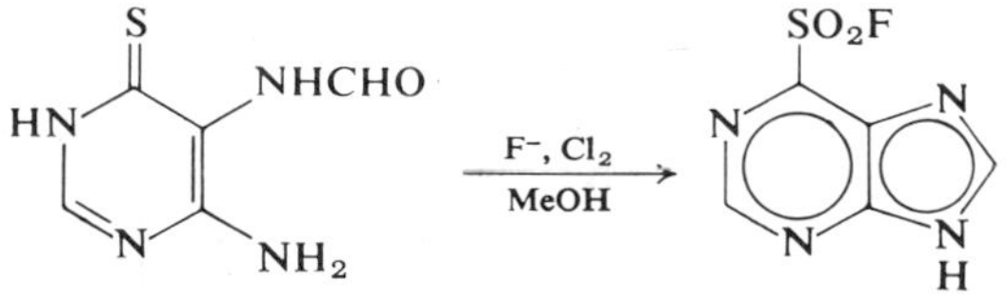

An unusual purine preparation, occurring at 0°, is of 6-fluorosulphonylpurine (15) on treating a methanol solution of the thiopyrimidine (14) with hydrofluoric acid and chlorine. Such a facile ring closure can only be the result of prior formation of the fluorosulphonyl group under the oxidative conditions present.[55]

Due to the labile nature of halogen atoms at $C_{(6)}$ and, to a less extent, at $C_{(2)}$ in the purine ring toward acid conditions, formic acid closures of chloropyrimidines give the corresponding 2- or 6-oxopurines. Hypoxanthine results on heating 4,5-diamino-6-chloropyrimidine,[26] and this reaction forms the basis for preparing a number of derivatives of hypoxanthine and xanthine from 6-chloro- and 2,6-dichlorodiaminopyrimidines, respectively.[41, 57] Although halogen atoms are only held to be labile after the purine ring has been formed, a case in which hydrolysis occurs under conditions permitting only formylation of the 5-aminopyrimidine is known.[57] The presence of deactivating groups can produce exceptions to the above generalisations. This can be illustrated by a formic acid cyclisation giving 9-benzyl-6-chloro-2-trifluoromethylpurine after 3 hours under reflux.[58] Other examples of 2-chloro- and 6-chloropurines formed in this way through deactivation due to trifluoromethyl groups are known.[58]

As a corollary to this section it is pertinent to note a reaction which is an extension of the Traube synthesis although this is not immediately apparent. A number of investigations have reported the conversion of uric acid (16) to xanthine (17) after strenuous heating with formic acid[59] or formic acid precursors which include mixtures of chloroform and sodium hydroxide[60, 61] and of oxalic acid and glycerol.[62] Such a replacement of an 8-oxo function can only take place through opening of the imidazole ring to give the 5-ureidopyrimidine (18) after which further degradation to the 4,5-diaminopyrimidine is followed by formylation and cyclisation of the 5-formamidopyrimidine to xanthine.

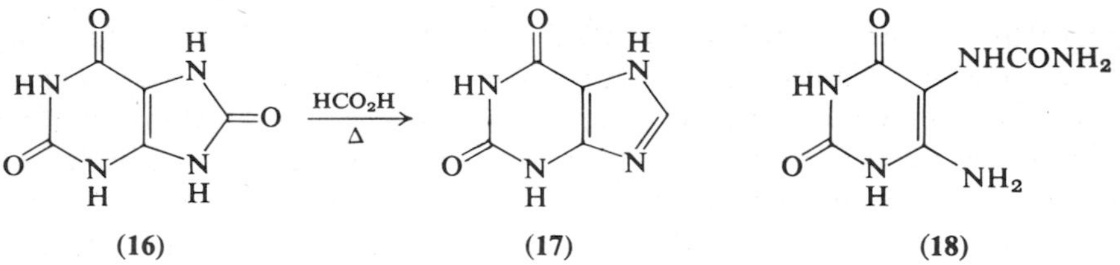

C. Cyclisation with Dithioformic Acid

The application of this derivative to purine syntheses arose from the need for a cyclising reagent capable of reacting under mild conditions, particularly in the presence of labile glycosyl residues.[63] 4,5-Diaminopyrimidines with an aqueous solution of the sodium or potassium salt of dithioformic acid (HCS_2H) are readily converted to the 4-amino-5-thioformamido derivatives.[64] Although ring closure to the purine was originally effected by heating alone as, for example, in forming 6-methyl-purine which was the first purine to be prepared by this route,[64] these conditions proved too vigorous for use in subsequent syntheses,[63] and were replaced by closures in pyridine or quinoline employing reflux times of only a few minutes. Exemplifying the latter procedure are 6-amino-2-methyl- (20, R = Me, R′ = H),[63] 6-amino-2-dimethylamino-(20, R = NMe$_2$, R′ = H),[65] and 6-amino-9-phenylpurines (20, R = H, R′ = C$_6$H$_5$)[66] derived from the corresponding 5-thioformamido-pyrimidines (19). Derivatives of both 2- and 6-oxopurines have been derived likewise.[63] In place of the above solvents water can be employed, in some cases, but longer heating times are necessary. Both 4,6-diamino-2-methylthio-5-thioformamidopyrimidine (19, R = MeS, R′ = H) and the 4-methylaminopyrimidine analogue (19, R = MeS, R′ = Me) were converted to the respective purines (20, R = MeS, R′ = H and Me)

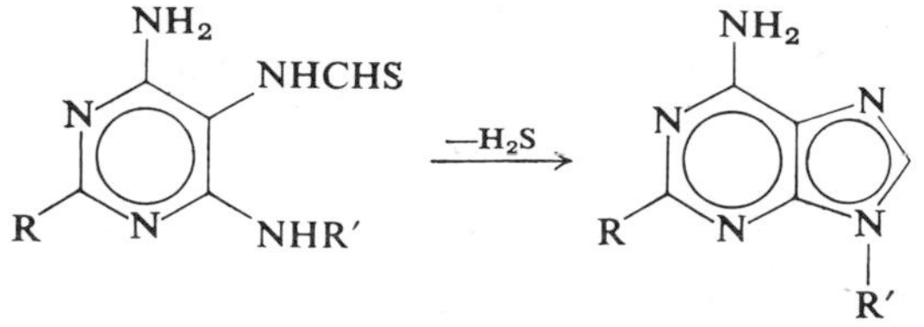

(19)   (20)

in this way, the former requiring 36 h in boiling water while the latter, involving the more basic methylamino group, was completed in 5 h.[63] Adenine has been claimed to be obtained in almost quantitative yield by this route (12 h)[67] but subsequent preparations[58] have given only 65% yields of the purine together with appreciable amounts of 4,6-diamino-5-formamidopyrimidine. A similar mixture is produced when the reaction is carried out in pyridine.[43] The presence of a formamido-pyrimidine in the reaction mixture is a common feature of thioform-amide ring closures under aqueous conditions; replacement of the sul-phur atom by oxygen appears to take place before cyclisation to the purine is completed. The other possible route to a formamidopyrimidine, through hydrolytic fission of the purine imidazole ring, is unlikely under the reaction conditions employed. With a 4,6-diamino-5-thioform-amidopyrimidine having a secondary amino group at either C$_{(4)}$ or C$_{(6)}$, condensation involves the substituted-amino group, as with a form-amidopyrimidine, giving a 9-substituted purine as the major product, as seen above in 20 and in the formation of 6-amino-2,9-dimethylpurine (20, R = R′ = Me).[69]

The foregoing work formed the basis for a nucleoside synthesis programme as it suggested that direct cyclisation of 4-amino-6-glycosyl-amino-5-thioformamidopyrimidines to 9-glycosylpurines could be carried out. Initially, using pyridine, and a nitrogen atmosphere for the cyclisation various adenine-9-glycosides have been prepared. These include the 9-β-D-xylopyranosyl derivatives of adenine,[70] 6-amino-2-methyl-,[69] and 6-amino-2-methylthiopurine. The 9-β-D-ribosyl analogue[71] of the last purine[72] is likewise derived. Where the hydroxy groups of the sugar moiety are protected by acetylation, cyclisation can in some cases give rise to a mixture of the 9-glycosyladenine[70, 72] and the isomeric 6-glycosylaminopurine (Eq. 4). The fact that only a

$$(4)$$

9-glycosyladenine is obtained with the unprotected glycosylamino-pyrimidine[72] may be due to a hydrogen bonded state existing, before cyclisation, between the glycosidic nitrogen atom and the acetyl group at $C_{(2)}$ of the xylose moiety leading to the amino groups at $C_{(4)}$ and $C_{(6)}$ possessing comparable activities.[70]

Other solvents tried in attempts to improve yields include borax in sodium hydroxide solution[72] and sodium methoxide in various alcohols.[73] Some improvement (giving up to 60% yield) was found using ethanol but higher boiling alcohols caused some decomposition to occur.[73] In all cases some glycosylpurine is obtained whether the acetylated sugar derivative or unprotected form is used. Furthermore, irrespective of the original sugar configuration, the sugar in the cyclised product is in the pyranose form with a β-glycosyl linkage with the purine.[74]

The majority of examples arising by the procedure are 9-glycosyl derivatives of adenine or of 6-amino-2-methylthiopurine in which the $C_{(9)}$ substituents (all in the pyranose form) are D-xylosyl,[73] D-ribosyl,[73] D-mannosyl,[75] L-arabinosyl,[76] D-galactosyl,[77] or D-glucosyl.[65, 78] The feeling that this methoxide treatment, with the accompanying concomitant deacetylation, was occasioning a furanose → pyranose transformation seems justified in the light of cyclisations performed in an aprotic solvent, like acetonitrile, containing fused potassium acetate, in which acetyl group loss was found to be minimal and 9-glycosylpurines with an unchanged furanose residue resulted.[74, 77] One unexpected result arises if acetic acid is present in this solvent mixture, the product being exclusively the 6-glycosylaminopurine.[73]

Cyclisation with alkoxides can give 9-furanosylpurines but suitable protecting groups must be present before ring closure. Preparation of the naturally occurring adenosine, 9-β-D-ribofuranosyladenine, has, however, been possible by this means.[79]

D. Cyclisation with Other Carboxylic Acids

Procedures parallel those used with formic acid, the initial reaction product being generally the 5-acylaminopyrimidine requiring further treatment to effect ring closure. The range of carboxylic acids available enables a variety of functional substituents to be placed at $C_{(8)}$ of the purine. Traube first utilised acetic acid in forming the 8-methyl derivatives of guanine and 3-methylhypoxanthine, cyclisation of the acetamidopyrimidine being effected in hot sodium hydroxide solution. Cases of direct preparations of 8-methylhypoxanthine derivatives, with alkyl[80] and aryl[81] substituents at $C_{(2)}$, are, however, reported by prolonged heating of the diamine in the acid solution. Trifluoroacetic acid forms 8-trifluoromethylpurines; the 5-trifluoromethylacetamido intermediates undergo ring closure either on fusion[82, 83] or in alkali solutions.[58] Higher acid homologues afford the appropriate 8-alkylpurine similarly, an early example being 8-ethylguanine from 2,4,5-triamino-1,6-dihydro-6-oxopyrimidine and propionic acid. In this instance ring closure occurred on heating the sodium salt of the propionamido derivative at 240°.[12]

Extensive use has been made of this route to prepare 8-substituted derivatives of the therapeutically important theophylline and related alkylated xanthines. Successful condensations have been carried out between acetic, propionic, butyric, isobutyric and n-valeric acid, and 1,3-dialkyl-4,5-diaminouracils,[84] the intermediate pyrimidines (21) being ring closed by heating in a metal bath[84] or treatment with aqueous alkali[85] to the respective theophylline (22, R = Me, Et, Pr^n, Pr^i, and Bu^n). Variations in the pyrimidine moiety have given the 1,3-diethyl and 1,3-dibutyl analogues by the same route.[84]

8-Alkenyltheophyllines of the type 22 (R = allyl, crotyl, β-pentenyl, and γ-pentenyl) arise similarly,[85, 86] and other variations of the substituent at $C_{(8)}$ of theophylline have been made with phenyl-, benzyl-cyclohexyl-, dibenzyl-, 1-naphthyl-, and phenoxyacetic acids.[86] A number of 8-substituted theophyllines have been formed directly by interaction of the acid and diaminopyrimidine in phosphoryl chloride.

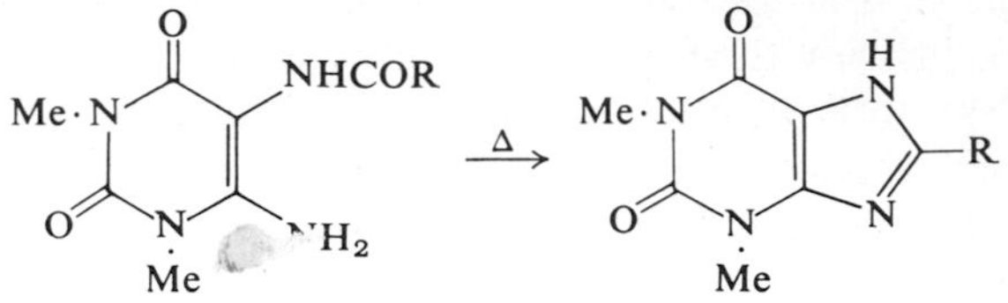

(21)

(22)

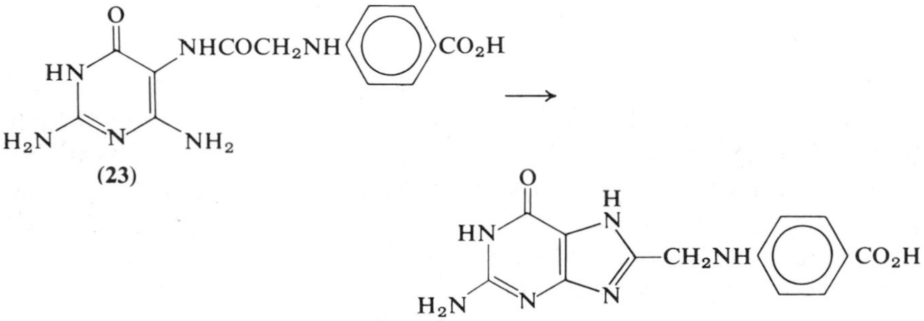

(23)

(24)

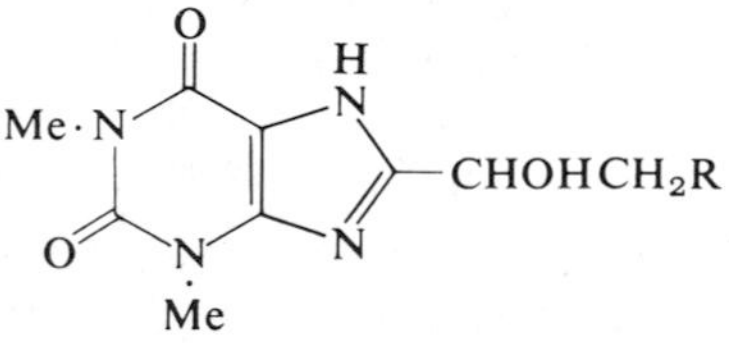

(25)

These include the 8-phenyl,[87] 8-benzyl,[97] and 8-styryl[86] derivatives (22, R = C_6H_5, $CH_2C_6H_5$, and $CH:CHC_6H_5$) respectively, among others.[86] Further examples of acidic conditions being employed in purine forma- tion are seen with 8-phenylguanine for which ring closure occurred in a phosphoric acid–phosphorus pentoxide medium at 180°[88] while the purine analogue of pteroic acid (24) resulted from cyclodehydration of the amide (23) in sulphuric acid on a water bath.[89] Cyclisations in acid media are the exception rather than the rule; almost all examples utilise alkaline solutions for the ultimate stage. In addition to the above other acids employed, but not necessarily affording theophylline derivatives, include 2- and 3-thienylacetic acid[90] and α-hydroxyacids of which glycollic acid provides a facile route to 8-hydroxymethylpurines, the parent member arising from a condensation with 4,5-diaminopyrimi- dine.[91] An interesting point is that whereas with glycollic acid the 5-hydroxymethylacetamidopyrimidine is obtained requiring alkali treatment to effect closure, use of ethyl glycollate gives the purine directly.[91] Although other applications of the ester are recorded[92] the majority of 8-hydroxymethylpurines has arisen by means of the two-stage procedure. The variety is exemplified by the 8-substituted analogues of hypoxanthines,[93] 1,6-dihydro-6-thiopurine,[94] and 3-methyl- and 1,3-dibenzylxanthine.[95] Related theophylline derivatives were similarly obtained, lactic acid and glyceric acid giving the respective 8-α-hydroxy-ethyl[96] (25, R = H) and 8-α,β-dihydroxyethyl[97] derivatives (25, R = OH). Gluconic acid likewise affords the appropriate 8-pentahydroxy-pentyl[97] derivative although the hypoxanthine analogue could not be obtained owing to failure to cyclise the gluconylaminopyrimidine.[93]

Dicarboxylic acids have had wide application where purines bearing carboxyl groups, directly or indirectly linked to $C_{(8)}$, are needed. The use of oxalic acid to prepare 8-carboxypurines must be approached with caution as condensation of 1,2-dicarbonyl compounds with 4,5-diaminopyrimidines constitutes a standard synthesis of 6,7-dihydroxy-

$$(5)$$

pteridines (the Isay reaction) (Eq. 5).* Provided too vigorous heat treatment is avoided, the oxalylamino derivative (26) which forms then undergoes preferential cyclisation to the purine (27) on heating if converted first to the sodium salt. Both 8-carboxyhypoxanthine (27) and the 3-methyl analogue[12, 101] were obtained by this route but when the preparation of the 6-thiopurine analogue (28, R = H) was attempted,[94] 7-aminothiazole[5,4-d]pyrimidine (30) resulted through decarboxylation and the alternative cyclisation involving the thio group occurring. The correct cyclisation mode is followed, giving 28 (R = Et) if diethyl oxalate replaces oxalic acid in the reaction with 29.[94]

8-Carboxymethylpurines result from condensations with cyanacetic ester [this reagent being used in preference to malonic acid], in which hydrolysis of the nitrile group occurs during alkaline cyclisation,[12, 101]. The carboxyethyl derivatives of 3-methylxanthine,[12] theophylline,[12] hypoxanthine,[93] 1,6-dihydro-6-thiopurine,[94] and guanine[12, 101] are derived similarly from succinic acid. Glycine derivatives are utilised to give 8-aminomethylpurines of general type (31).[101, 102]

Use of acids as cyclising reagents is extended to the conversion of existing purines to the appropriate 8-substituted analogues. One

---

* Fusion of the appropriate diaminopyrimidine with oxalic acid has given 6,7-di-hydroxypteridine (R = R' = H),[98] also the 2,4-dihydroxy (R = R' = OH)[99] and 4-hydroxy-2-mercapto (R = SH, R' = OH)[100] derivatives (Eq. 5).

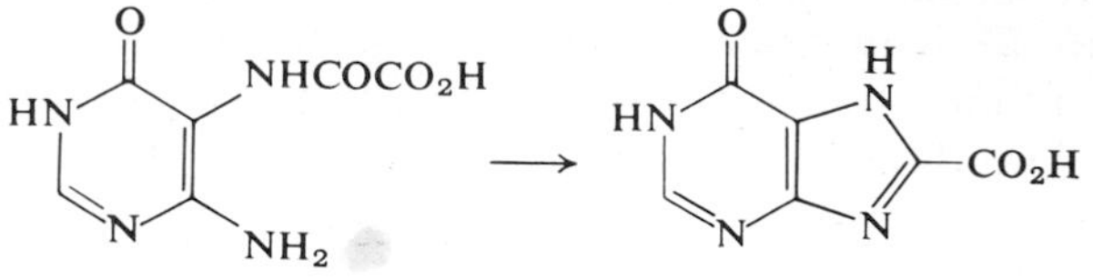

(26)   (27)

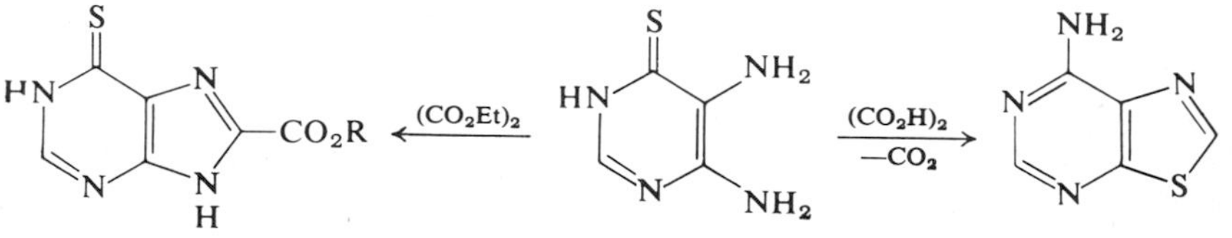

(28)   (29)   (30)

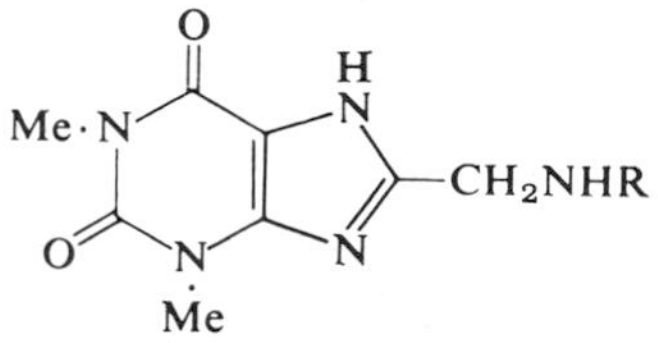

(31)

example is the formation of 8-benzyltheophylline on heating 8-bromo-theophylline with phenylacetic acid in sodium hydroxide at 250°. Other 8-benzylxanthines follow similarly by this modified Traube synthesis.[102a]

E. Cyclisation with Acid Anhydrides

The limited range of anhydrides at hand restricts the range of examples available. Acetic anhydride has been extensively used to prepare 8-methylpurines. Cyclisation may or may not go to completion in one stage, but conversion to the 8-methylpurine in the latter case is usually facile. Refluxing acetic anhydride was used by Isay[4] and others[8] to ring close 4,5-diaminopyrimidine directly to the parent member of the series, 8-methylpurine. A series of 2-oxo-8-methylpurines (32) prepared by Johns and his co-workers illustrates the techniques employed. The 8-methylpurine (32, R = R' = H)[31] resulted from fusion of the potassium salt of the intermediate acetamidopyrimidine whereas the 9-methyl homologue (32, R = H, R' = Me)[14] arose from direct reaction between the diaminopyrimidine and acetic anhydride. The latter preparation illustrates the greater facility for cyclisation of the more basic 4-alkylamino-5-aminopyrimidines over 4,5-diaminopyrimidines.[14]

As a general rule the more acidic nature of oxopyrimidines shows in the formation of only the 5-acetamidopyrimidines in acetic anhydride under reflux conditions. Procedures already elaborated in the sections devoted to formic acid (Sect. 1B) and other carboxylic acids (Sect. 1D) are utilised to cyclise such derivatives to the appropriate 8-methyl-purines. These include dry-state fusions (as for example, **32**, R = Me, R′ = Et)[103] or heating in aqueous alkali, 8-methylhypoxanthine and various analogues[104] being so prepared. Phosphoryl chloride effected ring closure to 1,6-dihydro-2,8-dimethyl-6-oxopurine[105] while in hydrochloric acid both 8-methyl- and 1,3,8,9-tetramethylxanthine have been prepared.[106] An attempt to cyclise 2,4,5-triamino-1,6-dihydro-6-oxopyrimidine directly to 8-methylguanine failed, but conversion of the 5-acetamido analogue was carried out in acetamide at reflux temperature.[29, 51]

By contrast adenine[104] derivatives and those of related amino-purines,[107] being derived from more basic pyrimidines, are usually obtained directly from acetic anhydride treatment.

Complications may ensue if thio groups or halogen atoms are also present; thus 7-amino-2-methylthiazolo[5,4-*d*]pyrimidine is the major product if 4,5-diamino-1,6-dihydro-6-thiopyrimidine is used.[104] A sur-

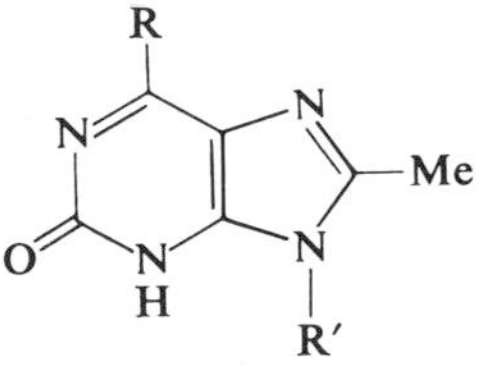

(32) 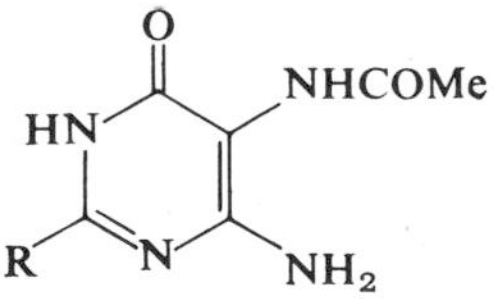   (33) 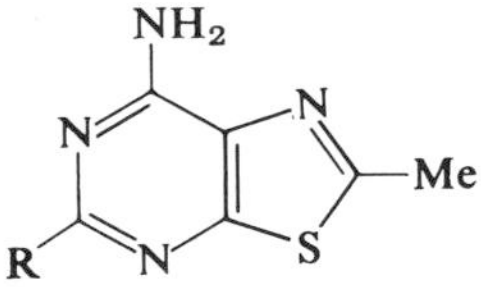   (34) 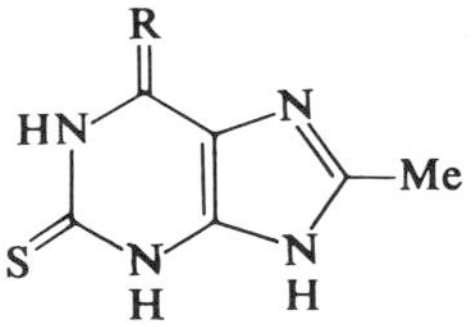   (35)

prising feature of this derivative was the stability in alkali, the expected isomerisation to 1,6-dihydro-8-methyl-6-thiopurine not being observed. In this connection it is noteworthy that, on heating the 5-acetamido-6-oxopyrimidines (**33**, R = NH_2 and Me) with phosphorus pentasulphide in pyridine, replacement of oxygen by sulphur at $C_{(6)}$ must precede cyclisation as the major products are the thiazolo[5,4-*d*]pyrimidines (**34**, R = NH_2 and Me).[108] If, however, the 2-thiopyrimidine analogue

(33, R = SH) is similarly treated, the product is a mixture of the 6-oxo-2-thio- (35, R = O) and 2,6-dithiopurine (35, R = S).[108]

The action of acetic anhydride on 4,5-diamino derivatives of 1,2,3,6-tetrahydro-1,3-dimethyl-2,6-dioxopyrimidine has been the subject of a number of investigations.[47, 109] Traube[12] obtained 1,3,8-trimethylxanthine using only the anhydride but the method is improved if pyridine is also present during reflux, as in the preparation of 3,8-dimethyl- and 1,3,8-trimethylxanthines.[47] When either the 4,5-dimethylamino- (36)[109] or 4-amino-5-dimethylamino-[110] (37) derivative are reacted, the product in either case is 1,3,7,8-trimethylxanthine (38). Mechanisms proposed for both reactions entail loss of methanol from the respective adducts, 39 and 40. It is interesting to note under fusion conditions that the related acetamidopyrimidine (41) suffers no loss of methyl groups and gives 7,8,9-trimethylxanthine betaine (42).[111]

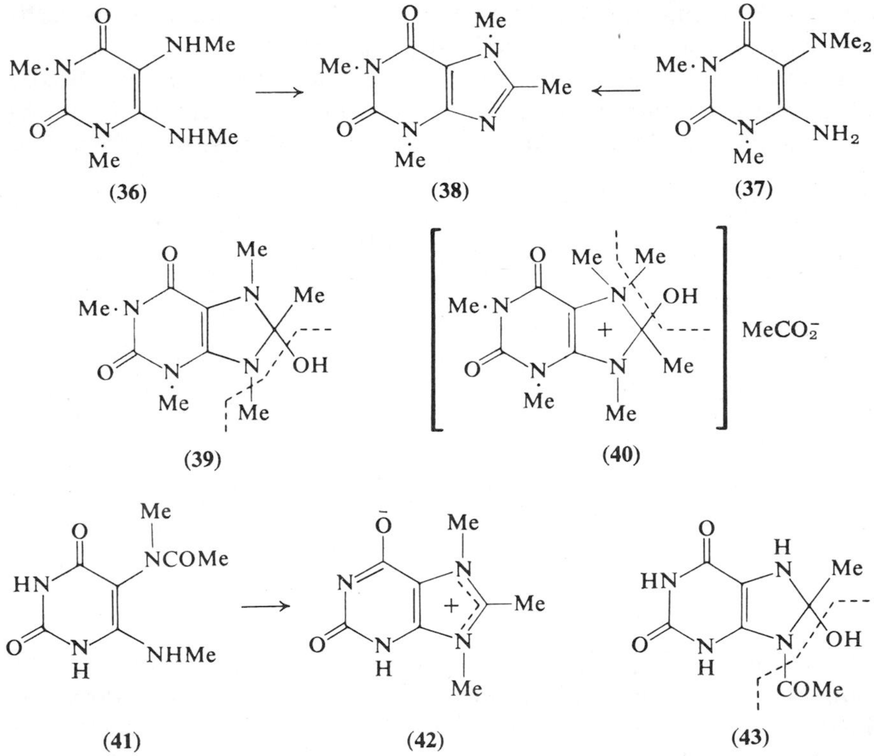

(36)   (38)   (37)

(39)   (40)

(41)   (42)   (43)

Trifluoroacetic anhydride, either alone or combined with trifluoroacetic acid, is used to prepare 8-trifluoromethylpurines. Direct cyclisation may occur but heating (210°) of the acetylated pyrimidine may be required. These techniques have given 8-trifluoromethylpurine,[82] and the 6-methyl homologue [112] as well as 2,6-diamino-8-trifluoromethyl-[82] and 2-amino-6,8-*bis*-trifluoromethylpurine.[82] In spite of the excellent dehydrating properties of anhydrides, the use of trifluoroacetic anhydride to cyclise 4,5-diamino-6-chloropyrimidine resulted in concomitant hydrolysis of the halogen atom, giving 8-trifluoromethylhypoxanthine.[82]

Anhydride-type reactions also include use of sugar lactones as cyclising agents. Both D-gluconolactone and D-ribonolactone have been fused at 140° with 4,5-diaminopyrimidines affording 8-D-glucono-pentahydroxypentyl- and 8-D-ribonotetrahydroxybutylpurines (Eq. 6, $n = 4$ and 3, respectively). Analogous compounds of 6-dimethylamino-2-methyl- and 2-methyl-6-morpholinopurine are known.[92]

$$
\begin{array}{c}
\text{NEt}_2 \\
\end{array}
\quad + \quad
\underset{\substack{| \\ \text{CH}_2\text{OH}}}{\overset{\text{OC}\!-\!-\!-}{\underset{\text{HC}\!-\!-\!-}{(\text{HCOH})_{n-1}\,\text{O}}}}
\quad \longrightarrow \quad
\begin{array}{c}
\text{NEt}_2 \\
\end{array}
\;-(\text{CHOH})_n\text{CH}_2\text{OH} \quad (6)
$$

Modification of the Traube synthesis by utilising existing purines rather than 4,5-diaminopyrimidines has been noted already in Section 1B. Applications of this are also found using anhydrides to replace the $C_{(8)}$-oxo function in uric acid derivatives by an alkyl group. Uric acid heated with acetic anhydride in pyridine or dimethylaniline gives 8-methylxanthine [47, 112–116] directly. The intermediate adduct (**43**) has been isolated and is converted to 8-methylxanthine on heating in benzyl alcohol.[115] *N*-Methylated analogues behave similarly, from 3-methyl- and 1,3-dimethyluric acid were obtained 3,8-dimethyl-[114] and 1,3,8-trimethylxanthine,[106] respectively. Propionic anhydride likewise affords 8-ethylxanthine derivatives.[117] Certain 8-alkoxypurines undergo this transformation as illustrated by the formation of 8-methyl- and 8-ethyl-caffeines from the 8-ethoxy analogues using acetic and propionic anhydride, respectively.[117] As noted previously (Sect. 1B), a mechanism for reactions of this type is considered to be through fission at the imidazole ring [47, 106] followed by recyclisation.

F. Cyclisation with Acid Chlorides

Because of the availability of more convenient alternatives the simple acyl chlorides have had little application. By contrast the higher molecular weight homologues have considerable utility in being highly reactive liquids whereas the corresponding more conventional reagent, for example, an acid or anhydride, is a solid or is not readily available. The formation of a series of 8-α-hydroxybenzylpurines by means of mandelyl chloride illustrates this, the intermediate 5-amido derivative being ring closed in potassium methoxide.[118] Dipurin-8-ylalkanes of the type **45** arise when chlorides of dicarboxylic acids are employed. Ring closure of the diamide derivative (**44**) in sodium hydroxide gives 1,4-di(hypoxanthin-8-yl)butane (**45**) but 1,4-di(6-chloropurin-8-yl)butane is formed if a mixture of phosphoryl chloride and dimethylaniline is employed.[119] Succinyl chloride affords the corresponding 1,2-dipurin-8-ylethanes.[119] The reaction between oxalyl chloride and the tris trimethylsilyl derivative of 2,4,5-triamino-6-methoxy(or ethoxy)pyrimidine in toluene at 0° in the presence of triethylamine is unusual in that the product after removal of the trimethylsilyl groups by alcoholysis is an 8-oxopurine (**46**, R = Me and Et) instead of the corresponding 6,7-dioxopteridine (**47**).[120] Use of this activated form of 4,5-diaminopyrimidines is referred to again later (Sects. 1Q and 1S).

In contrast to the few examples of acyl chloride utilisation the converse holds for aroyl chlorides which, by virtue of their greater reactivities compared with the corresponding benzoic acids, have enjoyed more extensive use for the preparation of 8-phenylpurines. Condensation of the benzoyl chloride derivative is made in dilute sodium hydroxide or pyridine[5] with subsequent cyclisation of the 5-benzamidopyrimidine being carried out in a number of ways. Albert cyclised 4-amino-5-benzamidopyrimidine to 8-phenylpurine by heating at 200° for some minutes,[8] like conditions affording 2-amino-8-phenyl- and 6-amino-8-(4-chlorophenyl)-purine[121] in addition to a number of 2,6-diamino-8-substituted-phenyl derivatives.[122] The acidic character of oxopyrimidines can lead to cyclisation difficulty unless prior conversion to the sodium salt is made, this procedure being necessary to prepare some 8-phenylguanines,[122] with a high temperature (280°) used. The corresponding theophylline derivative was prepared from the benzamidopyrimidine by heating in aqueous sodium hydroxide.[86] Some refractory pyrimidines, as for example 2,4-diamino-5-benzamido-1,6-dihydro-6-oxopyrimidine, were converted to purines by heating in benzamide.[122] A recent technique, possibly of wider general application, makes use of cyclohydration in polyphosphoric acid, the preparation of a representative range of 8-phenyl- and 8-p-tolylpurines being demonstrated.[123, 124] With this reagent in the presence of a 6-thio group the main product is 7-amino-2-phenylthiazolo[5,4-d]pyrimidine.[123] The thiazolopyrimidine is obtained in better yield using the sodium salt of the benzamidopyrimidine. This fact is surprising considering that the analogous 5-acetamido-6-thiopyrimidines form only purines under these conditions,[53] and 4-amino-5-benzamido-1,2,3,6-tetrahydro-2,6-dithiopyrimidine and similar derivatives give only the corresponding 8-phenyl-2,6-dithiopurine on cyclisation.[123] Phosphoryl chloride and, to a lesser extent, phosphoryl bromide have been employed to cyclise 5-benzamidopyrimidines. The general applicability of this medium is demonstrated in the fact that in one series nineteen out of twenty attempted ring closures were successful.[121] It was also useful in effecting conversion to the purine when other methods had failed as, for example, in the case of 2,6-diamino-8-phenylpurine which could not be obtained from 2,4,6-triamino-5-benzamidopyrimidine by prolonged heating with benzamide.[122] Yields are usually good but the presence of oxo groups in the pyrimidine leads to formation of the appropriate chloro- (or bromo-) purines.[121, 122] A number of 2-amino-6-halogeno-8-substituted phenyl purines[121, 122, 125, 126] have been formed in this manner. Halogenation has been established to take place before cyclisation and is borne out by the failure to chlorinate 2-amino-1,6-dihydro-6-oxo-8-phenylpurine which in this respect behaves like guanine itself, which is not chlorinated under these conditions.[127] Although chlorination of any oxo group present is not a serious drawback to the method, acid hydrolysis of the resulting chloropurine being relatively simple, other

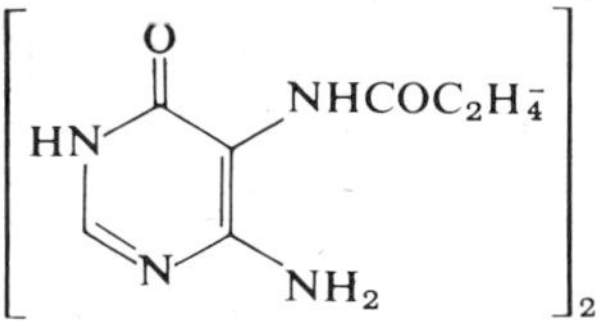

(44)

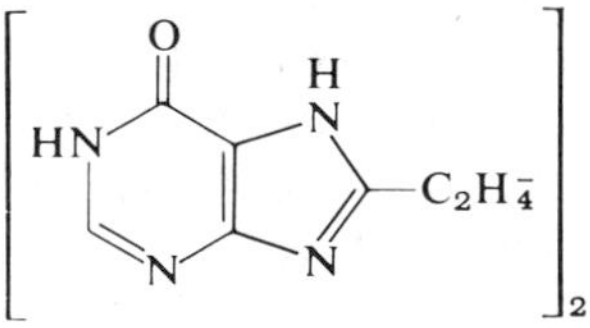

(45)

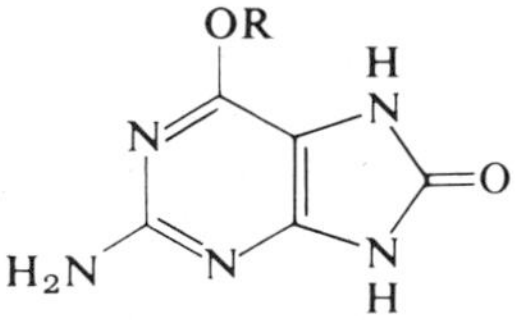

(46)

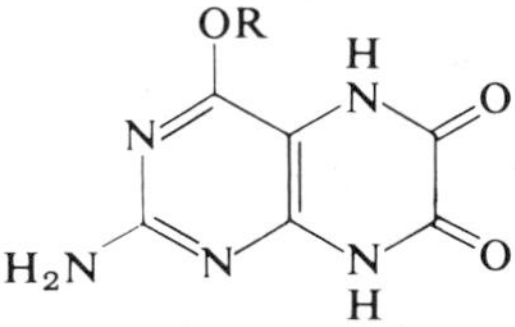

(47)

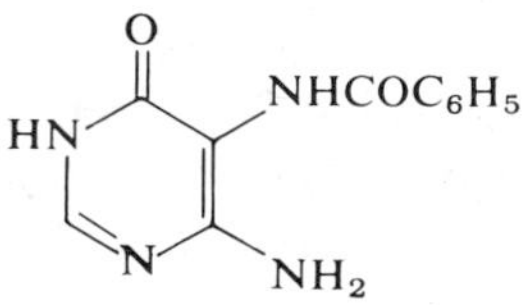

(48)

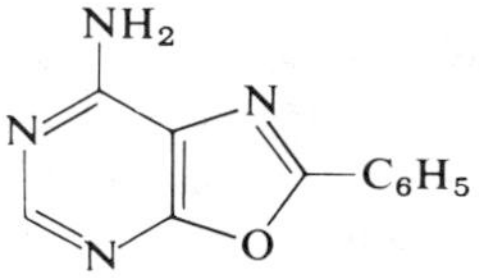

(49)

side effects may occur. These include chlorination of the amide group[121] and loss of *N*-methyl[128] groups but of more importance, if the reagent contains moisture, concurrent cyclisation of **48** may afford both 6-chloro-8-phenylpurine and the oxazolo[5,4-*d*]pyrimidine (**49**). The factors governing this cyclisation are known;[121] and, as in the case of the related thiazolo[5,4-*d*]pyrimidines, acid conditions are not found to be conducive to purine formation. Oxazole formation can be minimised by the use of freshly distilled phosphoryl chloride which gives almost exclusively 6-chloro-8-arylpurines.[122]

G. Cyclisations with Orthoesters and Diethoxymethyl Acetate

These reagents fall into the same category as those of the preceding sections in giving rise to 8-unsubstituted- or 8-alkylpurines. Trialkyl orthoesters by themselves have only limited use; direct cyclisation occurs in some cases, while in others only the intermediate product is formed and a further step is needed to close the ring. Instances of both happenings are found in the cases of hypoxanthine and 2,6-dichloropurine which arise from direct ring closures, while 4,5-diaminopyrimidine and 2- and 6-chlorodiaminopyrimidines only form a 5-ethoxymethyleneaminopyrimidine (**50**) under these conditions. Analogous products (**51**, R = Me or Et) arise with triethyl orthoacetate or triethyl orthopropionate and 4,5-diamino-6-chloropyrimidine. These, on heating under fusion conditions or in dimethylsulphoxide, give 8-alkyl-6-chloropurines (**52**, R = Me or Et).[129]

Mixtures of orthoesters and acetic anhydride provide more effective cyclising media, in which faster reaction times and milder conditions are possible. In almost every case a good yield of purine is isolated from the reaction mixture although a notable exception is 4,5-diaminopyrimidine which gives mainly 4,5-diacetamidopyrimidine (**53**) and only a trace of purine.[129] Acetic anhydride alone also produces **53**.[27] The outstanding advantage of this class of reagent over those discussed previously is that direct cyclisation of halogenopyrimidines to halogenopurines is possible with only a slight risk of concomitant hydrolysis of the halogen atom being present. Discovery of the mixed reagent was made simultaneously and independently by two schools,[130, 131] who demonstrated its applicability to the synthesis of a range of chloropurines. Subsequently it was shown[132] that purines containing a more reactive fluorine atom were also accessible although attempts to prepare 6-fluoropurine itself from 4,5-diamino-6-fluoropyrimidine have failed, only hypoxanthine being isolated.[132]

TABLE 2. Cyclisation of 4,5-diamino-6-dimethylamino-2-methylthiopyrimidine to 6-dimethylamino-2-methylthiopurine with orthoesters

Reagent Mixture	Yield (%)
Triethyl orthoformate/acetic anhydride	82
Triethyl orthoformate/propionic anhydride	76
Triethyl orthoformate/butyric anhydride	23
Trimethyl orthoformate/acetic anhydride	80
Tripropyl orthoformate/acetic anhydride	96

The anhydride concentration is not critical, varying in practice between 10% and 50%, and the reagent can also act as an acetylating agent at the purine stage. After cyclisation a mixture of the purine and 9- (or 7-) acetyl derivative may result, but production of the latter can be minimised by keeping the anhydride concentration low.[133] Such derivatives present no disadvantage, as deacetylation occurs readily in hot water or cold dilute alkali.[130, 131] Studies carried out with different anhydrides and orthoesters have not produced any combination significantly better then the original one[131] (see Table 2).

One important aspect in which the reagent differs from those discussed previously is its behaviour toward 6-alkyl(aryl)amino-4,5-diaminopyrimidines. Whereas previously ring closure, through involvement of the 5-amino group and the secondary amino group, has led to a 9-alkyl(aryl)purine, in this case a 6-substituted-aminopurine arises through reaction with both primary amino groups. This route has been

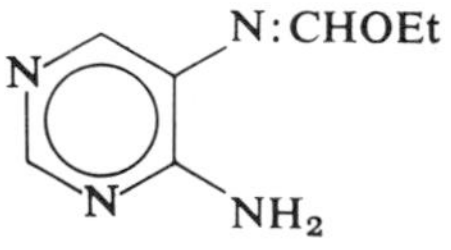

(50) 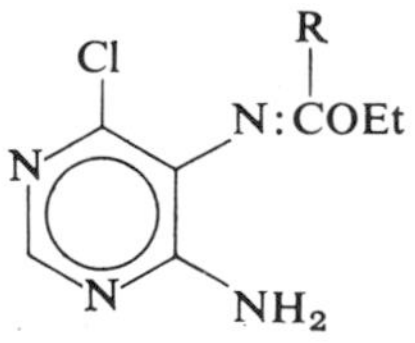   (51) 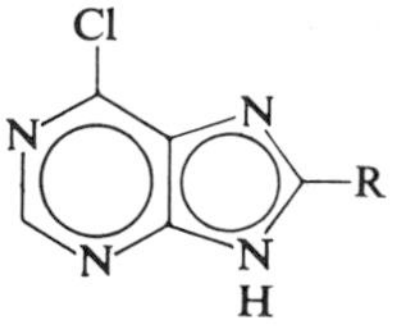   (52)

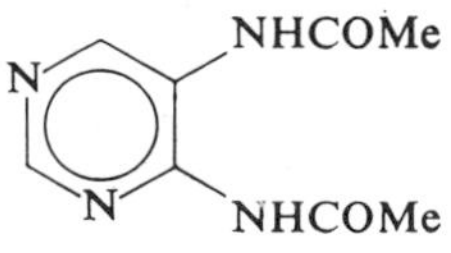

(53) 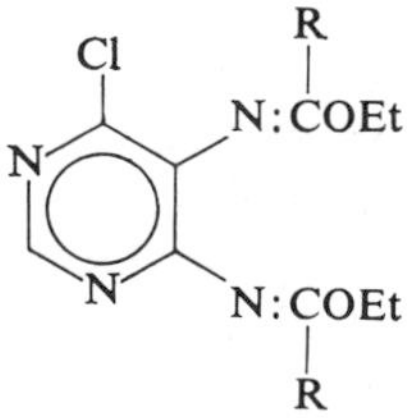   (54)

employed in the formation of 6-alkylaminopurines[134] including the plant cell division factor, kinetin, 6-furfurylaminopurine from 4,5-diamino-6-furfurylaminopyrimidine.[135]

Purines with methyl or ethyl substituents at $C_{(8)}$ have been derived with triethyl orthoacetate[104, 129, 136] and triethyl orthopropionate, respectively. Substitution of hydrochloric acid for acetic anhydride gives a medium in which cyclisation can be carried out at room temperature, a series of 9-alkyl-chloropurines being prepared[137] in better yields than when mixtures containing acetic anhydride were employed under reflux conditions. Slight elevation of the temperature, however, above that prevailing results in hydrolysis of halogen atoms occurring.[137] In the acid reagent 5-amino-4-chloro-6-hydrazinopyrimidine gives two products, the pyrimido[5,4-*e*]dihydrotriazine and 9-acetamido-6-chloropurine,[138] whereas only the latter is formed using the anhydride reagent and heating[139] (Eq. 7). Formic acid–orthoester combinations have been tried on 4,5-diaminopyrimidines. Both 2- and 6-chloro-4,5-diaminopyrimidine with acidified triethyl orthoformate give the appropriate chloropurine directly but only intermediates of the type **54** (R = Me or Et) are obtained with orthoacetate or orthopropionate,[129] requiring

(a) (EtO)$_3$CH/HCl   (b) (EtO)$_3$CH/(MeCO)$_2$O

submination techniques to convert them to 8-alkyl-6-chloropurines.

A refinement to the orthoester–anhydride cyclising technique was the introduction of diethoxymethyl acetate,[140] which is produced in 40% yield[141] when triethyl orthoformate and acetic anhydride stand together for some days at room temperature. Advantages of this reagent over the original orthoester–anhydride combination are that yields are generally improved[133, 142] and there is no tendency to form $N_{(7)}$- or $N_{(9)}$-acylated purines. In some cases cyclisations are possible which have not been successful in the anhydride mixture. Examples of this are the preparation of purine from 4,5-diaminopyrimidine,[129] and of 7-benzylguanine from the appropriate 4-amino-5-benzylaminopyrimidine.[143] In both cases other reagents give only uncyclised products. Comparative studies have been carried out with the two types of reagent and in one instance steric factors appear to emphasise the different modes of action.

With the *cis* isomer of 5-amino-6-chloro-4-(cyclopentan-2-olamino)-pyrimidine, cyclisation with the anhydride mixture was unsuccessful but the purine **55a** resulted when diethoxymethylacetate was used. The converse held for the *trans* isomer (**55b**), which gave a good yield of

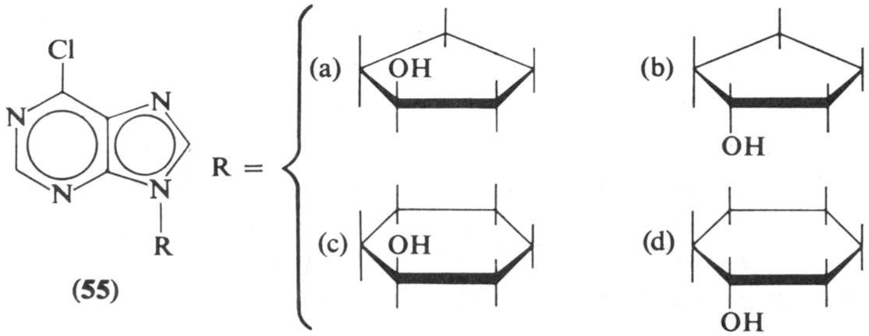

purine using the orthoester but a negligible one if diethoxymethyl acetate was employed.[144] On replacing the cyclopentanol residues with *cis* and *trans* cyclohexan-2-ol groups, both pyrimidines are readily cyclised to the corresponding purine, **55c** and **55d**, with diethoxymethyl acetate.[145] Application of any of the above procedures to the cyclisation of 5(4)-amino-4(5)-ribosylaminopyrimidines to the appropriate 9- or 7-ribosylpurine are unsuccessful, ring closure being concomitant with removal of the sugar moiety.[146] A purine with the β-D-ribofuranosyl group located at $N_{(1)}$ has, however, been prepared by this procedure.[147] Further extensions with this type of reagent to the preparation of 8-alkylpurines are not likely as diethoxyethyl acetate and its homologues cannot be prepared from acetic anhydride and the corresponding ethyl orthoesters.[141]

H. Cyclisation with Formamide

The historical background, in which the original application of formamide was purely as a vehicle to carry out thermal cyclisation of 4-amino-5-formamidopyrimidines, has been given (Sect. 1B).

The preparation of hypoxanthine[148] and isoguanine[17] on heating the appropriate 4,5-diaminopyrimidine with a mixture of formic acid and formamide appear to be the first examples of the use of formamide as a formylating agent. Initially, the reactions were carried out in sealed tubes but subsequently open-vessel techniques were developed. Although the presence of formic acid is not required, the reaction gives the best results if acidic conditions are initially present. In practice these are supplied by using either the sulphate or hydrochloride salt of the diaminopyrimidine[9] or, if the pyrimidine free base is employed, by the addition of hydrochloric acid or other mineral acid to the formamide. Commercially, these requirements have been met in the synthesis of xanthine by using N-(4-amino-1,6-dihydro-2,6-dioxopyrimidin-5-yl)-sulphamic acid (56).[149] Cyclisations attempted in an aprotic (neutral) environment have either failed or given extremely poor yields.[9] The superiority of this reagent over formic acid in forming purines unsubstituted at $C_{(8)}$ shows in a number of ways. Ring closure occurs in situ, without isolation of an intermediate formamidopyrimidine, and the product, in the majority of cases, crystallises from the reaction mixture on cooling or after addition of water. Yields tend to be higher—adenine, for example, being obtained in 95% yield whereas cyclisation with formic acid reduces this to 62%.[43, 150] A further advantage is that cases are known where the amide was successful when formic acid had failed to effect ring closure.[17] Not unexpectedly, halogenopurines are not formed in this way; attempts to do so invariably give the corresponding oxopurine. Less reactive groups may also suffer through prolonged contact with the reagent; replacement of a 2-methylthio group by amino group has been found,[151] due to ammonia liberated on breakdown of the formamide on heating.

Where the 4,5-diaminopyrimidine has a secondary amino group at $C_{(6)}$, the cyclisation pattern follows that of formic acid—the major product being a 9-substituted adenine with the isomeric 6-substituted-aminopurine present to a minor extent. Separation of the isomers is effected by digestion in dilute sodium hydroxide which removes the 6-substituted-amino derivative as the soluble sodium salt.[152] As a generalisation the isomeric mixture comprises 75%, often more, of the 9-substituted adenine. When kinetin, 6-furfurylaminopurine, was made from 4,5-diamino-6-furfurylaminopyrimidine the larger component of the resulting mixture was 9-furfuryladenine.[135] An interesting illustration of this competitive ring closure is the cyclisation of 1,2-di(4,5-diaminopyrimidin-6-ylamino)ethane (57)[153] which gave three isomeric products. As expected, the diadeninylethane (58) formed the largest component (80%) but, surprisingly, the second largest (18%) was the asymmetric isomer (59), while only a trace (1–2%) of the 1,2-di(purin-6-ylamino)ethane (60) was found. In terms of pyrimidine moieties present this cyclisation gives 90% of the 9-substituted adenine.

As cyclisations take place at fairly elevated temperatures (160–180°), side effects are commonly encountered. In one example the 6-(2-hydroxyethylamine)pyrimidine (61), undergoing ring closure in the presence of a trace of hydrochloric acid, gave not the expected 9-(2-hydroxyethyl)purine but the isomeric imidazo(1,2-c)pyrimidine (62)

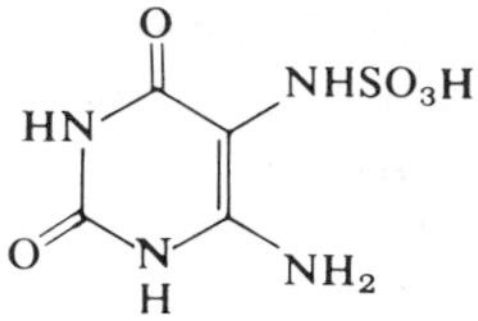

(56)

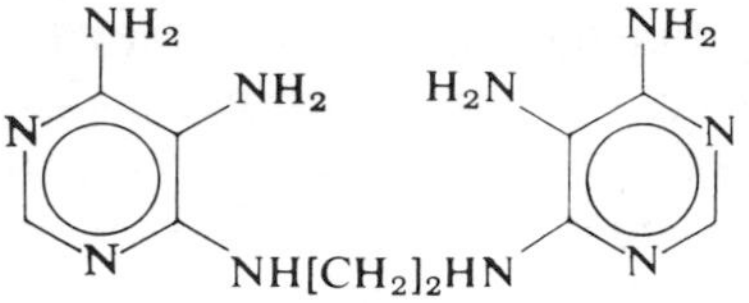

(57)

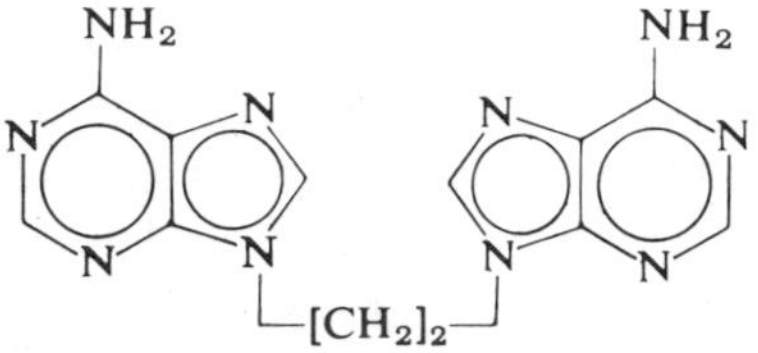

(58)

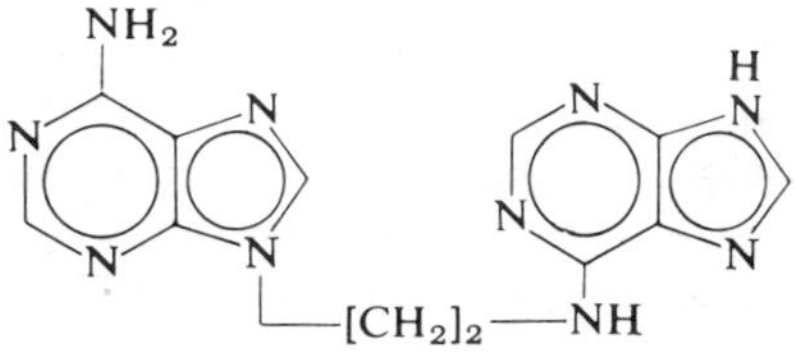

(59)

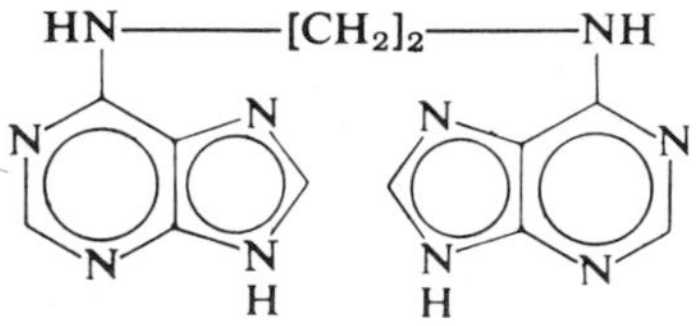

(60)

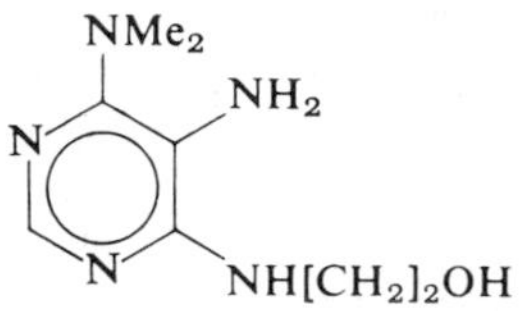

(61)

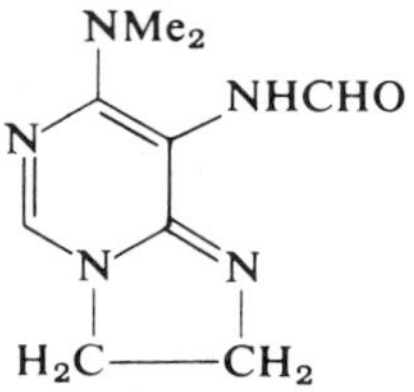

(62)

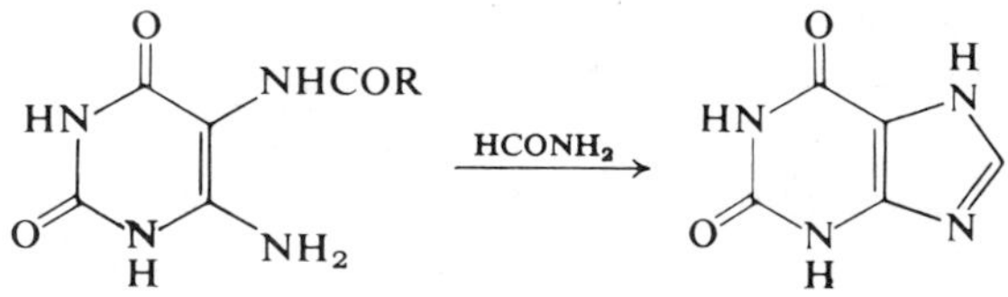

(63)

$\xrightarrow{\text{HCONH}_2}$

(64)

(Ring index No. 1181). An explanation for this reaction,[154] based on results of synthetic studies,[155] is that chlorination of the hydroxyl group by the acid is followed by internal quaternisation at $N_{(3)}$, the resulting salt undergoing subsequent conversion to the base due to the prevailing high pH of the medium. With 5-amido-4-aminopyrimidines (for example, 63) having either alkyl or aryl amide groups, the cyclising conditions are forcing enough to remove the acyl or aryl group and produce a $C_{(8)}$-unsubstituted purine. The preparation of xanthine (64) from either 63 (R = Me) or 63 (R = Et) illustrates this effect.[47, 105, 156] Various $N_{(1)}$- and $N_{(3)}$-alkylated xanthines were obtained in this way.[156, 157] Arylamidopyrimidines to some degree behave similarly; an attempted formamide ring closure of 2,4,6-triamino-5-(4-chlorobenz-amido)pyrimidine produced an appreciable quantity of 2,6-diamino-purine in addition to the required 8-(4-chlorophenyl) analogue.[122]

The likelihood of thiazolo[5,4-*d*]pyrimidine formation occurring when 4,5-diamino-1,6-dihydro-6-thiopyrimidines are treated with cyclising reagents is great in most cases. With formamide, however, the appropriate 6-thiopurine is obtained.[158] In any event any of the thiazolo derivative formed should be isomerised to the purine under the reaction conditions employed[13] (see Sect. 8A).

In previous sections (1B, 1E) the transformation of the $C_{(8)}$-group or atom in an existing purine by means of the cyclising agent has been noted. Formamide can react likewise; conversion of uric acid derivatives to the corresponding xanthines has been demonstrated with the parent member,[159, 160] also 1-methyl,[159, 161] 3-methyl-,[159] and 1,3-dimethyluric acid.[162] The 7- and 9-methyl analogues do not appear to react under these conditions. A related reaction is the preparation of xanthine by formamide treatment of the 8-thio analogue.[163]

I. Cyclisation with Other Amides

Under reflux conditions, homologues of formamide convert 4,5-diaminopyrimidines directly to the corresponding 8-alkyl (or -aryl) purines. Acetamide gives the 8-methyl derivative of guanine[51] and xanthine[47] while trifluoroacetamide affords 8-trifluoromethyladenine.[82] Propionamide[47, 164] and benzamide[164] find use in forming 8-ethyl- and 8-phenylxanthines. Condensation of the diamides of malonic and succinic acids with the diaminopyrimidine in glycol has given the bis theophylline derivatives (**65**, $n = 1$ and 2).[165]

The product from acetamide treatment of 4,5-diamino-1,6-dihydro-6-thiopyrimidine is 1,6-dihydro-8-methyl-6-thiopurine, none of the isomeric thiazolopyrimidine being formed[29] (*cf.* preceding section).

Ring closure of various 4-amino-5-acylaminopyrimidines in acetamide produces only 8-methylpurines; initial loss of the $C_{(5)}$-acyl group is followed by acetylation and cyclisation.[47, 136] This phenomenon is general when acyl groups of solvent amide and pyrimidine are not the same, the 8-alkyl group of the resulting purine being derived from the

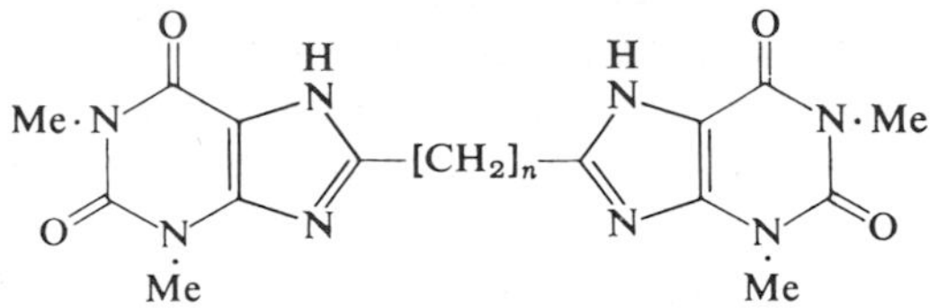

(**65**)

amide used as cyclising agent. Aryl- as well as alkylamides may also suffer this displacement; the required 8-phenyl-2,6-diaminopurine resulted on heating 2,4,6-triamino-5-benzamidopyrimidines with the corresponding benzamide derivative, but a mixture of 2,6-diamino-purine and 2,6-diamino-8-phenylpurines resulted when formamide was used.[122] This interchange reaction also applies to certain purines. With uric acid, for example, prolonged heating in acetamide gives a reasonable yield of 8-methylxanthine (Eq. 8).[159, 162]

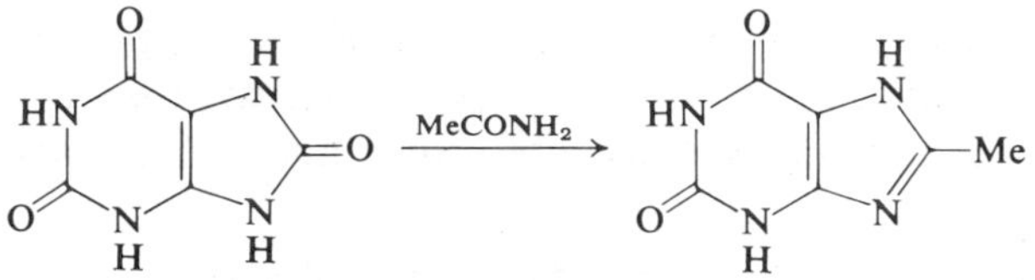

(8)

J. Cyclisation with NN-Dialkylamides and Phosphoryl Chloride

The formylating properties of dimethylformamide–phosphoryl chloride mixtures form the basis of the long-established Vilsmeier–Haack procedure.[166] A recent adaptation[167, 168] enables 4,5-diaminopyrimi-dines to be formylated and cyclised in situ, and has the advantage of employing low operating temperatures. Halogenopurines can be derived from the appropriate halogenopyrimidines; representative products include various 2-chloropurines[167, 168] and also the bis 6-chloropurine (66).[168] Formation of the latter by this route is of interest as when a triethyl orthoformate-acetic anhydride mixture was used only the bis 5-acetamido-4-aminopyrimidine resulted. Some difficulties are reported with the formation of unsubstituted halogeno-purines. The product from the attempted cyclisation of 4,5-diamino-6-

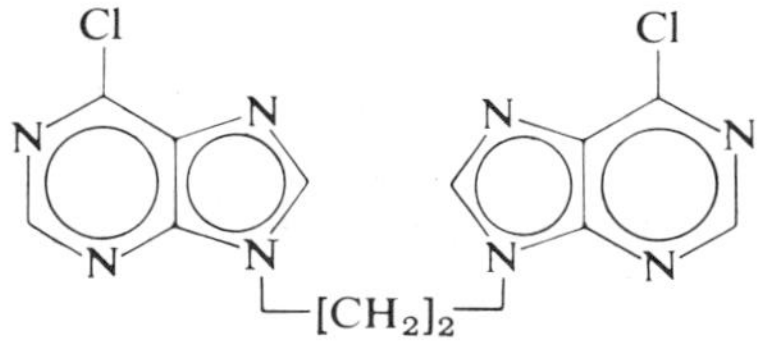

(66)

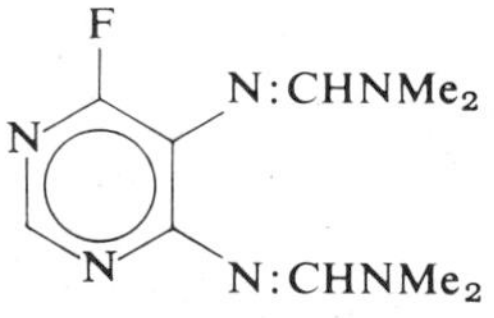

(67)

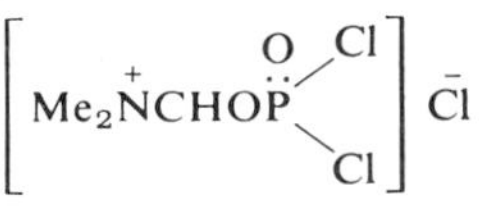

(68)

fluoropyrimidine was the disubstituted azamethine derivative (67), whereas the analogous 6-chloropyrimidine gave a mixture of 6-chloro-purine and the 6-chloro analogue of 67.[132] The reaction has given good results for aminopurines[168] and mono- and dioxopurines.[168] No thio-purines have been reported by this route but the reaction conditions appear more likely to favour thiazolopyrimidine formation.* When dimethyl acetamide–phosphoryl chloride is used, a 6-chloro-8-methyl purine is obtained[107] and the method appears capable of extension to the formation of other 8-substituted purines.

A mechanism for this reaction, which involves an intermediate aza-methine derivative such as 67, is presumed to operate through the agency of an amide-phosphoryl chloride adduct of the type 68.

K. Cyclisation with Amidines

As formamidine itself is unstable this reagent is not a means to forming $C_{(8)}$-unsubstituted purines. They can, nevertheless, be obtained by using the fact that sym-triazines in the presence of amines are degraded to NN'-disubstituted formamidines in situ. In this way purine itself and theophylline are obtained using the appropriate 4,5-diaminopyrimidine as the amine.[170] Owing to the inaccessibility of C-alkylformamidines generally cyclisations have been limited to the preparation of 8-methyl- and 8-phenylpurines. For the former deriva-tives acetamidine, usually as the hydrochloride, admixed with sodium acetate, is fused (160–190°) with the diaminopyrimidine for some minutes, the purine being isolated after suitable treatment of the aqueous solution obtained by dissolving the cooled melt in water.

Among various 8-methyl derivatives prepared by the procedure are amino-,[92, 171] oxo-,[171] $N_{(7)}$-alkyl,[172] and thiopurines.[171] A point of interest with the thiopurines is that, although both the 2-oxo-6-thio and 2,6-dithio derivatives of 4,5-diaminopyrimidine are converted to the appropriate 8-methylthiopurine in good yields, no products resulted with 4,5-diamino-1,6-dihydro-6-thiopyrimidine under the same con-ditions.[171] A mechanism proposed for these condensations involves a series of acid-catalysed transamidation reactions.[171]

With benzamidine similar procedures give rise to 8-phenylpurine,[173] also oxo,[171, 173, 174] aryl, and N-alkyl[172, 173] derivatives. Nicotinamidine gives 8-(2-pyridyl)-purines.[173, 175] Also known are the 8-(3-pyridyl) and 8-(4-pyridyl)[174] analogues.

---

* Evidence to support this is the facile preparation of benzothiazole from 2-mercapto-aniline with this mixed reagent.[169]

L. Cyclisation with Guanidines

Although this route appears to provide a valuable means of forming 8-aminopurines directly, this is not found to be the case in practice. Condensations are erratic and have been successful with only a few oxo- and thiopyrimidines. Initially guanidine salts were employed, the thiocyanate salt has successfully converted the diaminopyrimidine (69) to the 8-aminopurine (70) but has failed to give the $C_{(2)}$-thio analogue

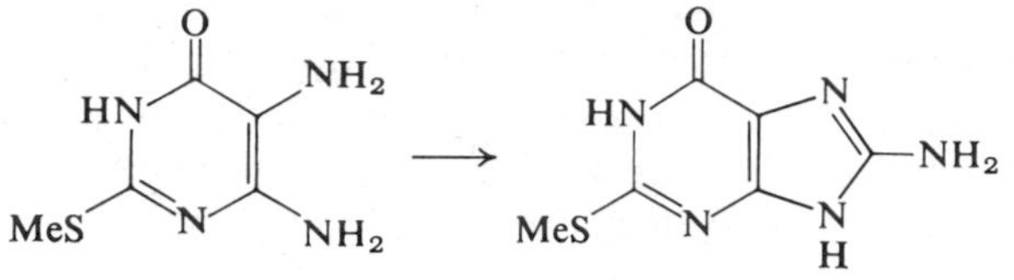

(69)   (70)

of 70 under the same conditions.[176] The results of condensations with guanidine itself are likewise unpredictable. Thus while 2-oxo-[177] and 6-oxo-dihydro-4,5-diaminopyrimidine[178] afford the appropriate 8-amino-oxodihydropurine with the 2-thio-[177] and 6-thiopyrimidine[177, 178] analogues, only the latter is converted to an 8-aminopurine. The reaction with 2,4,5-triaminopyrimidine also failed.[177]

M. Cyclisation with Urea

Condensations of this type afford 8-oxopurines directly, these being versatile derivatives in which the oxo group is readily replaced by chlorine which can then suffer displacement by various nucleophilic agents. The reaction is normally carried out under fusion conditions (160–180°), an excess of urea being present, the product remaining behind on trituration of the melt with water. Examples of the condensation being effected in a solvent are known. The mechanism has been the subject of a number of investigations the results of which show that the ammonia evolved is derived exclusively from the urea, the nitrogen atoms of the pyrimidine amino groups remaining attached to the ring. Two illustrations are the fact that uric acid, devoid of activity, results from interaction of 4,5-diaminouracil and urea which had been labelled with ^{15}N in both nitrogen atoms,[179, 180] and by the formation of 7,8-dihydro-8-oxopurine using 4,5-diaminopyrimidine and NN'-dimethylurea (Eq. 9).[181]

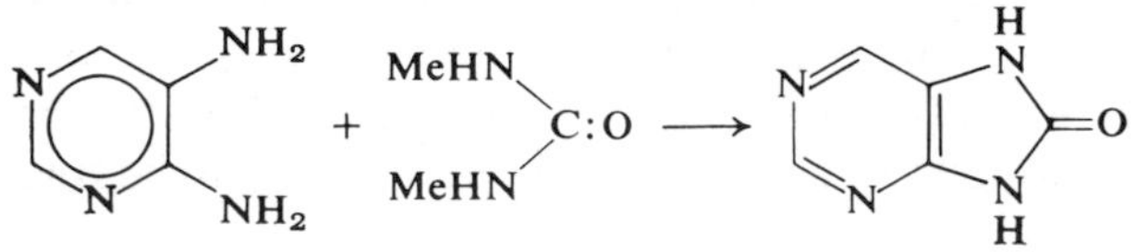

(9)

In view of the severity of the cyclising conditions the preparation of chloropurines by this procedure seems surprising. Two examples, however, illustrating this are 2-chloro-7,8-dihydro-8-oxopurine[182] and the 6-chloropurine (**71**)[183] from the appropriate chloro-4,5-diaminopyrimidine. In both cases the lack of reactivity shown by the halogen atom is attributed to fixation of the $C_{(4)}$-$C_{(5)}$ double bond and attendant mesomerism, and is comparable with that found in 2- or 4-chloro-4,5-diaminopyrimidines.

Thiopyrimidines give the appropriate 8-oxothiopurine; no evidence for the formation of isomeric thiazolopyrimidines has been presented. Some involvement of the ester group was found with a 4,5-diamino-6-ethoxycarbonylpyrimidine; under fusion conditions interaction with the ammonia liberated gave the 6-carbamoylpurine whereas with urea in

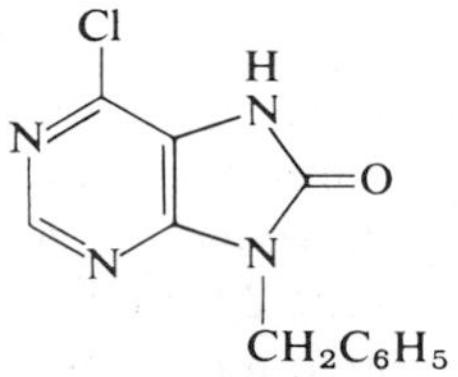

(**71**)

pyridine, under reflux, hydrolysis to the 6-carboxypurine ensued.[18]

N. Cyclisation with Thiourea

Thiourea cyclisation procedures parallel those used with urea but higher reaction temperatures (180–220)° are required. Although most condensations give the required 8-thiopurines directly, the pronounced electronegative character of the sulphur atom predisposes toward the likelihood of side reactions occurring. The earliest reported example of this is from a fusion with 4,5-diamino-1,6-dihydro-2-methylthio-6-oxopyrimidine (**72**) which gave the 8-aminopurine (**73**) and not the expected 1,6,7,8-tetrahydro-2-methylthio-6-oxo-8-thiopurine (**74**). This reaction is explicable if the effective cyclising reagent is guanidine thiocyanate, formed by thermal rearrangements of the thiourea. The preparation of **73** using guanidine thiocyanate (Sect. 1L) gives considerable support to this mechanism.[176] The course of the reaction may in part be influenced by the presence of the 2-methylthio group as the 2-thiopyrimidine analogue condenses normally with thiourea giving 1,2,3,6,7,8-hexahydro-6-oxo-2,8-dithiopurine.[184] An unusual loss of a methyl group follows reaction of the 5-amino-4-methylaminopyrimidine

(75), the product being the corresponding 8-thiopurine (76) unsubstituted at $C_{(9)}$.[165] Use of NN'-dimethylthiourea at moderate temperatures affords the same 8-thiopurines as are obtained with thiourea but with more elevated temperatures S-methylation through migration of methyl groups is possible. Both diaminopyrimidines 77 (R = O) and 77 (R = S) with this derivative at 190° gave the appropriate 8-thiopurines (78, R = O or S) but, on raising the melt temperature to 260°, the corresponding 8-methylthiopurines (79, R = O or S)[165] were obtained.

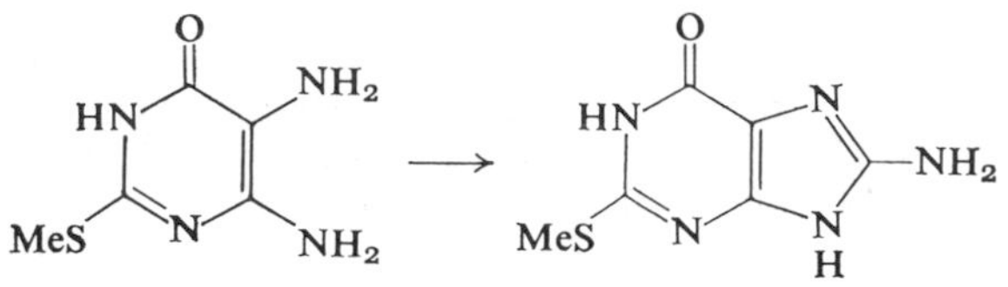

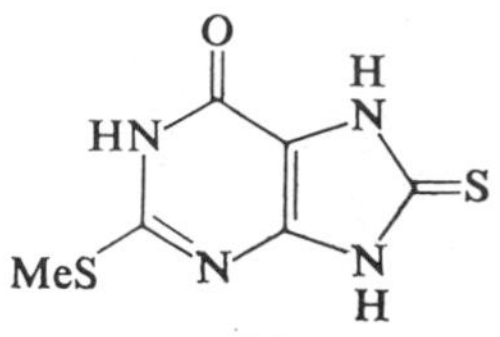

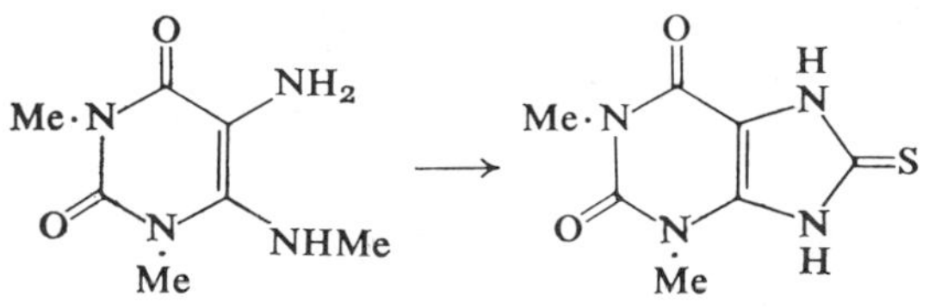

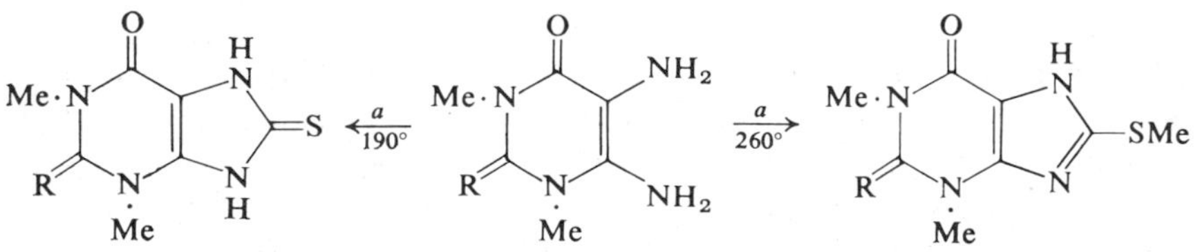

O. Cyclisation with Cyanates, Isocyanates and Derivatives

A synthesis of uric acid by Traube is an early example of this procedure, 4,5-diaminouracil being converted by potassium cyanate to the 5-ureido derivative which was initially cyclised by fusion,[12] but a later modification utilised heating with hydrochloric acid in sealed tubes for this purpose.[185] For preparing this and other 9-unsubstituted-8-oxopurines the urea fusion method (Sect. M) is undoubtedly superior with the advantage of the reaction taking place in one step. One point in

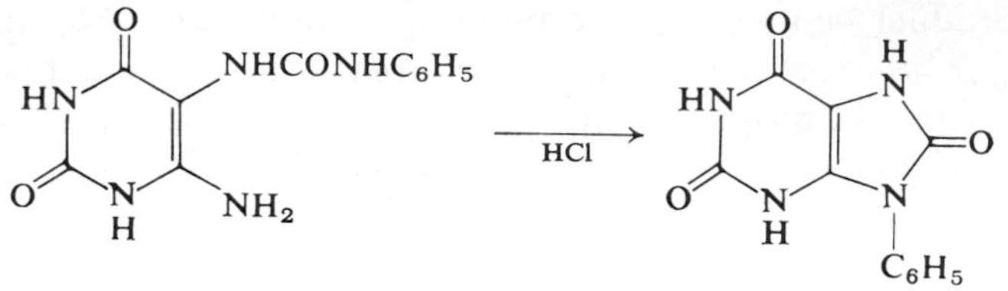

(80)   (81)

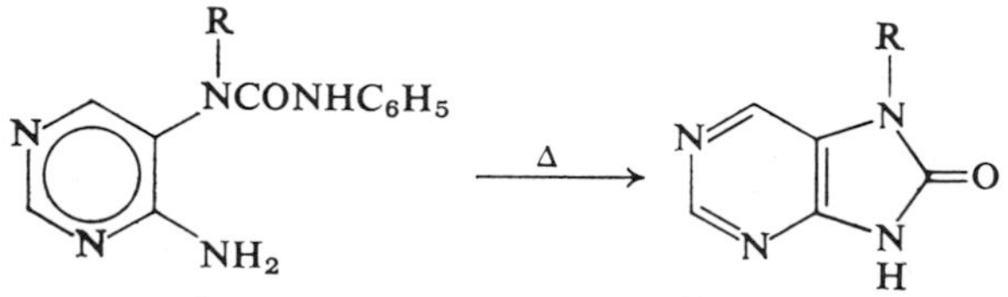

(82)   (83)

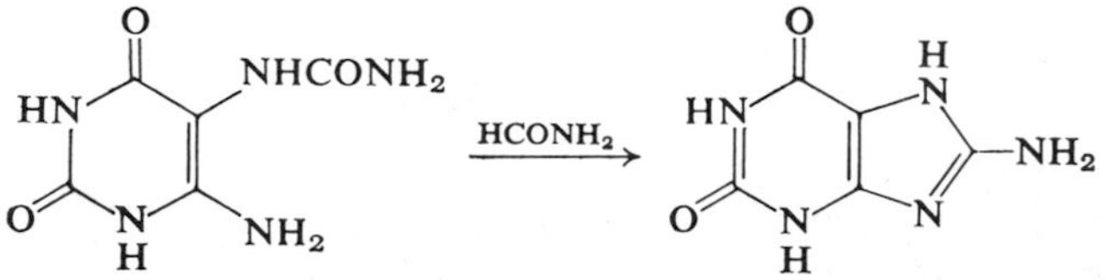

(84)   (85)

favour of the cyanate procedure worth noting is that by using organic isocyanates 9-alkyl- or arylpurines can be obtained from diamino-pyrimidines having only a primary amino group at $C_{(4)}$. In this way 9-phenyluric acid (81) is produced on cyclisation of 80, formed from the diaminopyrimidine and phenyl isocyanate, in hydrochloric acid.[12] Other modes of cyclisation of 5-ureidopyrimidines are observed using different reagents or conditions to the above. In the case of the phenyl-ureido derivatives (82, R = Me, Et, and C_6H_5) on heating to above their fusion points, appropriate 7-alkyl- or 7-aryl-8,9-dihydro-8-oxopurines (83) result, with evolution of aniline.[173] The action of formamide on the other hand converts the pyrimidine (84), through cyclodehydration, to the 8-aminopurine (85)[163] but the reaction is complicated as xanthine is also formed. The preparation of an 8-anilinopurine from a 5-phenyl-ureidopyrimidine is reported following treatment with phosphorus trichloride in toluene,[186] this example being more fully detailed in Section W.

P. Cyclisation with Isothiocyanates

Procedures generally follow those outlined above for isocyanates, but comparatively more use has been made of this class of reagent for preparing the appropriate 9-substituted purines, especially in the case of theophylline derivatives. The thioureidopyrimidine (**86**) from 4,5-diamino-1,2,3,6-tetrahydro-1,3-dimethyl-2,6-dioxopyrimidine and methyl isothiocyanate affords the 9-methylpurine (**87**) after heating for some hours in hydrochloric acid.[187] Homologous 9-alkyl derivatives have been prepared likewise.[187–189] Direct formation of 9-phenylpurines from 4,5-diaminopyrimidines has been demonstrated by prolonged heating with phenyl isothiocyanate (7 h/160°) in dimethylformamide

(**86**) (**87**)

(**88**) (**89**)

(**90**) (**91**)

(**92**) (**93**)

(**94**)

(**95**)

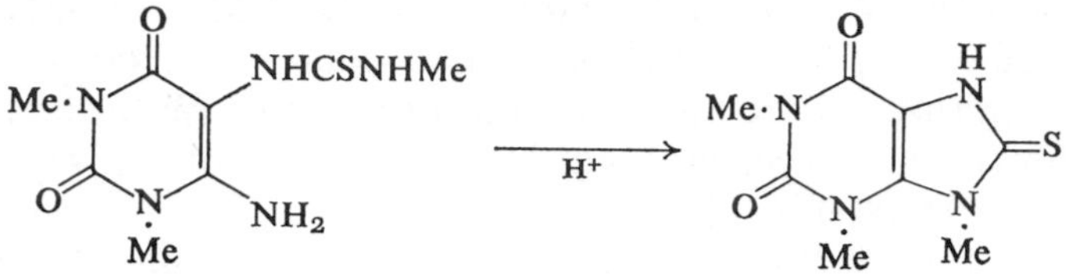

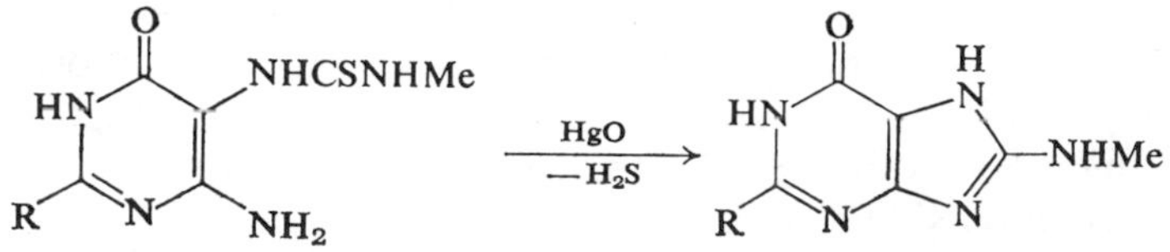

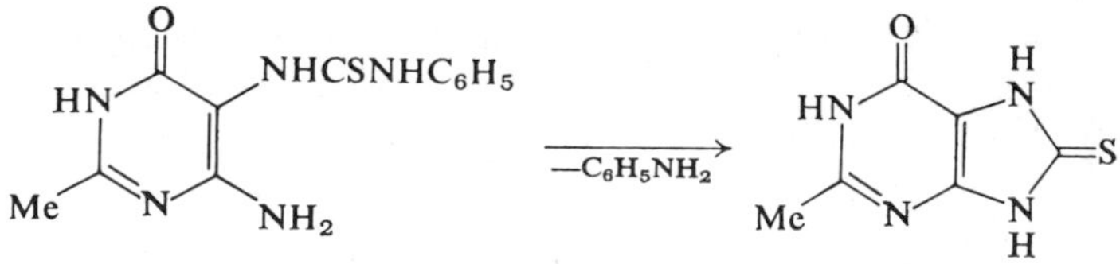

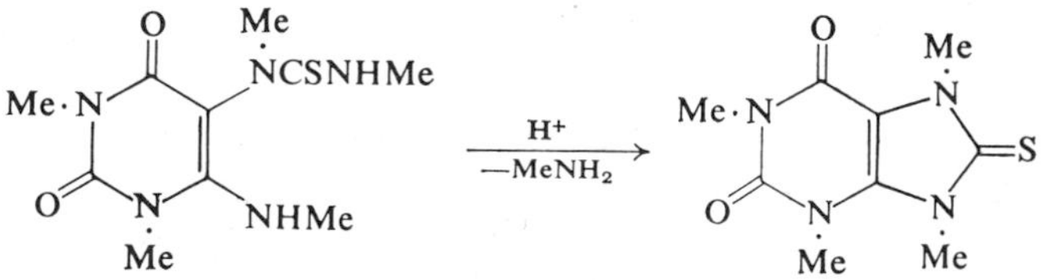

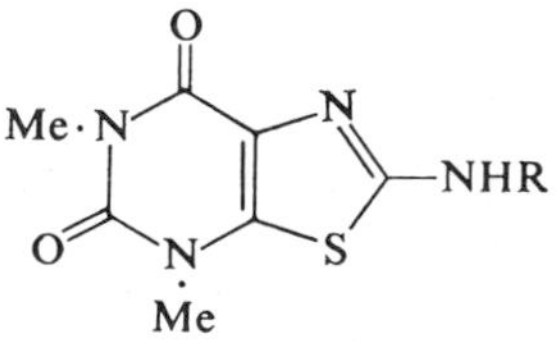

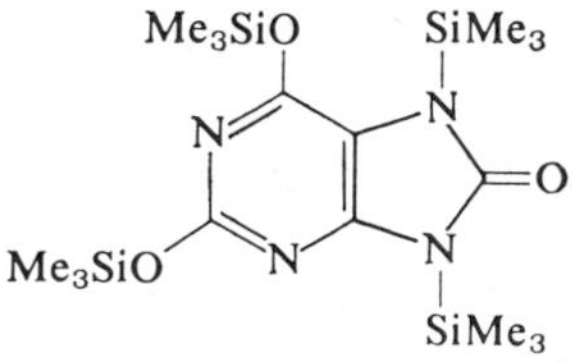

but if dioxan is used as solvent only the phenylthioureidopyrimidine is isolated.[173] In the preparation of phenylthioureides the use of phenyl-dithiocarbamic acid ($C_6H_5NHCS_2H$), as the ammonium salt, in place of phenyl isothiocyanate has been demonstrated.[186] As is found in the isocyanate cyclisations other pathways are possible. In the presence of sulphide-removing agents a reaction analogous to cyclodehydration can be effected as, for example, when methylthioureidopyrimidines of the type **88** are treated with alkaline mercuric oxide solution, the products being the 8-methylaminopurines (**89**, R = H, SMe, or OH).[93, 190] Removal of the terminal alkyl or aryl substituent of the thioureido group during ring closure also features in this series. With the phenyl-thioureidopyrimidine (**90**) prolonged heating in pyridine gives 1,6,7,8-tetrahydro-2-methyl-6-oxo-8-thiopurine (**91**).[186] Two examples of elimination of ethylamine from ethylthioureidopyrimidines are report-ed, in one case under fusion conditions (250°)[187] but in the other[191] conditions are unspecified. Ring closure of the methylthioureido derivative (**92**) in mineral acid to the purine (**93**)[189] is most likely a further example of this type as alkylamino groups attached to the pyrimidine nucleus are usually found to be less readily removed than those located on side chains.

A further cyclisation made from which thiazolo[5,4-*d*]pyrimidines can result is seen in the fact that a number of the above 9-substituted-8-thiotheophyllines were found to be contaminated with these isomeric derivatives (**94**). In the case of a closure involving cyclohexyl isothio-cyanate the thiazolo derivative (**94**, R = cyclohexyl) was the only product.[187]

Q. Cyclisation with Carbon Dioxide

The specialised conditions required make this reagent of only academic interest in purine synthesis. Reaction with 4,5-diamino-pyrimidines requires prior conversion of them to the trimethylsilylated "active form" and is carried out under pressure at elevated tempera-tures. Uric acid was obtained in good yield from 4,5-bis-trimethylsilyl-amino-2,6-bis-trimethylsilyloxypyrimidine under these conditions,[192] hydrolysis of the resulting trimethylsilylated purine (**95**) to uric acid being effected with aqueous ethanol.[192]

R. Cyclisation with Carbon Disulphide

In contrast to the preceding reagent the thio analogue has extensive application to the formation of 8-thiopurines. The first use, in 1903, was to convert 4,5-diaminouracil to 1,2,3,6,7,8-hexhydro-2,6-dioxo-8-thiopurine, in the absence of solvent, under pressure conditions.[193] By employing a basic solvent such as pyridine[194] the reaction will take place under reflux conditions. Derivatives having a low solubility, for example, oxopyrimidines, can be taken into solution in pyridine containing either solid potassium hydroxide[195] or an aqueous solution of the alkali.[56, 196, 197] More soluble pyrimidines have been reacted in a 50% ethanol–pyridine mixture.[198] Replacement of the pyridine by an aqueous sodium hydroxide–ethanol combination has been utilised to prepare 8-thiotheophylline derivatives[165] while triethylamine has found use as solvent in large-scale preparations.[199, 200] Improvements in yield and greater solubility of the reactants are obtained from carrying out the cyclisations in dimethylformamide.[201]

Compared with thiourea fusions, these procedures have the distinct advantage of needing less rigorous conditions and although heating for periods, varying from one-half to some hours, is required in most cases, some examples are found where cyclisation proceeds at room temperature.[199, 200] This fact is well illustrated by ring closure of a 5-amino-4-glycosylaminopyrimidine to the 9-glycosyl-8-thiopurine after prior acetylation of the sugar moiety.[202]

The preparation of 2-chloro-7,8-dihydro-8-thiopurine from 4,5-diamino-2-chloropyrimidine in refluxing pyridine[177] cannot be taken as an indication of the suitability of this route for obtaining other halogeno-8-thiopurines. In this instance the well known unreactive nature of a $C_{(2)}$-halogen on a purine with a strong electron-releasing group at $C_{(8)}$ is shown (as, for example, in the similar case of the 2-chloro-8-oxo analogue, see Sect. 1M). With halogen atoms at the $C_{(6)}$-position thiazolopyrimidine formation results through an initial replacement of halogen atom by sulphur, thus 7-amino-2-mercapto-thiazolo[5,4-d]pyrimidine (97) was obtained when 4,5-diamino-6-chloropyrimidine (96, R = Cl) was cyclised in dimethylformamide.[201] Other examples from 6-chloropyrimidines are known.[201] Obviously both thio and alkylthio groups at $C_{(6)}$ are suspect in this reaction as seen in the attempted synthesis[56] of 1,6,7,8-tetrahydro-6,8-dithiopurine (98) from 4,5-diamino-1,6-dihydro-6-thiopyrimidine which gave instead the thiazolopyrimidine (97). The latter was also obtained when the 6-benzylthio analogue (96, R = SCH$_2$C$_6$H$_5$) was used.[201] The presumption that ring closures of this kind occur through the agency of a pyrimidin-5-yldithiocarbamate ion of the type 99[201] is given indirect support from the fact that 4-amino-5-methylaminopyrimidines[191, 203] fail to cyclise to the respective 8-thiopurines under these conditions. Steric hindrance by the methyl group in preventing dithiocarbamoyl group formation is suggested to be a controlling factor. Isomerisation of the thiazolopyrimidine (97) to 1,6,7,8-tetrahydro-6,8-dithiopurine (98) is effectively carried out in hot dilute sodium hydroxide.[56] As with the majority of other reagents, if the 6-position of a 4,5-diamino-pyrimidine carries a secondary amino group then this group is the one involved in cyclisation rather than the $C_{(4)}$-primary-amino group, giving 9-alkyl(aryl)-8-thiopurines.[191, 194] Examples of the Dimroth rearrangement are found under these conditions. Using equal parts of pyridine and ethanol as solvent, the N-methylpyrimidine (100) was

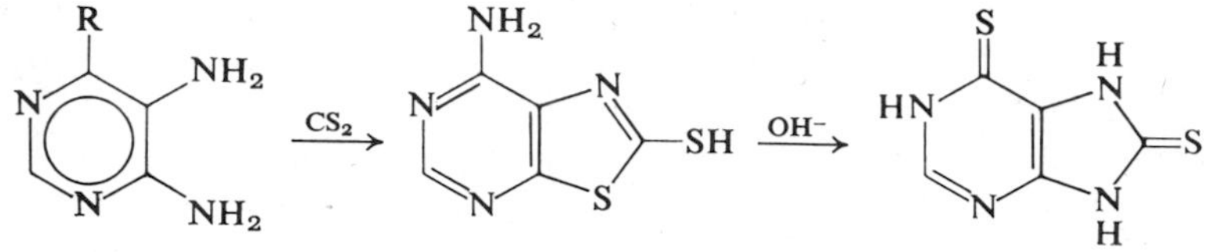

(96)   (97)   (98)

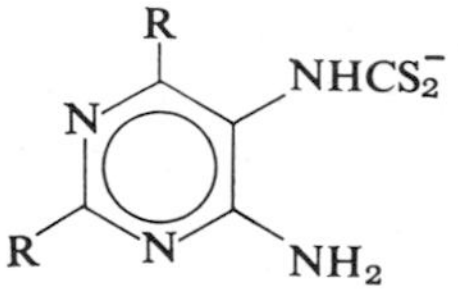

(99)

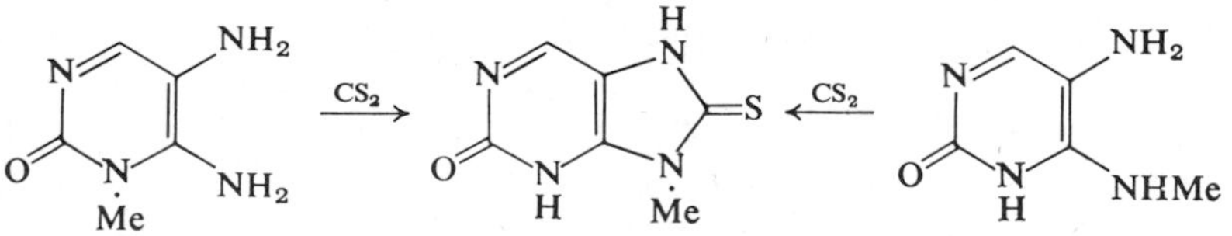

(100)   (101)   (102)

converted to 2,3,7,8-tetrahydro-9-methyl-2-oxo-8-thiopurine (101), the structure of which followed from its synthesis, using the same reagent from 5-amino-2,3-dihydro-4-methylamino-2-oxopyrimidine (102).[198] The N-butyl analogue of 100 rearranges in the same manner. In such cases alkyl group rearrangement occurs before cyclisation and involves pyrimidine ring fission, bond rotation, and recyclisation.[204] A number of the reagents dealt with already are capable of transforming the $C_{(8)}$-group of existing purines. In this respect carbon disulphide is no exception and with it both theophylline and 3-methyluric acid in aqueous potassium hydroxide at 150° have been converted to their 8-thio analogues.[205]

S. Cyclisation with Phosgene

This route permits cyclisation to 8-oxopurines to be made in aqueous acid or, more usually, alkaline media, the reaction being completed in one step, at room temperature, on passage of phosgene through the solution of the 4,5-diaminopyrimidine. In dilute (5–10%) sodium hydroxide solutions, 2-amino-7,8-dihydro-8-oxo-,6-amino-2,3,7,8-tetra-hydro-2,8-dioxo-,[206] 1,6,7,8-tetrahydro-6,8-dioxo-,[206] and 6-amino-7,8-dihydro-8-oxopurines[206] have been prepared. Some halogen-containing derivatives are accessible as, for example, 6-chloro-7,8-dihydro-8-oxopurine, under the same conditions.[207, 207a]

Uric acid derivatives by this route include the 9-(2-hydroxyethyl)-derivative[208] and the isomeric 3-(2-hydroxyethyl)analogue, preparation of the latter being carried out in aqueous bicarbonate.[209] Uric acid itself is not obtained from 4,5-diaminouracil, no reaction occurring with phosgene,[192] but a condensation in toluene follows if the trimethyl-silylated pyrimidine (103) is employed in the presence of triethyl-amine.[192] A noteworthy point arising from this is the considerable solubility of the trimethylsilylated pyrimidine in hydrocarbon solvents.

The derivative formed in dilute hydrochloric acid from 5-amino-4-chloro-6-hydrazinopyrimidine, originally formulated as the 6-chloro-purine (104, R = H)[207] is now known to be 8-amino-7-chloro-3-oxo-s-triazolo[4,3-c]pyrimidine (104a).[207a] An unambiguous synthesis of 104 (R = H) has been achieved by use of suitable blocking groups on the terminal nitrogen atom of the hydrazino moiety, for example, carbobenzoxy or benzhydryl, these being removed by acid treatment after ring closure to the purine 104 (R = $OCOCH_2C_6H_5$ or $CH(C_6H_5)_2$)[207a] The disadvantages associated with using the gaseous reagent have been in some cases overcome by using the solid phosgene-pyridine

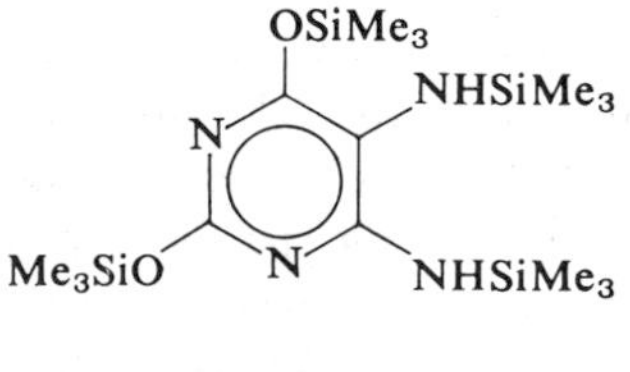

(103) 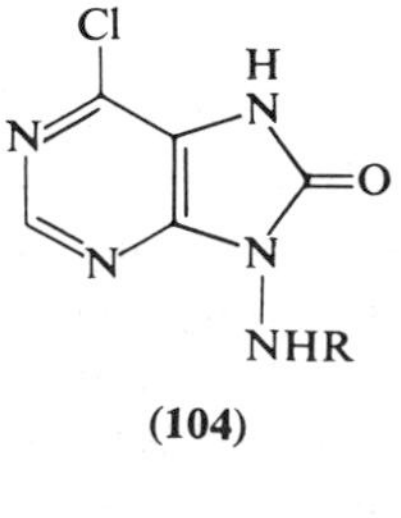   (104) 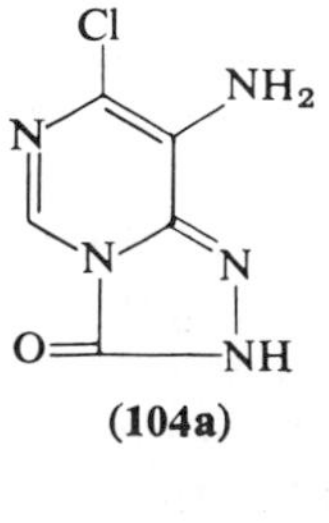   (104a)

complex of Scholtissek[210] and carrying out the reaction in benzene or dioxan under reflux conditions.[207a, 211]

T. Cyclisation with Thiophosgene

This route to 8-thiopurines requires conditions similar to those used for phosgene condensations. The reagent has had very limited use of which the formation of 9-amino-6-chloro-7,8-dihydro-8-thiopurine[207] and 1,6,7,8-tetrahydro-2-methylthio-6-oxo-8-thiopurine[103] are examples.

U. Cyclisation with Chlorocarbonic Esters

Although this type of ring closure, which usually requires two stages, was much favoured by the earlier workers for forming 8-oxopurines it is now largely, but not completely, replaced by the single-stage urea fusion or phosgene procedures. Condensation of ethyl (or methyl) chlorocarbonate to the pyrimidin-5-ylurethane (105) is carried out in alkaline solution, the product, after isolation, being converted to the purine under fusion conditions (200–250°). The majority of the purines thus formed have been derivatives of uric acid of which Traube prepared the parent member, also the 3-methyl[2] and 1,3-dimethyl derivatives,[2] the 3-ethyl[212] and 1,3-diethyl analogues[213] following later. Derivatives alkylated in the imidazole ring include the 7,9-dimethyl-[214] and 7-ethyl-9-methyluric acids,[215] also the fully N-methylated 1,3,7,9-tetramethyl-uric acid (106).[189] A noteworthy point is that interaction of the ester with 4,6-diamino-5-formamidopyrimidine in dimethylformamide gives the $N_{(7)}$-formylpurine (107) directly.[48] Cyclisation of the urethane intermediates does not in every case necessitate fusion conditions; the use of dilute sodium hydroxide for this purpose is reported.[211] In another case by carrying out the reaction in pyridine containing phosphorus pentasulphide the 6-oxopyrimidine (108) is converted to the 6-thiopurine (109).[108] More recently carbobenzoxy chloride under similar conditions has been shown to be a suitable ring-closing reagent for forming 8-oxopurines. In aqueous sodium acetate 4,5-diamino-6-benzylaminopyrimidine gives 110, which is converted to 6-amino-9-benzyl-7,8-dihydro-8-oxopurine by fusion.[183, 216] Under the more basic conditions obtained using aqueous sodium hydrogen carbonate, further Schotten-Baumann reactions occur giving the purine (111) directly.[216]

Treatment of 5-amino-4-chloro-6-hydrazinopyrimidine with ethyl chloroformate involves only the more basic hydrazine group, and the product (112) in hot diethoxymethyl acetate gives the purine (113) rather than the isomeric pyrimidodihydrotriazine (114).[217]

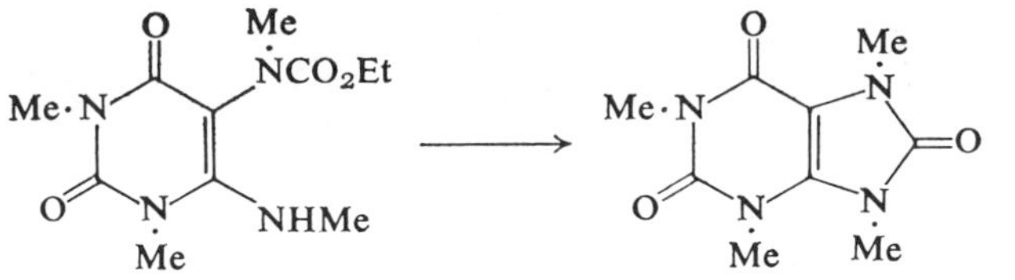

The structural formulas (105)–(114) are shown.

(105) → (106) 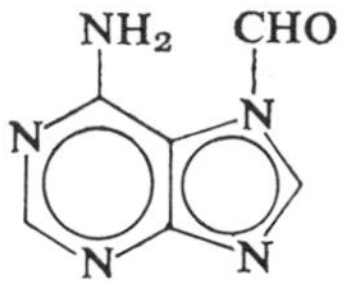 (107)

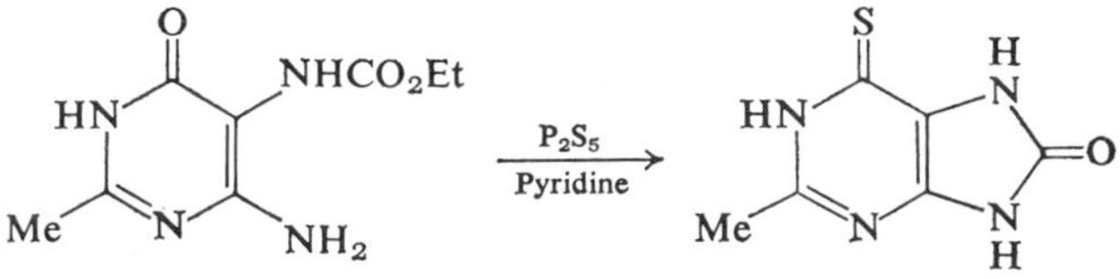

(108) $\xrightarrow[\text{Pyridine}]{P_2S_5}$ (109)

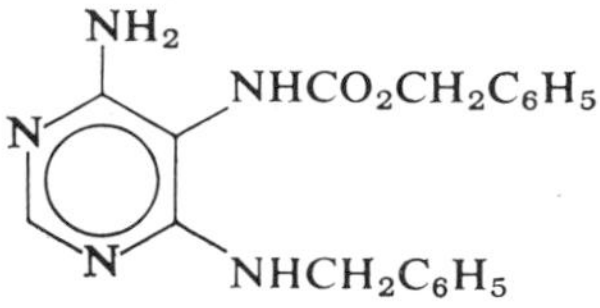

(110) 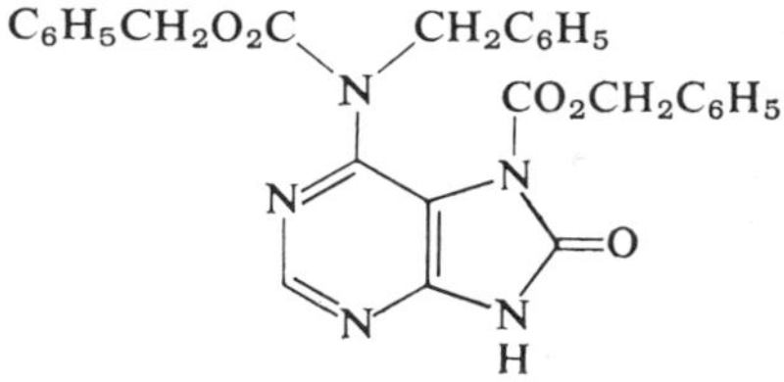 (111)

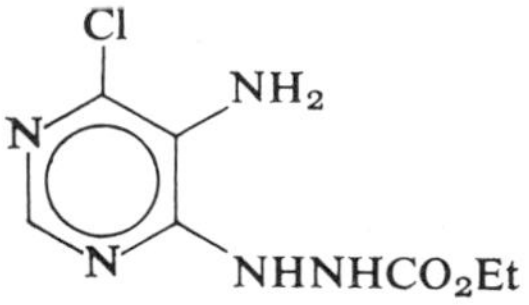

(112) 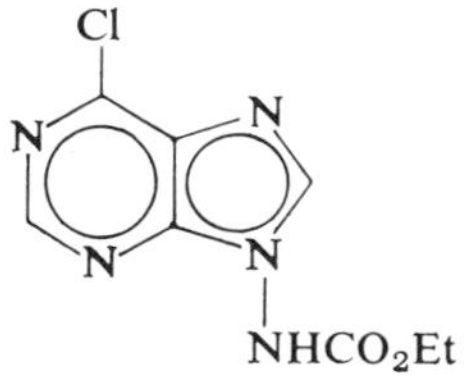 (113) 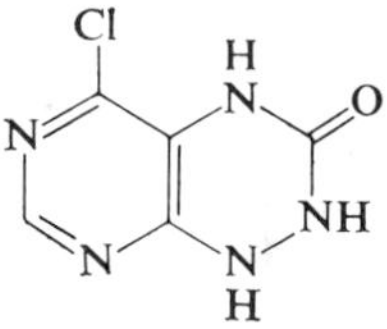 (114)

V. Cyclisation with Aldehydes and Ketones

The potentialities of this method of ring closure have not yet been fully exploited. They provide a facile route not only to 8-alkyl (and 8-aryl) purines but also to 8-unsubstituted derivatives using readily available reagents. Initially, a Schiff base, not isolated, is formed through the 5-amino group of the pyrimidine and conversion to the purine requires oxidative conditions (Eq. 10). A variety of reagents have been adopted for this purpose, ferric chloride being originally used,[218] but later prolonged heating of the reactants in nitrobenzene, in which the mild oxidising power of this solvent itself is utilised, was effective in forming a series of 8-phenylpurines.[219] The reaction is also possible in a

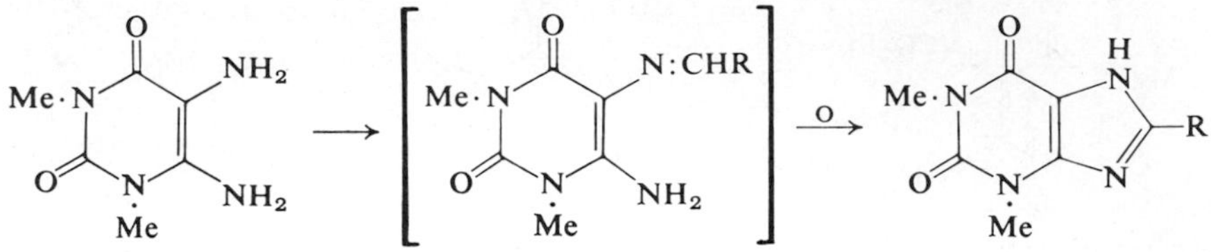

(10)

lower boiling solvent, such as benzene, provided a palladium catalyst is present and air is bubbled through the mixture.[219] 8-Phenyltheophylline is one example derived by this technique.[220] A recent approach[221] utilises the bisulphite adduct of the aldehyde, in ethanol with reflux times from 5 to 30 minutes. The reaction in this instance appears to proceed though the condensation product (115) rather than the azine (116), as the latter cannot be cyclised to 1,3,8-trimethylxanthine by

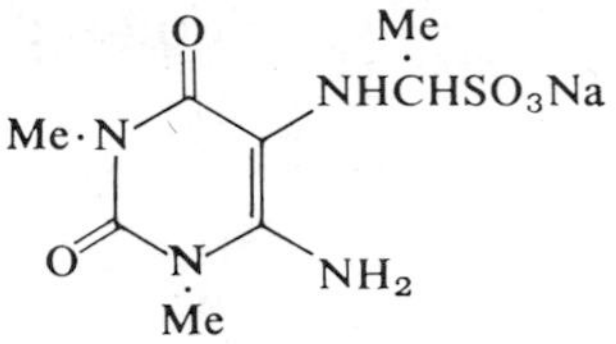

(115)

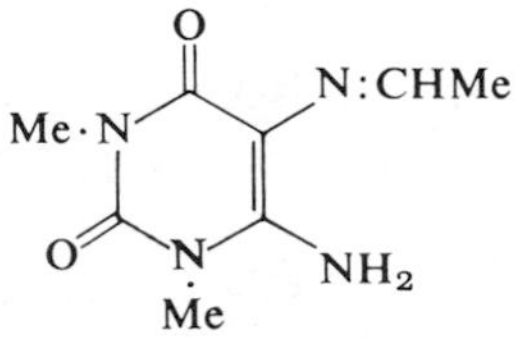

(116)

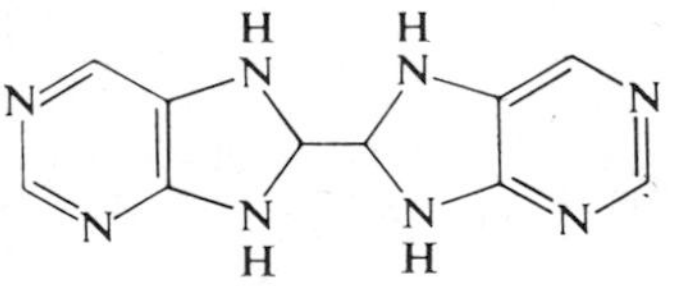

(117)

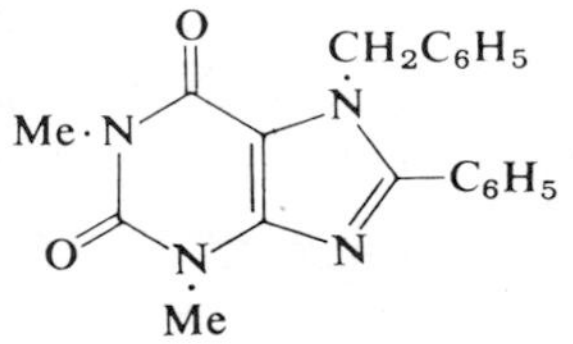

(118)

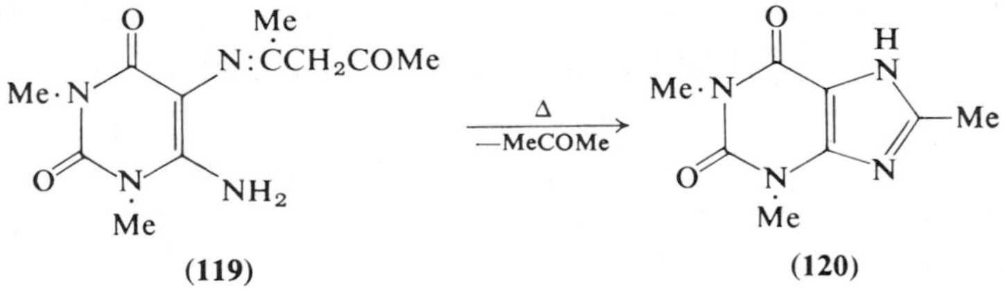

(119)

(120)

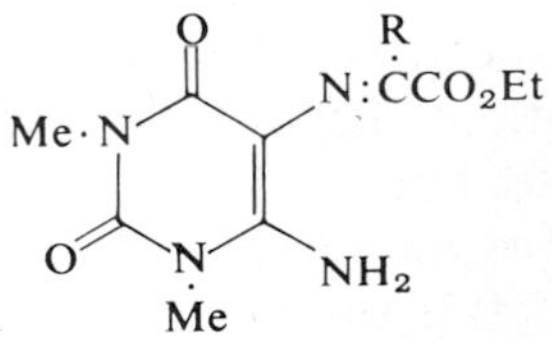

(121)

heating with ethanolic sodium bisulphite or with the aldehyde–bisulphite adduct. Oxidation of the intermediate dihydropurine is presumed to be effected by the liberated bisulphite anion.

Instead of the expected pteridine derivatives the $C_{(8)}$-linked bis 7,8-dihydropurines (117) result when aqueous glyoxal and the appropriate 4,5-diaminopyrimidine in molar excess interact.[119] Oxo-,[222] amino-,[222] and methyldihydropurines[222] are examples of the extension of this reaction. The suggestion that this mode of cyclisation is due to steric factors favouring purine rather than pteridine formation and proceeds via an ammonia–aldehyde adduct on the $C_{(4)}$- instead of the expected $C_{(5)}$-amino group has still to be substantiated.[223]

With an excess of aromatic aldehyde two molecules undergo condensation with the diamine, and 8-aryl-7-benzyl purines (118)[218] result.

So far very little use has been made of ketones for cyclising purposes and the few examples available employ derivatives of 1,3-diketones. Acetylacetone and the diamine give initially an azamethine of the type 119 which, by loss of acetone under fusion conditions ($\sim 270°$), is converted to the 8-methylpurine (120). Replacement of the acetylacetone by benzoylacetone in this condensation also gives 120 as the final product, acetophenone in this case being abstracted. Some 8-phenyltheophylline is also produced in the latter reaction due to ring closure of small amounts of the azamethine derivative formed through the benzoyl group.[224] Keto esters react likewise giving azamethine analogues such as 121 (R = Me or C_6H_5); these cyclise at lower temperatures ($\sim 230°$) to the 8-alkyl(aryl)theophylline with an accompanying deletion of the elements of ethyl formate.[224]

W. Cyclisation with Cyanogen Derivatives

Few applications are recorded, poor yields being given generally but a point in favour is that cyclisation is a single-stage reaction. Cyanogen bromide with 4,5-diaminopyrimidine gives only a minute yield of 8-aminopurine.[8] The same reagent in dilute alkali under reflux converts the corresponding 6-thiopyrimidine to 8-amino-1,6-dihydro-6-thiopurine more productively (31%).[225] These reactions are presumed to involve intermediate adducts of the type (122). Attention can be drawn to ring closure of the 5-phenylureidopyrimidine (123), in toluene containing phosphorus trichloride, to 8-anilino-1,6-dihydro-2-methyl-6-oxopurine, as the related chloro intermediate (124) may be involved.[186] Cyclisation of 4,5-diamino-1,6-dihydro-2-methyl-6-oxopyrimidine to the same 8-anilinopurine by means of phenyl cyanamide in butanol is successful but on using ethanol as solvent only the intermediate adduct (125) is obtained which has resisted all attempts at ring closure to the purine.[186] Formation of 6-amino-8-benzylpurine from heating 4,5,6-triaminopyrimidine with benzyl cyanide is of interest as in this case an intermediate of the type (125a) could be involved.[225a] For convenience the reaction of diphenyl carbodiimide ($C_6H_5N:C:NC_6H_5$) with a series of 5-alkyl(aryl)amino-4-aminopyrimidines in dimethylformamide (160°/6h) is noted here, the products being 8-anilino-7-alkyl(aryl)purines.[172]

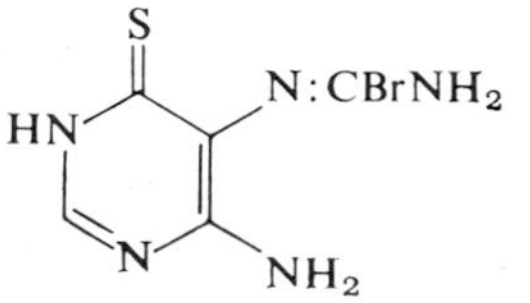

(122) 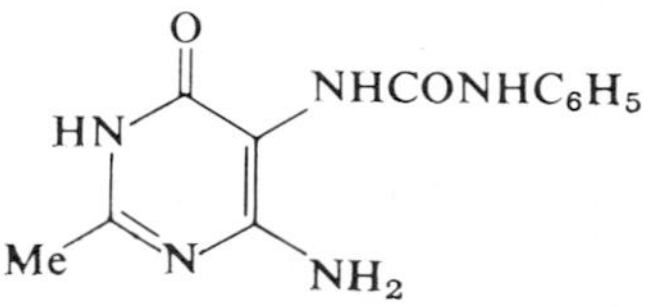   (123)

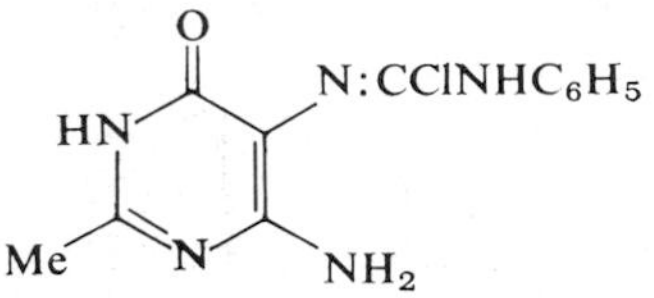

(124) 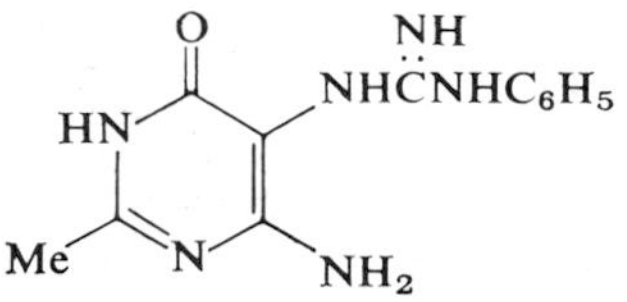   (125) 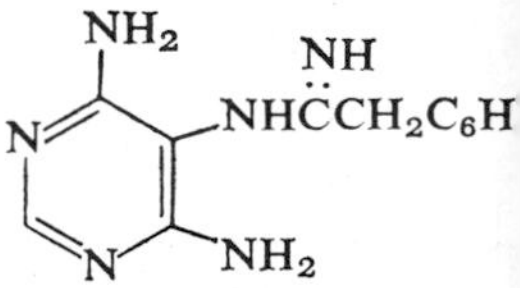   (125a)

2. Syntheses from 4-Amino-5-nitro- and 4-Amino-5-nitroso-pyrimidines

Two unrelated types of closure are grouped under this heading. The first which makes use of an *in situ* reduction of the nitro or nitroso group before cyclisation is, in effect, a modified Traube synthesis whereas the second proceeds by an internal cyclodehydration reaction, involving the oxygen atom of the nitro or nitroso group and a $C_{(4)}$-alkylamino group.

A. Cyclisations Involving Reduction of a Nitro, Nitroso, or Other Group at $C_{(5)}$

On heating 4-amino-5-nitrosopyrimidines in 90% formic acid in the presence of zinc dust or Raney nickel direct formation of the $C_{(8)}$-unsubstituted purine is possible.[226] Under these conditions if a thio group is present simultaneous dethiation and cyclisation can occur if

Raney nickel is the reducing agent used.[226] This route was successful in producing guanine, hypoxanthine, xanthine, and 3-methylxanthine. If more than one oxo group is present the addition of sodium acetate is beneficial and, in the case of xanthine, the reaction does not take place without it.[226] Theophylline could not be obtained in this way; only the intermediate 5-formamidopyrimidine was formed which gave the purine on brief treatment with hot dilute sodium hydroxide.[226] The majority of reductive cyclisations have taken place in formamide, containing small amounts of formic acid, using a variety of reducing agents. These include hydrogen sulphide,[227, 228] sulphur dioxide,[227] sodium sulphide,[228] sodium sulphite,[227] the latter having had extensive application. Finely divided iron and Raney nickel have also been used.[228] Formamide-dithionite reaction mixtures have given rise to xanthine[227–230] and other oxo-, thio-, and aminopurines[227, 229, 231] from the appropriate 5-nitrosopyrimidines. One dithionite reduction, giving a uric acid derivative, took place in aqueous bicarbonate solution into which phosgene was introduced.[209] Similar procedures with 5-nitropyrimidines have converted them to adenine[232] and 2-substituted purines.[233]

An analogous reductive cyclisation has been demonstrated with 5-phenylazopyrimidines; a low yield (24%) of 2-phenyladenine results on prolonged (6 h) passage of hydrogen sulphide through a solution of 4,6-diamino-2-phenyl-5-phenylazopyrimidine in a boiling mixture of dimethylformamide and triethyl orthoformate (Eq. 11).[234]

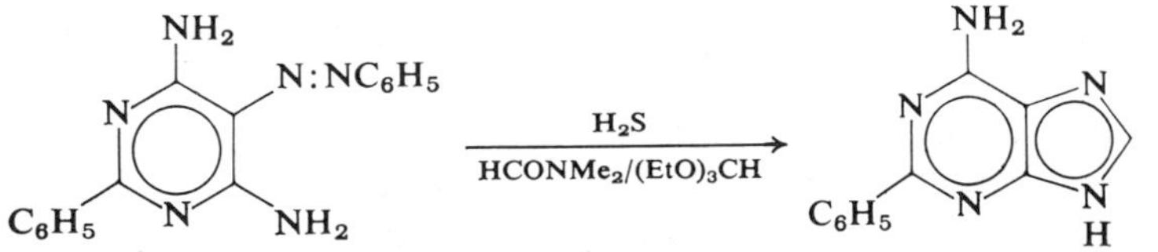

$$\text{(11)}$$

On catalytic reduction of the nitro group in the 4-acetamidopyrimidine (126, R = p-acetamidostyryl), the 8-methylpurine (127) is directly obtained due to the characteristic instability of the intermediate 4-acetamido-5-aminopyrimidine.[235] A series of 8-methylpurines has been produced in this way,[236] from both 4-acetamido-5-nitro- and 4-acetamido-5-nitrosopyrimidines. It should be noted that the fairly vigorous

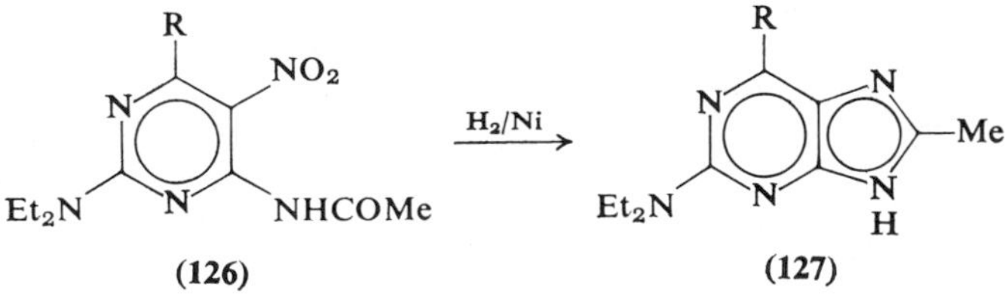

(126)   (127)

acylation procedures adopted for the nitropyrimidines must be moderated for the nitroso analogues in view of their observed rearrangement,[237, 238] at temperatures over 120°, to cyano-s-triazines in acetic anhydride.

B. Cyclisation through Cyclodehydration

This novel approach, which has been directed mainly to the preparation of N-alkyl (or aryl) derivatives of xanthine, utilises the reaction between a $C_{(5)}$-nitroso group and a $C_{(4)}$-alkylamino group, in which the elements of water are abstracted. Thus 1,2,3,4-tetrahydro-1,3-dimethyl-6-methylamino-5-nitroso-2,4-dioxopyrimidine (**128**, R = H) on heating in butanol or xylene gives theophylline (**129**, R = H);[239] likewise the 4-benzylaminopyrimidine (**128**, R = C_6H_5) affords the 8-phenyl derivative (**129**, R = C_6H_5).[240] The method appears general for derivatives of this type, thermal cyclisations being equally well carried out in dimethylformamide or N-methylpyrrolidone or by fusion.[241] The severity of the ring closure conditions were shown, in one example, to be unnecessary if acylation of the nitrosopyrimidine preceded cyclisation. Thus, theophylline (**129**, R = H) can be obtained from the O-acylated derivative (**130**) in ethanol under reflux conditions.[242]

Where only one hydrogen atom is available in the alkyl moiety, as for example in the 4-isopropylamino derivative (**131**), cyclisation is still possible by abstraction of the second atom from the amino group. The products in this case are the nonclassical 8,8-dialkylpurines,[241, 243] illustrated by **132**, which isomerise at their melting points to the corresponding 7,8-dialkylpurines,[244] for example, **133**.

A useful extension of this route allows the direct formation of purines alkylated in the imidazole by carrying out ring closures in the presence of alkylating agents. In a simple example caffeine (**136**) is formed from the nitrosopyrimidine (**134**) using diazomethane, methyl iodide, or dimethyl sulphate.[245] A noteworthy feature of this reaction is the ease with which cyclodehydration occurs. This can be adequately explained by noting that methylation of the nitroso pyrimidine gives rise to a

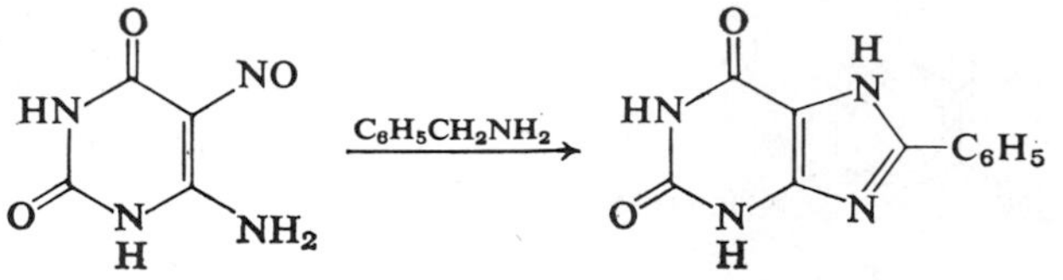

$$\text{(12)}$$

derivative (**135**) possessing pronounced *N*-oxide character and a facility to lose the oxygen atom.

Studies of the nitroso-alkylamine interaction reveal that the nitroso group and alkylamino group need not necessarily be located on the same nucleus. Thus, theophylline is obtained when the nitrosopyrimi-

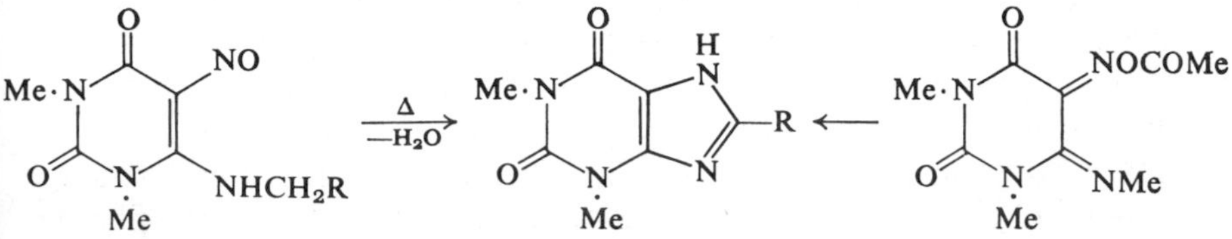

(128) (129) (130)

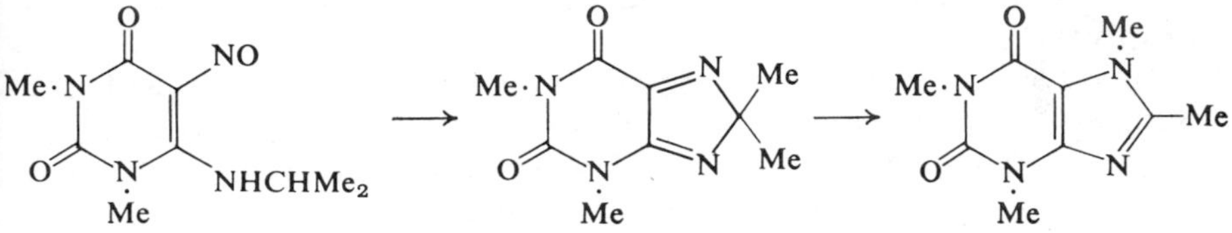

(131) (132) (133)

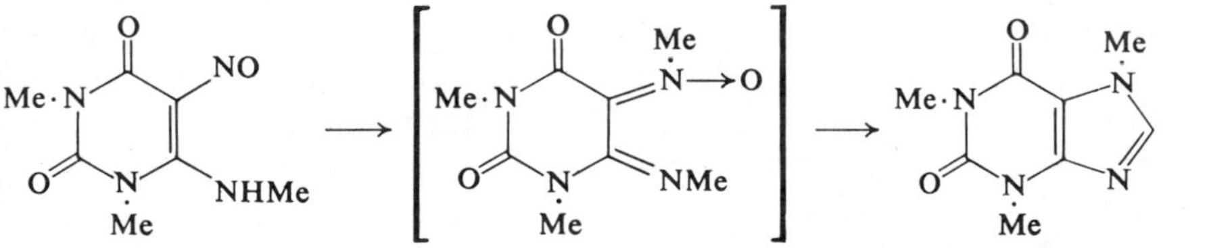

(134) (135) (136)

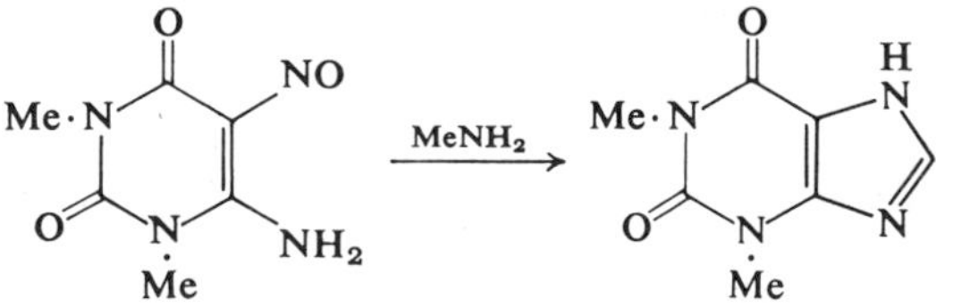

(137) (138)

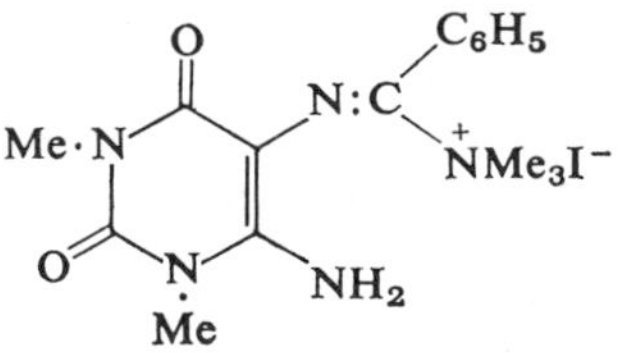

(139) 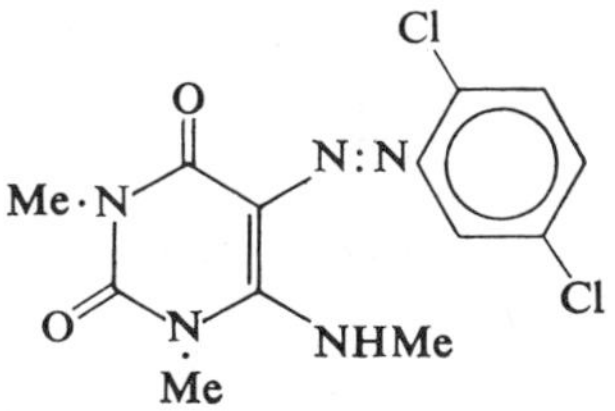 (140)

dine (137) and methylamine are reacted.[239] Benzylamine similarly gives
the 8-phenyl derivatives of 3-methyl-,[240] 1,3-dimethyl-,[239] and 3-benzyl-
xanthine.[240] The preparation of 8-phenylxanthine itself[240] (Eq. 12) is
the sole example to date of the formation in this way of a non-N-
alkylated purine.

A suggested mechanism assumes that the initial condensation is
through an exchange of the $C_{(4)}$-amino group with the alkylamine, such
amino group interchanges are known to occur with aminopurines[246]
and -pyrimidines. However, the possibility that interaction of the
methylene group with the nitroso group may represent the first stage
cannot be discounted, as an unconnected approach[247] to 8-phenyl-
theophylline, using benzyltrimethylammonium iodide in dimethylformamide under reflux conditions, postulates an intermediate (139)
of this type.

In place of the above 5-nitrosopyrimidines some success with the
5-phenylazo analogues has been obtained. A satisfactory yield (50%)
of theophylline, for example, is afforded on heating 140 in nitrobenzene.[248] Cyclisations of this type are also possible under fusion conditions.[248]

Examples in which benzaldehyde is utilised to prepare 8-phenyl-
purines are known. Reaction of 4,5-diamino-1,3-dimethyluracil with
the aldehyde in dimethylformamide gave 8-phenyltheophylline together
with the 7-oxide analogue. The addition of formic acid to the reaction
produces a reducing environment and the 8-phenylpurine is the sole
product.[249] The same effect has been obtained[249a] using the 1,1-di-
methylhydrazone of benzaldehyde and 137 in either dimethylformamide
or dimethylsulphoxide. Other related nitrosopyrimidine-benzaldehyde
interactions leading to purine-7-oxides are detailed in Chapter XI (Sect.
1C).

3. Syntheses from 5-Amino-4-oxopyrimidines

For the preparation of 8-oxopurines, as well as the 8-thio analogues,
use of these precursors presents an alternative route to the Traube
synthesis. Although this method has little current application it has
been widely used in the past.

A. Cyclisation with Cyanates and Isocyanates

The synthesis devised by Fischer and which finally established the
structure of uric acid was of this type.[250, 251] Later workers adapted the

method to produce a variety of uric acid and 2-amino-6,8-dioxopurine derivatives. For the former compounds 5-aminobarbituric acid is condensed with either potassium cyanate or an organic isocyanate, the latter if 9-alkyl[252] or -aryl[253, 254] derivatives are required; the resulting 5-ureidopyrimidine, the so-called "pseudouric acid" of Fischer, was originally cyclised in molten oxalic acid[250] but this was quickly superceded by boiling dilute or concentrated hydrochloric acid[251] (Eq. 13). Derivatives of 2-amino-6,8-dioxopurines have been similarly obtained.

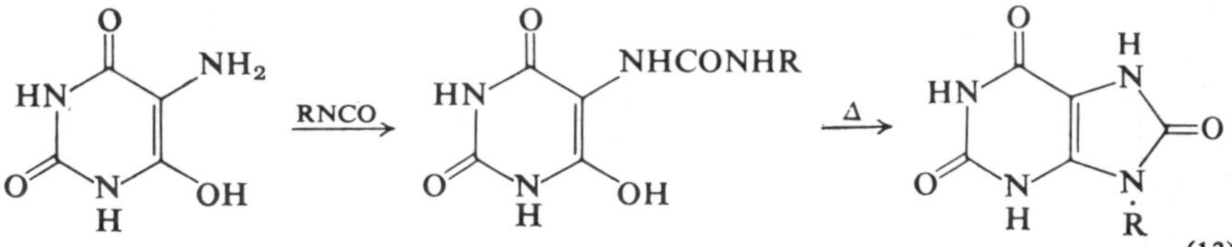

(13)

[251, 255] By using 5-alkylaminopyrimidines the corresponding 7-alkyl-purines follow.[251, 255–260] $N_{(1)}N_{(3)}$-Dimethylpyrimidines,[250, 251, 259, 261] other $N_{(1)}N_{(3)}$-dialkyl,[262] and -diaryl analogues[263] afford the respective $N_{(1)}N_{(3)}$-alkylated uric acids.

B. Cyclisation with Urea

Under fusion conditions 5-aminobarbituric acid and urea give pseudouric acid which is ring closed with acid as above.[179] A reaction with nitrourea differs from the above in that condensation is possible in a mildly alkaline aqueous solution on heating. Under these conditions the nitrourea reacts like potassium cyanate and the uric acid derivative is produced directly, emerging from the solution on cooling.[264]

C. Cyclisation with Alkyl- or Aryl Isothiocyanates

With isothiocyanates the conditions employed follow closely those of the cyanates and isocyanates above (Sect. 3A). The route is more useful, however, in that the 8-thiopurines obtained are readily desulphurised with Raney nickel to the $C_{(8)}$-unsubstituted derivative or after methylation give 8-methylthio derivatives which undergo facile reactions with various nucleophilic reagents. The preparation of 8-thioguanine derivatives is possible only if an alkyl isothiocyanate is used. Although aryl isothiocyanates can be used to prepare the intermediate thio-ureidopyrimidine, attempts at ring closure only lead to degradation.[265]

Reactions between the 5-aminopyrimidine and isothiocyanate are best carried out in dilute alkali with subsequent ring closure in hydrochloric acid of the "pseudothiouric acid" formed.[266-270] One novel approach by Fischer, giving 1,2,3,6,7,8-hexahydro-2,6-dioxo-8-thiopurine, resulted from a sulphide reduction of 5-cyanamidobarbituric acid **(141)** followed by acid treatment of the product.[271]

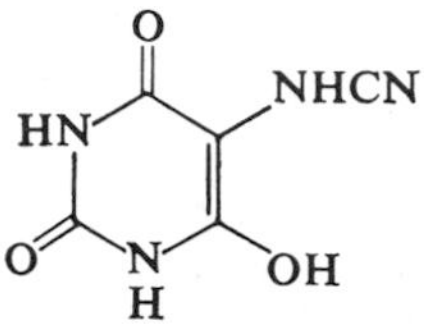

(141)

4. Syntheses from 4-Amino-5-unsubstituted-pyrimidines

Direct purine formation is possible; the two methods available both require *in situ* nitrosation as a first step, the reaction then following those already described for 5-nitroso-4-aminopyrimidines (Sects. 2A and 2B).

A. Using Traube Synthesis Conditions

Using a formamide–formic acid mixture as solvent concurrent nitrosation, reduction, and cyclisation occur on the addition of sodium nitrite at low temperature followed by sodium dithionite and heating.

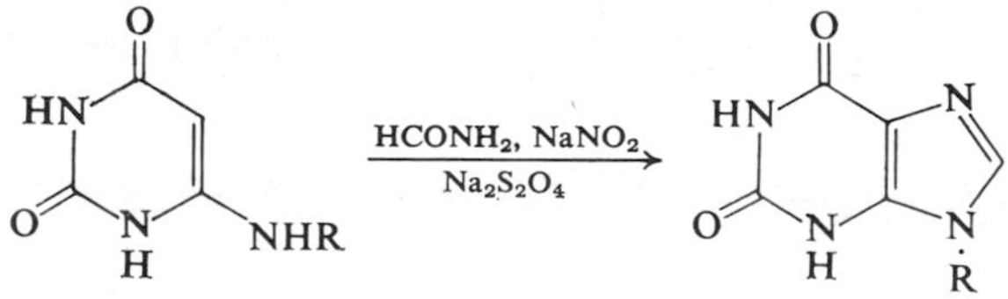

(14)

Xanthine and a number[229] of 9-substituted derivatives[272] (Eq. 14), and various 2-thio-,[229] 2-methylthio-,[229] and N-alkylated-oxopurines[229, 273] are directly formed in this way from the appropriate 4-aminopyrimidines.

B. Through Cyclodehydration of Nitrosopyrimidines

In contrast to those of the preceding section cyclisations of this type are carried out at room temperature. Nitrosation of the 4-aminopyrimidine is made with isoamyl nitrite and precipitation of the purine occurs on addition of cold ethanolic hydrogen chloride to the reaction mixture. The formation of 8-phenyltheophylline from the 4-benzylaminopurine (Eq. 15) is one example of a number of xanthine derivatives so obtained.[240]

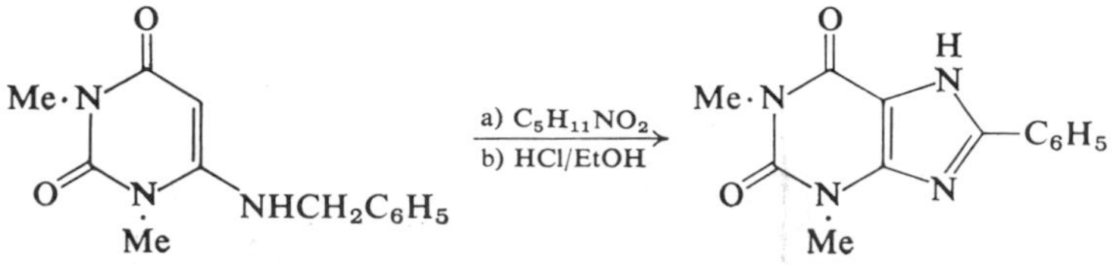

$$\text{a) } C_5H_{11}NO_2 \quad \text{b) } HCl/EtOH \qquad (15)$$

5. Syntheses from 5-Sulphoamino-4-unsubstituted-pyrimidines

2,4-Dioxo-5-sulphoaminopyrimidine derivatives with amides and urea afford the same products as when 4,5-diaminopyrimidines are employed. Thus, formamide at 180° converts both 142 (R = H)[274] and 142 (R = Me)[229] to xanthine (143, R = H)[274] and the 3-methyl homologue (143, R = Me),[229] respectively. The 8-methyl analogues follow using acetamide.[229, 274] Fusions with urea have produced uric acid[274] and the 3-methyl homologue.[229] Attempts to prepare theophylline failed as heating the $N_{(1)}N_{(3)}$-dimethyl analogue of 142 in formamide caused fission of the sulpho group and formylation to the corresponding 5-formamidopyrimidine.[229]

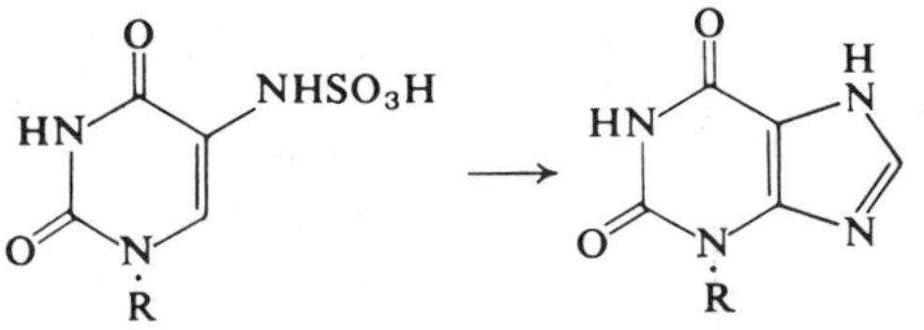

(142)   (143)

6. Syntheses from 4,5-Dioxopyrimidines

This approach has little to commend itself from a practical standpoint, the inaccessibility of suitable 4,5-dioxopyrimidines alone limits the usefulness of the method. It has, however, considerable historical significance in being the means of an early synthesis* of uric acid by heating isodialuric acid and urea together in dilute sulphuric acid.[275] Later Prusse[276] studied the reaction in detail using mono- and dimethylureas, obtaining methylated uric acids in very low yields, usually of the order of 10%. Isodialuric acid with N-methylurea in either sulphuric or hydrochloric acid gives 7-methyluric acid, and with NN'-dimethylurea

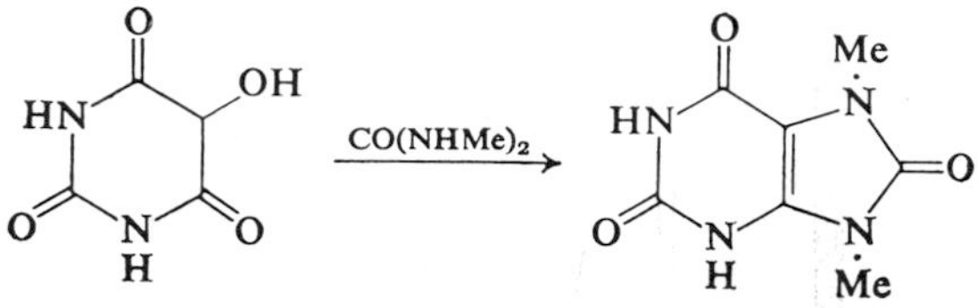

$$(16)$$

the 7,9-dimethyluric acid is obtained (Eq. 16). This latter case is noteworthy as it appears to be the only 9-methyluric acid derivative prepared by this route. Similar condensations of urea and N-methylurea were carried out with 1-methyl- and 3-methylisodialuric acids giving 1-methyl-, 1,7-dimethyl-, and 3,7-dimethyluric acids. No condensation could be effected with NN'-dimethylurea and preparations of 1,3-dimethyl-, 1,9-dimethyl-, 3,9-dimethyluric acids or any of the trimethyl or tetramethyl homologues were unsuccessful.

7. Syntheses from 4-Amino-5-carbamoylpyrimidines

These derivatives undergo the Hofmann rearrangement, forming 8-oxopurines, with alkaline solutions of hypochlorite or hypobromite. The parent member (145, R = H)[277] is obtained in 60% yield from 144, and in like manner the appropriate 2-substituted pyrimidines give the 2-methyl (145, R = Me),[278, 279] 2-trifluoromethyl (145, R = F_3C),[280] and 2-amino (145, R = NH_2)[281] analogues. From the last purine preparation some of the isomeric 6-amino-2,3-dihydro-3-oxopyrazolo-[5,4-d]pyrimidine (146) was isolated.[281]

The related Curtius reaction applied to 4-amino-5-hydrazinocarbonyl-pyrimidines (147) is also a route to 8-oxopurines. Although nitrous acid treatment of 147 (R = H) gave 149 (R = H) directly,[277] the formation of the 2-methyl (149, R = Me), 2-amino (149, R = NH_2), and 2-oxo derivatives required gentle heating of the intermediate carboxyazides (148) in xylene for some minutes.[278, 279]

---

* This synthesis preceded by nearly ten years Fischer's synthesis which confirmed the structure of uric acid.

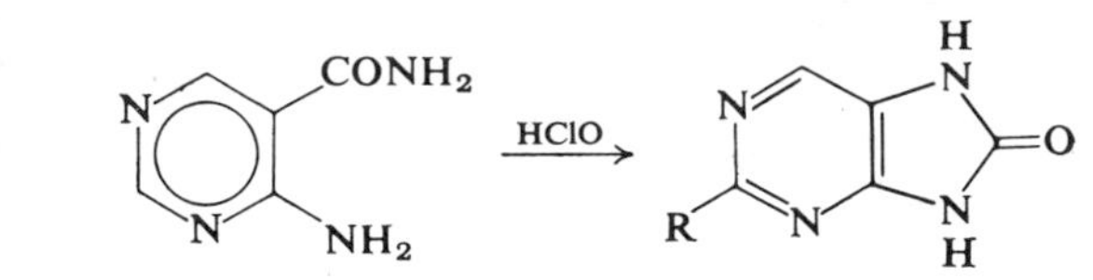

(144)   (145) 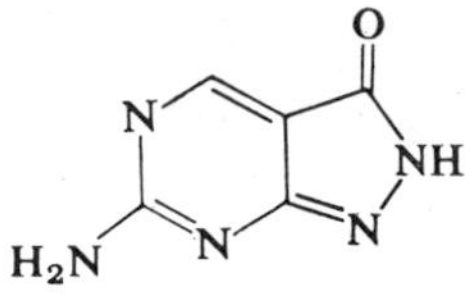   (146)

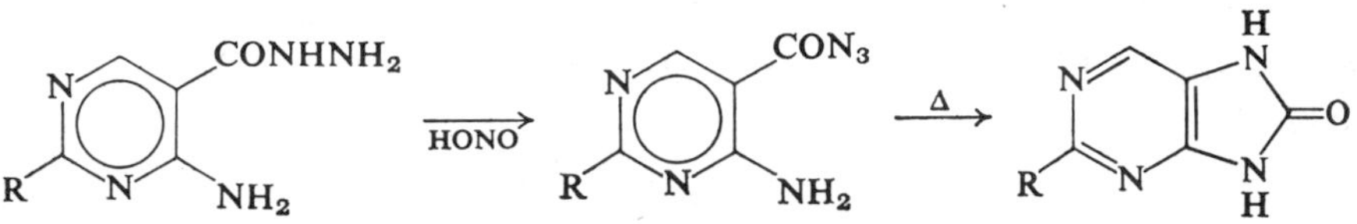

(147)   (148)   (149)

8. Purines derived from Pyrimidine-containing Heterocycles

Some bicyclic derivatives, which contain a pyrimidine ring as part of the structure, undergo rearrangement usually under alkaline conditions to purines.

A. From Thiazolo[5,4-*d*]pyrimidines

This rearrangement, which has already been encountered in previous sections, gives rise to 6-thiopurines on heating the thiazole derivative in formamide or dilute sodium hydroxide. Simple purines so derived include the parent 1,6-dihydro-6-thiopurine and the 2-methyl and 2-butyl analogues from the appropriate 7-aminothiazolo[5,4-*d*]pyrimidine.[13] By using the 7-alkylaminothiazole analogues, alkali treatment gives the 9-alkyl-1,6-dihydro-6-thiopurines.[13] The formation of 1,6,7,8-tetra-hydro-6,8-dithiopurine from the 2-mercaptothiazole derivative (Eq. 17, R = SH)[56] and of 8-amino-1,6-dihydro-6-thiopurine from the 2-aminothiazole analogue (Eq. 17, R = NH$_2$)[225] illustrate the preparation of more complex derivatives by this procedure.

One ingenious use of this rearrangement has been to prepare an *N*-oxide of 1,6-dihydro-6-thiopurine which could not be obtained by direct oxidation of the purine (see Ch. XI, Sect. 1B).

The mechanism postulated for these rearrangements assumes ring opening to a 4-amino-5-formamido-6-thiopyrimidine anion followed by recyclisation through the $C_{(4)}$-amino group.[13]

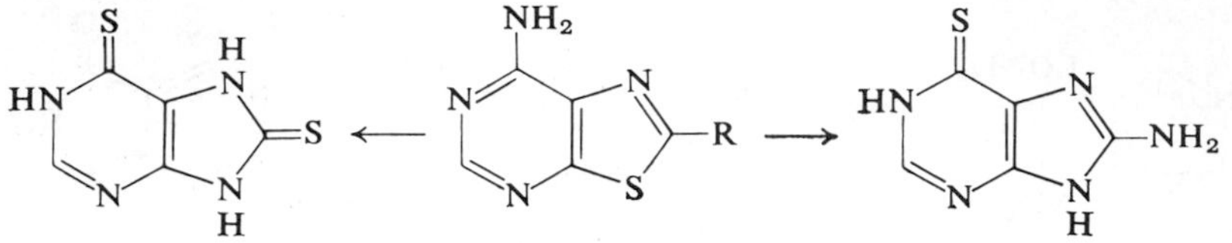

(17)

An apparently isolated example of thiazolopyrimidine rearrangement in which only the thiazole moiety itself is involved is seen in the isomerisation of **150** to the 8-thiopurine (**151**) on prolonged heating in alkali.[282]

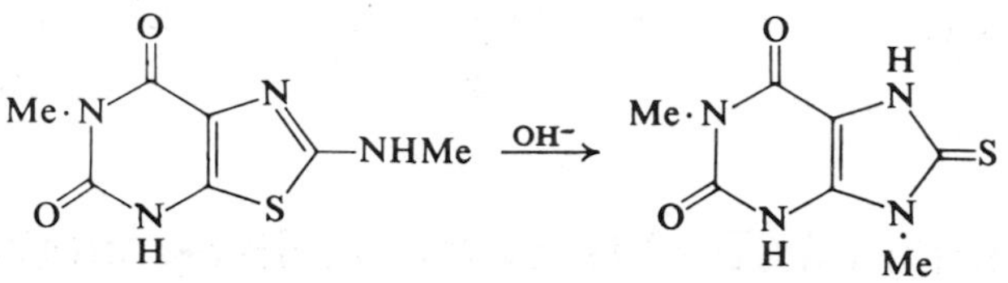

(150)   (151)

B. From Oxazolo[5,4-*d*]pyrimidines

On heating with ammonia or amines the oxygen atom is replaced by nitrogen with ensuing formation of the imidazole ring. Ammonia gives rise to $C_{(9)}$-unsubstituted derivatives but the 9-methyl and 9-propyl homologues result with methylamine and propylamine, respectively (Eq. 18, R = Me and Pr). No comparable conversions could be effected with benzylamine.[121] Isomerisation of 7-amino-oxazolopyrimidines to hypoxanthines occurs in hot alkaline solution[283] while with peracetic acid, in the cold, the corresponding 6,8-dioxopurine results, most likely through rearrangement of an *N*-oxide intermediate.[284]

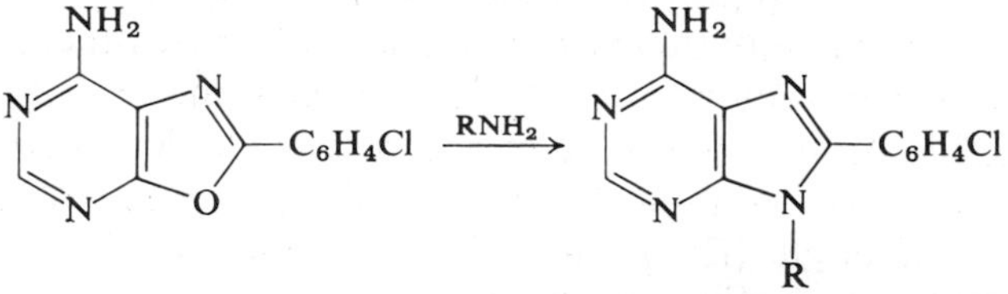

(18)

C. From Tetrazolo[1,5-*c*]pyrimidines

Heating under reflux conditions a solution of 8-amino-7-chloro-tetrazolo[1,5-*c*]pyrimidine in diethoxymethyl acetate affords 6-chloro-8-ethoxypurine directly (Eq. 19).[285]

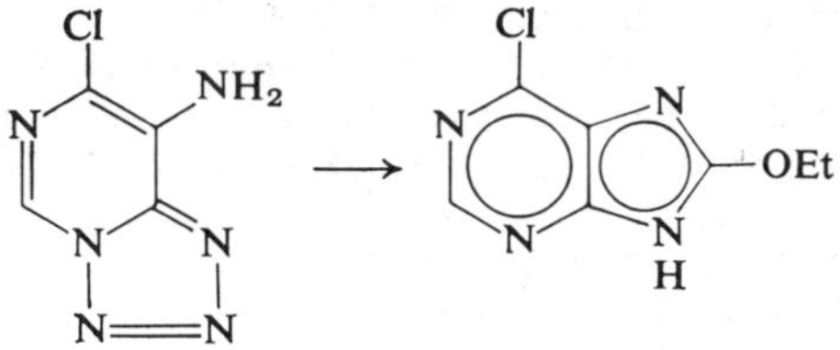

$$(19)$$

D. From Pyrimido[5,4-c]oxadiazines

Some pyrimido[5,4-c]oxadiazines when heated to 200° lose the elements of formaldehyde, and rearrangement to purines occurs (Eq. 20).[286]

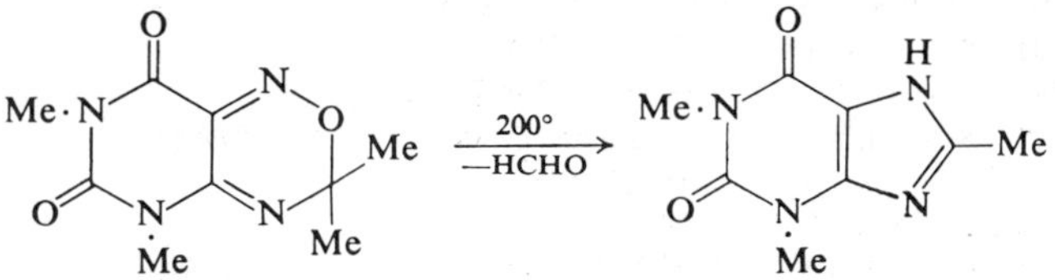

$$(20)$$

E. From Thiazolo[3,4,5-gh]purines

Acid or base hydrolyses of the tricyclic 2,2-diamino-2H-thiazolo-[3,4,5-gh]purines affords 8-amino-1,6-dihydro-6-thiopurines (Eq. 21).[225]

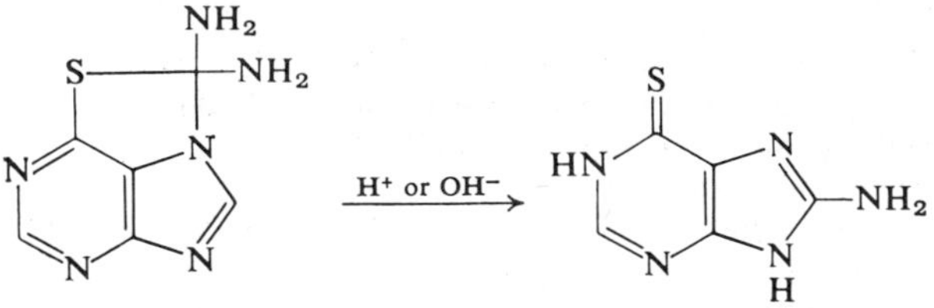

$$(21)$$

References

1. Traube, *Ber.*, **33**, 1371 (1900).
2. Traube, *Ber.*, **33**, 3035 (1900).
3. Gabriel and Colman, *Ber.*, **34**, 1234 (1901).
4. Isay, *Ber.*, **39**, 250 (1906).
5. Wilson, *J. Chem. Soc.*, **1948**, 1157.
6. Brown, *J. Appl. Chem.*, **1954**, 358.
7. Clark and Kalcker, *J. Chem. Soc.*, **1950**, 1029.

8. Albert and Brown, *J. Chem. Soc.*, **1954**, 2060.
9. Robins, Dille, Willits, and Christensen, *J. Amer. Chem. Soc.*, **75**, 263 (1953).
10. Elion, Burgi, and Hitchings, *J. Amer. Chem. Soc.*, **74**, 411 (1952).
11. Townsend and Robins, *J. Org. Chem.*, **27**, 990 (1962).
12. Traube and co-workers, *Annalen*, **432**, 266 (1923).
13. Brown and Mason, *J. Chem. Soc.*, **1957**, 682.
14. Johns, *J. Biol. Chem.*, **12**, 91 (1912).
15. Traube, *Annalen*, **331**, 64 (1904).
16. Hoffer, *Jubilee Volume Dedicated to Emil Christoph Barrel*, 1946, p. 428; through *Chem. Abstracts*, **41**, 4108 (1947).
17. Bendich, Tinker, and Brown, *J. Amer. Chem. Soc.*, **70**, 3109 (1948).
18. Clark, Kernick, and Layton, *J. Chem. Soc.*, **1964**, 3221.
19. Cresswell, Maurer, Strauss, and Brown, *J. Org. Chem.*, **30**, 408 (1965).
20. Baker, Joseph, and Schaub, *J. Org. Chem.*, **19**, 631 (1954).
21. Haley and Maitland, *J. Chem. Soc.*, **1951**, 3155.
22. Traube, *Ber.*, **37**, 4544 (1904).
23. Baker, Joseph, and Williams, *J. Org. Chem.*, **19**, 1793 (1954).
24. Ruttink, *Rec. Trav. chim.*, **65**, 751 (1946).
25. Matsuura and Goto, *J. Chem. Soc.*, **1965**, 623
26. Bendich, Russell, and Fox, *J. Amer. Chem. Soc.*, **76**, 6073 (1954).
27. Brown, *J. Appl. Chem.*, **7**, 109 (1957).
28. Engelmann, *Ber.*, **42**, 177 (1909).
29. Elion, Goodman, Lange, and Hitchings, *J. Amer. Chem. Soc.*, **81**, 1898 (1959).
30. Traube, *Annalen*, **331**, 77 (1904).
31. Johns, *J. Biol. Chem.*, **11**, 67 (1912).
32. Johns, *J. Biol. Chem.*, **11**, 73 (1912).
33. Bobransky and Synowiedski, *J. Amer. Pharm. Assoc. Sci. Edn.*, **37**, 62 (1948).
34. Comte, U.S. Pat., 2,542,396 (1951); through *Chem. Abstracts*, **45**, 6657 (1951).
35. Ballantyne, U.S. Pat., 2,564,351 (1951); through *Chem. Abstracts*, **46**, 2574 (1952).
36. Homeyer, U.S. Pat., 2,646,432 (1953); through *Chem. Abstracts*, **48**, 8819 (1954).
37. American Cyanamid Co., Brit. Pat., 744,866 (1956); through *Chem. Abstracts*, **51**, 2062 (1957).
38. Elion and Hitchings, *J. Amer. Chem. Soc.*, **76**, 4027 (1954).
39. Gepner and Krebs, *Zhur. obshchei Khim.*, **16**, 179 (1946).
40. Brown and Jacobsen, *J. Chem. Soc.*, **1965**, 3770.
41. Montgomery and Hewson, *J. Org. Chem.*, **26**, 4469 (1961).
42. Cavalieri, Tinker, and Bendich, *J. Amer. Chem. Soc.*, **71**, 533 (1949).
43. Freer and Sherman, *Amer. Chem. J.*, **20**, 223 (1898).
44. Cavalieri and Brown, *J. Amer. Chem. Soc.*, **71**, 2246 (1949).
45. Gordon, *J. Chem. Soc.*, **1954**, 757.
46. Abrams and Clark, *J. Amer. Chem. Soc.*, **73**, 4609 (1951).
47. Bredereck, Hennig, Pfleiderer, and Weber, *Chem. Ber.*, **86**, 333 (1953).
48. Denayer, *Bull. Soc. chim. (France)*, **1962**, 1358.
49. Weygand and Grosskinsky, *Chem. Ber.*, **84**, 839 (1951).
50. Rose, *J. Chem. Soc.*, **1952**, 3448.
51. Davies, Noell, Robins, Koppel, and Beaman, *J. Amer. Chem. Soc.*, **82**, 2633 (1960).
52. Noell, Smith, and Robins, *J. Medicin. Chem.*, **5**, 996 (1962).
53. Elion, Lange, and Hitchings, *J. Amer. Chem. Soc.*, **78**, 2858 (1956).
54. Robins and Lin, *J. Amer. Chem. Soc.*, **79**, 490 (1957).
55. Beaman and Robins, *J. Amer. Chem. Soc.*, **83**, 4038 (1961).

56. Beaman, Gerster, and Robins, *J. Org. Chem.*, **27**, 986 (1962).
57. Robins, Dille, and Christensen, *J. Org. Chem.*, **19**, 930 (1954).
58. Nagano, Inoue, Saggiomo, and Nodiff, *J. Medicin. Chem.*, **7**, 215 (1968).
59. Biltz, *J. prakt. Chem.*, **118**, 166 (1927).
60. Sundwik, *Z. physiol. Chem.*, **23**, 476 (1897).
61. Sundwik, *Z. physiol. Chem.*, **26**, 131 (1898).
62. Sundwik, *Z. physiol. Chem.*, **76**, 486 (1912).
63. Baddiley, Lythgoe, McNeil, and Todd, *J. Chem. Soc.*, **1943**, 383.
64. Todd, Bergel, and Karimullah, *J. Chem. Soc.*, **1936**, 1557.
65. Andrews, Anand, Todd, and Topham, *J. Chem. Soc.*, **1949**, 2490.
66. Daly and Christensen, *J. Org. Chem.*, **21**, 177 (1956).
67. Baddiley, Lythgoe, and Todd, *J. Chem. Soc.*, **1943**, 386.
68. Brown, Roll, Plentl, and Cavalieri, *J. Biol. Chem.*, **172**, 469 (1948).
69. Baddiley, Lythgoe, and Todd, *J. Chem. Soc.*, **1944**, 318.
70. Kenner, Lythgoe, and Todd, *J. Chem. Soc.*, **1944**, 652.
71. Baddiley, Kenner, Lythgoe, and Todd, *J. Chem. Soc.*, **1944**, 657.
72. Howard, Lythgoe, and Todd, *J. Chem. Soc.*, **1945**, 556.
73. Kenner and Todd, *J. Chem. Soc.*, **1946**, 852.
74. Kenner, Rodda, and Todd, *J. Chem. Soc.*, **1949**, 1613.
75. Lythgoe, Smith, and Todd, *J. Chem. Soc.*, **1947**, 355.
76. Kenner, Lythgoe, and Todd, *J. Chem. Soc.*, **1948**, 957.
77. Andrews, Kenner, and Todd, *J. Chem. Soc.*, **1949**, 2302.
78. Holland, Lythgoe, and Todd, *J. Chem. Soc.*, **1948**, 965.
79. Kenner, Taylor, and Todd, *J. Chem. Soc.*, **1949**, 1613.
80. Craveri and Zoni, *Chimica*, **33**, 473 (1957).
81. Pappelardo and Conderelli, *Ann. Chim.* (*Italy*), **43**, 727 (1953).
82. Giner-Sorolla and Bendich, *J. Amer. Chem. Soc.*, **80**, 5744 (1958).
83. Albert, *J. Chem. Soc.* (*B*), **1966**, 438.
84. Speer and Raymond, *J. Amer. Chem. Soc.*, **75**, 114 (1953).
85. Furst and Eburt, *Chem. Ber.*, **93**, 99 (1960).
86. Hager, Krantz, Harmond, and Burgison, *J. Amer. Pharm. Assoc. Sci. Edn.*, **43**, 152 (1954).
87. Hager and Kaiser, *J. Amer. Pharm. Assoc. Sci. Edn.*, **43**, 148 (1954).
88. Fu, Hargis, Chinoporos, and Malkiel, *J. Medicin. Chem.*, **10**, 109 (1967).
89. Caldwell and Cheng, *J. Amer. Chem. Soc.*, **77**, 6631 (1955).
90. Hager, Ichniowski, and Wisek, *J. Amer. Pharm. Assoc. Sci. Edn.*, **43**, 156 (1954).
91. Albert, *J. Chem. Soc.*, **1955**, 2690.
92. Hull, *J. Chem. Soc.*, **1958**, 4069.
93. Ishidate and Yuki, *Chem. and Pharm. Bull* (*Japan*), **5**, 240 (1957).
94. Ishidate and Yuki, *Chem. and Pharm. Bull.* (*Japan*), **5**, 244 (1957).
95. Bredereck, Siegel, and Föhlisch, *Chem. Ber.*, **95**, 403 (1962).
96. Erhart and Hennig, *Arch. Pharm.*, **289**, 453 (1956).
97. Satoda, Fukui, Matsuo, and Okumura, *Yakugaku Kenkyu*, **28**, 633 (1956); through *Chem. Abstracts*, **51**, 16495 (1957).
98. Albert, Brown, and Cheeseman, *J. Chem. Soc.*, **1952**, 1620.
99. Bertho and Bentler, *Annalen*, **570**, 127 (1950).
100. Elion and Hitchings, *J. Amer. Chem. Soc.*, **69**, 2553 (1947).
101. Bayer, Ger. Pat., 213,711 (1908); *Frdl.*, **9**, 1010 (1908–1910).
102. Bayer, Ger. Pat., 209,728 (1908); *Frdl.*, **9**, 1009 (1908–1910).
102a. Von Schuh, Ger. Pat., 1,091,570 (1960); through *Chem. Abstracts*, **56**, 14305 (1962).

103. Johns and Baumann, *J. Biol. Chem.*, **15**, 515 (1913).
104. Koppel and Robins, *J. Org. Chem.*, **23**, 1457 (1958); Prasad, Noell, and Robins, *J. Amer. Chem. Soc.*, **81**, 193 (1959).
105. Acker and Castle, *J. Org. Chem.*, **23**, 2010 (1958).
106. Golovchinskaya, *Zhur. obshchei. Khim.*, **24**, 146 (1954); through *Chem. Abstracts*, **49**, 3205 (1955).
107. Lister, *J. Chem. Soc.*, **1963**, 2228.
108. Ueda, Tsuji, Momona, *Chem. and Pharm. Bull.* (*Japan*), **11**, 912 (1963).
109. Bredereck, Hennig, Pfleiderer, and Deschler, *Chem. Ber.*, **86**, 845 (1953).
110. Bredereck, Herlinger, and Resemann, *Chem. Ber.*, **93**, 236 (1960).
111. Bredereck, Küpsch, and Wieland, *Chem. Ber.*, **92**, 566 (1959).
112. Cohen and Vincze, *Israel J. Chem.*, **2**, 1 (1964).
113. Khemelevski, *Zhur. obshchei. Khim.*, **31**, 3123 (1961); through *Chem. Abstracts*, **56**, 15505 (1962).
114. Biltz and Schmidt, *Annalen*, **431**, 70 (1923).
115. Strukov and Diskina, *Farmatsiya*, **8**, 16 (1945).
116. Golovchinskaya, *Zhur. obshchei Khim.*, **19**, 1173 (1946); through *Chem. Abstracts*, **42**, 2580 (1948).
117. Huston and Allen, *J. Amer. Chem. Soc.*, **56**, 1793 (1934).
118. Haggerty, Springer, and Cheng, *J. Medicin. Chem.*, **8**, 797 (1965).
119. Mautner, *J. Org. Chem.*, **26**, 1914 (1961).
120. Birkofer, Ritter, and Kühlthau, *Chem. Ber.*, **97**, 934 (1964).
121. Falco, Elion, Burgi, and Hitchings, *J. Amer. Chem. Soc.*, **74**, 4897 (1952).
122. Elion, Burgi, and Hitchings, *J. Amer. Chem. Soc.*, **73**, 5235 (1951).
123. Fu, Chinoporos, and Terzian, *J. Org. Chem.*, **30**, 1916 (1965).
124. Fu and Chinoporos, *J. Heterocyclic Chem.*, **3**, 476 (1966).
125. Hitchings and Elion, U.S. Pat., 2,628,235 (1953); through *Chem. Abstracts*, **48**, 713 (1954).
126. Wellcome Foundation Ltd., Brit. Pat., 691,427 (1953); through *Chem. Abstracts*, **48**, 4003 (1954).
127. Adams and Whitmore, *J. Amer. Chem. Soc.*, **67**, 1271 (1945).
128. Uretskaya, Rybkina, and Menshikov, *Zhur. obshchei Khim.*, **30**, 327 (1960).
129. Montgomery and Temple, *J. Org. Chem.*, **25**, 395 (1960).
130. Montgomery, *J. Amer. Chem. Soc.*, **78**, 1928 (1956).
131. Goldman, Marsico, and Gazzola, *J. Org. Chem.*, **21**, 599 (1956).
132. Beaman and Robins, *J. Medicin. Chem.*, **5**, 1067 (1962).
133. Montgomery and Holum, *J. Amer. Chem. Soc.*, **80**, 404 (1958).
134. Ikehara and Ohtsuka, *Chem. and Pharm. Bull.* (*Japan*), **9**, 27 (1961).
135. Hull, *J. Chem. Soc.*, **1958**, 2746.
136. Pfleiderer, *Annalen*, **647**, 161 (1961).
137. Temple, Kussner, and Montgomery, *J. Medicin. Chem.*, **5**, 866 (1962).
138. Temple, McKee, and Montgomery, *J. Org. Chem.*, **28**, 923 (1963).
139. Montgomery and Temple, *J. Amer. Chem. Soc.*, **82**, 4592 (1960).
140. Montgomery and Temple, *J. Amer. Chem. Soc.*, **79**, 5238 (1957).
141. Post and Erikson, *J. Org. Chem.*, **2**, 260 (1938).
142. Montgomery and Temple, *J. Amer. Chem. Soc.*, **80**, 409 (1958).
143. Ulbricht, Personal communication.
144. Schaeffer and Weimar, *J. Org. Chem.*, **25**, 774 (1960).
145. Schaeffer and Weimar, *J. Amer. Chem. Soc.*, **81**, 197 (1959).
146. Blackburn and Johnson, *J. Chem. Soc.*, **1960**, 4358.

147. Fox and Van Praag, *J. Org. Chem.*, **26**, 526 (1961).
148. Getler, Roll, Tinker, and Brown, *J. Biol. Chem.*, **178**, 259 (1949).
149. Boehringer, Ger. Pat., 854,953 (1952); through *Chem. Abstracts*, **47**, 4905 (1953).
150. Berezovski and Yurkevich, *Zhur. obshchei Khim.*, **32**, 1655 (1962).
151. Elion, *J. Org. Chem.*, **27**, 2478 (1962).
152. Leese and Timmis, *J. Chem. Soc.*, **1958**, 4107.
153. Lister, *J. Chem. Soc.*, **1960**, 3682.
154. Lister, *J. Chem. Soc.*, **1960**, 899.
155. Ramage and Trappe, *J. Chem. Soc.*, **1952**, 4410.
156. Bredereck, Hennig, and Weber, Brit. Pat., 706,424 (1954); through *Chem. Abstracts*, **49**, 9045 (1955).
157. Pfleiderer and Schundehütte, *Annalen*, **612**, 158 (1958).
158. Soemmer, Ger. Pat., 1,152,108 (1963); through *Chem. Abstracts*, **60**, 2974 (1964).
159. Bredereck, Von Schuh, and Martini, *Chem. Ber.*, **83**, 201 (1950).
160. Scheuing and Konz, Ger. Pat., 804,210 (1951); through *Chem. Abstracts*, **45**, 8037 (1951).
161. Dikstein, Bergmann, and Chaimovitz, *J. Biol. Chem.*, **221**, 239 (1956).
162. Bredereck and Von Schuh, Ger. Pat., 864,870 (1953); through *Chem. Abstracts*, **47**, 11237 (1953).
163. Saum, Dissertation, Technischen Hochschule, Stuttgart (1959).
164. Bredereck and Von Schuh, Ger. Pat., 864,868 (1953); through *Chem. Abstracts*, **47**, 11238 (1953).
165. Merz and Stahl, *Arzneim.-Forsch.*, **15**, 10 (1965).
166. Vilsmeier and Haack, *Ber.*, **60**, 119 (1927).
167. Clark and Ramage, *J. Chem. Soc.*, **1958**, 2821.
168. Clark and Lister, *J. Chem. Soc.*, **1961**, 5048.
169. Davis, Knevel, and Jenkins, *J. Org. Chem.*, **27**, 1919 (1962).
170. Grundmann and Kreutzberger, *J. Amer. Chem. Soc.*, **77**, 6559 (1955).
171. Bergmann and Tamari, *J. Chem. Soc.*, **1961**, 4468.
172. Bredereck, Effenberger, and Österlin, *Chem. Ber.*, **100**, 2280 (1967).
173. Bergmann, Neiman, and Kleiner, *J. Chem. Soc.* (C), **1966**, 10.
174. Bergmann, Kalmus, Ungar, and Kwietny-Govrin, *J. Chem. Soc.*, **1963**, 3729.
175. Bergmann, Rashi, Kleiner, and Knafo, *J. Chem. Soc.* (C), **1967**, 1254.
176. Johns and Bauman, *J. Biol. Chem.*, **14**, 381 (1913).
177. Lewis, Beaman, and Robins, *Canadian J. Chem.*, **41**, 1807 (1963).
178. Robins, *J. Amer. Chem. Soc.*, **80**, 6671 (1958).
179. Cavalieri, Blair, and Brown, *J. Amer. Chem. Soc.*, **70**, 1240 (1948).
180. Clusius and Vecchi, *Helv. Chim. Acta*, **36**, 1324 (1953).
181. Brown, *The Chemistry and Biology of Purines*, Eds., Wolstenholme and O'Connor J and A Churchill, London, 1957, p. 50.
182. Bergmann, Ungar, and Kalmus, *Biochim. Biophys. Acta*, **45**, 49 (1960).
183. Altman and Ben-Ishai, *J. Heterocyclic Chem.*, **5**, 679 (1968).
184. Johns and Hagan, *J. Biol. Chem.*, **14**, 299 (1913).
185. Levene and Senior, *J. Biol. Chem.*, **25**, 607 (1916).
186. King and King, *J. Chem. Soc.*, **1947**, 943.
187. Blicke and Schaaf, *J. Amer. Chem. Soc.*, **78**, 5857 (1956).
188. Golovchinskaya, Ovcharova, and Cherkasova, *Zhur. obshchei Khim.*, **30**, 3332 (1960).
189. Bredereck, Küpsch, and Wieland, *Chem. Ber.*, **92**, 583 (1959).
190. Cook and Thomas, *J. Chem. Soc.*, **1950**, 1888.

191. Blackburn and Johnson, *J. Chem. Soc.*, **1960**, 4347.
192. Birkofer, Kühlthau, and Ritter, *Chem. Ber.*, **93**, 2810 (1960).
193. Boehringer, Ger. Pat., 142,468 (1902); *Frdl.*, **7**, 668 (1902–1904).
194. Cook and Smith, *J. Chem. Soc.*, **1949**, 3001.
195. Noell and Robins, *J. Amer. Chem. Soc.*, **81**, 5997 (1959).
196. Cresswell, Strauss, and Brown, *J. Medicin. Chem.*, **6**, 817 (1963).
197. Dietz and Burgison, *J. Medicin. Chem.*, **9**, 160 (1966).
198. Brown, *J. Appl. Chem.*, **9**, 203 (1959).
199. American Cyanamid Co., Brit. Pat., 744,869 (1956); through *Chem. Abstracts*, **50**, 10136 (1956).
200. Baker and Schaub, U.S. Pat., 2,705,715 (1955); through *Chem. Abstracts*, **50**, 5041 (1956).
201. Balsiger, Fikes, Johnson, and Montgomery, *J. Org. Chem.*, **26**, 3386 (1961).
202. Pfleiderer and Bühler, *Chem. Ber.*, **99**, 3022 (1966).
203. Baker, Schaub, and Joseph, *J. Org. Chem.*, **19**, 638 (1954).
204. Brown and Harper, *J. Chem. Soc.*, **1963**, 1276.
205. Boehringer, Ger. Pat., 128,117 (1902); *Chem. Zentr.*, **2**, 314 (1902).
206. Cavalieri and Bendich, *J. Amer. Chem. Soc.*, **72**, 2587 (1950).
207. Krackov and Christensen, *J. Org. Chem.*, **28**, 2677 (1963).
207a. Temple, Smith, and Montgomery, *J. Org. Chem.*, **33**, 530 (1968).
208. Forrest, Hatfield, and Lagowski, *J. Chem. Soc.*, **1961**, 963.
209. Lohrmann, Lagowski, and Forrest, *J. Chem. Soc.*, **1964**, 451.
210. Scholtissek, *Chem. Ber.*, **89**, 2562 (1956).
211. Taylor, Barton, and Paudler, *J. Org. Chem.*, **26**, 4961 (1961).
212. Biltz and Peukert, *Ber.*, **58**, 2190 (1925).
213. Bergmann and Dikstein, *J. Amer. Chem. Soc.*, **77**, 691 (1955).
214. Biltz and Bülow, *Annalen*, **426**, 246 (1922).
215. Biltz and Heidrich, *Annalen*, **426**, 269 (1922).
216. Altman and Ben-Ishai, *Bull. Res. Council Israel*, **11A**, 4 (1962).
217. Temple, McKee, and Montgomery, *J. Org. Chem.*, **28**, 2257 (1963).
218. Traube and Nithack, *Ber.*, **39**, 227 (1906).
219. Jerchel, Kracht, and Krucker, *Annalen*, **590**, 232 (1954).
220. Jerchel, Ger. Pat., 955,861 (1957); through *Chem. Abstracts*, **53**, 4317 (1959).
221. Ridley, Spickett, and Timmis, *J. Heterocyclic Chem.*, **2**, 453 (1965).
222. Fidler and Wood, *J. Chem. Soc.*, **1956**, 3311.
223. Fidler and Wood, *J. Chem. Soc.*, **1957**, 3980.
224. Ried and Torinus, *Chem. Ber.*, **92**, 2902 (1959).
225. Temple and Montgomery, *J. Org. Chem.*, **31**, 1417 (1966).
225a. Giner-Sorolla and Brown, unpublished result.
226. Liau, Yamashita, and Matsui, *Agric. and Biol. Chem.* (*Japan*), **26**, 624 (1962).
227. Scheuing and Konz, Ger. Pat., 834,105 (1952); through *Chem. Abstracts*, **47**, 1733 (1953).
228. Ishidate, Sekiya, and Kurita, *J. Pharm. Soc. Japan*, **74**, 420 (1954).
229. Bredereck and Edenhofer, *Chem. Ber.*, **88**, 1306 (1955).
230. Vogl and Taylor, *J. Amer. Chem. Soc.*, **79**, 1518 (1957).
231. Taylor, Vogel, and Cheng, *J. Amer. Chem. Soc.*, **81**, 2442 (1959).
232. Fujimoto and Ono, *J. Pharm. Soc. Japan*, **86**, 3671 (1965).
233. Levin and Tamari, *J. Chem. Soc.*, **1960**, 2782.
234. Taylor and Morrison, *J. Amer. Chem. Soc.*, **87**, 1976 (1965).
235. Lister and Timmis, *J. Chem. Soc.*, **1960**, 1113.

236. Kempter, Rokos, and Pfleiderer, *Angew. Chem. Internat. Edn.*, **6**, 258 (1967).
237. Taylor, Jefford, and Cheng, *J. Amer. Chem. Soc.*, **83**, 1261 (1961).
238. Taylor and Jefford, *J. Amer. Chem. Soc.*, **84**, 3744 (1962).
239. Goldner, Dietz, and Carstens, *Naturwiss.*, **51**, 137 (1964).
240. Goldner, Dietz, and Carstens, *Annalen*, **691**, 142 (1966).
241. Bühler and Pfleiderer, *Angew. Chem.*, **77**, 129 (1965).
242. Pfleiderer and Kempter, *Angew. Chem. Internat. Edn.*, **6**, 258 (1967).
243. Goldner, Dietz, and Carstens, Fr. Pat., 1,367,786 (1964); through *Chem. Abstracts*, **62**, 575 (1965).
244. Goldner, Dietz, and Carstens, *Z. Chem.*, **4**, 454 (1964).
245. Goldner, Dietz, and Carstens, *Tetrahedron Letters*, **1965**, 2701; *Annalen*, **698**, 145 (1966).
246. Whitehead and Traverso, *J. Amer. Chem. Soc.*, **82**, 3971 (1960).
247. Taylor and Garcia, *J. Amer. Chem. Soc.*, **86**, 4720 (1964).
248. Pfleiderer and Blank, *Angew. Chem. Internat. Edn.*, **5**, 666 (1966).
249. Taylor and Garcia, *J. Amer. Chem. Soc.*, **86**, 4721 (1964).
249a. Yoneda, Ogiwara, Kanahori, and Nishigaki, *Chem. Comm.*, **1970**, 1068.
250. Fischer and Ach, *Ber.*, **28**, 2473 (1895).
251. Fischer, *Ber.*, **30**, 559 (1897).
252. Armstrong, *Ber.*, **30**, 2308 (1897).
253. Fischer, *Ber.*, **33**, 1701 (1900).
254. Biltz and Strufe, *Annalen*, **423**, 242 (1921).
255. Borowitz, Bloom, and Sprinson, *Biochemistry*, **4**, 650 (1965).
256. Fischer and Clemm, *Ber.*, **30**, 3089 (1897).
257. Biltz, Marwitzky, and Heyn, *Annalen*, **423**, 122 (1921).
258. Biltz, Marwitzky, and Heyn, *Annalen*, **423**, 147 (1921).
259. Biltz and Zellner, *Annalen*, **423**, 192 (1923).
260. Gatewood, *J. Amer. Chem. Soc.*, **45**, 3056 (1921).
261. Biltz and Heyn, *Annalen*, **423**, 185 (1921).
262. Sembritsky, *Ber.*, **30**, 1814 (1897).
263. Whitely, *J. Chem. Soc.*, **91**, 1330 (1907).
264. Miles and Bogert, *J. Amer. Chem. Soc.*, **62**, 1173 (1940).
265. Koppel and Robins, *J. Amer. Chem. Soc.*, **80**, 2751 (1958).
266. Biltz and Bülow, *Annalen*, **426**, 299 (1922).
267. Biltz and Strufe, *Annalen*, **423**, 200 (1921).
268. Gulland and Story, *J. Chem. Soc.*, **1938**, 692.
269. Biltz and Heidrich, *Annalen*, **426**, 290 (1922).
270. Biltz and Bülow, *Annalen*, **426**, 283 (1922).
271. Fischer and Tullner, *Ber.*, **35**, 2563 (1902).
272. Pfleiderer and Nübel, *Annalen*, **631**, 168 (1960).
273. Viout and Rumpf, *Bull. Soc. chim. France*, **1962**, 1250.
274. Fischer, Neumann, and Roch, *Chem. Ber.*, **85**, 752 (1952).
275. Behrend and Roosen, *Annalen*, **251**, 235 (1888).
276. Prusse, *Annalen*, **441**, 203 (1925).
277. Bredereck, Effenberger, and Schweizer, *Chem. Ber.*, **95**, 956 (1962).
278. Dornow and Hinz, *Chem. Ber.*, **91**, 1834 (1958).
279. Dornow, Ger. Pat., 1,064,950 (1959); through *Chem. Abstracts*, **55**, 10483 (1961).
280. Barone, *J. Medicin. Chem.*, **6**, 39 (1963).
281. Taylor and Knopf in *The Chemistry and Biology of Purines*, Eds., Wolstenholme and O'Connor, J and A Churchill, London, 1957, p. 36.

282. Cook, Downer, and Heilbron, *J. Chem. Soc.*, **1949**, 1069.
283. Ohtsuka, *Bull. Chem. Soc. Japan*, **43**, 954 (1970).
284. Ohtsuka and Sugimoto, *Bull. Chem. Soc. Japan*, **43**, 2281 (1970)
285. Temple, McKee, and Montgomery, *J. Amer. Chem. Soc.*, **27**, 1671 (1962).
286. Goldner, Dietz, and Carstens, *Annalen*, **692**, 134 (1966).

CHAPTER III

Purine Syntheses from Imidazoles and Other Precursors

Although the greatest number of purines synthesised to date have been derived from pyrimidine precursors, the use of alternative intermediates is of growing importance. Imidazole derivatives and, to a lesser degree, noncyclic intermediates are now being widely exploited for this purpose. Routes starting with the noncyclic intermediates can be considered to be extensions of standard imidazole syntheses as mechanistic studies show that in such cases purine formation proceeds by way of imidazole precursors. It has also been demonstrated that very simple compounds, such as hydrogen cyanide and ammonia, combine under mild conditions to form naturally occurring purines. This work, like the preceding, shows that imidazoles rather than pyrimidines are initially produced and provides a link between purely chemical syntheses on the one hand and biological syntheses, which also involve imidazole intermediates, on the other.

1. Using Imidazoles

The first purine prepared by this route was due to Sarasin and Wegmann in 1924[1] but the lack of other suitable imidazole derivatives at that time resulted in little subsequent attention being given to this approach, which remained dormant for nearly twenty years. A renewed interest was shown by workers in the nucleoside field for whom this approach offered a means of preparing purine glycosides under mild conditions. Although imidazole derivatives can be converted to purines under these conditions, in practice only a restricted number of nucleosides have been obtained in this way. Nevertheless, the use of imidazoles has greatly benefited the syntheses of purines and has enabled the preparation of some derivatives not possible by other methods.

A. Syntheses from 4,5-Dicarbamoylimidazoles

In alkaline potassium hypobromite solution at 0° 4,5-dicarbamoyl-imidazole (**1**) undergoes a Hofmann reaction from which xanthine (**2**, R = H) is obtained in good yield.[2] By using the *N*-methylated imida-zole 9-methylxanthine (**2**, R = Me) is formed.[3, 4] This result, which is confirmed by later examples, shows that the major cyclisation product of such *N*-alkylimidazoles is the appropriate 9-alkylxanthine and demonstrates that the carbamoyl group on the carbon atom adjacent to the substituted ring nitrogen is the one converted to an isocyanate group. A small amount of a 7-substituted xanthine has, however, been isolated along with the 9-isomer in one preparation.[5] For large-scale preparations the hypobromite may be replaced by sodium hypo-chlorite.[6] Although the procedure is limited to preparing xanthine derivatives, it has been explored as a possible route to nucleosides, a number of xanthine-9-glycosides being derived from the appropriate 4,5-dicarbamoyl-1-glycosylimidazoles. Pilot experiments were success-ful with the 9-D-mannopyranosyl-,[7] 9-D-ribopyranosyl-,[7] 9-D-xylo-pyranosyl-,[5, 8] and 9-D-glycopyranosylxanthines[5] but failed with the 9-D-arabinopyranosyl derivative.[8] The nucleoside, xanthosine, having the ribose moiety in the furanose form, was likewise obtained.[9] An interesting application of the Lossen rearrangement, which mechanisti-cally parallels the Hofmann reaction in that an essential part of the reaction entails conversion of a $-\overset{\overset{\textstyle O}{\cdots}}{C}-N$ group to an $-N{:}C{:}O$ group,

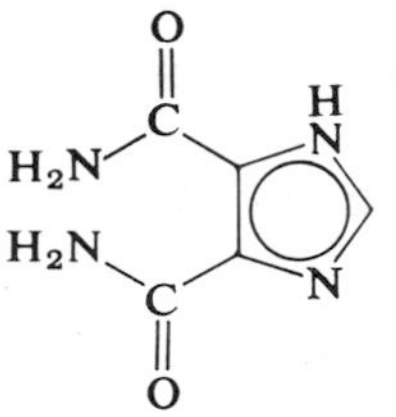

(1) 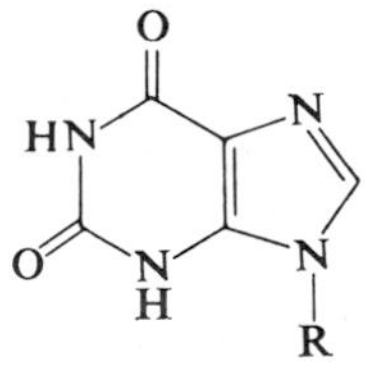   (2)

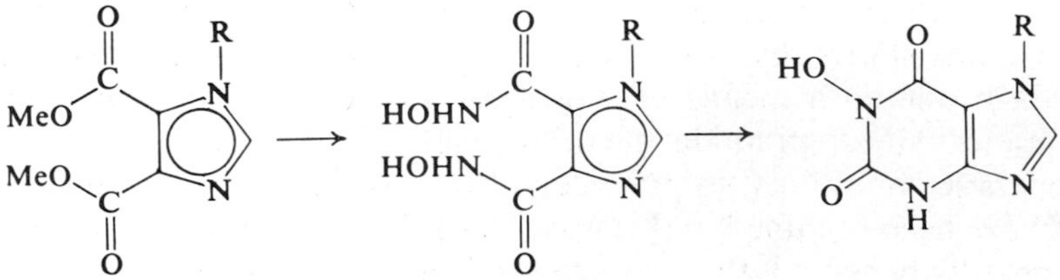

(3)   (4)   (5)

is to the formation of xanthine-1-oxides. Although this reaction is given more detailed treatment elsewhere (Ch. XI, Sect. 1A), it is exemplified, for comparison, by the cyclisation of the 4,5-di-(N-hydroxy-carbamoyl)imidazole (4), obtained from the reaction of 3 with hydroxyl-amine, to a xanthine-1-oxide (5) when treated with benzenesulphonyl chloride in tetrahydrofuran. In addition to the parent derivative (5, R = H),[10] this procedure has given the 7-methyl (5, R = Me) and 7-benzyl (5, R = $CH_2C_6H_5$) derivatives among others.[11] A noteworthy point is that in contrast to the Hofmann reaction Lossen rearrangement of an N-alkylimidazole gives the 7-alkylpurine.

B. Syntheses from 4(5)-Amino-5(4)-carbamoylimidazoles

Sarasin and Wegmann were the first to demonstrate a purine synthesis from an imidazole obtaining 7-methylxanthine (7) by heating 4-amino-5-carbamoyl-1-methylimidazole (6) with ethyl carbonate.[1] Development of this route allow formation of a wide range of purines. In many

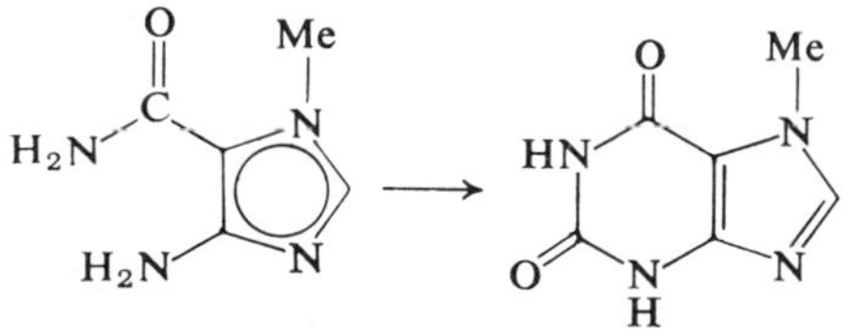

(6)   (7)

respects the procedure is analogous to the Traube synthesis in that both require insertion of a one-carbon moiety between two nitrogen atoms in order to form the corresponding ring. It is not surprising, therefore, to find that the majority of reagents and conditions are common to both types of cyclisation.

a. Cyclisation with Formic Acid

Formylation is best made with a mixture of formic acid and acetic anhydride, the intermediate formylated imidazole then being further treated to effect cyclisation. In some examples[12-15] prolonged refluxing alone suffices to give the hypoxanthine derivative directly. Hypoxanthine itself (10, R = H) follows conversion of 4(5)-amino-5(4)-carbamoyl-imidazole (8, R = H) to the formyl derivative (9, R = H), and sub-sequent closure is effected by heating in aqueous sodium bicarbonate

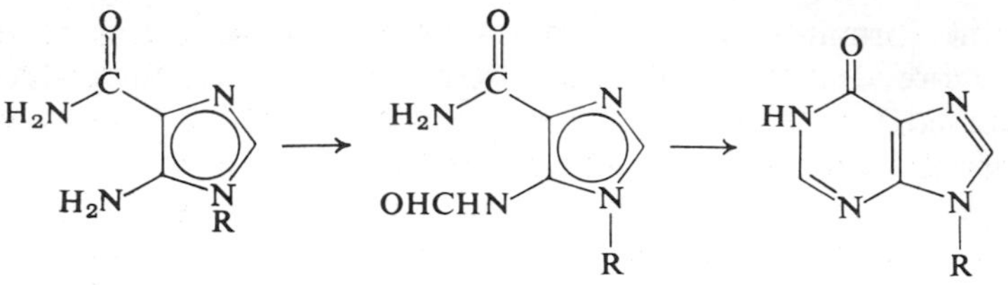

(8)　　　　　　　　(9)　　　　　　　　(10)

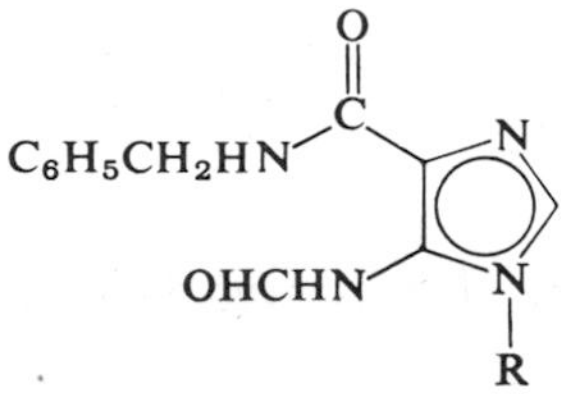

(11)

solution.[16, 17] These conditions are sufficiently mild to allow the formation of the nucleoside, inosine (**10**, R = β-D-ribofuranosyl).[18-20] The isomeric 7-riboside was likewise obtained[19] although a partial loss of the ribose moiety was noted. In some instances cold alkali alone will induce cyclisation as, for example, in the formation of 1-benzylinosine on allowing the imidazole (**11**, R = β-D-ribofuranosyl) to stand (48 h) in 3N-sodium hydroxide solution at room temperature.[21] Alkaline cyclisation also occurs under nonaqueous conditions, hypoxanthine resulting from a closure with sodium ethoxide in ethanol.[16] The formyl derivative of the imidazole may also be ring closed in a high-boiling solvent; formamide being employed in forming 1-benzylhypoxanthine.[21]

b. *Cyclisation with Esters and Orthoesters*

Both alkyl esters and trialkyl orthoesters are versatile reagents as they allow preparation of hypoxanthine derivatives containing alkyl (or aryl) groups at $C_{(2)}$. Alkyl esters have had only limited application, examples being the formation of hypoxanthine, using ethyl formate,[21a] and the 2-methyl[21a, 22] and 2-ethyl[21a] analogues from ethyl acetate and ethyl propionate, respectively, the ring closures being carried out in ethanolic sodium ethoxide under reflux. By contrast extensive use of trialkyl orthoesters is found in the literature, these being almost invariably employed in the form of a mixture with acetic anhydride under reflux conditions for one or two hours.[23-30] Omission of the anhydride gives rise to the alkoxymethyleneamino derivative (**12**) but cyclisation to the purine follows on heating in acetic anhydride[23, 31] or dimethylformamide.[32, 33] Direct ring closures under the latter conditions are possible using a mixture of trialkyl orthoester and dimethylformamide, exemplified by the preparation of 8-methylhypoxanthine from the appropriate 2-methylimidazole,[32] while hypoxanthine-1-oxide is similarly derived from 4-amino-5-(N-hydroxy)carbamoylimidazole (13)[33] (see Ch. XI, Sect. 1A). Where more vigorous treatment is necessary to cyclise the alkoxymethyleneaminoimidazole, fusion and vacuum sublimation have been successful—the former in converting 4-carbamoyl-5-methylethoxymethyleneaminoimidazole (12, R = Me) to 2-methylhypoxanthine (14, R = Me, R′ = H) while the latter technique

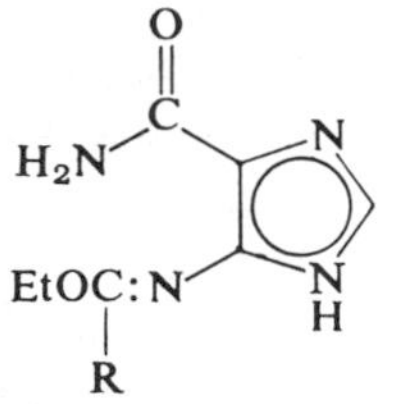

(12) 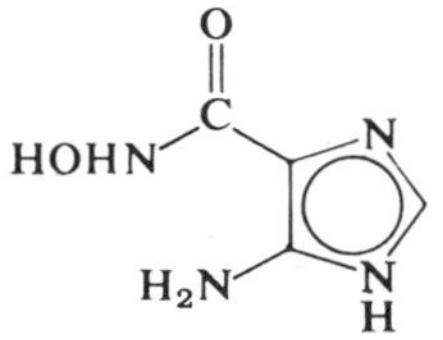   (13) 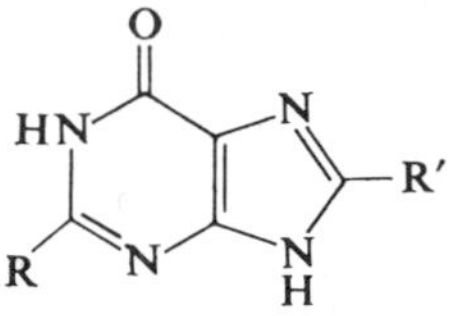   (14)

was adopted in forming 2,8-dimethyl- and 2,8-diphenyl-hypoxanthine (14, R = R′ = Me and R – R′ = C_6H_5) from the respective imidazoles.[32]

c. Cyclisation with Amides

Although wide use of these reagents is found in the pyrimidine series, only occasionally are they employed in imidazole closures. Hypoxanthine,[16] the 1-benzyl,[21] 9-benzyl,[15] and 7-methyl[34] derivatives

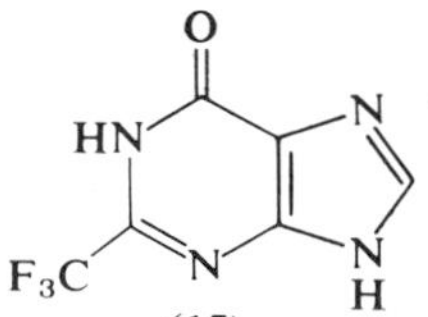

(15)

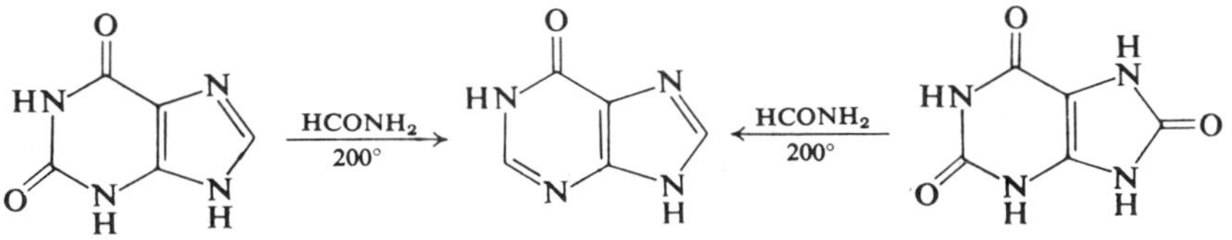

(16)   (17)   (18)

follow from using formamide. On heating for some hours 4-amino-5-carbamoylimidazole with trifluoroacetamide at the reflux point 2-trifluoromethylhypoxanthine (15) results.[35] The conversion of both xanthine (16)[36] and uric acid (18)[37] to hypoxanthine (17) on heating in formamide (200°) under pressure for some hours seems to be an extension of the above route. Opening of the pyrimidine ring in both purines to give a 4-amino-5-carbamoylimidazole must precede recyclisation by the formamide to hypoxanthine. In the case of uric acid (18) a concomitant opening and closing of the imidazole ring is also implicated.

d. Cyclisation with Urea

This provides a satisfactory route to xanthine derivatives. The parent member[16, 38] and 7-methyl homologues[29, 34] are formed in this way. 9-Amino-8-methylxanthine (20) is likewise obtained from 1,5-diamino-4-carbamoyl-2-methylimidazole (19).[23] These reactions require fusion temperatures (170–180°) but if lower temperatures are employed the intermediate ureidoimidazole can be isolated. One such derivative (21), from 1-methyl-4-methylamino-5-(N-methyl)carbamoylimidazole, is formed at 140° in the presence of hydrogen chloride. Conversion to theobromine (22) is effected by fusion or heating in dilute mineral acid with ensuing removal of the methyl group of the carbamoyl moiety.[39]

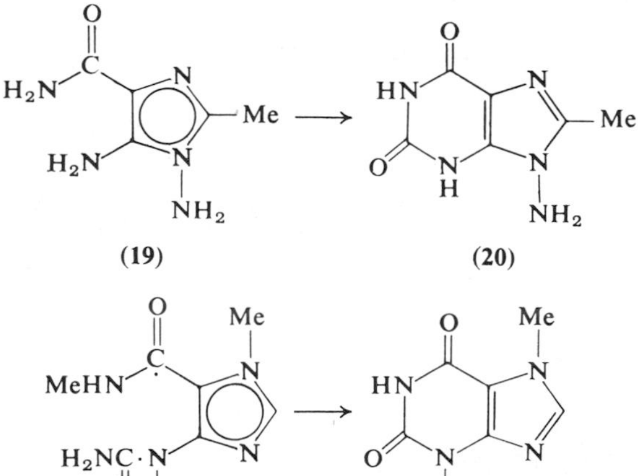

(19)   (20)

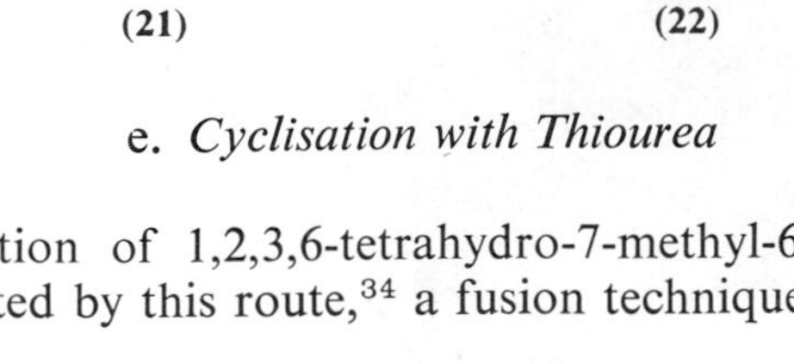

(21)   (22)

e. Cyclisation with Thiourea

The preparation of 1,2,3,6-tetrahydro-7-methyl-6-oxo-2-thiopurine has been reported by this route,[34] a fusion technique being employed.

f. *Cyclisation with Cyanates*

In aqueous solution, at room temperature, caffeidine (**23**) on addition of potassium cyanate is rapidly converted to caffeine (**24**).[39] Under these conditions 4-amino-5-carbamoyl-1-methyl-2-methylthioimidazole failed to react.[13]

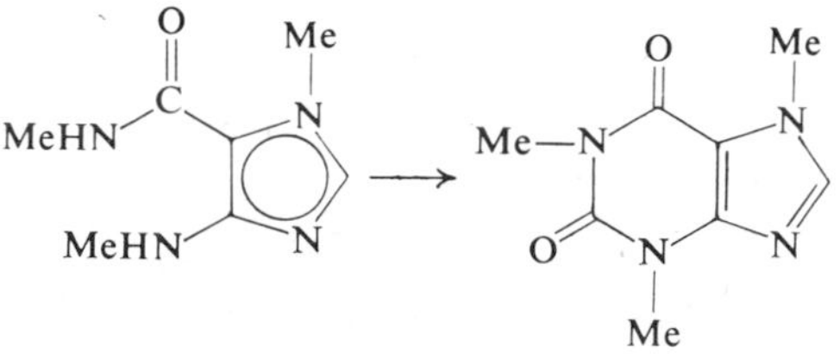

(23)   (24)

g. *Cyclisation with Isothiocyanates*

These reagents show an unusual versatility in that as well as leading to the expected 6-oxo-2-thiopurines they can be utilised for 6-amino-

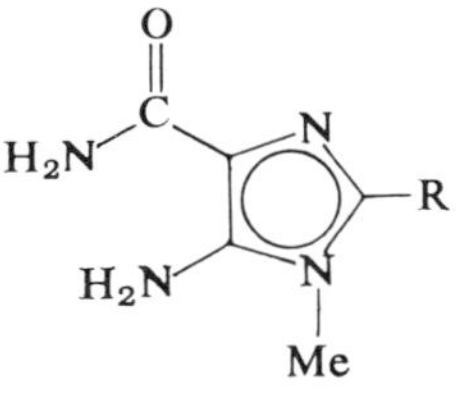

(25) 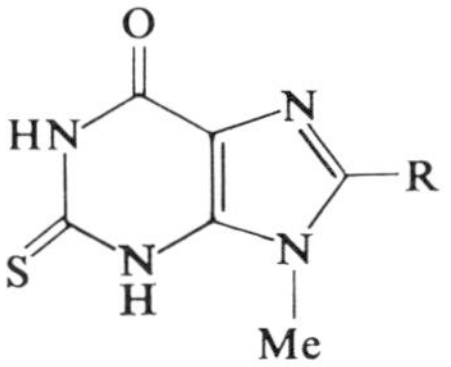   (26) 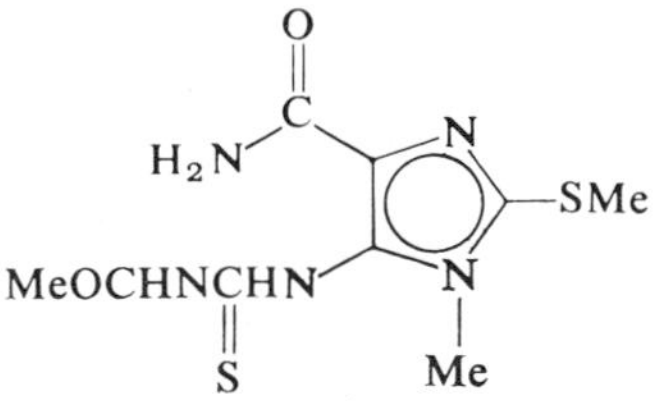   (27)

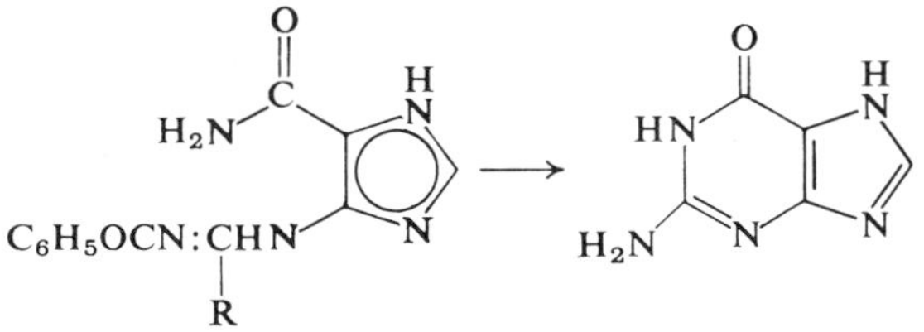

(28)   (29)

purine formation. In boiling pyridine containing methyl isothiocyanate, the 4-amino-5-carbamoylimidazoles (**25**, R = H and SMe) are directly converted to the purines (**26**, R = H and SMe).[13] The latter (**26**, R = SMe) was also obtained by the action of acetyl isothiocyanate on the imidazole (**25**, R = SMe) in ethyl acetate, the thioureido intermediate (**27**) in this case being isolated and subsequently cyclised by briefly heating in dilute sodium hydroxide.[13] This route has been extended to the preparation of guanine derivatives; guanine itself (29) results from the action of benzoyl isothiocyanate on 4(5)-amino-5(4)-carbamoylimidazole, followed by *S*-methylation of the product (28 R = SH) to 28 (R = SMe) and then amination to the guanidinyl derivative (28, R = NH$_2$) before cyclisation to 29 in hot alkali.[40] If thioureidoimidazoles such as 27 are treated before cyclisation with hot phosphoryl chloride, dichloro derivatives of 4-cyano-5-thioureido-imidazoles result which undergo ring closure in alkali affording 2-thioadenines (see Sect. 1Eb).

h. *Cyclisation with Chlorocarbonic Esters*

The milder conditions required with this type of reagent favour its use over that of urea for xanthine syntheses. In some instances isolation of the intermediate ethoxycarbonylamino derivative has been made but this is not essential as direct cyclisation is possible. The procedure is illustrated by the preparation of xanthine; condensation of 4(5)-amino-5(4)-carbamoylimidazole with ethyl chlorocarbonate is carried out at 0° being followed by thermal cyclisation for which various conditions are used including fusion and heating in concentrated ammonia solution or nitrobenzene.[16] In like fashion the imidazole (30), prepared from 5-amino-4-carbamoyl-1-methyl-2-methylthioimidazole and phenylthio-

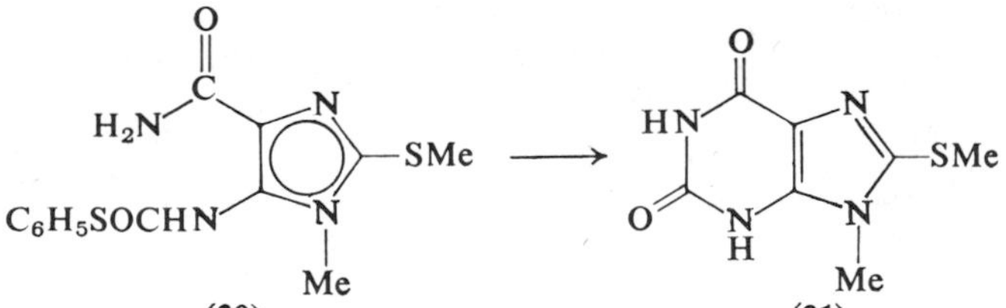

(30)　　　　　(31)

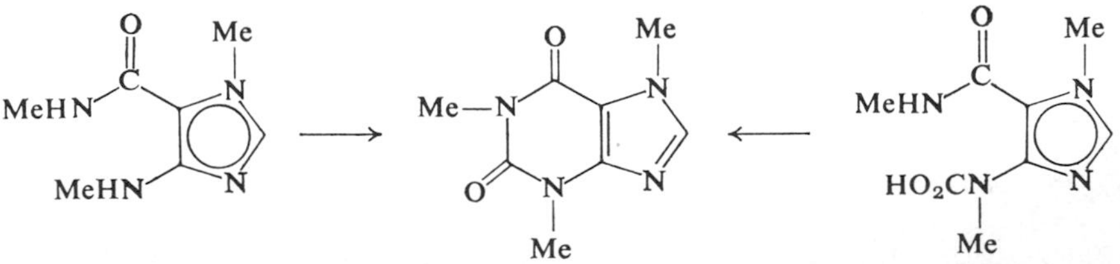

(32)　　　　　(33)　　　　　(34)

chlorocarbonate (C$_6$H$_5$SCOCl) gives 9-methyl-8-methylthioxanthine (31) after a short period in boiling dioxan.[12] The same solvent is a good medium in which to effect direct cyclisation, the 4-amino-5-carbonyl imidazole and ethyl chlorocarbonate being heated together for some hours in the presence of potassium carbonate. Examples by this route are 1,7-dimethyl-,[41, 42] 1-ethyl-7-methyl-[41] and 1,7-diethyl-8-methyl-,[41] 7,8-dialkyl-,[29] and 1,3,8,9-tetramethylxanthine.[43] The ring closure of caffeidine (32), as nitrate, to caffeine (33) takes place likewise in aqueous bicarbonate,[39] but conversion of the intermediate caffeidine carboxylic acid (34) to 33 is made in phosphoryl chloride (115°/3 h).[44]

i. Cyclisation with Diethyl Carbonate

Little use has been made of this volatile reagent for preparing xanthine derivatives. The formation of 7-methyl[1] and 7-ethyl-8-methyl[45] derivatives by early workers employed sealed-tube procedures but a recent synthesis of xanthine[21a] itself was effected in ethanolic sodium ethoxide under reflux.

j. Cyclisation with Carbon Disulphide

This procedure provides a workable route to 6-oxo-2-thiopurine derivatives, the reaction being carried out in pyridine under reflux conditions for some hours. The 9-methyl-2-thio- (36, R = H) and 2,8-

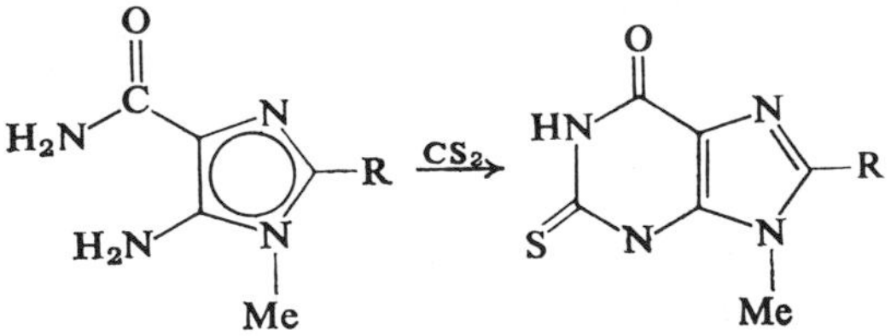

(35)   (36)

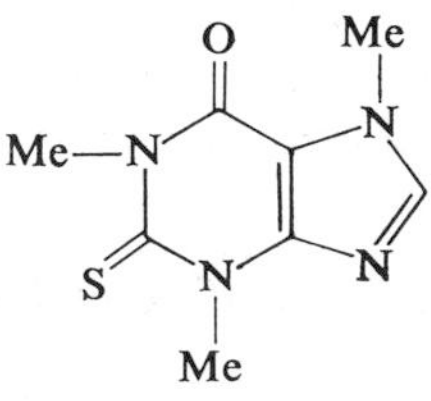

(37) 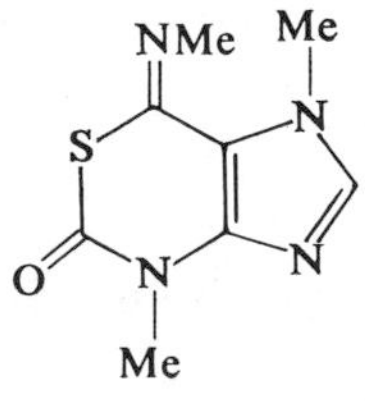   (37a)

dithio-9-methyl-oxopurines (36, R = SH) are thus derived from the imidazoles (35, R = H and SH).[12]

k. *Cyclisation with Thiophosgene and Thioesters*

The product obtained by shaking caffeidine nitrate (**32**) with thio-phosgene in aqueous sodium carbonate, originally formulated[39] as the 2-thio analogue (**37**) of caffeine, has now been shown[45a] to be the isomeric 1,4,5,7-tetrahydro-1,4-dimethyl-7-methylimino-5-oxoimidazo-[4,5-*d*][1,3]thiazine (**37a**). Conversion of 4(5)-amino-5(4)-carbamoylimi-dazole hydrochloride to 2-thiohypoxanthine is effected on heating in dimethylformamide with sodium methyl xanthate ($MeOCS_2Na$) under reflux conditions.[46]

C. Syntheses from 4(5)-Amino-5(4)-thiocarbamoylimidazoles

Thiocarbamoylimidazoles are derived from the appropriate car-bamoyl derivative by interaction with phosphorus pentasulphide in pyridine[47, 48] or by direct synthesis.[49] Cyclisation procedures, which follow those described for the carbamoyl compounds, give the 6-thio-purine analogues. Closure with formic acid–acetic anhydride mixtures affords initially a formamido derivative (**38**), which is converted to the

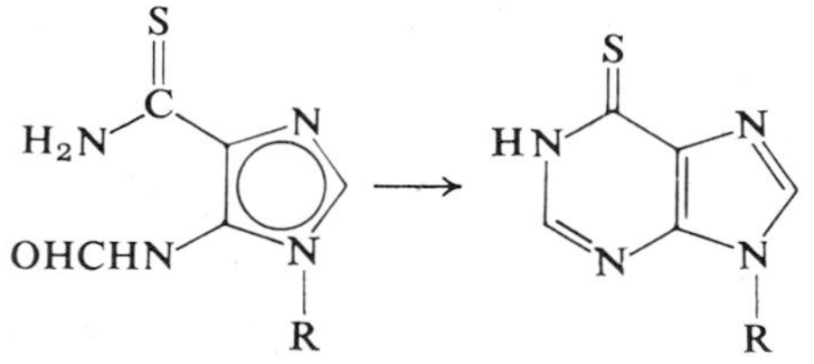

(**38**)   (**39**)

purine in boiling aqueous bicarbonate solution,[49] examples being 1,6-dihydro-9-methyl- (**39**, R = Me) and 1,6-dihydro-8,9-dimethyl-6-thiopurine. An attempt to prepare the parent member (**39**, R = H) by this route gives only hypoxanthine[48] but successful closures with form-amide[47] or ethyl formate in sodium ethoxide[49a] can be made. Fusion with urea gives 1,2,3,6-tetrahydro-2-oxo-6-thiopurine.[49a]

D. Syntheses from 4(5)-Amino-5(4)-amidinoimidazoles

These imidazoles provide useful intermediates for the synthesis of adenine derivatives, the parent member (**41**) being produced by formyla-tion of **40** (R = H) using a formic acid–acetic anhydride mixture followed by aqueous alkali treatment of the product (**40**, R = CHO).[16] Cyclisation of the formyl derivative occurs more readily than it does with the corresponding formylated carbamoyl compound; even on crystallisation from hot water some adenine is formed.[16] Using isotopically labelled formic acid, adenine marked at the 2-position is obtained.[50] Reflux conditions at 170° with trifluoroacetamide has given 6-amino-2-trifluoromethylpurine (**42**).[35] Urea fusion gives isoguanine (**43**),[16, 51] which is also formed by the action of phosgene on the imidazole in cold sodium hydroxide solution.[51] The route can be adapted to prepare adenine-1-oxide by cyclisation of 4-amino-5-(*N*-hydroxy)amidinoimidazole (**44**) with triethyl orthoformate in dimethyl-

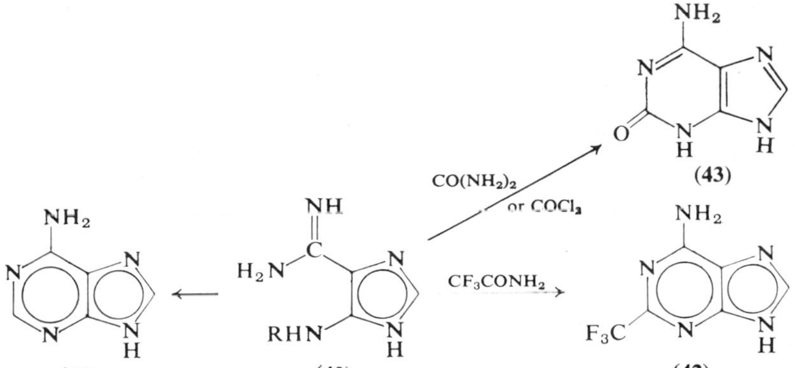

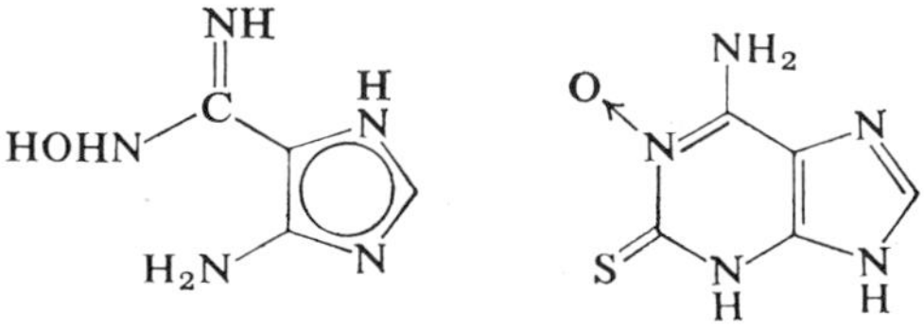

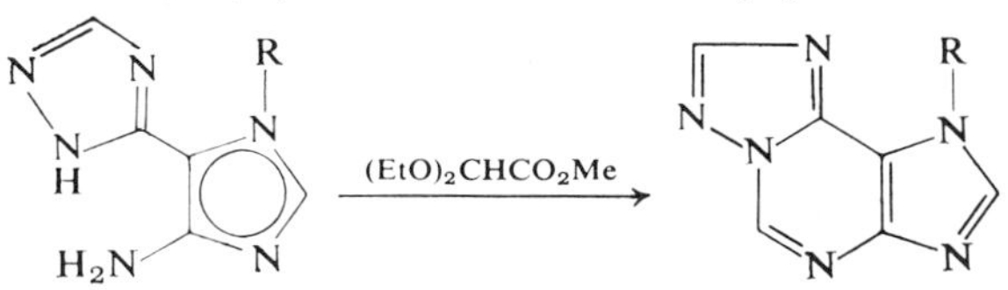

formamide (see Ch. XI, Sect. 1A). The corresponding 2-thio analogue (**45**) is notable for its ease of formation, being produced on allowing the imidazole (**44**, R = H) to stand at room temperature (2 days) in a solution of pyridine containing carbon disulphide.[52] Similarly, no heating is needed to cyclise the cyclic amidino derivatives (**46**, R = H or $CH_2C_6H_5$) to the S-triazolo[5,1-i]purines (**47**, R = H or $CH_2C_6H_5$), standing for some hours in diethoxymethyl acetate being sufficient.[53] The 9-benzyl isomer of **47** is likewise derived.[53]

E. Syntheses from 4(5)-Amino-5(4)-cyanoimidazoles

Although this approach gives adenine derivatives also it possesses distinct advantages over the above amidinoimidazole syntheses (Sect. 1D), the most important of these being that with the much wider range of cyanoimidazoles available considerably more synthetic scope is allowed.

a. *Cyclisation Involving Orthoesters*

Triethyl orthoformate readily forms the ethoxymethyleneaminoimidazole (**48**) from the corresponding 5-aminoimidazole and ring closure to 9-methyladenine (**49**) follows on heating with ethanolic ammonia in a sealed tube.[49] By contrast, a modified preparation of the 7-methyl analogue entails cyclisation of the isomeric N-methyl ethoxymethyleneamino intermediate (**50**) with ethanolic ammonia at room temperature.[54]

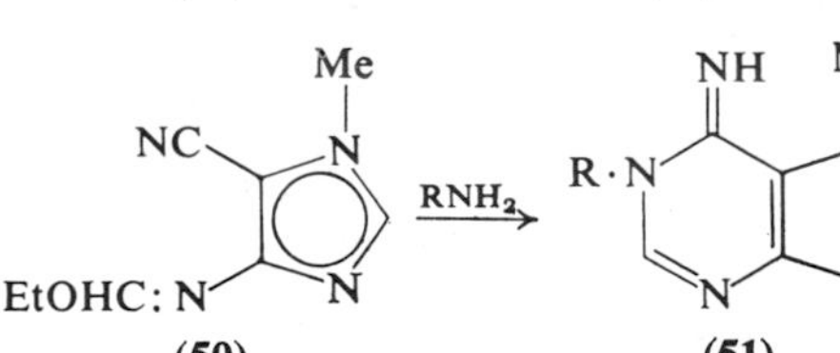

(**48**)   (**49**)

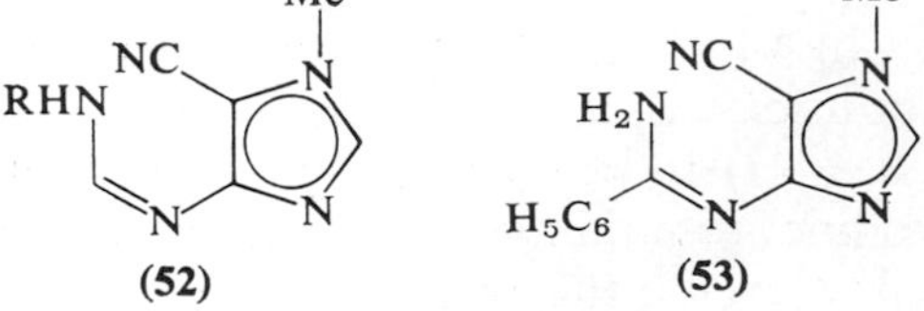

(**50**)   (**51**)

(**52**)   (**53**)

Replacing ammonia by amines and carrying out the reaction in boiling benzene produced the 1-methyl (**51**, R = Me) and 1-butyl (**51**, R = Bu) homologues,[54] while hydrazine affords the 1-amino derivative (**51**, R = NH$_2$).[54] The corresponding N-benzylimidazole as the ethoxy-methyleneamino derivative affords 7-benzyladenine after a few hours in ethanolic ammonia at room temperature while the 1,7-dibenzyl homologue results with benzylamine in hot ethyl acetate.[55] A formamidine intermediate of the type **52** is thought to be involved in such closures.[54] The synthesis of 6-amino-7-methyl-2-phenylpurine from 4-amino-5-cyano-1-methylimidazole and benzonitrile in methanolic ammonia under pressure[56] probably proceeds through the related derivative (**53**).

b. *Cyclisation with Amide-like Reagents*

Direct cyclisation to the purine is possible with these reagents. With formamide, under reflux, adenine[34] and the 8-methylthio analogue[57] both result from the appropriate 4-amino-5-cyanoimidazole. Although the same reagent failed to convert 5-cyano-4-hydroxyamino-1-methyl-imidazole into adenine-3-oxide, this purine was obtained using form-amidine acetate in ethanol[58] (Ch. XI, Sect. 1B). A good yield of adenine is claimed from reaction of formamidine acetate with 4(5)-amino-5(4)-cyanoimidazole in 2-methoxyethanol.[59, 60] Fusion with urea of the imidazole (**54**) gave the 2-oxopurine (**55**) directly.[61] Probably the same compound (**55**) was obtained by a synthesis,[62] now only of historic interest, using urethane in a sealed tube. Cyanoimidazoles are involved,

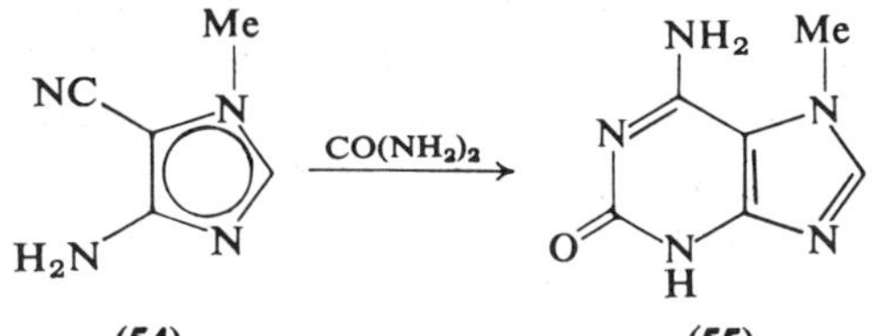

(**54**)   (**55**)

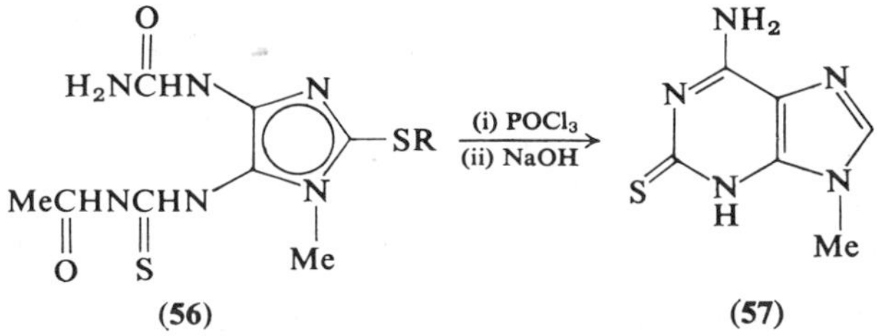

(**56**)   (**57**)

but not usually isolated, in the consecutive treatment of carbamoyl-imidazoles with phosphoryl chloride and dilute alkali. Examples of this procedure are provided by conversion of the imidazoles (**56**, R = Me and C_6H_5) to the purines (**57**, R = Me and C_6H_5). The intermediates, following the phosphoryl chloride treatment, are characterised as dichloro derivatives of 5-(N-acetyl)thioureido-4-cyanoimidazoles but no structures are presented.[13] The same route affords 6-amino-8-methyl-thio-9-phenyl-2-thiopurine.[13]

F. Syntheses from 4(5)-Amino-5(4)-alkoxycarbonyl-imidazoles

Cyclising agents, which include cyanates, isothiocyanates, and ureas, lead to derivatives of xanthine or the 6-oxo-2-thiopurine analogues.

a. *Cyclisation with Cyanates and Isocyanates*

Xanthine itself arose from a three-stage reaction, the initial step involving condensation of 4-amino-5-methoxycarbonylimidazole (**58**) with aqueous potassium cyanate following which the resulting ureido derivative (**59**, R = Me) is hydrolysed with dilute alkali to the acid (**59**, R = H)—the purine (**60**) being isolated on treatment of the latter with boiling 50% hydrochloric acid.[63] These conditions for ring closure are comparable with those used for conversion of 4-amino-5-ureido-pyrimidines to 8-oxopurines (Ch. II, Sect. 10). Subsequently, direct cyclisation of the ester form of the imidazole to the purine by means of dilute sodium hydroxide has been demonstrated.[64] With methyl iso-cyanate replacing the potassium cyanate 1-methylxanthines[64-67] are obtained. Uric acid and the 1,7-dimethyl analogue were similarly

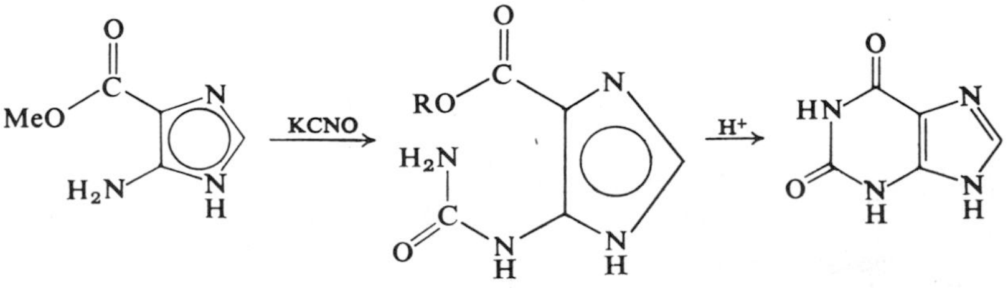

| (58)   (59)   (60) |

derived from the appropriate 2-hydroxyimidazoles.[68]

b. *Cyclisation with Urea*

The action of urea on the aminoimidazole in hot pyridine gives a ureidoimidazole which is then closed to the xanthine in the usual manner. In this way 8-phenyl- and 8-benzylxanthine, and 7-methyl-8-phenylxanthine, have been formed but the yields are poor.[66]

c. *Cyclisation with Isothiocyanates*

This forms a useful means of preparing 6-oxo-2-thiopurine derivatives, the procedures used following those previously employed for cyanates

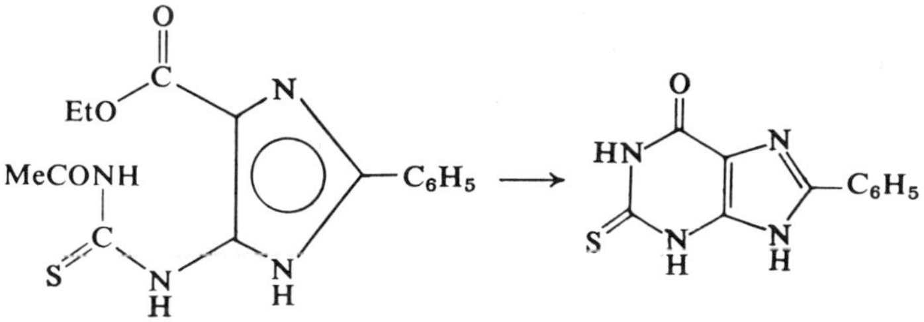

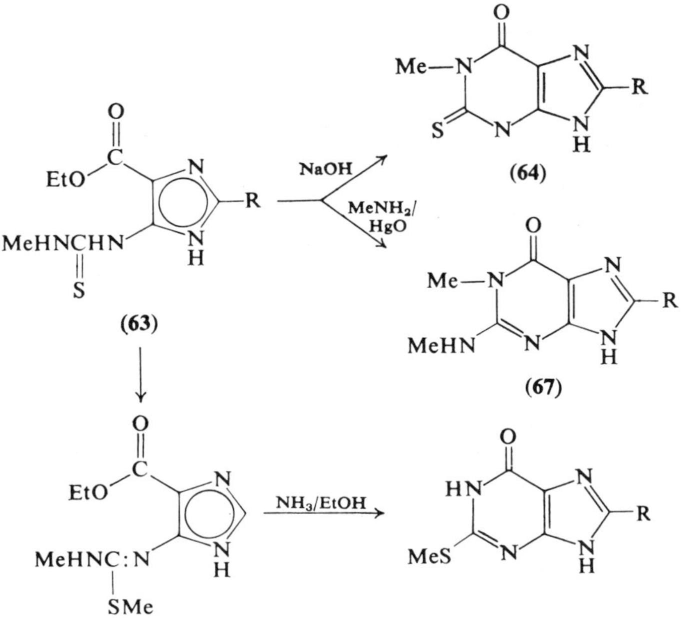

and isocyanates. With acetyl isothiocyanate an acetylthioureidoimid-
azole such as **61** is formed from which the 2-thiopurine (**62**), unsubstituted at $N_{(1)}$ obtains following deacetylation and cyclisation using
2N-hydrochloric acid and dilute alkali, respectively.[66] Methyl isothiocyanate and, to a much lesser extent, phenyl isothiocyanate are employed under similar conditions for forming the 1-methyl-[57, 64–66, 69–71]
(**64**) and 1-phenyl-6-oxo-2-thiotetrahydropurines.[57] Methylation of the
thioureido intermediate (**63**) with diazomethane gives the S-methylated
form (**65**) converted to the corresponding 2-methylthiopurine (**66**) in
ethanolic ammonia under pressure.[71]

As in Traube synthesis (see Ch. II, Sect. 1P) cyclisation of a thioureido
derivative in the presence of mercuric compounds favours production
of amino- rather than the expected thiopurine. A series of methylated
guanine derivatives (**67**, R = H, C_6H_5, and $CH_2C_6H_5$) have been
obtained from ring closures in aqueous methylamine containing
mercuric oxide.[71]

G. Synthesis from 4-Amino-5-aroylimidazoles

A 6-phenylpurine (**69**) results when 4-amino-5-benzoyl-1-methyl-2-

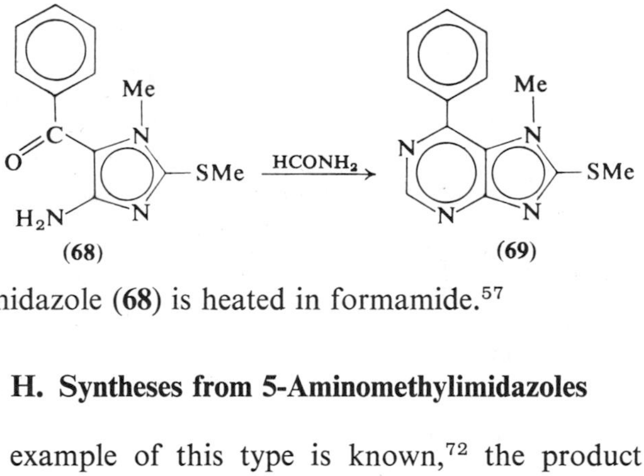

methylthioimidazole (**68**) is heated in formamide.[57]

H. Syntheses from 5-Aminomethylimidazoles

Only one example of this type is known,[72] the product being a
dihydropurine (**72**). The 5-aminomethyl group in **70** is converted to

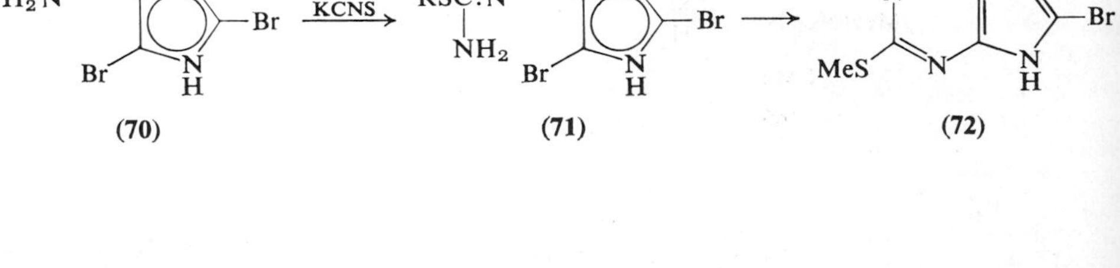

thioureidomethyl (**71**, R = H) by means of potassium thiocyanate and, after methylation to **71** (R = Me), cyclisation to 8-bromo-1,6-dihydro-2-methylthiopurine (**72**) results on heating in ethanolic pyridine. An attempted extension of this work using related aminomethylimidazoles was unsuccessful due to instability of these derivatives.[73]

2. Using Acyclic Precursors

Under this heading are grouped various approaches which have in common the direct formation of a purine by interaction of aliphatic components. The term "one-step syntheses" is generally applied to such preparations which, on the one hand, may only involve heating together two or more very simple compounds, usually at high temperature and pressure or, on the other hand, may utilise more complex aliphatic derivatives which undergo a bicyclisation step in which both imidazole and pyrimidine rings are concurrently formed under relatively mild conditions. An offshoot of the former type is the so-called "abiotic synthesis" by which the formation of the naturally occurring purines can be explained under the conditions existing on a primitive earth—utilising only the simplest of molecules and without entailing enzymic intervention.

A. The "One-Step Synthesis"

This is by no means a new concept; an early preparation of uric acid[74, 74a] of this type, due to Horbaczewski,[74] consisted of fusing two molecular equivalents of urea with trichlorolactic acid amide (Eq. 1).

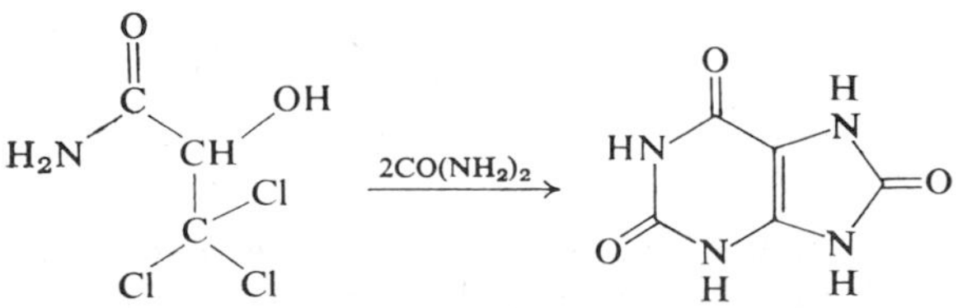

$$\text{Eq. (1)}$$

Synthetic studies using simpler derivatives than these have given varying yields of the purines. Of more theoretical than practical interest was a synthesis of purine, in minute yield, on heating formamide and ammonia together at 180° under pressure for some hours.[75] By contrast good yields (40–50%) of adenine are obtained with formamide and phosphoryl chloride, in molar excess, using a lower (120°) temperature but prolonged (15 h) heating.[76, 77] The necessity for pressure conditions in this type of preparation is seen in the absence of any adenine when the reaction was repeated at atmospheric pressure. A mixture of products resulted when triethyl orthoformate and liquid ammonia (1:12) react under vigorous conditions (200°, 16 h) the major ones being adenine (21%), purine (14%), and 8-aminopurine (3%). If chloroform replaces the triethyl orthoformate, the product yields drop to 5%, 7%, and 1%, respectively. Poorer yields of the three purines result from subjecting formamidine–ammonia or formiminoethyl ether–ammonia mixtures to the above conditions. In either case it is found that purine rather than adenine is now the major product.[77a] The significance of temperature in these reactions is shown by the absence of 8-aminopurine if the temperature falls below 160°, while with a further decrease to 120° no purines can be detected in the reaction mixture.[77a]

Examples of the more sophisticated approach in which bicyclisation is involved utilise derivatives of malonamide as the main precursor fragment and an excess of the cyclising reagent as solvent. A disadvantage of this route is that it is limited to the formation of purines having the same substituent at the 2- and 8-position. Hypoxanthine was the first example, prepared in 62% yield by heating formamidomalonamidamidine hydrochloride in formamide[16] (Eq. 2). Later workers,[32]

$$\text{(2)}$$

using the unformylated amidine salt, converted this to a number of 2,8-dialkylhypoxanthines by means of alkyl orthoesters, the cyclisation being carried out in boiling dimethylformamide. Reaction times varied with the nature of the alkyl groups, thus hypoxanthine itself resulted from 5 minutes heating with triethyl orthoformate whereas the 2,8-dimethyl (Eq. 2, R' = R'' = Me) and 2,8-diethyl (Eq. 2, R' = R'' = Et) homologues required 10 hours and 60 hours heating, respectively with the appropriate orthoester. Shorter reaction times or using one equivalent of the orthoester gives the appropriate 4-carbamoyl-5-ethoxymethyleneaminoimidazole. Hypoxanthine is obtainable, in 85% yield, on heating aminomalonamidamidine hydrochloride in a mixture of triethyl orthoformate and acetic anhydride.[78]

The corresponding 2,8-dialkyl adenine derivatives result when aminomalondiamidine hydrochloride is reacted under the above conditions but employing shorter reaction times (Eq. 3, R = Me and Et).[32]

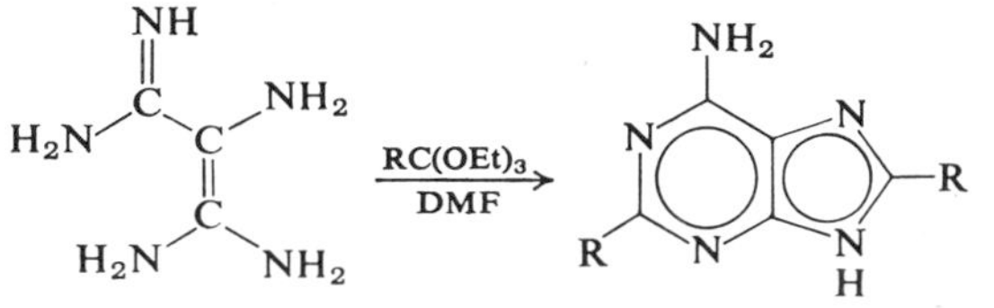

$$\text{(3)}$$

Acetonitrile can be used as solvent in place of the dimethylformamide. When malonodinitrile is treated with nitrous acid followed by hydroxylamine, the resulting 3-amino-4-(N-hydroxyamidino)furazan (73) being structurally similar to aminomalondiamidine, gives adenine on heating in formic acid containing Raney nickel. Lower yields were recorded using zinc dust as reducing agent.[17]

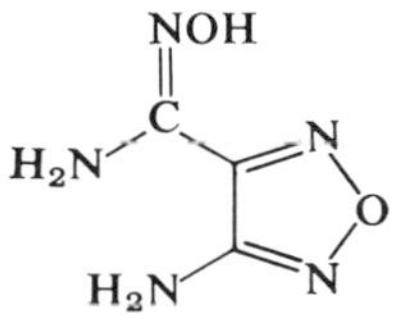

(73) 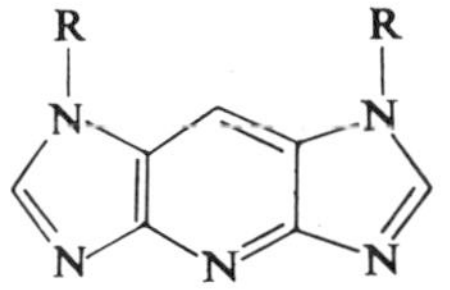   (74) 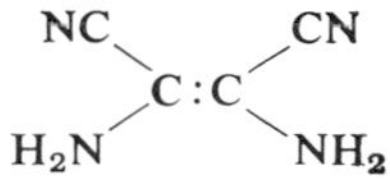   (75)

Adenine derivatives also result from reaction between isonitrosomalononitrile and formamidine salts in formamide containing a reducing agent. Unlike the preceding examples, which pass through an imidazole stage, the intermediates in this case are 4-amino-5-nitrosopyrimidines. Guanidine carbonate and the potassium salt of isonitrosomalononitrile, when heated in formamide in the presence of sodium dithionite, give 2,6-diaminopurine (Eq. 4) directly.[79] Modification of the method, using aryl amidines, allows the preparation of 2-phenyl- and 2-β-pyridyladenines.[80]

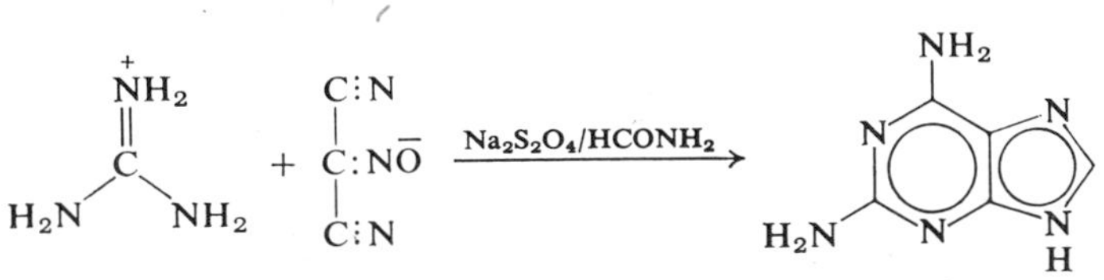

$$\text{(4)}$$

One school has investigated the reaction of N-substituted aminoacetonitriles with formamidine derivatives, purine arising in 40% yield when N-cyanomethylphthalimide and tris formamidomethane were heated in formamide[81] (Eq. 5).

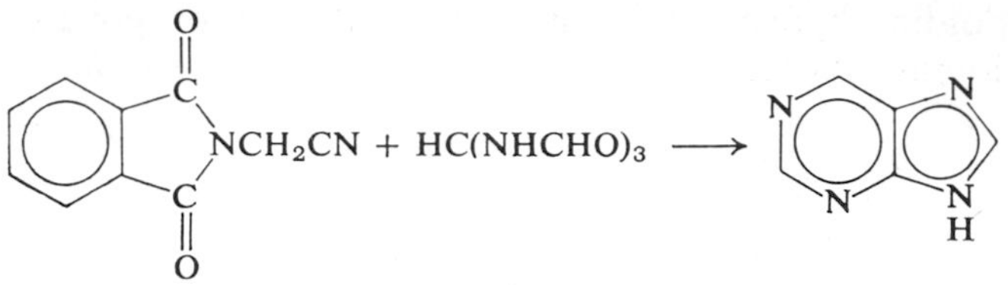

(5)

Subsequent condensations using either tris formamidomethane or formamidine acetate and carried out in a butanol or formamide solvent gave improved yields.[82, 83] By employing alkyl- and arylaminoaceto-nitriles the method is extended to produce 7-substituted purines (Eq. 6, R = Me, Et, C_2H_4OH, and C_6H_5), but other products are also found, notably di-imidazolo[4,5-b:5',4'-e]pyridines of the type **74**.[84] Separation

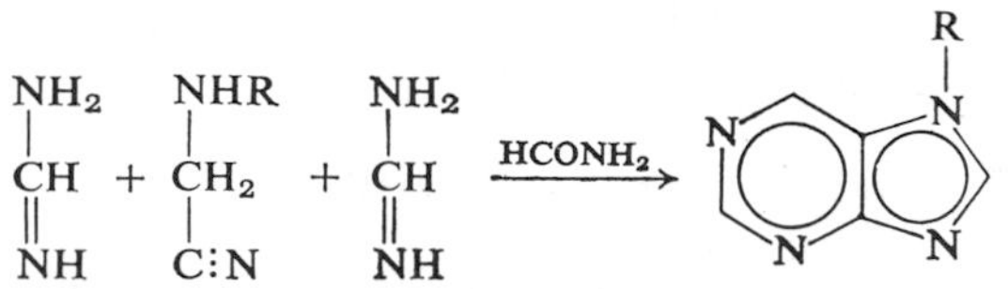

(6)

of the more volatile 7-alkylpurines is made by distillation or sublimation.

A related synthesis,[85] giving a 75% yield of hypoxanthine, is heating under pressure ethyl acetamidocyanoacetate in ethanolic ammonia containing ammonium acetate and triethyl orthoformate (Eq. 7). The known facility with which triethyl orthoformate reacts with ammonia

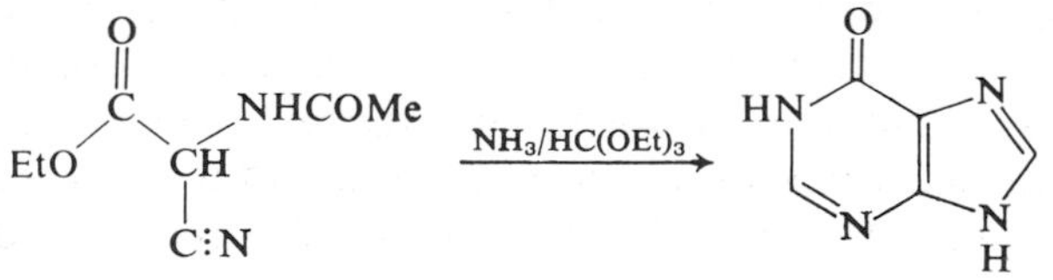

(7)

under these conditions to form formamidine[86] suggests that the latter is the actual cyclising reagent involved.

B. Abiotic Syntheses

An ingenious hypothesis, partly verified by experimental evidence, has been developed to explain the formation of vital organic compounds, such as purines and amino acids, when the earth was in a prebiological state. All chemical requirements are met with using only ammonia, water, and hydrogen cyanide; the latter, by virtue of its ready polymerisation, is the key material.

Adenine is produced on passing hydrogen cyanide into aqueous ammonia solution, the mixture then being kept at below 70° for some time. Two days is sufficient for small amounts of adenine to accumulate providing that the maximum temperature is adopted but at ambient temperatures periods up to 19 days are required.[87] By carrying out the reaction under anhydrous conditions, using hydrogen cyanide in liquid ammonia, a thirtyfold increase in yield of adenine is obtained in a few hours.[88, 88a] The overall reaction may be represented by Equation 8, but a stepwise pathway has been proposed[89] by which an initial

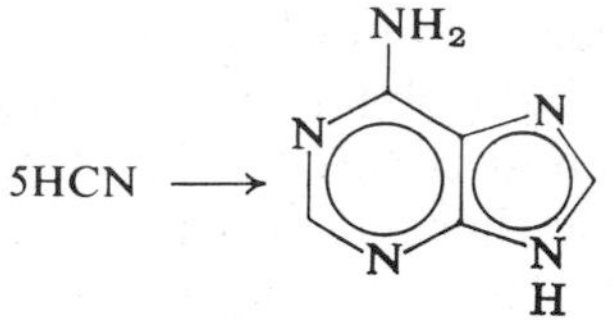

$$5HCN \longrightarrow \qquad\qquad\qquad (8)$$

hydrogen cyanide polymerisation gives rise to the trimeric derivative, aminomalononitrile (Eq. 9). Interaction with ammonia leads to amino-malondiamidine (Eq. 10), while a concurrent reaction produces formamidine (Eq. 11). Cyclisation to adenine follows interaction of the products

$$3HCN \longrightarrow H_2NCH(CN)_2 \qquad\qquad (9)$$

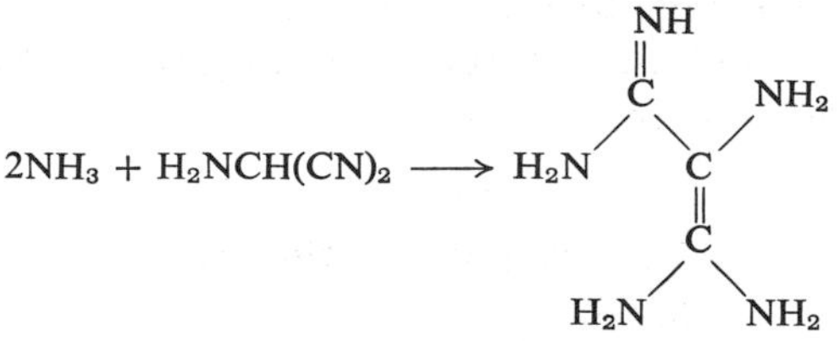

$$2NH_3 + H_2NCH(CN)_2 \longrightarrow \qquad\qquad (10)$$

$$HCN + NH_3 \longrightarrow HN:CHNH_2 \qquad\qquad (11)$$

of Equations 10 and 11 through the intermediate formation of 4-amino-5-amidinoimidazole, which requires the addition of a further molecule of formamidine to give the purine (Eq. 12). Detailed experimental

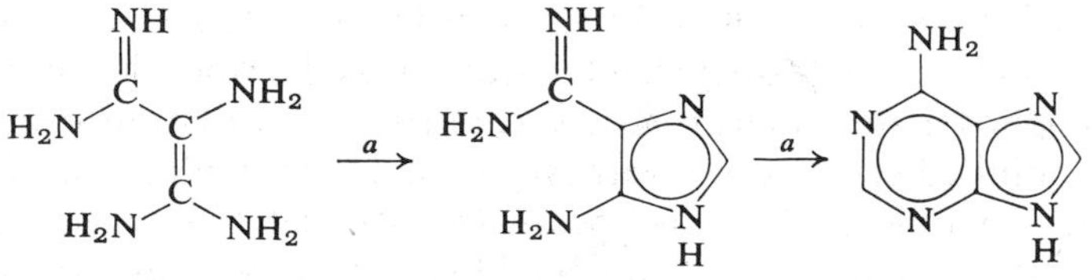

$$\qquad\qquad\qquad\qquad (12)$$

$$a = HN:CHNH_2$$

studies have resulted in the identification of most of the intermediates and have shown that the above reaction scheme is possible under these conditions[90] and that other purines, or their precursors, may also be involved. Some indications of the presence of small amounts of hypoxanthine have been found[91] but under acid conditions cyclisation of 4(5)-amino-5(4)-carbamoylimidazole, which is present in reasonable amounts in the reaction mixture, is not favoured.[16] Further products identified from the reaction include 4(5)-amino-5(4)-cyanoimidazole and 1,2-diaminomalononitrile (75); the former has been prepared by the action of aqueous formamidine acetate on aminomalononitrile and the latter arises when hydrogen cyanide replaces the amidine acetate.[59] Both substances are possibly obligatory intermediates in the reaction. No guanine has so far been discovered but its formation is not unlikely as it has been obtained by thermal polymerisation of the amino acid mixtures[92] which can be isolated from aged ammonia–hydrocyanic acid solutions, conditions similar to those used for adenine being employed. Extensions of the liquid ammonia procedure in which hydrogen cyanide is replaced by acetamidine (120°, 6 h) give rise to a mixture of 2-methyl-, 8-methyl-, and 2,8-dimethyladenines. If liquid methylamine and hydrogen cyanide interact, the products are 7- and 9-methyl-6-dimethyl-aminopurines.[88a] In passing it should be noted that the idea of abiotic syntheses is not new as over eighty years ago uric acid was obtained by Horbaczewski on heating the amino acid, glycine,[93] alone. None of the stages of the above abiotic synthetic studies give any evidence of the formation of pyrimidine intermediates.[90]

C. Biological Syntheses

Reference to the *in vivo* synthesis of purines is made here for the sake of completeness as imidazole derivatives are involved. A number of detailed and comprehensive reviews by workers in the field are available.[94–96] The realisation that intermediates of this kind, not pyrimidines, were implicated arose from the observation that cultures of *E. coli* metabolically inhibited by a sulphonamide[38] contained appreciable amounts of 4(5)-amino-5(4)-carbamoylimidazole.[97] It was subsequently realised[18] that the imidazole occurs naturally as the ribotide from which the sugar-phosphate moiety is cleaved during the isolation procedure. With uninhibited systems biosynthesis is carried to completion through the agency of folic acid coenzymes with formylation and subsequent closure of the imidazole to inosinic acid followed by transformation of this nucleotide by divergent pathways, to either adenylic or guanylic acids.[98] It is a surprising fact, as these investigations reveal, that purine formation does not involve a pyrimidine stage considering that over half the complement of bases in nucleic acids are pyrimidines. An assumption of some common initial pathway would seem reasonable but is wrong as pyrimidine biosynthesis is known to follow an independent route.[99]

References

1. Sarasin and Wegmann, *Helv. Chim. Acta*, **7**, 713 (1924).
2. Baxter and Spring, *Nature*, **154**, 462 (1944).
3. Baxter, Gowenlock, Newbold, Woods, and Spring, *Chem. and Ind.*, **23**, 77 (1945).
4. Baxter and Spring, *J. Chem. Soc.*, **1945**, 232.
5. Baddiley, Buchanan, and Osborne, *J. Chem. Soc.*, **1958**, 3606.
6. Woodward, U.S. Pat., 2,534,331 (1950); through *Chem. Abstracts*, **45**, 5191 (1951).
7. Baxter, McLean, and Spring, *J. Chem. Soc.*, **1948**, 523.
8. Baxter and Spring, *J. Chem. Soc.*, **1947**, 378.
9. Howard, McLean, Newbold, Spring, and Todd, *J. Chem. Soc.*, **1949**, 232.
10. Bauer and Dhawan, *J. Heterocyclic Chem.*, **2**, 220 (1965).
11. Bauer, Nambury, and Dhawan, *J. Heterocyclic Chem.*, **1**, 275 (1964).
12. Cook and Smith, *J. Chem. Soc.*, **1949**, 2329.
13. Cook and Smith, *J. Chem. Soc.*, **1949**, 3001.
14. Shaw, Warrener, Butler, and Ralph, *J. Chem. Soc.*, **1959**, 1648.
15. Shaw, *J. Org. Chem.*, **30**, 3371 (1965).
16. Shaw, *J. Biol. Chem.*, **185**, 439 (1950).
17. Ichikawa, Kato, and Takenishi, *J. Heterocyclic Chem.*, **2**, 253 (1965).
18. Greenberg and Spilman, *J. Biol. Chem.*, **219**, 411 (1956).
19. Baddiley, Buchanan, Hardy, and Stewart, *J. Chem. Soc.*, **1959**, 2893.
20. Shaw and Wilson, *J. Chem. Soc.*, **1962**, 2937.
21. Shaw, *J. Amer. Chem. Soc.*, **80**, 3899 (1958).
21a. Yamazaki, Kumashiro, and Takenishi, *J. Org. Chem.*, **32**, 3258 (1967).
22. Rousseau, Robins, and Townsend, *J. Amer. Chem. Soc.*, **90**, 2661 (1968).
23. Naylor, Shaw, Wilson, and Butler, *J. Chem. Soc.*, **1961**, 4845.
24. Leese and Timmis, *J. Chem. Soc.*, **1961**, 3818.
25. Glushkov and Magidson, *Doklady Akad. Nauk. S.S.S.R.*, **133**, 585 (1960).
26. Glushkov and Magidson, *Zhur. obshchei Khim.*, **31**, 1173 (1961).
27. Glushkov and Magidson, *Zhur. obshchei Khim.*, **31**, 1906 (1961).
28. Glushkov and Magidson, *Khim. geterotsikl. Soedinenii*, **1965**, 85.
29. Trout and Levy, *Rec. Trav. chim.*, **85**, 1254 (1966).
30. Nakata, *Meiji Yakka Daigaku Kenkyu Kiyo*, **2**, 66 (1963); through *Chem. Abstracts*, **61**, 1864 (1964).
31. Yamazaki, Kumashiro, and Takenishi, *Chem. and Pharm. Bull.* (*Japan*), **16**, 1561 (1968).
32. Richter, Loeffler, and Taylor, *J. Amer. Chem. Soc.*, **82**, 3144 (1960).
33. Taylor, Cheng, and Vogel, *J. Org. Chem.*, **24**, 2019 (1959).
34. Prasad and Robins, *J. Amer. Chem. Soc.*, **79**, 6401 (1957).
35. Giner-Sorolla and Bendich, *J. Amer. Chem. Soc.*, **80**, 5744 (1958).

36. Scheuing and Konz, Ger. Pat., 806,670 (1951); through *Chem. Abstracts*, **46**, 1035 (1952).
37. Scheuing and Konz, Ger. Pat., 804,210 (1951); through *Chem. Abstracts*, **45**, 8037 (1951).
38. Stetton and Fox, *J. Biol. Chem.*, **161**, 333 (1945).
39. Biltz and Rakett, *Ber.*, **61**, 1409 (1928).
40. Yamakazi, Kumashiro, and Takenishi, *J. Org. Chem.*, **32**, 1825 (1967).
41. Mann and Porter, *J. Chem. Soc.*, **1945**, 751.
42. Blicke and Godt, *J. Amer. Chem. Soc.*, **76**, 3653 (1954).
43. Golovchinskaya, Kolganova, Nikolaeva, and Chaman, *Zhur. obshchei Khim.*, **33**, 1650 (1963).
44. Fischer and Bromberg, *Ber.*, **30**, 219 (1897).
45. Montequi, *Anales real Soc. espan. Fis. Quim.*, **24**, 731 (1926).
45a. Walentowski and Wanzlick, *Chem. Ber.*, **102**, 3000 (1969).
46. Yamazaki, Kumashiro, and Takenishi, *J. Org. Chem.*, **32**, 3032 (1967).
47. Hitchings and Elion, U.S. Pat., 2,756,228 (1956); through *Chem. Abstracts*, **51**, 2887 (1957).
48. Ikehara, Nakazawa, and Nakayama, *Chem. and Pharm. Bull. (Japan)*, **10**, 660 (1962).
49. Shaw and Butler, *J. Chem. Soc.*, **1959**, 4040.
49a. Yamazaki, Kumashiro, Takenishi, and Ikehara, *Chem. and Pharm. Bull. (Japan)*, **16**, 2172 (1968).
50. Paterson and Zbarsky, *J. Amer. Chem. Soc.*, **75**, 5753 (1953).
51. Cavalieri, Tinker, and Brown, *J. Amer. Chem. Soc.*, **71**, 3973 (1949).
52. Cresswell and Brown, *J. Org. Chem.*, **28**, 2560 (1963).
53. Temple, Kussner, and Montgomery, *J. Org. Chem.*, **30**, 3601 (1965).
54. Taylor and Loeffler, *J. Amer. Chem. Soc.*, **82**, 3147 (1960).
55. Leonard, Carraway, and Helgeson, *J. Heterocyclic Chem.*, **2**, 291 (1965).
56. Taylor and Borror, *J. Org. Chem.*, **26**, 4967 (1961).
57. Gompper, Gäng, and Saygin, *Tetrahedron Letters*, **1966**, 1885.
58. Taylor and Loeffler, *J. Org. Chem.*, **24**, 2035 (1959).
59. Ferris and Orgel, *J. Amer. Chem. Soc.*, **87**, 4976 (1965).
60. Ferris and Orgel, *J. Amer. Chem. Soc.*, **88**, 3829 (1966).
61. Shaw, *J. Org. Chem.*, **27**, 883 (1962).
62. Montequi, *Anales real Soc. espan. Fis. Quim.*, **25**, 182 (1927).
63. Allsebrook, Gulland, and Story, *J. Chem. Soc.*, **1942**, 232.
64. Heilbron and Cook, Brit. Pat., 683,523 (1952); through *Chem. Abstracts*, **48**, 2093 (1954).
65. Cook, Davis, Heilbron, and Thomas, *J. Chem. Soc.*, **1949**, 1071.
66. Cook and Thomas, *J. Chem. Soc.*, **1950**, 1884.
67. Bader and Downer, *J. Chem. Soc.*, **1953**, 1636.
68. Bills, Gebura, Meek, and Sweeting, *J. Org. Chem.*, **27**, 4633 (1962).
69. Bader and Downer, *J. Chem. Soc.*, **1953**, 1641.
70. Cook, Downer, and Heilbron, *J. Chem. Soc.*, **1949**, 1069.
71. Cook and Thomas, *J. Chem. Soc.*, **1950**, 1888.
72. Mitter and Chatterjee, *J. Indian Chem. Soc.*, **11**, 867 (1934).
73. Hoskinson, *Austral. J. Chem.*, **21**, 1913 (1968).
74. Horbaczewski, *Monatsh.*, **8**, 201, 584 (1887).
74a. Behrend, *Annalen*, **441**, 215 (1925).
75. Bredereck, Ulmer, and Waldmann, *Chem. Ber.*, **89**, 12 (1956).

76. Morita, Ochiai, and Marumuto, *Chem. and Ind.*, **1968**, 1117.
77. Ochiai, Marumuto, Kobayashi, Shimazu, and Morita, *Tetrahedron*, **24**, 5731 (1968).
77a. Kobayashi and Honjo, *Chem. and Pharm. Bull. (Japan)*, **17**, 703 (1969).
78. Richter and Taylor, *Angew. Chem.*, **67**, 303 (1955).
79. Vogel and Taylor, *J. Amer. Chem. Soc.*, **79**, 1518 (1957).
80. Taylor, Vogel, and Cheng, *J. Amer. Chem. Soc.*, **81**, 2442 (1959).
81. Bredereck, Effenberger, and Rainer, *Angew. Chem.*, **73**, 63 (1961).
82. Bredereck, Effenberger, Rainer, and Schosser, *Annalen*, **659**, 133 (1962).
83. Bredereck, Effenberger, and Rainer, Ger. Pat., 1,150,988 (1963); through *Chem. Abstracts*, **60**, 2975 (1964).
84. Bredereck, Effenberger, and Rainer, *Annalen*, **673**, 82 (1964).
85. Taylor and Cheng, *Tetrahedron Letters*, **1959**, 9.
86. Taylor and Erhart, *J. Amer. Chem. Soc.*, **82**, 3138 (1960).
87. Oró and Kimball, *Arch. Biochem. Biophys.*, **94**, 217 (1961).
88. Wakamatsu, Yamada, Saito, Kumashiro, and Takenishi, *J. Org. Chem.*, **31**, 2035 (1966).
88a. Yamada, Kumashiro, and Takenishi, *J. Org. Chem.*, **33**, 642 (1968).
89. Oró, *Nature*, **191**, 1193 (1961).
90. Oró and Kimball, *Arch. Biochem. Biophys.*, **96**, 293 (1962).
91. Lowe, Rees, and Markham, *Nature*, **199**, 219 (1963).
92. Ponnamperuma, Young, and Munoz, *Fed. Proc.*, **22**, 479 (1963).
93. Horbaczewski, *Monatsh.*, **3**, 796 (1882).
94. Buchanan and Hartmann, *Adv. Enzymol.*, **21**, 199 (1959).
95. Buchanan, *Harvey Lectures*, **54**, 104 (1958–1959).
96. Greenberg and Jaenicke, *Chemistry and Biology of Purines*, Eds., Wolstenholme and O'Connor, J and A Churchill, London, 1957, p. 204.
97. Shive, Ackermann, Gordon, Getzendaner, and Eakin, *J. Amer. Chem. Soc.*, **69**, 725 (1947).
98. Magasanik and Karibian, *J. Biol. Chem.*, **235**, 2672 (1960).
99. Baddiley and Buchanan, *Ann. Reports*, **54**, 331 (1957).

CHAPTER IV

Purine and the *C*-Alkyl and *C*-Aryl Derivatives

The preparation of purine itself, by Emil Fischer in 1898, occurred well over a century after the first derivative of the ring system, uric acid, had been synthesised. All mono- and di-*C*-methyl purines and the trimethyl derivative have been prepared, but the higher alkyl homologues are not well represented. Although purine occurs naturally as the nucleoside "nebularine," no simple alkyl purine homologues have been isolated from biological systems. It is of interest, therefore, to contrast the lack of *C*-methyl derivatives with the abundance of *N*-methyl purines which exists in nature.

1. Purine (Unsubstituted)

It is only lately that the parent member has become readily available. This is due largely to the introduction of the two techniques of desulphurisation with Raney nickel and halogen removal by catalytic hydrogenation. Both procedures are valuable for abstraction of the appropriate element either from a pyrimidine intermediate or from a purine.

A. Preparation of Purine

The first direct synthesis of Isay,[1] involving a Traube condensation of 4,5-diaminopyrimidine with formic acid, was later much improved (83% yield) by using a carbon dioxide atmosphere.[2-4] Pyrimidines labelled with deuterium at the 2- and 4-positions have given the corresponding 2- and 6-deuteropurines.[5] Other cyclising reagents employed are diethoxymethyl acetate (82%)[6] and *s*-triazine, which is degraded by the conditions used and reacts like formamidine.[7] Purine preparations involving removal of groups from preformed purines include Fischer's original purine synthesis (65%) entailing dehalogenation of 2,6-di-iodopurine by zinc dust in aqueous solution[8] and catalytic hydrogenations, with a palladium catalyst, in the presence of a proton acceptor, of 6-chloro-,[4, 9] 2,6-dichloro-,[9, 10] and 2,6,8-trichloropurine.[9]

Similar reductions carried out in a deuterium atmosphere with 2-chloro- and 6-chloropurine afford 2-deutero- and 6-deuteropurine, respectively.[11] The latter purine is also derived from 6-iodopurine.[12]

Dethiolation of 6-thio-[4, 13] and 2,6-dithiopurine[13] with Raney nickel in aqueous solution gives purine in 40% yield. Purine also results through concurrent reduction of both oxygen and sulphur when 1,6-dihydro-6-thiopurine-3-oxide is similarly treated.[14, 15] Using deuterated Raney nickel 6-deutero- and 8-deuteropurine are obtained from the respective 6-thio- and 8-thiodihydropurine.[12] Removal of thiocyanato groups from 6-thiocyanato- and 2,6-dithiocyanatopurine can also be effected with Raney nickel.[15a] Decarboxylation of 8-carboxypurine occurs in boiling water[16] but the isomeric 6-carboxylic acid must be heated to 190° to accomplish this.[17, 18] Removal of a $C_{(6)}$ acid function also occurs on heating purine-6-sulphinic acid in formic acid.[19] Purine also results from treatment of 6-hydrazinopurine with dilute sodium hydroxide.[20] The work of one school shows that acyclic precursors can be utilised in purine synthesis, and an almost 60% yield is obtained when derivatives of aminoacetonitrile and either formamidine or a mixture of tris formamidomethane and formamide are reacted together at 130°.[21] Just over a 1% yield of purine has been obtained from rigorous heating of a mixture of ammonia and formamide under pressure.[22]

B. Properties of Purine

Purine is a colourless solid, m.p. 216°, readily subliming *in vacuo* and crystallising from toluene or ethanol as needles. Solubility in water is high (1 g in 2 ml at 20°) giving rise to an alkaline solution. A moderate solubility is shown in alcohol and benzene, less so in ethyl acetate or acetone, while in ether or chloroform it is almost insoluble. A weakly basic nature is shown, anionic and cationic pK_a values are 8.9 and 2.4, respectively.[2] Stability shown toward acids and alkali is pronounced[2, 4] when compared with that of similar heterocycles, such as pteridine, and vigorous conditions are needed before degradation ensues.[23] This property can be correlated with the high overall electron density; both

5- and 6-membered rings possess an aromatic sextet of π-electrons. In addition, ionization introduces a further stabilising factor through the associated resonance forms. Monobasic salts are formed with acids exemplified by the hydrochloride (m.p. 208°),[24] hydrobromide (m.p. 232°),[22] nitrate [m.p. 205° (dec.)],[8] and picrate (m.p. 208°).[4, 8] Sodium, potassium, and silver salts are derived by replacement of the acidic hydrogen at $N_{(9)}$ while with divalent metals, such as zinc, copper, and nickel (2:1) base-metal complexes arise.[25]

C. Reactions of Purine

The limited range of reactions undergone by the parent compound reflects the largely aromatic character of the molecule. Gentle warming with acetic anhydride gives a mixture of 7- and 9-acetyl derivatives[6] while methylation procedures (Sect. 3Aa) give only 9-methylpurine.[22] Other electrophilic reagents that react include bromine but only if purine hydrobromide is used; the unstable dibromo adduct originally formed reverts again in boiling acetone to purine hydrobromide. Sulphur under fusion conditions (245°) gives 7,8-dihydro-8-thiopurine in 75% yield directly.[26] In spite of intensive studies[27] evidence for the addition of water across the peripheral double bond system as occurs for example, in the related pteridines, is lacking. However, adducts

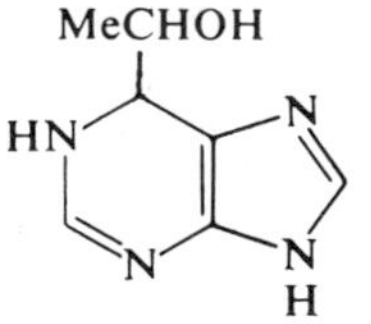

(1) 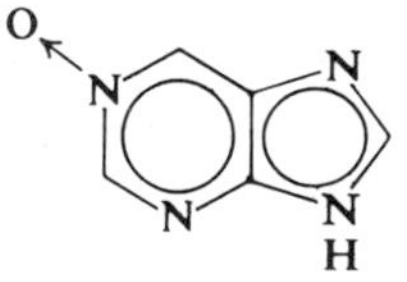   (2) 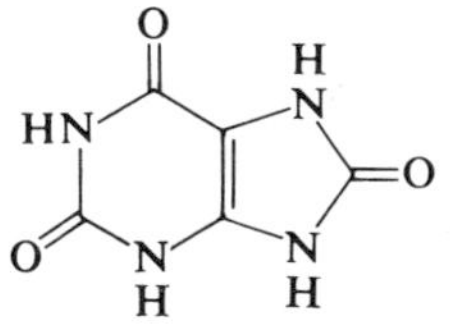   (3)

are formed with methanol, ethanol, and higher homologues on irradiation of the deoxygenated solutions.[28] Products are 1,6-dihydro-6-α-hydroxyalkylpurines, such as **1**, obtained with ethanol. Oxidation with an acetic anhydride–hydrogen peroxide mixture or perbenzoic acid gives mixtures of the 1-oxide (**2**) and the 3-oxide in poor yield,[29] the majority of the purine being degraded by this treatment. Enzymic oxidation, in the presence of xanthine oxidase, converts purine by a three-stage oxidation process to uric acid (**3**).[30] Polarographic studies[31] show that reduction occurs in two stages in acid solution but not at all in neutral or alkaline solution. Hydrogenation over palladium catalysts has given hydrogen uptakes corresponding to the formation of di-[4]

and tetrahydropurine[32] but the products appear to be unstable. With a platinum oxide catalyst in acetic anhydride[32a], however, stepwise reduction with isolatable, reduced intermediates is possible. The fully reduced octahydro (perhydro) purine is finally obtained as the fairly stable 1,3,7,9-tetra-acetyl derivative (See Ch. XII, Sect. 1Ab).

2. *C*-Alkyl- and -Arylpurines

A. Preparation of Alkylpurines

As the introduction of an alkyl group into a purine is possible in only a minority of cases, direct synthesis from intermediates having the necessary alkyl groups *in situ* is the usual route. Frequently, such intermediates contain other groups as well but these are usually removed after the purine ring is formed.

a. *By Direct Synthesis*

Detailed aspects of the Traube synthesis are given in Chapter II. By using 2- or 6-alkyl-4,5-diaminopyrimidines and appropriate cyclising agents 2-, 6-, and 8-alkylpurines are possible with this procedure. Examples are 2-methyl-,[5] 6-methyl-,[5, 31, 33] 8-methyl-,[1, 2, 5] 2,6-dimethyl-,[34] 6,8-dimethyl-,[34] and 2,6,8-trimethylpurine.[34] Other examples by this route include the 2-butyl-,[35] 2-phenyl-,[36–38] 2,8-diphenyl-,[36] 6-methyl-8-phenyl,[37] and 8-phenylpurine.[2] The latter purine is also derived by thermal cyclodehydration of 4-benzylamino-5-nitrosopyrimidine.[39] An unspecified synthesis,[40] presumably of the Traube type, has given 6-propylpurine.

b. *By Dehalogenation of Halogenopurines*

Examples of removal of nuclear halogen atoms by hydrogenation using a palladium catalyst have been cited previously (Sect. 1A). Others of this type are formation of 2-methyl- and 2,8-dimethylpurine from the corresponding 6-chloro analogues.[34] Exocyclic halogen atoms are displaced reductively, exemplified in the preparation of 6-methylpurine by prolonged heating, under reflux conditions, of 6-chloromethyl-, 6-bromomethyl-, or 6-tribromomethylpurine with thiolacetic acid.[41]

Surprisingly, no 6-methylpurine results from the 6-dibromomethyl analogue when treated in this way.[41] A number of 8-alkylpurines have been derived by catalytic reduction of the appropriate 8-chloroalkyl-purines.[18, 42]

c. *By Removal of Thio and Methylthio Groups*

Almost invariably the technique for displacement of sulphur-con-taining groups by hydrogen atoms is by heating the purine with Raney nickel catalyst, usually in a dilute alkaline or alcoholic medium. Some application of nitrous and nitric acid for dethiolation of substituted thiopurines is found elsewhere but no simple alkylpurines have been obtained with these acid reagents.

All monomethylpurines have resulted from Raney nickel desulphur-isations; 2-methyl-[34] and 8-methylpurine[43] result from the correspond-ing 6-thiopurine; while 6-methylpurine is obtained from either 2-thio- or 8-thio-dihydro-6-methylpurine.[26] Simultaneous removal of more than one group can be achieved as in the formation of 8-methylpurine (**5**) from the 2,6-dithiopurine (**4**)[44] and 1,4-di(8-purinyl)butane (**6**) from the analogous 6-thiopurinyl derivative.[45] Although most 6-thio or 6-methyl-thio groups are readily removed by this treatment, difficulty in abstract-ing the sulphur from either 6-thio- or 6-methylthio-2-phenylpurine is

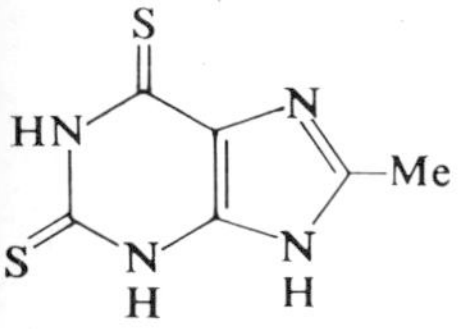

(**4**) 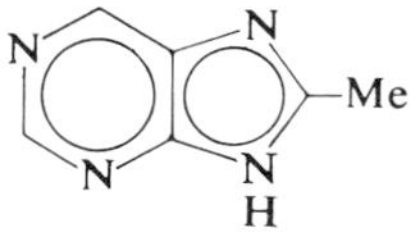   (**5**) 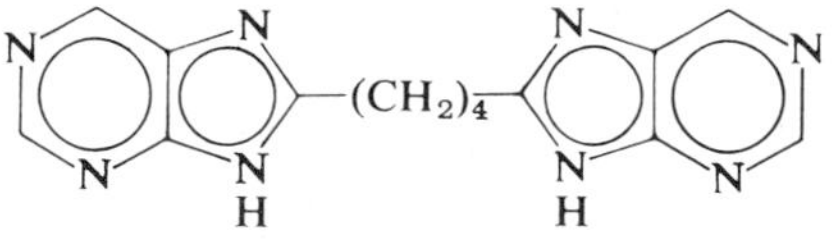   (**6**)

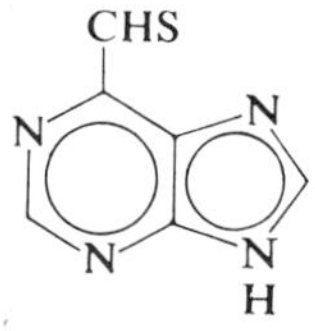

(**7**) 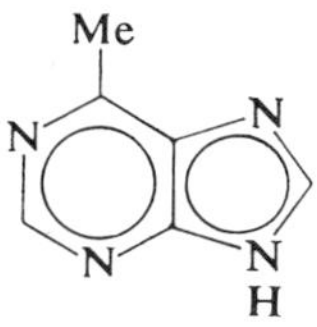   (**8**) 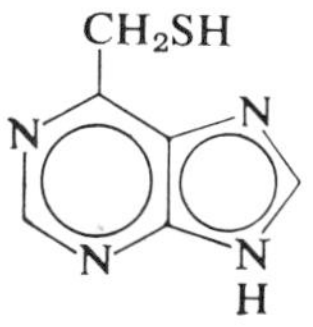   (**9**)

reported.[36] Exocyclic sulphur atoms may be similarly replaced as illustrated by the preparation of 6-methylpurine (**8**) from either purine-6-thioaldehyde (**7**)[26] or 6-mercaptomethylpurine (**9**).[46]

d. *By Conversion of Chloro to Alkyl Groups*

Extensive use has been made of the sodium derivative of malonic ester for this purpose. In a simple example, 6-chloropurine after conversion to 6-di(ethoxycarbonyl)methylpurine (**10**, R = H) undergoes facile decarboxylation to 6-methylpurine (**11**, R = H),[47] the 9-methyl analogue (**11**, R = Me) being derived likewise.[48] Apart from 6-chloro-

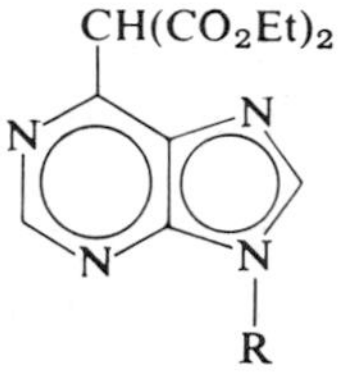

(10) 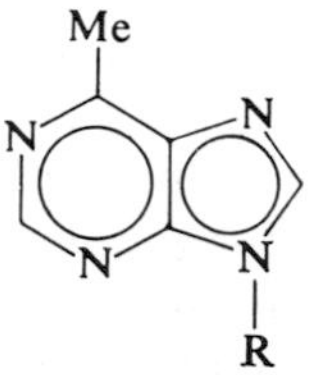   (11)

purines the method has been successfully applied to 2-chloro-[49] and 8-chloropurines also.[50] The versatility of the method is illustrated by the reaction of 8-chlorocaffeine with malonic ester, the product (**12**) then being treated with alkyl iodides and the alkyl derivatives (**13**, R = alkyl) subsequently decarboxyled in hot dilute acid. Derivatives so obtained are 8-methyl- (**14**, R = H),[51] 8-propyl- (**14**, R = Et),[52] 8-butyl- (**14**, R = Pr),[52] and 8-pentyl-caffeine (**14**, R = Bu).[52]

Application of Grignard and similar reagents to halogenopurines is rare. Some 2,7-dimethylpurine (**16**) results from the action of methylmagnesium iodide on 2-chloro-7-methylpurine (**15**).[53] Although 6-

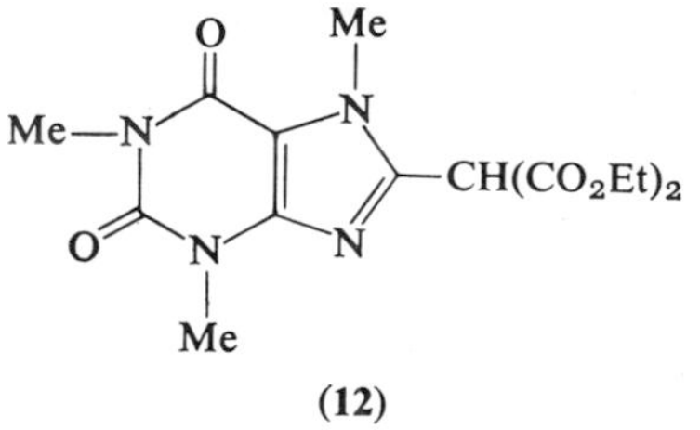

(12) 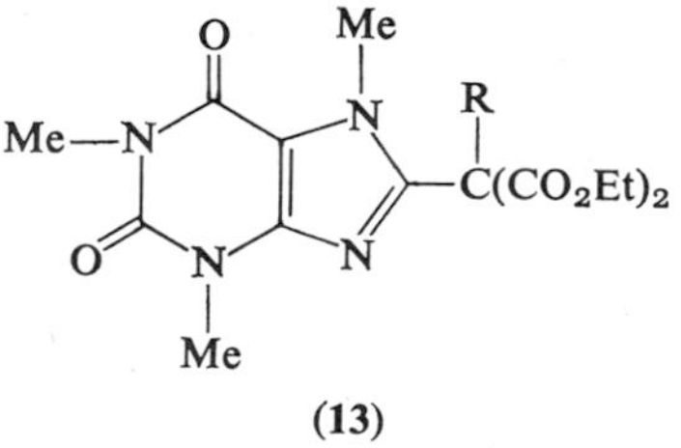   (13)

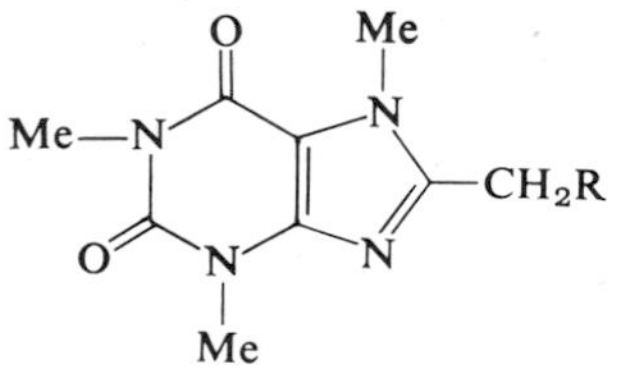

(14) 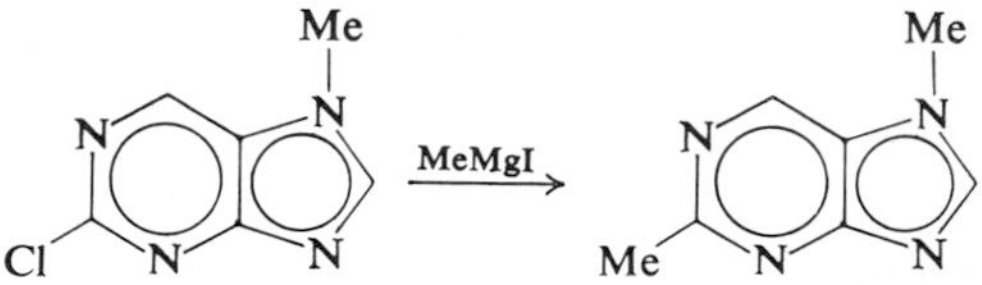   (15)   (16)

chloropurine was unaffected by Grignard reagents[47] with phenyl lithium, some 6-phenylpurine was formed.[47]

e. *By Other Methods*

Examples of direct *C*-alkylation are almost unknown. A claim[54] in the older literature to have obtained 8-methylpurines from 8-unsub-stituted derivatives by the action of methylating agents has been shown[55] to be erroneous—the products, in fact, being formed by methylation of an imidazole ring nitrogen atom.

An unusual but authentic case of $C_{(8)}$-alkylation arises when sodium theophyllinate, in aqueous solution, and but-2-enyl bromide react at room temperature, the crystalline product being 8-(but-2-enyl)theo-phylline (**17**). Confirmation of this assignment follows hydrogenation (Raney nickel) to the known 8-butyltheophylline.[56] The absence of any $N_{(7)}$-alkylation is surprising in view of the facility with which theophylline forms 7-alkyl derivatives (Ch. VI, Sect. 6Db). Other $C_{(8)}$-alkylations of this type may occur more widely than is presently appreciated. Supporting this is the isolation of both 7-benzyltheophylline and an isomeric product, in 20% yield, presumed to be 8-benzyl-

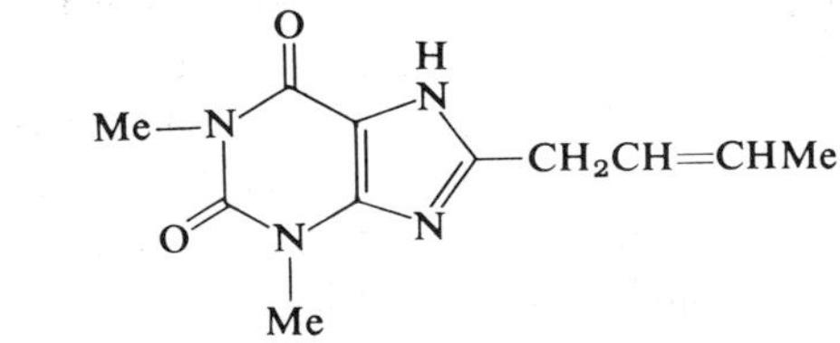

(**17**)

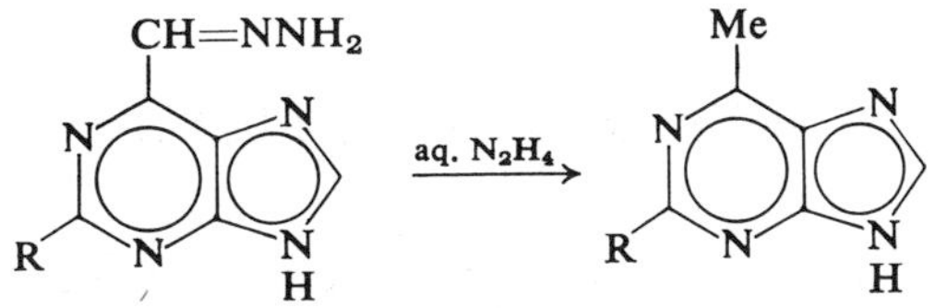

(**18**)   aq. N$_2$H$_4$ →   (**19**)

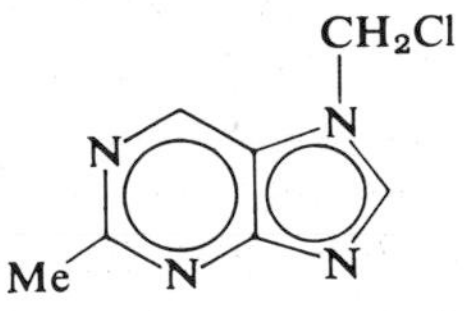

(**20**) 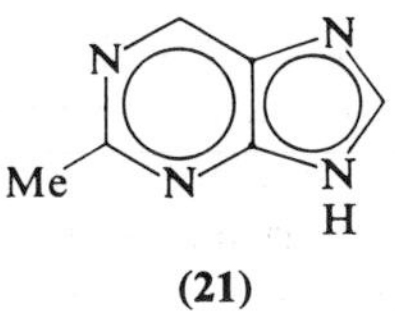   (**21**)

theophylline from a benzyl chloride alkylation, in alkali, of theophylline.[57]

Certain 8-unsubstituted, as well as 8-oxo- and 8-alkoxy-, purines are
converted to 8-alkylpurines on fusion or vigorous heating with amides
or acid anhydrides.[58, 59] These reactions involving ring-opened intermediates are dealt with fully in Chapter II, Sections 1E and 1I.

Reduction of 6-diethylamino-8-hydroxymethyl-2-methylpurine with
phosphorus and hydriodic acid gives 6-diethylamino-2,8-dimethyl-
purine.[60] Another reductive method is hydrazinolysis of the hydrazones
of C-formylpurines. Both hydrazones of purine-6-aldehyde (**18**, R =
H)[26] and the 2-amino derivative (**18**, R = NH$_2$)[61] give the respective
6-methylpurine (**19**, R = H and NH$_2$) on prolonged heating with
aqueous hydrazine. Removal of the 7-chloromethyl group occurs on
boiling an aqueous solution of the purine (**20**), affording 2-methylpurine
(**21**) in good yield.[62]

B. Properties of Alkyl- and Arylpurines

The mono- and dimethylpurines and 2,6,8-trimethylpurine are solids,
the melting points of which lie in the 219–286° range, the lower figure
representing 2,8-dimethyl- and the upper 2-methylpurine. Melting
points for the known mono- and diphenylpurines are also within the
above range. Purification of all the methyl derivatives can be effected
through vacuum sublimation. As with the pyrimidines introduction of
methyl groups causes a slight increase in basicity although in the purines
this is somewhat less marked (6-methyl, pK_a 2.6; 8-methyl, pK_a 2.9).
The presence of a phenyl group likewise produces little effect (8-phenyl,
pK_a 2.7).

C. Reactions of Alkyl Purines

The presence of alkyl groups does not significantly alter the general
reactions of the ring system, which behaves like that of purine itself.
With respect to the alkyl groups (in this context methyl groups are
understood, unless otherwise indicated), some associated reactivity is
found. Due to the paucity of methylpurines, which existed until fairly
recently, reactions undergone by methyl groups are largely unexplored.
In general they are less reactive than those attached to a pyrimidine
nucleus but, nevertheless, in their behaviour some common analogies
toward a number of reagents can be drawn. In the absence of other

substituents in the ring a $C_{(6)}$-methyl group is generally the most active. The presence of electron-donating groups can, however, lead to considerable enhancement in the reactivity of a $C_{(8)}$-methyl group. The 2-position shows a generally insignificant reactivity. In broadest terms the activity can be correlated with the location of the groups on the ring and the degree of conjugation that is present with a ring nitrogen atom. The methylazine (MeĊ:N) system so formed can, in effect, be likened to the methylketo (MeĊ:O) system.

a. *Chloral with Methylpurines*

Condensation of chloral with 6-methylpurine in the presence of acetic acid gives the addition compound, 6-(3,3,3-trichloro-2-hydroxypropyl)purine (**22**),[63] which alkaline hydrolysis converts to the lactic acid derivative (**23**), and which loses water on treatment with sodium

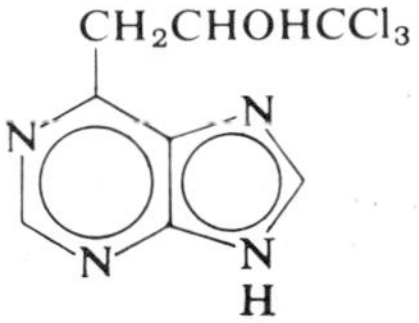

$CH_2CHOHCCl_3$ $CH_2CHOHCO_2H$ $CH{=}CHCO_2H$

(**22**) 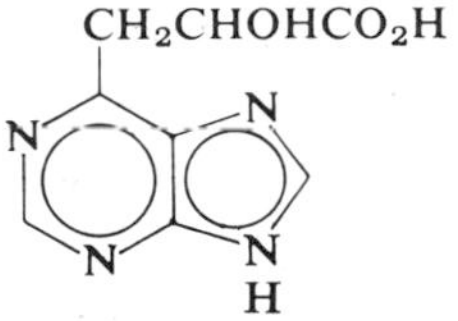   (**23**) 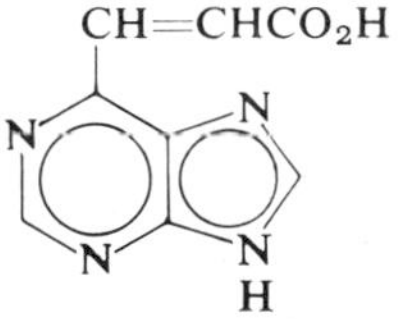   (**24**)

ethoxide, thus forming 6-β-carboxyvinylpurine (**24**).[63]

b. *Methyl- to Carboxypurines*

Where conversion of a methyl to a carboxyl group is possible, a useful means for methyl group removal is available, as decarboxylation is usually a facile procedure.

Direct oxidation of a methyl group does not appear to be a good route to carboxypurines. The action of alkaline permanganate at 80° on 6-methylpurine gives only a poor yield (8%) of 6-carboxypurine.[61] With selenium dioxide in dimethylformamide a mixture of the acid and 6-formylpurine is obtained.[61] By using 6-hydroxymethylpurine (**25**) in place of 6-methylpurine, cold permanganate oxidation gives a 51% yield of the acid.[61] A further modification involving conversion of 6-methylpurine to purin-6-ylmethylpyridinium iodide (**26**) gives a better yield (77%) of the acid on oxidation.[64] Conversion of a $C_{(8)}$-methyl group to the pyridinium salt is known but oxidation to the 8-carboxy derivative does not seem to have been attempted.[65] Chloromethyl groups are readily oxidised; the preparation of a 6-carboxypurine with aqueous permanganate from the 6-chloromethyl analogue has been carried out.[48] On complete chlorination the resulting trichloromethyl-purine can be hydrolysed to the carboxypurine, sodium bicarbonate solution being employed to obtain 6-carboxypurine.[18] If methanolic sodium methoxide is used instead the methyl ester is obtained.[18]

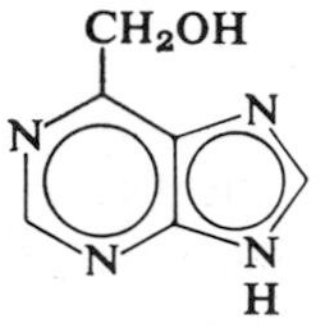

(25) 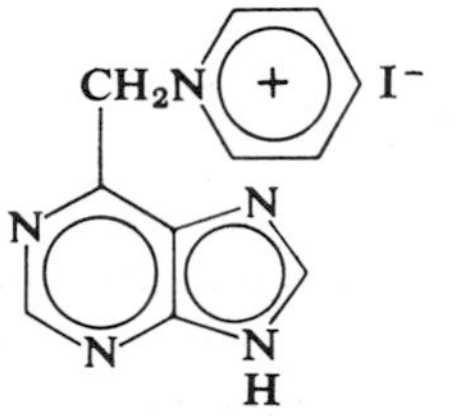   (26) 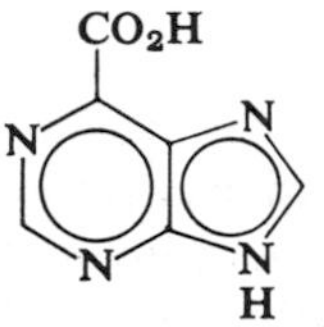   (27)

c. *Other Oxidations of Alkylpurines*

Sodium nitrite in acetic acid at 3° converts 6-methylpurine to the oxime of 6-formylpurine.[61] Peroxyacids or acetic acid–hydrogen per-oxide mixtures at 80° afford 6-methylpurine-1-oxide (28).[29] In this derivative enhancement of the methyl group reactivity, due to the adjacent positively charged nitrogen atom, is demonstrable by the formation of 6-chloromethylpurine (29) following treatment with methanesulphonyl chloride.[18] With hot acetic anhydride 28 is con-verted to 6-methoxycarbonylmethylpurine (30), which arises from re-

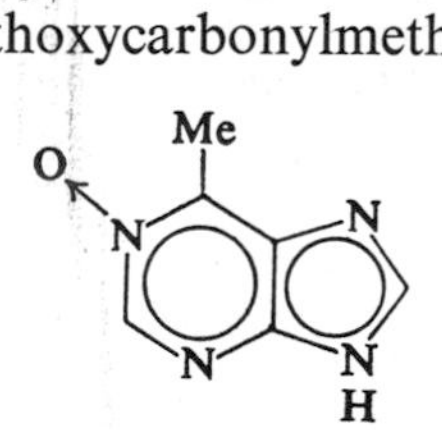

(28) 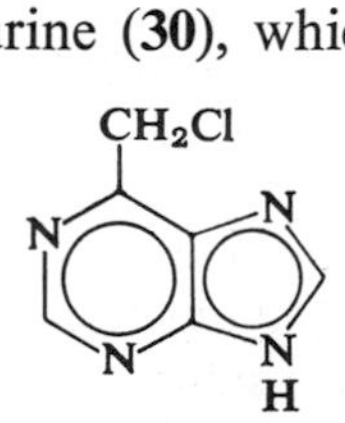   (29)

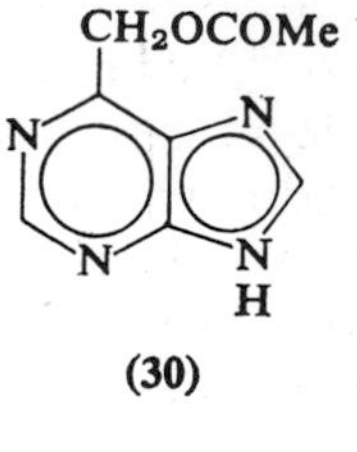

(30) 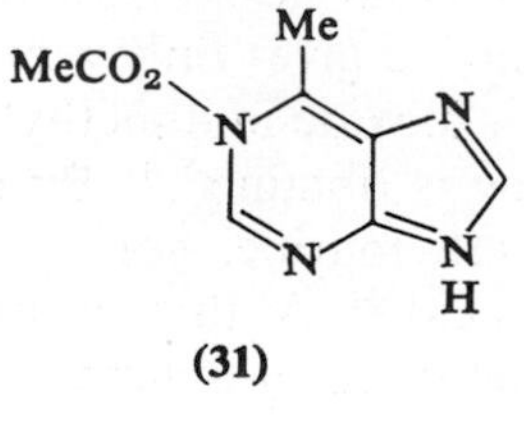   (31)

arrangement of the initially formed *O*-acylated isomer (**31**).[29] Reaction of **28** with thioacetic anhydride, or thioacetic acid, does not involve the methyl group but through rearrangement a mixture of the 2- and 8-thio derivatives of 6-methylpurine results.[26] The ratio of thiopurines obtained is time dependent; less than 4 hours reflux gives mainly the 2-thio isomer, while with longer heating periods the 8-isomer comprises the major component.[26] Such results are in contrast with the findings that neither anhydride nor acid have any effect on 6-methylpurine itself.[41] These and other reactions of *N*-oxides are further elaborated in Chapter XI, Section 3.

d. *Halogenation of Alkylpurines*

Successful direct halogenation of the purine nucleus requires the presence of suitable electron-releasing groups, the products being formed by reaction at the nucleophilic 8-carbon atom. With simple alkylpurines the poor electron-donor properties of alkyl groups does not permit substitutions of this type to take place but instead halogenation of the alkyl groups occurs. 6-Methylpurine undergoes an 80% conversion to 6-trichloromethylpurine (**32**, R = R′ = H) on using an excess of sulphuryl chloride in trifluoroacetic acid.[18] On omitting the trifluoroacetic acid or replacing it with acetic acid no reaction ensues.[18] The 6-trichloromethyl analogues of 2-amino- (**32**, R = NH_2, R′ = H), 8-trifluoromethyl- (**33**), and 9-methyl-2-methylaminopurine are prepared with this reagent.[66] Although *N*-chlorosuccinimide is a more practical form of chlorinating agent, the application in trifluoroacetic acid to 6-methylpurine gave only a poor yield (36%) of trichloromethyl derivative. In contrast while *N*-bromosuccinimide, in excess, affords a near theoretical yield of 6-tribromomethylpurine (**34**)[18] with equimolar amounts of purine and reagent, the 6-dibromomethyl analogue (**35**) is obtained.

The preparation of derivatives having halogen substituted methyl groups at $C_{(2)}$ and $C_{(8)}$ appears confined to polysubstituted purines, these being mainly derivatives of *N*-alkylated oxopurines. The more reactive methyl group of the two toward halogenation is that located at $C_{(8)}$. Reactions of this type form the bases of a number of patents.[67-69] One, two, or all hydrogen atoms may be replaced under a variety of conditions. Passing chlorine through a solution of 1,7-diethyl-8-methylxanthine in phosphoryl chloride at 50° gives the 8-trichloromethyl derivative.[70] Aluminium trichloride, in dichloromethane at 60° was used to prepare 8-trichloromethyltheobromine[71] in over 80° yield. On

changing the solvent to chlorobenzene a mixture of 8-dichloromethyl-
and 8-trichloromethyl-theobromine resulted.[72]

8-Methylcaffein (36) provides an illuminating example of the effect
produced by a change in reaction conditions. In chlorobenzene below 7°
the product is the 8-trichloromethylpurine (37)[72,73] but with higher
temperature further chlorination at the $N_{(7)}$-methyl group giving 38[73]
occurs. Under Wohl–Ziegler conditions the action of N-bromosuccinimide in the presence of a benzoyl peroxide catalyst and ultraviolet light
gives only 8-bromomethylcaffeine (39),[74] but the isocaffeine analogue
with N-chlorosuccinimide in chlorobenzene is converted to the 8-tri-
chloromethyl derivative.[72,75] Further reaction at higher temperatures
with this reagent produces the 3-chloromethyl-8-trichloromethylpurine
(40).[75] The failure of 8-methyltheophylline to react under the above conditions is most probably due to ionisation of the imidazole ring causing
a reduction in the nucleophilicity of the $C_{(8)}$-methyl group.[74]

Under less forcing halogenating conditions, as are obtained using
sulphuryl chloride in chloroform, only the mono- or dichloromethyl
derivatives of the above purines result.[76]

Although trifluoromethylpurines with fluoromethyl groups located
on the 2-, 6-, or 8-carbon atoms are known, they have not resulted

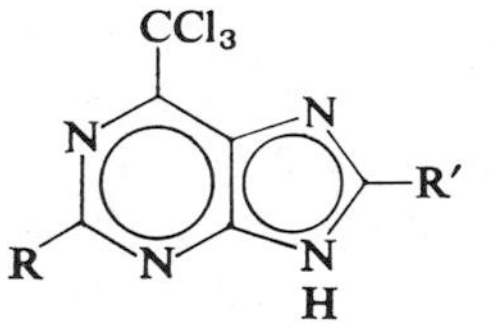

(32) 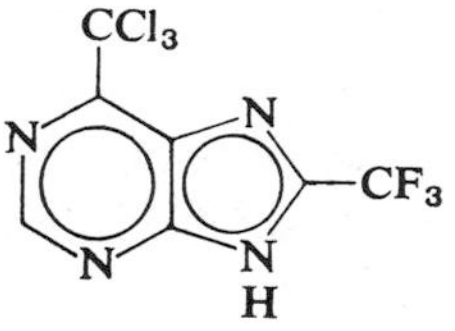   (33) 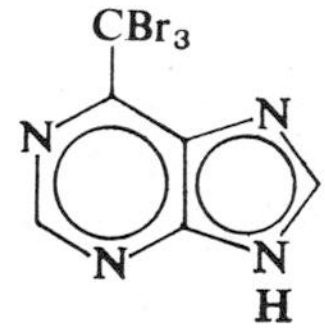   (34)

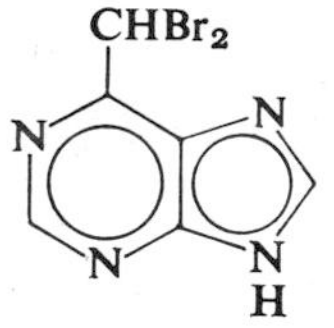

(35) 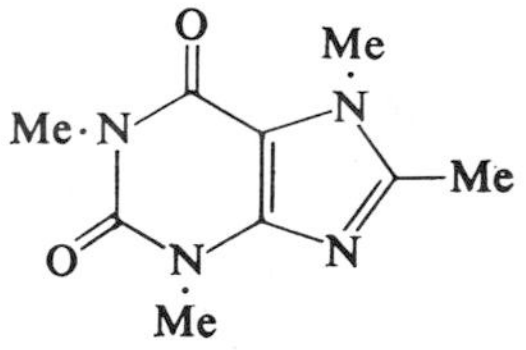   (36) 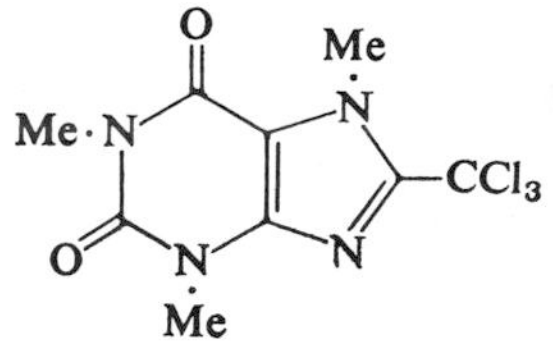   (37)

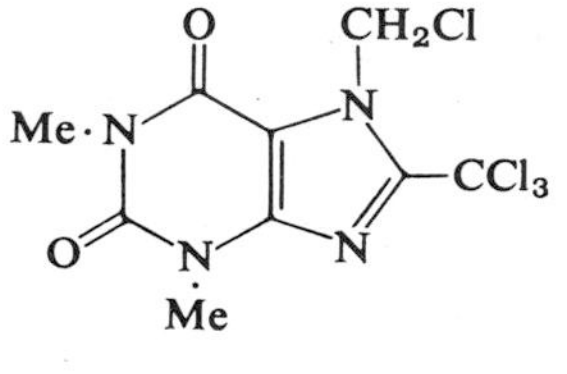

(38) 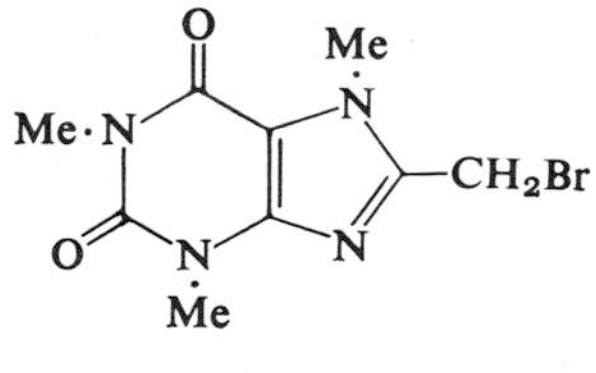   (39) 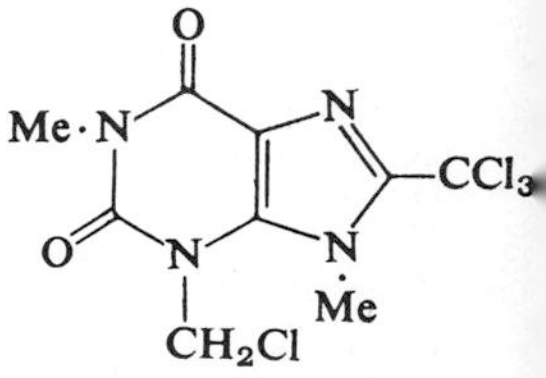   (40)

from fluorination of methylpurines but by synthesis from pyrimidine, imidazole, or other precursors[66,77,78] containing fluoromethyl groups (suitable examples of these are found in Ch. II, Sect. 1E and Ch. III, Sect. 1Bc).

3. The *N*-Alkylpurines

All four mono-*N*-methylpurines are known, as are three of the *N*-ethyl derivatives. Preparation of these by direct synthesis from *N*-alkylpyrimidines or by conversion of appropriately *N*-alkylated purines is usual. Direct alkylation is only of restricted value in *N*-alkylpurine formation. In both 1- and 3-alkylpurines the adoption of a quinonoid structure is necessary to preserve aromaticity.

A. Preparation of *N*-Alkylpurines

a. *By Direct Alkylation*

Purine with dimethyl sulphate in aqueous solution gives only 9-methylpurine,[22] no trace of the $N_{(7)}$-isomer being found. Other routes to this derivative are the action of diazomethane in an ether-alcohol medium[22] or by methyl iodide in dimethylformamide at room temperature on the thallium salt of purine.[78a] An interesting feature of the latter preparation is the accompanying formation of some 7,9-dimethylpurinium iodide (**41**). This is better formed from purine using an excess of methyl iodide and dimethylformamide as solvent[78a] or by the original route[8] involving heating 7-methylpurine with methanolic methyl iodide. From reaction with dimethyl sulphate in methanolic potassium hydroxide, 6-methylpurine gives two products—the expected 6,9-dimethylpurine (**42**) comprising the major one (50%) with a lower yield (15%) of 3,6-dimethylpurine (**43**) as the other product.[79]

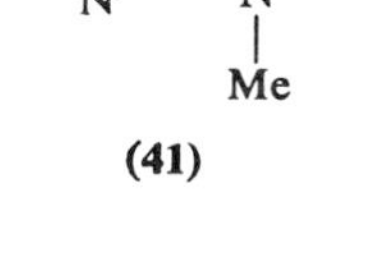

(**41**) 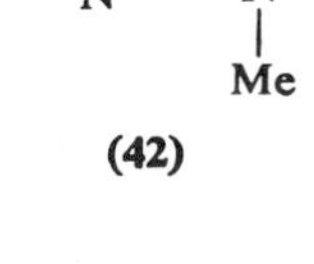   (**42**) 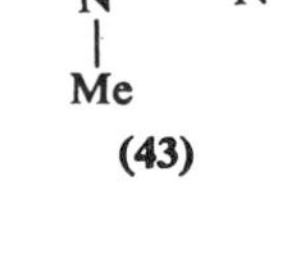   (**43**)

b. *By Other Means*

Of the *N*-methyl purines the $N_{(9)}$-isomer can be prepared by various routes which include reductive dehalogenation over palladium of either 2- or 8-chloro-9-methylpurine,[80] heating 2-iodo-9-methylpurine with zinc dust and water,[8] or formic acid closure of 5-amino-4-methylamino-pyrimidine.[4,81] Preparation of the 7-methyl isomer follows from catalytic reduction of either 2-chloro-[80] or 2,6-dichloro-7-methylpurine,[4] the latter also being reduced satisfactorily with phosphonium iodide and hydriodic acid.[8] The action of zinc dust in water on 2-iodo-7-methylpurine is a further route to 7-methylpurine.[8] A novel preparation of the same purine consists of heating together methylaminoacetonitrile and formamidine acetate in formamide.[82] Representative examples of 7- and

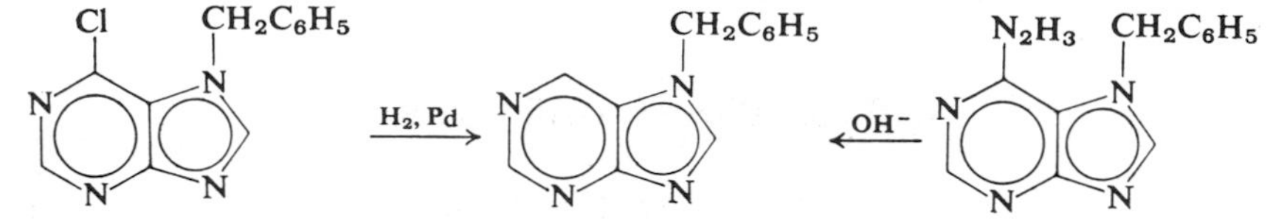

(44)   (45)   (46)

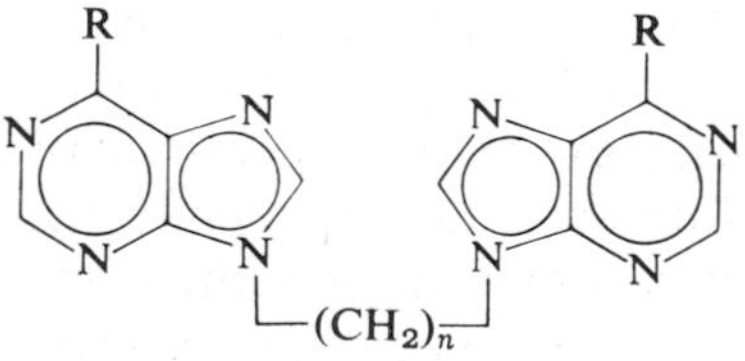

(47) 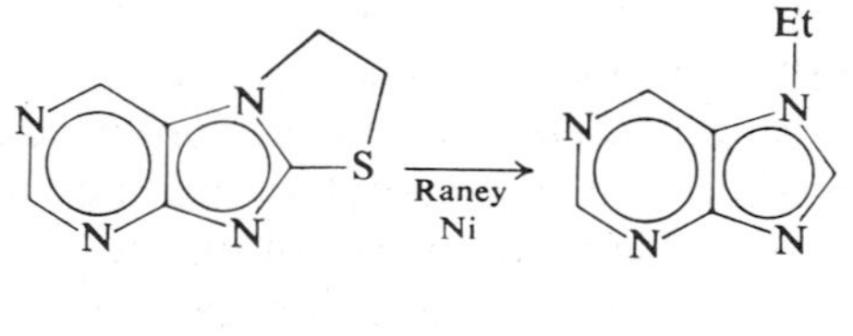   (48)   (49)

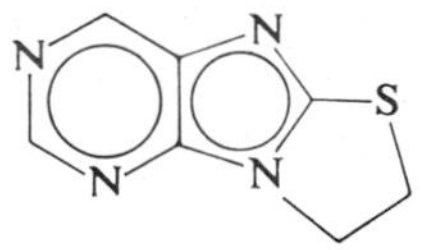

(50) 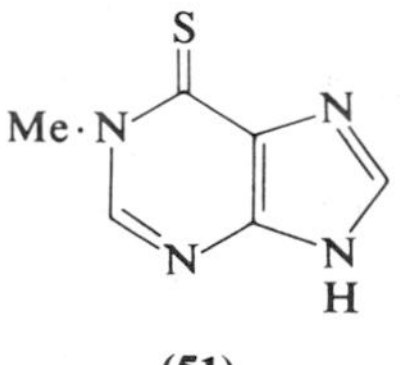   (51) 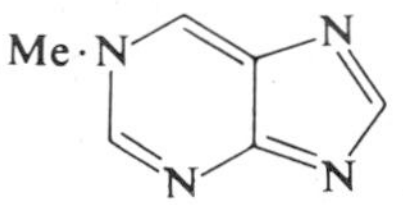   (52)

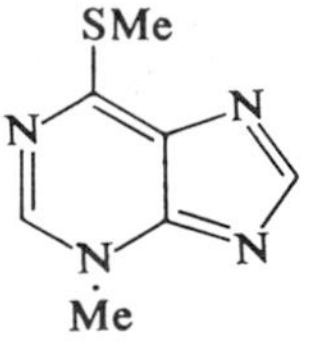

(53) 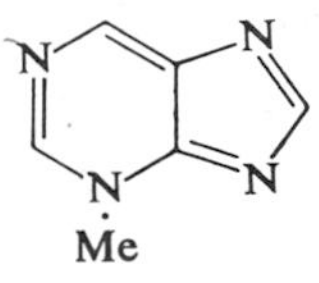   (54) 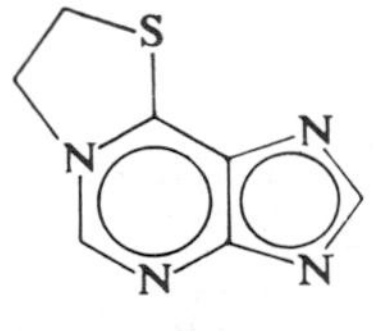   (55)

9-alkyl(or aryl)purines obtained using one or more of the above routes are 7-ethyl-,[82] 9-ethyl-,[83] 9-butyl-,[84] 7-phenyl-,[82] 9-phenyl-,[85] 7-benzyl-(45),[86] and 9-benzylpurine.[86] Reduction over a palladium catalyst of either 2-chloro- or 2,6-dichloro-9-vinylpurine, giving 9-ethylpurine, is an extension of the dehalogenation procedures.[86a] Apart from catalytic hydrogenation of 7-benzyl-6-chloropurine (44), formation of 45 occurs on mild alkaline treatment of 7-benzyl-6-hydrazinopurine (46).[20] Bis-purines of the type 47 ($n = 2$, R = H, Me, or Pr)[87] and 47 ($n = 6$, R = H),[88] derived by Traube syntheses, are other examples of 9-alkyl-purines. An unusual route to 7- or 9-ethylpurines is by desulphurisation of dihydrothiazolopurines with Raney nickel[89] in refluxing ethanol. In this manner 6,7-dihydrothiazolo[2,3-f]purine (48) [RRI 2341] gives 7-ethylpurine (49), while the isomeric 7,8-dihydrothiazolo[3,2-e]purine (50) affords 9-ethylpurine.[89] Both 1-methyl- (52) and 3-methylpurine (54) result from Raney nickel treatment, the former from 1,6-dihydro-1-methyl-6-thiopurine (51) in boiling water,[90] while the latter prepara-tion from 3-methyl-6-methylthiopurine (53) is carried out in methanol under reflux.[91] Similar conditions also provide a route to 1-ethylpurine from 7,8-dihydrothiazolo[2,3-i]purine (55).[89]

B. Properties of N-Alkylpurines

All N-alkylpurines are crystalline, well-defined, colourless solids that possess moderate storage stability, the least stable being compounds alkylated in the pyrimidine ring. This fact, together with the more pronounced basic nature found in the latter compounds, can be correlated with the quinonoid structures required. The case of 1-ethyl purine (pK_a 5.08) exemplifies this enhanced basicity when contrasted with that of 7- or 9-ethylpurine (pK_a 2.67)[89] or of purine itself (pK_a 2.39).[2]

C. Reactions of N-Alkylpurines

Few examples of reactions are known; chlorination of an N-methyl group is possible as shown by the formation of 7-chloromethyl-2-methylpurine (57) when chlorine was bubbled through a solution of 2,7-dimethylpurine (56) in trichloroethane. The inability to form anions, and hence stable structures, leads to disruption of N-alkylpurines in alkaline solution. 9-Methylpurine, for example, in 10N-sodium hydroxide for one hour at 100°, is converted into 5-amino-4-methylaminopyrimi-dine (58),[2] this behaviour contrasting sharply with the pronounced

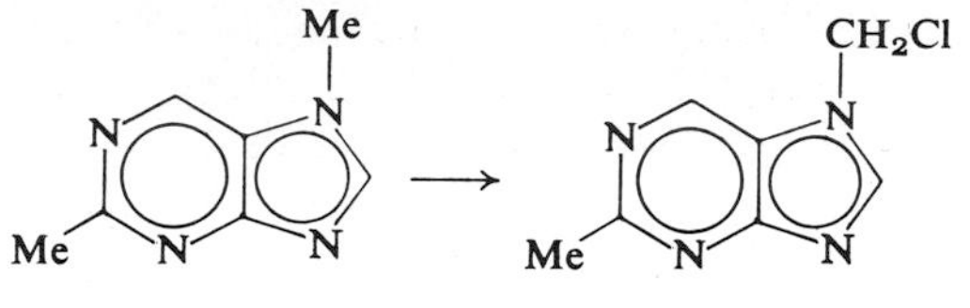

(56)   (57) 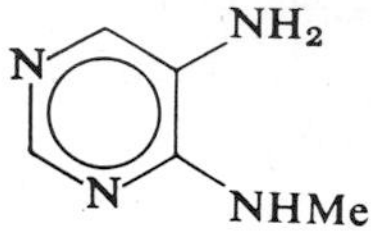   (58)

stability shown by purine itself toward alkali. Degradations to 5-amino-4-alkylaminopyrimidines are noted with other 9-alkylpurines[86] and 9-ribofuranosylpurine,[23] the latter taking place at room temperature in 0.05N-alkali. A stable methiodide is formed on treating 7-methylpurine with methyl iodide in methanol.[8]

4. Natural Occurrence

Purine can be isolated, in the form of the 9-β-D-ribofuranosyl derivative "nebularine," from the mushroom *Agaricus nebulais*[92] and the mould *Streptomyces yokosukanensis*.[93] Although even weak alkali rapidly breaks nebularine down to 5-amino-4-ribosylaminopyrimidine, hydrolysis with 3% hydrochloric acid affords purine itself.[92] Various syntheses of nebularine have been carried out.[15a,94]

References

1. Isay, *Ber.*, **39**, 250 (1906).
2. Albert and Brown, *J. Chem. Soc.*, **1954**, 2060.
3. Brown, *J. Appl. Chem.*, **5**, 358 (1955).
4. Bendich, Russell, and Fox, *J. Amer. Chem. Soc.*, **76**, 6073 (1954).
5. Matsuura and Goto, *J. Chem. Soc.*, **1965**, 623.
6. Montgomery and Temple, *J. Org. Chem.*, **25**, 395 (1960).
7. Grundmann and Kreutzberger, *J. Amer. Chem. Soc.*, **77**, 6559 (1955).
8. Fischer, *Ber.*, **31**, 2550 (1898).
9. Bredereck, Herlinger, and Graudums, *Chem. Ber.*, **95**, 54 (1962).
10. Bredereck, Herlinger, and Graudums, *Angew. Chem.*, **71**, 524 (1959).
11. Coburn, Thorpe, Montgomery, and Hewson, *J. Org. Chem.*, **30**, 1110 (1965).
12. Schweizer, Chan, Helmkamp, and Ts'o, *J. Amer. Chem. Soc.*, **86**, 696 (1964).
13. Beaman, *J. Amer. Chem. Soc.*, **76**, 5633 (1954).
14. Brown, Levin, Murphy, Sele, Reilly, Tarnowski, Schmid, Teller, and Stock, *J. Medicin. Chem.*, **8**, 190 (1965).
15. Levin and Brown, *J. Medicin. Chem.*, **6**, 825 (1963).
15a. Saneyoshi and Chihari, *Chem. and Pharm. Bull.* (*Japan*), **15**, 909 (1967).
16. Albert, *J. Chem. Soc.*, **1960**, 4705.

17. Mackay and Hitchings, *J. Amer. Chem. Soc.*, **78**, 3511 (1956).
18. Cohen, Thom, and Bendich, *J. Org. Chem.*, **27**, 3545 (1962).
19. Doerr, Wempen, Clarke, and Fox, *J. Org. Chem.*, **26**, 3401 (1961).
20. Temple, Kussner, and Montgomery, *J. Org. Chem.*, **30**, 3601 (1965).
21. Bredereck, Effenberger, Rainer, and Schosser, *Annalen*, **659**, 133 (1962).
22. Bredereck, Ulmer, and Waldmann, *Chem. Ber.*, **89**, 12 (1956).
23. Gordon, Weliky, and Brown, *J. Amer. Chem. Soc.*, **79**, 3245 (1957).
24. Chan, Schweizer, Ts'o, and Helmkamp, *J. Amer. Chem. Soc.*, **86**, 4182 (1964).
25. Cheney, Freiser, and Fernando, *J. Amer. Chem. Soc.*, **81**, 2611 (1959).
26. Giner-Sorolla, Thom, and Bendich, *J. Org. Chem.*, **29**, 3209 (1964).
27. Albert, *J. Chem. Soc.* (*B*), **1966**, 438.
28. Linschitz and Connolly, *J. Amer. Chem. Soc.*, **90**, 2979 (1968); Connolly and Linschitz, *Photochem and Photobiol.*, **7**, 791 (1968).
29. Stevens, Giner-Sorolla, Smith, and Brown, *J. Org. Chem.*, **27**, 567 (1962); Giner-Sorolla, Gryte, Cox, and Parham, unpublished results.
30. Bergmann and Dikstein, *J. Biol. Chem.*, **223**, 765 (1956).
31. Smith and Elving, *J. Amer. Chem. Soc.*, **84**, 1412 (1962).
32. Breshears, Wang, Bechtolt, and Christensen, *J. Amer. Chem. Soc.*, **81**, 3789 (1959).
32a Betula, *Annalen*, **729**, 73 (1969).
33. Gabriel and Colman, *Ber.*, **34**, 1234 (1901).
34. Prasad, Noell, and Robins, *J. Amer. Chem. Soc.*, **81**, 193 (1959).
35. Brown and Mason, *J. Chem. Soc.*, **1957**, 682.
36. Bergmann, Kalmus, Ungar-Waron, and Kwietny-Govrin, *J. Chem. Soc.*, **1963**, 3729.
37. Fu, Chinoporos, and Terzian, *J. Org. Chem.*, **30**, 1916 (1965).
38. Bergmann, Neiman, and Kleiner, *J. Chem. Soc.* (*C*), **1966**, 10.
39. Goldner, Dietz, and Carstens, *Naturwiss.*, **51**, 137 (1964).
40. Clark, Elion, Hitchings, and Stock, *Cancer Res.*, **18**, 445 (1958).
41. Giner-Sorolla and Bendich, *J. Medicin. Chem.*, **8**, 667 (1965).
42. Golovchinskaya and Chaman, *Zhur. obshchei Khim.*, **22**, 2220 (1952).
43. Koppel and Robins, *J. Org. Chem.*, **23**, 1457 (1958).
44. Bergmann and Tamari, *J. Chem. Soc.*, **1961**, 4468.
45. Mautner, *J. Org. Chem.*, **26**, 1914 (1961).
46. Dyer and Bender, *J. Medicin. Chem.*, **7**, 10 (1964).
47. Lettré, Ballweg, Maurer, and Rehberger, *Naturwiss.*, **50**, 224 (1963).
48. Chaman and Golovchinskaya, *Zhur. obshchei Khim.*, **33**, 3342 (1963).
49. Ovcharova and Golovchinskaya, *Zhur. obshchei Khim.*, **34**, 3254 (1964).
50. Nikolaeva and Golovchinskaya, *Zhur. obshchei Khim.*, **34**, 1137 (1964).
51. Golovchinskaya, *Sbornik Statei obshchei Khim., Akad. Nauk. S.S.S.R.*, **1**, 692 (1953).
52. Golovchinskaya, *Sbornik Statei obshchei Khim., Akad. Nauk. S.S.S.R.*, **1**, 702 (1953).
53. Shioi, Jap. Pat., 177,356 (1949); through *Chem. Abstracts*, **45**, 7590 (1951).
54. Biltz and Sauer, *Ber.*, **64**, 752 (1931).
55. Jones and Robins, *J. Amer. Chem. Soc.*, **84**, 1914 (1962).
56. Donat and Carstens, *Chem. Ber.*, **92**, 1500 (1959).
57. Serchi, Sancio, and Bichi, *Farmaco. Ed. Sci.*, **10**, 733 (1955).
58. Bredereck, von Schuh, and Martini, *Chem. Ber.*, **83**, 201 (1950).
59. Huston and Allen, *J. Amer. Chem. Soc.*, **56**, 1793 (1934).
60. Hull, *J. Chem. Soc.*, **1958**, 4069.

61. Giner-Sorolla, *Chem. Ber.*, **101**, 611 (1968).
62. Shioi, Jap. Pat., 177,355 (1949); through *Chem. Abstracts*, **45**, 7590 (1951).
63. Lettré and Woenckhaus, *Annalen*, **649**, 131 (1961).
64. Giner-Sorolla, Zimmerman, and Bendich, *J. Amer. Chem. Soc.*, **81**, 2515 (1959).
65. Bredereck, Siegel, and Föhlisch, *Chem. Ber.*, **95**, 403 (1962).
66. Cohen and Vincze, *Israel J. Chem.*, **2**, 1 (1964).
67. Boehringer, Ger. Pat., 146,714 (1902); *Frdl.*, **7**, 670 (1902–1904).
68. Boehringer, Ger. Pat., 151,133 (1902); *Frdl.*, **7**, 672 (1902–1904).
69. Boehringer, Ger. Pat., 153,121 (1902); *Frdl.*, **7**, 674 (1902–1904).
70. Mann and Porter, *J. Chem. Soc.*, **1945**, 751.
71. Golovchinskaya, Fedosova, and Cherkasova, *Zhur. priklad. Khim.*, **31**, 1241 (1958).
72. Chaman, Cherkasova, and Golovchinskaya, *Zhur. obshchei Khim.*, **30**, 1878 (1960).
73. Golovchinskaya, *Zhur. priklad. Khim.*, **31**, 918 (1958).
74. Zimmer and Atchley, *Arzneim.-Forsch.*, **16**, 541 (1966).
75. Golovchinskaya, Ovcharova, and Cherkasova, *Zhur. obshchei Khim.*, **30**, 3332 (1960).
76. Chaman and Golovchinskaya, *Zhur. obshchei Khim.*, **32**, 2015 (1962).
77. Giner-Sorolla and Bendich, *J. Amer. Chem. Soc.*, **80**, 5744 (1958).
78. Kaiser and Burger, *J. Org. Chem.*, **24**, 113 (1959).
78a. Taylor, Maki, and McKillop, *J. Org. Chem.*, **34**, 1170 (1969).
79. Vincze and Cohen, *Israel J. Chem.*, **4**, 23 (1966).
80. Beaman and Robins, *J. Org. Chem.*, **28**, 2310 (1963).
81. Brown, *J. Appl. Chem.*, **7**, 109 (1957).
82. Bredereck, Effenberger, and Rainer, *Annalen*, **673**, 82 (1964).
83. Montgomery and Temple, *J. Amer. Chem. Soc.*, **79**, 5238 (1957).
84. Montgomery and Temple, *J. Amer. Chem. Soc.*, **80**, 409 (1958).
85. Fischer and Loeben, *Ber.*, **33**, 2278 (1900).
86. Montgomery and Temple, *J. Amer. Chem. Soc.*, **83**, 630 (1961).
86a. Pitha and Ts'o, *J. Org. Chem.*, **33**, 1341 (1968).
87. Lister, *J. Chem. Soc.*, **1963**, 2228.
88. Lister, *J. Chem. Soc.*, **1960**, 3394.
89. Balsiger, Fikes, Johnston, and Montgomery, *J. Org. Chem.*, **26**, 3386 (1961).
90. Townsend and Robins, *J. Org. Chem.*, **27**, 990 (1962).
91. Townsend and Robins, *J. Heterocyclic Chem.*, **3**, 241 (1966).
92. Löfgren, Lüning, and Hedstrom, *Acta Chem. Scand.*, **8**, 670 (1954).
93. Nakamura, *J. Antibiotics (Japan)*, Ser. A, **14**, 94 (1961).
94. Brown and Weliky, *J. Biol. Chem.*, **204**, 1019 (1953).

CHAPTER V

Halogenopurines

The exceptional reactivity toward nucleophilic reagents shown by halogenopurines makes them invaluable compounds for purine inter-conversion. Halogen atoms located at the 2-, 6-, or 8-position are all replaceable but the ease or difficulty of displacement is modified if electron-releasing groups are present, this being most profound in the case of a 2-halogenopurine.

1. The Preparation of 2-, 6-, and 8-Halogenopurines

Purine chemistry is fortunate in affording a number of ways of synthesising the versatile halogenopurines. While on the one hand 2- and 6-halogeno derivatives can be obtained by cyclisation of appropriate halogen-containing pyrimidines, they are also obtainable from purines by halogen replacement of existing groups. The latter route is also suitable for preparing 8-halogenopurines but, due to the unique proper-ties of the 8-carbon atom, these purines can, in many instances, be formed by direct halogenation of an 8-unsubstituted purine.

A. Chloropurines by Cyclisation of Chloro-4,5-diaminopyrimidines

This aspect, which is limited to the formation of 2- and 6-chloro-purines, has been fully covered in the appropriate sections in Chapter II.

B. By Chlorination of Oxopurines

This, the oldest approach, is still extensively used and gives 2-, 6-, or 8-chloropurines from the appropriate oxo analogues.

135

a. *In Simple Purines*

Phosphoryl chloride alone may be employed but more usually an adjunct, which may be another chlorinating agent or a base catalyst, is also present. Successful chlorinations with the neat reagent usually require sealed-tube conditions; 6-chloro-2-phenyl-,[1] 6-chloro-7-methyl-,[2] 6-chloro2,8-dimethyl-,[3] and 2,6-dichloro-9-methylpurine[4,5] are examples obtained in this way, although cases of 6-chloropurine formation occurring under reflux conditions are also known.[6] The outstanding application, now only of historical importance, was Fischer's attempt to prepare 2,6,8-trichloropurines[7-10] from uric acid derivatives from which, in all cases, he obtained initially 2,6-dichloro-8-oxopurines needing further treatment[8,11] with the same chlorinating agent under pressure to insert the third chlorine atom. Reexamination of this synthesis more recently has confirmed the need for a two-stage reaction but has shown that yields of up to 90% of trichloropurine are possible by using a large excess of chlorinating agent under less rigorous conditions.[12] Xanthine is unusual in that it will not chlorinate with anhydrous phosphoryl chloride even under pressure but the required 2,6-dichloropurine results if the reagent is first pretreated with a half mole equivalent of water. The effective reagent in this instance is claimed to be the pyrophosphoryl chloride formed *in situ*.[13]

Phosphoryl chloride–phosphorus pentachloride combinations obviate the need for sealed-tube procedures, reflux conditions sufficing. Examples derived from the appropriate di- or trioxo derivatives by this means include 2,6-dichloro-7-methyl-,[2*] 2,6-dichloro-9-phenyl-,[14] 2,6,8-trichloro-7-methyl-,[8] and 2,6,8-trichloro-9-phenylpurine.[9] The reaction is temperature dependent as with insufficient heating only partially halogenated derivatives may be formed.[15] More efficacious and extensively employed are mixtures of phosphoryl chloride with tertiary bases, usually *NN*-dimethyl- or diethylaniline, under reflux conditions. Hypoxanthine was first converted to 6-chloropurine[16-18] using dimethylaniline, but replacing this by *N*-ethylpiperidine is claimed to afford better yields.[19] The use of toluene as solvent for such chlorinations has recently been described.[20] Homologous 6-chloropurines prepared in the presence of dialkylanilines include the 2-methyl,[21] 8-methyl,[22] and related 2,8-dialkyl derivatives.[23] Xanthine does not react under these conditions[13] although the isomeric 6,8-dioxo derivatives are readily converted to 6,8-dichloropurines.[24,25] Uric acid is transformed to

---

* Surprisingly, chlorination of 9-methylxanthine under these conditions fails[14] but succeeds when phosphoryl chloride is used alone under pressure.[4]

2,6,8-trichloropurine directly but the yield is low (25%).[26] A more successful route to the latter is to obtain 2,6-dichloro-7,8-dihydro-8-oxopurine from uric acid by means of phosphoryl chloride alone and then complete the chlorination using the base-catalysed reagent.[27] The procedure has been adapted for nucleosides, a prior requirement being protection of the ribose moiety by acetylation.

Interaction between the halogen atom of the purine and the tertiary aromatic base is rare, but the attempted chlorination of the hypo-xanthine derivative (1) in the presence of dimethylaniline gave the 6-N-methylanilinopurine (2), while with phosphoryl chloride alone normal chlorination gave 6-chloro-2-methyl-8-methylthiopurine.[28] A more general base involvement occurs when aliphatic tertiary bases are substituted for the aromatic bases in the reaction. Both hypoxanthine and xanthine in phosphoryl chloride containing triethylamine suffer replacement of the oxo-groups by diethylamino groups affording 6-diethylamino- and 2,6-bisdiethylaminopurine (3), respectively.[29] Corre-sponding displacements occur with trimethylamine but with higher tertiary amines only the 6-position is involved; both tripropylamine and tributylamine with xanthine give the corresponding 2-oxo-6-dialkyl-aminopurine. Amination of the 6-position also occurs with the mono-potassium salt of uric acid, 2,8-dichloro-6-diethylaminopurine being formed.[29] On limiting the amount of triethylamine the product, sur-prisingly, is 8-chloroxanthine[29] obtained in 85% yield. As noted above simple uric acid derivatives usually afford 2,6-dichloro-8-oxo derivatives. Formation of 8-chloro-2,6-dioxo derivatives is a feature of some N-alkylated uric acids (see Sect. 1Bd).

(1)   (2)   (3)

The nucleoside inosine, with the sugar moiety protected by acetyla-tion, is smoothly converted to 6-chloropurine-9-β-D-riboside in phos-phoryl chloride-dimethylaniline under reflux[30] but omission of the tertiary base results in cleavage of the riboside-purine linkage.[30] Other successful chlorinations of this nucleoside under similar conditions have been carried out with Bosshard reagent using either thionyl chloride in dimethylformamide[31] or NN-diethylchloromethylene ammonium chlor-ide[32] in chloroform.

Phenylphosphonic dichloride, which has successfully chlorinated other oxo heterocyclic compounds, failed to give recognisable products when applied to uric acid or theobromine.[33]

b. *In the Presence of Amino Groups*

Guanine itself shows resistance to chlorination,[34] as is also found with the 8-phenyl analogue.[35] However, most substituted guanines and other amino-oxopurines are more reactive and form corresponding aminochloropurines. With phosphoryl chloride containing phosphorus pentachloride only the $C_{(6)}$-oxo group reacts in the 6,8-dioxopurine (**4**) giving 2-amino-6-chloro-7,8-dihydro-8-oxopurine.[36] Under sealed-tube conditions the 8-oxo group in 6-amino-2-chloro-7,8-dihydro-8-oxo-purine was converted to a chlorine atom.[37] In the same way the 7- and 9-methyl analogues give the appropriate 6-amino-2,8-dichloro-7(or 9)-methylpurine.[37] Amino groups located at $C_{(8)}$ appear to deactivate oxo groups toward halogenation. Using various chlorinating agents no reaction with 8-anilino-1,6-dihydro-6-oxo-2-methylpurine was observed.[28] Phosphoryl chloride alone, under reflux conditions, smoothly converts the 9-aminohypoxanthine (**5**) to 6-chloro-9-dimethylamino-8-methylpurine.[38]

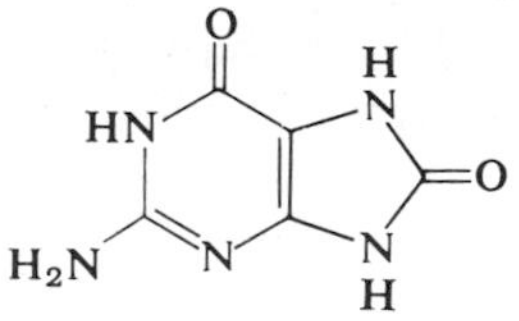

(**4**) 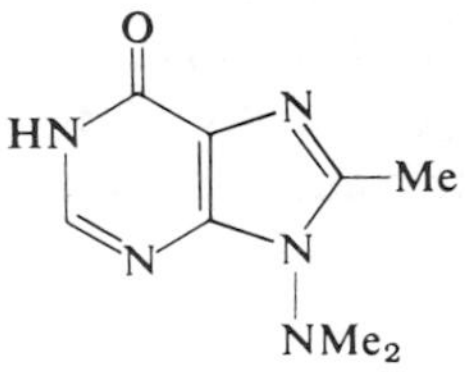   (**5**)

The concomitant chlorination and cyclisation of 2,4-diamino-5-benzamido-1,6-dihydro-6-oxopyrimidines to the corresponding 2-amino-6-chloro-8-phenylpurines[35] is discussed in Chapter II, Section 1F.

c. *In the Presence of Thio (and Alkylthio) Groups*

Provided the thio group is in the alkylated form no untoward effects are apparent. Successful chlorinations with phosphoryl chloride alone or with a dialkylaniline catalyst have, with the appropriate oxo derivative, given 6-chloro-2-methylthio-,[25] 8-chloro-2-methylthio-,[25] 6-chloro-8-methyl-2-methylthio-,[22] 6,8-dichloro-2-methylthio-,[25] 8-chloro-6-

methylthio-,[24] 6-chloro-8-methylthio-,[24] 6-chloro-2-methyl-8-methyl-thio-,[28] 6-chloro-2,8-bismethylthio-,[39] and 8-chloro-2,6-bismethylthiopurine.[39]

d. *In the Presence of N-Alkyl Groups*

Chlorination of *N*-alkyl oxopurines is, as a general rule, only possible in those cases where the oxo function is capable of existing in the enol form also. The reaction can be complicated by the occurrence of subsidiary effects, the most likely of these being removal of an *N*-methyl group. In addition where an unsubstituted 8-position is present nuclear chlorination at this site is a further possibility.

Mono-oxopurines with *N*-alkyl groups in the imidazole ring behave normally; the example of 6-chloro-7-methylpurine has been stated.[2] Of the methylated forms of xanthine theophylline does not react with phosphoryl chloride alone but theobromine (**6**) suffers loss of the $N_{(3)}$-methyl group, thus allowing chlorination of both 2- and 6-positions (**7**) to occur.[40] In the presence of dimethylaniline the same purine (**7**) arises[41] but using phosphoryl chloride–phosphorus pentachloride mixtures a concurrent nuclear chlorination at $C_{(8)}$ occurs giving 2,6,8-trichloro-7-methylpurine (**8**).[40, 42] Removal of a pyrimidine-sited *N*-methyl group is also found when 1,9-dimethylxanthine (**9**), in phosphoryl chloride at 140°[5, 43] for some hours, gives 2,6-dichloro-9-methylpurine (**10**). The isolation of the intermediate 2-chloro-1,6-dihydro-1,9-dimethyl-6-oxopurine (**11**) under milder conditions indicates that halogenation of the $C_{(2)}$-oxo group precedes removal of the $N_{(1)}$-methyl group.[5] Caffeine, through chlorination of the 8-position and removal of both pyrimidine *N*-methyl groups, is converted to 2,6,8-trichloro-7-methylpurine when treated with a mixture of phosphorus pentachloride and phosphoryl chloride at 180°.[42] Chlorination of a doubly bound oxygen atom without loss of the associated *N*-methyl group does appear to have been demonstrated.[44] The product from phosphoryl chloride treatment of 1,9-dimethylhypoxanthine (**12**) was not isolated but condensed with an amine with the formation of the iminopurine (**13**).

Uric acid derivatives behave somewhat differently under these conditions owing to the special nature of the oxo group located at $C_{(8)}$. Those alkylated only in the imidazole ring behave as the parent compound in that, with phosphoryl chloride alone, the appropriate 7-(or 9)-alkyl-2,6-dichloro-8-oxodihydropurine[8, 15] results. If phosphorus pentachloride is also present the fully chlorinated 2,6,8-trichloropurine[8, 15] is obtained. A similar course is taken by polymethylated uric acid derivatives which can, by loss of a methyl group, afford compounds methylated only in the imidazole ring. This is exemplified by the chlorination of the 3,9-dimethyl (**14**, R = H)[8] and 3,7,9-trimethyl (**14**, R = Me)[42] homologues of uric acid both of which lose the 3-methyl group affording the respective 9-methyl- (**15**, R = H) and 7,9-dimethyl- (**15**, R = Me) 2,6-dichloropurines.

The majority of methylated uric acid derivatives undergo primary chlorination of the 8-oxo group only and exhibit little tendency toward displacement of *N*-methyl groups. An exception is the 1,3,7,9-tetra-

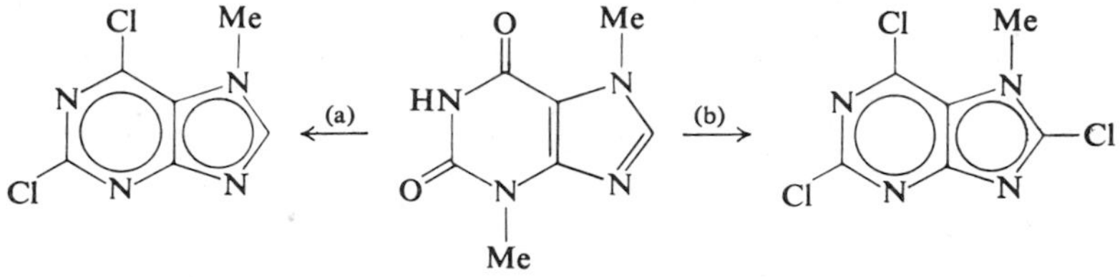

(**7**)   (**6**)   (**8**)

(*a*) POCl₃   (*b*) POCl₃—PCl₅

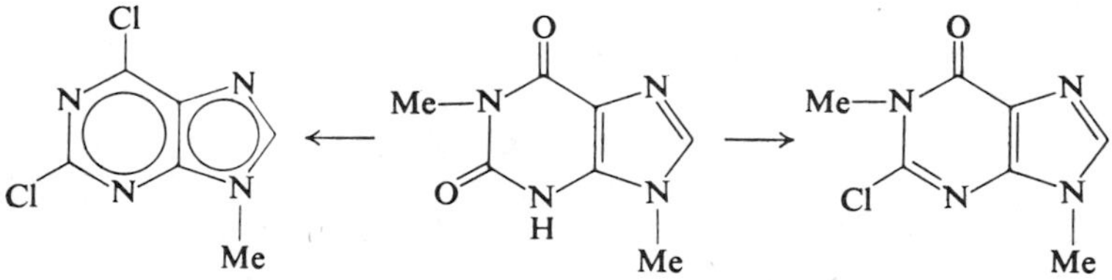

(**10**)   (**9**)   (**11**)

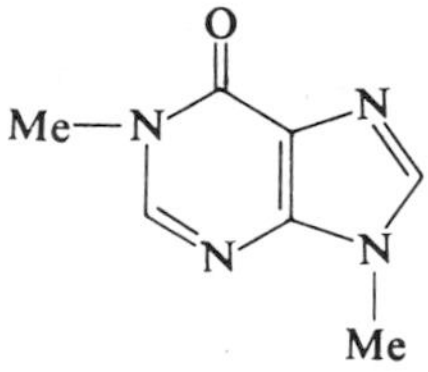

(**12**) 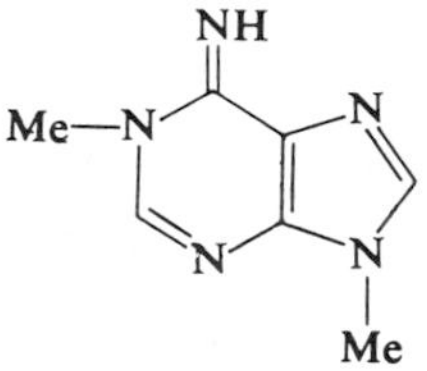   (**13**) 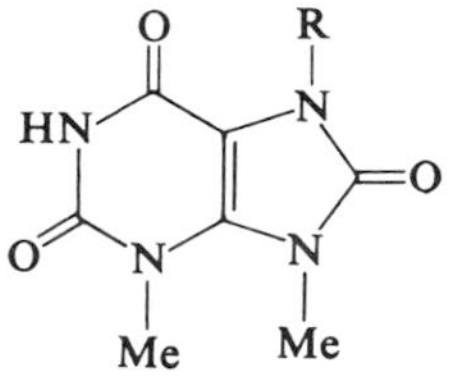   (**14**)

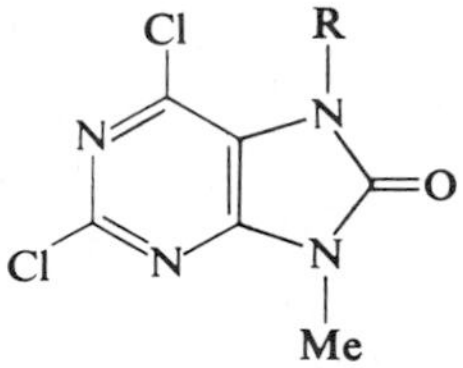

(**15**) 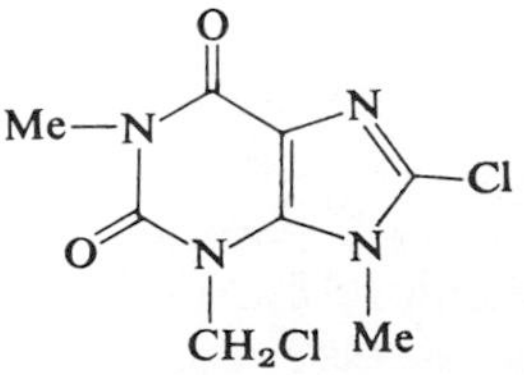   (**16**)

methyluric acid which loses $N_{(9)}$-methyl group forming 8-chloro-1,3,7-trimethylxanthine.[45] Generally, use of phosphoryl chloride alone is effective; in this way the 3-methyl,[46,47] 1,3-dimethyl,[48] 1,7-dimethyl,[49,50] 3,7-dimethyl,[46,47] and 1,3,7-trimethyl[51] derivatives of 8-chloroxanthine are formed from the uric acid analogues. Under rigorous conditions, in phosphoryl chloride at 170° containing phosphorus pentachloride, 3,7-dimethyluric acid suffers loss of the $N_{(3)}$-methyl group, being converted to 2,6,8-trichloro-7-methylpurine.[42]

Demethylations of the above kind proceed by initial conversion of the methyl group to a chloromethyl form, in which one or all hydrogen atoms are replaced by halogen, after which hydrolytic fission of the nitrogen–carbon bond occurs. The point is illustrated by the conversion of the $N_{(3)}$-chloromethyl derivative (16) of 8-chlorocaffeine to 8-chloro-1,9-dimethylxanthine simply on heating in water.[52] Other examples of this type are recorded,[53] including the removal of either the 3- or 7- or all three methyl groups of 8-chlorocaffeine.

C. By Replacement of Thio or Methylthio Groups

The oxidative removal of a thio group by chlorine can result in production of a chloropurine, the halogen being located at the site of the deposed sulphur group. This approach was based on an original observation that when 1,6-dihydro-6-thiopurine in cold ethanol was subjected to the passage of chlorine gas 6-chloropurine was obtained.[54,55]

Subsequently it has been demonstrated that methylthiopurines, because of their greater solubility, are more suitable. Some difficulties encountered with certain derivatives are overcome either by the addition of concentrated hydrochloric acid or saturation of the methanolic solution with hydrogen chloride. The use of a combination of both these reagents is occasionally encountered. Chloropurines, which have been prepared in yields superior to those obtained with phosphoryl chloride, include 2-chloro- (95%),[17] 6-chloro- (88%),*[17,39] 8-chloro- (81%),[17] 2,6-dichloro- (92%),[17] 6,8-dichloro- (74%),[17] and 2,6,8-trichloropurine,[56] also 6-chloro-3-methylpurine and the 8-phenyl analogue[57] were derived from the corresponding thio- or methylthiopurines. One advantage of the method is that it allows the preparation of 2- and 8-chloropurines not obtainable by chlorination of the corresponding oxopurines.[58]

* The comparable yield of 6-chloropurine obtained from hypoxanthine using phosphoryl chloride was 74%.[17]

In the absence of hydrogen chloride, some instances of partial chlorination are found. With 2,3,7,8-tetrahydro-2,8-dithiopurine the product is 2-chloro-7,8-dihydro-8-oxopurine,[17] whereas the 2,6,8-trithio analogue of uric acid is converted to 6-chloro-7,8-dihydro-8-oxo-2-sulphopurine.[56]

The reaction is also suitable for thiopurines carrying amino and other electron-releasing groups. Initially preparations[54, 55, 59] of 2-amino-6-chloropurine were made by this route from 6-thioguanine but later the 6-methylthio analogue was employed.[60] Further examples include 6-amino-8-chloropurine,[61] 2-amino-6-chloro-3-methylpurine,[62] and the 1-oxide of 6-amino-2-chloropurine.[63]

Oxothiopurines give the appropriate chloro-oxopurines, examples being 8-chloro-2-oxo-,[64] 8-chloro-6-oxo-,[56] 6-chloro-2,8-dioxo-,[56] and 2,8-dichloro-6-oxopurines.[56] The procedure has also been employed for the conversion of thiopurine nucleosides without protection of the ribose moiety. Thus the riboside of either 6-thio- or 6-methylthiopurine gives 6-chloropurine riboside in over 82% yield using only chlorine in methanol at $-10°$.[65] These conditions also successfully chlorinate the riboside of 2-amino-6-methylthiopurine[65] but, if the corresponding 6-thiopurine riboside is used, the addition of hydrogen chloride is required for this reaction to succeed.[30]

D. By Other Means

Conversion of amino groups to chloro by diazotisation is a rare occurrence but 2-chloropurine, in low yield, results from the action of concentrated hydrochloric acid on the diazonium salt of 2-amino-purine[66] and the 9-β-D-ribosides of 2-amino-6-chloro- and 2-amino-1,6-dihydro-6-thiopurine are likewise transformed to the 2,6-dichloro and 2-chloro-6-thio analogues.[67] Although adenine derivatives are not affected by these conditions, 6,8-diaminopurine gives 6-amino-8-chloropurine.[68] The preparation of 6-chloropurine from 6-hydrazino-purine has been accomplished with acidified ferric chloride at room temperature[69] and, in the same manner, 8-chlorocaffeine follows from the 8-hydrazino analogue.[70] Another route to 6-chloropurines is by heating the 6-sulphopurine with thionyl chloride at 50°, as used in preparations of 6-chloro- and 2-amino-6-chloropurines.[71] A related example of the conversion of a 9-alkyl-6-thio- to a 9-alkyl-6-chloro-purine, employing thionyl chloride under reflux conditions, is known.[72]

E. Preparation of 2-, 6-, and 8-Bromopurines

Procedures analogous to those described for the preparation of chloropurines are employed, the exception being that no cyclisation of a bromo-4,5-diaminopyrimidine is recorded.

a. *By Bromination of Oxopurines*

Prolonged heating of hypoxanthine in phosphoryl bromide gives the 6-bromo derivative in moderate yield (40%),[13, 20] with xanthine only a poor yield (10%) of 2,6-dibromopurine results.[13] Ring closures of 4-amino-5-benzamido-1,6-dihydro-6-oxopyrimidines with phosphoryl bromide to 6-bromo-8-phenylpurines[35] are described in Chapter II, Section 1F.

b. *By Replacement of Thio or Methylthio Groups*

Bromine water treatment of 8-thio-9-methylxanthine affords 8-bromo-9-methylxanthine.[73] This appears to be the first example of an oxidative halogenation of a thiopurine. Present procedures are patterned on those followed with the chloropurines, the reactions being carried out with a mixture of 47% hydrobromic acid and bromine in methanol below 10°. Like the chloropurine preparations the presence of amino or oxo groups is not found to hinder the reaction. Bromopurines so derived are exemplified by 2-bromo-,[74] 6-bromo-,[74] 8-bromo-,[74] 2,6-dibromo-,[74] 6,8-dibromo-, and 2,6,8-tribromopurine.[74] Although 2,3,7,8-tetrahydro-2,8-dithiopurine readily converts to the 2,8-dibromo derivative, hydrolysis of the highly reactive $C_{(8)}$-halogen atom occurs during isolation giving the 2-bromo-8-oxo derivative as product.[74] Amino derivatives prepared include 2-amino-6-bromo-,[74] 6-amino-8-bromo-, and 2,6-diamino-8-bromopurine.[74] Analogous oxo derivatives are 8-bromo-6-oxo-,[74] 2-bromo-6-oxo-,[74] and 8-bromo-2,6-dioxopurines.[74] The same conditions also permit the formation of 6-bromo- and 2-amino-6-bromopurine ribosides from the 6-thio analogues.[30]

F. Preparation of 2-, 6-, and 8-Iodopurines

Due to the restricted procedures available for forming iodopurines and their generally high reactivity toward nucleophilic reagents, relatively few derivatives have been made. With the exception of some

8-iodopurines, which can result from displacement of 8-thio groups (Sect. 2B) or by direct halogenation (Sect. 2A), the usual procedure is by exchange of an existing halogen atom, usually chlorine. Although 2- and 8-iodopurine are unknown, the action of strong hydriodic acid (47%) at 0° on 6-chloropurine gives 6-iodopurine.[13] In like manner 2-iodo-9-methyl-,[75] 2-iodo-9-phenyl-,[9] 6-iodo-2-methyl-,[76] and 6-iodo-9-methylpurine[76] arise. With 2,6-dichloro-[13] and 2,8-dichloropurine[64] the $C_{(2)}$-chlorine atom is resistant to displacement giving, respectively, 2-chloro-6-iodo-[13] and 2-chloro-8-iodopurine.[64] 2,6,8-Trichloropurine is converted to 2-chloro-6,8-diiodopurine similarly.[77] Incorporation of phosphonium iodide or a mixture of red phosphorus and iodine with the hydriodic acid effects reductive removal of some chlorine atoms and iodination of the remaining ones. This procedure was first used to obtain 2-iodo-7-methylpurine from 2,6-dichloro-7-methylpurine[75] and affords a preparation of 2,6-diiodopurine from 2,6,8-trichloropurine (*cf.* above).[75, 78] With the same reagent 2-amino-7,8-dihydro-6-iodo-8-oxopurine was formed from the 6-chloro analogue[36] although transformation of both 2-amino-6-chloro- and 2-amino-8-chloropurine to the corresponding 2-amino-iodopurine can be effected with cold hydriodic acid (− 10°, 1.5 h) alone.[78a]

G. Preparation of 2-, 6-, and 8-Fluoropurines

Of the unsubstituted fluoropurines only 2-fluoro- and 6-fluoropurine are known, whereas a substantial number of substituted 2-, 6-, and 8-fluoropurines have been prepared. Although direct syntheses from fluoropyrimidines have been carried out,[78, 79] the two usual methods are displacement of chlorine atoms or diazotisation of aminopurines by a modified Schiemann reaction. Attempts either to fluorinate an oxopurine with phosphoryl fluoride[80] or to replace a hydrazino group with ferric fluoride[81] were unsuccessful.

a. *By Cyclisation of Fluoro-4,5-diaminopyrimidines*

This approach, which has had very restricted application, is noted in Chapter II, Section 1G.

b. *By Replacement of Chlorine Atoms*

6-Fluoropurine is obtained by a two-stage reaction from 6-chloropurine by treating the latter with trimethylamine and heating the quaternised product with potassium hydrogen fluoride in ethanol at 50°.[81a] Various chloropurines have been successfully converted to the corresponding fluoropurine by heating with silver fluoride in toluene. In the case of the less reactive 2-chloropurines the higher boiling xylene is required. None of the required products were obtained with unsubstituted 2-chloro-,[82] 6-chloro-,[81] or 8-chloropurine[52] but various C- or N-alkyl derivatives will react, the products so obtained including 2-fluoro-9-methyl-,[82] 6-fluoro-9-methyl-,[82] 8-fluoro-9-methyl-,[82] and 2,6-difluoro-7-methylpurine.[82] When 2,6,8-trichloro-7-methylpurine was likewise treated, 2,6-difluoro-8,9-dihydro-7-methyl-8-oxopurine was formed.[82]

No replacement occurred with 2-amino-6-chloropurine[57] although the corresponding riboside, as triacetate, was fluorinated in 25% yield.[83]

c. *By Replacement of Amino Groups*

Under diazotisation conditions a modified Schiemann procedure on 2-aminopurine, involving successive treatment with sodium nitrite solution and 48% fluoboric acid,[66, 84] gives a 41% yield of 2-fluoropurine. In contrast 2-fluoro-6-methylpurine is obtained only in 6% yield.[66] No adverse effect on the reaction is found due to the presence of oxo or thio groups or halogen atoms, as shown by the ready formation of 2-fluoro-6-oxo-,[66] 2-fluoro-6-thio-,[66] 2-fluoro-6-methylthio-,[66, 85] and 6-chloro-2-fluoropurine[66] from the appropriate 2-amino derivatives. Neither adenine[81] nor 8-aminopurine[66] give products under these conditions. With 2,6-diaminopurine only the $C_{(2)}$-amino group undergoes diazotisation, giving a poor yield (6%) of 6-amino-2-fluoropurine,[66] this being improved (22%) if the reaction is carried out in liquid hydrogen fluoride containing sodium nitrite.[85a] More realistic yields were obtained from the preparation of 6-benzylamino-2-fluoro-,[85] 6-amino-9-benzyl-2-fluoro-,[66] 6-dimethylamino-2-fluoro-,[66] and 2-fluoro-6-[2-(imidazol-2-yl)ethylamino]purine.[77]

The low reaction temperatures adopted permit some aminopurine ribosides to undergo fluorination without prior protection of the sugar moiety being necessary; exemplified in the preparation of 2-fluoro-1,6-dihydro-6-thiopurine riboside.[66] Adenosine, like adenine, remains unchanged[86] and the 9-β-D-riboside of 2,6-diaminopurine also behaves like the parent purine in only giving the 6-amino-2-fluoro analogue.[66, 84, 86, 87] In contrast, the 2′,3′,5′-tri-O-acetyl derivatives of these nucleosides give the respective 6-fluoro- and 2,6-difluoro derivatives under the same conditions[86] but the 6,8-diaminopurine analogue only affords 8-fluoroadenosine triacetate.[87a]

2. The Preparation of 8-Halogenopurines by Direct Halogenation and Other Specific Routes

Owing to the conjunction of an electron-rich imidazole ring with an electron-deficient pyrimidine ring, the 8-position of the purine ring has unique character, being available for attack by both nucleophilic and electrophilic reagents. The insertion of electron-releasing groups may partly or completely rectify the electron deficit in the pyrimidine moiety thereby enhancing the negative nature of the 8-carbon atom. Direct halogenation of this position is therefore possible in purines containing one or more electron-releasing groups.

A. By Direct Halogenation

One aspect of this, in which phosphoryl chloride–phosphorus penta-chloride combinations concurrently chlorinate oxo functions and the unsubstituted 8-position of N-methylated derivatives has been already dealt with (Sect. 1Bc).

Chlorine is employed for selective chlorination of the 8-position, by passing it through a suspension of the purine in an inert solvent, which is usually, but not always, of a nonpolar type, at a fairly low temperature. Chloroform is widely used, and in it have been prepared the 8-chloro derivatives of theophylline,[88, 89] theobromine,[90, 91] and caffeine.[92, 93] Tetrachloroethane, dichlorobenzene, nitrobenzene, and nitromethane[94] are other solvents used. More polar media include water, in which caffeine was successfully chlorinated,[95] and warm acetic acid, as in the preparation of 8-chlorotheophylline[94] and isocaffeine.[52, 96]*

Sulphur analogues of the N-methylxanthines likewise give the corresponding 8-chloropurines under these conditions.[99] Sulphuryl chloride, serving both as chlorinating agent and solvent, effectively converts 1,9-dimethylxanthines to the 8-chloro analogues[100–102] at room temperature over a long (100 h) period. Older techniques described for 8-chlorocaffeine preparations are the direct action of chlorine on

---

* Some involvement of N-methyl groups may result from prolonged treatment. In one example 8-chlorocaffeine suspended in o-dichlorobenzene, containing small amounts of iodine, at reflux point, first suffered loss of the $N_{(3)}$-methyl group[97, 98] after 1.5 hours and then loss of the $N_{(7)}$-methyl group after 3 hours' reaction.[97]

powdered caffeine[51] and heating an acidified solution of the purine with potassium chlorate.[51, 103]

A wider range of derivatives is available from bromination, the resulting more reactive 8-bromopurines having been extensively utilised for $C_{(8)}$ nucleophilic substitution reactions. Bromination is a more versatile procedure than chlorination as even simple monosubstituted purines can be directly brominated. Adenine with bromine alone affords a bromoadenine adduct breaking down readily to 8-bromoadenine.[104] Hypoxanthine [104, 105] and xanthine [82, 93] form their 8-bromo analogues likewise. A similar procedure gives 8-bromoguanine,[93] which has been claimed to be formed, but not isolated, by bromination in dimethyl-formamide.[106] With N-methylated xanthine derivatives brominations with or without solvent have been effected. Heating with bromine alone affords 8-bromo-1-methylxanthine[107] but preparations of 8-bromo-9-methylxanthine,[108] 8-bromo-1,9-dimethylxanthine,[109] and 8-bromo-1,3,9-trimethylxanthine [110] were carried out in an acetic acid media. Various procedures for forming the 8-bromo derivatives of theophylline, theobromine, and caffeine have been described. 8-Bromotheophylline results from the use of bromine alone, either hot [48] or cold,[93] in hot ethanol,[111, 112] in hot acetic acid,[111, 113] and in a carbon tetrachloride–nitrobenzene mixture under reflux conditions.[114, 115] 8-Bromotheobromine is formed in bromine alone,[51, 113] in chloroform,[34] and in carbon tetrachloride–nitrobenzene,[92] this last reaction being adapted to large-scale preparation.[116] Originally, with bromine alone, poor yields [51, 117, 118] of 8-bromocaffeine were encountered but better results are obtained in carbon tetrachloride–nitrobenzene mixtures.[92, 113, 119, 120]

Suitable adaptations permit purine nucleosides to undergo $C_{(8)}$-bromination after protection of the ribose moiety, either as the triacetyl or diisopropylidene derivative, is made. Adenosine [121, 122] and 2'-deoxyadenosine [121] with N-bromoacetamide in chloroform under reflux affords the corresponding 8-bromonucleoside. Guanosine triacetate is similarly brominated in acetic acid at 50°.[121]

Examples of direct iodination of the 8-position are few; 8-iodocaffeine is formed when caffeine and iodine are heated in a sealed tube at 150° for 6 hours.[119] Both guanosine and xanthosine are converted to the 8-iodo derivatives in good yield in dimethylsulphoxide by N-iodo-succinimide, the reactions being catalysed by n-butyl disulphide.[123]

B. By Other Means

The stable diazo compounds given by some 8-aminopurines are transformed to the corresponding 8-chloropurines on treatment with hydrochloric acid.[68] Reference to 8-fluoropurine formation using fluoboric acid instead has been made previously (Sect. 1Gc). Hydrochloric acid also converts 8-nitropurines to the appropriate 8-chloro analogues, the 8-bromo derivatives arise with 48% hydrobromic acid likewise, reflux conditions in both cases being necessary.[124] A rearrangement common to both 3-hydroxyxanthine and guanine-3-oxide results in their conversion to 8-chloroxanthine and -guanine, respectively, on reaction with inorganic or organic chlorides in a dipolar aprotic solvent (see Ch. XI, Sect. 3C). A variety of 8-iodopurines has arisen from displacements carried out on 8-thiopurines in cold aqueous sodium bicarbonate with potassium iodide–iodine mixtures. The reaction appears specific for thio groups at $C_{(8)}$ as failure to displace $C_{(2)}$- or $C_{(6)}$-thio groups is found. The best results obtain if one electron-releasing group is present, the ready formation of 2-amino-, 6-amino-, 2-oxo-, and 6-oxo-8-iodopurines exemplifies this.[78a] Some more highly substituted 8-thiopurines also respond to this reagent as seen in the formation of the 8-iodo derivatives of xanthine,[125] and the 9-methyl,[126] 1,3-, 1,9-, and 3,7-dimethyl,[125] and 1,3,7- and 1,3,9-trimethyl[125] homologues. The conversion of 8-thioisoguanine is likewise carried out although the isomeric 8-thioguanine does not react.[78a] A route to 8-iodoguanine is, however, available by utilising the fact that the analogous nucleoside, 8-thioguanosine, is converted to the 8-iodo analogue under the above conditions after which acid cleavage of the ribosyl moiety affords the required 8-iodopurine.[78a]

3. The Preparation of Extranuclear Halogenopurines

Examples of purines with *C*-halogenoalkyl groups are relatively few compared with the large number of existing examples that contain *N*-halogenoalkyl groups.

A. By Direct Halogenation

A methyl group at $C_{(6)}$ can be chlorinated in stages from the mono- to the trichloromethyl state, some 8-methyl purines react under these conditions. This aspect has already been fully dealt with (Ch. IV, Sect. 2Cd).

B. From Hydroxyalkylpurines

The 8-hydroxyalkylpurines, which are more accessible than the corresponding 6-hydroxyalkyl analogues, are readily chlorinated with thionyl chloride containing pyridine in catalytic amounts. A solvent such as chloroform is sometimes used. Due to the important therapeutic applications of derivatives of N-methylated xanthines, the majority of 8-hydroxyalkyl group chlorinations are found with purines of this type. Illustrative of this method are the 8-chloromethyl derivatives of 3-methylxanthine,[127] theophylline,[127, 128] theobromine,[127-129] caffeine,[130] isocaffeine,[131] also the 8-chloroethyl analogues of theophylline,[128] theobromine,[132] 3-methylxanthine,[128] and 1-benzyl-3,7-dimethylxanthine.[132]

Chlorination of hydroxyalkyl groups at $C_{(6)}$ has been carried out with phosphoryl chloride in boiling tetrahydrofuran, exemplified by the preparation of 6-(3-chloropropyl)purine.[133] Many examples exist of halogenation of hydroxyalkyl groups attached to nuclear nitrogen atoms. Theophylline, through alkylation with chloro- or bromo-alkanols, gives 7-hydroxyalkyl derivatives which thionyl chloride either alone or in benzene converts to 7-chloroalkyl derivatives.[134-138] A more direct alternative route consists of direct alkylation of a sodium or silver theophylline salt with an $\alpha\omega$-dihalogenoalkane. The preparation of the 7-(2-bromoethyl),[136] 7-(4-chlorobutyl),[139] or 7-(5-chloropentyl)[139] derivative from 1,2-dibromethane, 1-chloro-4-iodobutane, and 1-chloro-5-iodopentane, respectively, illustrate the procedure which is carried out in boiling aqueous alkali. Theobromine with a stoicheometric amount of 1,2-dibromoethane gives 1-(2-bromoethyl)theobromine.[136] Representative examples of $N_{(9)}$-chloroalkyl purines arising from thionyl chloride treatment are 6-amino-9-(2-chloroethyl)-,[140] 6-amino-9-(3-chloropropyl)-,[140] 2-amino-6-benzylthio-9-(2'-chloroethyl),[72] and 9-(2-chloroethyl)-6-methylthiopurine.[72] In all cases heating with the reagent itself or in benzene on a steam bath for 30 minutes suffices. Direct alkylation, using 1-bromo-2-chloroethane, giving 2-amino-9-(2-chloroethyl)-6-methylthiopurine from 2-amino-6-methylthiopurine, has been carried out in dimethylsulphoxide containing potassium carbonate at room temperature.[72]

Owing to the therapeutic importance of derivatives of NN-di-2-chloroethylamine (the "nitrogen mustards"), the purine nucleus has been incorporated into this type of compound. Considerable effort has been directed toward chlorination of 6-(2-hydroxyethylamino)- and 6-NN-di(2-hydroxyethylamino)purine. On treating the former (17) with hot thionyl chloride the 6-(2-chloroethylamino)purine initially formed undergoes spontaneous cyclisation to the hydrochloride of 7,8-dihydroimidazo[2,1-i]purine (18, R = H).[141] The NN-dihydroxy-ethylaminopurine (19) is also found to give a cyclised derivative[141–144] the structure of which (18, R = CH_2CH_2Cl) has been proved by an unambiguous synthesis.[141] No comparable cyclisation is noted during the preparation of 8-NN-di(2-chloroethylamino)purines, demonstrated by the isolation of the xanthine derivative (20).[145] Analogous 8-sub-

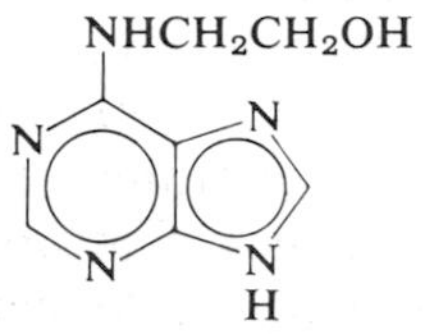

(17) 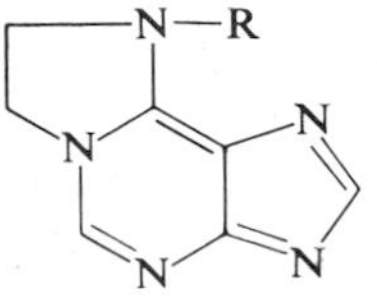   (18)

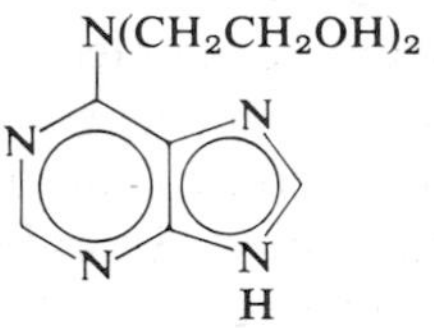

(19) 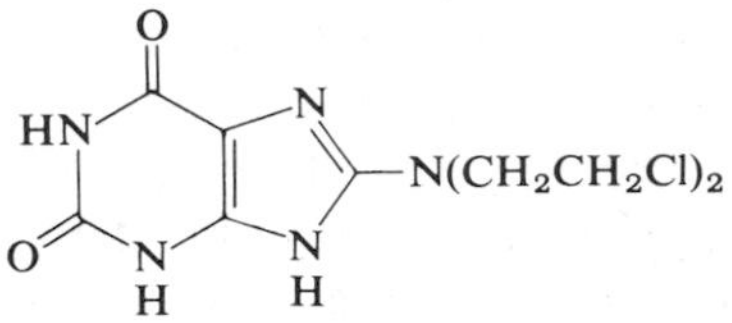   (20)

stituted compounds of adenine,[145] hypoxanthine,[145] theobromine,[116] theophylline,[116] and caffeine[116] are known. No identifiable product was obtained from the attempted chlorination of 8-NN-di(2-hydroxyethyl-amino)guanine.[145] Other purines with mustard groups are known.[146,147]

C. From Halogenopurines

Substitution of a $C_{(2)}$-, $C_{(6)}$-, or $C_{(8)}$-halogen atom by a chloromethyl (or bromomethyl) group is possible by condensation of the halogeno-purine with sodium diethyl malonate followed by chlorination (or bromination) of the malonyl moiety and subsequent decarboxylation. This route provides examples of the formation of 2-chloromethyl-,[147a] 6-chloromethyl-,[147b] and 8-bromomethylpurines.[131] The major value of this route lies in the means it provides of converting a halogen atom to a methyl group, for which see Chapter IV, Section 2Ad.

D. By Synthesis

As purine methyl groups are not amenable to direct fluorination procedures, fluoromethylpurines are prepared from suitable precursors

containing preformed fluoromethyl groups. Examples are found of
2-trifluoromethyl-[148, 81] and 6-trifluoromethylpurines,[81, 148–150] derived
by cyclisation of the appropriate trifluoromethyl-4,5-diaminopyrimi-
dine, while 8-trifluoromethylpurines arise using the requisite 4,5-
diaminopyrimidine and trifluoroacetic anhydride,[81, 148, 151] acid[81, 152]
or amide,[81] as cyclising reagent.

E. By Other Means

Heating 6-methylpurine with chloral in the presence of acetic acid
below 100° gives a good yield of 6-(3,3,3-trichloropropyl)purine.[133]
Theophylline forms an $N_{(7)}$-chloromethyl derivative with aqueous
formaldehyde and hydrogen chloride.[134] Reference to the direct halogenation of N-methyl groups can be found elsewhere (Sect. 1Bd).

4. Properties of Halogenopurines

Unsubstituted halogenopurines are colourless, well-defined solids
with a tendency toward decomposition rather than true melting. For
comparison Table 3 gives the melting or decomposition point data for
all the simple derivatives. Although unsubstituted fluoro- and iodopurines are not well represented a number of stable mono- and di-
halogenated $N_{(9)}$-alkyl derivatives of this type are known including
2-fluoro-9-methyl-, 6-fluoro-9-methyl-, 8-fluoro-9-methyl-, 2,6-difluoro-
9-methyl-, and 2-iodo-9-ethylpurine.

TABLE 3.   Melting Points of Simple Halogenopurines.

Position	Fluoro	Chloro	Bromo	Iodo
2-	216°(dec)	231°	243°(dec)	—
6-	126°	177°	194°(dec)	167°(dec)
8-	—	200°(dec)	—	—
2,6-	—	181°(dec)	207°(dec)	224°(dec)
2,8-	—	148°	(unstable)	—
6,8-	—	178°(dec)	223°	—
2,6,8-	—	181°	223°(dec)	—

5. Reactions of 2-, 6-, and 8-Halogenopurines

The high reactivity of a halogen atom, in particular of chlorine, is the dominant feature of halogenopurine chemistry and is, with the possible exception of the methylsulphonyl group, not equalled by any other atom or group. Fischer,[153] whose prodigious synthetic efforts included the preparation of all the naturally occurring purines then known, recognised the invaluable part played by halogeno derivatives in purine metatheses and singled out 2,6,8-trichloropurine as being the foremost in importance.*

It would be no exaggeration to say that starting with a 2,6,8-trichloropurine derivative nearly all the known purines could be prepared. A valuable feature of halogen atoms is their ready removal reductively. This, therefore, affords a facile route for removal of an oxo group, provided this is capable of being halogenated, thereby leaving an unsubstituted carbon atom in its place.

A. Reaction Studies with Halogenopurines

Although the ease of substitution with any particular nucleophilic reagent depends largely upon the position of the halogen atom in the ring, the influence exerted by other groups present must also be considered. Among the monohalogeno derivatives the 2-halogenopurines show the lowest activity while the 6- and 8-halogeno isomers exhibit greater and more similar tendencies toward nucleophilic substitution.[154] Kinetic studies of the action of alkaline reagents on various chloropurines indicate the reactions to be second order and of a bimolecular type.[154, 155]

With polyhalogenopurines the order of replacement of halogen atoms has been rationalised as a result of studies carried out on 2,6,8-trichloropurines.[156] The results of the action of strong aqueous alkali on 2,6,8-trichloropurine have been compared with those obtained with 7- or 9-alkyl-2,6,8-trichloropurines. With the 7- or 9-unsubstituted chloropurine (21, R = H) the initial attack at $C_{(6)}$, giving 2,8-dichloro-1,6-dihydro-6-oxopurine (22) is followed by substitution of the halogen at $C_{(2)}$ and then at $C_{(8)}$. In the case of a 7- or 9-alkyl-2,6,8-trichloropurine (21, R = Me) the first product is the 8-oxopurine (23) with succeeding halogen replacement at $C_{(6)}$, that at $C_{(2)}$ being the last to react. From

---

* The words used were "*An die Spitze derselben stelle ich das Trichloropurin welches bei der Synthese der natürlichen Purinkörper die Hauptrolle gespielt hat.*"

these findings has arisen the postulation that at the moment of attack with **21** (R = H) the active purine species involved is a resonance-stabilised anion of the type **24** which possesses an induced electron-rich centre in the imidazole ring. This precludes any nucleophilic attack at $C_{(8)}$ which is therefore directed to $C_{(6)}$ instead. In the cases of the 7- or 9-alkylated purines, in which no anionic forms comparable with **24** are possible, $C_{(8)}$ is the first site of reaction by virtue of the inherent electron-depleted character of the imidazole ring extant in the neutral molecule. This effect is also demonstrated with 6,8-dichloropurine which gives 8-chlorohypoxanthine in strong alkali.[17] The near equality of reactivity of the 6- and 8-positions is shown by the formation of a 2-chloro-6,8-disubstituted purine when 2,6,8-trichloropurine is treated with a strong but nonbasic type of nucleophilic reagent [77] such as thiourea. Behaviour, contrary to that expected from the "anion formation theory" is shown by 2,8-dichloropurine which, with basic nucleophilic reagents, under-goes replacement of the $C_{(8)}$-rather than the $C_{(2)}$-chlorine atom.[17, 64] The lack of reactivity of a $C_{(2)}$-chlorine atom is best explained in terms of a partial fixation of the $C_{(4)}-C_{(5)}$ double bond producing an increased electron density, the effect of which is associated at the 2-position with the mesomeric nature of the halogen atom. The "active ion" theory has been extended along the same lines to include substitution under acid conditions, in this case a protonated form, such as **25**, is presumed to be involved and the reverse order of attack would be expected. This

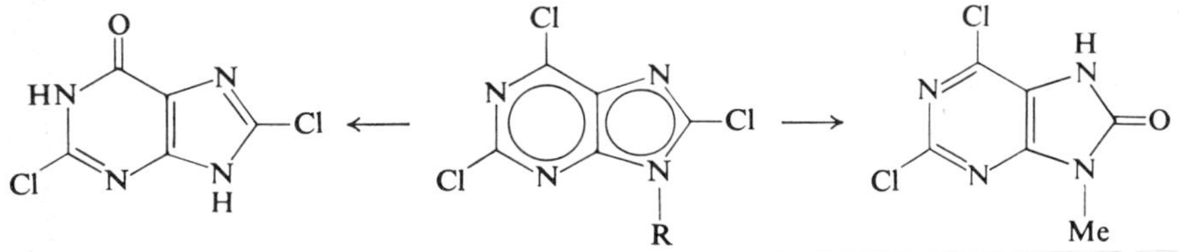

(22)   (21)   (23)

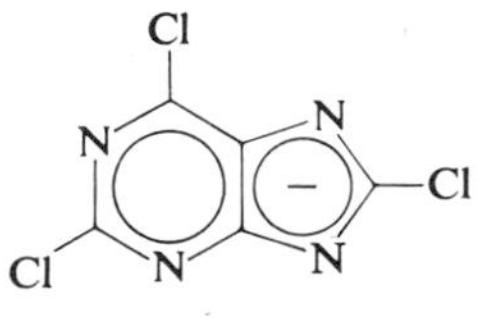

(24) 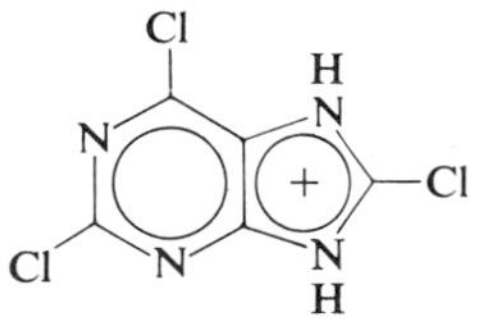   (25)

effect does in practice obtain; in hot acid 6,8-dichloro-[24] and 2,6,8-trichloropurine [11] are converted to the respective 8-oxo-6-chloro- and 8-oxo-2,6-dichloropurines but the anomalous behaviour shown by other chloropurines is an indication that other parameters, as yet unidentified, may operate, among which protonation at more than one site is a possibility.

B. Removal of Halogen Atoms

Various reduction procedures are available: chemical, catalytic, and electrochemical. By carefully controlling the conditions stepwise dehalogenation of polyhalogenopurines is possible. The most favoured chemical means has been the use of hydriodic acid and phosphonium iodide, variations of this use a mixture of hydriodic acid, red phosphorus and iodine[157] or hydriodic acid through which phosphine is passing.[158] For dehalogenation in neutral media finely divided zinc in water is often successful. Both of these older methods have now been largely superseded by the easily controlled catalytic hydrogenolysis which is carried out in aqueous or alcoholic media, usually in the presence of an inorganic base, such as magnesium oxide. The latter, which serves to prevent hydrochloride formation and so predispose toward reduction of the purine nucleus, can be dispensed with by carrying out the reaction in dilute alkali. The third procedure, entailing electrolytic reduction, is now only of historic importance.

a. In the Presence of only Alkyl Groups

The removal of halogen atoms from C-alkyl and N-alkyl purines has been considered already (Ch. IV, Sects. 2Ab, 3Ab). Dehalogenation of 2,6,8-trichloropurine (26) gives first 2,8-dichloro- (27) and on further reduction 2-chloropurine (28),[77] the reaction being carried out in N-sodium hydroxide with a palladium catalyst. The latter purine (28) arises from 2-chloro-6,8-diiodopurine[77] or 2,6-dichloropurine[159, 160] in the same way. Examples from the older literature include 2-chloro-7-methylpurine, from either 2,6-dichloro- or 2,6,8-trichloro-7-methyl-purine[75] on heating under reflux in aqueous solution with zinc powder, also 2-chloro-9-methyl-[75] and 2-chloro-9-phenylpurine[9] similarly from the respective trichloropurines (21, R = Me and C_6H_5). Catalytic

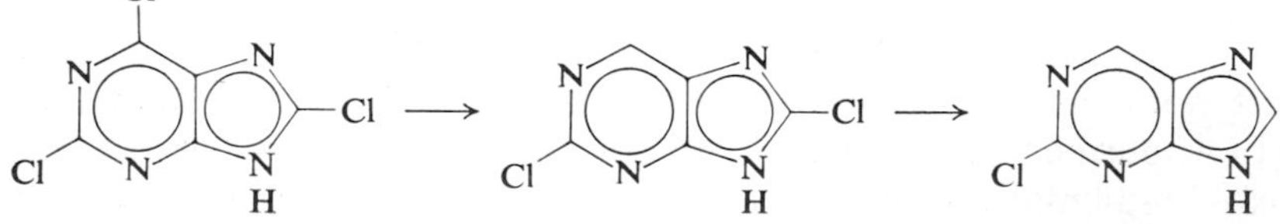

(26)   (27)   (28)

reduction of 2,6-dichloro-9-methylpurine[161] also gives 2-chloro-9-methylpurine. Fluoropurines show a marked inertness towards hydrogenolytic removal, an example of this is the conversion of 2-chloro-6-fluoropurine to 6-fluoropurine using a palladium catalyst.[161a]

b. *In the Presence of Amino Groups*

Procedures adopted for halogen removal are not usually complicated by the presence of amino groups; the older use of hydriodic acid and phosphonium iodide or catalytic hydrogenation on a palladium catalyst are the two most favoured. Either reagent is capable of removing one or two halogen atoms from an aminochloropurine, the preparation of adenine provides a good example, both chlorine atoms of 6-amino-2,8-dichloropurine being removed equally efficiently by either hydriodic acid[162, 37, 77] or catalytic means.[163] The naturally occurring 6-substituted-aminopurine derivatives "kinetin,"[159] the purine base of "puromycin",[159] and 6-succinoaminopurine[158] follow from hydriodic acid reduction of the respective 2,8-dichloropurine. The same reagent affords 7-methyladenine from either the 2-chloro-[37] or 2,8-dichloro-[37] analogues, and 9-methyladenine from 6-amino-2,8-dichloro-9-methylpurine.[37, 162] The isomeric 8-amino-7-methylpurine follows from the 2,6-dichloropurine,[164] and a further example is 2-dimethylamino-9-methylpurine from 6-chloro-2-dimethylamino-9-methylpurine.[165] With warm hydriodic acid alone 2,8-dichloro-6-diethylaminopurine gives 6-diethylaminopurine.[29] The versatility of catalytic over acid reduction is seen in the formation either of adenine[163] or 6-amino-2-chloropurine[77, 166] from 6-amino-2,8-dichloropurine by control of the hydrogen intake. A like reduction of 2-chloro-8-methylaminopurine gives 8-methylaminopurine.[64] An interesting dechlorination arises during the attempted debenzylation in liquid ammonia containing

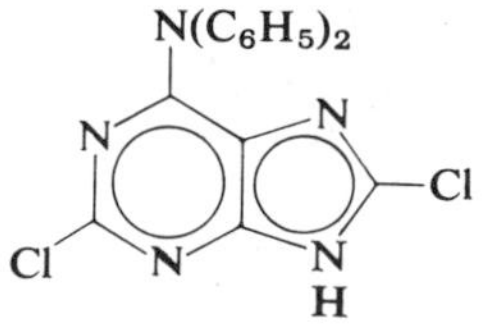

(29) 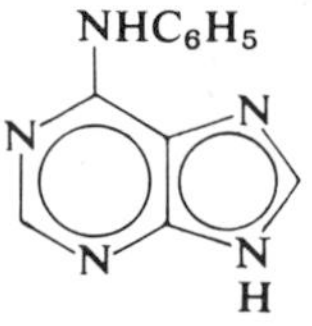   (30)

sodium of 2,8-dichloro-6-dibenzylaminopurine (29) the product, 6-benzylaminopurine (30), resulting from removal of only one of the benzyl groups but both halogen atoms.[167]

c. *In the Presence of Oxo (and Methoxy) Groups*

Halogen atoms attached to oxopurines are usually removed under conditions similar to those described for aminopurines. Examples of dehalogenation with hydriodic acid–phosphorus combinations include the formation of hypoxanthine from the 2,8-dichloro analogue,[162] 7,8-dihydro-8-oxopurine from the 2,6-dichloro derivative,[7] also 7-methylhypoxanthine by removal of a 2-chlorine[40] and 7,8-dihydro-9-methyl-8-oxopurine from 2,6-dichloro-7,8-dihydro-9-methyl-8-oxopurine.[15] Xanthine results from removal of a chlorine at $C_{(8)}$ using hydriodic acid alone.[29] This route serves also for a preparation of guanine and isomeric forms, the same purine arising from reduction of 2-amino-8-chloro-1,6-dihydro-6-oxopurine.[162] Similarly derived are isoguanine[162] from the 8-chloro analogue and 6-amino-7,8-dihydro-8-oxopurine[7] by removal of a chlorine atom at $C_{(2)}$. Dehalogenations with this reagent in the presence of alkoxy groups induces concomitant dealkylation, exemplified by formation of isoguanine from 6-amino-8-chloro-2-ethoxypurine (31),[162] also xanthine from 8-chloro-2,6-diethoxypurine[162] and 6-dibutylamino-2,3-dihydro-2-oxopurine from 6-dibutylamino-8-chloro-2-ethoxypurine.[29]

The conversion of 2-chloro-6,8-diethoxy-7-methylpurine to 1,6,8,9-tetrahydro-7-methyl-6,8-dioxopurine is a further example.[164] A more recent modification, using hydriodic acid alone, allows not only the removal, *in situ*, of chlorine atoms and a tetrahydro-2-pyranyl group but also allows hydrolysis of the ether group. This occurs in the conversion of 2,6-dichloro-8-ethoxy-9-(tetrahydropyran-2-yl)purine (32) to 7,8-dihydro-8-oxopurine (33) in one step.[156] Other examples of this type are available.[156] Apart from its controllability, catalytic reduction over palladium is useful in that it can be performed in alkaline media in

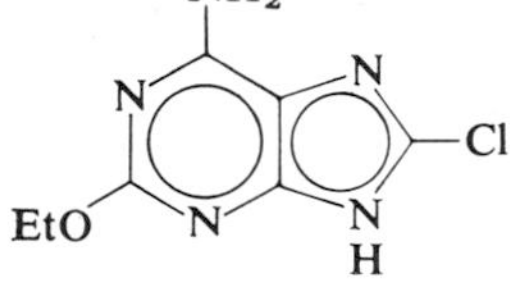

(31)   (33)

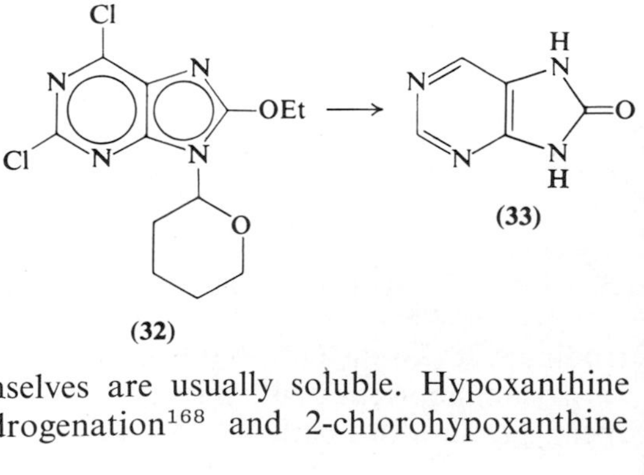

(32)

which the oxopurines themselves are usually soluble. Hypoxanthine results from complete hydrogenation[168] and 2-chlorohypoxanthine from partial hydrogenation[77] of 2,8-dichlorohypoxanthine. Xanthine is similarly obtained from 8-chloroxanthine.[168]

d. *In the Presence of Thio Groups*

Hydriodic acid treatment of thiopurines has successfully removed halogen atoms in such derivatives. As examples may be cited the conversion of 1,6-dihydro-2-iodo-6-thiopurine to 1,6-dihydro-6-thiopurine[157] and 2-chloro-1,6-dihydro-7-methyl-6-thio- to 1,6-dihydro-7-methyl-6-thiopurine.[169]

e. *From N-Alkylpurines*

The older literature is replete with examples of removal of halogen atoms from *N*-methylated oxopurines. Chlorine atoms have been removed from 2-chloro[8, 164] and 6-chloro[42] derivatives of this type using hydriodic acid reductions. Most attention has been focussed on reduction of 8-halogeno derivatives of *N*-alkylated xanthines, especially those of theophylline, theobromine, and caffeine. With hydriodic acid and phosphorus the 8-chloro-*N*-methylxanthine was reduced to the corresponding *N*-methylxanthine in the case of 1-methyl,[97] 1,3-dimethyl,[53] 1,7-dimethyl,[97] 1,9-dimethyl,[52] 1,3,9-trimethyl,[52] 3-ethyl,[170] 3,7-diethyl,[170, 171] 1,3,7-triethyl,[171] 3,7-dibutyl,[171] and 1,3,7-tributyl[171] derivatives. Hydriodic acid alone converted 8-iodotheophylline and 8-iodotheobromine to theophylline and theobromine, respectively.[125] An early variant of this method gave theophylline and 1,7-dimethylxanthine from the appropriate 8-chloro analogue on heating in tetrahydronaphthalene containing less than one molecular equivalent of iodine.[172] In this procedure hydriodic acid is presumed to be formed through interaction between iodine and protons derived from the solvent. To a lesser degree zinc dust in aqueous solution has been used in forming theophylline,[173] caffeine,[51, 118] and similar derivatives from their 8-chloro analogues, but more use has been made of hydrogenolysis over colloidal palladium catalysts for this purpose.[119, 174–176]

Electrolysis in dilute sulphuric acid at a lead cathode successfully converted 8-chlorotheophylline to the parent purine.[177] Under this treatment 8-chloro-7-chloromethyltheophylline loses first the exocyclic halogen atom, forming 8-chlorocaffeine, but with longer reaction times (> 2 h) further reduction to caffeine ensues.[175]

f. *By Indirect Means*

In most examples removal of chlorine atoms is best made directly. Occasionally prior conversion to a thio group is made for which removal can be made reductively, with Raney nickel, or oxidatively, by means of hydrogen peroxide or nitric acid (Ch. VII, Sect. 1Ca).

Treatment of 6-chloropurines with sodium sulphite solution affords purine-6-sulphinates which are themselves readily transformed to the corresponding 6-unsubstituted purine on gentle warming in formic acid.[71] A more tedious removal of halogen involves first replacement by a nitrile group which is hydrolysed to carboxyl and finally thermally decarboxylated to the halogen-free compound. This route has been used to prepare purine itself from 6-iodopurine.[178]

C. Replacement of 2-, 6-, and 8-Halogens by Amino Groups

Great variations are found in the conditions required for substitution of halogen by amino groups. In general terms, the halogen sited at $C_{(6)}$ is the most reactive with those at the remaining positions showing a lesser degree of reactivity. However, pH changes in the reaction medium or the presence of other groups in the purine molecule can profoundly alter the above order.

a. *With One Halogen Atom*

Moderately vigorous conditions appear necessary to replace the halogen atom of 2-chloropurine itself. With a primary amine, such as histamine, prolonged boiling in butanol gives the 2-imidazolylethylamino derivative.[77] Hydrazine, a stronger nucleophile, forms 2-hydrazinopurine on heating the chloropurine in it at 80° for 12 hours.[179] More forceful treatment is needed with the 7- and 9-alkyl analogues, in the case of the 2-chloro-7-methyl-[75] and 2-chloro-9-methylpurine,[75, 161] or the corresponding 2-iodopurines,[75] sealed-tube procedures employing methanolic ammonia are necessary to form the appropriate 2-amino derivatives. Similar conditions are required for piperidine and 2-chloro-9-methylpurine to interact.[154] Butylamine at 100° at atmospheric pressure gives 2-butylamino-6,9-dimethylpurine from the 2-chloro analogue.[180]

With 6-chloropurine, the most reactive and useful of the monochloropurines, conversion to adenine requires a temperature of 150°

with ammonia-saturated butanol in a sealed vessel.[16] Numerous adenine derivatives have been prepared using ethanolic or methanolic ammonia; examples are the 2-phenyl,[1] 7-methyl,[2] 9-methyl,[181] 9-furyl,[182] and other 9-alkyl[183-185] and 9-aryl[186] derivatives. An interesting side-light pertaining to the preparation of 6-amino-7-methylpurine is that whereas ethanolic ammonia with 6-chloro-7-methylpurine at 160° in a closed tube gives the required 6-amino-7-methylpurine, the dipurinyl-amine (34) is produced on heating the chloropurine in the open on a steam bath.[2] This derivative most likely arises through interaction of some of the initially formed 6-amino-7-methylpurine with the starting material, the feasibility of this being shown by the preparation of 7-methyl-6-(7-methylpurin-6-yl)aminopurine (34) on heating 6-amino-7-methylpurine with 6-chloro-7-methylpurine in ethanol.[2] Liquid ammonia under pressure at 70° converts 6-chloro-7-(3-hydroxypropyl)purine to the corresponding 6-aminopurine.[187] In contrast to the rigorous condi-tions used with 6-chloropurine the conversion of 6-fluoro-9-methyl-purine to the 6-amino analogue occurs with aqueous ammonia on a water bath.[79] The greater nucleophilicity of primary and secondary amines allows less severe reaction conditions to be employed; even the more volatile members, for example, methylamine or dimethylamine, in the form of their aqueous or ethanolic solutions, readily react at atmos-pheric pressure at temperatures below 100°. A number of biologically significant 6-substituted-aminopurines have arisen from 6-chloropurine in this way. Amongst these are "puromycin" base, 6-dimethylamino-purine, using methanolic solutions of dimethylamine,[188] "kinetin," 6-furfurylaminopurine[19, 189, 190] and 6-succinoaminopurine in which the amine derivative used was L-aspartic acid.[191] Many other examples are known of amines interacting with 6-chloro-[21, 192, 193] and the 6-chloro-7-[2, 194] or -9-alkyl (or aryl) purines.[181, 185]. With alkyldi-amines, bispurines of the type 35 are formed.[195, 196]

Under similar mild conditions 6-chloropurines have reacted with a variety of amino compounds including amino acids[191, 197, 198] (Ch. IX, Sect. 9A), hydroxylamine,[69, 199] hydrazine[179, 181] (see Sect. D), and ethyleneimine.[200]

Tertiary amines give rise to quaternary halides in the cold on addition of trialkylamine to 6-chloropurine in an anhydrous solvent. In dimethyl-formamide with trimethylamine a good yield of trimethylpurin-6-ylammonium chloride (36) precipitates almost immediately.[201] The stable betaine form (37) is commercially available[202] under the name "Alpurine." A number of 9-substituted analogues[203, 204] including the 9-riboside[30] are known.

In contrast to the large number of reactions known to occur with

8-halogeno derivatives of *N*-alkylated oxopurines, little study has been made of simple 8-halogenopurines. The expected reluctance to react with nucleophilic reagents is shown by 8-chloropurine[17] but the 9-methyl analogue after treatment with piperidine at 45° for some days affords 9-methyl-8-piperidinopurine.[154]

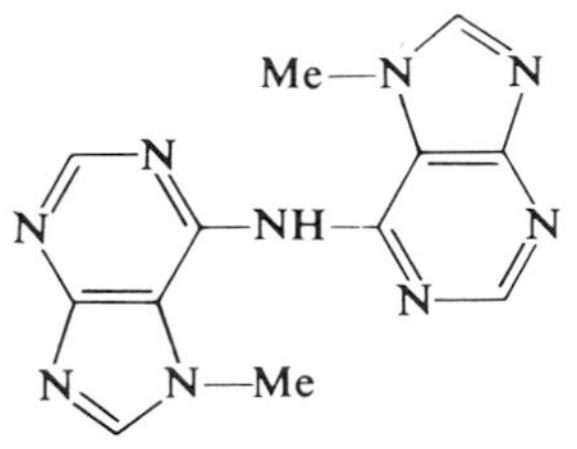

(34) 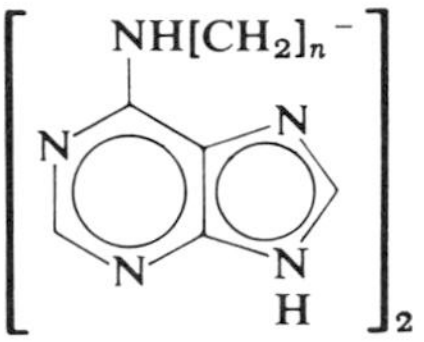   (35)

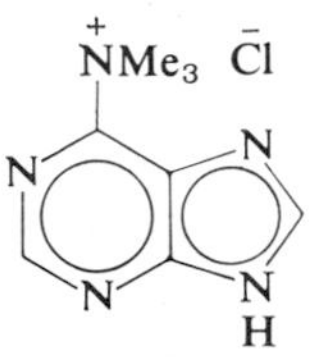

(36) 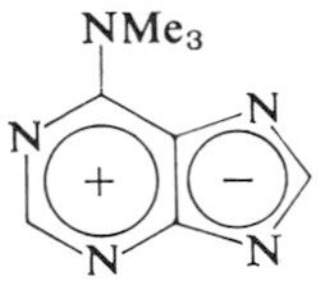   (37)

b. *With Two Halogen Atoms*

Stepwise or simultaneous replacement of the chlorine atoms in 2,6-dichloropurine (38) is possible. With alcoholic ammonia under sealed-tube conditions 6-amino-2-chloropurine (39) is formed[205] but in aqueous ammonia total amination results giving 2,6-diaminopurine (40).[205] The presence of an *N*-alkyl group in the imidazole ring increases the electrophilic nature of $C_{(6)}$—this being reflected in the facile formation of 6-amino-2-chloro-7-methylpurine from the 2,6-dichloro analogue in alcoholic ammonia on a steam bath.[2] These same conditions also convert 2,6-dichloro-9-substituted-purines to the corresponding 6-amino-2-chloro derivative.[204] Previously[34, 37] pressure conditions were thought necessary to form 6-amino-2-chloro-9-methylpurine.[206] 2,6-Diamino-7-methylpurine[37] like the above $N_{(7)}$-unsubstituted analogue is obtained under pressure using aqueous ammonia. The more basic primary or secondary amines react with less vigorous treatment, aqueous or alcoholic solutions at the reflux point suffice to replace 6-chlorine atoms but pressure conditions are required to complete amination[77, 207] of the 2-position. The unreactive nature generally shown by a $C_{(2)}$-chlorine is in these cases reinforced by the inductive effect of the amino group now inserted at $C_{(6)}$. Many examples are to be found of the formation of 2-chloro-6-substituted amino purines in this way.[14, 34, 37, 159]

Both chlorine atoms can be replaced under reflux conditions using an excess of amine as solvent. Amines employed include morpholine, furfurylamine, and piperidine from which the corresponding 2,6-di-aminopurines are derived.[159, 160] Aqueous solutions of these amines give only the respective 6-amino-2-chloropurines.[159] The order of substitution of a 2,6-dihalogenopurine is reversed in the case of the riboside of 6-chloro-2-fluoropurine. On standing in methanolic ammonia at room temperature for 24 hours preferential amination of the more reactive fluorine atom leads to the formation of the 2-amino-6-chloro-purine analogue.[67]

The pronounced reactivity shown by a halogen at $C_{(6)}$ is a feature common to both 2,6- and 6,8-dihalogenopurines, as is also the relative inactivity of the remaining halogen in each case.

Pressure conditions replace one or both halogens by amino groups, concentrated ammonia at 100° for 12 hours provides 6-amino-8-chloro-purine but at 135° half this time suffices to form 6,8-diaminopurine.[24] Primary and secondary amines, in aqueous solution, give rise to the corresponding 6-substituted-amino-8-chloropurine[24, 25] on heating below 100° in the open but replacement of the remaining chlorine atom in such a compound requires elevated temperatures and pressure, the conversion of 8-chloro-6-methylaminopurine to 8-dimethylamino-6-methylaminopurine being made by heating at 125° for 5 hours.[24] Pressure conditions are also necessary to replace both halogen atoms by

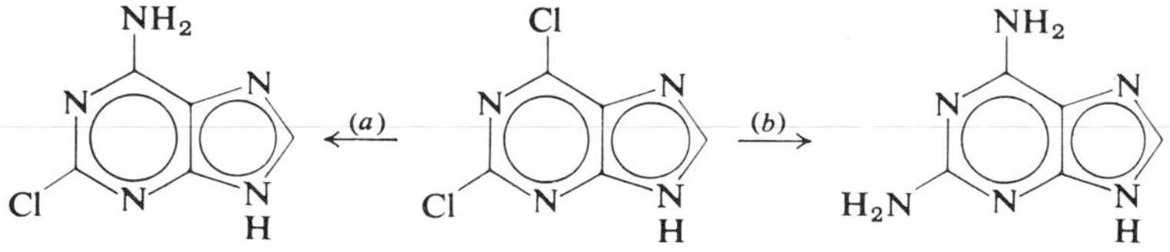

(39)   (38)   (40)

(a) EtOH—NH₃   (b) NH₄OH

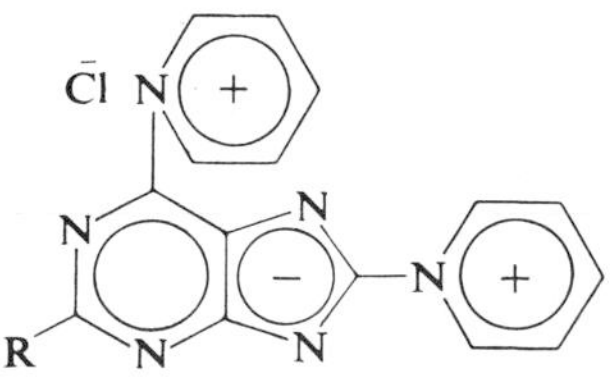

(41)

substituted amino groups in one step.[24] Alkylation of an imidazole nitrogen atom in the case of 6,8-dichloropurine does not, surprisingly, reverse the order of reactivity, with 25% aqueous dimethylamine under reflux 9-benzyl-6,8-dichloropurine gives 9-benzyl-8-chloro-6-dimethyl-aminopurine in good yield.[200]

Unexpected behaviour is shown by 2,8-dichloropurine which in warm aqueous methylamine solution suffers displacement of the $C_{(8)}$-halogen atom in preference to the one at $C_{(2)}$.[64] Aromatic tertiary amines react with 6,8-dichloropurine, thus with pyridine alone or in alcoholic solution, the dipyridinium chloride (**41**, R = H) is formed.[208]

c. *With Three Halogen Atoms*

Of all halogenopurines 2,6,8-trichloropurine (and the *N*-alkylated forms) has been the most studied with respect to reactions with nucleophilic reagents. If no *N*-alkyl group is present in the imidazole ring, initial substitution occurs at $C_{(6)}$. Thus, in aqueous ammonia, at 100° for 6 hours, in a sealed tube, 2,6,8-trichloropurine gives 6-amino-2,8-dichloropurine.[12, 162] With primary and secondary amines heating an aqueous solution of the amine and chloropurine under reflux suffices to form 2,8-dichloro-6-dibutylamino-,[29] 2,8-dichloro-6-diethylamino-,[29] 2,8-dichloro-6-dimethylamino-,[159] 2,8-dichloro-6-furylamino-,[159] 2,8-dichloro-6-piperidino-,[159] and other 6-substituted-aminopurines.[158, 159, 197] If anhydrous conditions are employed replacement of all chlorine atoms is possible either by heating the purine with the amine alone at boiling point or at 150–170° in a sealed tube.[159]

In ethanolic solution containing pyridine, trichloropurine gives rise to the 6-pyridinium betaine form (**42**) but if the reaction is carried out in pyridine alone 2-chloro-6,8-dipyridiniumpurinyl chloride (**41**, R = Cl) is obtained.[208]

On replacing the imidazole proton by an alkyl group at either $N_{(7)}$ or $N_{(9)}$ anionic forms are precluded and initial amination is now favoured at $C_{(8)}$. Illustrating this is the formation of 8-amino-2,6-dichloro-7-methylpurine (**43**) from the 7-methyltrichloropurine (**44**) when alcoholic ammonia is used under pressure[164] but with aqueous ammonia the product is 6,8-diamino-2-chloro-7-methylpurine (**45**).[209] With the isomeric 9-methyltrichloropurine an anomalous finding is reported[8] in that both 8-amino-2,6-dichloro- and 6-amino-2,8-dichloro-9-methyl-purine are produced with ethanolic ammonia but in the mixture the 6-amino isomer predominates. The expected 8-amino isomer (69%) results from this treatment of 2,6,8-trichloro-9-(tetrahydropyran-2-yl)-

purine at room temperature,[156] with only 25% of 6-amino-2,6-dichloro-9-(tetrahydropyran-2-yl)purine being obtained.

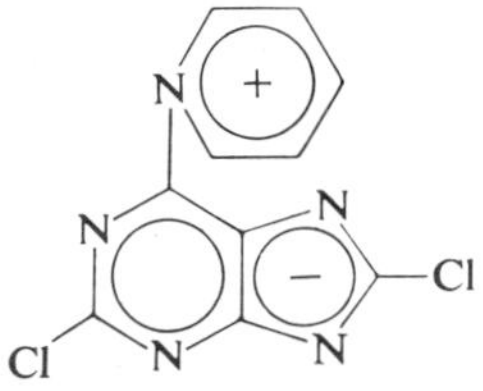

(42)

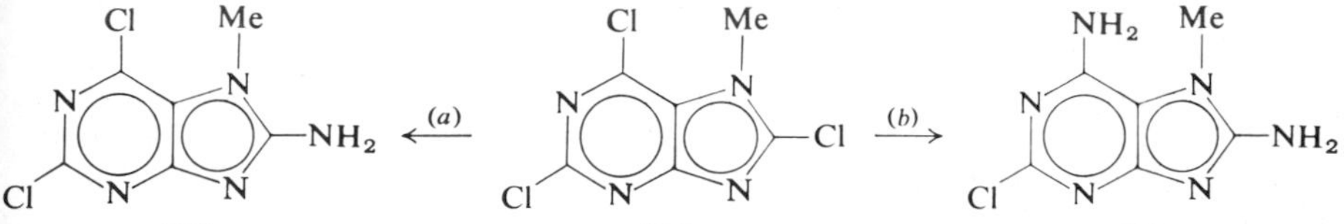

(43)   **(44)**   **(45)**

(a) EtOH—NH$_3$ (b) NH$_4$OH

d. *With Chloro-aminopurines*

As previously noted insertion of an amino group may induce partial or total deactivation of the remaining halogen atoms toward further amination. The difficulty in converting 2-chloro-6-aminopurine to 2,6-diaminopurine with ammonia has been long appreciated[34] but with aqueous methylamine, under pressure for 16 hours at 130°, 6-amino-2-methylaminopurine is obtained.[207] Although aqueous ammonia at 130° does not react with 6-alkylamino-2-chloropurines[207] with ammonia saturated solutions of butanol at 160° 2-amino-6-substituted-amino-purines[77] are obtained. Aqueous solutions of primary and secondary amines likewise react under pressure.[207] With 6-amino-2-chloro-7-methylpurine, similarly, ammonia is ineffective and high temperature and pressure are required to insert an alkylamine group at C$_{(2)}$[34] and 9-substituted 6-amino-2-chloropurines also need forcing conditions to effect reaction with amine. With either methylamine or dimethylamine the 9-β-ribofuranoside gives the 2-substituted-amine only after some hours under pressure at 100°.[210] By contrast the high reactivity of a fluorine atom at C$_{(2)}$ shows in the fact that 6-amino-2-fluoropurine-9-β-D-ribofuranoside affords the 2-butylamino analogue with butylamine in ethanol under reflux.[66]

In comparison with the 2-halogeno compounds a slight increase in reactivity is shown by halogen atoms in 2-amino-6-halogenopurines.

Passage of ammonia through a boiling ethanolic solution of 2-amino-6-chloropurine is without effect, but with 2-amino-6-chloro-3-methyl-purine transformation into 2,6-diamino-3-methylpurine is found after some hours.[62] The activation induced by the $N_{(3)}$-methyl group is further seen in the formation of 2-amino-6-butylamino-3-methylpurine on interaction with butylamine at room temperature.[62] Alkylation at other sites in some cases aids replacement, 2-amino-9-benzyl-6-chloro-purine, for example, reacts with cyclic amines under mild conditions[200] while the 9-ribosides of 2-amino-6-fluoropurine[83] and 2-amino-6-iodopurine,[30] with methanolic ammonia and aqueous dimethylamine give rise, respectively, to the 2,6-diamino and 2-amino-6-dimethylamino derivatives on the water bath.

The conditions required for amination of halogen atoms at $C_{(8)}$ are generally similar to those for replacing $C_{(2)}$-halogen atoms. The rigorous procedures (160–170° for 16 h) used to convert either 6-amino-8-chloro-[61] or 6-amino-8-bromopurine[211] to 6,8-diaminopurine illustrate this. Pressure conditions, but at a lower temperature, have been used to insert amino groups at $C_{(8)}$[211] although 6-amino-8-bromopurine and ethanolamine will react in boiling 2-methoxyethanol.[212] Typical examples are 8-chloro-6-methylamino- and 8-chloro-6-dimethylamino-purine which are transformed to 6,8-disubstituted-aminopurines under pressure.[24]

Amination of amino-dichloropurines is virtually confined to examples using derivatives of 6-amino-2,8-dichloropurine. With morpholine, furfurylamine, and hexylamine, using an excess of the amine as solvent, under reflux or under sealed-tube conditions, simultaneous replacement of both 2- and 8-chlorine atoms occurs.[159]

Investigation of N-amino-6-chloropurines reveals that no special properties can be attributed to the halogen atom. Methanolic ammonia at 100° gives 6-amino-9-dimethylamino-8-methylpurine (**46**) from the 6-chloro analogue,[38] but replacement of the chlorine by a dimethyl-amino-, diethylamino-, or furfurylamino group took place in ethanolic solution of the amine under reflux. No comparable reaction was found with 9-amino-6-chloro-7,8-dihydro-8-oxopurine (**47**), in which the halogen remained untouched even in boiling morpholine.[213]

(46)   (47)

e. With Chloro-oxopurines

The generally inert character shown by halogen atoms in amino-purines is paralleled by the behaviour of their counterparts in the oxopurines. Some activation may be encountered, however, where the purine is N-alkylated.

Owing to the lack of reactivity with ammonia amination of 2-chloro-hypoxanthine is not a practical route to guanine. With alkylamines corresponding 2-alkylaminohypoxanthines form under moderate conditions, for example, ethanolic methylamine reacts at 130° over some hours[214] whereas histamine is successfully condensed in hot butanol.[77] Alkylation in the imidazole ring slightly increases the ease of substitution, both the 7-[4, 40] and 9-methyl[4] derivatives of 2-chlorohypoxanthine on prolonged treatment with aqueous ammonia at 150° give the respective 7- and 9-methylguanines. Aqueous dimethylamine in dioxan at 100° was employed in forming 9-benzyl-1,6-dihydro-2-dimethyl-amino-6-oxopurine from the corresponding 2-chloropurine.[200] Tertiary amines react to form betaines, pyridine gives the 2-pyridinium derivative (48).[208] Methylation in the pyrimidine ring facilitates amination of 2-chloropurines, seen in the reaction of the 1,9-dimethylpurine (49) with aqueous ammonia at 80°.[5] This enhancement of activity does not seem to have been appreciated by earlier workers, Fischer having used pressure conditions to aminate the 2-chloro-1,7-dimethyl analogue.[40] Aqueous amines—for example, benzylamine—react below 100° but neat benzylamine, under reflux conditions, is required to replace the chlorine atom in 2-chloro-1,6,8,9-tetrahydro-7-methyl-6,8-dioxopurine.[215] Few examples occur in the literature of 6-chlorooxopurines undergoing amination, the resulting 6-aminopurines are better derived by hydrolysis of 6-aminochloropurines. Aqueous ammonia at elevated temperatures converts the 3-methyl and 3,7-dimethyl derivatives of 6-chloro-2,8-dioxopurines to the corresponding 6-aminopurines 50 (R = H)[216] and 50 (R = Me).[217, 218]

8-Oxo derivatives of 2,6-dichloropurines require vigorous treatment to effect replacement of both halogen atoms and an overall similarity of reaction with that of 2,6-dichloropurine itself is found. Thus, in alcoholic ammonia at 150° only the $C_{(6)}$-halogen in 2,6-dichloro-7,8-dihydro-8-oxopurine is replaced whereas in aqueous ammonia at this temperature the 2,6-diamino analogue is formed.[7] Corresponding replacements with 7-methyl- and 9-methyl-2,6-dichloro-8-oxo-dihydropurine have been carried out.[37] Formation of 2-chloro-6-diethylamino-7,8-dihydro-8-oxopurine requires prolonged heating of the 2,6-dichloropurine in diethylamine at 110°; more elevated temperatures are needed to insert the second amino group.[34]

Whereas 2,6-dichloro-7,8-dihydro-8-oxopurine reacts like 2,6-di-chloropurine, the isomeric 2,8-dichloro-1,6-dihydro-6-oxopurine be-haves differently from 2,8-dichloropurine. In alcoholic ammonia at 150° for some hours 2-amino-8-chloro-1,6-dihydro-6-oxopurine is formed[162] in contrast to 8-amino-2-chloropurine obtained from 2,8-dichloropurine under the same conditions.[64] Much attention has been directed toward amination of 8-halogeno-N-methylxanthines, especially with derivatives of caffeine, theobromine, and theophylline. Vigorous conditions are mandatory in all cases for conversion to the 8-amino derivative, pressure conditions over a temperature range of 140–180° being usual. Because of the pharmacological significance caffeine derivatives are well repre-sented, 8-aminocaffeine being originally obtained from alcoholic ammonia treatment, at 130–150°, of the 8-bromo[51] or 8-chloro ana-logue.[219] Two more recent routes to this derivative, using nonpressure conditions, both employ 8-bromocaffeine as starting material. In the first procedure condensation with phthalimide in boiling dimethyl-formamide (18 h) is followed by dilute acid hydrolysis of the 8-phthali-midocaffeine.[220] The alternative route entails conversion to 8-hydrazino-caffeine, in aqueous hydrazine under reflux, which gives 8-aminocaffeine when heated in dimethylformamide.[220] Of the two, the latter procedure gives the better (43%) yield.

Reactions of an 8-halogenocaffeine with alcoholic solutions of aliphatic amines are exemplified in formation of the 8-methylamino,[221] -dimethylamino,[222] -ethylamine,[221] -diethylamino,[222] -benzylamino,[222] -dibenzylamino,[222] and phenylethylamino[223] derivatives. Other more complex amines have been used.[224] 1,2-Diaminoethane and homologues and an excess of 8-chlorocaffeine afford NN'-di(caffein-8-yl)amino-alkanes of the type 51.[89] Aromatic and cycloaliphatic amines giving appropriate 8-aminocaffeines include aniline,[219] p-toluidine,[221] m-xyli-dine,[221] 2-aminopyridine,[225] morpholine,[223] piperidine,[222, 223] and pyrrolidine.[223] Similar condensations are recorded with various higher alkyl homologues of caffeine. The 3-ethyl-1,7-dimethyl, 3-butyl-1,7-dimethyl, and 3-alkyl-1,7-dimethyl derivatives of 8-chloroxanthine give products with ammonia and amines.[96] Under the pressure conditions employed other groups present may react. Thus, 8-bromo-7-carboxy-methyltheophylline gives the 8-amino-7-carbamoylmethyl derivative (52),[112] and diethylamine similarly converts 7-(2-bromoethyl)-8-chlorotheophylline to the 8-diethylamino-7-(2-diethylaminoethyl)purine (53).[226] Other examples of this type are known.[227]

Amino derivatives of theobromine are similarly prepared, usually from the 8-bromopurine, characteristic examples being the 8-amino-,[228] from alcoholic ammonia at 180°, various 8-alkylamino-,[34] 8-anilino,[228] and 8-amphetaminotheobromines.[113] The isomeric 8-chloro-1,2,3,6-tetrahydro-1,7-dimethyl-2,6-dioxopurine (8-chloroparaxanthine) gives 8-aminopurines similarly.[229]

Surprisingly, the halogen at $C_{(8)}$ of theophylline is reactive, contrasting with the fact that other 8-halogenopurines not alkylated in the imidazole ring are inert. Reactions occur with alcoholic ammonia,[230] alkyl-amines,[89, 113, 226] hydroxyalkylamines,[231] and aniline[230] under the same conditions as for theobromine and caffeine derivatives.

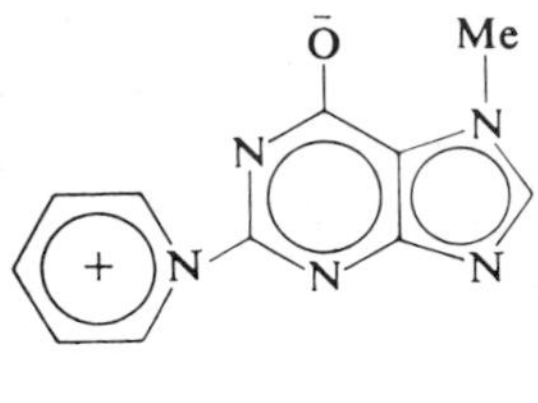

(48) 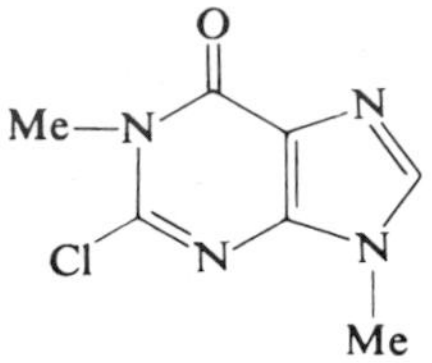   (49) 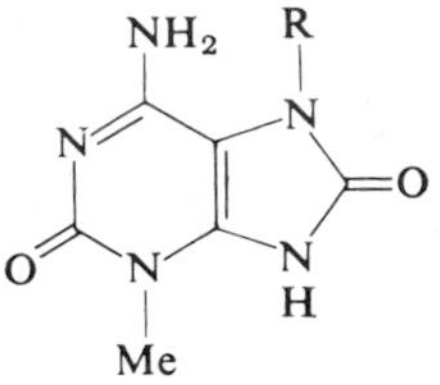   (50)

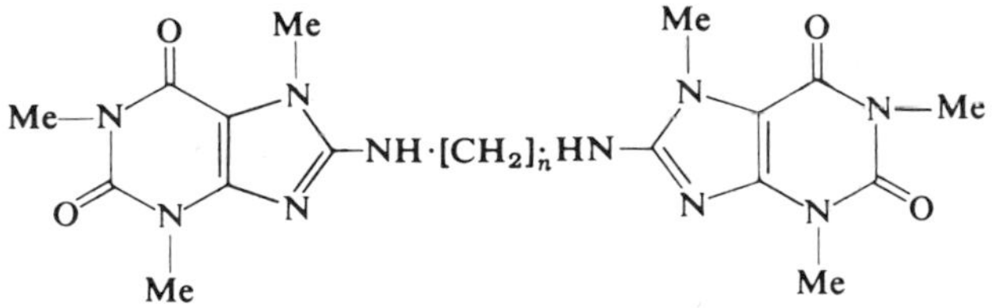

(51)

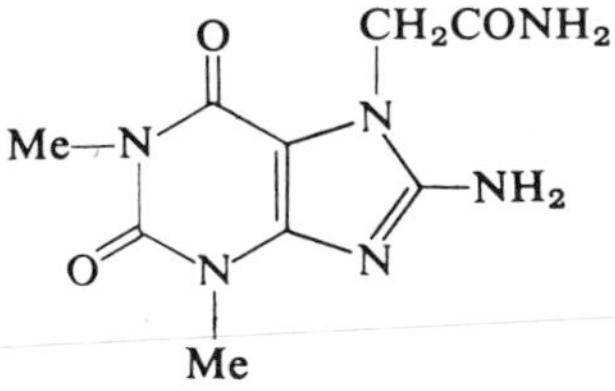

(52) 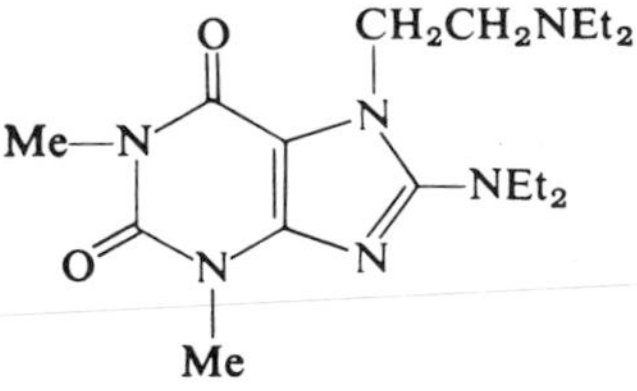   (53)

f. *With Chloro-thio (and-methylthio)purines*

Owing to the danger of replacement of both halogen atom and thio group by the amine, it follows that suitable purines are restricted to those with readily displaced halogen atoms or with unreactive thio groups. The action of aqueous dimethylamine on 6-chloro-2-methyl-thio-[25] and 6-chloro-8-methyl-2-methylthio-purine[22] affords the respective 6-dimethylaminopurine on heating for 1 or 2 hours on a steam bath. With 6,8-dichloro-2-methylthiopurine, by analogy with 6,8-dichloropurine itself, only 8-chloro-6-dimethylamino-2-methylthiopurine is formed.[25] That halogen replacement, using ethanolic alkylamines, of the riboside of 2-fluoro-1,6-dihydro-6-thiopurine to the 2-methylamino- or 2-dimethylamino-1,6-dihydro-6-thiopurine analogue can be carried out at room temperature,[67] reflects the reactivity of the $C_{(2)}$-fluorine rather than the suitability of the reaction to halogenopurine ribosides in general.

The characteristic inertness of 8-methylthio groups shows in 6-chloro-8-methylthiopurine which with aqueous ammonia at 100° gives 6-amino-8-methylthiopurine and with alkylamines and aniline corresponding derivatives below this temperature.[24] The preparation of 6-(3-diethylaminopropylamino)-2-methyl-8-methylthiopurine from the 6-chloro analogue was effected in refluxing toluene.[28]

g. With Chloropurines Having Other Groups Present

Purines with acid functions may show some slight enhancement of halogen activity. For example, 6-carboxy-2-chloro-9-methylpurine is transformed into the corresponding 2-dimethylamino derivative on heating with aqueous dimethylamine for some hours.[165] By contrast the usually more reactive halogen found at $C_{(6)}$ is not present in 6-chloro-7,8-dihydro-8-oxo-2-sulphopurine (54) which is unaffected by dimethylamine under these conditions.[56] The inert halogen in this case is more likely to be an effect of the presence of the 8-oxo function rather than the sulphonic acid group as amination of halogens is possible in the presence of methylsulphonyl groups. Aqueous dimethylamine, for example, on a water bath transforms 6-chloro-2-methylsulphonylpurine (55) to 6-dimethylamino-2-methylsulphonylpurine. However, care must be exercised in the conditions employed due to the ease of displacement of a methylsulphonyl group, claimed to be approaching that of a chlorine atom, by nucleophilic reagents.[39] Thus, use of higher temperatures in the above reaction produces 2,6-bisdimethylaminopurine (56).[39]

6-Dimethylamino-2,8-dimethylsulphonylpurine likewise followed from the 6-chloropurine and 8-chloro-6-dimethylamino-2-methylsulphonylpurine (58) from the 6,8-dichloro-2-methylsulphonyl derivative (57).[33] The expected inert character shown by the $C_{(8)}$-halogen is further highlighted by the attempted amination of 8-chloro-2,6-dimethylsulphonylpurine (59) which, through a preferential displacement of the group at $C_{(6)}$, gives 58 as product.[39]

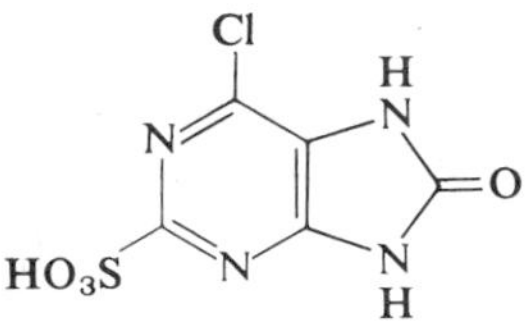

Activation of otherwise poorly reactive halogen atoms often follows oxide formation at an adjacent nitrogen atom, as exemplified by the preparation of the 2-morpholino derivative from 2-chloroadenine-1-oxide and morpholine in dimethylformamide under reflux[63] (see Ch. XI, Sect. 3D).

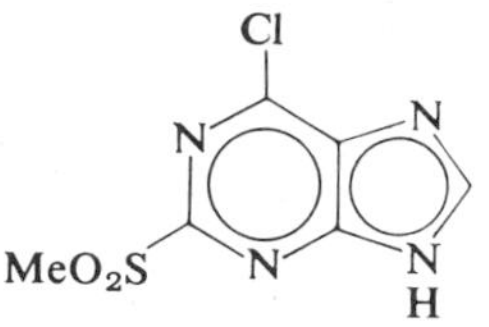

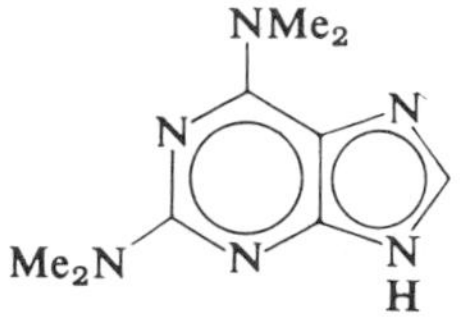

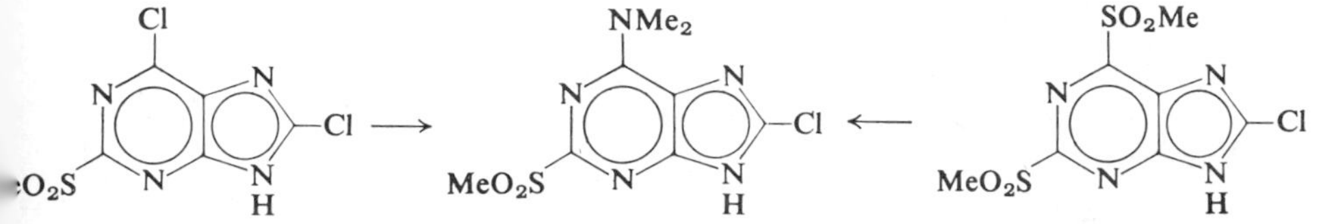

D. Replacement of 2-, 6-, and 8-Halogen Atoms by Hydrazino, Hydroxyamino, Azido, and Related Groups

Prolonged heating (12 h) at 80° with hydrazine is needed to prepare 2-hydrazinopurine from 2-chloropurine,[179] and to convert 6-amino-2-chloropurine to 6-amino-2-hydrazinopurine.[179] The 6-substituted-amino-2-chloropurines react with hydrazine hydrate under reflux conditions.[159] Although 2-chlorohypoxanthine forms the 2-hydrazino-purine at water bath temperature,[179] neither 2-chloro-6-ethoxy-7-methyl-nor 2-chloro-6,8-diethoxy-7-methylpurine do so[209] under this treatment.

Hydrazine converts 6-chloropurine to 6-hydrazinopurine[179] at room temperature. Similarly derived are the 6-hydrazino derivatives of 7-methyl-,[2] 9-methyl-,[181] 9-ethyl-,[185] 7-benzyl-,[232] and 9-benzylpurine.[232] No reaction occurred with 2-amino-6-chloropurine, under a variety of conditions,[179] but 6-chloro-8-methyl-2-methylthiopurine affords the 6-unsymmetrical dimethyl hydrazide with ethanolic NN-dimethyl-hydrazine at 100°.[22]

Simple 8-hydrazinopurines are few. The attempted formation of 6-amino-8-hydrazinopurine by heating 6-amino-8-bromopurine with hydrazine gave 6,8-diaminopurine.[211] An analogous reaction in the nucleoside series converts 8-bromoguanosine to the 8-amino analogue.[233]

In both examples initial formation of the 8-hydrazino derivative is likely as thermal breakdown of an 8-hydrazino- to an 8-aminopurine is known to occur.[220] The 8-halogeno derivatives of theobromine,[70] caffeine,[120, 221] and 1,7-dimethylxanthine[70] give the 8-hydrazino compounds but the corresponding theophylline derivative is not formed under atmospheric conditions.[70, 120] A route to 8-hydrazinotheophylline exists by hydrazinolysis of 8-chloro-7-chloromethyltheophylline (60),[70, 120] which reacts like 8-chlorocaffeine before a reductive removal of the chloromethyl group ensues. Under pressure conditions at 150° 8-bromo-theophylline and hydrazine hydrate affords NN-di(theophyllin-8-yl)-hydrazine (61).[120] A similar preparation of an NN-di(purin-6-yl)-hydrazine has been described by Fischer.[37] Owing to the presence of two functional centres, 7-acetonyl-8-bromotheophylline (62) forms the triazinopurine (63) with an excess of alcoholic hydrazine.[234]

Stepwise replacement occurs with dichloropurines, 2,6-dichloropurine being converted to 2-chloro-6-hydrazinopurine[179] at room temperature. The corresponding 2-chloro-6-hydrazino-7-methylpurine is obtained using warm ethanolic[209] or aqueous[37] hydrazine. With anhydrous hydrazine at 80° or hydrazine hydrate under reflux both chlorine atoms are replaced giving the respective 2,6-dihydrazino-[179] or 2,6-dihydrazino-7-methylpurine[209] (65). The latter purine also arises by simultaneous replacement of ethoxy group and halogen in 2-chloro-6-ethoxy-7-methylpurine (64) with boiling hydrazine hydrate.[209] Both halogens in 6,8-dichloropurine are replaced on prolonged heating on a steam bath with aqueous hydrazine.[24]

An example of a 2,8-dichloropurine undergoing hydrazination is the conversion of 2,8-dichloro-6-morpholinopurine to the 2,8-dihydrazino derivative in hydrazine at reflux point.[159]

Boiling hydrazine hydrate replaces all halogen atoms in both 2,6,8-trichloropurine[159] and 2,6,8-trichloro-7-methylpurine.[209] The latter trihydrazinopurine can be obtained using 2,6-dichloro-8-ethoxy-7-methylpurine also.[209] In hot 80% hydrazine 2,6,8-trichloro-7-methyl-purine gives only the 2-chloro-6,8-dihydrazinopurine.[209]

Reactions with hydroxylamine are few, 6-hydroxyaminopurine is formed on heating 6-chloropurine in ethanolic hydroxylamine for some hours, the 9-β-D-riboside being similarly derived.[235, 236] Others by this route include 2-amino-6-hydroxyamino-9-methyl- and 6-hydroxyamino-9-methoxymethylpurine from the corresponding 6-chloropurine. Little or no reaction is found if the 6-iodopurines are used.[199] The activity of a fluorine atom at $C_{(2)}$ is demonstrated by conversion of 2-fluorohypo-xanthine to the 2-hydroxyamino analogue[67] and formation of 2,6-dihydroxyaminopurine from 2-fluoro-6-chloropurine,[237] both replacement reactions being carried out in refluxing ethanol. Further details of these derivatives are given in Chapter VIII, Section 7B.

The formation of azidopurines from halogenopurines is not common, the alternative route entailing nitrous acid treatment of hydrazino-purines is the one more usually followed (Ch. VIII, Sect. 7Ca).

The synthesis of 6-azidopurine (66) from 6-chloropurine has been reported without detail.[80] Heating 2,6-dichloro- and 2,6,8-trichloro-purine with ethanolic sodium azide for a few minutes gives, respectively, 2,6-diazido- and 2,6,8-triazidopurine.[238] An example of an 8-azido derivative is 6-amino-8-azidopurine riboside from 8-bromoadenosine by means of sodium azide in dimethylsulphoxide.[233]

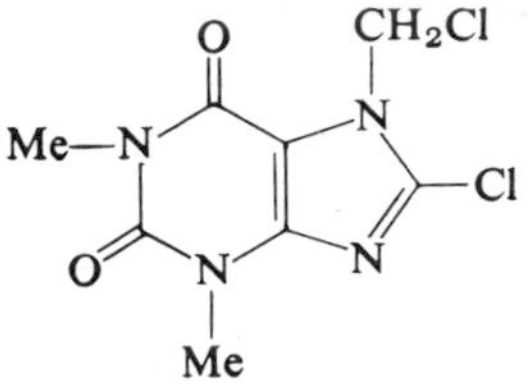

(60) 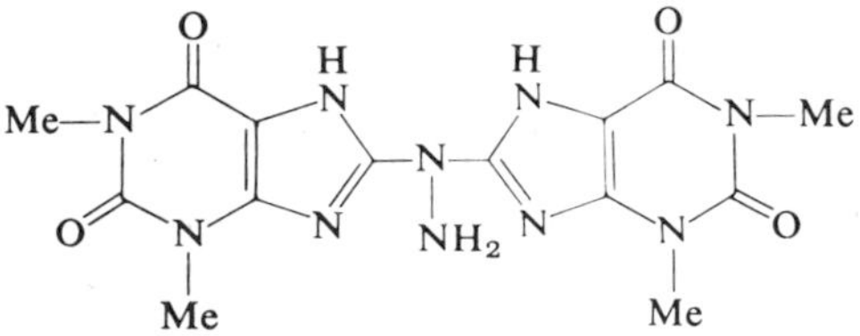   (61)

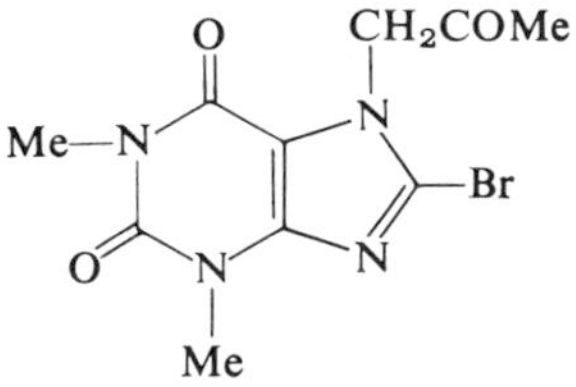

(62) 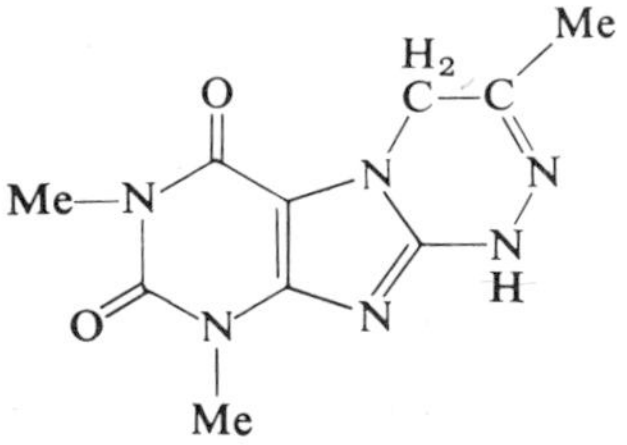   (63)

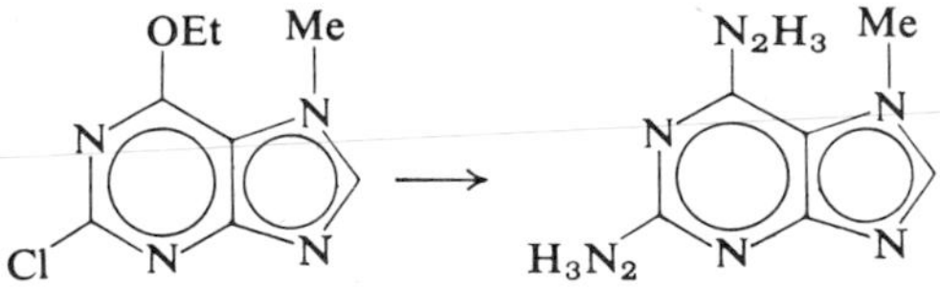

(64)   (65) 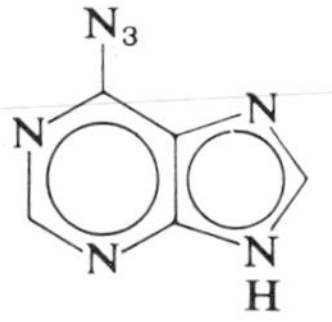   (66)

E. Replacement of 2-, 6-, and 8-Halogen Atoms by Alkoxy Groups

As a general rule alkoxylation of a $C_{(2)}$-halogen atom requires equally as vigorous conditions as are used for carrying out the corresponding amination. Furthermore, where the presence of an unsubstituted $N_{(7)}$ or $N_{(9)}$ allows anion formation to occur, even more resistance to replacement may be found due to coulombic repulsion. In the simplest case, while 2-chloropurine and sodium ethoxide require a temperature of 150° for conversion to 2-ethoxypurine, the 9-methyl analogue affords 2-ethoxy-9-methylpurine in ethanol under reflux.[155] Sodium methoxide or ethoxide in boiling toluene give 2-methoxy- or 2-ethoxy-6-methylpurine from 2-chloro-6-methylpurine.[3]

Conditions for the reaction between alkoxides and 6-chloropurines are usually to heat the purine in an alcoholic solution of the sodium alkoxide, either on a steam bath or under reflux. As well as 6-methoxy-purine,[239] others similarly prepared are the 6-ethoxy, -propoxy, iso-propoxy, -butoxy,[239] and -benzyloxy[204] derivatives. Both 6-(furfur-2-yloxy)purine (**67**),[19] an analogue of the plant growth factor, kinetin, and the reduced form 6-(tetrahydrofurfur-2-yloxy)purine[239] are further examples of the route. Although 7- and 9-alkyl-6-chloropurines readily form the 6-methoxy[2, 181, 232] or 6-ethoxy derivatives[182] below 100°, exceptional reactivity is shown by 6-chloro-9-ethyl-[232] and 6-chloro-9-furfurylpurine, which are rapidly converted to the respective 6-ethoxy-purines in dilute sodium hydroxide containing small amounts of ethanol at just above room temperature. 6-Aryloxy analogues, for example, both 7-[2] and 9-methyl[181] isomers of 6-p-bromophenoxypurine, result from treatment of the appropriate chloropurine with the phenol in hot potassium hydroxide solution.

Enhanced activity due to an alkyl group at $N_{(3)}$ is shown by the ease of formation of 6-methoxy and -ethoxy derivatives from 6-chloro-3-methylpurine simply on warming a methanolic or ethanolic solution of the purine hydrochloride.[57]

Both 8-chloropurine and 8-chloro-9-methylpurine react with sodium ethoxide, the former at 150° over 1 hour and the latter on leaving at room temperature for some hours.[155] As in the case of 2-chloropurines the sluggish nature of the halogenopurine anion is beautifully demonstrated.

Unsubstituted di- and trichloropurines show the same substitution patterns with alkoxides as with other nucleophiles. Prolonged reflux conditions (20 h) only convert 2,6-dichloropurine to 2-chloro-6-methoxypurine[240]—replacement of the second halogen atom being rendered more difficult by the purine being in the anionic form. This can be overcome by precluding ionisation by means of an easily removed blocking group, such as tetrahydropyran-2-yl, at $N_{(9)}$. This device enables 2,6-diethoxypurine to be prepared under reflux conditions.[240a] Stepwise replacement of halogens in 2,6-dichloro-7-methyl-purine is readily achieved, exemplified by the formation of 2-chloro-6-methoxy- and 2,6-dimethoxy-7-methylpurine.[241] Higher molecular weight alkoxides require more vigorous treatment; both 2,6-diprop-oxy-[242] and 2,6-diallyloxy-7-methylpurine[242] are formed at higher temperatures (100–140°) under sealed conditions.

Both halogens, surprisingly, are replaced in 2,8-dichloropurine in sodium methoxide under reflux conditions,[64] but similar treatment of 6,8-dichloropurine gives only 8-chloro-6-methoxypurine.[24] The 6-ethoxy analogue follows likewise.[24] At room temperature ethanolic sodium ethoxide converts 2,6,8-trichloropurine (**68**, R = H) to 2,8-dichloro-6-ethoxypurine (**69**) but at 100° further substitution at the 2-position gives (**70**).[162, 171] Following alkylation at $N_{(7)}$ or $N_{(9)}$ the expected activity of the $C_{(8)}$-halogen is shown. Thus ethanolic potassium hydroxide at 3° with the 7-methyl homologue (**68**, R = Me) affords 2,6-dichloro-8-ethoxy-7-methylpurine (**71**) while at 40° 2-chloro-6,8-diethoxy-7-methylpurine (**72**) results.[164] The same order of substitution is found with 9-methyl[164] and other 9-alkyl-2,6,8-trichloropurines.[156] In one example using an excess of sodium ethoxide both reactive chlorine atoms can be replaced at room temperature.[156] Although the expected forcing conditions (150°, 3–5 h) are needed to methoxylate 6-amino-2-chloropurine,[243, 244] their use with the 9-methyl homologue[206] seems unnecessary as other related 9-substituted purines give the 2-methoxy derivative by reflux procedures.[204] It should be noted that attempted ethoxylation of the 7-methyl isomer does not give 6-amino-2-ethoxy-7-methylpurine but 7-methylguanine. This alkali-induced rearrangement is described subsequently (Sect. 5Fa).

Prolonged heating (18 h) in the appropriate alcohol at the boiling point with an excess of one equivalent of sodium gives the 6-methoxy-, -ethoxy, -propoxy, -butoxy,[59] and -benzyloxy[204] derivatives with 2-amino-6-chloropurine, and also converts 2-amino-6-chloro-3-methylpurine to the 6-methoxy- or 6-ethoxypurine.[62] The expected activity of the halogen atom in this case is not shown due to the deactivating influence of the $C_{(2)}$-amino group. Formation of 6-amino-8-methoxy- and 8-methoxy-6-methylaminopurine from the respective 6-amino-8-chloropurine requires sealed-tube conditions and elevated temperatures.[24] Methoxylation of 6-amino-2,8-dichloropurines also necessitates pressure and prolonged heating, the products being the appropriate 2-alkoxy-6-amino-8-chloropurines.[29, 162]

Few reactions are recorded for chlorothiopurines but 2-chloro-1,6-dihydro-7-methyl-6-thiopurine in methanolic sodium methoxide at 100° for 3 hours gives the 2-methoxy analogue.[169] The inert nature of an 8-methylthio group is shown by the formation of 6-methoxy-8-methylthiopurine from the corresponding 6-chloropurine.[24]

Of the oxopurines studied attention has been mainly focussed on reactions with 8-bromo or 8-chloro derivatives of the *N*-alkylated forms.

Among the 2-halogenopurines the behaviour of the highly reactive 2-chloro-1,9-dimethylhypoxanthine (**73**), in forming the 2-ethoxy analogue at room temperature,[5] contrasts strikingly with the conditions needed (sealed tube at 140°) for converting 2-chloro-7-methylhypoxanthine (**74**) to the 2-methoxypurine.[245] Anion formation in the latter example would explain the difference in $C_{(2)}$-halogen activity toward alkoxylation. Both halogen atoms are active in 2,6-dichloro-7,8-,dihydro-7,9-dimethyl-8-oxopurine which, under mild conditions, gives the 2,6-diethoxypurine (**75**).[15]

The 8-halogeno derivatives of *N*-alkylated xanthines react with alkoxides at reflux temperatures with the notable exception of theophylline which does not react even under pressure.[246] Sluggishness is

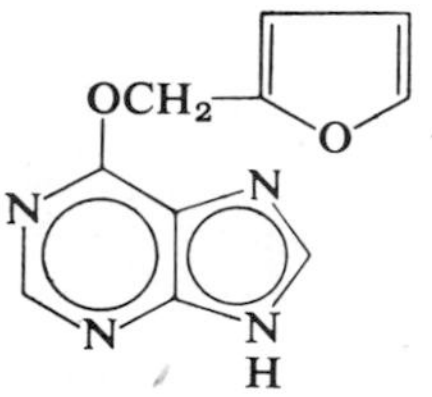

(67) 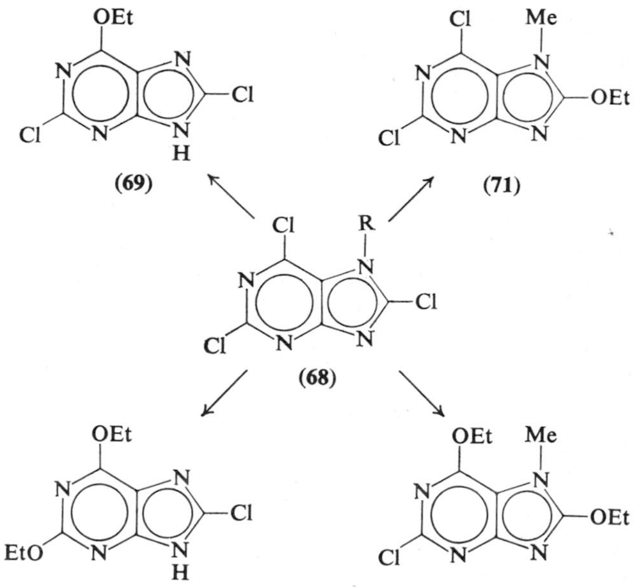   (69)   (71)

(68)

(70)   (72)

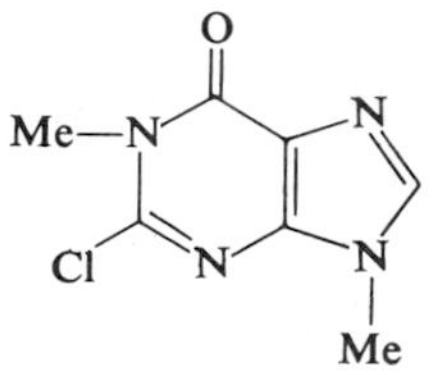

(73) 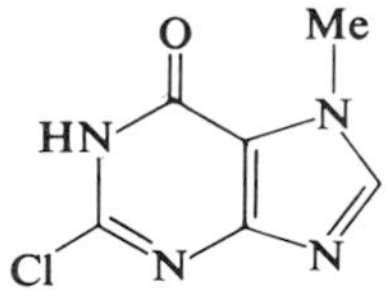   (74) 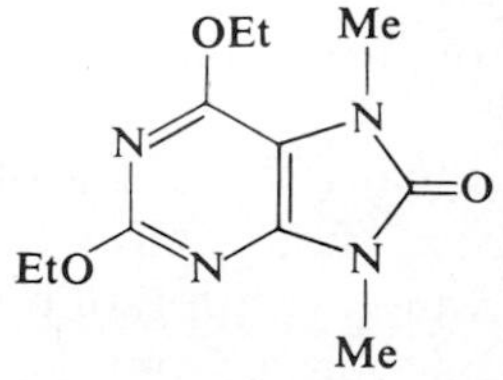   (75)

shown by the 8-chloro-1,9-dimethylxanthine, ethoxylation only being successful with autoclave conditions. Higher molecular weight alkoxy groups can, however, be inserted under reflux conditions.[100] In contrast 8-chlorotheobromine[246] and the $N_{(3)}N_{(7)}$-diethyl homologue[171] form the 8-methoxy- and 8-ethoxypurines at water bath temperatures.

Using either the 8-chloro- or 8-bromocaffeine, Fischer[247] and others[248-251] converted these to the 8-methoxy- or 8-ethoxypurine by means of methanolic or ethanolic potassium hydroxide[247] at 100°. Similarly derived were 8-propoxy,[250] -butoxy,[250] -allyloxy,[250] -benzyloxy,[250] -phenylethyloxy,[250] -methoxyethyloxy,[252] -ethoxyethyloxy,[253] and other alkoxyethyloxy groups,[253] -dialkylaminoethyloxy[254, 255] (reaction in benzene or toluene) and -phenoxy[250, 256] derivatives of caffeine. A novel route to 8-(2-bromoethoxy)caffeine entails heating the 8-chloropurine with bromohydrin and sodium acetate at 190° for 48 hours.[89] Related examples of 8-methoxy and 8-ethoxypurines are found in the 1-ethyl-3,7-dimethyl[257, 258] and 3,7-diethyl-1-methyl[171] homologues of caffeine. 8-Chloroisocaffeine gives 8-alkoxypurines similarly.[100]

F. Replacement of 2-, 6-, and 8-Halogen Atoms by Oxo Groups

Either aqueous alkali or dilute or concentrated acids are used for hydrolyses.* As the purine molecule is capable of anion and cation formation, respectively, in these media the halogen atom first replaced, where more than one is present, may not be the same in both cases.

a. *With Alkali*

Both 2-chloro- and 2-iodo-7-methylpurine afford 2,3-dihydro-7-methyl-2-oxopurine in hot N-potassium hydroxide,[75] the ease of removal of the halogens contrasting sharply with the almost inert nature exhibited by them toward replacement by other nucleophiles. With 6-chloropurine conversion to hypoxanthine requires some hours heating in 0.1N-sodium hydroxide,[16] but if electron-demanding groups are also present in the ring the halogen becomes more labile,[259] requiring briefer exposure to the alkali. Hydrolyses of 6-chloro-9-alkylpurines may give rise to a mixture of required 9-alkylhypoxanthine and 4,5-diamino-6-

---

* Due to the exceptional reactivity of the halogen in 6-fluoro-9-methylpurine,[79] formation of 9-methylhypoxanthine occurs when an aqueous solution of the fluoropurine is allowed to stand at room temperature for a few days.

chloropyrimidine. A recent study of this reaction[260] indicates that alkali instability is a feature of purines devoid of strong electron-releasing groups. Stabilisation, however, is possible if the molecule can assume an ionic state, as happens when hydroxyl ion attack is directed first toward the halogen-bearing carbon atom. If this is not the case nucleophilic substitution occurs at the alternative site, that is, at $C_{(8)}$ with subsequent fission of the imidazole ring and formation of a 4,5-diamino-6-chloropyrimidine.

Of the dihalogenopurines the 2,6-dichloro derivative after heating 1 hour in N-alkali gives 2-chlorohypoxanthine[179]; the corresponding 7-methyl,[40] 9-methyl,[4] and 9-phenyl analogues arise in the same way.[14] Similar but more vigorous conditions were used to obtain 8-chlorohypoxanthine from 6,8-dichloropurine,[24] the 2-methyl analogue following likewise.[25]

Hot dilute alkali replaces only the 6-halogen in 2,6,8-trichloropurine (76, R = H)[162, 168] giving the 6-oxopurine (77). With 9- or 7-alkyl-2,6,8-trichloropurine (76, R = Alkyl) the appropriate 2,6-dichloro-8-oxopurine (78) results[156, 164] on leaving an alkaline solution for some hours at room temperature.

In boiling 0.1N-sodium hydroxide 2-amino-6-chloropurine is recovered largely unchanged.[59] The attempted conversion by Fischer of 6-amino-2-chloro-7-methylpurine (79) to 7-methylisoguanine (80) in this way gave instead 7-methylguanine (81).[261] This rearrangement, probably the first recorded for purines, has subsequently been investigated and a partial mechanism proposed.[262] Although some 80 was detected in trace amounts it does not appear to arise by simple hydrolysis of the $C_{(2)}$-chlorine atom, a more complex route is envisaged. The above conditions also isomerise 2-chloro-7-methyl-6-methylamino-purine to 1,7-dimethylguanine.[261]

Halogenothiopurines have been little studied but the 2-methylthio group in 6,8-dichloro-2-methylthiopurine remains intact after heating for 2 hours in N-sodium hydroxide, the product being 8-chloro-2-methylthiohypoxanthine.[25]

Alkaline hydrolysis converts 6-chloro-7,8-dihydro-8-oxopurine to 1,6,7,8-tetrahydro-6,8-dioxopurine[263] under unspecified conditions. The majority of examples relating to chloro-oxo purines are of N-alkylated derivatives. With 2,6-dichloro-7,8-dihydro-7,9-dimethyl-8-oxopurine, the 2-chloro-6-oxo analogue is formed with some fission of the imidazole ring taking place in addition.[264] Although 8-halogeno-xanthines[42] and -theophyllines[111] are stable, alkaline conditions readily remove halogens in the corresponding 7- (or 9-)alkylated derivatives, exemplified by the conversion of 8-bromo-[42] or 8-chlorocaffeine[51] to 1,3,7-trimethyluric acid. Other caffeine homologues are similarly transformed to the uric acid analogues.[170]

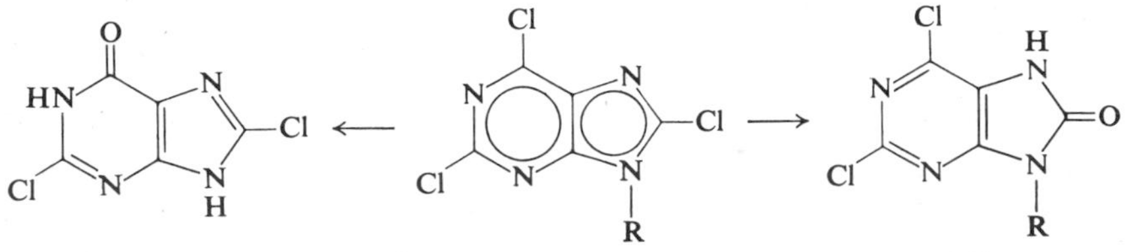

(77)   (76)   (78)

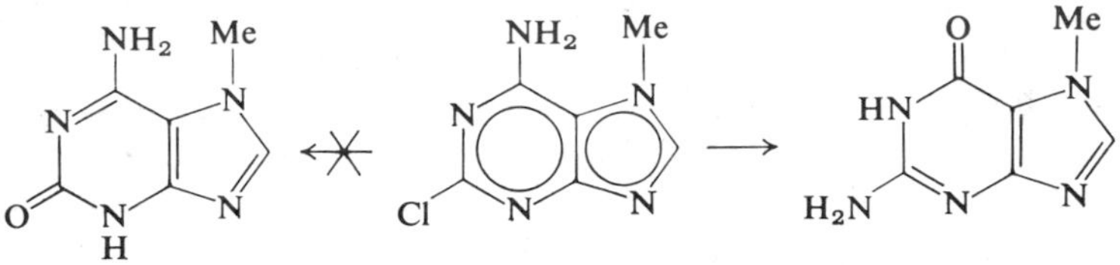

(80)   (79)   (81)

b. With Acid

The purine ring is more stable to hydrolysis in acid media and in many cases replacement of halogen by oxo occurs much faster than in alkaline solution. Hypoxanthine, for example, results from heating 6-chloropurine in 0.1 N-hydrochloric acid under reflux for 1 hour, compared with the 4-hour heating period required if 0.1 N-sodium hydroxide is used.[16]

Because of the reduced risk of imidazole ring fission the route is preferable for hydrolysis of 6-chloro-7-alkyl-[187] and 9-alkylpurines[184, 185, 265, 266] to the hypoxanthine derivatives. The more reactive 6-chloro-3-methyl-8-phenylpurine is converted to the 6-oxo purine on treatment with an acetic acid–sodium acetate mixture.[57]

Under sealed-tube conditions 2,6-diiodopurine gives xanthine directly,[75] 7-methylxanthine arising similarly from 2,6-dichloro-7-methylpurine.[40] With 6,8-dichloropurine the reverse order of substitution to that found with alkali occurs. In 50% aqueous hydrochloric acid on a steam bath the product is 6-chloro-7,8-dihydro-8-oxopurine (82, R = H)[24]; the 2-methyl (82, R = Me)[25] and 2-methylsulphonyl derivatives (82, R = SO$_2$Me)[39] being derived likewise. By analogy with 6-substituted-2,8-dichloropurines the halogen at C$_{(2)}$ in 2-chloro-8-iodopurine would be expected to undergo hydrolysis whereas in fact the normally less reactive 8-position is the site of attack giving 2-chloro-7,8-dihydro-8-oxopurine.[64]

2,6,8-Trichloropurines (**83**, R = H or Me) with[215] or without[11] 7- or 9-alkyl groups, on warming with strong acid in the open undergo replacement of the $C_{(8)}$-halogen only (**84**), pressure conditions and elevated temperatures being needed to convert the trichloropurine directly to the uric acid derivatives (**85**).[7, 42]

After 20 minutes in boiling 30% hydrochloric acid, isoguanine (**87**) is formed from 6-amino-2-chloropurine (**86**, R = H).[166] Under more drastic conditions, involving fuming hydrochloric acid at 120°, the 7-methyl analogue (**86**, R = Me) suffers hydrolysis of both halogen atom and amino group giving 7-methylxanthine (**88**).[37] In the related 6-amino-2-chloro-8,9-dihydro-7-methyl-8-oxopurine, however, halogen replacement necessitates closed vessel techniques.[37]

Deactivation of a halogen atom associated with a $C_{(8)}$-oxo group is again seen with 2-amino-6-chloro-7,8-dihydro-8-oxopurine requiring the above conditions for conversion to 2-amino-1,6,7,8-tetrahydro-6,8-dioxopurine.[36] By comparison 2-amino-6-chloro-[59, 60] and 2-amino-6-chloro-3-methylpurine[62] react under reflux conditions alone. The effect of a $C_{(8)}$-oxo group on the activity of a $C_{(6)}$-halogen is noted above, and a parallel situation exists between an oxo group at $C_{(6)}$ and a $C_{(8)}$-halogen atom. Thus, while the corresponding 8-oxo derivative follows hydrochloric acid hydrolysis of 6-amino-8-chloro-,[24] 8-chloro-6-methylamino-,[24] and 8-chloro-6-dimethylamino-2-methylsulphonylpurine,[39] in the case of 8-bromoguanine conversion to 2-amino-1,6,7,8-tetrahydro-6,8-dioxopurine requires heat and pressure.[267] Concentrated hydrochloric acid at 120° affords the 2,6-dioxo derivative from 8-amino-2,6-dichloro-7-methylpurine[164] and the 2,8-dioxopurine from 6-amino-2,8-dichloropurine. Under reflux in aqueous acid the latter dichloropurine gives only the corresponding 2-oxopurine.[162]

A rearrangement follows ethanolic hydrogen chloride treatment of 9-amino-6-chloro-7,8-dihydro-8-oxopurine (**89**) giving 5-chloro-1,2,3,4-tetrahydro-3-oxo[5,4-*e*]pyrimidotriazine (**90**) as product.[213]

Among the chloro-oxopurines various *N*-methylated derivatives of 6,8-dioxo-2-chloropurine are converted to the analogous uric acid derivatives in concentrated acid at 110°.[264] A related example of this type is the formation of 1,7-dimethylxanthine from 2-chloro-1,6-dihydro-1,7-dimethyl-6-oxopurine.[40]

Hydrolysis of 2,8-dichlorohypoxanthine in this way provides a route to 8-chloroxanthine. The structure of the latter was confirmed by reductive dehalogenation to xanthine[168] in view of a previous claim[56] that the product of the hydrolysis was the isomeric 2-chloro-1,6,7,8-tetrahydro-6,8-dioxopurine.

The inert nature of an 8-halogen toward acid conditions in the above

purine is also reflected in the lack of reactivity found in 8-chloro derivatives of caffeine and 1,7-dimethylxanthine.[45]

That certain chloromethylthiopurines can be safely hydrolysed is
exemplified by the preparation of 6-chloro-7,8-dihydro-2-methylthio-8-
oxopurine from 6,8-dichloro-2-methylthiopurine by heating in dilute
acid.[25]

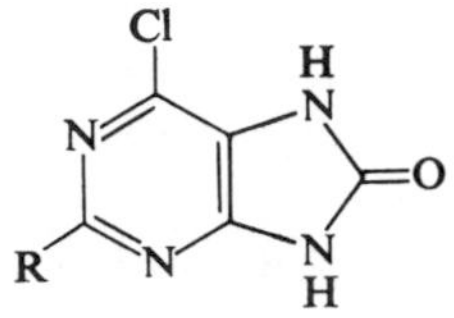

(82)

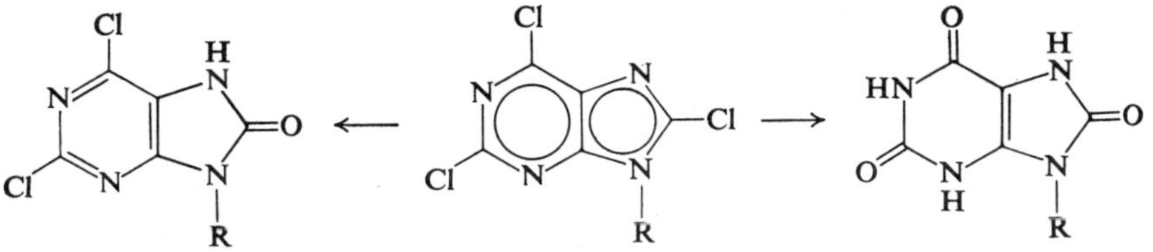

(84)   (83)   (85)

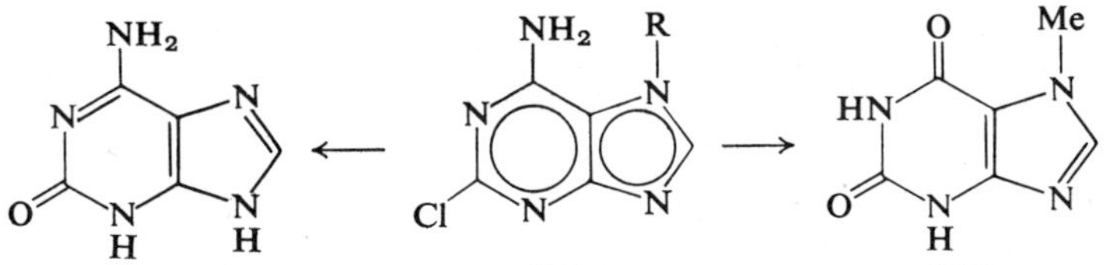

(87)   (86)   (88)

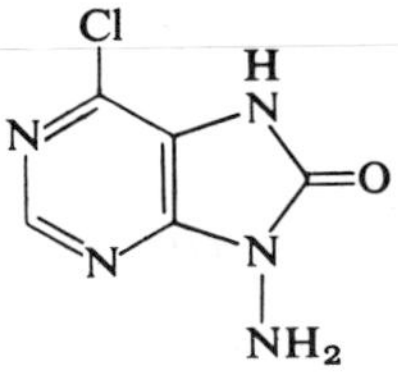

(89) 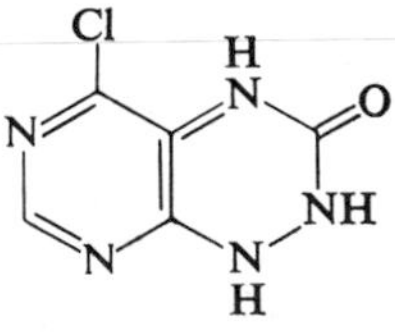   (90)

G. Replacement of 2-, 6-, and 8-Halogen Atoms by Alkyl(and aryl)thio Groups

Although most simple alkylthiopurines are most conveniently prepared by alkylation of thiopurines, thereby obviating the use of the

noxious alkyl mercaptans, direct replacement of a halogen by an alkylthio group is, nevertheless, a valuable route to purine thioethers. Due to the clinical importance of "6-mercaptopurine" in neoplastic disease therapy the majority of alkylthiopurines that have been made are derivatives of this compound. Condensation of the halogenopurine with the sodium or potassium salt of the mercaptan is made in aqueous or alcoholic solution, or other suitable solvent, temperatures below 100° usually being adequate. In practice the mercaptan in sodium or potassium hydroxide solution is utilised in place of the salt. 6-Chloropurine reacts in this way to give 6-ethylthio-, 6-propylthio-, 6-isopropylthio-, 6-butylthio-, and higher alkylthiopurine homologues.[268] Derivatives with cyclic substituents include the 6-phenylthio,[269, 270] and -substituted-phenylthio,[270] -benzylthio,[190] -fur-2-ylthio,[190] -naphth-2-ylthio,[270] and -imidazol-2-ylthio.[270a] With 1,2-dimercaptoethane, 1,2-di(purinyl-6-thio)ethane (91) is formed.[271] Reaction of aryl mercaptans and chloropurine will take place in dilute alkali at room temperature.[269] Similar treatment with alkyl mercaptans have been carried out successfully with 6-chloro-2-ethyl-,[272] 6-chloro-8-methyl-,[22] and 6-chloro-9-methyl-purine.[181] Dichloropurines react with mercaptans in the same order as with alkoxides, 2,8-dichloropurine affords 2-chloro-8-methylthio-purine,[64] while the 6,8-dichloro isomer gives 8-chloro-6-methylthio-purine[24] and the 8-chloro-6-ethylthio analogue.[24] The reaction can be utilised to insert methylthio groups into existing chloromethylthio-purines, as in the conversion of 6-chloro-2-methylthiopurine to 2,6-dimethylthiopurine,[39] 6,8-dimethylthiopurine likewise arising from 6-chloro-8-methylthiopurine.[24] The preparation of 2,6,8-trimethylthio-purine from 6-chloro-2,8-dimethylthiopurine can be accomplished in this way[39] but only 8-chloro-2,6-dimethylthiopurine results if 6,8-dichloro-2-methylthiopurine is employed as starting material.[39]

Simple oxopurines are little represented in these replacements but 6-chloro-7,8-dihydro-2-methylthio-8-oxopurine has been converted to 7,8-dihydro-2,6-dimethylthio-8-oxopurine.[39] Some of the earliest examples of interaction between a halogenopurine and alkanethiol are found in the formation of 8-thioethers of caffeine,[88] which include the

(91)   (92)

methylthio, ethylthio, propylthio, butylthio; also the phenylthio and benzylthio analogues. With cysteine 8-(2-amino-2-carboxyethylthio)-caffeine (92) resulted.[88]

H. Replacement of 2-, 6-, and 8-Halogen Atoms by Thio Groups

Two principle routes are available, the older one which involves heating the halogenopurine with alkaline solutions of a sulphide or the more agreeable method in which reaction with thiourea in alcoholic media gives rise to an unstable thiouronium salt, not usually isolated, which breaks down readily to the thiopurine in alkaline solution.

a. *With Sulphides*

Alkaline solutions of sulphides are conveniently prepared by saturating sodium or potassium hydroxide with hydrogen sulphide. Ammonium sulphide, similarly obtained, is used to a lesser degree. Reaction conditions vary from room temperature to heating under reflux, only occasionally are pressure conditions needed.

With this procedure ^{35}S-labelled 1,6-dihydro-6-thiopurine was obtained in excellent yield from 6-chloropurine using labelled barium sulphide.[273] In place of sulphide solutions thiolacetic acid has been successfully employed with 6-chloro- and 6-iodopurine.[274] The sulphide technique also converts other 6-chloropurines,[57] 6-chloro-7-alkyl-, and 6-chloro-9-alkylpurines[203, 275] to the 6-thio analogues. Cold ammonium sulphide was sufficient to transform 6-chloro-3-methyl-8-phenylpurine to the 6-thio analogue.[57] A solution of sodium ethoxide saturated with hydrogen selenide under reflux was used to prepare 6-selenopurine (93).[276]

At room temperature in potassium hydrosulphide only replacement of the $C_{(6)}$-halogen in 2,6-dichloro-7-methylpurine takes place but at 100° disubstitution, giving 1,2,3,6-tetrahydro-7-methyl-2,6-dithiopurine (94, R = Me) occurs.[169] Under mild conditions 6-chloro-2-sulphamoyl-purine gives the 6-thio analogue (95) as major product (48%) but some dithiopurine (94, R = H) is found present (21%), arising through conversion of the sulphamoyl group at $C_{(2)}$.[259] Both 6,8-dichloro-purine[24] and the 2-methyl homologue[25] behave normally, only the 6-chlorine atom being replaced under reflux conditions. Attempts to react 2,6,8-trichloropurine with alcoholic sodium hydrosulphide at room temperature give mixed products[169, 277] but at 100° in aqueous

potassium hydrosulphide the 2,6,8-trithio derivative results.[169] 2,6,8-
Trichloro-7-methylpurine similarly gives the corresponding trithio
derivative.[169] Partial halogen replacement was found with one 9-alkyl-
2,6,8-trichloropurine at room temperature in ethanolic sulphide, the
product being the corresponding 2-chloro-6,8-dithiopurine.[156] Examples
of thiolation conditions used with aminopurines are the preparation of
thioguanine by heating 2-amino-6-chloropurine under reflux with
sodium hydrosulphide solution for 2 hours.[60] In contrast is the sealed-
tube procedure required with the 8-oxo derivative (96).[269] Prolonged
reflux temperatures (19 h) are required to convert 2-amino-6-chloro-
purine to 6-selenoguanine (97), using ethanolic sodium ethoxide
saturated with hydrogen selenide.[278, 279] The more strenuous conditions
necessary with 8-halogeno-aminopurines are exemplified by the formation of 7,8-dihydro-6-methylamino-8-thiopurine (98) requiring 3 hours
in aqueous hydrosulphide at 125°.[24] The relatively inert character of a
methylthio group at $C_{(2)}$ toward these reagents is demonstrated by
8-chloro-1,6-dihydro-2-methylthio-6-thiopurine resulting from heating
under reflux (4 h) 6,8-dichloro-2-methylthiopurine in a solution of
potassium hydrosulphide.[25]

The repeated failure of 2,6-dichloro-7,8-dihydro-8-oxopurine to react
with alcoholic sulphide solutions may be a solubility phenomenon as,

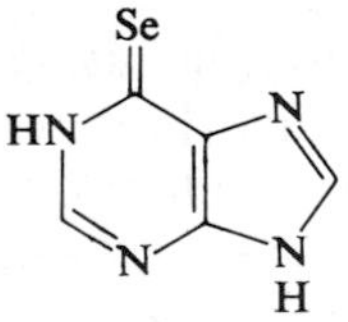

(93) 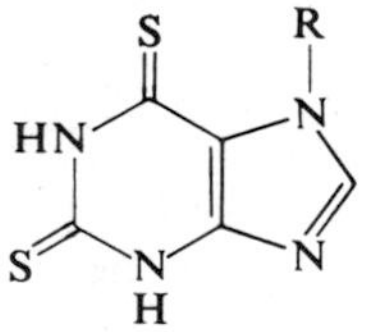   (94)

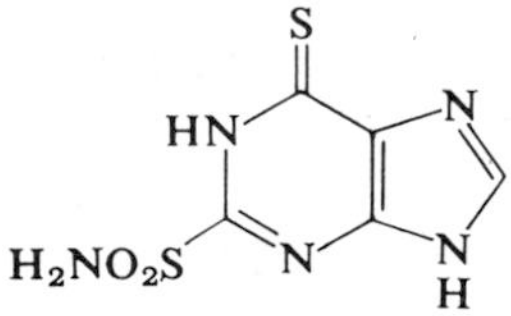

(95) 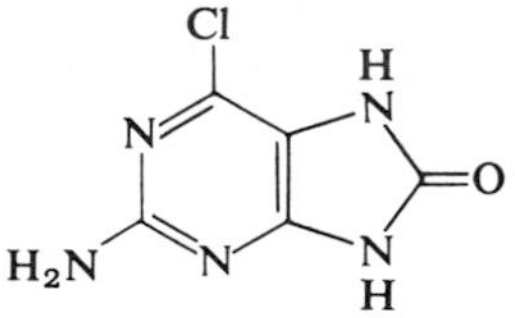   (96) 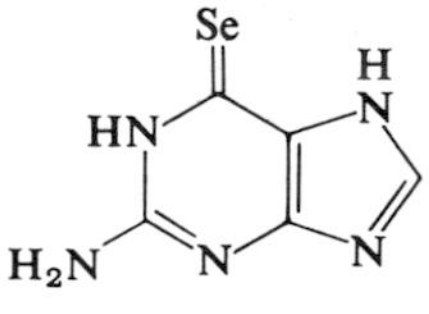   (97)

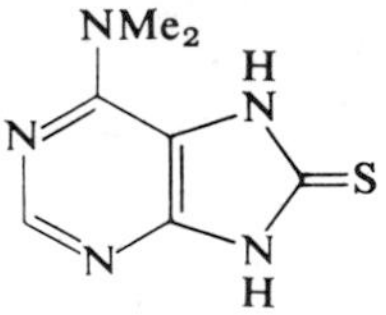

(98) 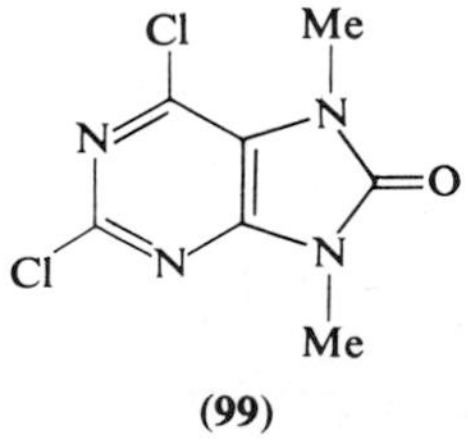   (99)

on standing at room temperature in aqueous ammonium sulphide, mainly the 2-chloro-8-oxo-6-thio analogue is formed together with traces of the 2,6-dithiopurine.[277] The latter purine arises directly from the dichloropurine by using pressure conditions.[280] Similarly, the $N_{(7)}N_{(9)}$-dimethyl homologue (**99**) on heating with alcoholic sulphide solution gives the 2-chloro-6-thiopurine but under pressure forms the 2,6-thio derivative.[280] In view of the reluctance shown by 8-halogeno-xanthines to react with nucleophilic reagents generally it is surprising to find that 8-bromoxanthine forms the 8-thio analogue with potassium hydrosulphide at 120°.[169] The N-alkyl derivatives of 8-halogeno-xanthines, on the other hand, react under even milder treatment; suitable examples of 8-thioxanthines prepared on a water bath include the 3-ethyl,[170] 1,7-dimethyl-3-ethyl,[170] and 1,3,9-trimethyl[153] derivatives.

b. *With Thiourea*

Advantages gained using this procedure include the use of more moderate reaction temperatures, the process being generally carried out in boiling ethanol or propanol, and the removal of the risk of any susceptible groups present being reduced, a possibility always present when hydrosulphides are used. Additionally, as thiourea is a nonbasic nucleophilic reagent, interaction occurs largely with the purine as a neutral molecule, rather than as an ionic species, resulting in a more facile halogen replacement.

In ethanol under reflux for 1 hour, equimolar quantities of thiourea and 6-chloropurine give 1,6-dihydro-6-thiopurine in 67% yield.[16] This increases to near quantitative amounts if thioacetamide replaces thiourea.[281] Homologues likewise prepared are the 2-methyl,[21] 8-methyl,[22] and 2,8-dimethyl[3] derivatives. Although 6-selenopurine can be obtained by sodium hydroselenide treatment, the use of selenourea in ethanol is reported to increase the yield to 92%.[276] Both 7-[2] and 9-methyl-6-chloropurine[181] and related 9-ethyl-,[185] 9-substituted-ethyl-,[184, 282] 9-cyclopentyl-,[183, 265] and 9-phenyl-[186] analogues afford the appropriate 6-thiopurines with thiourea. With a larger alkyl group than ethyl at $N_{(9)}$ ethanol is no longer a suitable solvent and is replaced by the higher boiling propanol.

The only disadvantage associated with this reaction is the formation of various side products. Studies made[272] with 6-chloropurine and thiourea in boiling propanol show that although the 6-thiopurine is the major product, 2,2-diamino-2H-thiazolo[3,4,5-gh]purine (**100**, R = H) is also formed, in up to 13% yield. Three products were obtained with 2-ethyl-6-chloropurine, 2-ethyl-1,6-dihydro-6-thiopurine (38%), the corresponding thiazolopurine (**100**, R = Et) (18%) and 2-ethyl-6-propylthiopurine which is assumed to arise through interaction between solvent and thiouronium derivative (**101**, R = Et).[272] The corresponding thiouronium purine (**101**, R = H) can be isolated by heating 6-chloropurine with thiourea in acetonitrile, conversion to 1,6-dihydro-6-thiopurine being accomplished by a short reflux period in ethanol.[283] Thiourea and 6-chloro-2,9-diethylpurine gives both the 6-thio- and 6-propylthiopurine as products; substitution of the imidazole proton by the ethyl group precludes any thiazolopurine formation.[272] It is pertinent to note that the above thiazolopurines (**100**) in hot acid or alkali rearrange to the appropriate 8-amino-1,6-dihydro-6-thiopurines (**102**).[272] See Chapter II, Section 8E.

Both halogen atoms are readily replaced in 2,6-dichloropurines, 2,6-dithio derivatives of 7-methyl-[2] and 9-phenylpurines[14] arising in this way. The weak activity of the $C_{(2)}$-halogen in 2,8-dichloropurine is again shown by the formation of only 2-chloro-7,8-dihydro-8-thiopurine.[64] Selective replacement is possible with 6,8-dichloropurines, one equivalent of thiourea in hot methanol gives the 8-chloro-6-thio derivative whereas an excess of the reagent, in ethanol as solvent, affords the 6,8-dithiopurine.[24,25] The mixture of products arising with 2,6,8-trichloropurine[277] has been shown to comprise 2-chloro-1,6,7,8-tetrahydro-6,8-dithio- and 1,2,3,6,7,8-hexahydro-2,6,8-trithiopurine.[77]

The appropriate 6-thio derivatives arise with 6-chloro-2-methyl-thio-,[25] 6-chloro-8-methyl-2-methylthio-,[22] 6-chloro-8-methylthio-,[24]

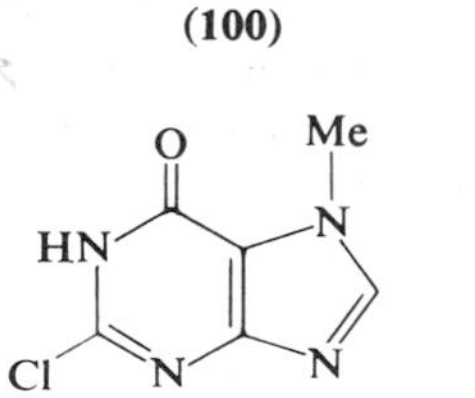

(100) 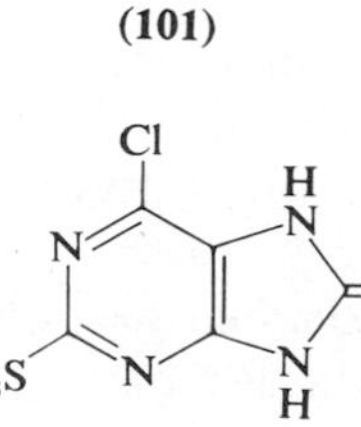   (101) 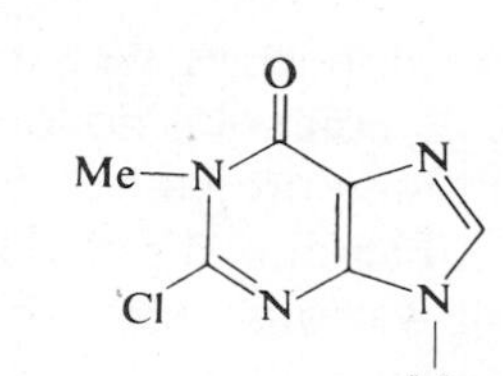   (102)

(103)   (104)   (105)

and 6-chloro-2,8-dimethylthiopurines.[39] Similarly prepared from the chloro derivatives are 6-methylthio-8-thio-[24] and 2-methylthio-6,8-dithiopurines.[25] Although the $C_{(2)}$-halogen atom in the 6-oxopurine (**103**)[2] and that at $C_{(6)}$ in the 8-oxopurine (**104**)[56] require vigorous treatment for replacement by basic nucleophilic reagents, they react with thiourea under more moderate conditions. Alkylation at $N_{(1)}$ in the 2-chlorohypoxanthine (**105**) activates the halogen sufficiently to allow reaction with thiourea in ethanol at room temperature.[5]

I. Replacement of Halogen Atoms by Sulpho and Thiocyanato Groups

Purine sulphonic acids are formed under mild conditions by reacting chloropurines with aqueous sodium sulphite (Ch. VII, Sect. 4A).

Under reflux conditions 6-chloropurine in methanolic potassium thiocyanate affords 6-thiocyanatopurine (**106**).[269] Although this compound is extremely unstable in alkali, being converted rapidly to 1,6-dihydro-6-thiopurine, isomerisation to the isothiocyanate form does not occur in hot alcohol as has been observed in related heterocyclic systems.[284]

J. Replacement of Halogen Atoms by Other Groups

For direct replacement by alkyl (or aryl) groups a number of successful methods have been devised. Of paramount importance is the reaction of halogenopurines with the sodium derivative of diethyl malonate in boiling ethanol. Examples of such condensations are found with derivatives of 2-chloro-,[285] 6-chloro-,[286] 8-chloro-,[102] and 2,6-dichloropurines.[165] The further conversion of such compounds to methyl- and higher alkylpurines is dealt with in Chapter IV (Sect. 2Ad). The condensation of 8-chlorocaffeine with 1-cyano-3-dimethylamino-1-phenylpropane in the presence of sodamide is also of this type, the product being 8-(1-cyano-3-dimethylamino-1-phenylpropyl)caffeine (**107**).[225] Although at elevated temperatures (250°) 8-bromotheophylline, and other 8-bromo-N-alkylxanthines, react with phenylacetic acid in sodium hydroxide solution giving good yields of 8-benzylpurines,[287] these reactions are best considered as being Traube-type cyclisations rather than as reactions of the $C_{(8)}$-halogen atom. Although the successful application of Grignard reagents to halogenopurines is still awaited, related reactions are known with lithium-alkyl and -aryl compounds.[286]

The formation of 6-phenylpurine from 6-chloropurine by means of lithium phenyl is reported[286] but side reactions also occur leading to mixed products.

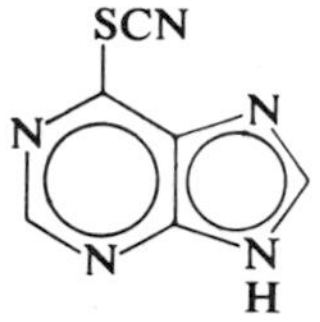

(106) 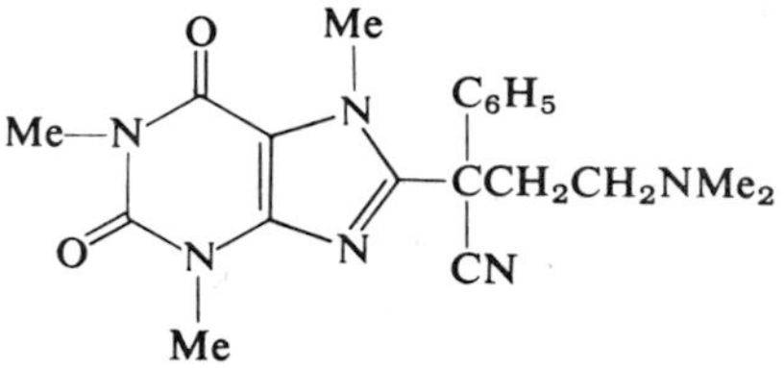   (107)

6. Reactions with Extranuclear Halogen Atoms

Where the halogen is separated from the nucleus by two or more carbon atoms the behaviour toward nucleophilic reagents resembles that of the halogen in the corresponding alkyl halide itself. However, whereas the volatile nature of the latter necessitates closed-vessel reaction conditions the halogenoalkyl purines, which are generally solids, can usually, but not always, be reacted under reflux conditions. Halogenomethyl derivatives, on the other hand, are highly reactive and provide a valuable means of replacing a methyl group by various functional groups.

A. Replacement by Amino Groups

Representative examples of amination of chloromethyl groups at $C_{(2)}$ and $C_{(6)}$ are seen in the formation of 2-dimethylaminomethyl-1,6-dihydro-1,9-dimethyl-6-oxopurine (108)[285] and 2-chloro-6-diethylamino-methyl-9-methylpurine (109)[165] from the respective 2- and 6-chloromethyl derivatives under mild treatment. By using an aqueous in place of alcoholic solution of dialkylamine in the latter example the $C_{(2)}$-chlorine is also replaced.[165] The 6-aminomethyl analogue of 109 results from a multistage reaction in which after heating the 6-chloromethyl-purine with hexamethylene tetramine, the product is acetylated and then hydrolysed with acid to 6-aminomethyl-2-chloro-9-methylpurine.[165] This route compares unfavourably with use of phthalimide for the amination of the 8-chloromethyl derivatives of caffeine,[130] theobromine,[129] and isocaffeine.[288] These were carried out with potassium phthalimide in a high boiling solvent, such as p-chlorotoluene, the resulting 8-phthalimidomethylpurine (**110**) being transformed to the 8-aminomethylpurine on heating with hydrazine hydrate. Reaction with alkyl(and aryl)amines proceeds under reflux conditions, typical 8-alkylaminomethyl derivatives of theobromine,[129] theophylline,[128, 289] caffeine,[130] and related methylated xanthines[128, 288] being obtained. Side reactions leading to di(theophyllin-8-ylmethyl)amines may sometimes occur.[128] Heating 8-chloromethylpurines with pyridine on a water bath in the absence of solvent gives the quaternary chloride; examples include the 8-pyridiniummethyl chlorides of theophylline (**111**),[128] theobromine,[127] caffeine,[127] and 3-methylxanthine.[127] Pressure conditions are used to aminate 6-(3-chloropropyl)purine with aqueous dimethylamine[133] but

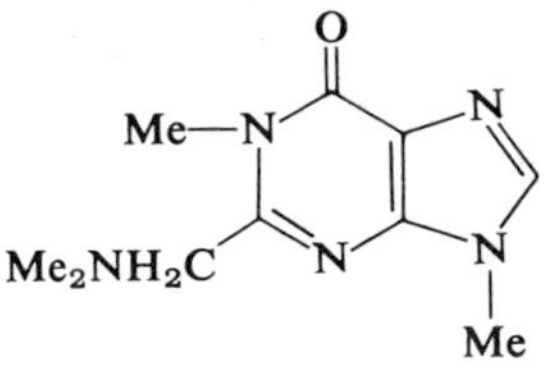

(108)

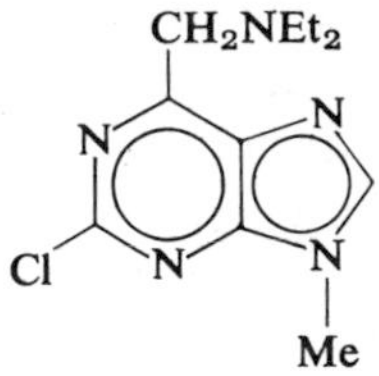

(109)

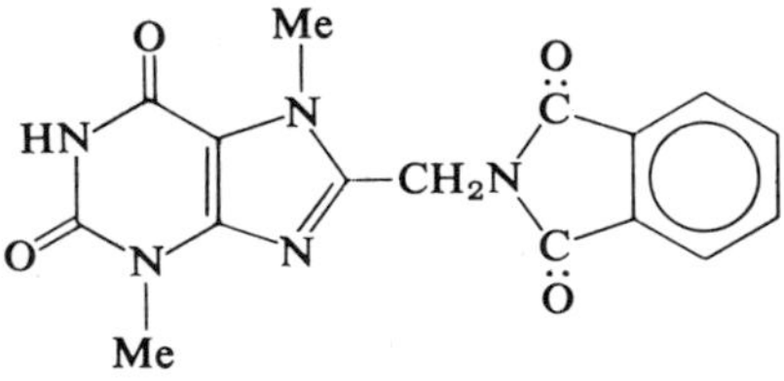

(110)

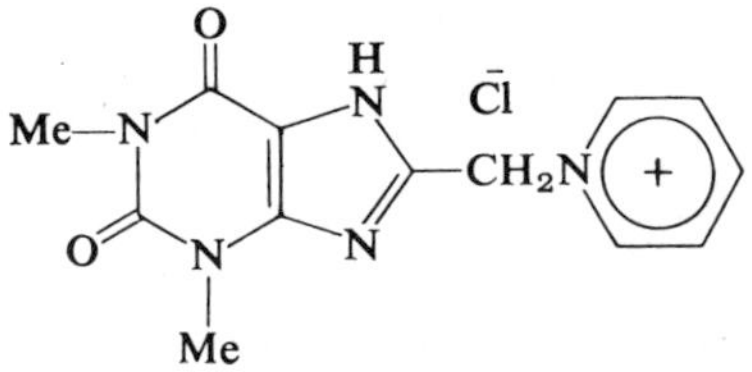

(111)

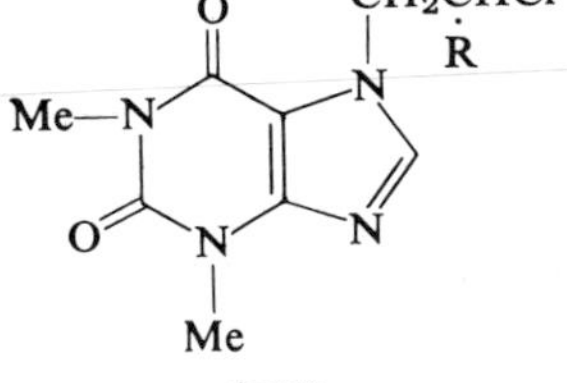

(112)

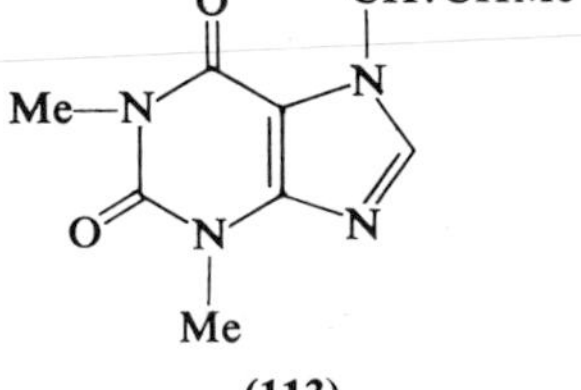

(113)

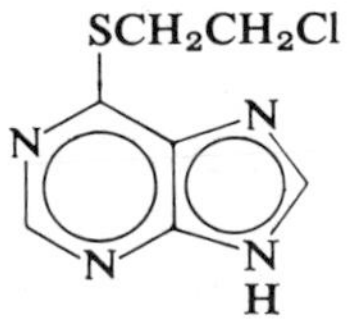

(114)

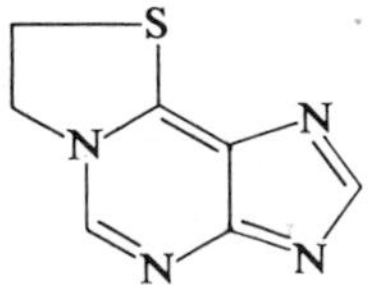

(115)

8-(2-chloroethyl)theobromine reacts with aqueous diethylamine below the reflux point.[132] The conversion of 6-trichloromethylpurine by means of aqueous ammonia at room temperature to 6-carbamoylpurine[290] is noted elsewhere (Ch. IX, Sect. 3Ab).

Halogen atoms in N-halogenoalkylpurines show typically low activities, 7-(2-chloroethyl)theophylline (**112**, R = H) only reacts with ammonia and amines[226, 291] under pressure. Diethylamine in benzene with 7-(2-chloropropyl)theophylline (**112**, R = Me), however, does not give expected 7-(2-diethylaminopropyl)purine but through base-catalysed removal of hydrogen chloride 7-propenyltheophylline[137] (**113**) is produced).

Amination of 9-chloroethyl-6-alkylthiopurines is carried out in a high boiling solvent such as 2-methoxyethanol.[72] Chloroethylthio groups located at $C_{(2)}$, $C_{(6)}$, and $C_{(8)}$ have been observed to internally alkylate an adjacent ring nitrogen atom on gentle heating (60°) in ethanol. Illustrated is the formation of 7,8-dihydrothiazolo[2,3-i]purine (**115**) from 6-(2-chloroethylthio)purine[141, 270] (**114**) but corresponding dihydrothiazolopurines are formed analogously by 2-(2-chloroethyl)- and 8-(2-chloroethyl)purine.[292] Other related examples involving quaternary chloride formation are known.[293]

B. Replacement by Hydroxy and Alkoxy Groups

Heating with dilute hydrochloric acid converts a 2-chloromethyl-purine to the 2-hydroxymethyl analogue.[285] The appropriate 8-hydroxy-methyl derivatives of caffeine[294] and isocaffeine[288] are similarly derived. Longer chain alkyl halides can be suitably hydroxylated by way of the acetoxy intermediates. Representative examples are the conversion of 7-iodobutyl- and 7-iodopentyltheophyllines to the 7-acetoxyalkyl derivatives from which the 7-alkanols are derived in hot ethanolic potassium hydroxide.[139]

Dichloromethyl groups in boiling water alone are converted to formyl groups, illustrative examples resulting from this treatment are given in Chapter IX, Section 5Aa. It should be noted that under these conditions some N-chloromethyl groups may be removed.[52]

Hydrolysis of trichloromethyl groups provides a valuable route to some carboxypurines (Ch. IX, Sect. 1Aa) but instances where this procedure leads to removal of the groups, presumably through a subse-quent decarboxylation, are known.[109, 295] A similar loss of an 8-tri-fluoromethyl group, giving adenine, occurs when the purine is treated with 5N-sodium hydroxide at 25°.[295a]

Alkoxy derivatives of chloromethylpurines are usually prepared by heating with alcoholic sodium alkoxide; this provides a range of examples of 2-ethoxymethyl-,[285] 6-ethoxymethyl-,[165] and 8-ethoxy-methyl-[129, 131, 294] and other 8-alkoxymethylpurines.[294] Methanolic potassium hydroxide reacts with 6-trichloromethylpurine giving 6-trimethoxymethylpurine converted by acid hydrolysis to 6-methoxy-carbonylpurine.[290] The carboxylic esters are more usually prepared directly by heating the trichloromethylpurine with alcohol. This aspect is fully detailed in Chapter IX, Section 2Ab.

C. Replacement by Other Groups

Thio and alkyl(and aryl)thio groups readily replace halogen atoms in chloromethyl groups. A poor yield of 6-mercaptomethylpurine (**116**), "homomercaptopurine," results from treating 6-chloromethylpurine with hydrogen sulphide-saturated ethanol containing free sulphur.[281] More productive routes to this derivative are degradation of 6-dithio-carbamoylmethylpurine (**117**) with aqueous ammonia at 55° or of 6-acetylthiomethylpurine (**118**) in 0.88 ammonia at room temperature. Both **117** and **118** arise from 6-chloromethylpurine following condensa-tions with ammonium dithiocarbamate and thiolacetic acid, respec-tively.[281] Surprisingly, 6-chloromethylpurine appears not to react with thiourea but 6-bromomethylpurine gives a good yield of the isothio-uronium (**119**) salt.[281] Although attempts to convert **119** to 6-mercapto-methylpurine were unsuccessful, other mercaptomethylpurines have resulted through the isothiouronium salts of a number of 8-chloromethyl purines[129, 288]; also 7-(2-bromoethyl)-,[296] 7-(4-iodobutyl)-,[139] and 7-(5-iodopentyl)purines[139] likewise form the 7-mercaptoalkyl analogues.

Reactions of halogenoalkylpurines with alkylthiols in ethanol under reflux conditions has given 6-methylthio-, 6-ethylthio-, 6-benzylthio-, and 6-phenylthioalkylpurines in good yield.[281] A comprehensive range of examples of 7-(2-alkylthioethyl)purines has been made.[296]

Thiolacetic acid under appropriate conditions exerts a pronounced reducing action. As noted above, brief treatment of 6-chloromethyl-purine gives the 6-acetylthiomethyl derivative (**118**) but prolonged reflux effects complete reduction to 6-methylpurine. 6-Bromomethylpurine behaves likewise.[281] Differences are observed between the behaviour of 6-trichloromethylpurine and the 6-tribromomethyl analogue with the above reagent, the former being reduced only to the stable 6-dichloro-methylpurine after 2 hours heating whereas the bromo derivative rapidly undergoes complete reduction to 6-methylpurine.[281] An interesting aspect of the latter reaction is that it does not appear to pass through an intermediate 6-dibromomethyl stage as this purine, like the 6-dichloromethyl analogue, is not attacked by thiolacetic acid.[281]

A methanolic suspension of 6-bromomethylpurine with potassium thiocyanate at 60° gives a 70% yield of 6-thiocyanatomethylpurine (**120**), raising to over 80% if the chloromethylpurine is used.[281] Preparation of 7-(2-thiocyanatoethyl)theophylline from the 7-bromoethyl analogue is made in acetone under reflux.[296]

Exocyclic halogen atoms react with the sodium derivative of malonic ester or its derivatives in a similar manner to that of nuclear halogens. Examples of this type, which are found with 2-chloromethyl-,[285] 6-chloromethyl-,[165] and 8-chloromethylpurines,[288] are given later (Ch. IX, Sect. 1Aa). Application of this reaction in the case of 7-(2-chloroethyl)theophylline (**112**, R = H) led to the unexpected formation of the 7-vinyl analogue (**121**).[136] In this case the basic reaction conditions are responsible for this effect as other examples of base-induced hydrogen chloride abstraction are known.[133, 137]

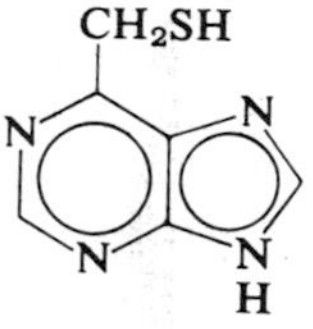

(116) 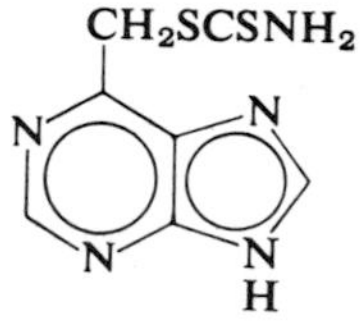   (117) 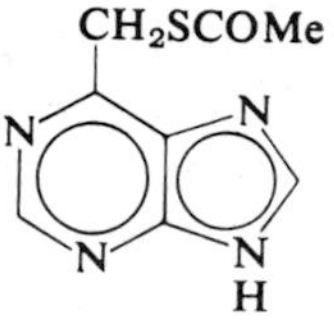   (118)

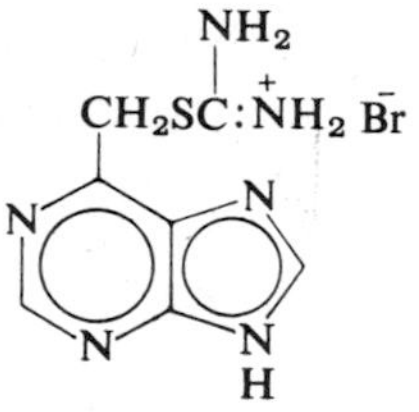

(119) 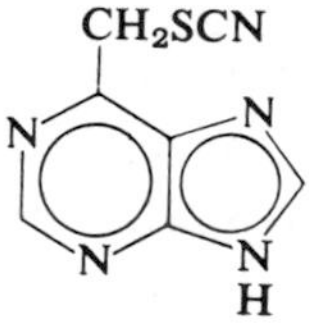   (120) 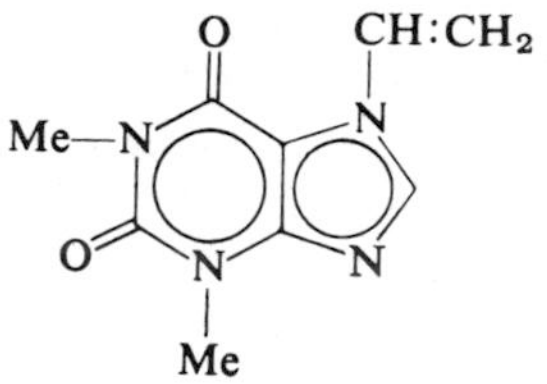   (121)

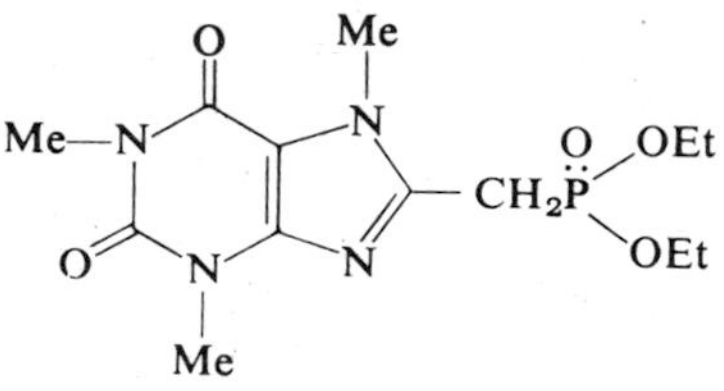

(122) 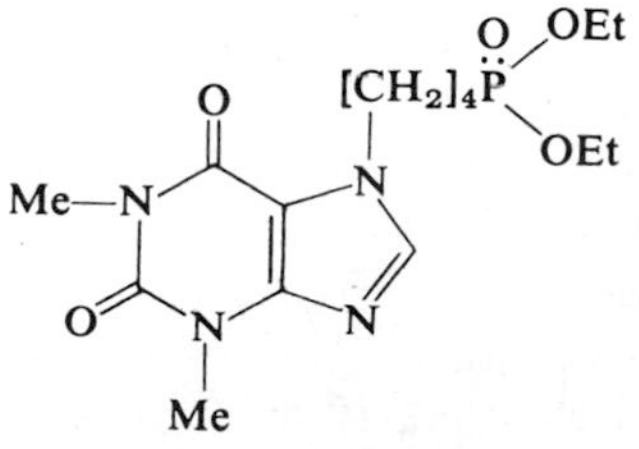   (123)

Both *C*- and *N*-halogenoalkylpurines with triethyl phosphite undergo the Arbuzov rearrangement, 8-chloromethylcaffeine is converted to 8-diethylphosphonylmethylcaffeine (**122**).[297] Corresponding esters arise with tripropyl- and tributylphosphites.[297] The 7-(4-iodobutyl) and 7-(5-iodopentyl) derivatives of theophylline afford 7-diethylphosphonyl-alkylpurines of the type **123**.[139]

7. *N*-Alkylhalogenopurines

Few examples of simple halogenopurines alkylated in the pyrimidine moiety are known but derivatives with $N_{(7)}$- or $N_{(9)}$-alkyl groups are easily obtained.

A. Preparation of *N*-Alkylhalogenopurines

Two routes of general synthetic applicability are available, these being supplemented by the more limited approach of direct alkylation.

a. *By Direct Synthesis*

Only purines with halogen atoms located at the 2- and 6-position are possible by this means. By employing appropriately *N*-alkylated halogeno-4,5-diaminopyrimidines the Traube reaction can be applied to the synthesis of alkylhalogenopurines with one or more of the ring nitrogen atoms carrying an alkyl group.

b. *By Replacement of Groups by Halogen*

Halogenation of existing *N*-alkylated oxopurines has been described earlier. This route, however, in the case of *N*-methyl derivatives, does run the risk of removal of the methyl group if vigorous conditions are used (see Sects. 1Bd, 1Ea, and 2). Under this heading comes also the conversion of thio and methylthio groups, by means of elementary halogen in methanolic solution (Sects. 1C, 1Eb). Both types of replacement reaction are equally suitable for preparing halogenopurines *N*-alkylated in either pyrimidine or imidazole moieties.

c. *By Alkylation*

Alkylating agents with mono-, di-, and trichloropurines may give rise to isomeric mixtures of 7- and 9- alkylpurines, in which the latter purine is usually the major component. Initially, methylation of 6-chloropurine with dimethyl sulphate was effected in dilute sodium hydroxide[275] but more recent studies favour dimethyl sulphoxide or dimethylformamide containing anhydrous potassium carbonate. In many preparations the 7-isomer is formed in only trace amounts so that recrystallisation affords the pure 9-alkylpurine. Where appreciable contamination by the 7-isomer exists solubility differences between the two isomers are exploited to separate them,[82, 232] the $C_{(9)}$-isomer being generally much more soluble. A route claimed to give specifically the 9-alkylpurine is by alkylation of the thallium salt of the chloropurine in an aprotic solvent.[297a] A survey of alkylation studies with chloropurines is given in Table 4. A noteworthy point is that while treatment of 6,8-dichloro-

TABLE 4.   Alkylation of Chloropurines.

Purine	Alkylating agent	Reaction medium	Mixture (%) product 7-Isomer	9-Isomer	Ref.
2-chloro	methyl iodide	DMSO/K$_2$CO$_3$	9	85	82, 154
6-chloro	dimethyl sulphate	N-NaOH	not separated		275
6-chloro	ethyl iodide	DMSO/K$_2$CO$_3$	5	50	232
6-chloro	2-bromoethanol	DMSO/K$_2$CO$_3$	14	61	298
6-chloro	2-bromocyclohexene	DMF/K$_2$CO$_3$	trace	26	299
6-chloro	benzyl chloride	DMSO/K$_2$CO$_3$	15	38	232
6-chloro	p-nitrobenzyl bromide	DMF/Et$_3$N	18	62	300
6-chloro	2-bromoethyl chloride	DMSO/K$_2$CO$_3$	—	19	232
6-chloro	chloroacetonitrile	DMSO/K$_2$CO$_3$	—	35	232
6-chloro	2-bromoethyl acetate	DMSO/K$_2$CO$_3$	—	46	232
6-chloro	ethyl bromoacetate	DMSO/K$_2$CO$_3$	—	35	232
8-chloro	methyl iodide	DMSO/K$_2$CO$_3$	4	8	82
8-chloro	diazomethane	ether	—	38	154
2,6-dichloro	diazomethane	ether	11	20	82
2,6-dichloro	benzyl chloride	ether	30	12	200
6,8-dichloro	benzyl chloride	DMF/K$_2$CO$_3$	trace	57	200
2,6,8-trichloro	methyl iodide	N-KOH	not separated		11
2,6,8-trichloro	methyl iodide	DMSO/K$_2$CO$_3$	—	50	156
2-amino-6-chloro	benzyl chloride	DMSO/K$_2$CO$_3$	24	40	200
6-amino-2,8-dichloro	methyl iodide	KOH	—	90 (Mixture?)	162

purine with benzyl chloride in hot dimethylformamide (80°) containing potassium carbonate gives the expected 9-benzyl-6,8-dichloropurine,[200] with methyl iodide in dimethylsulphoxide containing potassium carbonate, at low temperature ($< 40°$) 6-chloro-7,8-dihydro-7,9-dimethyl-8-oxopurine (124)[82] is obtained. The foregoing result is of interest as two reaction pathways are feasible. In the first an initial methylation gives (presumably) the $N_{(9)}$-methylpurine as a result of which the $C_{(8)}$-halogen now becomes the more labile of the two toward basic nucleophiles. As the reaction was carried out in a notoriously hygroscopic solvent a subsequent hydrolysis to the 8-oxopurine is possible, in which event the remaining imidazole nitrogen atom, now protonated, is available for methylation. In the alternative scheme hydrolysis of the $C_{(8)}$-halogen would also be greatly facilitated by formation of the dimethylated quaternary intermediate (125); charged structures of this type have been demonstrated with related purines.[301] 6-Fluoropurine with methyl iodide behaves normally under like conditions (DMSO/K_2CO_3) giving 7-(40%) and 9-methyl (60%) derivatives. [161a]

Under similar reaction conditions, alkylation specifically at $N_{(9)}$ using unsaturated compounds and involving base-catalysed Michael additions is possible. After some hours at room temperature a solution of 6-chloropurine in dimethylsulphoxide containing acrylonitrile and potassium carbonate affords, in 60% yield, 6-chloro-9-(2-cyanoethyl)-purine (126).[184]

Alkylation, also at $N_{(9)}$, occurs with unsaturated cyclic ethers in the presence of catalytic amounts of acid. A reaction of this type is considered mechanistically in terms of the alkylating agent being a proton-

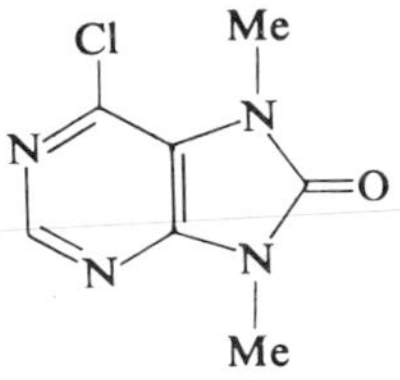

(124) 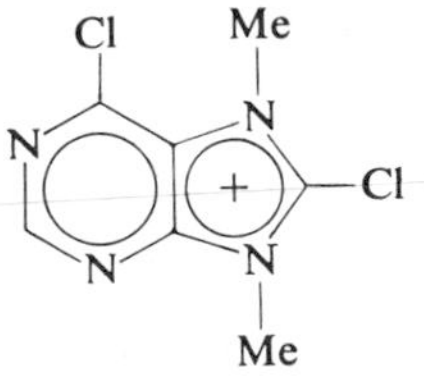   (125)

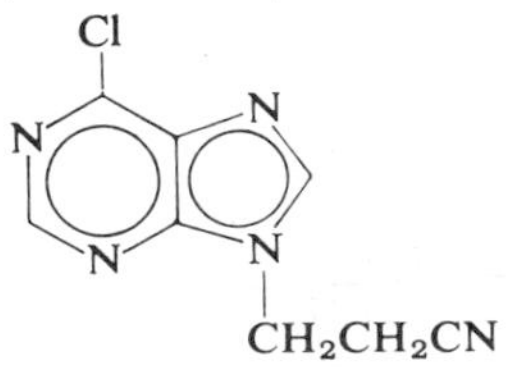

(126) 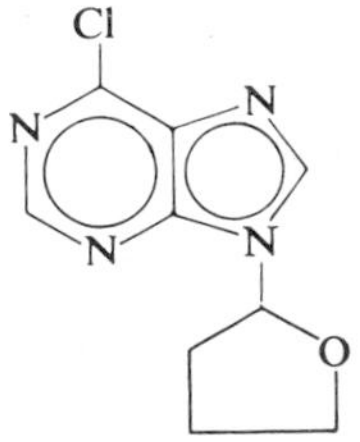   (127)

ated form of the cyclic ether which interacts with an anionic form of the purine. In practice the reaction is carried out in ethyl acetate, acid catalysed with *p*-toluenesulphonic acid, at moderate temperatures (ca. 70°). In this way, 6-chloropurine and 2,3-dihydrofuran gives 6-chloro-9-(tetrahydrofuran-2-yl)purine (**127**) in good yield.[203] Analogous derivatives from 6-chloro-, 6-bromo-, and 6-iodopurine arise with 2,3-dihydro-2-methylfuran,[204] 2,3-dihydropyran,[302] 2,3-dihydro-2-hydroxymethyl-pyran,[302] and 2,3-dihydrothiophene.[203] With this last thioether reflux temperatures are required to form 6-chloro-9-(tetrahydrothien-2-yl)-purine.[203] Corresponding 9-(tetrahydropyran-2-yl) derivatives are given by 2,6-dichloro-[204] and 2,6,8-trichloropurine.[156]

Both 6-chloro- and 2,6-dichloropurine are converted to the 9-vinyl analogues, also by an acid-catalysed procedure, using vinyl acetate in ethyl acetate containing mercuric chloride at 50°, on leaving the mixture to stand for 5 days.[303]

References

1. Traube and Herrman, *Ber.*, **37**, 2267 (1904).
2. Prasad and Robins, *J. Amer. Chem. Soc.*, **79**, 6401 (1957).
3. Prasad, Noell, and Robins, *J. Amer. Chem. Soc.*, **81**, 193 (1959).
4. Gulland and Story, *J. Chem. Soc.*, **1938**, 692.
5. Ovcharova, Nikolaeva, Chaman, and Golovchinskaya, *Zhur. obshchei Khim.*, **32**, 2010 (1962).
6. Lin and Price, *J. Org. Chem.*, **26**, 108 (1961).
7. Fischer and Ach, *Ber.*, **30**, 2208 (1897).
8. Fischer, *Ber.*, **32**, 267 (1899).
9. Fischer and Loeben, *Ber.*, **33**, 2278 (1900).
10. Fischer, *Ber.*, **33**, 1701 (1900).
11. Fischer, *Ber.*, **30**, 2220 (1897).
12. Boldyrev and Makita, *J. Appl. Chem. U.S.S.R.* (Eng. Ed.), **28**, 399 (1955).
13. Elion and Hitchings, *J. Amer. Chem. Soc.*, **78**, 3508 (1956).
14. Koppel and Robins, *J. Amer. Chem. Soc.*, **80**, 2751 (1958).
15. Fischer, *Ber.*, **17**, 328 (1884).
16. Bendich, Russell, and Fox, *J. Amer. Chem. Soc.*, **76**, 6073 (1954).
17. Beaman and Robins, *J. Appl. Chem.*, **12**, 432 (1962).
18. Hitchings and Elion, U.S. Pat., 2,746,961 (1956); through *Chem. Abstracts*, **51**, 1258 (1957).
19. Schütte, Schaaf, Liebisch, Benes, Kozel, and Veres, *Z. Chem.*, **4**, 430 (1964).
20. Fujimoto, Jap. Pat., 6918 (1967); through *Chem. Abstracts*, **67**, 82223 (1967).
21. Robins, Jones, and Lin, *J. Org. Chem.*, **21**, 695 (1956).
22. Koppel and Robins, *J. Org. Chem.*, **23**, 1457 (1958).
23. Craveri and Zoni, *Boll. chim. farm.*, **97**, 393 (1958).
24. Robins, *J. Amer. Chem. Soc.*, **80**, 6671 (1958).
25. Noell and Robins, *J. Org. Chem.*, **24**, 320 (1959).

26. Davoll and Lowy, *J. Amer. Chem. Soc.*, **73**, 2936 (1951).
27. Davoll, Lythgoe, and Todd, *J. Chem. Soc.*, **1946**, 833.
28. King and King, *J. Chem. Soc.*, **1947**, 943.
29. Robins and Christensen, *J. Amer. Chem. Soc.*, **74**, 3624 (1952).
30. Gerster, Jones, and Robins, *J. Org. Chem.*, **28**, 945 (1963).
31. Ikehara, Uno, and Ishikawa, *Chem. and Pharm. Bull. (Japan)*, **12**, 267 (1964).
32. Sörm and Zemlicka, Czech. Pat., 110,944 (1964); through *Chem. Abstracts*, **61**, 13406 (1964).
33. Robison, *J. Amer. Chem. Soc.*, **80**, 5481 (1958).
34. Adams and Whitmore, *J. Amer. Chem. Soc.*, **67**, 1271 (1945).
35. Elion, Burgi, and Hitchings, *J. Amer. Chem. Soc.*, **73**, 5235 (1951).
36. Fischer, *Ber.*, **31**, 2619 (1898).
37. Fischer, *Ber.*, **31**, 104 (1898).
38. Leese and Timmis, *J. Chem. Soc.*, **1961**, 3818.
39. Noell and Robins, *J. Amer. Chem. Soc.*, **81**, 5997 (1959).
40. Fischer, *Ber.*, **30**, 2400 (1897).
41. Urestskaya, Rybkin, and Menshikov, *Zhur. obshchei Khim.*, **30**, 327 (1960).
42. Fischer, *Ber.*, **28**, 2480 (1895).
43. Golovchinskaya, Nikolaeva, Ovcharova, and Chaman, U.S.S.R. Pat., 141,870 (1961); through *Chem. Abstracts*, **57**, 3459 (1962).
44. Ovcharova and Golovchinskaya, *Zhur. obshchei Khim.*, **34**, 2472 (1964).
45. Fischer, *Ber.*, **30**, 3009 (1897).
46. Fischer and Ach, *Ber.*, **31**, 1980 (1898).
47. Fischer and Ach, Ger. Pat., 99,122 (1897); *Frdl.*, **5**, 852 (1897–1900).
48. Fischer and Ach, *Ber.*, **28**, 3135 (1895).
49. Fischer and Clemm, *Ber.*, **31**, 2622 (1898).
50. Boehringer, Ger. Pat., 107,507 (1898); *Frdl.*, **5**, 859 (1897–1900).
51. Fischer, *Annalen*, **215**, 253 (1882).
52. Golovchinskaya and Chaman, *Zhur. obshchei Khim.*, **30**, 1873 (1960).
53. Fischer and Ach, *Ber.*, **39**, 423 (1906).
54. Wellcome Foundation Ltd., Brit. Pat., 767,216 (1957); through *Chem. Abstracts*, **51**, 14796 (1957).
55. Hitchings and Elion, U.S. Pat., 2,815,346 (1957); through *Chem. Abstracts*, **52**, 6417 (1958).
56. Robins, *J. Org. Chem.*, **26**, 447 (1961).
57. Bergmann, Neiman, and Kleiner, *J. Chem. Soc. (C)*, **1966**, 10.
58. Beaman and Robins, unpublished results.
59. Balsiger and Montgomery, *J. Org. Chem.*, **25**, 1573 (1960).
60. Daves, Noell, Robins, Koppel, and Beaman, *J. Amer. Chem. Soc.*, **82**, 2633 (1960).
61. Usbeck, Jones, and Robins, *J. Amer. Chem. Soc.*, **83**, 1113 (1961).
62. Townsend and Robins, *J. Amer. Chem. Soc.*, **84**, 3008 (1962).
63. Cresswell and Brown, *J. Org. Chem.*, **28**, 2560 (1963).
64. Lewis, Beaman, and Robins, *Canad. J. Chem.*, **41**, 1807 (1963).
65. Robins, *J. Amer. Chem. Soc.*, **82**, 2654 (1960).
66. Montgomery and Hewson, *J. Amer. Chem. Soc.*, **82**, 463 (1960).
67. Gerster and Robins, *J. Org. Chem.*, **31**, 3258 (1966).
68. Jones and Robins, *J. Amer. Chem. Soc.*, **82**, 3773 (1960).
69. Giner-Sorolla and Bendich, *J. Amer. Chem. Soc.*, **80**, 3932 (1958).
70. Priewe and Poljack, *Chem. Ber.*, **88**, 1932 (1955).
71. Doerr, Wempen, Clarke, and Fox, *J. Org. Chem.*, **26**, 3401 (1961).

72. O'Brien, Westover, Robins, and Cheng, *J. Medicin. Chem.*, **8**, 182 (1965).
73. Biltz and Sauer, *Ber.*, **64**, 752 (1931).
74. Beaman, Gerster, and Robins, *J. Org. Chem.*, **27**, 986 (1962).
75. Fischer, *Ber.*, **31**, 2550 (1898).
76. Dyer, Reitz, and Farris, *J. Medicin. Chem.*, **6**, 289 (1963).
77. Ballweg, *Annalen*, **649**, 114 (1961).
78. Garkusa, *Zhur. obshchei Khim.*, **27**, 1712 (1957).
78a. Koda, Biles, and Wolf, *J. Pharm. Sci.*, **57**, 2056 (1968).
79. Beaman and Robins, *J. Medicin. Chem.*, **5**, 1067 (1962).
80. Bendich, Giner-Sorolla, and Fox, *Chemistry and Biology of Purines*, Eds., Wolsten-holme and O'Connor, J. and A. Churchill Ltd., London, 1957, p. 7.
81. Giner-Sorolla and Bendich, *J. Amer. Chem. Soc.*, **80**, 5744 (1958).
81a. Kiburis and Lister, *Chem. Comm.*, **1969**, 381.
82. Beaman and Robins, *J. Org. Chem.*, **28**, 2310 (1963).
83. Gerster, Beaman, and Robins, *J. Medicin. Chem.*, **6**, 340 (1963).
84. Montgomery and Hewson, *J. Amer. Chem. Soc.*, **79**, 4559 (1957).
85. Leonard, Skinner, and Shive, *Arch. Biochem. Biophys.*, **92**, 33 (1961).
85a. Eaton and Denny, *J. Org. Chem.*, **34**, 747 (1969).
86. Montgomery and Hewson, *J. Heterocyclic Chem.*, **4**, 463 (1967); *J. Org. Chem.*, **34**, 1396 (1969).
87. Dickinson, Holly, Walton, and Zimmerman, *J. Medicin. Chem.*, **10**, 1165 (1967).
87a. Ikehara and Yamada, *Chem. Comm.*, **1968**, 1509.
88. Long, *J. Amer. Chem. Soc.*, **69**, 2939 (1947).
89. Burkhalter and Dill, *J. Amer. Pharm. Assoc. Sci. Edn.*, **48**, 190 (1959).
90. Biltz and Damm, *Annalen*, **406**, 22 (1914).
91. Biltz and Topp, *Ber.*, **44**, 1524 (1911).
92. Klosa, *Arch. Pharm.*, **289**, 211 (1956).
93. Fischer and Reese, *Annalen*, **221**, 336 (1883).
94. Homeyer and Delamater, U.S. Pat., 2,614,105 (1952); through *Chem. Abstracts*, **48**, 3389 (1954).
95. Rochleder, *Jahresber. fur Chem.*, **1850**, 435.
96. Golovchinskaya, Ovcharova, and Chaman, U.S.S.R. Pat., 135,085 (1961); through *Chem. Abstracts*, **55**, 16576 (1961).
97. Mann and Porter, *J. Chem. Soc.*, **1945**, 751.
98. Blicke and Godt, *J. Amer. Chem. Soc.*, **76**, 3655 (1954).
99. Biltz and Rakett, *Ber.*, **61**, 1409 (1928).
100. Ovcharova and Golovchinskaya, *Zhur. obshchei Khim.*, **30**, 3339 (1960).
101. Ovcharova and Golovchinskaya, U.S.S.R. Pat., 135,084 (1961); through *Chem. Abstracts*, **55**, 16596 (1961).
102. Nikolaeva and Golovchinskaya, *Zhur. obshchei Khim.*, **34**, 1137 (1964).
103. Biltz, *Ber.*, **43**, 1618 (1910).
104. Brühne, *Ber.*, **23**, 225 (1890).
105. Krüger, *Z. physiol. Chem.*, **18**, 434 (1894).
106. Duval and Ebel, *Bull. Soc. Chim. biol.*, **46**, 1059 (1964).
107. Krüger and Saloman, *Z. physiol. Chem.*, **26**, 369 (1898).
108. Biltz, Strufe, Topp, Heyn, and Robl, *Annalen*, **423**, 200 (1921).
109. Golovchinskaya, Ovcharova, and Cherkasova, *Zhur. obshchei Khim.*, **30**, 3332 (1960).
110. Biltz and Strufe, *Annalen*, **423**, 242 (1921).
111. Biltz and Strufe, *Annalen*, **404**, 131 (1914).

112. Cacace, Crisera, and Zifferero, *Ann. Chim. (Italy)*, **46**, 99 (1956).
113. Lespagnol and Gaumeton, *Bull. Soc. chim. France*, **1961**, 253.
114. Yoshitomi, *J. Pharm. Soc. Japan*, **524**, 884 (1925).
115. Serchi, Sancio, and Bichi, *Farmaco. Ed. Sci.*, **10**, 733 (1955).
116. Koppel, Springer, Robins, Schneider, and Cheng, *J. Org. Chem.*, **27**, 2173 (1962).
117. Schultzen, *Z. Chem.*, **1867**, 614.
118. Maly and Hinteregger, *Monatsh.*, **3**, 86 (1882).
119. Yoshitomi, *J. Pharm. Soc. Japan*, **508**, 460 (1924).
120. Libermann and Rouaix, *Bull. Soc. chim. France*, **1959**, 1793.
121. Holmes and Robins, *J. Amer. Chem. Soc.*, **86**, 1242 (1964).
122. Ikehara, Muneyama, and Kaneko, *J. Amer. Chem. Soc.*, **88**, 3165 (1966).
123. Lipkin, Howard, Nowotny, and Sano, *J. Biol. Chem.*, **238**, 2249 (1963).
124. Cacace and Masironi, *Ann. Chim. (Italy)*, **47**, 366 (1957).
125. Biltz and Beck, *J. prakt. Chem.*, **118**, 149 (1928).
126. Biltz and Bulow, *Annalen*, **426**, 306 (1922).
127. Bredereck, Siegel, and Föhlisch, *Chem. Ber.*, **95**, 403 (1962).
128. Kallischnigg, U.S. Pat., 2,879,271 (1959); through *Chem. Abstracts*, **54**, 591 (1960).
129. Golovchinskaya, Ebed, and Chaman, *Zhur. obshchei Khim.*, **32**, 4097 (1962).
130. Golovchinskaya and Chaman, *Zhur. obshchei Khim.*, **22**, 535 (1952).
131. Chaman and Golovchinskaya, *Zhur. obshchei Khim.*, **31**, 2645 (1961).
132. Ebed, Chaman, and Golovchinskaya, *Zhur. obshchei Khim.*, **36**, 816 (1966).
133. Lettré and Woenkhaus, *Annalen*, **649**, 131 (1961).
134. Burkhalter and Dill, *J. Org. Chem.*, **24**, 562 (1959).
135. DiPaco and Tauro, *Ann. Chim. (Italy)*, **47**, 698 (1957).
136. Cacace, Fabrizi, and Zifferero, *Ann. Chim. (Italy)*, **46**, 91 (1956).
137. Zelnik, *Bull. Soc. chim. France*, **1960**, 1917.
138. Richter, Rubinstein, and Elming, *J. Medicin. Chem.*, **6**, 192 (1963).
139. Parikh and Burger, *J. Amer. Chem. Soc.*, **77**, 2386 (1955).
140. Lister and Timmis, *J. Chem. Soc.*, **1960**, 327.
141. Johnston, Fikes, and Montgomery, *J. Org. Chem.*, **27**, 973 (1962).
142. Huber, *Angew Chem.*, **68**, 706 (1956).
143. Lyttle and Petering, U.S. Pat., 2,957,875 (1960); through *Chem. Abstracts*, **55**, 8445 (1961).
144. Burstein and Ringold, *Canad. J. Chem.*, **40**, 561 (1962).
145. Chu, Harris, and Mautner, *J. Org. Chem.*, **25**, 1759 (1960).
146. Lin and Price, *J. Org. Chem.*, **26**, 264 (1961).
147. Lin and Price, *J. Org. Chem.*, **26**, 266 (1961).
147a. Ovcharova and Golovchinskaya, *Zhur. obshchei Khim.*, **34**, 3254 (1964).
147b. A good example of this type is found in reference 165.
148. Nagano, Inoue, Saggiomo, and Nodiff, *J. Medicin. Chem.*, **7**, 215 (1964).
149. Kaiser and Burger, *J. Org. Chem.*, **24**, 113 (1959).
150. Lister, *J. Chem. Soc.*, **1963**, 2228.
151. Cohen and Vincze, *Israel J. Chem.*, **2**, 1 (1964).
152. Albert, *J. Chem. Soc. (C)*, **1966**, 438.
153. Fischer, *Ber.*, **32**, 435 (1899).
154. Barlin and Chapman, *J. Chem. Soc.*, **1965**, 3017.
155. Barlin, *J. Chem. Soc. (B)*, **1967**, 954.
156. Sutcliffe and Robins, *J. Org. Chem.*, **28**, 1662 (1963).
157. Garkusa, U.S.S.R. Pat., 104,281 (1956); through *Chem. Abstracts*, **51**, 5847 (1957).
158. Baddiley, Buchanan, Hawker, and Stephenson, *J. Chem. Soc.*, **1956**, 4659.

198 Chapter V

159. Breshears, Wang, Bechtolt, and Christensen, *J. Amer. Chem. Soc.*, **81**, 3789 (1959).
160. Panagapoulos and co-workers, *Arzneim.-Forsch.*, **15**, 204 (1965).
161. Beaman, Tautz, Duschinsky, and Grundberg, *J. Medicin. Chem.*, **9**, 373 (1966).
161a. Kiburis and Lister, unpublished results.
162. Fischer, *Ber.*, **30**, 2226 (1897).
163. Bredereck, Herlinger, and Graudums, *Angew Chem.*, **71**, 524 (1959).
164. Fischer, *Ber.*, **30**, 1846 (1897).
165. Chaman and Golovchinskaya, *Zhur. obshchei Khim.*, **33**, 3342 (1963).
166. Davoll and Lowy, *J. Amer. Chem. Soc.*, **74**, 1563 (1952).
167. Blackburn and Johnson, *J. Chem. Soc.*, **1960**, 4347.
168. Lloyd, *Chem. and Ind.*, **1963**, 953.
169. Fischer, *Ber.*, **31**, 431 (1898).
170. Biltz and Peukert, *Ber.*, **58**, 2190 (1925).
171. MacCorquodale, *J. Amer. Chem. Soc.*, **51**, 2245 (1929).
172. Gottler, Ger. Pat., 576,604 (1933); through *Chem. Abstracts*, **27**, 5757 (1933).
173. Comte, U.S. Pat., 2,142,935 (1939); through *Chem. Abstracts*, **33**, 2916 (1939).
174. Scheuing and Gottler, Ger. Pat., 582,435 (1933); through *Chem. Abstracts*, **27**, 5754 (1933).
175. Yoshitomi, *J. Pharm. Soc. Japan*, **512**, 839 (1924).
176. Rosenmund and Zetsche, *Ber.*, **51**, 578 (1918).
177. Yoshitomi, *J. Pharm. Soc. Japan*, **510**, 649 (1924).
178. Mackay and Hitchings, *J. Amer. Chem. Soc.*, **78**, 3511 (1956).
179. Montgomery and Holum, *J. Amer. Chem. Soc.*, **79**, 2185 (1957).
180. Brown, England and Lyall, *J. Chem. Soc. (C)*, **1966**, 266.
181. Robins and Lin, *J. Amer. Chem. Soc.*, **79**, 490 (1957).
182. Hull, *J. Chem. Soc.*, **1958**, 2746.
183. Montgomery and Temple, *J. Amer. Chem. Soc.*, **80**, 409 (1958).
184. Baker and Tanna, *J. Org. Chem.*, **30**, 2857 (1965).
185. Montgomery and Temple, *J. Amer. Chem. Soc.*, **79**, 5238 (1957).
186. Greenberg, Ross, and Robins, *J. Org. Chem.*, **24**, 1314 (1959).
187. Schaeffer and Vince, *J. Medicin. Chem.*, **8**, 710 (1965).
188. Albert and Brown, *J. Chem. Soc.*, **1954**, 2060.
189. Daly and Christensen, *J. Org. Chem.*, **21**, 177 (1956).
190. Bullock, Hand, and Stokstad, *J. Amer. Chem. Soc.*, **78**, 3693 (1956).
191. Carter, *J. Biol. Chem.*, **223**, 139 (1956).
192. Sutherland and Christensen, *J. Amer. Chem. Soc.*, **79**, 2251 (1957).
193. Skinner, Shive, Ham, Fitzgerald, and Eakin, *J. Amer. Chem. Soc.*, **78**, 5097 (1956).
194. Townsend, Robins, Loeppky, and Leonard, *J. Amer. Chem. Soc.*, **86**, 5320 (1964).
195. Lettré and Ballweg, *Annalen*, **649**, 124 (1961).
196. Lister, *J. Chem. Soc.*, **1960**, 3682.
197. Lettré and Ballweg, *Annalen*, **633**, 171 (1960).
198. Ballio and DiVittorio, *Gazzetta*, **90**, 501 (1960).
199. Giner-Sorolla, O'Bryant, Burchenal, and Bendich, *Biochemistry*, **5**, 3057 (1966).
200. Montgomery, Hewson, and Temple, *J. Medicin. Chem.*, **5**, 15 (1962).
201. Horwitz and Vaitkevicius, *Experentia*, **17**, 552 (1961).
202. Aldrich Chemical Company Inc. Catalogue, Milwaukee, Wisconsin, U.S.A., 1967.
203. Lewis, Schneider, and Robins, *J. Org. Chem.*, **26**, 3837 (1961).
204. Bowles, Schneider, Lewis, and Robins, *J. Medicin. Chem.*, **6**, 471 (1963).
205. Brown and Weliky, *J. Org. Chem.*, **23**, 125 (1958).
206. Falconer, Gulland, and Story, *J. Chem. Soc.*, **1939**, 1784.

207. Montgomery and Holum, *J. Amer. Chem. Soc.*, **80**, 404 (1958).
208. Bredereck, Christmann, Graudums, and Koser, *Angew. Chem.*, **72**, 708 (1960).
209. Itai and Ito, *Chem. and Pharm. Bull.* (*Japan*), **10**, 1141 (1962).
210. Schaeffer and Thomas, *J. Amer. Chem. Soc.*, **80**, 3738 (1958).
211. Burgison, 133rd A.C.S. Meeting, San Francisco, Calif., 1958, p. 14 M.
212. Chu and Mautner, *J. Org. Chem.*, **26**, 4498 (1961).
213. Krackov and Christensen, *J. Org. Chem.*, **28**, 2677 (1963).
214. Gerster and Robins, *J. Amer. Chem. Soc.*, **87**, 3752 (1965).
215. Borowitz, Bloom, Rothschild, and Sprinson, *Biochemistry*, **4**, 650 (1965).
216. Fischer and Ach, *Ber.*, **32**, 2721 (1899).
217. Fischer, *Ber.*, **30**, 1839 (1897).
218. Boehinger, Ger. Pat., 96,926 (1897); *Frdl.*, **5**, 834 (1897–1900).
219. Gomberg, *Amer. Chem. J.*, **23**, 51 (1900).
220. Zimmer and Mettalia, *J. Org. Chem.*, **24**, 1813 (1959).
221. Cramer, *Ber.*, **27**, 3089 (1894).
222. Einhorn and Baumeister, *Ber.*, **31**, 1138 (1898).
223. Blicke and Godt, *J. Amer. Chem. Soc.*, **76**, 2835 (1954).
224. Klosa, *Naturwiss.*, **46**, 401 (1959).
225. Ehrhart, *Arch. Pharm.*, **290**, 16 (1957).
226. Damiens and Delaby, *Bull. Soc. chim. France*, **1955**, 888.
227. Merz and Stähle, *Arch. Pharm.*, **293**, 801 (1965).
228. Boehringer, Ger. Pat., 164,425 (1903); *Frdl.*, **8**, 1146 (1905–1907).
229. Boehringer, Ger. Pat., 156,901 (1903); *Frdl.*, **7**, 678 (1902–1904).
230. Boehringer, Ger. Pat., 156,900 (1903); *Frdl.*, **7**, 677 (1902–1904).
231. Satoda, Fukui, Matsuo, and Okumura, *Yakugaku Kenkyu*, **28**, 621 (1956).
232. Montgomery and Temple, *J. Amer. Chem. Soc.*, **83**, 630 (1961).
233. Holmes and Robins, *J. Amer. Chem. Soc.*, **87**, 1772 (1965).
234. Polonovski, Pesson, and Zelnik, *Compt. rend.*, **236**, 2519 (1953).
235. Chang, *J. Medicin. Chem.*, **8**, 884 (1965).
236. Giner-Sorolla, Medrek, and Bendich, *J. Medicin. Chem.*, **9**, 143 (1966).
237. Giner-Sorolla, Nanos, Burchenal, Dollinger, and Bendich, *J. Medicin. Chem.*, **11**, 521 (1968).
238. Smirnova and Postovski, *Zhur. Vsesoyuz Khim. obshch. im D.I. Mendeleeva*, **9**, 711 (1964).
239. Huber, *Chem. Ber.*, **90**, 698 (1957).
240. Coburn, Thorpe, Montgomery, and Hewson, *J. Org. Chem.*, **30**, 1110 (1965).
240a. Cassidy, Olsen, and Robins, *J. Heterocyclic Chem.*, **5**, 461 (1968).
241. Bergmann and Heimhold, *J. Chem. Soc.*, **1935**, 955.
242. Bergmann and Heimhold, *J. Chem. Soc.*, **1935**, 1365.
243. Bergmann and Burke, *J. Org. Chem.*, **21**, 226 (1956).
244. Bergmann and Stempien, *J. Org. Chem.*, **22**, 1575 (1957).
245. Bredereck, Küpsch, and Wieland, *Chem. Ber.*, **92**, 566 (1959).
246. Biltz and Pardon, *Ber.*, **63**, 2876 (1930).
247. Fischer, *Ber.*, **17**, 1785 (1884).
248. Wislicenus and Körber, *Ber.*, **35**, 1991 (1902).
249. Biltz and Bergius, *Annalen*, **414**, 54 (1917).
250. Huston and Allen, *J. Amer. Chem. Soc.*, **56**, 1356 (1934).
251. Balaban, *J. Chem. Soc.*, **1926**, 569.
252. Bahner, Di Paolo, and Jones, *J. Org. Chem.*, **23**, 1816 (1958).
253. Klosa, *J. prakt. Chem.*, **18**, 117 (1962).

254. Chakravarty and Jones, *J. Amer. Pharm. Assoc. Sci. Edn.*, **47**, 233 (1958).
255. Cooper and Cheney, U.S. Pat., 2,688,618 (1954); through *Chem. Abstracts*, **49**, 11729 (1955).
256. Baumann, *Arch. Pharm.*, **10**, 127 (1913).
257. Biltz and Max, *Annalen*, **414**, 68 (1917).
258. Biltz and Max, *Annalen*, **414**, 79 (1917).
259. Beaman and Robins, *J. Amer. Chem. Soc.*, **83**, 4038 (1961).
260. Jones, Mian, and Walker, *J. Chem. Soc. (C)*, **1966**, 692.
261. Fischer, *Ber.*, **31**, 542 (1898).
262. Shaw, *J. Org. Chem.*, **27**, 883 (1962).
263. Temple, McKee, and Montgomery, *J. Org. Chem.*, **27**, 1671 (1962).
264. Fischer, *Ber.*, **32**, 250 (1899).
265. Schaeffer and Weimar, *J. Org. Chem.*, **25**, 774 (1960).
266. Schaeffer, Vogel, and Vince, *J. Medicin. Chem.*, **8**, 502 (1965).
267. Fischer, *Ber.*, **30**, 559 (1897).
268. Koppel, O'Brien, and Robins, *J. Org. Chem.*, **24**, 259 (1959).
269. Elion, Goodman, Lange, and Hitchings, *J. Amer. Chem. Soc.*, **81**, 1898 (1959).
270. Johnston, Holum, and Montgomery, *J. Amer. Chem. Soc.*, **80**, 6265 (1958).
270a. Schneider, Niemers, and Warnecke, *Z. physiol. Chem.*, **349**, 1739 (1968).
271. Lewis, Noell, Beaman, and Robins, *J. Medicin. Chem.*, **5**, 607 (1962).
272. Temple and Montgomery, *J. Org. Chem.*, **31**, 1417 (1966).
273. Elion and Hitchings, *J. Amer. Chem. Soc.*, **76**, 4027 (1954).
274. Giner-Sorolla, Thom, and Bendich, *J. Org. Chem.*, **29**, 3209 (1964).
275. Elion, *J. Org. Chem.*, **27**, 2478 (1962).
276. Mautner, *J. Amer. Chem. Soc.*, **78**, 5292 (1956).
277. Elion, Mueller, and Hitchings, *J. Amer. Chem. Soc.*, **81**, 3042 (1959).
278. Mautner and Jaffe, *Biochem. Pharmacol.*, **5**, 343 (1961).
279. Mautner, Chu, Jaffe, and Sartorelli, *J. Medicin. Chem.*, **6**, 36 (1963).
280. Ray, Chakravarti, and Bose, *J. Chem. Soc.*, **1923**, 1957.
281. Giner-Sorolla and Bendich, *J. Medicin. Chem.*, **8**, 667 (1965).
282. Lin and Price, *J. Org. Chem.*, **26**, 108 (1961).
283. McInerney and Kupchik, *J. Medicin. Chem.*, **10**, 741 (1967).
284. Johnson and McCollum, *Amer. Chem. J.*, **36**, 136 (1906).
285. Chaman and Golovchinskaya, *Zhur. obshchei Khim.*, **34**, 3254 (1964).
286. Lettré, Ballweg, Maurer, and Rehberger, *Naturwiss.*, **50**, 224 (1963).
287. Von Schuh, Ger. Pat., 1,091,570 (1960); through *Chem. Abstracts*, **56**, 14305 (1962).
288. Golovchinskaya and Chaman, *Zhur. obshchei Khim.*, **32**, 3245 (1962).
289. Kallischnigg, Ger. Pat., 1,001,273 (1957); through *Chem. Abstracts*, **54**, 1569 (1960).
290. Cohen, Thom, and Bendich, *J. Org. Chem.*, **27**, 3545 (1962).
291. Klinger and Kohlstadt, Ger. Pat., 1,011,424 (1957); through *Chem. Abstracts*, **53**, 18071 (1959).
292. Balsiger, Fikes, Johnston, and Montgomery, *J. Org. Chem.*, **26**, 3446 (1961).
293. Zirm and Pongratz, Austrian Pat., 205,041 (1959); through *Chem. Abstracts*, **54**, 1571 (1960).
294. Golovchinskaya, *Zhur. obshchei Khim.*, **18**, 2129 (1948).
295. Golovchinskaya, Ovcharova, Chaman, and Cherkasova, U.S.S.R. Pat., 129,658 (1960); through *Chem. Abstracts*, **55**, 4549 (1961).
295a. Giner-Sorolla and Brown, *J. Chem. Soc. (C)*, in press.

296. Eckstein, Gorczyka, Kocwa, and Zejc, *Dissertationes Pharm.*, **10**, 239 (1958).

297. Lugovkin, *Zhur. obshchei Khim.*, **27**, 1524 (1957).

297a. Taylor, Maki, and McKillop, *J. Org. Chem.*, **34**, 1170 (1969).

298. Schaeffer and Vince, *J. Medicin. Chem.*, **8**, 33 (1965).

299. Schaeffer and Weimar, *J. Amer. Chem. Soc.*, **81**, 197 (1959).

300. Schaeffer and Odin, *J. Medicin. Chem.*, **9**, 576 (1966).

301. Broom and Robins, *J. Heterocyclic Chem.*, **1**, 113 (1964).

302. Robins, Godefroi, Taylor, Lewis, and Jackson, *J. Amer. Chem. Soc.*, **83**, 2574 (1961).

303. Pitha and Ts'o, *J. Org. Chem.*, **33**, 1341 (1968).

CHAPTER VI

The Oxo-(Hydroxy-) and Alkoxypurines

By far the most widespread of any class of purines, examples of mono-, di-, and trioxopurines are to be found in diverse biological systems in which they are present as intermediate or end products of metabolism. Simple derivatives encountered are hypoxanthine, xanthine, or uric acid, usually from the animal kingdom, while N-methylated xanthines, notably theophylline, theobromine, or caffeine, are metabolites associated with botanical sources. Derivatives of the latter category have had wide application in medicine as stimulants and diuretics.

Nomenclature problems abound with these purines and a satisfactory general textual presentation is difficult. While on the one hand the use of the long established but misleading terminology "hydroxypurines" is unacceptable, as it infers that a hydroxylated form is the dominant specie present, to call them "purinones" by which, for example, xanthine becomes purine-2,6-dione, is equally invalid as the implied ketonic character does not actually exist. The position can be further complicated when considering N-alkylated oxopurines in which derivatives containing both "oxo" and "hydroxy" forms can be formulated. While the systematic nomenclature, with which xanthine becomes 1,2,3,6-tetrahydro-2,6-dioxopurine, is precise and leads to no confusion instant recognition of the compound is virtually impossible. In an effort to reduce this difficulty the policy adopted in this text is to name, where possible, compounds as simple derivatives of hypoxanthine, xanthine, uric acid, and to a lesser degree, those of theophylline, theobromine, or caffeine. Such names will be used in the additive sense rather than indicating substitution of a function or group. For example, 2-thiohypoxanthine would be the preferred form to 2-thioxanthine.* Where derivatives of 2- or 8-oxopurine are encountered the above remarks are

---

* The one exception to this ruling is thioguanine (2-amino-1,6-dihydro-6-thiopurine) but the name is retained because of common usage.

inapplicable and systematic nomenclature is adopted as in the case of the multisubstituted-oxopurines. Diagrammatic representation will be consistently the oxo tautomer and location of the associated proton is arbitrarily fixed as follows: with the oxygen atom at $C_{(2)}$ the proton is shown at $N_{(3)}$, with hypoxanthine derivatives the proton is at $N_{(1)}$, while with 8-oxopurines the proton may be shown either at $N_{(7)}$ or $N_{(9)}$, whichever is appropriate. These same general considerations are applied in the succeeding chapter on the thiopurines.

1. Preparation of 2-, 6-, and 8-Oxopurines

The ease with which oxopurines can be converted into the corresponding chloropurines by phosphoryl chloride makes them of prime importance as starting materials in purine transformations.

A. By the Traube Synthesis

The many facets of this approach are extensively treated in Chapter II, but they may be generally summarised as follows: a 2- or 6-oxopurine is derived from the corresponding 4,5-diamino-oxopyrimidine on cyclisation with an acid, ester, amide, or amidine. If ring closure is effected with reagents such as urea, alkyl chloroformates, urethanes, phosgene, and alkyl isocyanates the 8-oxopurine results. An extension of this enables formation of an 8-oxopurine to be made directly from a $C_{(8)}$-unsubstituted purine by fusion of the latter with urea.

B. From Imidazoles

Use of these intermediates gives an approach almost as versatile as the Traube synthesis; 2-, 6-, and 8-oxopurines can be prepared. Ring closure of 4(5)-amino-5-(4)-carbamoylimidazoles is the most extensively used, by using the appropriate cyclising reagent either 6-oxo-(hypoxanthine) or 2,6-dioxo(xanthine)purines are formed. With 2-hydroxyimidazole analogues either 6,8-dioxo- or 2,6,8-trioxopurines are possible. Reagents leading to hypoxanthine derivatives include formic acid, formamide, and triethyl orthoformate, the corresponding xanthines arising from closure with ethyl chloroformate, ethyl carbonate, or urea fusion. Xanthine derivatives also result from a Hofmann reaction on 4,5-dicarbamoylimidazoles, closure of 4(5)-amino-5(4)-cyanoimidazoles with urethane and the action of isocyanates on alkyl 4(5)-amino-5(4)-carboxyimidazoles. Full treatment of these methods is given in the appropriate sections in Chapter III.

C. From Halogenopurines

The conversion of halogeno- to oxopurines is given in Chapter V, Section 5F.

D. From Aminopurines and Related Derivatives

a. *In the Absence of Other Groups*

With simple 2- and 6-aminopurines conversion to the corresponding 2- or 6-oxo analogue can be effected with warm nitrous acid, one of the earliest applications[1] being the formation of hypoxanthine from adenine. Related successful transformations occur with C-alkyl-adenines[2] and adenine-1-oxide[3] but, due to the acid labile nature of the starting material in the latter case, cold nitrous acid is employed for the hydrolysis. Although 2-aminopurine reacts readily,[4, 5] the effect of nitrous acid on the 8-amino isomer does not appear to have been reported although the preparation of 8-oxopurines from extensively substituted 8-aminopurines is well documented.[6] In contrast to the case of adenine the $C_{(6)}$-amino group in both 2,6- and 6,8-diaminopurine is inert and a 2-[4] or 8-oxo[6] derivative of adenine results. Mechanistic studies[6, 7] of the replacement of amino group by oxo functions indicate that, whereas at $C_{(2)}$ and $C_{(6)}$ replacement may take place through addition of water to the purine cation,[7] the 8-oxopurine is formed via a diazonium intermediate that is capable of isolation.[6]

b. *In the Presence of Other Groups*

An amino group at $C_{(6)}$ usually resists hydrolysis by nitrous acid if a $C_{(2)}$-oxo group is present. Thus, isoguanine,[7, 8] 7-methylisoguanine,[9] and isoguanine-1-N-oxide[10] are inert to this reagent. The failure to hydrolyse both groups in 2,6-diaminopurine, noted in Section a above, is due to the initial formation of isoguanine. Isoguanine derivatives can, however, be converted to the xanthine analogues by dilute mineral acid hydrolyses.[7, 9, 11–13] The contrasting behaviour toward nitrous acid exhibited by the 3,7-[14] and 3,9-dimethyl[15] derivatives of isoguanine, seen in their facile conversion to the corresponding dimethylxanthines, appears to be the result of activation of a $C_{(6)}$-group through $N_{(3)}$-methylation. The isomeric guanine derivatives are readily transformed to xanthines with nitrous acid, good examples occur with guanine itself,[16-20] also the 1-methyl,[21] 7-methyl,[9, 21] 9-benzyl,[22] 1,7-dimethyl,[21] and 7,9-dimethyl[23] analogues. Hydrochloric acid hydrolyses give the same products but prolonged heating is required, as for the formation of xanthine from guanine (32 h) in 25% acid.[24] Studies carried out with 2,6-diaminopurine show dissimilarity between nitrous and hydrochloric acid hydrolyses, the latter, initially gives guanine which is then further hydrolysed to xanthine.[7] Like guanine, the isomeric 6-amino-7,8-dihydro-8-oxopurine with nitrous acid gives 1,6,7,8-tetrahydro-6,8-dioxopurine.[25] A synthesis of uric acid from 2-amino-1,6,7,8-tetrahydro-6,8-dioxopurine in this way is reported.[26]

Although amino groups in N-methylated purines are normally readily displaced by nitrous acid treatment, this is not the case with 6-amino-3-alkylpurines, for example, 3-methylguanine is unchanged in hot nitrous or hydrochloric acid but affords 3-methylxanthine when heated under reflux with 2N-sodium hydroxide[27] or in a sealed tube with ammonia solution.[28] Although likewise inert to mineral acids,[23] 3-alkyladenines are converted to 3-alkylhypoxanthines by the action of nitrosyl chloride on the purine in an acetic acid–acetic anhydride mixture containing a small amount of pyridine.[29] Hydrolyses with sodium hydroxide are effective but yields are reduced.[29] Both 7-[30] and 9-alkyladenines[30, 31] react normally with nitrous acid giving the appropriate alkylhypoxanthines.

Halogen substituents of aminopurines show predictable behaviour toward mineral acid treatment. At temperatures below 60° the chlorine atom in 6-amino-2-chloro-7,8-dihydro-9-methyl-8-oxopurine remains intact during a nitrous acid conversion to the corresponding 6,8-dioxopurine.[32] More vigorous hydrolyses, with hydrochloric acid at 120°, removes both amino group and halogen illustrated by the formation of 7-methylxanthine from 6-amino-2-chloro-7-methylpurine.[30] The more reactive $C_{(6)}$-halogen atom in 2-amino-6-chloro-7,8-dihydro-8-oxopurine (1) is demonstrated by the formation of uric acid (2) on leaving in nitrous acid at room temperature for some hours.[33] A halogen at $C_{(8)}$ shows the expected reluctance to undergo nucleophilic attack; both 2-[34] and 6-amino-8-bromopurine[34, 35] give the respective 8-bromo-oxopurine with hot nitrous acid. The same reagent in the cold converts a 2-methylthioadenine to the 2-methylthiohypoxanthine analogue,[36] and a 2-aminopurine possessing a trifluoromethyl group at $C_{(8)}$ to 2,3-dihydro-2-oxo-8-trifluoromethylpurine.[37]

Treatment of *N*-aminopurines with cold nitrous acid does not give *N*-hydroxypurines but effectively replaces the amino group with hydrogen. Hypoxanthine,[38] 8-methylhypoxanthine,[39] and 8-methyl-xanthine[39] have been obtained from the corresponding 9-aminopurine by this means. Hydrolysis of the 9-aminopurine with concentrated hydrochloric acid was unsuccessful.[38]

Dilute mineral acids convert hydroxyaminopurines to the oxo analogue. Hypoxanthine was obtained from 6-hydroxyaminopurine using nitric acid[40] while xanthine results from dilute hydrochloric acid treatment of either 2,3-dihydro-6-hydroxyamino-2-oxopurine[13] or 2,6-dihydroxyam-inopurine.[41] These conversions can be effected with trifluoroacetic acid–hydrogen peroxide mixtures in place of mineral acids, as in the formation of guanine from 2-amino-6-hydroxyaminopurine.[42] Hypoxanthine

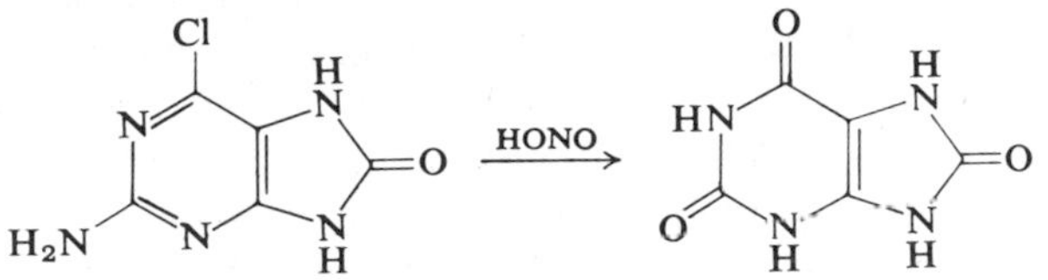

(1)   HONO →   (2)

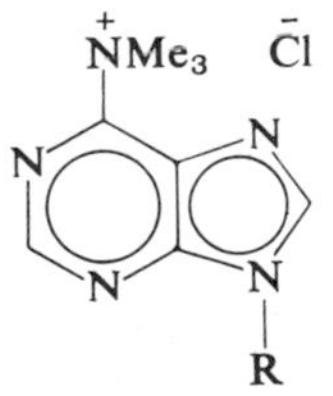

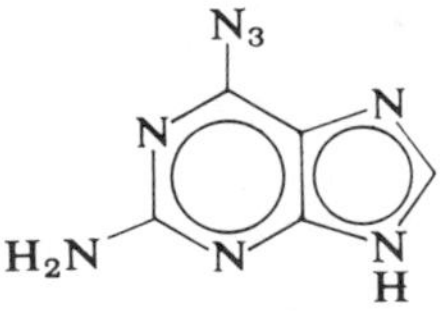

(3)   (4)

derivatives result when trimethylpurin-6-yl ammonium chlorides (3) are heated at 60° in dilute sodium hydroxide.[43] The indirect conversion of hydrazinopurines is possible in some instances by first formation and then acid hydrolysis of the corresponding azidopurine. Thus, hot concentrated hydrochloric acid transforms 2-amino-6-azidopurine (4) to guanine while the same conditions afford xanthine from 6-azido-2-chloropurine. Other 6-azidopurines do not hydrolyse in this way[44] (see Ch. VIII, Sect. 7Cb).

E. From Alkoxypurines

These transformations are subsequently discussed in Sections 8A, 8B, and 8D.

F. From Thio- and Alkylthiopurines

Various oxidising agents will bring about conversions of this kind. Gentle warming in dilute nitric acid (25–50%) rapidly produces the hypoxanthine derivative from the corresponding 6-thiopurine.[45, 46] Xanthine is similarly obtained[47] from 1,2,3,6-tetrahydro-2-oxo-6-thiopurine using either manganese dioxide or hydrogen peroxide as oxidant.[48] Hydrogen peroxide, in ammoniacal solution, converts the isomeric 6-oxo-2-thiopurines to their xanthine analogues.[49, 50] More recently an elegant oxidation procedure using chlorine in methanolic or ethanolic solution converts 6-thiopurines to the hypoxanthine analogue[51, 52] but under these conditions 2,3,7,8-tetrahydro-2,8-dithiopurine is converted to the 2-chloro-8-oxopurine analogue.[53]

After undergoing S-alkylation thiopurines become more easily converted to the corresponding oxopurines. Hydrolyses with strong hydrochloric acid (6N) is extensively used to replace methylthio groups at $C_{(2)}$,[27, 31] some hours heating being usually employed. Although 6-oxo-2-methylthiopurines can be likewise hydrolysed,[54] the presence of an amino group at $C_{(6)}$ is found to inhibit removal of the $C_{(2)}$-methylthio group by this means.[55] Hypoxanthine derivatives are derived from 6-alkylthiopurines by dilute acid hydrolysis[56] and 8-methylthiopurines are similarly hydrolysed.[4] The incorporation of hydrogen peroxide in the acid medium has also been successful[57, 58] but hydrogen peroxide[24, 59] in water or ethanol[60] has been widely used, mild conditions, including reactions at room temperature, often being sufficient. The preparation of isoguanine from 6-amino-2-methylthiopurine in this way[60] should be compared with the failure to hydrolyse the latter with acid, noted above. The facility with which 3-methyl-6-methylthiopurine is converted to 3-methylhypoxanthine at room temperature by either aqueous peroxide or 2N-sodium hydroxide is ascribed to the electron-depleted nature of the pyrimidine ring.[23] The isomeric 1-methylhypoxanthine results from chlorine oxidation of the 1-methyl-6-benzylthio analogue.[51]

The formation of carboxymethylthiopurines from thiopurines is noted elsewhere (Ch. IX, Sect. 9Ba), these derivatives affording useful intermediates in the conversion of thio- to oxopurines. Although 2-carboxymethylthiopurine itself is not hydrolysed by acid,[4] the 6-oxo- and 6,8-dioxo analogues are readily converted to xanthine and uric acid,[61] respectively. A noteworthy point is that 2-carboxymethylthioadenine shows comparable behaviour with 2-methylthioadenine in not undergoing acid hydrolysis.[12] Hypoxanthine derivatives are derived from 6-carboxymethylthiopurines[31, 56] while 8-oxopurines result from rigorous acid treatment of 8-carboxymethylthiopurines.[62, 63] If both $C_{(2)}$ and $C_{(6)}$ are oxo functions hydrolysis of the thio group at $C_{(8)}$ is only possible if $N_{(9)}$ is substituted. This is seen in the failure to hydrolyse 8-carboxymethylthioxanthine[62] and by the fact that this derivative is obtained on acid treatment of the 2,8-dicarboxymethylthio analogue.[61] In contrast the conversion to 9-alkyluric acids of the corresponding 9-alkyl-8-carboxymethylthioxanthines is readily effected.[62] Hydrolysis of hydroxyethylthio groups occurs more readily. Illustrating this are the formation of 6-oxo- and 8-oxopurines from the 6-[56] or 8-(2-hydroxyethyl)thiopurines[64] on treatment in boiling water or on heating alone.

G. By Reduction of Benzyloxy Groups

This aspect is reviewed in Section 8B of this chapter.

H. By Conversion of Other Groups

In addition to the foregoing, a variety of other groups undergo acid or alkali hydrolysis to give the corresponding oxopurine. Although these reactions are dealt with in the appropriate chapters, the more important groups that can be utilised include sulphinic and sulpho (Ch. VII, Sect. 4B), halogenosulphonyl (Ch. VII, Sect. 5B), sulphamoyl (Ch. VII, Sect. 6B), and alkylsulphonyl (Ch. VII, Sect. 7B). The most facile hydrolyses take place with these groups sited at $C_{(6)}$ but preparations of 2- and 8-oxopurines by these procedures are available. Less obvious routes to hypoxanthine include the hydrolysis of 6-cyanopurine with warm $2N$-sulphuric acid[65] and treatment of 6-phenylcarbonylpurine with a cold hydrogen peroxide–trifluoroacetic acid mixture.[66] Conversion of an 8-nitropurine to the 8-oxo analogue occurs in boiling dilute sulphuric acid.[67]

I. By Rearrangement of Purine-N-oxides

Isomerisation of purine-N-oxides to oxopurines can occur on treatment with acetic anhydride or on exposure to ultraviolet radiation. The isolation of isoguanine after irradiation of an aqueous solution of adenine-1-oxide illustrates the reaction.[68] A complete study of these rearrangements is given in Chapter XI, Section 3C.

J. By Other Routes

A specific means of forming 8-oxopurines is provided either by Hofmann reaction on derivatives of 4-amino-5-carbamoylpyrimidines or, alternatively, by the Curtius reaction on 4-amino-5-hydrazino-carbonylpyrimidines. Examples of these reactions are given in Chapter II, Section 7.

K. By Free Radical Attack

Purines react with hydroxyl radicals generated either by use of Fenton's reagent[69] or *in situ* by subjecting aqueous solutions of the purine to high-energy radiation[70] under anaerobic conditions. Initially the hypoxanthine analogue is formed if $C_{(6)}$ is unsubstituted. Succeeding attack is directed to $C_{(8)}$ but no reaction at $C_{(2)}$ has been found. Illustrative of these reactions are the formation of guanine from 2-amino-purine[69] and the 8-oxo analogues of adenine[69, 71] and hypoxanthine.[69]

L. By Enzymic Oxidations

The biological oxidation of purines presents a complex and, as yet, incomplete picture. No comprehensive review is available but a summary of some of the main features has been made.[72] Using *in vitro* techniques purine has been shown to be successively oxidised by means of xanthine oxidase, a milk enzyme, first to hypoxanthine, then to xanthine, and finally to uric acid.[73] The reaction, presumed to occur through the agency of hydroxyl ions,[74] is seen to follow the usual order of nucleophilic displacement of a $N_{(9)}$-unsubstituted purine, that is, $C_{(6)}$, $C_{(2)}$, then $C_{(8)}$. With other enzyme systems the order of oxidation can be completely reversed[75] and evidence is also available which indicates that simultaneous oxidation to the same product along two different pathways is possible with the one enzyme.[74] Substituted purines with xanthine oxidase may undergo different initial sites of attack, most 6-substituted purines, for example, unlike hypoxanthine, are first oxidised to the 8-oxo analogue.[76-78] The majority of N-methyl-ated purines seem resistant to oxidation by enzymic means, but some have been transformed with difficulty, the conversion of 3-methylhypo-xanthine to 3-methyluric acid, for instance, could only be achieved by using unrelated enzyme systems for the two oxidation stages.[79]

2. Preparation of *N*-Alkylated Oxopurines

Both direct synthesis or alkylation of an oxopurine precursor are available routes, the choice being governed largely by the particular alkylated derivative required. Methods of preparing methylated xanthines from uric acid[80] and xanthine derivatives[81] have been reviewed.

A. By the Traube Synthesis

This permits the formation of derivatives alkylated at any of the four nitrogen atoms. While purines with alkyl groups at $N_{(1)}$, $N_{(3)}$, and $N_{(9)}$ are readily accessible, the difficulties in the preparation of 4-amino-5-alkylamino-oxopyrimidine limit the value of the method for forming $N_{(7)}$-alkylated oxopurines. The variations and scope of the route are fully covered in Chapter II.

B. From Imidazoles

Although of less general utility than the previous type of synthesis it is principally useful where $N_{(7)}$- or $N_{(9)}$-alkyl derivatives are required; these follow when the appropriate $N_{(1)}$- or $N_{(3)}$-alkylimidazole derivative is cyclised. Variations in the method allow for its extension to the synthesis of some $N_{(1)}$-alkylated purines. The application to the formation of $N_{(3)}$-alkylpurines is limited to a few examples only, of which the cyclisation of caffeidine with various reagents that give rise to derivatives of 3,7-dimethyl- or 1-alkyl-3,7-dimethylxanthine is the most noteworthy.[82] Other illustrations are to be found in Chapter III.

C. By Direct Alkylation

Direct alkylation of an oxopurine may give rise to a mixture of *N*-alkylated analogues; furthermore, by changing the reaction medium, alkylation may be induced to take place at different nitrogen atoms. Modes of alkylation of mono-, di-, and trioxopurines are described later in the chapter (Sects. 6Da, b, and c).

D. By Other Methods

The rearrangement of alkoxypurines to the isomeric *N*-alkylated oxopurines is discussed later (Sect. 8D). Reduction of *N*-hydroxymethylpurines gives the corresponding *N*-methylpurines (see Sect. 6E).

3. Preparation of Extranuclear Hydroxypurines

Hydroxymethylpurines, by reason of their unusual reactivity compared with the higher hydroxyalkyl homologues, are represented by a number of examples. Although derivatives of 2- and 6-hydroxymethylpurines are few, the 2-hydroxymethyl derivative (5) followed from dilute hydrochloric acid hydrolysis of the 2-chloromethyl analogue.[83] 6-Hydroxymethylpurine, "homohypoxanthine," is formed by reduction of purine-6-aldehyde[84] with a platinum catalyst in ethanolic solution or from an acetic anhydride rearrangement of 6-methylpurine-1-oxide[85] (Ch. XI, Sect. 3C). Both 7- and 9-hydroxymethyl derivatives are formed by the action of formaldehyde on the parent purine, with theophylline, in the presence of hydrogen chloride, the 7-chloromethyl derivative first produced is converted to the hydroxymethyl analogue by means of hot ethanol.[86] Detailed treatment of N-hydroxymethylation is given later (Sect. 6E). 6-Alkylthio- but not 6-thiopurines form the 9-hydroxymethylpurines directly in sodium carbonate solution at room temperature.[87] A number of approaches to the stable 8-hydroxymethylpurines are available. The Traube synthesis using glycollic acid as cyclising reagent is well represented[88–91] and has been extended to thermal cyclisation of 4-(2-hydroxymethylamino)-5-nitrosopyrimidines[92] (Eq. 1). Reduction of the 8-aldehyde has given the respective 8-hydroxymethyl derivative of theobromine and caffeine[93, 94]; related derivatives of this type were formed on acid hydrolysis of 8-chloromethylpurines.[95, 96] An unusual

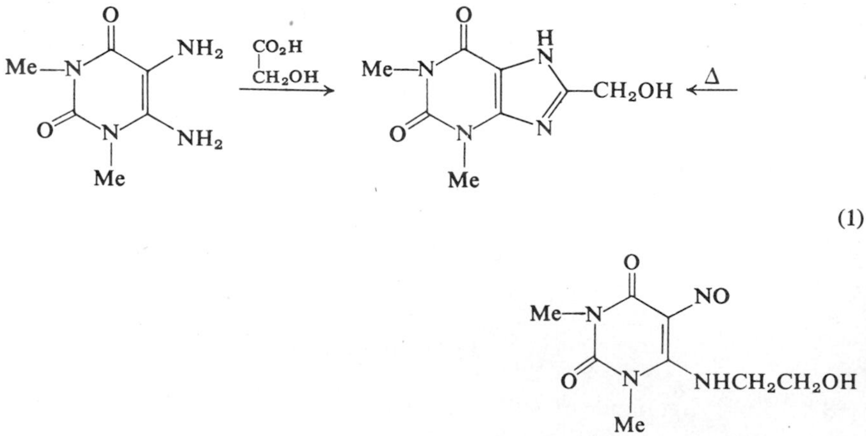

(1)

route to 8-hydroxymethylcaffeine consists of heating caffeine with paraformaldehyde at 170° for some hours.[89]

Homologous hydroxyalkylpurines result from the Traube synthesis, as in the formation of 9-(2-hydroxyethyl)adenine from formamide cyclisation of 4,5-diamino-6-(2-hydroxyethylamino)pyrimidine.[97, 98] Direct alkylation permits the location of hydroxyalkyl groups at $N_{(1)}$, $N_{(7)}$, and $N_{(9)}$ according to the purine used. Both $N_{(7)}$- and $N_{(9)}$-hydroxyethyl derivatives are usually formed with 2-bromo- or 2-chloroethanol,[99, 100] but theophylline gives only the expected 7-hydroxyalkyl analogue.[101] Replacement of the halogenated alcohol with either bromo-[102] or chlorohydrins[103] or an epoxide[103–105] produces the same derivatives. These reagents with theobromine form the corresponding 1-hydroxyalkyl derivatives.[106]

Modifications to exocyclic groups in some cases provide a useful means of preparing hydroxyalkyl analogues. Suitable examples are the reduction of 9-(2-ethoxycarbonylethyl)adenine[107] (6) to the 9-(3-hydroxypropyl) analogue with lithium aluminium hydride[107] or the formation of 7-(2-hydroxypropyl)theophylline by nickel reduction of the 7-acetonyl precursor (7).[108] The isomeric 1-(2-hydroxypropyl)-theobromine arose from the action of sulphuric acid at 80° on 1-allyl-theobromine (8).[109]

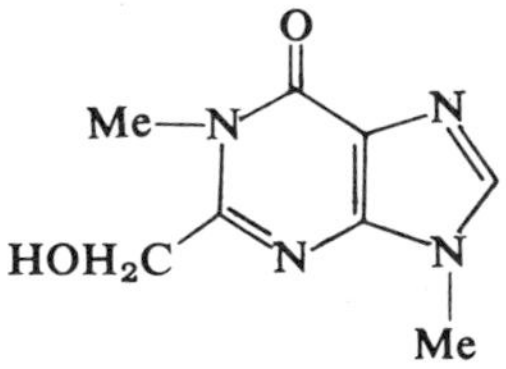

(5) 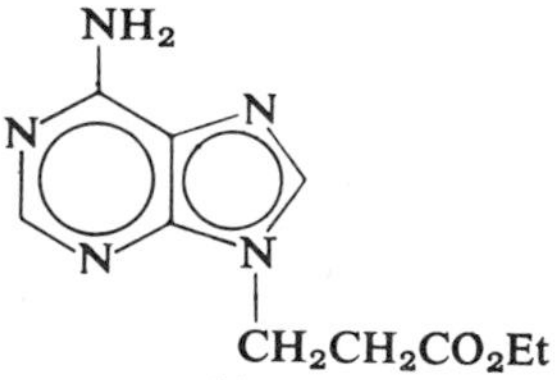　　　　　　(6)

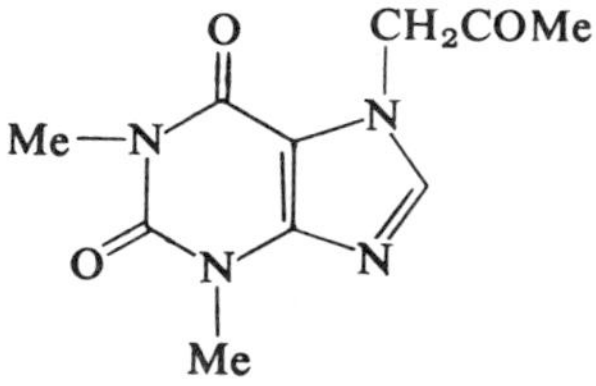

(7) 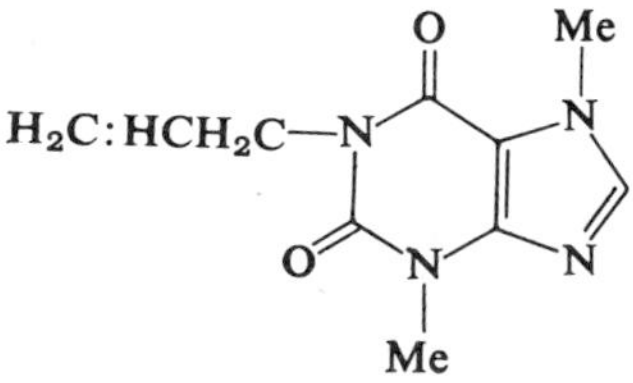　　　　　　(8)

Purines, in which the alkanol moiety is separated from the purine by a heteroatom such as sulphur or nitrogen, are usually formed through

interaction between the thiopurine and alkanol halide and the halogeno-
or methylthiopurine with an amine alcohol. Reactions of these types are
specialised forms of general nucleophilic displacement, details of which
are given in Chapters V and VII.

4. Preparation of Alkoxypurines

Nearly all alkoxypurines have resulted from nucleophilic displacement of a halogen atom or some suitable group. Direct alkylation of
oxopurines, as a general rule, is not a practical route as *N*- rather than
O-alkylation is favoured. However, exceptions to this rule are to be
found.

A. By the Traube Synthesis

The possibility of displacement of alkoxy groups under the conditions
used for cyclisation inhibits the general use of this route. Careful
choice of the cyclising reagents has enabled the synthesis of 2-methoxy-
(acetic formic anhydride/50°,1 h),[4] 2-ethoxy-6-methyl- (triethyl orthoformate-acetic anhydride/reflux,1 h),[110] and 2,6-dimethoxypurine (form-
amide/180°,20 min)[111] to be carried out from the appropriate
4,5-diaminopyrimidine. Some 6-alkoxy-8-oxopurines result from ring
closure of trimethylsilylated pyrimidines with oxalyl chloride.[112]

B. By Displacement of Halogen Atom or Other Substituent

A comprehensive account of the formation of alkoxypurines from the
corresponding halogenopurines has been given elsewhere (Ch. V, Sect.
5E).
Methylsulphonyl group displacements afford a few examples. In the
case of 6-amino-9-methyl-2-methylsulphonylpurine, heating under

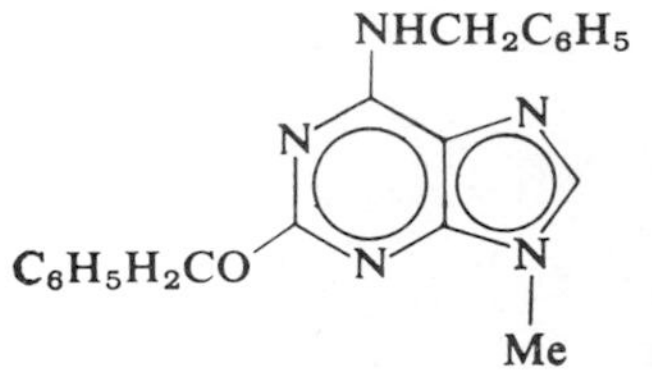

(9) 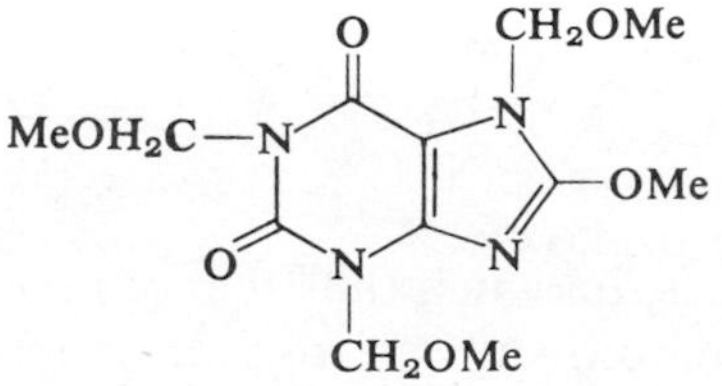   (10)

reflux for 5 hours with equimolar amounts of the sodium alkoxide is a route to the 2-ethoxy and 2-benzyloxy analogues. With an excess of sodium benzyloxide further benzylation involving the amino group to give 9 is observed.[55]

The conversion of 8-methylsulphonylpurine to the 8-methoxy analogue required heating at 180° for some hours.[113] Extranuclear methoxypurines are exemplified by the tetramethoxyxanthine (10) from the 8-chloro-1,3,7-tris-trichloromethyl derivative[114] and 6-trimethoxy-methyl- from 6-trichloromethylpurine.[115]

C. By *O*-Alkylation

Except for certain uric acid derivatives direct *O*-alkylation of free oxopurines seems to be unknown.* The best examples of this procedure

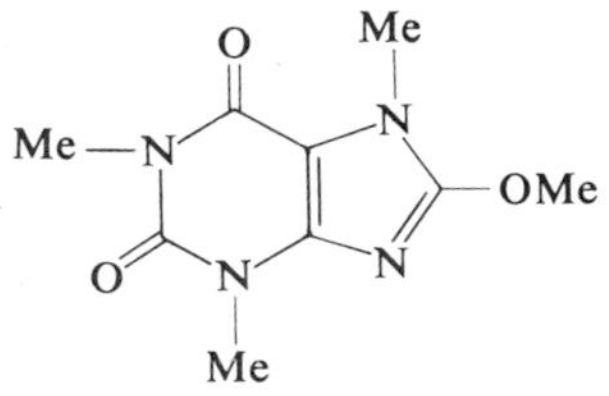

(11) 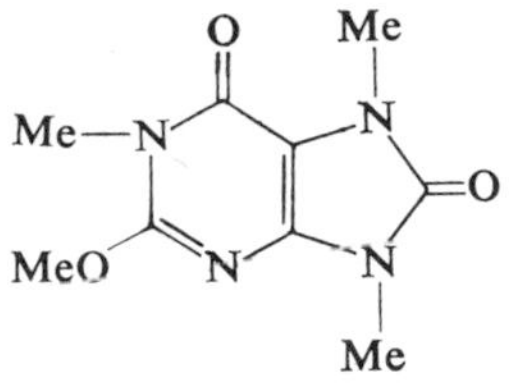   (12)

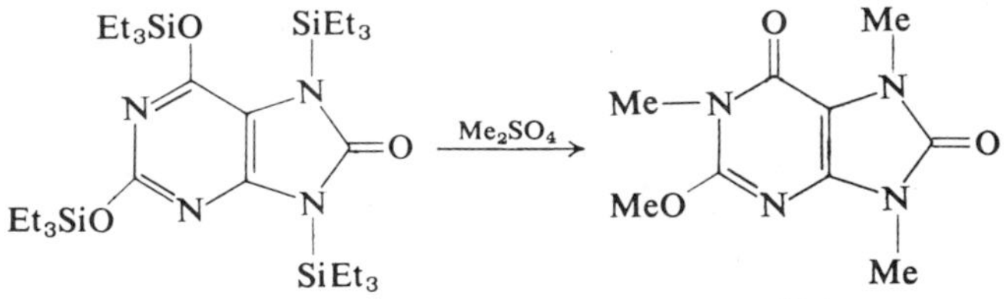

(13) $\xrightarrow{\text{Me}_2\text{SO}_4}$ (14)

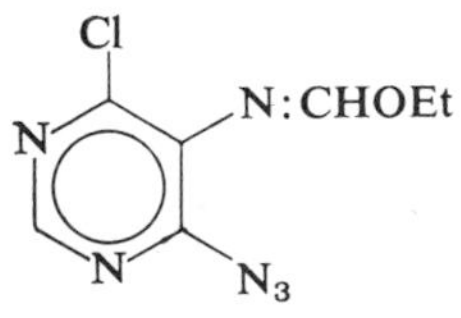

(15)

---

* A claim to have converted 3-methylhypoxanthine to the 6-methoxy analogue,[116] using methyl iodide in dimethyl formamide, has been shown to be erroneous,[117] the product appearing to be 1,3-dimethylhypoxanthinium iodide.[118]

The nucleoside derivatives, however, are exceptions as isolation of the $C_{(6)}$-methoxy analogues of inosine[119,120] and guanosine[121] is reported following diazomethane treatment of the appropriate nucleoside.

occur with *N*-alkylated derivatives of uric acid. Diazomethane in ethereal solution at room temperature converts 1,3,7-trimethyluric acid to the 8-methoxy analogue (**11**) and 1,7,9-trimethyluric acid to the isomeric 2-methoxy derivative (**12**).[122] Concomitant *N*- and *O*-alkylation occurs with mono- and di-*N*-alkylated uric acids under the above treatment affording the trialkyl homologues of **11** and **12**.[123] Diazoethane gives the corresponding 2- and 8-ethoxy derivatives.[122, 123] Comparable *O*-alkylations have been effected at $C_{(8)}$ by using the silver salt of 1,3,7-trimethyluric acid with ethyl iodide at 100°, which gives 8-ethoxycaffeine in good yield,[124] and at $C_{(2)}$ on treating 1,2,3,6,7,8-hexahydro-8-oxo-7,9-bis-triethylsilyl-2,6-bis-triethylsilyloxypurine (**13**) with dimethyl sulphate in toluene, which gave the 2-methoxypurine (**14**) as product.[112]

D. By Other Routes

Two 8-alkoxypurines have resulted from novel preparations. Treatment of 5-amino-4-azido-6-chloropyrimidine with diethoxyethylacetate at room temperature for some time gives the azine derivative (**15**), which in anisole, under reflux, is rapidly converted to 6-chloro-8-ethoxypurine.[125] Thermal isomerisation of an 8-isopropyl-7-methoxypurine to the more stable 8-methoxy-7-isopropylpurine[126] is noted in Chapter XI, Section 4B.

5. Properties of Oxo- and Alkoxypurines

The very high degree of intermolecular hydrogen bonding exhibited by all simple oxopurines shows itself in the absence of true melting points, the recorded figures being essentially the temperatures at which decomposition is observed, and in the very low order of solubility of the derivatives in water. For this reason these derivatives are best obtained in a pure form by repeated precipitation from alkaline solution with acid. *N*-Methylation substantially reduces the possibility of hydrogen bonding, as seen in the more soluble nature of theophylline (1:120) and caffeine (1:46) compared with the parent xanthine (1:15,000). The rather high figure (1:2000) for theobromine may be an indication that $N_{(1)}$, when unsubstituted, is one of the more important bonding sites.

That oxopurines show little evidence for existing to any degree in the hydroxy tautomer is well substantiated,[113, 127] although this form must be an intermediate specie in some reactions. The ability to form anions generally confers stability toward nucleophilic attack by hydroxyl ions. Surprisingly, in oxopurines this only holds good for the mono-oxopurines. The presence of two or more oxo groups leads to some instability toward alkali.[4] With acids the converse is the case, with mono-oxopurines being the most readily attacked.[4] An extreme example is 2,3-dihydro-2-oxopurine which is converted to 4,5-diamino-2,3-dihydro-2-oxopyrimidine by N-sulphuric acid at room temperature.[128] It would seem from inspection of the melting points of the methyl ethers of the mono-oxopurines (2-OMe, m.p. 206°; 6-OMe, m.p. 195°; 8-OMe, m.p. 154°) that the above considerations do not apply but the absence of a melting point below 300° for either 2,6- or 2,8-dimethoxypurine suggests that a high lattice energy is inherent and therefore some degree of hydrogen interbonding exists in these derivatives.

6. Reactions of Oxopurines

The two most useful reactions of the oxo function are halogenation and thiation as, in both cases, the oxo function is replaced by an atom or group capable of ready substitution by nucleophiles or of being removed reductively.

A. Oxo Group to Halogen

An extensive account of this important reaction is given in Chapter V, Section 1B.

B. Oxo to Thio Group

Thiation of oxopurines usually follows the normal order of nucleophilic substitution, an oxo group at $C_{(6)}$ being the most easily replaced in all cases. With simple purines thiation of groups at $C_{(2)}$ and $C_{(8)}$ does not occur[129] unless other strong electron-releasing groups are present. The most widely used procedure entails heating the purine, under reflux, in pyridine containing phosphorus pentasulphide for some hours. In early studies tetralin was favoured as solvent but others, including kerosine and β-picoline, have also been tried. Thiation in the absence of solvent has been accomplished in some cases.

Hypoxanthine was originally converted to 1,6-dihydro-6-thiopurine ("6-mercaptopurine") in 40% yield, on heating (8 h) with phosphorus pentasulphide in tetralin,[130] but improved yields (73%), on a kilogram scale, are achieved with shorter heating times (3 h) in the lower boiling pyridine.[131] Hypoxanthine derivatives similarly converted to the corresponding 6-thiopurine are (reflux times in pyridine) the 1-methyl (3 h),[31] 2-methyl (tetralin, 190°, 4 h),[132] 3-methyl (2.5–4 h),[31, 133] 7-methyl (3 h),[134] 8-methyl (4 h),[135] or (tetralin, 210°, 3 h),[136] 9-methyl (2 h),[137] and 1,9-dimethyl (β-picoline, 1.5 h)[46] homologues. Similarly obtained are the 6-thio derivatives of 2-phenyl-,[138] 8-phenyl-,[139, 140] 1-benzyl-,[52] 3-benzyl,[52] and 1,9-dibenzylpurine,[141] in the latter example an excess of phosphorus pentasulphide was employed as solvent, a low yield (10%) being obtained.

The resistance to thiation found in 2- and 8-oxopurines is reflected in similar behaviour with the dioxopurines. Xanthine[47] and derivatives,[31, 133, 134, 142] and the isomeric 6,8-dioxopurine[143, 144] and its derivatives[138] give only the respective 2- or 8-oxo-6-thiotetrahydropurines. The exception is the case of 1,6,7,8-tetrahydro-2-methyl-6,8-dioxopurine which is reported to have been converted to the 6,8-dithio analogue under these conditions.[129] In view of the general findings the inert character of 2,8-dioxopurines toward this reaction is not unexpected.[144] This is further exemplified by the isolation of only the 6-thio derivative (16) on prolonged heating (24 h) of uric acid with phosphorus pentasulphide in pyridine.[145]

(16)

Routes to 2,6- and 6,8-dithiopurines are possible by the above procedures if the starting materials are the appropriate 2- or 8-thio-6-oxopurines. Suitable examples of the former are 1,2,3,6-tetrahydro-2,6-dithiopurine itself, from 2-thiohypoxanthine,[47] and the 1-methyl,[31] 3-methyl,[31, 133] and other homologues.[146] Apart from the parent derivative[136] other 6,8-dithiopurine derivatives so obtained include the 2-phenylpurine.[138] Replacing either the $C_{(2)}$- or $C_{(8)}$-oxo function of uric acid by sulphur does not affect the reactivity of the other oxygen atoms toward replacement by sulphur, in either case the derivative undergoes replacement only at $C_{(6)}$ as in uric acid itself. Thus, 2,6-dioxo-8-thio- is converted to 2-oxo-6,8-dithiohexahydropurine,[129] the 8-oxo-

2,6-dithio isomer is derived from 6,8-dioxo-2-thiohexahydropurine,[147] while the 2,6,8-trithio analogue of uric acid results from 6-oxo-2,8-dithio-1,2,3,6,7,8-hexahydropurine.[147]

The presence of amino groups does not appear to inhibit replacement of an oxygen at $C_{(6)}$ but longer heating times seem usual. The original conversion of guanine to 2-amino-1,6-dihydro-6-thiopurine (thioguanine) employed pentasulphide in tetralin at 200° for 8 hours,[148] but the solvent was later changed to pyridine and a longer reflux period (18 h) applied.[149] Substantial heating times are likewise adopted with the 1-methyl (6 h),[150] 3-methyl (16 h),[27] 7-methyl (10 h),[134] 8-methyl (8 h),[149] 9-methyl (8 h),[142] and other 9-alkyl and -aryl (4–24 h)[151] homologues. With 2-substituted-amino-1,6-dihydro-6-oxopurines reduced heating times (3 h) suffice for conversion to the 6-thio derivatives.[152] A comparison can also be made between the heating times required for preparing 8-amino-1,6-dihydro-6-thiopurine (10 h)[143] and 1,6-dihydro-8-methylamino-6-thiopurine (3.5 h).[90] Although 8-thioguanine does not react the 8-methylthio analogue, through concurrent thiation and S-demethylation, is converted to 2-amino-1,6,7,8-tetrahydro-6,8-dithiopurine.[136] No loss of methylthio groups occurred in the formation of 1,6-dihydro-2,8-dimethylthio-6-thiopurine from the corresponding 6-oxopurine.[147]

Phosphorus pentasulphide in pyridine, or other solvent, will thiate some oxo groups fixed in the doubly bound form. Both theophylline (pyridine, 8 h)[153] or (decalin, 1.5 h)[154] and theobromine (6 h)[153, 155]

Reagent P₂S₃/kerosine

afford only the appropriate 2-oxo-6-thio derivative. Caffeine does not react in pyridine, even on prolonged heating, but gives the 6-thio analogue with phosphorus trisulphide in kerosine at 150°.[156] At higher temperatures both $C_{(2)}$- and $C_{(6)}$-groups are replaced in caffeine (210°, 1 h)[156] (Eq. 2) and theobromine (180°, 1 hr).[157] The 2,6-dithio analogue of theophylline is obtained with phosphorus pentasulphide in the absence of solvent at 275° for a few minutes.[154] Much milder conditions (P₂S₅, pyridine, 2 h) are reported to replace both oxo groups in 3-methylxanthine derivatives.[117]

Although uric acid[158] and its *N*-alkyl homologues[159] are converted to the 8-thio analogues on prolonged heating with carbon disulphide, these are not true examples of 8-oxo group thiation but rather special cases of Traube syntheses involving pyrimidine intermediates (Ch. II, Sect. 1R).

C. Replacement of Oxo Group by Hydrogen

Ways to accomplish this replacement are described in other chapters, the reduction products depending upon whether the oxopurine is considered to be reacting as the "oxo" or "hydroxy" tautomer. Direct reduction, electrochemically, of oxopurines gives dihydropurines but the reaction is successful only with 6-oxopurine derivatives (Ch. XII, Sect. 1Aa). The "hydroxy" forms are replaced indirectly, the usual means being by conversion to a halogen atom, usually chlorine, which is then removed by hydrogenation over a palladium catalyst. Alternatively the thio analogue is prepared and then reduced by heating with Raney nickel in solution. Both procedures preserve the aromatic character of the purine ring system, but the latter reaction is noteworthy in that with some fixed oxo forms thiation to a doubly bound thio analogue is possible (see Sect. 6B), which, with Raney nickel, is converted to a dihydro derivative (see Ch. VII, Sect. 1Ca).

Special cases of group removal at $C_{(8)}$ are found with uric acid derivatives which give the corresponding xanthines when heated with cyclising reagents of the type used in Traube syntheses. Xanthine, for example, results from treating uric acid with formic acid,[160] oxalic acid in glycerol,[161] or formamide.[20] The last reagent also converts 1,3-dimethyluric acid to theophylline.[20] Removal of the 2-oxo group, giving hypoxanthine in unspecified yield, is claimed to occur when a solution of xanthine in sodium hydroxide is heated with chloroform at 70° for 2 hours.[161] This sole example of oxygen elimination at $C_{(2)}$ by a ring-opening reaction would profit from further investigation.

D. *N*-Alkylation of Oxopurines

With the notable exception of diazoalkanes, alkylating agents give only *N*-alkylated purines, even under alkaline conditions which would be expected to favour *O*-alkylation. This generalisation, however, cannot be applied to uric acid derivatives for which reference to Section 4C should be made.

The pattern of alkylation varies according to the number of oxo groups present and, furthermore, in the same purine, sites of alkylation differ depending on whether alkylation is carried out in a basic, acid or neutral medium. Not included in this section is N-alkylation by re-arrangement of alkoxypurines, details of which are noted in Section 8D.

a. *Mono-Oxopurines*

The few studies that have been made indicate that in the majority of cases more than one nitrogen atom may be alkylated during the reaction. With 2,3-dihydro-2-oxopurine the location of the most basic nitrogen is uncertain but it would not be unreasonable to presume that $N_{(3)}$, as in xanthine, is the most likely site. The formation of the 3,7-dimethyl derivative from 2,3-dihydro-7-methyl-2-oxopurine with methyl iodide in alkaline solution at 80° lends support to this assumption.[162] Hypo-xanthine, under similar conditions, affords the 1,7-dimethyl homo-logue[35, 163] which also obtains from methyl iodide treatment of 7-methylhypoxanthine (Eq. 3).[30] Using isopentyl iodide unspecified

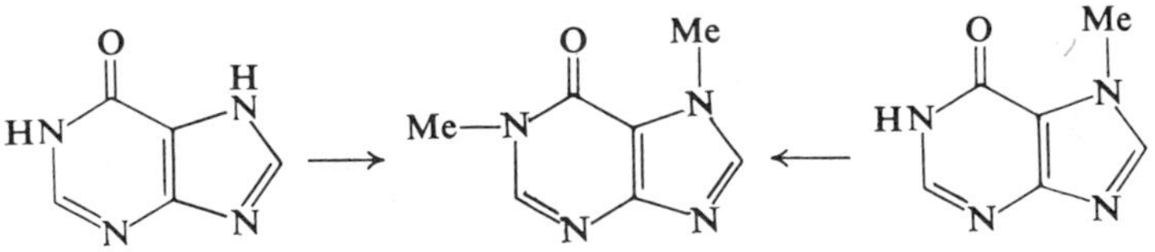

$$(3)$$

mono- and dialkylated hypoxanthines have been claimed.[164] With 9-methylhypoxanthine conversion to 1,9-dimethylhypoxanthine occurs rapidly at room temperature with alkaline dimethyl sulphate.[31] A comprehensive alkylation study of hypoxanthine made with benzyl halides confirmed the earlier methylation results but revealed that com-plex product mixtures are present. The reactions were carried out in nonaqueous media, usually dimethylformamide or dimethylacetamide, in the presence of anhydrous potassium carbonate. Under these condi-tions (95°, 22 h) 1-benzylhypoxanthine gives, in addition to the expected 1,7-dibenzyl analogue (42%), half as much (21%) of the 1,9-dibenzyl isomer.[118] Benzylation of 3-benzylhypoxanthine (105°, 16 h) gives a good yield (74%) of the 3,7-dibenzyl analogue.[52] Benzyl bromide converts 7-benzylhypoxanthine into 1,7-dibenzylhypoxanthine in fair yield (57%).[118] Some of the 3,7-dibenzyl isomer is also formed in the above reaction.[118] The 1,9-dibenzyl analogue is similarly derived (95°, 2.5 h) from 9-benzylhypoxanthine.[52]

In nonbasic conditions alkylation follows a different pathway. Hypoxanthine itself with either dimethyl sulphate or methyl p-toluene-
sulphonate in dimethylacetamide (125°),[23] or with methyl p-toluene
sulphonate alone (250°),[165] affords the appropriate 7,9-dimethyl
hypoxanthinium salt which methanolic ammonia converts to the betaine
(17). Both 7- and 9-methylhypoxanthine are transformed to 17 using
methyl p-toluenesulphonate as alkylating agent.[23] The 7,9-dibenzyl
analogue results from heating hypoxanthine in dimethylformamide
(100°, 2 h) with benzyl bromide or the 7- or 9-benzyl purine with the
same reagent in acetonitrile for prolonged periods (16–18 h) under
reflux.[118] In the complex mixture that arises on heating 1-benzylhypo-
xanthine with benzyl chloride in acetonitrile (48 h), has been identified
1,3-dibenzyl- (8%), 1,7-dibenzyl- (17%), and 1,9-dibenzylhypoxanthines
(8%) in addition to 1,7,9-tribenzylhypoxanthinium bromide (18) as a

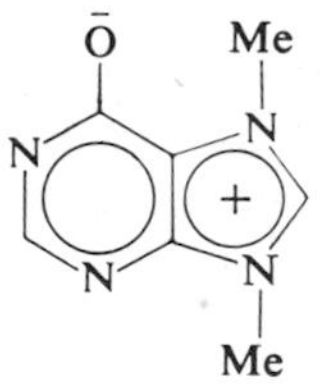

(17) 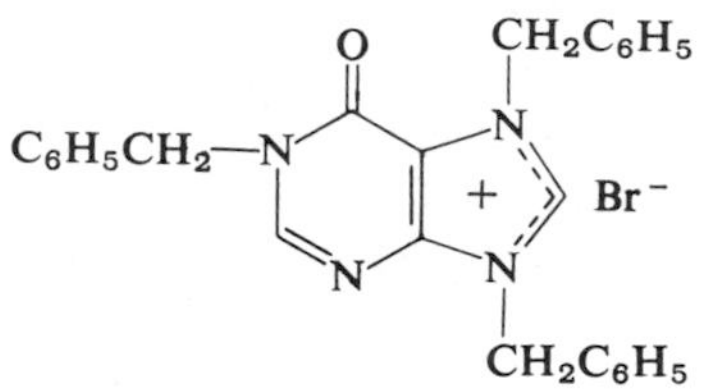   (18)

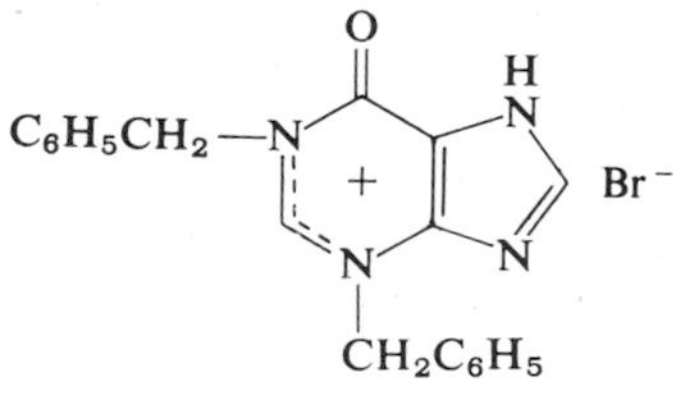

(19) 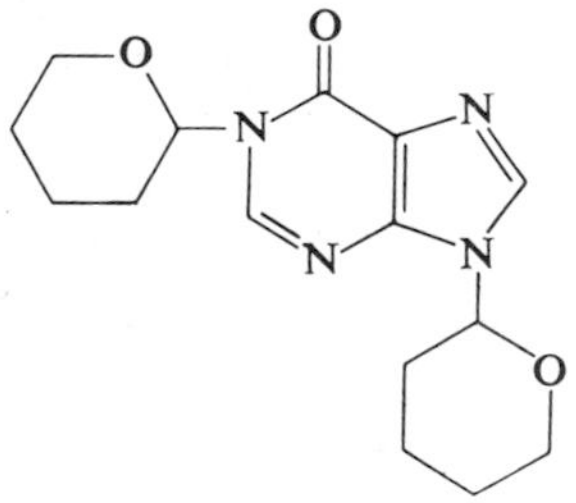   (20)

major (17%) product.[118] The corresponding reaction with 3-benzyl-
hypoxanthine is complicated by benzyl group migration occurring at
elevated temperatures. With benzyl bromide in acetonitrile at 80° (16 h),
1,3-dibenzylhypoxanthinium bromide (19) is formed but, on raising the
reaction temperature to 110°, the two main products are 1,7- and 1,9-
dibenzylhypoxanthine (this result is further discussed in Sect. 7B). Both
1,7- and 1,9-dibenzylhypoxanthine give the 1,7,9-tribenzyl derivative
(18) on further benzylation.[118]

Under more acidic conditions alkylation with 2,3-dihydropyran gives
mainly 1,9-bis-tetrahydropyran-2-ylhypoxanthine (20) but also some

1,7-dialkyl isomer as well.[166] This reaction, which succeeds in dimethyl sulphoxide at 60° (15 h) with a hydrogen chloride catalyst, fails[167] in ethyl acetate containing p-toluene sulphonic acid. In these reactions the alkylating agent is assumed to be the tetrahydropyranyl cation.[168]

Unsubstituted 7,8-dihydro-8-oxopurine has not been alkylated but the course of the reaction is seen in the formation of the 7,9-dimethyl analogue on treating 2,6-dichloro-7,8-dihydro-8-oxopurine with di-methyl sulphate in cold alkaline solution.[169]

b. Di-Oxopurines

The alkylation patterns of xanthine are well defined, under alkaline conditions replacement of protons by alkyl groups is in order of decreasing acidity—that is $N_{(3)}$, $N_{(7)}$, then $N_{(1)}$—whereas in neutral or mildly acid media both imidazole ring nitrogens are involved, giving quaternised or betaine forms of the purine.

Like hypoxanthine under alkaline conditions xanthine gives di- rather than monomethylated derivatives. Theobromine results by dimethylation at $N_{(3)}$ and $N_{(7)}$, by dimethyl sulphate, in potassium hydroxide at 60°.[20, 170] These conditions also convert 3-methylxanthine to the same product.[20, 171]. Similar preparations of theobromine derivatives are known.[172–174] Theobromine is readily alkylated further to caffeine[175–177]; using dimethyl sulphate a near theoretical yield is claimed.[175] Both mono- and di-N-methylxanthines can be transformed to caffeine; with the former, the intermediate dimethylated xanthine is obtainable if the methylating agent is restricted to one equivalent. Thus, 1-methyl-xanthine[177–180] gives initially the 1,3-dimethyl derivative (theophylline), which is further alkylated to caffeine.[175, 176, 178, 180] Likewise the 3-methyl[20, 172] and 7-methyl[181] derivatives pass through theobromine. The 1,7-dimethyl analogue (paraxanthine) also gives caffeine[182] and various 3-alkyl homologues[183] in this way. Direct conversion of a xanthine to a caffeine derivative occurs using an excess of methylating agent[20, 174, 176, 184]; commercially, methyl acetate in acetone has been employed.[185] A summary of xanthine alkylation is given in Scheme A. Treatment of the triethylsilylated xanthine (21) with methyl iodide at room temperature gives 7-methylxanthine or theobromine on warming at 60°.[106]

Because of the pharmacological properties of N-alkylxanthines, many derivatives of this type, including N-ethyl[176, 177, 186] and -benzyl[163, 187] analogues, are frequently encountered. The most numerous examples are derived from alkylation of theophylline and theobromine as detailed in Tables 5 and 6.

TABLE 5.   Alkylation of Theophylline.

Alkylating Agent	Purine Salt/Conditions	$N_{(7)}$-Alkyl group	Reference
EtI	Na or Ag/EtOH/water bath	Et	224
PrI	K/EtOH/water bath	Pr	223
$CH_2:CHCH_2Br$	Na/100°/closed vessel	$CH_2CH:CH_2$	188
$CH:CCH_2Br$	Na/EtOH, 120°	$CH_2C:CH$	189
$C_6H_5CH_2Cl$	K/100°/closed vessel	$CH_2C_6H_5$	224
$HO(CH_2)_2Cl$	Na/H$_2$O, 100°	$(CH_2)_2OH$	190
$HO(CH_2)_3Cl$	Na/H$_2$O, 100°	$(CH_2)_3OH$	101, 191
$HO(CH_2)_5I$	Ag/DMF, 135°	$(CH_2)_5OH$	192
$MeCHOHCH_2Cl$	Na/H$_2$O, 100°	$CH_2CHOHMe$	101, 108
$HOCH_2CHOHCH_2Cl$	Na/H$_2$O, 110°	$CH_2CHOHCH_2OH$	193
$\overset{O}{\overset{/\backslash}{CH_2-CHCH_2Cl}}$	K/125°/closed vessel	$CH_2CHOHCH_2OH$	194, 195
HCHO/HCl	Base/H$_2$O, 20°	CH_2Cl	86
$Br(CH_2)_2Br^a$	Na/Pr^{i}OH, reflux	$(CH_2)_2Br$	196
$Cl(CH_2)_4I^a$	Ag/xylene, reflux	$(CH_2)_4Cl$	192
$Cl(CH_2)_5I^a$	Ag/xylene, reflux	$(CH_2)_5Cl$	192
$\overset{NH}{\overset{/\backslash}{CH_2-CH_2}}$	DMF/60°	$(CH_2)_2NH_2$	197
$Me_2N(CH_2)_2Cl$	Na/Pr^{i}OH, reflux	$(CH_2)_2NMe_2$	198
$Et_2N(CH_2)_2Cl$	Na/Pr^{i}OH, reflux	$(CH_2)_2NEt_2$	198, 199
$Me_2N(CH_2)_3Cl$	Na/Toluene, reflux	$(CH_2)_3NMe_2$	200
EtO_2CCl	Na/120°/closed vessel	CO_2Et	191
HO_2CCH_2Cl	Na/H$_2$O, 100°	CH_2CO_2H	201, 202
$EtO_2C(CH_2)_2Br$	Na/H$_2$O, 100°	$(CH_2)_2CO_2Et$	191
H_2NOCCH_2Cl	Na/H$_2$O, 90°	CH_2CONH_2	203
Et_2NOCCH_2Cl	Na/H$_2$O, 90°	CH_2CONEt_2	203, 204
$C_6NHOCCH_2Cl$	Na/MeOH, reflux	$CH_2CONHC_6H_5$	191, 203
$MeOCCH_2Br(Cl)$	Na/H$_2$O or EtOH, reflux	CH_2COMe	198, 205–207
$MeOCCHMeCl(Br)$	Na/MeOH, reflux	$CHMeCOMe$	208, 209
$C_6H_5OCCH_2Br$	Na/EtOH, reflux	$CH_2COC_6H_5$	198, 206
$C_6H_5OC(CH_2)_2Br$	Na/EtOH, reflux	$(CH_2)_2COC_6H_5$	226
$\overset{SO_3}{\overset{/\backslash}{CH_2-CHCH_2Cl^b}}$	K/H$_2$O, 170°	$\overset{CH_2-O}{CHSO}\atop{CH_2-O}$	210
$(HO)_2PO(CH_2)_3I$	Ag/DMF, 140°	$(CH_2)_3PO(OH)_2$	192
Me_2SO	Base/P$_2$O$_5$, 75°	CH_2SMe	211

^{a} With difunctional alkylating agents and an excess of theophylline, $N_{(7)}$-linked bis-theophyllinyl alkanes are formed.[190, 192, 196, 207, 215–217]

^{b} Rearrangement from a pentacyclic to a hexacyclic sulphite moiety occurs during alkylation.

TABLE 6. Alkylation of Theobromine.

Alkylating Agent	Purine Salt/Conditions	$N_{(1)}$-Alkyl Group	Reference
PrI	K/EtOH/water bath	Pr	223
$C_6H_{13}Cl$	Na/Pr^{i}OH, 100°/closed vessel	C_6H_{13}	212
EtO_2CCl	K/Toluene, reflux	CO_2Et	213
Me_2NOCCH_2Cl	Na/EtOH, reflux	CH_2CONMe_2	214
Et_2NOCCH_2Cl	Na/EtOH, reflux	CH_2CONEt_2	204
$C_6H_5NHOCCH_2Cl$	Na/EtOH, reflux	$CH_2CONHC_6H_5$	214
$C_6H_5CH_2NHOCCH_2Cl$	Na/EtOH, reflux	$CH_2CONHCH_2C_6H_5$	214

Although alkylations of the above types are usually carried out with the sodium or potassium salt of the xanthine in aqueous solution, the early literature is replete with examples employing the silver or lead salt. In these reactions no solvent was normally used although occasional interactions in methanol or xylene are encountered. Generally the products are the same as those obtained using aqueous conditions but slight differences are noted. Both lead[17, 124, 218] and silver salts[219] of

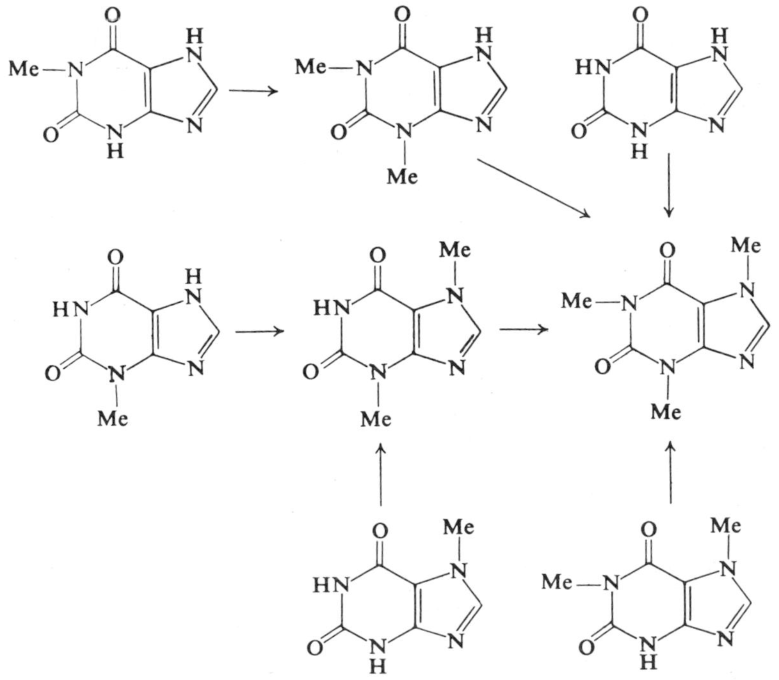

SCHEME A

xanthine give mainly theobromine on methylation whereas caffeine is obtained using the aqueous method. The silver salts of 3-methyl-xanthine[220] and theobromine[16, 221] give caffeine on heating with methyl iodide. This reaction has also produced 7- or 1-alkyl homologues of caffeine on alkylation of the silver salts of theophylline[222–224] and theobromine,[221, 223] respectively.

Alkylation in the dry state is effected by heating to 200° the salt formed by the purine with a tetra-alkyl ammonium hydroxide. Caffeine is obtained in excellent yield from theophylline or theobromine using tetramethylammonium hydroxide.[225] Formation of 7-(2-benzoylethyl)-theophylline from theophylline by heating with NN-dimethylamino-propiophenone hydrochloride may take place by a similar mechanism.[226] Diazomethane has been little used in the xanthine series. Xanthine itself is converted to the 1,3,7-trimethyl analogue.[184] Methylation at $N_{(3)}$ occurs in 1,9-dimethyl-[227] and derivatives of 1,7-dimethylxanthine[49, 50] with this reagent. The perchlorate of 3,7,9-trimethylxanthine gives likewise the 1,3,7,9-tetramethyl derivative.[228]

Addition reactions of the Michael type occur at $N_{(1)}$ in theobromine and $N_{(7)}$ in theophylline, with acrylonitrile, ethyl acrylate, and 2-vinyl-pyridine giving the appropriate 1- or 7-(2-cyanoethyl), -(2-ethoxy-carbonylethyl), or -(2-pyridinylethyl) derivative.[229] Condensations are carried out in dioxan, ethyl acetate, or other neutral solvent incorporating a basic catalyst, benzyltrimethylammonium hydroxide (Triton B)[229, 230] or sodium methoxide in pyridine[231] being usually employed.

Neutral or acid media favour alkylation at $N_{(7)}$ or $N_{(9)}$, interaction with the alkylating agent occurring either in the absence or the presence of a solvent such as dimethylacetamide or dimethyl sulphoxide. An illustration is the formation of 7,9-dimethylxanthinium betaine (22) by heating xanthine with methyl iodide (150°, 5 h),[232]* dimethyl sulphate in dimethylacetamide (140°, 2 h),[23] or methyl p-toluenesulphonate (170°, 1.5 h).[228] The best yields seem to result from reactions effected in the solvent. Both 7-[23] and 9-methylxanthine[23, 227] are converted to 22 under any of the above conditions. It is interesting to note that the product obtained from methylation of 9-methylxanthine, under apparently alkaline conditions, originally claimed to be 3,9-dimethyl-xanthine[234] but later amended to the more improbable 8,9-dimethyl isomer,[62] has now been shown to be the betaine[23, 235] (22). Formation of this derivative is found to be due to the reaction medium becoming acid during the alkylation procedure.[23]

---

* Preparation of 7,9-dimethylxanthinium betaine by methyl iodide treatment of xanthine was actually described some 26 years before this but the product was incorrectly designated as 3,9-dimethylxanthine.[233]

Further alkylation of 7,9-dimethylxanthine affords 1,7,9-trimethyl-xanthine[236] which also arises from reaction with methyl p-toluene-sulphonate of 1-methyl-[228] or 1,9-dimethylxanthine.[228, 234] Similarly the 3,7,9-trimethyl analogue is obtained from either 3-methyl-[228] or 3,7-dimethylxanthine.[228, 234] Other studies of this type have been made.[234] Caffeine is quaternised to the 1,3,7,9-tetramethyl metho-sulphate on heating in nitrobenzene (12 h) with dimethyl sulphate on a

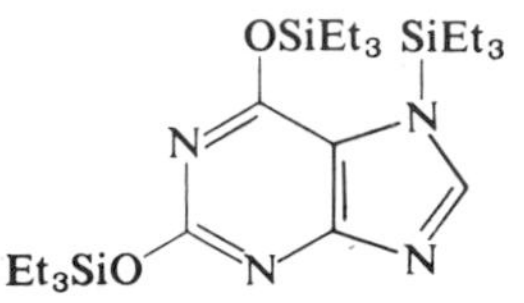

(21) 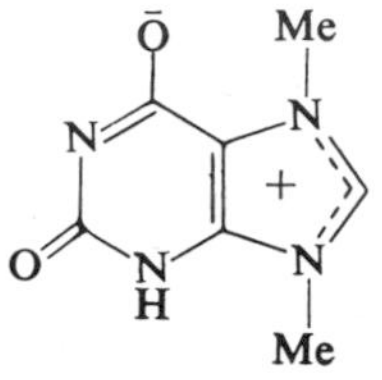   (22)

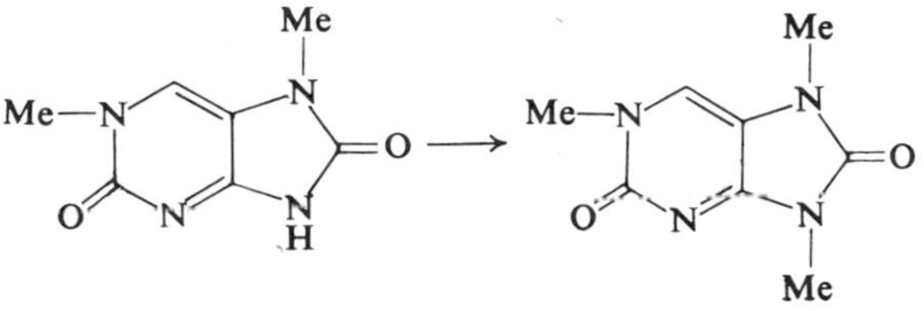

(23)   (24)

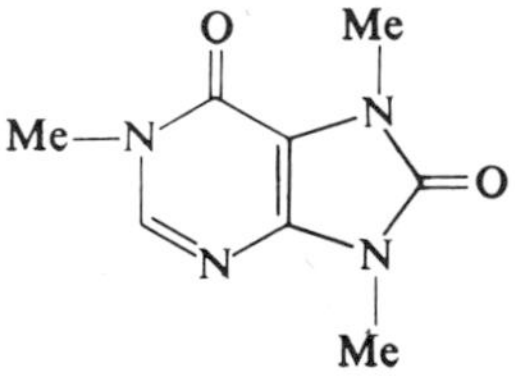

(25)

water bath.[234] When xanthine in warm dimethyl sulphoxide containing hydrogen chloride is treated with 2,3-dihydropyran, the 7-tetrahydro-pyran-2-yl derivative results.[166]

Few details are available of alkylation of the isomeric 2,8- and 6,8-dioxopurines. Because both nitrogen atoms in the imidazole ring are formally protonated, betaine structures of the above type are not possible. Dimethyl sulphate with 1,2,7,8-tetrahydro-1-methyl-2,8-dioxopurine in sodium hydroxide at room temperature rapidly (10 min) gives the 1,7-dimethyl analogue (23) as major product, some of the 1,9-dimethyl isomer also occurring.[237] Further alkylation of either product under the same conditions gives the 1,7,9-trimethyl derivative (24).[237] With methyl iodide in place of dimethyl sulphate, sealed tubes are required (100°, 1 h), this procedure also converting the isomeric 2,3,7,8-tetrahydro-3-methyl-2,8-dioxopurine to the 3,7,9-trimethylated form.[238]

Heating 1,6,7,8-tetrahydro-6,8-dioxopurine[25] or the 7-methyl[239] or 7,9-dimethyl[239] derivatives with methyl iodide (alkali, 100°) affords 1,6,7,8-tetrahydro-1,7,9-trimethyl-6,8-dioxopurine (25). The 3,7-dimethyl analogue is similarly converted to the 3,7,9-trimethylated form.[239]

c. Tri-Oxopurines

Uric acid shows well defined dibasic acid character. Acid and neutral metal salts are formed, those of lead, silver, and potassium being the most commonly encountered. Acidity due to the associated protons decreases in the order $N_{(3)}$, $N_{(9)}$, $N_{(1)}$, and $N_{(7)}$.

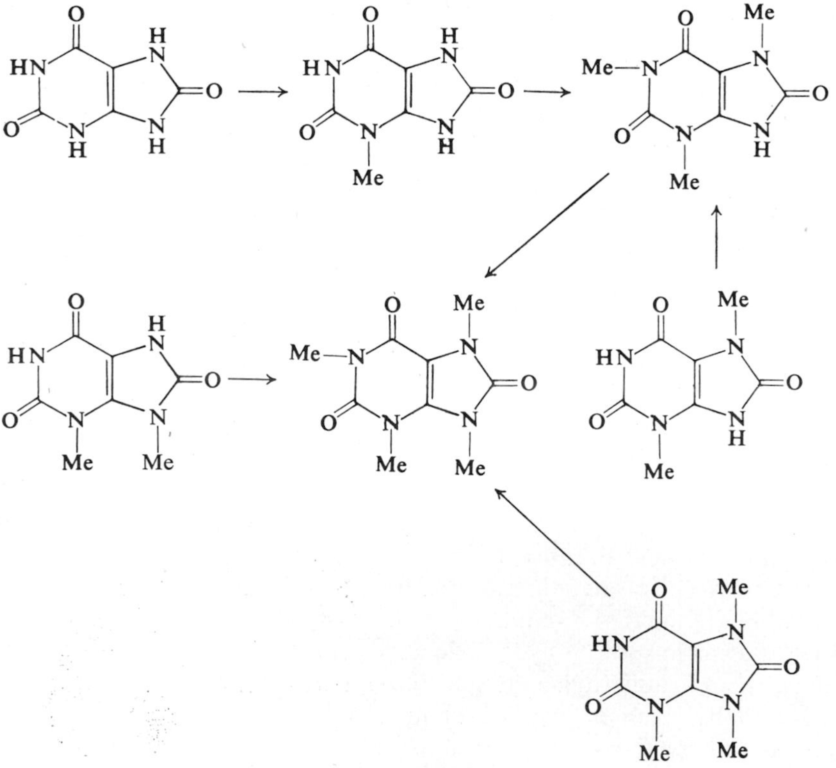

SCHEME B

In aqueous alkali, descriptively referred to by Fischer as the "wet way," alkylation shows some parallels with that of xanthine. With methyl[240] or ethyl iodide[241] in sodium hydroxide at 80° uric acid gives the 3-alkyl derivative which further alkylates to the 1,3,7-trialkyluric acid.[172, 238] Complete alkylation to 1,3,7,9-tetramethyluric acid (methyl iodide, potassium hydroxide, 100°) readily follows.[242] These and other results are summarised in the methylation of uric acid, Scheme B.

Examples of trialkyluric acid formation include 7-hydroxymethyl-1,3-dimethyl- from 7-hydroxymethyluric acid,[243] and also 1,3,7-trimethyl-uric acid from the 3,7-dimethyl analogue.[244] Methylation with an excess of reagent gives the corresponding 1,3,7,9-tetra-alkyluric acid from 3-methyl-,[245] 9-phenyl-,[246] and 3,9-dimethyluric acid.[247] 1,3,7,9-Tetra-methyluric acid can also be obtained from the 3,7,9-trimethyl ana-logue.[248] Other alkylations of this type are the formation of 1,3-di-methyluric acid from the 1-methyl derivative[249] and the 1,7,9-trimethyl homologue from 7,9-dimethyluric acid.[169]

Alkylation of uric acid by heating a metal salt with an alkyl halide or similar reagent (Fischer's "dry way") probably constitutes the earliest alkylation study made in the purine series.[250] Of the four available protons those at $N_{(3)}$ and $N_{(9)}$ show the most acidity, with the former being considered the more acidic. Alkylation by this route, therefore, involves first $N_{(3)}$, but if this is substituted then $N_{(9)}$ becomes the initial reaction site. If the monometal (acid) salt is used only one nitrogen atom is alkylated, whereas with the dimetal (neutral) salt two or more centres may be concerned. The acid lead salt of uric acid gives mainly the 3-methyl derivative with some of the 9-methyl isomer,[251–253] whereas the neutral lead or potassium salt gives 3,9-dimethyluric acid.[254] Improved yields result from using an inert solvent; carbon tetrachloride at 140° containing dimethyl sulphate was employed with dipotassium urate.[247] With ethyl bromide in the same solvent (120°, 12 h) 3,7,9-triethyluric acid was obtained,[255] none of the 3,9-diethyl derivative being formed. Further examples of the interrelationship between the 3- and 9-positions of uric acid derivatives toward alkylation are the formation of 3,9-dimethyl- from 3-methyl- (MeI, 130°, 16 h),[256] 3,7-dimethyl- from 7-methyl- (MeI, 170°, 12 h),[242] 1,3,9-trimethyl- from 1,3-dimethyl- (MeI in ether, 170°, 15 h),[257] 3,7,9-trimethyl- from 3,7-dimethyl- (MeI in ether, 170°, 15 h)[258] uric acids lead salts. Conversion of either 1,3,9-[257] or 3,7,9-trimethyluric acid[258] salts to the 1,3,7,9-tetramethyl derivative occurs under similar conditions. The production of 8-ethoxycaffeine in good yield rather than 9-ethyl-1,3,7-trimethyluric acid on heating ethyl iodide with the silver salt of 1,3,7-trimethyluric acid at 100°[124] is probably due to the lower temperature employed favouring O-alkylation. With methyl iodide these conditions give a mixture of 8-methoxycaffeine and 1,3,7,9-tetramethyluric acid.[242]

The application of diazoalkanes to the formation of 2- or 8-alkoxy-N-methyluric acids has been noted earlier (Sect. 4C). In some cases no O-alkylation is observed; uric acid itself is transformed to 1,3,7,9-tetramethyl derivative in ethereal diazomethane[259] while the 3-ethyl analogue gives 3-ethyl-1,7,9-trimethyluric acid.[260] It must, however, be pointed out that such examples appear to be the exception and might possibly result through the initial formation and subsequent thermal rearrangement of the 8-alkoxy derivative. The occurrence, in one case, of nearly equal amounts of 1,3,7-trialkyl-8-alkoxy- and tetra-alkyl derivative[122] supports this possibility. A smooth conversion of 3,7,9-triethyluric acid to the 3,7,9-triethyl-1-methyl and 1,3,7,9-tetraethyl analogues follows with diazomethane and diazoethane, respectively.[255] With methyl iodide in the cold the tetrakis-triethylsilyl derivative of uric acid gives, surprisingly, 1-methyluric acid but, if methylation is carried out in toluene at 150° (dimethyl sulphate), the two main products are 1,7,9-trimethyl- and 1,3,7,9-tetramethyluric acids. The occurrence of some $O_{(2)}$-methylation, to a minor extent, is also reported under these conditions.[112]

E. Hydroxymethylation and the Mannich Reaction

In alkaline solution formaldehyde can give N-hydroxymethyl derivatives, usually unstable to heat or prolonged treatment in aqueous solutions. Hypoxanthine, as the riboside, inosine, forms the 1-hydroxymethyl analogue, stable only if an excess of formaldehyde is present.[261] In warm 40% formaldehyde 3-methylxanthine gives an unidentified product, most probably 7-hydroxymethyl-3-methylxanthine.[262] A 7-hydroxymethyl derivative is produced with 2-chloro-1,6,7,8-tetrahydro-9-methyl-6,8-dioxopurine,[240] also by uric acid[263] and the 3-methyl[264] and 1,3-dimethyl[265] homologues, at room temperature. These derivatives are reduced to the 7-methyl analogue with acid stannous chloride, an example being 7-methyluric acid.[263] Theophylline and theobromine hydroxymethylate at $N_{(7)}$ and $N_{(1)}$, respectively.[262] If hydrogen chloride is present, theophylline gives the 7-chloroethyl derivative.[86]

An unusual instance of $C_{(8)}$-hydroxymethylation is reported on heating caffeine with paraformaldehyde and acid at 170° for some hours.[89]

If amines are also present and more neutral conditions obtain, aminomethyl derivatives (Mannich bases) result. Hypoxanthine with formaldehyde and either morpholine or piperidine affords appropriately the 1,9-diaminomethyl derivative (26, R = morpholino or piperidino)[266] whereas xanthine gives the 1,3,9-trisubstituted base.[266]

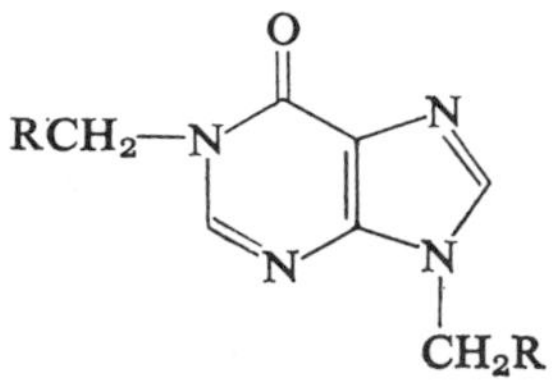

(26)

Uric acid, surprisingly, does not appear to react[266] but theophylline and theobromine readily form the respective 7-aminomethyl[86, 266–269] and 1-aminomethyl[266, 269] derivatives. Quaternisation of the exocyclic amino group has been effected with methyl iodide.[267]

F. Acylation of Oxopurines

Examples of such procedures are not common with either mono- or dioxopurines. When acylation does occur it is usually restricted to N-acylation, either at $N_{(1)}$ or $N_{(7)}$. With some uric acid derivatives both N- and O-acylation takes place, the latter event, however, being limited to the formation of 8-acetoxy derivatives.

Hypoxanthine with acetic anhydride in dimethylformamide under reflux affords 1-acetylhypoxanthine.[270] Dioxopurine examples are found with the N-methylated derivatives, theophylline gives the 7-acetyl analogue with acetic anhydride,[271] while 1-acetyltheobromine is obtained from sodium theobromine and acetyl chloride in chloroform or xylene.[272] The respective 7- and 1-benzoyl derivatives arise with benzoyl chloride. Other acyl derivatives of this type are known.[273, 274] By contrast uric acid and its N-acetylated forms have been extensively investigated. A significant feature of these derivatives is that all acetylate at $N_{(7)}$ unless this position is already substituted. In boiling acetic anhydride preparations of the 7-acetyl derivatives of 3-methyl-,[123] 9-methyl-,[123] 1,3-dimethyl-,[123] 1,9-dimethyl-,[123] 3,9-dimethyl-,[247] and 1,3,9-trimethyluric acid[123] are reported. Both 7-methyl-[123, 275] and 1,7-dimethyluric[123] acid under this treatment only undergo acylation of the 8-oxo group, as in 27[123] whereas uric acid itself and the 1-methyl homologues are claimed to give the corresponding 7-acetyl-8-acetoxy derivative.[123] Prolonged boiling (30 h) of uric acid in acetic anhydride in the presence of pyridine gives 8-methylxanthine.[276–278] Other sources claim[279] more vigorous conditions are required and heat for 80 h using dimethylaniline as base. This reaction has been studied in detail[280–282] and appears to proceed by way of mono-, di-, and triacetylated derivatives of 4,5-diaminouracil. Both open (28)[122] and cyclic (29)[280] forms have been proposed. Illustrated is the diacetate which further acetylates to the triacetate (30).[280] Conversion of the "acetates" to 8-methylxanthine requires heating with dilute acid or methanolic hydrogen chloride.[280]

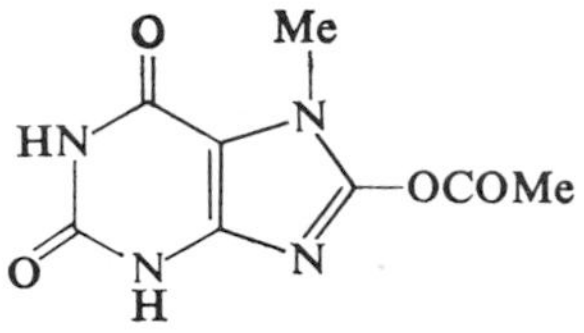

(27)

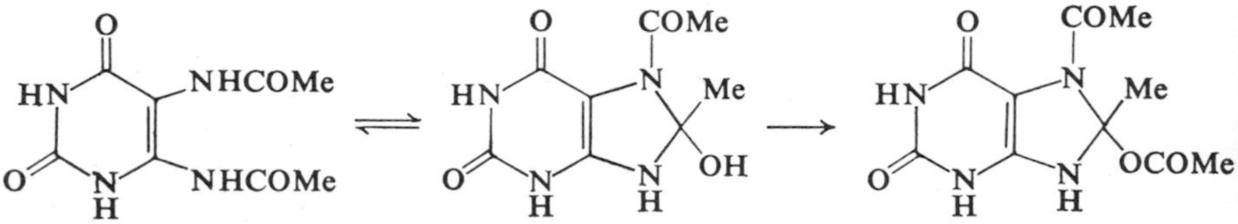

(28)   (29)   (30)

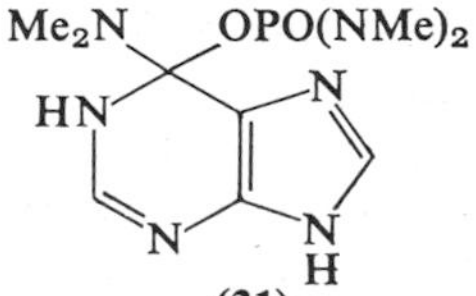

(31)

Under the above conditions no acetylation occurs with 3,7-dimethyl-, 7,9-dimethyl-, 1,3,7-trimethyl-, 1,7,9-trimethyl-, and 3,7,9-trimethyluric acids.[123]

G. Oxo to Amino Group

This reaction, which is of recent origin,[282a] has had only limited application so far to the purine series. There are, however, good grounds for supposing that the reaction can be extended to formation of a range of aminopurines. The examples reported derive from hypoxanthine which gives 6-dimethylamino- and 6-benzylaminopurines on being heated (230–260°, 5–10 min) with $NN'N''$-trisdimethyl- and $NN'N''$-

tribenzylphosphoramide, respectively. Amination, operating by way of an adduct intermediate of the type **31** which then loses phosphorodiamidic acid to generate the 6-alkylaminopurine, is postulated.[282a]

It is pertinent to draw attention here to examples of apparent oxo replacement by amino group, noted earlier (Ch. V, Sect. 1Ba). These dialkylaminopurines, which are obtained as side products from chlorination reactions using phosphoryl chloride–trialkylamine combinations, arise by quaternary salt formation through interaction of the chloropurine first formed and the trialkylamine. A subsequent thermally induced Martius–Hofmann degradation of the salt then affords the 6-dialkylaminopurine.

7. Reactions of *N*-Alkylated Oxopurines

A. Replacement of *N*-Alkyl Groups by Hydrogen

The use of phosphoryl chloride and other chlorinating agents to convert *N*-methyl groups to *N*-chloromethyl groups has been described (Ch. V, Sect. 3A), these groups being sometimes removed by heating in water, the conversion of 8-chloro-1,3,9-trimethylxanthine to 8-chloro-1,9-dimethylxanthine being made in this way.[227] In other examples the methyl group is lost during heating in the phosphoryl chloride; with 1,3,7,9-tetramethyluric acid this treatment gives 8-chlorocaffeine by concomitant chlorination at $C_{(8)}$- and $N_{(9)}$-demethylation.[240] Corresponding removals of an ethyl group at $N_{(9)}$ are recorded from 3,7,9-triethyl- and 1,3,7,9-tetraethyluric acids on heating with concentrated hydrochloric acid.[255] Aluminium tribromide in toluene is an effective reagent for $N_{(9)}$-demethylation.[165]

Because of their general ease of removal by reductive means, benzyl groups are useful as protecting agents for ring nitrogen atoms. Sodium in liquid ammonia is a widely favoured reagent for debenzylation. Successful results are obtained with a number of 8-substituted-1-benzyl-3,7-dimethylxanthines[283] and with 1-benzyl-3,9-dimethylxanthine.[187] Both benzyl groups in the case of 1,3-dibenzyl-8-hydroxymethyl-xanthine[89] are removed.

Hydrogenation over a palladium or charcoal catalyst (5%) in ethanol converts 3-benzylinosine to inosine.[52] With the methosulphate of 1,7-dibenzyl-3,9-dimethylxanthine this treatment removes only the benzyl group at $N_{(7)}$, that at $N_{(1)}$ requiring displacement by sodium in liquid ammonia.[187] Acid reagents, such as hydrogen bromide in acetic acid, are reported to remove an $N_{(7)}$-benzyl group when catalytic hydrogenation failed.[284] A more general application of hydrobromic acid alone, has been demonstrated[285] for debenzylation at $N_{(3)}$ in the case of 3-benzyl-, 3,7-dibenzyl-, and 1,3-dibenzylhypoxanthine. Group displacement is also noted with 3-benzyl-1,6,7,8-tetrahydro-6,8-dioxo-purine.[285] Oxidative removal of an alkenyl group is demonstrated by the formation of 1-methylhypoxanthine from the 1-methyl-9-propenyl analogue (32) in alkaline permanganate solution at room temperature.[286]

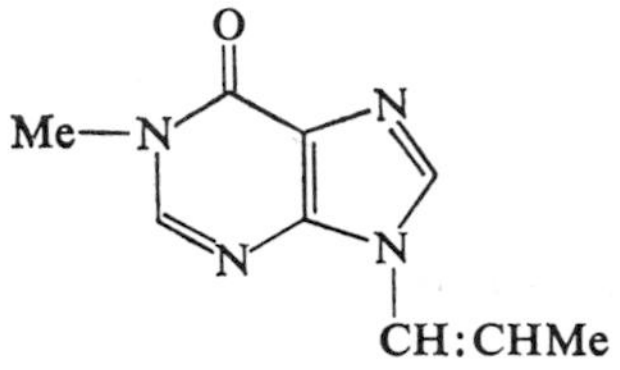

(32)

B. Rearrangement of *N*-Alkyl Groups

An apparent case of true alkyl group migration occurs on heating 1,3-dibenzylhypoxanthinium bromide in dimethylacetamide at 110°, the two main products, obtained in equal amounts, being 1,7- and 1,9-dibenzylhypoxanthine.[118, 287] The $N_{(3)}$-group migration is further demonstrated by heating 1,3-dibenzylhypoxanthinium bromide with adenine in dimethylacetamide at 120° for 4 h when some 3-benzyl-adenine is to be found in the reaction mixture.[118] As either 1,7- or 1,9-dibenzylhypoxanthine, or both, are the ultimate products of prolonged benzylation of hypoxanthine or its 1-, 3-, 7-, or 9-benzyl analogues, other examples of this migration are to be found.

C. Degradation of *N*-Alkylpurines

a. *With Alkali*

Alkaline stability can be largely correlated with the ability to form the anion. With the mono-oxopurines, where *N*-alkylation precludes this, degradation to an imidazole is found to occur readily. Thus 3-methyl-hypoxanthine in *N*-sodium hydroxide at 100° gives some 4(5)-carbamoyl-5(4)-methylaminoimidazole[288] while 1,9-dibenzylhypoxanthine with

6N-alkali is converted to 5-amino-1-benzyl-4-(N-benzyl)carbamoyl-imidazole also under like reflux conditions.[118, 289] The appropriate alkylated imidazoles result similarly with 1-alkylinosines,[290, 291] 1,7-dimethyl-,[292] and 1,7-dibenzylhypoxanthine,[289] also 1,3-dibenzyl-hypoxanthinium bromide giving the N-formylimidazole on addition of ammonia to an aqueous solution at ambient temperature[118] (Eq. 4).

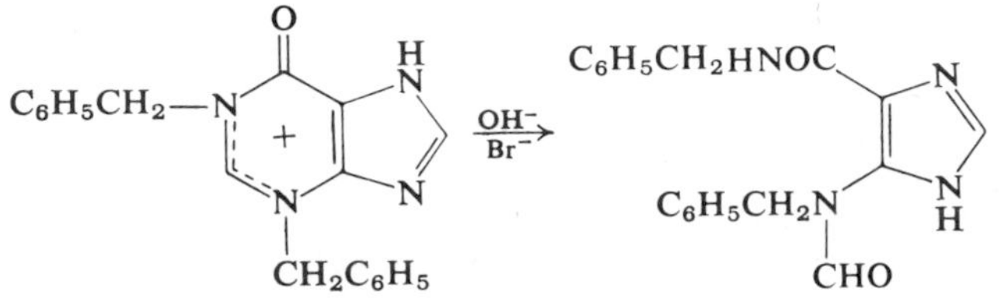

The association of the positive charge with the imidazole ring in 1,7,9-trimethylhypoxanthinium bromide induces nucleophilic attack at $C_{(8)}$ and formation of a pyrimidine derivative (Eq. 5). In contrast to the above the stability of simple 9-alkylhypoxanthines in alkali reflects the

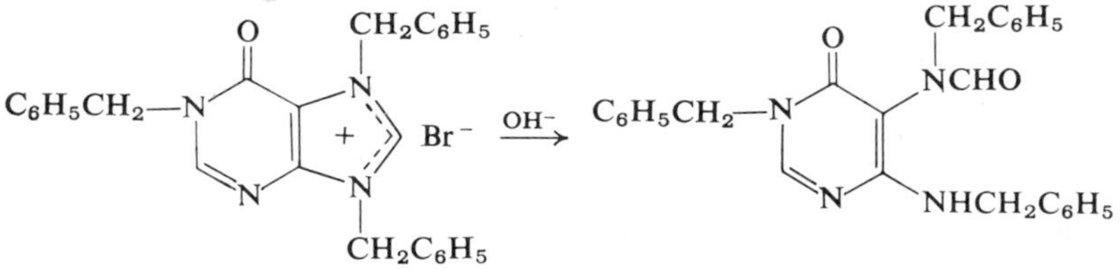

ability of these compounds to form the anion.[293] Mono- and di-N-alkylated dioxopurines for the same reason show the expected stability to alkaline conditions although vigorous treatment with strong alkali leads to decomposition, theobromine being converted to ammonia, methylamine, and sarcosine (N-methylglycine).[294] Trialkyl derivatives, precluded from anion formation, degrade readily. Caffeine illustrates this in affording caffeidine carboxylic acid on prolonged contact with cold N-alkali (3 days)[82, 295, 296] or for a brief period (15 min) at 100°. Decarboxylation by means of hot mineral acid gives 1-methyl-4-methyl-amino-5-(N-methyl)carbamoylimidazole.[82] The isomeric 3-methyl-imidazole analogue arises from dilute alkali treatment of 1,3,9-trimethyl-xanthine (isocaffeine).[297]

With tetra-alkylated purines the presence of the charged centre in the imidazole ring leads to fission of this moiety, illustrated by the rapid degradation of 1,3,7,9-tetramethylxanthine in warm 3% sodium hydroxide (Eq. 6).[298]

Stability studies of mono-, di-, and trimethyluric acids under prolonged heating (100°, 15–33 h) in N-potassium hydroxide reveal that in

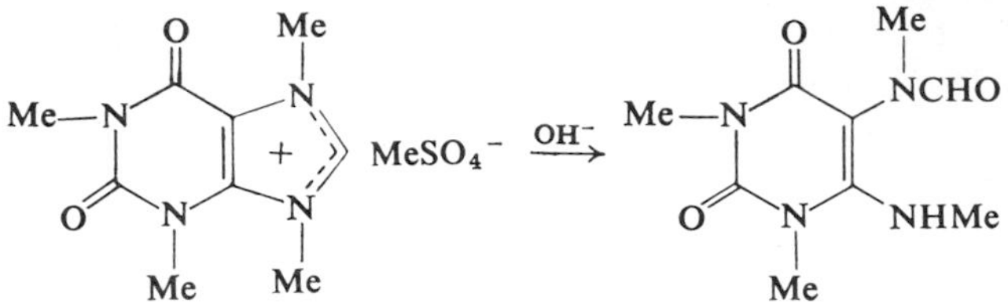

$$ \text{(6)} $$

the majority of cases only slight breakdown occurs. The exception is 1,3,9-trimethyluric acid, of which only 20% remains after 1 hour.[292] As a generalisation the greatest measure of resistance to alkaline degradation is found in $N_{(7)}$-alkylated derivatives. In line with previous observations on the instability of bases incapable of anion formation the complete decomposition of 1,3,7,9-tetramethyluric acid in cold N-alkali after a short time is not surprising.[292]

b. *With Acid*

Stability in dilute mineral acid solution is a feature of N-alkyl-oxopurines but in concentrated hydrochloric acid at elevated temperatures (ca. 200°) degradation ensues. The usual products are carbon monoxide, carbon dioxide, ammonia, and glycine,[299] but if methyl groups are attached to pyrimidine ring nitrogen atoms or to $N_{(9)}$ some methylamine is also produced. Additionally, if $N_{(7)}$ is methylated, as in 7-methylxanthine,[181] theobromine,[300] caffeine,[301] or 7-methyluric acid,[258] hydrolysis gives N-methylglycine (sarcosine) rather than glycine, showing the origin of the nitrogen atom in glycine as being that of $N_{(7)}$ in the purine.

8. Reactions of Alkoxypurines

Little use has been made of alkoxypurines in purine metatheses. Benzyloxy derivatives have some application as protected forms of the corresponding oxopurines, to which they are readily converted by hydrogenolysis. The most notable reaction shown by alkoxypurines is their facility in isomerising to N-alkyl-oxopurines on heating.

A. Alkoxy to Oxopurine

Acid conditions are invariably used for hydrolysis; alkali seems to be ineffective for this purpose. Mild conditions usually suffice, the reaction often being carried out at room temperature. Examples illustrating the ease of formation of the oxopurine are the preparation of 3-methyl-hypoxanthine from 3-methyl-6-methoxypurine (5N-HCl, 20°, 12 h),[117] isoguanine from 6-amino-2-ethoxypurine (HBr/HOAc, 20°, 48 h),[55] or 1,3,7-trimethyluric acid from 8-ethoxycaffeine (H$_2$O/HCl, 100°/10 min).[124] The hydrolysis of a benzyloxy group at C$_{(2)}$ has been effected by heating under reflux (1 h) in a mixture of hydrochloric and acetic acid.[55] With 8-chloro-2,6-diethoxypurine hydrochloric acid treatment involves both groups giving 8-chloroxanthine. Similar conditions with 6,8-diethoxy-7-methylpurine only hydrolyses the group at C$_{(6)}$[239] but complete conversion to 1,6,8,9-tetrahydro-7-methyl-6,8-dioxopurine is possible if cold hydriodic acid is employed.[239] It is sometimes advantageous to use the latter acid, either alone or with phosphorus, to hydrolyse the alkoxy group and remove a halogen atom simultaneously. The conversion of 8-chloro-2,6-diethoxypurine to xanthine[302] (Eq. 7)

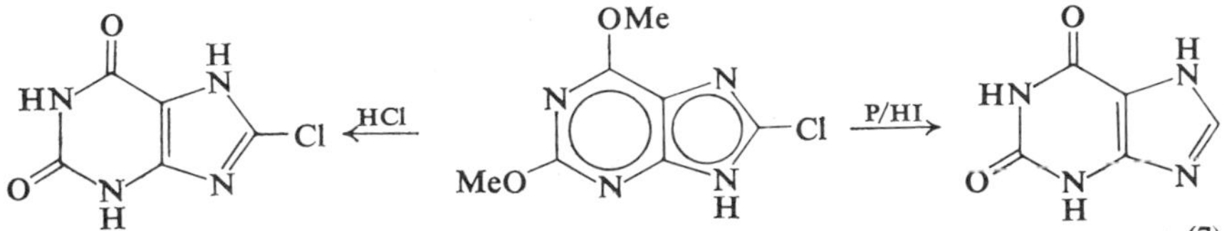

$$(7)$$

and the formation of 1,6,8,9-tetrahydro-7-methyl-6,8-dioxopurine from 2-chloro-6,8-diethoxy-7-methylpurine illustrates this.[239] Additional effects are removal of acid labile groups; examples involving all three types of replacement are seen in the preparation of 7,8-dihydro-8-oxopurine (**34**) and 1,6,7,8-tetrahydro-6,8-dioxopurine from 2,6-dichloro-8-ethoxy-9-(tetrahydropyran-2-yl)- (**33**) and 2-chloro-6,8-diethoxy-9-(tetrahydropyran-2-yl)purine, respectively.[303]

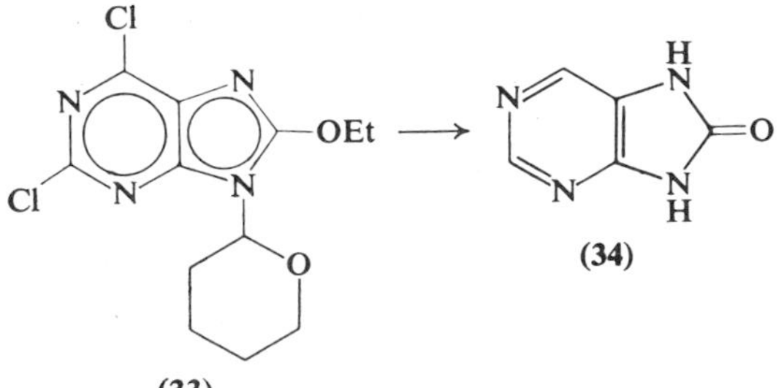

(33)   (34)

B. Hydrogenolysis of Alkoxy Groups

Virtually all conversions of this type relate to the reductive cleavage of benzyloxy groups, these having found some use in nucleoside chemistry as protected forms of oxo groups. Conversion of 6-amino-2-benzyloxy-9-methylpurine to 9-methylisoguanine was carried out in dilute hydrochloric acid over a palladous oxide catalyst.[55] Standard hydrogenation procedures now obtaining use ethanolic solutions of the purine and a 5% palladium on charcoal catalyst, the hydrogen being supplied at atmospheric pressure or slightly above this. Successful oxopurines have resulted from hydrogenolysis of 6-benzyloxy-2-fluoro-,[304] and 6-benzyloxy-9-(tetrahydrofur-2-yl)purine,[43] also the 9-β-D-ribofuranosides of 6-benzyloxy-2-methoxy-,[304] 6-benzyloxy-2-methylamino-,[58] 6-benzyloxy-2-dimethylamino-[58] purines and the 8-benzyloxy analogues of guaninosine and xanthosine.[305] The susceptibility of the benzyloxy group toward reductive cleavage is shared with the allyloxy group. In boiling propanol containing palladised barium sulphate 2,6-diallyloxy-7-methylpurine is hydrogenated to 7-methylxanthine.[306]

C. Alkoxy to Amino Group

Although alkoxy groups, especially the methoxy group, are extensively utilised as leaving groups in amination reactions in the pyrimidine series, few instances of their similar application with the purines are recorded. One such instance illustrating the facile replacement of a $C_{(6)}$-methoxy group is the conversion of the 9-β-D-riboside of 2-chloro-6-methoxypurine into that of 6-amino-2-chloropurine with methanolic ammonia at 80° for some hours.[307]

D. Rearrangement

Thermal isomerisation has been observed in purines with alkoxy groups at $C_{(2)}$, $C_{(6)}$, and $C_{(8)}$, an apparent migration of the alkyl group from the oxygen to an adjacent ring nitrogen atom occurs on heating them for a short period, usually at a temperature just above the melting

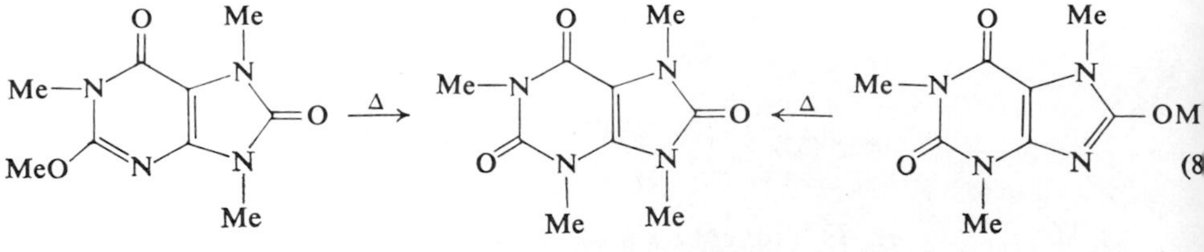

point. The formation of 1,3,7,9-tetramethyluric acid by heating either 1,6,7,8-tetrahydro-2-methoxy-1,7,9-trimethyl-6,8-dioxopurine[122] or 8-

methoxycaffeine[122, 308] to 200° for 30 min (Eq. 8) ably illustrates the scope of the reaction. A historical point is that the latter rearrangement was the first example to be studied of this type of isomerisation.[309]

Caffeine is obtained from 2,6-dimethoxy-7-methylpurine through rearrangement of both methoxy groups (210°, 15 min)[310] or alternatively from 2,3-dihydro-6-methoxy-3,7-dimethyl-2-oxopurine at a higher (300°, 1 h) temperature.[311] Other 2,6-dialkoxypurine rearrangements are noted with 2,6-diallyloxy-7-methyl-[306] and 2,6-dimethoxy-purine[111] although in the latter case the product (theophylline?) was not identified. Under similar temperatures (150°) and conditions the 2,6-dipropoxy analogue does not rearrange.[310] Analogous migrations are observed with 2-ethoxy-,[122] 8-ethoxy-,[308, 309, 312] and higher 8-alkoxypurine homologues[308]; 8-benzyloxycaffeine, for example, gives 9-benzyl-1,3,7-trimethyluric acid.[308] Others of this type are known[82, 123, 313] including a migration to $N_{(9)}$ of a dimethylaminoethyloxy group.[314]

E. Other Reactions

Benzylation of 6-methoxypurine with benzyl bromide in acetonitrile under reflux conditions (4 h) gives a poor yield (25%) of hypoxanthine containing smaller amounts of 1,3-dibenzyl- and 7,9-dibenzylhypo-xanthinium bromides. With dimethylformamide as solvent under milder conditions the products are 4(5)-benzylamino-5(4)-(N-benzyl)carba-moylimidazole (25%) and 7,9-dibenzylhypoxanthinium bromide.[118]

Conversion of an 8-alkoxy group to an 8-methyl or -ethyl group occurs on heating at an elevated temperature (270°) 8-alkoxycaffeines with acetic or propionic anhydride, the best yields arising from the compounds with the higher molecular weight alkoxy groups.[314a] These reactions cannot be considered to be true displacements of alkoxy groups as they are, in effect, extended Traube cyclisations (see Ch. II, Sect. 1E). Reactions of N-alkoxypurines are described in Chapter XI, Section 4B.

9. Oxidation of Oxopurines

Many aspects fall under this heading, some of which are dealt with elsewhere. Of these the most notable are the formation of N-oxides (Ch. XI, Sects. 1A, 1B) and enzymatic oxidation (this chapter, Sect. 1L).

As susceptibility to oxidation is highest in trioxopurines the major studies have been carried out with uric acid derivatives, but some contribution from xanthine chemistry has been made. In both the uric acid and xanthine series the same generalisation holds good in that under alkaline conditions fission of the pyrimidine ring gives rise to allantoin derivatives whereas the imidazole ring is degraded in acid media and the products are alloxans. A clear and concise account of the early studies with uric acid is available.[315] In this section both non-alkylated and alkylated purines will be treated together; many reactions are applicable to either series. Of the various oxidising agents chlorine has been the most extensively used and, because of the variety of the resulting products, is accorded separate treatment in the succeeding section devoted to acid oxidations.

A. Under Acid Conditions

a. *With Chlorine in Acid or Aqueous Media*

Various factors determine the oxidation product, of which the most important are the nature of the solvent, the reaction temperature, and the pH of the medium. Theobromine provides a good illustration in that chlorination in aqueous solution gives first a mixture of methyl-

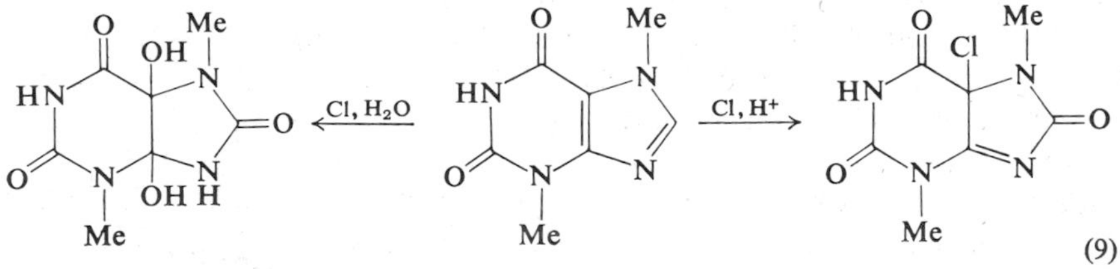

$$(9)$$

alloxan (**35**) and methylurea[222] which then interact to give 3,7-dimethyl-uric acid glycol (Eq. 9)[316]; the latter being produced also by direct oxidation of theobromine with potassium chlorate in hydrochloric acid.[317] By contrast chlorination in acetic acid affords 5-chloro-3,7-dimethylisouric acid (Eq. 9).[318] Theophylline gives, on short exposure, the chlorine in aqueous solution, 1,3-dimethyluric acid glycol[319] while longer times produce dimethylalloxan, which is also obtained on like oxidation of caffeine.[320] Xanthine is correspondingly converted into alloxan.[17]

In either concentrated or dilute acetic acid most alkylated uric acids, except 9-alkyl derivatives, give the corresponding 5-chloroisouric acids though loss of hydrogen chloride from the intermediate 4,5-dichloro-4,5-dihydrouric acid (36), as demonstrated by the conversion of 1,3,7-trimethyluric acid to 5-chloro-1,3,7-trimethylisouric acid (37).[321] The presence of an alkyl group at $N_{(9)}$ with $N_{(7)}$ unsubstituted leads to

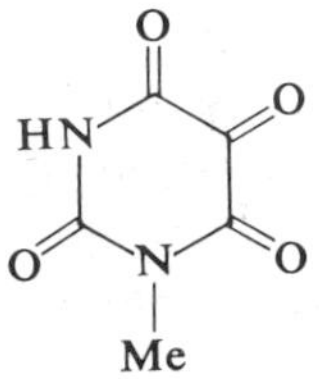

(35)

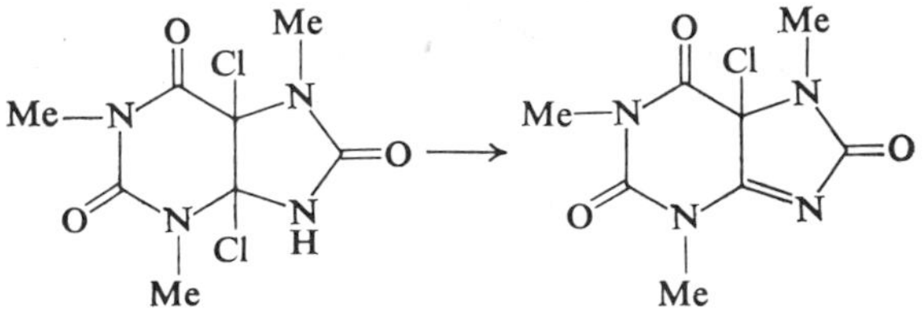

(36)   (37)

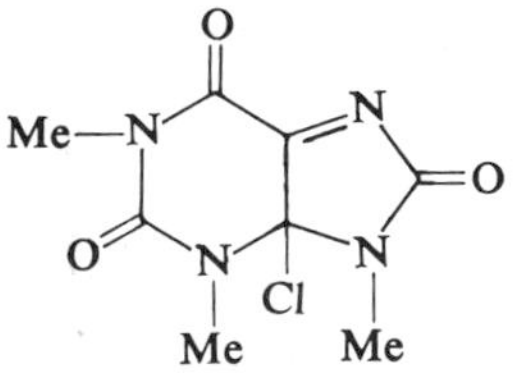

(38) 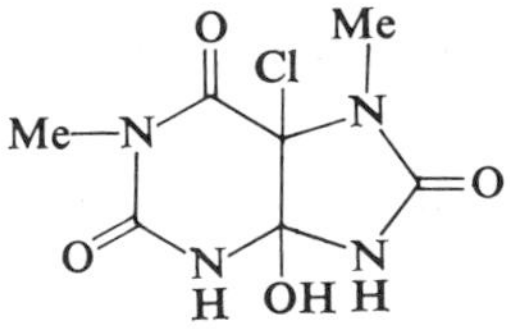   (39) 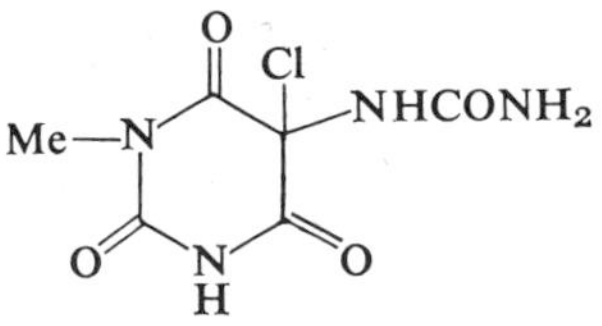   (40)

the isomeric 4-chloroisouric acids such as 38 derived from 1,3,9-trimethyluric acid.[322] Corresponding derivatives are formed by 9-methyl-,[323] 1,9-dimethyl-,[324] and 3,9-dimethyluric acid.[325]

Stable 4,5-dichloro-4,5-dihydrouric acids result from some uric acids having both imidazole nitrogen atoms alkylated; examples occur with 1,3,7,9-tetramethyluric acid[323] and other tetra-alkyl analogues.[312, 313, 326] Corresponding dichloro forms of 1,7-,[323] 3,7-[318] and 3,9-dimethyluric acids[323] are formed if hydrochloric acid is added to the acetic acid medium before chlorination proceeds. If water is present, even in small amounts, certain di- and trialkyluric acids give 5-chloro-4,5-dihydro-4-hydroxy derivatives of the type 39, these include 1,7-[323] and 3,7-dimethyl-[318] and 3,7,9-trimethyluric acid.[323] The exceptions to the above, which exclude ring fission, are the chlorination of uric acid[327] itself and the 1-[328] and 9-methyl[253] homologues in acetic acid. Through oxidative cleavage of the $C_{(4)}$—$N_{(9)}$ double bond, the corresponding

5-chloropseudouric acid results, as for example, **40**, from 1-methyluric acid.[328] In more dilute acid solutions the main product is the appropriate alloxan derivative as occurs with 1-methyl-[329] and 3-methyluric acid[330] and the parent acid, which is transformed to alloxan[331, 332] in one minute.

Alloxans are also formed by chlorination of uric acid and the monoalkylated analogues in water only but glycols are obtained with 1,3- and 3,7-dimethyluric acid. In those derivatives with alkyl groups at both $N_{(1)}$ and $N_{(7)}$ rearrangement to caffolides occurs. Thus, 1,3,7-trimethyluric acid gives 1,7-dimethylcaffolide[333] **(A)** the "apocaffeine" of

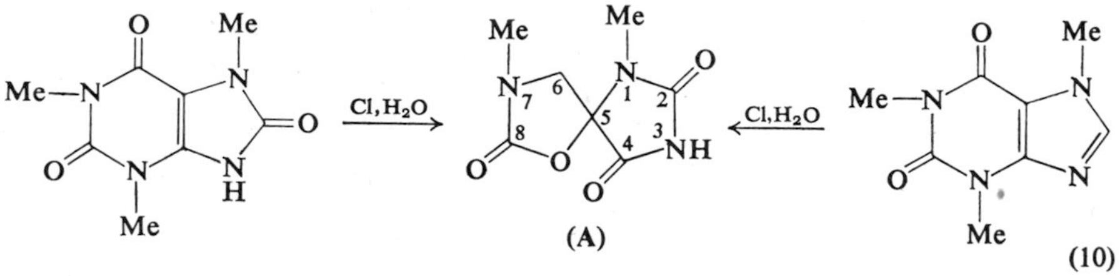

(A)

(10)

Fischer, which is also produced on similar oxidations of caffeine[333] (Eq. 10). Corresponding caffolides derived from 1,7-dimethyl-,[334] 1,7,9-trimethyl-,[333] and 1,3,7,9-tetramethyluric acid[335] are known.

b. *With Chlorine in Nonaqueous Media*

The majority of oxopurine chlorinations carried out in chloroform or other organic solvents have been made with uric acid derivatives. However, examples of this technique in the xanthine series are seen in the conversion of theobromine in boiling chloroform to the unusual pentachloroimidazole[336] **(41)** and imidazole ring fission in the case of

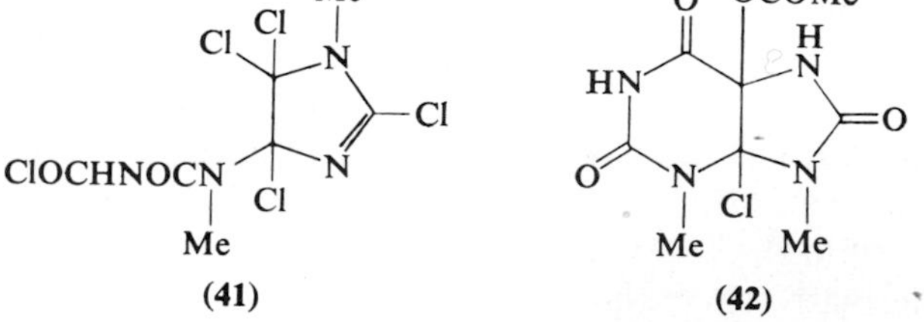

(41)

(42)

theophylline giving a 5-hydroxypseudouric acid[337] derivative. This medium is excellent for preparing 4,5-dichloro-4,5-dihydrouric acids being successful with 9-methyl-,[323] 7,9-dimethyl-,[169] 1,3,9-, 1,7,9-, and

3,7,9-trimethyl-[323] and 1,3,7,9-tetramethyluric acid.[323] Exceptions to this are 1,7-dimethyl-[332] and 1,3,7-trimethyluric acid[338] which give 5-chloroisouric acids.

In a few cases chlorinations in acetic anhydride have given 5-acetoxy-4-chloro-4,5-dihydrouric acids,[323, 325] such as **42** from 3,9-dimethyluric acid.

c. *With Other Acid Reagents*

Oxidations with nitric and nitrous acid have been reviewed by Fischer[339] and Blitz.[340] Chromic acid has been used with caffeine.[341, 342] Both xanthine and uric acid derivatives give the appropriate alloxan and a urea as primary products but further oxidation to parabanic acids or purpuric acids is possible. The purple colour that the latter compounds give on addition of ammonia solution constitutes the "murexide" test, which is positive for all purines capable of oxidation to an alloxan derivative. Studies made recently[343] have resulted in establishing the structures of the murexides derived from theophylline, theobromine, caffeine, and uric acid. Both caffeine and theophylline in a mixture of trifluoroacetic acid and peroxide at room temperature give NN'-dimethylparabanic acid.[344] Under mildly acid conditions electrochemical oxidation of uric acid gives an unstable product which degrades to equal amounts of alloxan and allantoin.[345] In strong acid (75% sulphuric acid) solution at a lead dioxide anode mainly urea together with some alloxan and parabanic acid[346] is formed. Theobromine affords methylalloxan under this treatment.[346]

B. Under Alkaline or Neutral Conditions

The isolation of allantoin (**A**), in 1838, after lead peroxide oxidation of uric acid[347] was followed by the recognition of other oxidation products,[348, 349] the chief ones, depending on the conditions obtaining, being oxonic acid (**B**), allantoxaidine (**C**), and oxaluric acid ($H_2NOCHNCOCO_2H$). The structures of these derivatives and the reaction pathways involved, having been subjects for great speculation over many years, were clarified by comprehensive studies[350–356] carried out with uric acid, all carbon atoms of which in turn were labelled with ^{14}C and, in addition, ^{15}N was employed at $N_{(7)}$ (Scheme C). Under similar conditions both 1- and 7-methyluric acid gave 3-methyl-allantoin while 1-methylallantoin is obtained from either 3- or 9-methyluric acid.[238] The ease of oxidation is demonstrated by the complete conversion of uric acid into allantoin after a few minutes in dilute

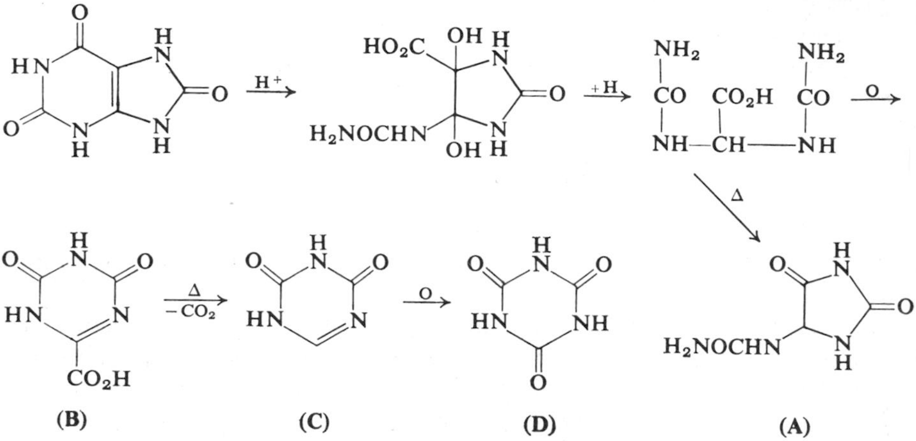

(B)   **(C)**   **(D)**   **(A)**

Scheme C

alkaline permanganate solution at 30°.[332] Oxidising agents used in neutral or alkaline solution include oxygen, ozone, iodine, potassium ferricyanide, and hydrogen peroxide. Allantoin or its derivatives also arise from oxidations in dilute acetic acid with potassium permanganate, potassium persulphate, or hydrogen peroxide.[349] With this last oxidant products also include cyanuric acid (**D**) and triuret [$(H_2NOCHN)_2CO$], in addition to allantoin.[357] Triuret also results from short exposure of an aqueous solution of uric acid to ultraviolet irradiation[358] or from electrochemical oxidation of the lithium salt in lithium carbonate solution at a lead dioxide anode.[346]

10. 4,5-Dihydrouric Acid Derivatives

Although these derivatives are presently of little practical interest, they form an integral part of the chemistry of uric acid and the most important are annotated below. Their formation involves at some stage an addition at the $C_{(4)}$–$C_{(5)}$ double bond. Such reactions are specific for uric acid derivatives and reflect the largely aliphatic, rather than aromatic, character of the bond. This property is in the main due to the isolated nature of the double bond which is not part of a formal resonance system.

The interrelationship of the various dihydrouric acids is summarised in the schematic form D.

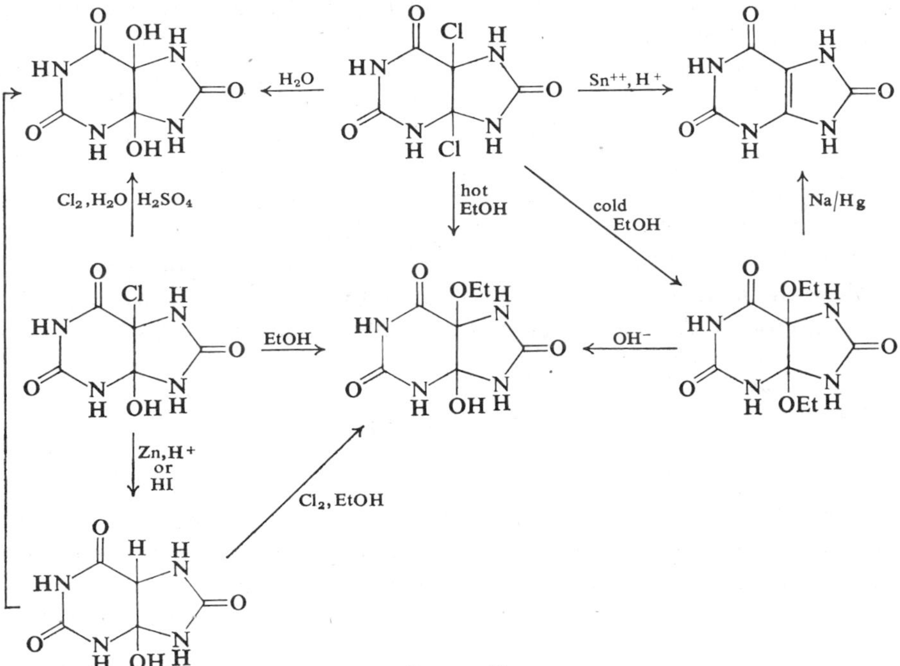

SCHEME D

A. 4,5-Dihydro-4,5-dihydroxyuric Acids (Uric Acid Glycols)

a. *Preparation*

Direct syntheses entail condensation of an alloxan with a urea using fusion conditions, as in the formation of 9-methyluric acid glycol from alloxan and methylurea[359] (Eq. 11). It seems likely that 5-hydroxy-pseudouric acids are intermediates in the above as their cyclisation to glycols has been demonstrated.[327]

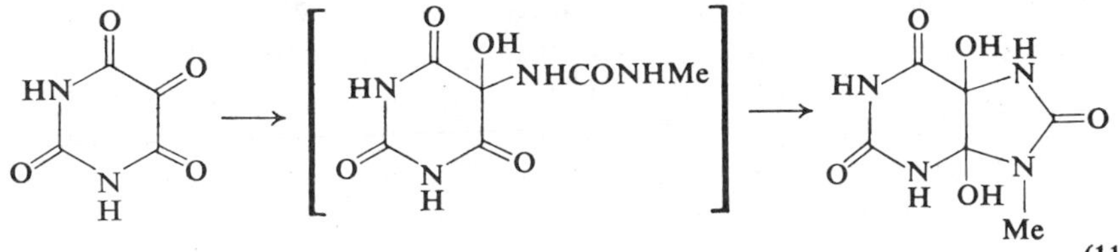

(11)

Although the product was not at the time identified the first example of this synthesis recorded was of uric acid glycol itself from alloxan and urea.[360] Oxidative preparations have used N-alkylated xanthine and uric acid derivatives (see Sect. 9Aa), also certain 4,5-dihydro-4-hydroxy-uric acids, notably the 3,7-dimethyl[318] and 3,7,9-trimethyl derivatives.[244] Under hydrolysis conditions 9-methyl-,[361] 1,3-dimethyl-,[323, 362] and 3,7-dimethyl-5-chloroisouric acids[318] give the appropriate glycol as do also certain 5-alkoxy-4,5-dihydrouric acids[169, 318] (uric acid hemiethers, Sect. 10Eb) in the same way. The formation of uric acid glycol following nitrous acid treatment of 5-amino-4,5-dihydro-4-hydroxyuric acid appears to be the only example of this kind to be described.[363]

b. *Properties and Reactions*

The glycols, generally, show more acidic character than the parent acids.[364] Although fairly stable toward dilute mineral acids the reduced pyrimidine ring is cleaved by warm aqueous nitric acid giving 5-hydroxy-hydantoylureas (**43**).[349, 365] Similar degradations are recorded on

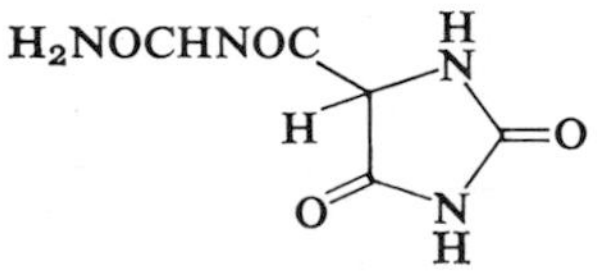

(**43**) 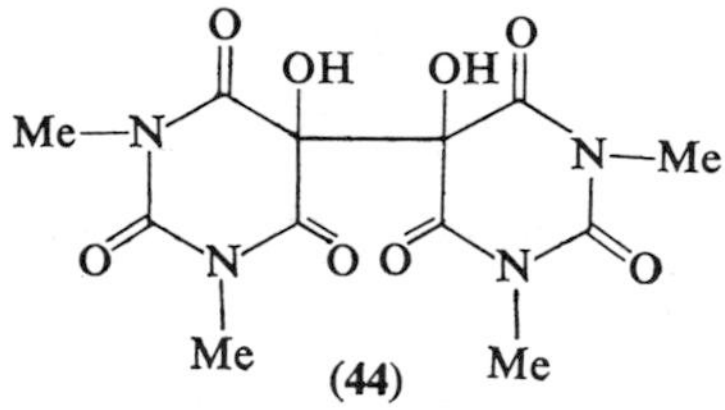   (**44**) 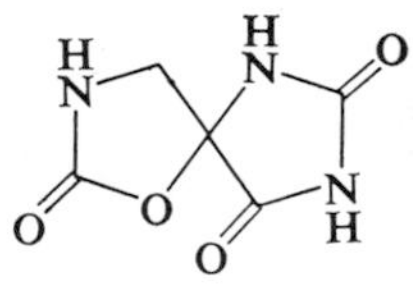   (**45**)

heating glycols with water, acetic acid, or ethanol or simply on leaving methanol or pyridine solutions to stand at room temperature.[316, 365–368] The exception is 1,3-dimethyluric acid glycol which undergoes imidazole ring fission in warm dilute nitric acid forming tetramethyl-alloxanthine (**44**).[364] Alkali converts some glycols to the caffolide notably those derived from uric acid (**45**)[327] and the 9-methyl[247] and 3,9-dimethyl derivatives.[247] With dehydrating agents such as sulphuric acid or acetic anhydride the corresponding spyrohydantoin results.[327] Reducing agents do not give the parent uric acid but degradation to a hydantoin derivative occurs. Only the hydroxyl group at $C_{(5)}$ can be alkylated to form the hemiether[340] and, although inert to this treatment, the $C_{(4)}$-hydroxyl group can, however, be acetylated.

B. 4,5-Dihydro-4-hydroxyuric Acids

a. *Preparation and Properties*

Only four examples of this class exist. Treatment of either 7,9-dimethyl- or 7,9-diethyluric acid glycol with phosphorus tribromide affords the appropriate required derivative.[369] This reaction most likely proceeds by reduction of the intermediate 5-bromo-4,5-dihydro-4-hydroxy derivative. A related preparation of 4,5-dihydro-4-hydroxy-3,7-dimethyluric acid (**47**, R = H) involves hydriodic acid or zinc-acetic acid reduction of the 5-chloro-4-hydroxy analogue (**46**, R = H).[318, 369] With methyl sulphate **47** (R = H) is converted to 4,5-dihydro-4-hydroxy-3,7,9-trimethyluric acid (**47**, R = Me).[244] Oxidation to the

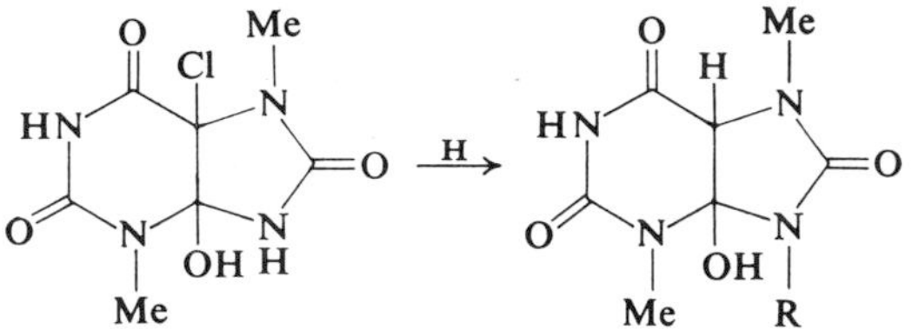

(46)   (47)

glycol is effected on passing chlorine through an aqueous solution whereas the hemiether results [240] using ethanolic solutions. Heating the hydroxyuric acids alone or in acid solution leads to fission of the pyrimidine ring and production of hydantoins.[318, 369]

C. 4,5-Dichloro-4,5-dihydrouric Acids

a. *Preparation*

Chlorination of uric acids in chloroform or an acetic acid–hydrochloric acid mixture has been elaborated already (Sects. 9Aa and 9Ab). An alternative route is the addition of hydrogen chloride to 5-chloroisouric acids in glacial acetic acid.[318, 323, 325] An unstable 4,5-dibromide results [124] on cold bromine treatment of 1,3,7-trimethyluric acid.

b. *Properties and Reactions*

A hygroscopic nature is shown. With stannous chloride reduction to the parent uric acid takes place. In cold aqueous solutions rapid hydrolysis to glycols occurs. Correspondingly glycol ethers are formed in cold ethanol but partial dealkylation, giving rise to hemiethers (Scheme D), is characteristic on heating the solution.

D. 4,5-Dialkoxy-4,5-dihydrouric Acids (Uric Acid Glycol Ethers)

a. *Preparation*

In addition to the action of cold alcohol on the 4,5-dichloro derivative noted above, the direct formation from a uric acid derivative is possible by passing chlorine through an alcoholic solution. Examples are the 4,5-dimethoxy and 4,5-diethoxy derivatives of 9-ethyl-1,3,7-trimethyluric acid formed in methanol and ethanol, respectively.[312] Methanolic solutions of xanthine derivatives undergo photosensitised oxidation in the presence of the dye stuff, rose bengal, to the corresponding 4,5-dimethoxyuric acid derivative.[370]

b. *Properties and Reactions*

These derivatives are obtainable in crystalline form, the most stable of which are those alkylated at $N_{(1)}$ or $N_{(7)}$. Reductive studies[169, 247, 263, 318, 321, 327, 328, 334, 361, 371] carried out with hydriodic acid or stannous chloride show that the intermediate pseudouric acid initially formed recyclises to give the appropriate uric acid as final product, whereas if sodium amalgam is the reducing agent the uric acid is formed through direct reduction of the ether groups. The nature of the product resulting from dilute acid hydrolysis depends on the particular uric acid derivative employed. Thus, aqueous hydrochloric acid on the 4,5-dihydro-4,5-dimethoxy derivatives of 7-methyl-, 3,7-, 3,9-, and 7,9-dimethyluric acids is the corresponding 4,5-dihydro-4-hydroxy-5-methoxy(hemiether)[169, 263, 318, 323] analogue. If an alkyl group is present at $N_{(1)}$, as with 1-methyl, 1,3-, and 1,9-dimethyluric acid ethers, fission of the imidazole ring gives a 5-alkoxypseudouric acid derivative.[321, 327, 328, 334, 372] More extensively alkylated ethers such as the 1,3,7- and 1,7,9-trimethyl and 1,3,7,9-tetramethyl analogues, having both $N_{(1)}$- and $N_{(7)}$-alkylated, undergo rearrangement[321, 327, 371] to caffolides, with dilute acid or spirohydantoins in lower pH solutions.[124, 373]

Aqueous or alkaline hydrolyses have been studied[374, 375]; the products are usually allantoins or related derivatives arising by opening of the pyrimidine moiety.

E. 5-Alkoxy-4,5-dihydro-4-hydroxyuric Acids (Uric Acid Glycol Hemiethers)

a. *Preparation*

The preceding section outlines their formation by acid hydrolyses of the glycol ethers, other routes are cyclisation of 5-alkoxypseudouric acids in alkaline solution[324, 327, 328, 334, 340, 361] and alcohol treatment of a 4,5-dihydro-5-chloro-4-hydroxyuric acid.[318, 323] As 1-alkyluric acid glycols, with the exception of the 1,3-dimethyl derivative, do not afford stable hemiethers the product from the action of ethanol on 4,5-dihydro-5-chloro-4-hydroxy-1,7-dimethyluric acid (**48**) through rearrangement, is the 5-ethoxy hemiether of 3,7-dimethyluric acid glycol[323] (**49**). This same derivative also arises when 1,7-dimethyluric acid glycol ether is acid hydrolysed.[328, 334] A further example of a rearrangement involving an apparent methyl group migration from $N_{(1)}$ to $N_{(3)}$ is found in the transformation of 1,9-dimethyl-5-chloro-isouric acid into the hemiether of 3,9-dimethyluric acid glycol.[247, 376]

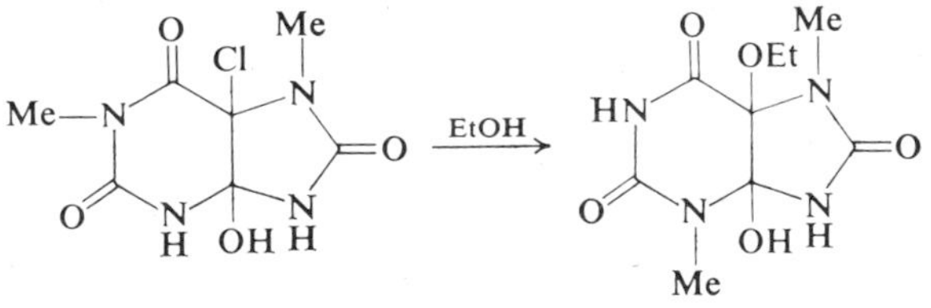

(**48**)   (**49**)

b. *Properties and Reactions*

The hemiethers are well-defined stable compounds, not readily attacked by dilute acids but converted to the parent glycols with warm concentrated sulphuric acid. More vigorous treatment with the latter results in conversion to the corresponding spyrohydantoin. Hydriodic acid degrades the majority of hemiethers to hydantoins,[247, 253, 263, 318, 327, 361] the exception being the 1,3-dimethyl analogue which is resistant to all but extreme conditions.[361] With ammonia and amines on gentle warming 5-alkoxyhydantoylamides result.[377] The corresponding acetylated 5-alkoxyhydantoylamides are produced on heating the hemiethers with acetic anhydride.[378] No alkylation to the glycol ether has been possible.

11. Hypoxanthine

This is the only mono-oxopurine of any biological significance, being widely distributed in the animal and plant kingdoms. Although not one of the component bases of nucleic acid it is, as the 9-β-D-ribofuranoside, inosine, of importance in nucleic acid biosynthesis as the precursor of both adenosine and guanosine. First isolated by Scherer, in 1850, from cattle spleen,[379] the name hypoxanthine was coined to denote the lower oxygen content of the compound compared with xanthine. Sarkin, a purine isolated from muscle tissue by Strecker[380] a few years later, is identical with hypoxanthine. The elaborate synthesis from 2,6,8-trichloropurine, by Fischer,[302] established the structure as 1,6-dihydro-6-oxopurine. Diverse synthetic routes now available, include *inter alia* variations of the Traube cyclisation (Ch. II) and ring closure of 5(4)-amino-4(5)-carbamoylimidazole (Ch. III). Of these the best yields (95%) result from Traube cyclisations, using triethyl orthoformate–acetic anhydride reagent[381] or from a one-stage synthesis (85%) comprising brief heating, under reflux (5 min) of aminomalonamidamidine hydrochloride in a mixture of triethyl orthoformate and dimethylformamide.[382] Hydrolysis of the amino group of adenine by nitrous acid[5] was an important step in demonstrating the relationship between the two bases.

Although hypoxanthine exhibits no melting point the picrate melts at 250–254°.[383] Crystallisation from boiling water, in which it is 70 parts soluble,[384] gives a colourless micro-crystalline anhydrous product. The various values quoted in the older literature for the solubility in cold water (1 in 1400) are due to the pronounced tendency toward formation of supersaturated solutions. In 95% ethanol at 17° solubility is 1 in 3700 parts rising to 1 in 900 at the boiling point.[380, 385] Hypoxanthine readily dissolves in cold mineral acids and alkali forming stable solutions. It can be reprecipitated from alkaline solution by carbon dioxide.[380, 385] An aqueous solution is degraded into ammonia, carbon dioxide, and formic acid on heating to 200°[386]; with concentrated hydrochloric acid at this temperature glycine is an additional product.[387] Partial decomposition occurs in hot aqueous alkali but fusion with potassium hydroxide at 200° gives ammonia and hydrogen cyanide.[386]

The most important reaction of hypoxanthine is the conversion to "6-mercaptopurine" on heating with phosphorus pentasulphide in pyridine,[130] this forming the basis of commercial preparations.[388, 389] With phosphoryl chloride under reflux in the presence of dialkylaniline, 6-chloropurine[390] is produced while phosphoryl bromide affords the 6-bromo analogue.[391] The various alkylated derivatives are described in Section 6Da. With methyl iodide in sodium methoxide the product is a hydrated complex of 1,7-dimethylhypoxanthine and sodium iodide, this appearing to be a feature of hypoxanthine derivatives only. Coupling with benzene diazonium salts, which takes place at $C_{(8)}$, is discussed in Chapter X, Section 3Aa. Opening of the imidazole ring occurs on heating with ethyl glyoxal hemiacetal in the presence of N-hydrochloric acid with the formation of 4,6-dihydroxypteridine, the latter arising through cyclisation of the degradation product with the acetal.[128]

The amphoteric nature (basic pK_a 1.98, acidic pK_a 8.94 and 12.10) shows in salt formation, usually as hydrates, for example, hydrochloride (monohydrate),[392] nitrate (monohydrate), sulphate (anhydrous),[393] or picrate (monohydrate)[392, 394] with the appropriate acid. Metal salts, such as the easily soluble sodium derivative[392] or the trihydrated silver salt, which partially dehydrates to the hemihydrate on heating to 120°,[385, 394] are produced on treatment with a basic aqueous solution of the metal ion, isolation being through precipitation, if insoluble, or on evaporation to dryness. Ligands are formed with many metal salts and bases including silver nitrate,[380, 385, 394, 395] perchlorates of copper^{II}, nickel, lead and zinc,[396] the chloride[394] and acetate[397] of mercury^{II}, barium hydroxide and cuprous oxide.[380, 385] Potentiometric titration studies show that two molecules of the purine can complex with each divalent metal ion but, if the latter is in excess, the reverse may occur; seen in the formation of 1,7-diacetoxymercurihypoxanthine using mercuric acetate.[397]

12. Xanthine

This usually is present in nature as a product of the later stages of metabolism, arising by further oxidation of hypoxanthine or hydrolytic conversion of guanine. Although not then identified, its isolation from urinary calculi was first reported[398] in 1817, by Marcet who named it "xanthic oxide" on account of the yellow colour it gave on nitric acid treatment. From this has been derived the "xanthine" of current usage. The subsequent efforts of Wohler and Liebig[399] produced the correct empirical formula but it was Fischer's synthesis from 2,6,8-trichloro-purine,[302] 60 years later, which resolved the controversy over the correct structure. Synthetic routes subsequently developed are mainly variations of the synthesis by Traube[18] involving cyclisation of 4,5-diamino-1,2,3,6-tetrahydro-2,6-dioxopurine, one such requiring only two stages, starting with urea and ethyl isonitrosocyanoacetate, gives xanthine in 60% overall yield.[400] Of less importance are Hofmann degradation of 4,5-dicarbamoylimidazole,[401] cyclisations of 4(5)-amino-5(4)-carbamoylimidazole with ethyl chloroformate[402] or urea,[402, 403] or related imidazole ring closures.[404] It is often more convenient to prepare xanthine by transformation of an existing purine, the most common of which is acid hydrolysis, by either hydrochloric or, better, nitrous acid, of guanine (Sect. 1Db) or mineral acid treatment of isoguanine (Sect. 1Db).

Xanthine, as the monohydrate, is obtained in a microcrystalline form on careful acidification of a dilute alkaline solution of the purine with acetic acid. On heating the hydrate to above 125° the anhydrous base results, which, having no melting point, is best characterised as the perchlorate (m.p. 262–264°). Although virtually insoluble in cold water (1 part in 14,000 at 16°), a tenfold increase in solubility is found with boiling water (1 part in 1500). The ready solubility in dilute acids and alkalis reflects the amphoteric character (acidic pK_a 7.44 and 11.12). Such solutions are stable on heating but more vigorous conditions, such as hydrochloric acid at 200°, degrades xanthine into carbon dioxide, ammonia, glycine, and formic acid. Although not readily oxidised conversion to alloxan follows treatment with potassium chlorate in hydrochloric acid (Sect. 9Aa). The enzymatic conversion of xanthine to uric acid is brought about with xanthine oxidase (Sect. 1L). Catalytic hydrogenation is ineffectual but reduction of the oxygen function at $C_{(6)}$ occurs with zinc and hydrochloric acid or sodium amalgam or by electrochemical means (Ch. XII, Sects. 1Aa and c). Electrophilic reagents will substitute at $C_{(8)}$, halogens give the 8-halogenoxanthine[405] while alkaline diazonium solutions give the appropriate 8-azo derivative which may be reduced to 8-aminoxanthine or hydrolysed to uric acid (Ch. X, Sects. 3Aa and b). With phosphorus pentasulphide in pyridine only the oxygen at $C_{(6)}$ is replaced by sulphur (Sect. 6B) but both oxo functions are replaced by chlorine using phosphoryl chloride containing small amounts of water (Ch. V, Sect. 1Ba). The various alkylation modes have been discussed earlier (Sect. 6Db). An unspecified di-*N*-hydroxymethyl derivative results on heating with formaldehyde in the presence of hydrochloric acid[406] which reverts to xanthine on treatment with dilute sodium hydroxide.

The usual salts are formed with hydrochloric, sulphuric, and nitric acids, all being rapidly hydrolysed in water.[380] Among the salts formed by bases the very soluble sodium derivative and insoluble silver,[124] lead,[124] and barium[380] derivatives are mainly encountered. If acid silver nitrate is used in place of the ammoniacal reagent the silver nitrate complex instead of the silver salt[380, 407] is formed. Like hypoxanthine with an excess of mercuric acetate the 1,7-diacetoxymercury derivative results.[397] The complex precipitated on addition of ammoniacal copper sulphate forms the basis of an assay method for xanthine.[408]

13. The *N*-Methylated Xanthines

Of the eleven possible *N*-methylxanthines seven have been identified in living matter. The most important commercially of these are theophylline, theobromine, and caffeine, brief monographs of which appear below. Of the remaining purines 1- and 7-methylxanthine (heteroxanthine) together with 1,7-dimethylxanthine (paraxanthine) are present in human urine, representing breakdown products of ingested theobromine, theophylline, and caffeine.

General features of the chemistry of *N*-methylxanthines parallel those of the parent xanthine, on acidic oxidation the appropriate alloxan is formed (Sect. 9Aa) while reduction of the $C_{(6)}$-oxo function gives the appropriate 6-deoxyxanthine (Ch. XII, Sects. 1Aa and c). Susceptibility to electrophilic attack at $C_{(8)}$ is retained as seen in the facile formation of 8-halogeno derivatives and of 8-nitropurines by direct nitration (Ch. X, Sect. 1Aa). Chromatographic and electrophoretic techniques for separating and identifying mixtures of *N*-alkylxanthines from natural sources have been developed.[409] The interconversion of derivatives by methylation is described in Section 6Db.

A. Theophylline (1,3-Dimethylxanthine, 1,2,3,6-Tetrahydro-1,3-dimethyl-2,6-dioxopurine, "Theocin")

Present to a small extent in the leaves of the tea plant, it was from this source that the first isolation in 1888 was reported by Kossel.[410, 411] For preparative purposes adaptations[20, 180, 400, 412] of the original Traube synthesis[173] are recommended.

Theophylline, as monohydrate, m.p. 272–274°, is obtained as needles or tablets from water. Although freely soluble in hot water, at 15° the solubility is 1 part in 226. The aqueous solution has a pK value of 8.7. It is soluble in dilute acids and bases, and forms a monohydrated hydrochloride and hydrobromide which become anhydrous above 100°. A soluble sodium salt can be prepared with sodium ethoxide but the insoluble silver salt is precipitated from ammoniacal solution by silver nitrate.[410, 411] With mercurous nitrate a 1:1 complex results[413] but mercuric nitrate[413] or mercuric acetate[397] gives di(theophyllin-7-yl)-mercury. Other complexes with heavy metal salts are known.[392, 414] Theophylline, as the sodium or calcium salt, forms many double salts with the corresponding metal salts of organic acids, used medicinally as diuretics and cardiac stimulants. Commercially available drugs are derived in this way from calcium or sodium salicylate, sodium acetate, or sodium glycinate. Also of therapeutic importance are derivatives of theophylline containing one or two molecules of an alkylamine, the most important of these adducts being "aminophylline" prepared from ethylene diamine.

B. Theobromine (3,7-Dimethylxanthine, 1,2,3,6-Tetrahydro-3,7-dimethyl-2,6-dioxopurine)

This was the first N-dimethylated xanthine to be recognised, being extracted by Woskresensky[415] from the cocoa plant. Although it was long believed to be the only purine of this type present recent studies[416] have identified theophylline, in small amounts, as a co-constituent. Other sources of theobromine include, *inter alia*, plants of the *Cola*[417] and *Cascarilla* families.[418] Synthesis is by methylation of 3-methyl-xanthine, various modifications[20, 180, 400] of this route are available. On heating, theobromine sublimes at 290° and finally melts at 351°. In cold water (17°) solubility is 1 part in 1600 (pK_a 9.9), at 100° a twenty-fold increase is noted. It is very poorly soluble in cold ethanol, ether, and chloroform, even less so in benzene, petroleum ether, and carbon tetrachloride. Being amphoteric it dissolves readily in dilute mineral acids and in solutions of bases. The hydrochloride and hydrobromide hydrolyse in water and on heating to 100° theobromine is regenerated. Salts formed with bases show greater stability but the very soluble derivative reverts back to theobromine on passing carbon dioxide through an aqueous solution.[295] Like the isomeric theophylline a 1:1 adduct forms with mercurous nitrate[413] but di(theobromin-1-yl)-mercury (?) is obtained[397] using mercuric salts. Double salts of medi-cinal value form between the alkali metal theobrominate and the corresponding metal salts of an organic acid. Such combinations arise with sodium acetate, sodium formate, sodium salicylate, and with calcium salts also. Main usages are as diuretics, cardiac stimulants, and vasodilators.

C. Caffeine (1,3,7-Trimethylxanthine, 1,2,3,6-Tetrahydro-1,3,7-trimethyl-2,6-dioxopurine, Theine, Guaranine)

This derivative, present in the leaves or fruit of a number of plants, mainly those of the tea, coffee, and cola species, was originally reported by Runge in 1820.[419] Commercially it is obtained by extraction of tea dust, which can contain up to 3.5% of the base. This process is the subject of a student's practical experiment,[420] using tea leaves, from which a caffeine yield of 1.5% can be demonstrated. Synthetically it is conveniently prepared by Traube cyclisation of 4-amino-5-formamido-1,3-dimethyluracil in ethanolic sodium ethoxide containing methyl iodide, concomitant ring closure to theophylline and methylation at

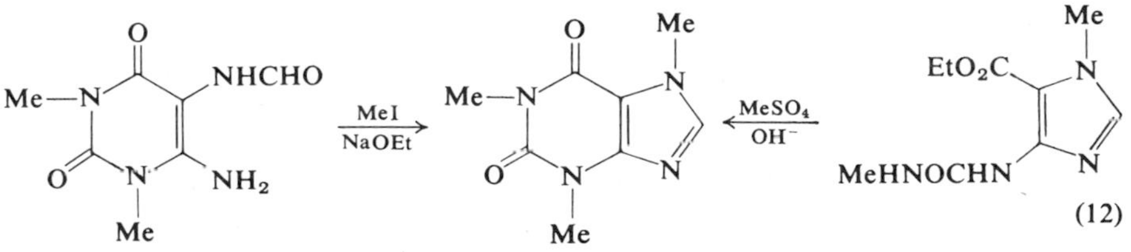

(12)

$N_{(7)}$ taking place in one stage.[173] The route from 5-ethoxycarbonyl-1-methyl-4-(N-methyl)ureidoimidazole involving ring closure and subsequent methylation in alkali is a practical alternative.[421] Both approaches are illustrated in Equation 12. Various conversions of N-methylxanthines to caffeine are elaborated in Section 6Db.

From water caffeine crystallises as the monohydrate which effloresces in air to give a fractionally hydrated derivative. Sublimation readily affords the anhydrous base, m.p. 238°. Cold water solubility is 1 part in 46 but in boiling water 1 part in 1.5 parts. Solutions of moderate strength are obtained in cold ether, ethanol, acetone, benzene, and ethyl acetate but poor solubility is found in petroleum ether. The absence of acid character shows in the neutral aqueous solution obtained; basic character is, however, present, the usual salts being formed with mineral acids but these are unstable in water, or on prolonged exposure to air. Hydrolytically unstable complexes, of therapeutic importance, are formed with organic acids, including acetic, benzoic, citric, and salicylic acids. Although generally inert towards boiling mineral acids, complete degradation occurs at more elevated temperatures (170–250°)[422] with hydrochloric and hydriodic acids. The exceptional instability towards alkali, shown in the formation of caffeidine carboxylic acid[295] at room temperature, can be related with the inability to form an anion. Ethanolic alkali causes a near-quantitative conversion to caffeidine while boiling solutions of alkali metal hydroxides result in complete breakdown to sarcosine, formic acid, methylamine, ammonia, and carbon dioxide.[301, 423]

With heavy metal salts 1:1 adducts are possible as with mercurous nitrate[413] or mercuric chloride. Although mercuric acetate is reported not to complex in similar fashion,[397] the formation of 8-acetoxy-mercuricaffeine (50) is claimed in acetic acid solution.[424]

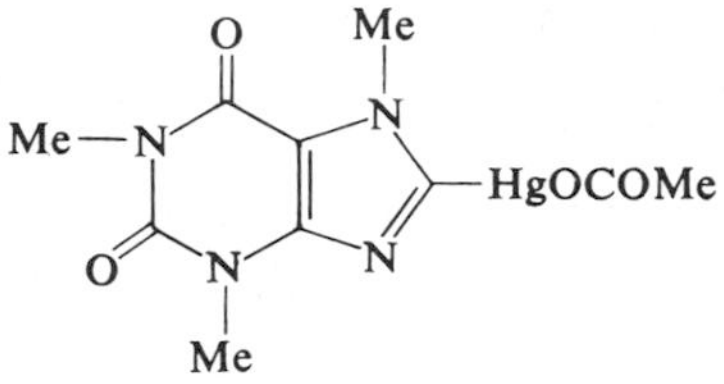

(50)

14. Uric Acid
(1,2,3,6,7,8-Hexahydro-2,6,8-trioxopurine)

Apart from the historical distinction of being the first purine to be isolated from a natural source, it is also notable in being the most extensively investigated purine derivative. It is present in many living systems as an end product of metabolism and it was from such a source, in this case urinary calculi, that Scheele first extracted it in 1776. An account of the elucidation of the structure is available,[425] the subject being treated in historical sequence culminating with Fischer's confirmatory synthesis[242] from pseudouric acid. Uric acid ribosides have been found in beef blood,[426] liver,[427] and the bacillus[428] *L. plantarum*. Although on ultraviolet spectral evidence the beef blood riboside was originally formulated as the $N_{(9)}$-riboside[427] later investigation revealed that this was actually the $N_{(3)}$-riboside,[429] this being supported by comparison with unambiguously synthesised material.[430] As a corresponding synthesis of the $N_{(9)}$-riboside confirms[305] its presence in the other sources, it is concluded that coexistence of $N_{(3)}$- and $N_{(9)}$-ribosides is possible through enzymatic interconversion.

Large-scale preparation of uric acid is by cyclisation of 4,5-diamino-1,2,3,6-tetrahydro-2,6-dioxopurine with urea under fusion conditions,[431, 432] this being utilised to prepare various carbon and nitrogen labelled uric acids.[350] On heating, uric acid shows no melting point but decomposes above 400° with evolution of hydrogen cyanide. Purification is best effected practically by addition of dilute mineral acid to an alkaline solution, the acid precipitating in a crystalline form. Direct crystallisation from water, although possible, is inconvenient due to the generally low solubility (1 part in 39,000 at 18°, 1 part in 2,000 at 100°) being, therefore, limited to small-scale purification. A further drawback is the ready tendency of such solutions to supersaturation in the presence of trace impurities. Dibasic acid character is shown (pK_a 5.75 and 10.3)[433] and alkaline solutions are readily obtained. Although acid and alkali stable to moderate conditions it is degraded to ammonia, carbon dioxide, and glycine on vigorous hydrochloric acid treatment.[350] With water under pressure and heat (200°), through a series of complex ring fissions and closures, a variety of products including 2,4,6- and 2,4,7-trihydroxypteridine have been identified.[434] Replacement of the oxo function at $C_{(8)}$ by another atom or group is possible, the reaction being essentially of the Traube type following from an initial opening of the imidazole ring. Thus, with formamide xanthine is formed[20] or 8-methylxanthine if acetamide or acetic anhydride[281] are employed. Conversion to the 8-thio analogue (51) is brought about by vigorous treatment with carbon disulphide.[158] A rapid degradation to the 4-mer-

(51)   (52)   (53)

capto analogue (52) of uramil ensues on heating potassium urate with ammonium sulphide at 160°.[435, 436] Reactions with halogenating agents are described in Chapter V and the partial conversion of the oxo functions to thio are noted in Section 6B. Although inert towards chemical reduction the various products of electrochemical reduction are described elsewhere (Ch. XII, Sect. 1Aa). Details have already been given of the acid oxidation to alloxan (Sect. 9A) or with alkaline reagents to allantoin (Sect. 9B). An excess of disylazane under rigorous conditions (200°, 14 h) gives a 70% yield of 1,2,3,6,7,8-hexahydro-8-oxo-7,9-bis-trimethylsilyl-2,6-bis-trimethylsilyloxypurine (53).[437]

References

1. Kossel, *Z. physiol. Chem.*, **10**, 248 (1886).
2. Baddiley, Lythgoe, and Todd, *J. Chem. Soc.*, **1944**, 318.

3. Parham, Fissekis, and Brown, *J. Org. Chem.*, **31**, 966 (1966).
4. Albert and Brown, *J. Chem. Soc.*, **1954**, 2060.
5. Tafel and Ach, *Ber.*, **34**, 1170 (1901).
6. Jones and Robins, *J. Amer. Chem. Soc.*, **82**, 3773 (1960).
7. Trattner, Elion, Hitchings, and Sharefkin, *J. Org. Chem.*, **29**, 2674 (1964).
8. Cherbuliez and Bernhard, *Helv. Chim. Acta*, **15**, 464 (1932).
9. Shaw, *J. Org. Chem.*, **27**, 883 (1962).
10. Cresswell and Brown, *J. Org. Chem.*, **28**, 2560 (1963).
11. Spies, *J. Amer. Chem. Soc.*, **61**, 350 (1939).
12. Bendich, Tinker, and Brown, *J. Amer. Chem. Soc.*, **70**, 3109 (1948).
13. Parham, Fissekis, and Brown, *J. Org. Chem.*, **32**, 1151 (1967).
14. Fischer, *Ber.*, **30**, 1839 (1897).
15. Okano, Goya, and Kaizu, *J. Pharm. Soc. Japan*, **87**, 469 (1967).
16. Strecker, *Annalen*, **118**, 166 (1861).
17. Fischer, *Annalen*, **215**, 309 (1882).
18. Traube, *Ber.*, **33**, 1371 (1900).
19. Mann and Porter, *J. Chem. Soc.*, **1945**, 751.
20. Bredereck, Von Schuh, and Martini, *Chem. Ber.*, **83**, 201 (1950).
21. Traube and Dudley, *Ber.*, **46**, 3839 (1913).
22. Koppel, O'Brien, and Robins, *J. Amer. Chem. Soc.*, **81**, 3046 (1959).
23. Jones and Robins, *J. Amer. Chem. Soc.*, **84**, 1914 (1962).
24. Fischer, *Ber.*, **43**, 805 (1910).
25. Fischer and Ach, *Ber.*, **30**, 2208 (1897).
26. Karrer, Manunta, and Schwyzer, *Helv. Chim. Acta*, **31**, 1214 (1948).
27. Townsend and Robins, *J. Amer. Chem. Soc.*, **84**, 3008 (1962).
28. Borowitz, Bloom, Rothschild, and Sprinson, *Biochemistry*, **4**, 650 (1965).
29. Thomas and Montgomery, *J. Org. Chem.*, **31**, 1413 (1966).
30. Fischer, *Ber.*, **31**, 104 (1898).
31. Elion, *J. Org. Chem.*, **27**, 2478 (1962).
32. Fischer and Ach, *Ber.*, **32**, 250 (1899).
33. Elion, Mueller, and Hitchings, *J. Amer. Chem. Soc.*, **81**, 3042 (1959).
34. Holmes and Robins, *J. Amer. Chem. Soc.*, **86**, 1242 (1964).
35. Kruger, *Ber.*, **26**, 1914 (1893).
36. Baddiley, Lythgoe, McNeil, and Todd, *J. Chem. Soc.*, **1943**, 383.
37. Albert, *J. Chem. Soc. (B)*, **1966**, 438.
38. Montgomery and Temple, *J. Amer. Chem. Soc.*, **82**, 4592 (1960).
39. Naylor, Shaw, Wilson, and Butler, *J. Chem. Soc.*, **1961**, 4845.
40. Giner-Sorolla and Bendich, *J. Amer. Chem. Soc.*, **80**, 3932 (1958).
41. Giner-Sorolla, Nanos, Burchenal, Dollinger, and Bendich, *J. Medicin. Chem.*, **11**, 521 (1968).
42. Giner-Sorolla, O'Bryant, Burchenal, and Bendich, *Biochemistry*, **5**, 3057 (1966).
43. Bowles, Schneider, Lewis, and Robins, *J. Medicin. Chem.*, **6**, 471 (1963).
44. Temple, Kussner, and Montgomery, *J. Org. Chem.*, **31**, 2210 (1966).
45. Fischer, *Ber.*, **31**, 431 (1898).
46. Neiman and Bergmann, *Israel J. Chem.*, **3**, 161 (1965).
47. Beaman, *J. Amer. Chem. Soc.*, **76**, 5633 (1954).
48. Boehringer, Ger. Pat., 143,725 (1902); *Frdl.*, **7**, 669 (1902–1904).
49. Bader and Downer, *J. Chem. Soc.*, **1953**, 1641.
50. Cook and Thomas, *J. Chem. Soc.*, **1950**, 1884.
51. Townsend and Robins, *J. Org. Chem.*, **27**, 990 (1962).

52. Montgomery and Thomas, *J. Org. Chem.*, **28**, 2304 (1963).
53. Beaman and Robins, *J. Appl. Chem.*, **1962**, 432.
54. Johns and Baumann, *J. Biol. Chem.*, **14**, 381 (1913).
55. Andrews, Anand, Todd, and Topham, *J. Chem. Soc.*, **1941**, 2490.
56. Johnston, Holum, and Montgomery, *J. Amer. Chem. Soc.*, **80**, 6265 (1958).
57. Haggerty, Springer, and Cheng, *J. Medicin. Chem.*, **8**, 797 (1965).
58. Gerster and Robins, *J. Amer. Chem. Soc.*, **87**, 3752 (1965).
59. Levin, Suguira, and Brown, *J. Medicin. Chem.*, **7**, 357 (1964).
60. Taylor, Vogel, and Cheng, *J. Amer. Chem. Soc.*, **81**, 2442 (1959).
61. Johns and Hogan, *J. Biol. Chem.*, **14**, 299 (1913).
62. Biltz and Sauer, *Ber.*, **64**, 752 (1931).
63. Johns and Baumann, *J. Biol. Chem.*, **15**, 515 (1913).
64. Blicke and Schaaf, *J. Amer. Chem. Soc.*, **78**, 5857 (1956).
65. Mackay and Hitchings, *J. Amer. Chem. Soc.*, **78**, 3511 (1956).
66. Cohen, Thom, and Bendich, *J. Org. Chem.*, **28**, 1379 (1963).
67. Cacace and Masironi, *Ann. Chim. (Italy)*, **47**, 366 (1957).
68. Brown, Levin, and Murphy, *Biochemistry*, **3**, 880 (1964).
69. Nofre, Lefier, and Cier, *Compt. rend.*, **253**, 687 (1961).
70. Weiss, *Nature*, **153**, 748 (1944).
71. Ponnamperuma, Lemmon, and Calvin, *Radiation Res.*, **18**, 540 (1963).
72. Lister, *Adv. Heterocyclic Chem.*, **6**, 1 (1966).
73. Bergmann and Dikstein, *J. Biol. Chem.*, **223**, 765 (1956).
74. Bergmann, Ungar, and Kalmus, *Biochim. Biophys. Acta*, **45**, 49 (1960).
75. Jamison and Gordon, *Biochim. Biophys. Acta*, **72**, 106 (1963).
76. Bergmann and Ungar, *J. Amer. Chem. Soc.*, **82**, 3957 (1960).
77. Wyngaarden and Dunn, *Arch. Biochem. Biophys.*, **70**, 150 (1957).
78. Bergmann, Ungar-Waron, Goldberg, and Kalmus, *Arch. Biochem. Biophys.*, **94**, 94 (1961).
79. Dikstein, Bergmann, and Henis, *J. Biol. Chem.*, **224**, 67 (1959).
80. Golovchinskaya, *Osnovnye Napravleniya Rabot Vsesoyuz. Nauk. Issledovatel. Khim. Farm. Inst. im S. Ordzhonikidze 1920–1957*, **1959**, 180.
81. Chkhikvadze, preceding reference, p. 213.
82. Biltz and Rakett, *Ber.*, **61**, 1409 (1928).
83. Ovcharova and Golovchinskaya, *Zhur. obshchei Khim.*, **34**, 3254 (1964).
84. Giner-Sorolla, *Chem. Ber.*, **101**, 611 (1968).
85. Stevens, Giner-Sorolla, Smith, and Brown, *J. Org. Chem.*, **27**, 567 (1962).
86. Burckhalter and Dill, *J. Org. Chem.*, **24**, 562 (1959).
87. Bryant and Harmon, *J. Medicin. Chem.*, **10**, 104 (1967).
88. Albert, *Chem. and Ind.*, **1955**, 202.
89. Bredereck, Siegel, and Föhlisch, *Chem. Ber.*, **95**, 403 (1962).
90. Ishidate and Yuki, *Chem. and Pharm. Bull. (Japan)*, **5**, 240,244 (1957).
91. Bayer, *Ger. Pat.*, 213,711 (1908); *Frdl.*, **9**, 1010 (1908–1910).
92. Goldner, Dietz, and Carstens, *Annalen*, **691**, 142 (1965).
93. Golovchinskaya and Chaman, *Zhur. obshchei Khim.*, **22**, 2225 (1952).
94. Golovchinskaya, Ebed, and Chaman, *Zhur. obshchei Khim.*, **32**, 4097 (1962).
95. Golovchinskaya and Chaman, *Zhur. obshchei Khim.*, **32**, 3245 (1962).
96. Golovchinskaya, *Zhur. obshchei Khim.*, **18**, 2129 (1948).
97. Lister and Timmis, *J. Chem. Soc.*, **1960**, 327.
98. Ikehara and Ohtsuka, *Chem. and Pharm. Bull. (Japan)*, **9**, 27 (1961).
99. Kazmirowsky, Dietz, and Carstens, *J. prakt. Chem.*, **19**, 162 (1963).

100. Montgomery and Temple, *J. Amer. Chem. Soc.*, **83**, 630 (1961).
101. Rice, U.S. Pat., 2,715,125 (1955).
102. O'Brien, Westover, Robins, and Cheng, *J. Medicin. Chem.*, **8**, 182 (1965).
103. Stoll and Schmidt, U.S. Pat., 2,756,229 (1956).
104. Roth, *Arch. Pharm.*, **292**, 234 (1959).
105. Seyden-Penne, Le Thi Minh, and Chabrier, *Bull. Soc. chim. France*, **1966**, 3934.
106. Gräfe, *Arch. Pharm.*, **300**, 111 (1967).
107. Lira and Huffman, *J. Org. Chem.*, **31**, 2188 (1966).
108. Zelnik, Pesson, and Polonovski, *Bull. Soc. chim. France*, **1956**, 1773.
109. Chemiewerk Homberg, A. G. Brit. Pat., 816,299 (1959); through *Chem. Abstracts*, **54**, 1571 (1960).
110. Prasad, Noell, and Robins, *J. Amer. Chem. Soc.*, **81**, 193 (1959).
111. Dille and Christensen, *J. Amer. Chem. Soc.*, **76**, 5087 (1954).
112. Birkofer, Ritter, and Kühlthau, *Chem. Ber.*, **97**, 934 (1964).
113. Brown and Mason, *J. Chem. Soc.*, **1957**, 682.
114. Fischer, *Ber.*, **39**, 423 (1906).
115. Cohen, Thom, and Bendich, *J. Org. Chem.*, **27**, 3545 (1962).
116. Bergmann, Kleiner, Neiman, and Rashi, *Israel J. Chem.*, **2**, 185 (1964).
117. Bergmann, Neiman, and Kleiner, *J. Chem. Soc. (C)*, **1966**, 10.
118. Montgomery, Hewson, Clayton, and Thomas, *J. Org. Chem.*, **31**, 2202 (1966).
119. Miles, *J. Org. Chem.*, **26**, 4761 (1961).
120. Scheit and Holy, *Biochim. Biophys. Acta*, **149**, 344 (1967).
121. Friedmann, Mahapatra, Dash, and Stevenson, *Biochim. Biophys. Acta*, **103**, 286 (1965).
122. Biltz and Max, *Ber.*, **53**, 2327 (1920).
123. Biltz and Pardon, *J. prakt. Chem.*, **134**, 310 (1932).
124. Fischer, *Annalen*, **215**, 253 (1882).
125. Temple, McKee, and Montgomery, *J. Org. Chem.*, **27**, 1671 (1962).
126. Goldner, Dietz, and Carstens, *Z. Chem.*, **4**, 454 (1964).
127. Miles, Howard, and Frazier, *Science*, **142**, 1458 (1963).
128. Albert, *Biochem. J.*, **65**, 124 (1957).
129. Bergmann and Kalmus, *J. Chem. Soc.*, **1962**, 860.
130. Elion, Burgi, and Hitchings, *J. Amer. Chem. Soc.*, **74**, 411 (1952).
131. Beaman and Robins, *J. Amer. Chem. Soc.*, **83**, 4038 (1961).
132. Robins, Jones, and Lin, *J. Org. Chem.*, **21**, 695 (1956).
133. Bergmann, Levin, Kalmus, and Kwietny-Govrin, *J. Org. Chem.*, **26**, 1504 (1961)
134. Prasad and Robins, *J. Amer. Chem. Soc.*, **79**, 6401 (1957).
135. Bergmann and Tamari, *J. Chem. Soc.*, **1961**, 4468.
136. Elion, Goodman, Lange, and Hitchings, *J. Amer. Chem. Soc.*, **81**, 1898 (1959).
137. Robins and Lin, *J. Amer. Chem. Soc.*, **79**, 490 (1957).
138. Bergmann, Kalmus, Ungar-Waron, and Kwietny-Govrin, *J. Chem. Soc.*, **1963**, 3729.
139. Neiman and Bergmann, *Israel J. Chem.*, **3**, 85 (1961).
140. Fu, Chinoporos, and Terzian, *J. Org. Chem.*, **30**, 1916 (1965).
141. Montgomery and Thomas, *J. Org. Chem.*, **31**, 1411 (1966).
142. Koppel and Robins, *J. Amer. Chem. Soc.*, **80**, 2751 (1958).
143. Robins, *J. Amer. Chem. Soc.*, **80**, 6671 (1958).
144. Bergmann and Kalmus, *J. Org. Chem.*, **26**, 1660 (1961).
145. Loo, Michael, Garceau, and Reid, *J. Amer. Chem. Soc.*, **81**, 3039 (1959).
146. Koppel and Robins, *J. Org. Chem.*, **13**, 1457 (1958).

147. Noell and Robins, *J. Amer. Chem. Soc.*, **81**, 5997 (1959).
148. Elion and Hitchings, *J. Amer. Chem. Soc.*, **77**, 1676 (1955).
149. Daves, Noell, Koppel, Robins, and Beaman, *J. Amer. Chem. Soc.*, **82**, 2633 (1960).
150. Noell, Smith, and Robins, *J. Medicin. Chem.*, **5**, 996 (1962).
151. Koppel, O'Brien, and Robins, *J. Amer. Chem. Soc.*, **81**, 3046 (1959).
152. Elion, Lange, and Hitchings, *J. Amer. Chem. Soc.*, **78**, 217 (1956).
153. Wooldridge and Slack, *J. Chem. Soc.*, **1962**, 1862.
154. Merz and Stähl, *Beitrage zur Biochemie und Physiologie von Naturstoffen,* Fischer Verlag, Jena, 1965, p. 285.
155. Kalmus and Bergmann, *J. Chem. Soc.*, **1960**, 3679.
156. Khaletski and Eshmann, *Zhur. obshchei Khim.*, **18**, 2116 (1948).
157. Khaletski and Eshmann, *Zhur. obshchei Khim.*, **20**, 1246 (1950).
158. Boehringer, Ger. Pat., 128,117 (1902); through *Chem. Zentr.*, I, 548 (1902).
159. Boehringer, Ger. Pat., 133,300 (1902); through *Chem. Zentr.*, II, 314 (1902).
160. Biltz and Beck, *J. prakt. Chem.*, **118**, 166 (1928).
161. Sundwik, *Z. physiol. Chem.*, **76**, 486 (1912).
162. Tafel, *Ber.*, **32**, 3194 (1899).
163. Traube et al., *Annalen*, **432**, 266 (1923).
164. Kruger, *Z. physiol. Chem.*, **18**, 422 (1894).
165. Bredereck, Schellenberg, Nast, Heise, and Christmann, *Chem. Ber.*, **99**, 944 (1966).
166. Nagasawa, Kumashiro, and Takenishi, *J. Org. Chem.*, **31**, 2685 (1966).
167. Robins, Godefroi, Taylor, Lewis, and Jackson, *J. Amer. Chem. Soc.*, **83**, 2574 (1961).
168. Lewis, Schneider, and Robins, *J. Org. Chem.*, **26**, 3837 (1961).
169. Biltz and Bulow, *Annalen*, **423**, 159 (1921).
170. Bredereck and Von Schuh, Ger. Pat., 864,869 (1953); through *Chem. Abstracts*, **47**, 11237 (1953).
171. Bredereck, Hennig, Pfleiderer, and Weber, *Chem. Ber.*, **86**, 333 (1953).
172. Fischer and Ach, *Ber.*, **31**, 1980 (1898).
173. Traube, *Ber.*, **33**, 3035 (1901).
174. Golovchinskaya, *Zhur. priklad. Khim.*, **30**, 1374 (1957).
175. Self and Rankin, *Quart. J. Pharmacol.*, **4**, 346 (1931).
176. Frydman and Troparesky, *Anales. Asoc. quim. argentina*, **45**, 79 (1957).
177. Rodionov, *Bull. Soc. chim. France*, **39**, 305 (1926).
178. Kruger and Saloman, *Z. physiol. Chem.*, **26**, 350 (1899).
179. Engelmann, *Ber.*, **42**, 177 (1909).
180. Gepner and Krebs, *Zhur. obshchei Khim.*, **16**, 179 (1946).
181. Kruger and Saloman, *Z. physiol. Chem.*, **21**, 169 (1895).
182. Fischer, *Ber.*, **30**, 2400 (1897).
183. Blicke and Godt, *J. Amer. Chem. Soc.*, **76**, 3653 (1954).
184. Biltz and Beck, *J. prakt. Chem.*, **118**, 198 (1928).
185. Decker, U.S. Pat., 2,509,084 (1950); through *Chem. Abstracts*, **44**, 8366 (1950).
186. Alexander and Marienthal, *J. Pharm. Sci.*, **53**, 962 (1964).
187. Vel'kina, Chaman, and Ebed, *Zhur. obshchei Khim.*, **37**, 508 (1967).
188. Zelnik, *Bull. Soc. chim. France*, **1960**, 1917.
189. Roche Products Ltd. Brit. Pat. 750,588 (1956); through *Chem. Abstracts*, **51**, 2888 (1957).
190. Di Paco and Tauro, *Ann. Chim. (Italy)*, **47**, 698 (1957).
191. Cacace, Fabrizi, and Zifferero, *Ann. Chim. (Italy)*, **45**, 983 (1955).
192. Parikh and Burger, *J. Amer. Chem. Soc.*, **77**, 2386 (1955).

193. Maney, Jones, and Korns, *J. Amer. Pharm. Assoc., Sci. Edn.*, **35**, 266 (1946).
194. Ishay, *J. Chem. Soc.*, **1956**, 3975.
195. Serchi and Bichi, *Farmaco, Ed. Sci.*, **12**, 594 (1957).
196. Cacace, Fabrizi, and Zifferero, *Ann. Chim. (Italy)*, **46**, 91 (1956).
197. Stieglitz and Stamm, Ger. Pat., 1,122,534 (1962); through *Chem. Abstracts*, **56**, 14306 (1962).
198. Klosa, *Arch. Pharm.*, **288**, 301 (1955).
199. Quevauviller, Chabrier, and Morin, *Bull. Soc. Chim. biol.*, **31**, 532 (1949).
200. Richter, Rubinstein, and Elming, *J. Medicin. Chem.*, **6**, 192 (1963).
201. Klosa, *Arch. Pharm.*, **288**, 114 (1955).
202. Baisse, *Bull. Soc. chim. France*, **1949**, 769.
203. Weissenburger, *Arch. Pharm.*, **288**, 532 (1955).
204. McMillan and Wuest, *J. Amer. Chem. Soc.*, **75**, 1998 (1953).
205. Serchi and Bichi, *Farmaco Ed. Sci.*, **11**, 501 (1956).
206. Zelnik, Pesson, and Polonovski, *Bull. Soc. chim. France*, **1956**, 888.
207. Polonovski, Pesson, and Zelnik, *Compt. rend.*, **236**, 2519 (1953).
208. Doebel and Spiegelberg, U.S. Pat., 2,761,863 (1956); through *Chem. Abstracts*, **51**, 3676 (1957).
209. Moroshita, Nakano, Satoda, Yoshita, and Fukuda, Jap. Pat., 6473 (1958); through *Chem. Abstracts*, **54**, 1571 (1960).
210. Ishay, *Arch. Pharm.*, **292**, 98 (1959).
211. Onodera, Hirano, Kashimura, and Yajima, *Tetrahedron Letters*, **1965**, 4327.
212. Eidenbenz and Von Schuh, Ger. Pat., 860,217 (1952); through *Chem. Abstracts*, **47**, 11238 (1953).
213. Merck, Ger. Pat., 290,910 (1914); *Frdl.*, **12**, 783 (1914–1916).
214. Cacace and Zifferero, *Ann. Chim. (Italy)*, **45**, 1026 (1955).
215. Merze and Stähle, *Arch. Pharm.*, **293**, 801 (1960).
216. Merz and Stähle, *Arznein.-Forsch.*, **15**, 10 (1965).
217. Damiens and Delaby, *Bull. Soc. chim. France*, **1955**, 888.
218. MacCorquodale, *J. Amer. Chem. Soc.*, **51**, 2245 (1929).
219. Pommerehne, *Arch. Pharm.*, **234**, 367 (1896); **236**, 105 (1898).
220. Bondzynski and Gottlieb, *Ber.*, **28**, 1113 (1895).
221. Biltz and Max, *Annalen*, **423**, 318 (1921).
222. Kossel, *Z. physiol. Chem.*, **13**, 298 (1889).
223. Van der Slooten, *Arch. Pharm.*, **235**, 469 (1897).
224. Schmidt and Schwabe, *Arch. Pharm.*, **245**, 312 (1907).
225. Myers and Zeleznick, *J. Org. Chem.*, **28**, 2087 (1963).
226. Polonovski, Pesson, and Zelnik, *Compt. rend.*, **240**, 2079 (1955).
227. Golovchinskaya and Chaman, *Zhur. obshchei Khim.*, **30**, 1873 (1960).
228. Bredereck, Christmann, Koser, Schellenberg, and Nast, *Chem. Ber.*, **95**, 1812 (1962).
229. Polonovski, Pesson, and Zelnik, *Compt. rend.*, **241**, 215 (1955).
230. Eckstein and Jajackowska, *Dissertationes Pharm. (Poland)*, **19**, 647 (1967).
231. Doebel and Spiegelberg, U.S. Pat., 2,761,862 (1956); through *Chem. Abstracts*, **51**, 3676 (1957).
232. Bredereck, Küpsch, and Wieland, *Chem. Ber.*, **92**, 566 (1959).
233. Gulland and MacRae, *J. Chem. Soc.*, **1933**, 662.
234. Biltz, Strufe, Topp, Heyn, and Robl, *Annalen*, **423**, 200 (1921).
235. Pfleiderer, *Annalen*, **647**, 161 (1961).
236. Pfleiderer and Sagi, *Annalen*, **673**, 78 (1964).

237. Johns, *J. Biol. Chem.*, **17**, 1 (1914).
238. Fischer and Ach, *Ber.*, **32**, 2721 (1899).
239. Fischer, *Ber.*, **30**, 1846 (1897).
240. Fischer, *Ber.*, **32**, 435 (1899).
241. Biilman and Bjerrum, *Ber.*, **50**, 837 (1917).
242. Fischer, *Ber.*, **30**, 559 (1897).
243. Boehringer, Ger. Pat., 106,493 (1897); *Frdl.*, **5**, 829 (1897–1900).
244. Biltz and Damm, *Annalen*, **413**, 186 (1916).
245. Fischer, Ger. Pat., 91811 (1896); *Frdl.*, **4**, 1252 (1897–1900).
246. Fischer, *Ber.*, **33**, 1701 (1900).
247. Biltz and Krzikalla, *Annalen*, **423**, 255 (1921).
248. Biltz and Pardon, *Ber.*, **63**, 2876 (1930).
249. Fischer and Clemm, *Ber.*, **30**, 3089 (1897).
250. Drygin, *Jahresber. Chem.*, **1864**, 629.
251. Hill and Mabery, *Ber.*, **9**, 370 (1876); *Amer. Chem. J.*, **2**, 305 (1880).
252. Fischer, *Ber.*, **17**, 328 (1884).
253. Biltz and Heyn, *Annalen*, **413**, 98 (1916); *J. Chem. Soc.*, **112**, 293 (1917).
254. Mabery and Hill, *Ber.*, **11**, 1329 (1878).
255. Biltz and Sedlatschek, *Ber.*, **57**, 175 (1924).
256. Fischer, *Ber.*, **32**, 267 (1899).
257. Fischer and Ach, *Ber.*, **28**, 2473 (1895).
258. Fischer, *Ber.*, **28**, 2480 (1895).
259. Herzig, *Z. physiol. Chem.*, **117**, 23 (1921).
260. Biltz and Peukert, *Ber.*, **58**, 2190 (1925).
261. Eyring and Ofengand, *Biochemistry*, **6**, 2500 (1967).
262. Bayer, Ger. Pat., 254,488 (1911); through *Chem. Zentr.*, I, 197 (1913).
263. Biltz, Markwitzky, and Heyn, *Annalen*, **423**, 122 (1921).
264. Boehringer, Ger. Pat., 102,158 (1897); *Frdl.*, **5**, 827 (1897–1900).
265. Boehringer, Ger. Pat., 106,503 (1898); *Frdl.*, **5**, 827 (1897–1900).
266. Brandes and Roth, *Arch. Pharm.*, **300**, 1000 (1967).
267. Okuda, *J. Pharm. Soc. Japan*, **80**, 205 (1960).
268. Mauverney, Brit. Pat., 881,827 (1960).
269. Roth and Brandes, *Arch. Pharm.*, **298**, 765 (1965).
270. Hashizume and Iwamura, *Tetrahedron Letters*, **1965**, 3095.
271. Biltz and Strufe, *Annalen*, **404**, 170 (1914).
272. Knoll, Ger. Pat., 252,641 (1911); *Frdl.*, **11**, 962 (1912–1914).
273. Vieth and Leube, *Biochem. Z.*, **163**, 13 (1925).
274. Ishido, Hosono, Isome, Maruyama, and Sato, *Bull. Chem. Soc. Japan*, **37**, 1389 (1964).
275. Biltz, *Annalen*, **423**, 119 (1921).
276. Boehringer, Ger. Pat., 121,224 (1900); *Frdl.*, **6**, 1182 (1900).
277. Biltz and Schmidt, *Annalen*, **431**, 70 (1923).
278. Khmelevskii, *Zhur. obshchei Khim.*, **31**, 3123 (1961).
279. Golovchinskaya, *Zhur. obshchei Khim.*, **19**, 1173 (1946).
280. Bredereck, Hennig, and Pfleiderer, *Chem. Ber.*, **86**, 321 (1953).
281. Golovchinskaya, *Zhur. obshchei Khim.*, **29**, 1213 (1959).
282. Golovchinskaya and Kolodkin, *Zhur. obshchei Khim.*, **29**, 1650 (1959).
282a. Aroutyunyan, Gounar, Gratcheva, and Zavialov, *Izvest. Akad. Nauk S.S.S.R.*, *Ser. khim.*, **1969**, 655.
283. Ebed, Chaman, and Golovchinskaya, *Zhur. obshchei Khim.*, **36**, 816 (1966).

284. Dolman, Van der Goot, Mos, and Moed, *Rec. Trav. Chim.*, **83**, 1215 (1964).
285. Neiman and Bergmann, *Israel J. Chem.*, **6**, 9 (1968).
286. Montgomery and Thomas, *J. Org. Chem.*, **30**, 3235 (1965).
287. Montgomery, Thomas, and Hewson, *Chem. and Ind.*, **1965**, 1596.
288. Pal and Horton, *J. Chem. Soc.*, **1964**, 400.
289. Rogers and Ulbricht, *J. Chem. Soc.*, (C) **1968**, 1929.
290. Shaw, *J. Amer. Chem. Soc.*, **83**, 4770 (1961).
291. Shaw, *J. Amer. Chem. Soc.*, **80**, 3899 (1958).
292. Fischer, *Ber.*, **31**, 3266 (1898).
293. Montgomery and Temple, *J. Amer. Chem. Soc.*, **79**, 5238 (1957).
294. Schmidt, *Annalen*, **295**, 299 (1883).
295. Maly and Andreasch, *Monatsh.*, **4**, 369 (1883).
296. Fischer and Bromberg, *Ber.*, **30**, 219 (1897).
297. Golovchinskaya, Kolganova, Nikolaeva, and Chaman, *Zhur. obshchei Khim.*, **33**, 1650 (1963).
298. Bredereck, Küpsch, and Wieland, *Chem. Ber.*, **92**, 583 (1959).
299. Lindsay, Paik, and Cohen, *Biochim. Biophys. Acta*, **58**, 585 (1962).
300. Schmidt and Pressler, *Annalen*, **217**, 287 (1883).
301. Schmidt, *Annalen*, **217**, 270 (1883).
302. Fischer, *Ber.*, **30**, 2226 (1897).
303. Sutcliffe and Robins, *J. Org. Chem.*, **28**, 1662 (1963).
304. Gerster and Robins, *J. Org. Chem.*, **61**, 3258 (1964).
305. Holmes and Robins, *J. Amer. Chem. Soc.*, **87**, 1772 (1965).
306. Bergmann and Heimhold, *J. Chem. Soc.*, **1935**, 1365.
307. Schaeffer and Thomas, *J. Amer. Chem. Soc.*, **80**, 3738 (1958).
308. Huston and Allen, *J. Amer. Chem. Soc.*, **56**, 1358 (1934).
309. Wislicenus and Körber, *Ber.*, **35**, 1991 (1902).
310. Bergmann and Heimhold, *J. Chem. Soc.*, **1935**, 955.
311. Ballon and Link, *J. Amer. Chem. Soc.*, **71**, 3743 (1949).
312. Biltz and Bergius, *Annalen*, **414**, 54 (1917); *J. Chem. Soc.*, **112**, 589 (1917).
313. Biltz and Max, *Annalen*, **414**, 68, 79 (1917); *J. Chem. Soc.*, **112**, 589, 590 (1917).
314. Klosa, *J. prakt. Chem.*, **6**, 8 (1958).
314a. Huston and Allen, *J. Amer. Chem. Soc.*, **56**, 1793 (1934).
315. Levene and Bass, "Nucleic Acids," American Chemical Society Monograph Series, New York, 1931.
316. Biltz and Topp, *Ber.*, **44**, 1524 (1911).
317. Biltz and Loewe, *Ber.*, **64**, 1014 (1931).
318. Biltz and Damm, *Annalen*, **406**, 22 (1914).
319. Biltz and Strufe, *Annalen*, **404**, 131 (1914).
320. Fischer, *Annalen*, **215**, 257 (1882).
321. Biltz and Heyn, *Annalen*, **413**, 179 (1916).
322. Biltz and Strufe, *Annalen*, **423**, 242 (1921).
323. Biltz and Pardon, *Annalen*, **515**, 201 (1935).
324. Biltz and Strufe, *Annalen*, **423**, 227 (1921).
325. Biltz and Krzikalla, *Annalen*, **457**, 131 (1927).
326. See reference 312.
327. Biltz and Heyn, *Annalen*, **413**, 7 (1916).
328. Biltz and Strufe, *Annalen*, **413**, 124 (1916).
329. Biltz, *Annalen*, **423**, 282 (1921).
330. Biltz and Heyn, *Ber.*, **52**, 768 (1919).

331. Biltz and Heyn, *Annalen*, **413**, 60 (1916).
332. Bils, Gebura, Meek, and Sweeting, *J. Org. Chem.*, **27**, 4633 (1962).
333. Biltz, *Ber.*, **43**, 1600, 1618 (1910).
334. Biltz and Damm, *Annalen*, **413**, 137 (1916).
335. Fischer, *Ber.*, **30**, 3009 (1897).
336. Todd and Whittaker, *J. Chem. Soc.*, **1946**, 628.
337. Biltz and Strufe, *Annalen*, **404**, 137 (1914).
338. Biltz, *Ber.*, **43**, 3553 (1910).
339. Fischer, *Ber.*, **32**, 447 (1899).
340. Biltz, *J. prakt. Chem.*, **145**, 65 (1936).
341. Maly and Hinteregger, *Monatsheft.*, **2**, 87 (1881).
342. Maly and Hinteregger, *Ber.*, **14**, 723 (1881).
343. Auterhoff and Bohle, *Arch. Pharm.*, **301**, 73 (1968).
344. Delia and Brown, *J. Org. Chem.*, **31**, 178 (1966).
345. Struck and Elving, *Biochemistry*, **4**, 1343 (1965).
346. Fichter and Kern, *Helv. Chim. Acta*, **9**, 429 (1926).
347. Wöhler and Liebig, *Annalen*, **26**, 241 (1838).
348. Biltz and Robl, *Ber.*, **53**, 1950, 1967 (1920).
349. Biltz and Schauder, *J. prakt. Chem.*, **106**, 108 (1923).
350. Dalgleish and Neuberger, *J. Chem. Soc.*, **1954**, 3407.
351. Brandenberger, *Biochim. Biophys. Acta*, **15**, 108 (1954).
352. Brandenberger, *Helv. Chim. Acta*, **37**, 641 (1954).
353. Brandenberger and Brandenberger, *Helv. Chim. Acta*, **37**, 2207 (1954).
354. Canellakis and Cohen, *J. Biol. Chem.*, **213**, 379 (1955).
355. Brandenberger, *Experientia*, **12**, 208 (1956).
356. Hartmann and Fellig, *J. Amer. Chem. Soc.*, **77**, 1051 (1955).
357. Schittenhelm and Warnat, *Z. physiol. Chem.*, **171**, 174 (1927).
358. Fellig, *Science*, **119**, 129 (1954).
359. Biltz, *Ber.*, **43**, 1511 (1910).
360. Mulden, *Ber.*, **6**, 1012 (1873).
361. Biltz and Heyn, *Annalen*, **413**, 89 (1916).
362. Biltz and Strufe, *Annalen*, **413**, 155, 197 (1916).
363. Biltz and Klemm, *Annalen*, **448**, 134 (1926).
364. Biltz and Heyn, *Ber.*, **45**, 1666 (1912).
365. Biltz and Heyn, *Ber.*, **45**, 1677 (1912).
366. Clemm, *Ber.*, **31**, 1450 (1898).
367. Biltz, *Ber.*, **43**, 1589 (1910).
368. Biltz and Topp, *Ber.*, **44**, 1511 (1911).
369. Biltz and Lemberg, *Annalen*, **432**, 137 (1923).
370. Matsuura and Saito, *Tetrahedron Letters*, **1968**, 3273.
371. Biltz and Krzikalla, *Annalen*, **423**, 177 (1921).
372. Biltz and Max, *Ber.*, **54**, 2477 (1921).
373. Biltz, *Ber.*, **44**, 282 (1911).
374. Biltz and Max, *Ber.*, **54**, 2451 (1921).
375. Biltz and Klein, *Ber.*, **58**, 2740 (1925).
376. Golovchinskaya, *Zhur. obshchei Khim.*, **24**, 136 (1954).
377. Biltz and Max, *Annalen*, **423**, 295 (1921).
378. Biltz and Loewe, *J. prakt. Chem.*, **141**, 246, 268 (1934).
379. Scherer, *Annalen*, **73**, 328 (1850).
380. Strecker, *Annalen*, **108**, 129, 141 (1858).

381. Taylor and Cheng, *J. Org. Chem.*, **25**, 148 (1960).
382. Richter, Loeffler, and Taylor, *J. Amer. Chem. Soc.*, **82**, 3144 (1960).
383. Yoshimura, *Biochem. Z.*, **37**, 480 (1911).
384. Traube, *Annalen*, **331**, 64 (1904).
385. Strecker, *Annalen*, **102**, 204 (1857).
386. Kossel, *Z. physiol. Chem.*, **6**, 421 (1882).
387. Kruger, *Z. physiol. Chem.*, **16**, 160 (1892).
388. Wellcome Foundation Ltd., Brit. Pat., 713,286 (1954).
389. Elion Hitchings, U.S. Pat., 2,691,654 (1955).
390. Bendich, Russell, and Fox, *J. Amer. Chem. Soc.*, **76**, 6073 (1954).
391. Elion and Hitchings, *J. Amer. Chem. Soc.*, **78**, 3508 (1956).
392. Kruger and Saloman, *Z. physiol. Chem.*, **24**, 364 (1898); **26**, 367 (1899).
393. Wulff, *Z. physiol. Chem.*, **17**, 468 (1893).
394. Bruhns, *Z. physiol. Chem.*, **14**, 536 (1890).
395. Hitchings, *J. Biol. Chem.*, **143**, 43 (1942).
396. Cheney, Freiser, and Fernando, *J. Amer. Chem. Soc.*, **81**, 2611 (1959).
397. Bayer, Posgay, and Majlat, *Pharm. Zentralhalle*, **101**, 476 (1962).
398. Marcet, "An Essay on the Chemical History and Medical Treatment of Calcul Disorders," London, 1817.
399. Wohler and Liebig, *Annalen*, **26**, 340 (1838).
400. Liau, Yamashita, and Matsui, *Agric. and Biol. Chem. (Japan)*, **26**, 624 (1962).
401. Baxter and Spring, *J. Chem. Soc.*, **1945**, 232.
402. Shaw, *J. Biol. Chem.*, **185**, 439 (1950).
403. Stetton and Fox, *J. Biol. Chem.*, **161**, 333 (1945).
404. Allsebrook, Gulland, and Story, *J. Chem. Soc.*, **1942**, 232.
405. Fischer and Reese, *Annalen*, **221**, 336 (1883).
406. Bayer, Ger. Pat., 254,488 (1911); *Frdl.*, **11**, 963 (1912–1914).
407. Lippmann, *Ber.*, **29**, 2645 (1896).
408. Williams, *J. Biol. Chem.*, **184**, 627 (1950).
409. Wieland and Bauer, *Angew. Chem.*, **63**, 511 (1951).
410. Kossel, *Ber.*, **21**, 2164 (1888).
411. Kossel, *Z. physiol. Chem.*, **13**, 298 (1889).
412. Grinberg, *Zhur. priklad. Khim.*, **13**, 1461 (1940).
413. Rosenthaler and Abelmann, *Ber. Pharm.*, **33**, 186 (1923).
414. Kruger, *Ber.*, **32**, 2818 (1899).
415. Woskresensky, *Annalen*, **41**, 125 (1842).
416. Franzke, Griehl, Grunert, and Hollstein, *Naturwiss.*, **55**, 299 (1968).
417. *Biochemisches Handlexicon*, Springer, Berlin, 1911, Vol. IV, p. 1060.
418. Farb, *Rev. Fac. Cienc. quim.*, **22**, 285 (1947).
419. Runge, *Neuste Phytochemische Entdeckungen*, Berlin, 1820, Vol. I, p. 144.
420. Cohen, *Practical Organic Chemistry*, Macmillan and Company, Ltd., London, 1933, p. 337.
421. Cook and Thomas, *J. Chem. Soc.*, **1950**, 1884.
422. Schmidt, *Ber.*, **14**, 814 (1881).
423. Schultzen, *Z. Chem.*, **1867**, 614.
424. Rosenthaler, *Arch. Pharm.*, **266**, 695 (1929).
425. Levene and Bass, "Nucleic Acids," American Chemical Society Monograph Series, New York, 1931, Ch. 4.
426. Davis, Newton, and Benedict, *J. Biol. Chem.*, **54**, 595 (1922).
427. Falconer and Gulland, *J. Chem. Soc.*, **1939**, 1369.

428. Hatfield, Greenland, Stewart, and Wyngaarden, *Biochim. Biophys. Acta*, **91**, 163 (1964).
429. Forrest, Hatfield, and Lagowski, *J. Chem. Soc.*, **1961**, 963.
430. Lohrmann, Lagowski, and Forrest, *J. Chem. Soc.*, **1964**, 451.
431. Johnson and Johns, *J. Amer. Chem. Soc.*, **36**, 545, 970 (1914).
432. Levene and Senior, *J. Biol. Chem.*, **25**, 607 (1916).
433. Bergmann and Dikstein, *J. Amer. Chem. Soc.*, **77**, 691 (1955).
434. Pfleiderer, *Chem. Ber.*, **92**, 2468 (1959).
435. Fischer and Ach, *Annalen*, **288**, 157 (1895).
436. Hager and Kaiser, *J. Amer. Pharm. Assoc., Sci. Edn.*, **44**, 193 (1955).
437. Birkofer, Külthau, and Ritter, *Chem. Ber.*, **93**, 2810 (1960).

CHAPTER VII

Thiopurines and Derivatives

This chapter embraces thiopurines and derived forms including thioethers, easily obtained by alkylation, which in turn can be oxidised to alkylsulphonyl derivatives. Oxidation of thiopurines has in a few cases given the disulphide analogue but usually higher oxidation states are reached leading to the formation of purine sulphinic and sulphonic acids.

The displacement reactions undergone by thio- and alkylthiopurines makes them, after the halogenopurines, the most useful compounds for purine transformations.

Thiocyanatopurines are fully described in Chapter IX, Section 7.

1. The 2-, 6-, and 8-Thiopurines

A. Preparation

The most important synthetic routes have been elaborated in previous chapters but, for completeness, are outlined below together with less well known preparative methods.

a. *By Synthesis from Pyrimidines*

Purines with thio groups at all three positions are possible using the Traube synthesis (Ch. II). Thus, cyclisation of a 4,5-diamino-2(or 6)-thiopyrimidine gives the corresponding 2-thio-[1] or 6-thiopurine.[2] If the cyclising reagent contains a thiocarbonyl group, as in thiophosgene,[3] thiourea,[1, 4] carbon disulphide,[5] or an isothiocyanate,[6] an 8-thiopurine results. Preparations of this type are especially valuable where a 2-thiopurine is required in view of the general difficulty experienced in replacing a 2-chlorine atom by a thio group or in thiation of a 2-oxo-purine.

A refinement allowing the direct formation of 6-thiopurines from 6-oxo-4,5-diaminopyrimidines, using phosphorus pentasulphide, is described elsewhere (Ch. II, Sect. 1B).

b. *By Synthesis from Imidazoles*

As with the pyrimidines use of a suitable imidazole and appropriate cyclising agent can produce a 2-, 6-, or 8-thiopurine. Ring closures of 4-amino-5-carbamoylimidazoles with thiophosgene,[7] carbon disulphide,[8] and related derivatives give 2-thiopurines while cyclisations of 4-amino-5-thiocarbamoyl-[9] and 4-amino-5-carbamoyl-2-mercaptoimidazoles[8] are routes to 6- and 8-thiopurines, respectively.

c. *By Replacement of Halogen Atoms*

The thiopurine arises on treatment of the halogeno derivative (usually the chloro) with sodium hydrosulphide or thiourea in ethanol under reflux conditions. Thiolacetic acid converts both 6-iodo- and 6-chloropurine to 1,6-dihydro-6-thiopurine in theoretical yield.[10] Thioacetamide in ethanol is also a means of transforming 6-halogenopurines to the 6-thio analogues.[11] These aspects are fully treated in Chapter V (Sects. 5Ha and b).

d. *From Oxopurines*

Conversion to a thio group occurs most readily with derivatives of 6-oxopurine on heating with phosphorus pentasulphide for some hours under reflux in a high boiling solvent, usually pyridine, but β-picoline, tetralin, and kerosene have also been employed. Occasionally, the reaction required heating with phosphorus pentasulphide or trisulphide in the absence of solvent to be effective. Most simple 2- and 8-oxopurines appear to be refractory towards this treatment, but the presence of other groups in the ring can assist the replacement of oxygen by sulphur at these positions. Some oxo functions, which cannot enolise, through the absence of an ionisable hydrogen atom, are directly converted to the thio form in this way.[12, 13] These aspects are more fully discussed in Chapter VI, Section 6B. The use of carbon disulphide in alkali at elevated temperatures to convert uric acid derivatives to the corresponding 8-thiouric acid analogues[14] is described in Chapter II, Section 1R.

e. *From Alkylthio- and Thiocyanatopurines*

Unlike the corresponding pyrimidine compounds thioethers of purines have not been widely exploited as a source of thiopurines, mainly due to the failure to effect a successful thiohydrolysis with conventional reagents, such as hydriodic acid.[15] Some success has been obtained using hydrogen sulphide, as in the conversion of the 3-methyl-6-methylthiopurines (**1**, R = Me and C_6H_5) to the 6-thiopurines (**2**, R = Me and C_6H_5) when the gas is bubbled through an aqueous ammoniacal solution or one of dimethylformamide containing sodium hydroxide.[13, 16]

The 6-thio analogue (**3**) of theophylline likewise arises from the corresponding 6-methylthiopurine.[16] An adaptation of this method converts the cyclic thioether, 7,8-dihydrothiazolo[2,3-*i*]purine (**4**), to 1,6-dihydro-1-(2-mercaptoethyl)-6-thiopurine (**5**).[17]

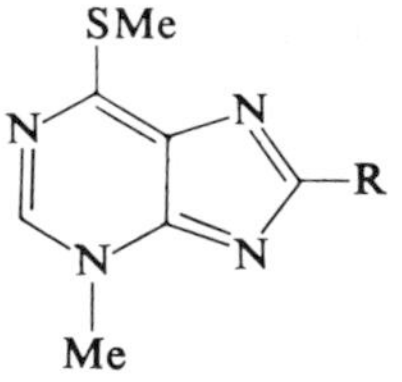

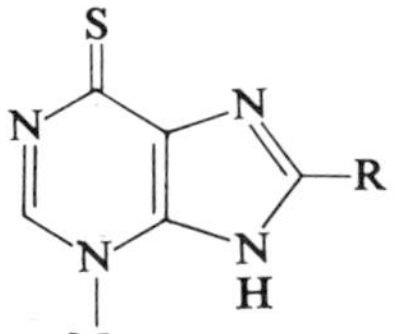

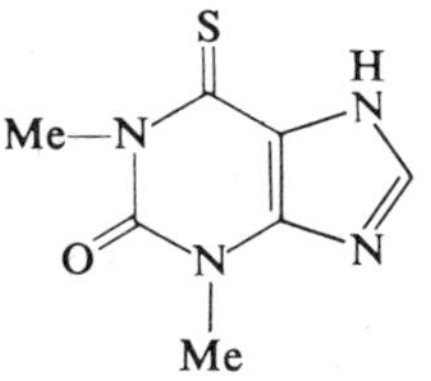

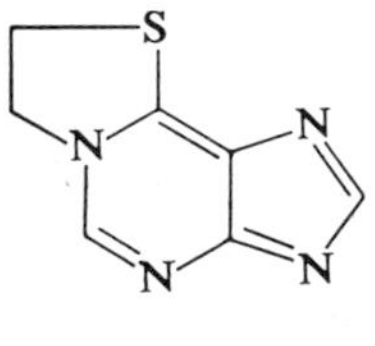

Although phosphorus pentasulphide has general application for demethylation of methylthiopyrimidines,[18] only occasional instances are to be found of use of this reagent in the purine series. One example is noteworthy in that concurrent thiation of an oxo group and thiohydrolysis of the thioether group occurred under the reaction conditions employed: 8-methylthioguanine (**6**) gives 2-amino-1,6,7,8-tetrahydro-

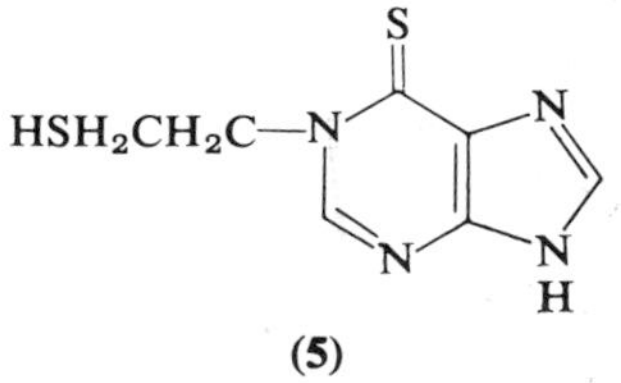

6,8-thiopurine (7).[19] Benzylthio derivatives are better subjects for thiohydrolysis than the methylthio analogues. Using sodium in liquid
ammonia the 2-thio derivatives of 6-methylamino-,[20] 6-butylamino-,[20]
6-dimethylaminopurine,[20] and 1-methylhypoxanthine[15] were prepared
from the corresponding 2-benzylthio analogue. Although formation of a
6-thiopurine by this procedure is known,[21] rapid heating in dimethylformamide was sufficient to debenzylate 3,7-dibenzyl-6-benzylthio-
purinium bromide (8).[22] A novel series of debenzylations have been
effected by heating benzylthiopurines in toluene at 80° in the presence
of aluminium bromide.[23] The p-toluenesulphonate of 6-amino-2-
benzylthio-7,9-dimethylpurine in this way affords 6-amino-7,9-dimethyl-
2-thiopurinyl betaine.[23] Both benzyl groups are removed from the salt
of 2,6-dibenzylthio-7,9-dimethylpurine giving the 7,9-dimethyl-2,6-
dithiopurinyl betaine (9).[23]

Under this section heading it is convenient to include thiocyanatopurines which in dilute sodium hydroxide at room temperature are
rapidly converted to the corresponding thiopurine. Examples of this
transformation are thioguanine from 2-amino-6-thiocyanatopurine and
1,2,3,6-tetrahydro-2,6-dithiopurine from the 2,6-dithiocyanato derivative.[24]

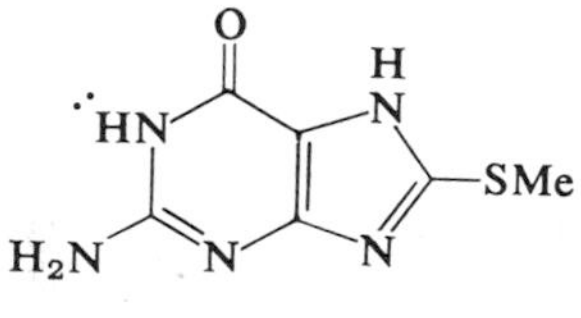

(6) 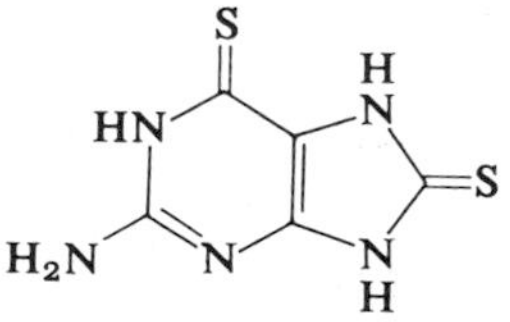   (7)

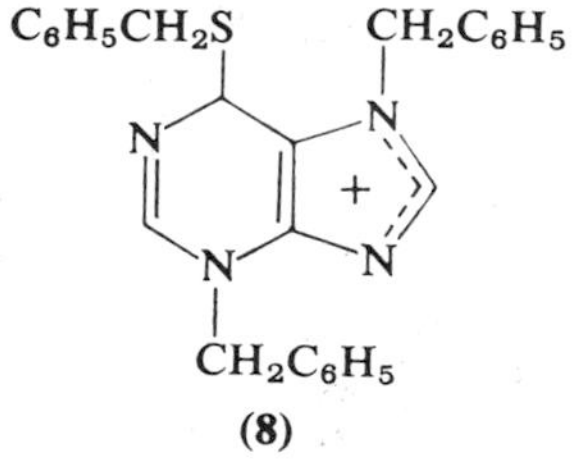

(8) 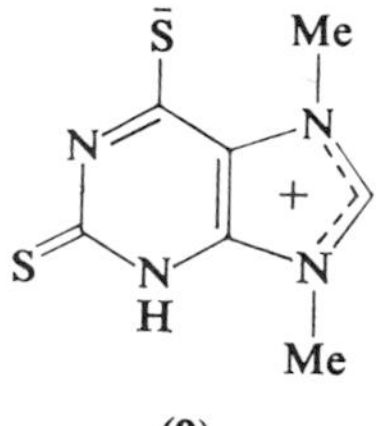   (9) 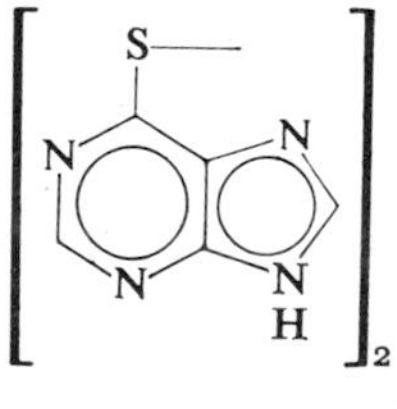   (10)

f. *By Reduction of Disulphides*

Few purine disulphides are known, but dipurin-6-yl disulphide (10)
and other purinyl disulphides are both acid and base labile. On standing

for 24 hours in N-hydrochloric acid the disulphide (10) gives a 79%
yield of 1,6-dihydro-6-thiopurine. An improved yield is obtained in
N-sodium hydroxide in a nitrogen atmosphere.[25]

g. By Direct Introduction of Sulphur

Fusion of purine with an equivalent of sulphur at 245° gives 7,8-
dihydro-8-thiopurine in good yield (75%)[10] but this is reduced if lower
temperatures or a hexachlorobenzene solvent is employed. This procedure also has afforded the 8-thio analogues of 2-amino-[10] and 6-
methylpurines.[10] Although hypoxanthine fails to react under these
conditions, the 8-thio derivatives of 1,9-dimethyl-, 7,9-dibenzyl-, and
7,9-dimethylhypoxanthine are formed in this way.[26] By heating 7,9-
dimethylhypoxanthine in β-picoline with phosphorus pentasulphide, in
addition to displacement of the oxo function by sulphur, insertion at
$C_{(8)}$ also occurs.[26] A solitary example of the formation of a 2-thio
derivative under these conditions is provided by reaction of 1,3,7-
trimethylhypoxanthinium nitrate with sulphur in pyridine.[26a] Sulphur
fusions are unsuccessful with guanine, 6-chloropurine, and 6-methyl-
purine-1-oxide[10] but, on heating the last derivative with thiolacetic acid
or diacetyl sulphide, a mixture of 2- and 8-thiodihydro-6-methylpurine
is formed.[10] Further details of the latter reaction are noted in Chapter
XI, Section 3C.

h. By Rearrangement of Thiazolo[5,4-d]pyrimidines

This route, which is specific for 6-thiopurine formation, is com-
prehensively treated in Chapter II, Section 8A.

i. By Other Methods

Although replacement of amino groups by sulphur is established
practice in pyrimidine chemistry, comparable examples involving
purines are rare. The formation of 1,6-dihydro-6-thiopurine by heating
6-hydrazino- or 6-hydroxyaminopurine with thiolacetic acid[10] are
reactions of this type. Certain 8-aminopurines which can be diazotised
give 8-thiopurines on treating the diazonium derivative with a sulphur-
containing reagent. This route converts 8-diazoxanthine to 8-thio-
xanthine, using thiourea in aqueous solution at room temperature,

while 8-thiohypoxanthine was formed from the diazo derivative in boiling sodium hydrosulphide.[27]

B. Preparation of Extranuclear Mercaptopurines

These compounds usually arise by variations of the standard techniques used for preparing nuclear thio groups. Although 6-mercaptomethylpurine (11) cannot be obtained from 6-chloromethylpurine by treatment with thiourea,[11] the use of thiolacetic acid under reflux conditions gives 6-acetylthiomethylpurine (12), from which 11 is obtained on standing in concentrated ammonia solution under nitrogen at room temperature for one hour. Ammonolysis under these conditions of 6-dithiocarbamylmethylpurine (13), prepared from 6-chloromethylpurine and ammonium dithiocarbamate, provides an alternative synthesis of 11.[11] Substituted 6-thiomethylpurines, in contrast to the parent number, can be prepared by way of thiourea on the 6-chloromethyl derivative and hydrolysis of the resulting isothiouronium compound with 2N-sodium hydroxide. An example from use of this approach is 2-chloro-9-methyl-6-thiomethylpurine.[28] Conversion of 8-chloro- to 8-thiomethylpurines by this approach is also known.[29] Reduction of disulphide groups in side chains affords extranuclear mercaptans, as with di[2-(purin-6-ylamino)ethyl] disulphide (14) which on hydrogenation in sodium hydroxide over palladium gives 6-(2-mercaptoethylamino)purine (15). An 8-(2-mercaptoethylamino)purine arises in the same way.[30] Dealkylation of thioethers is not a reliable means of forming the corresponding mercaptan owing to the likelihood of disulphide formation occurring; examples of attempted debenzylation of extranuclear benzylthioethers producing disulphides are

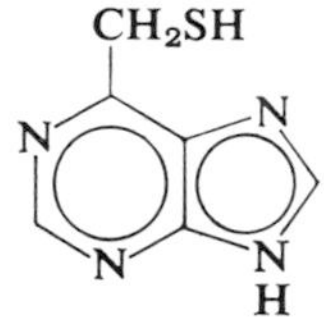

(11) 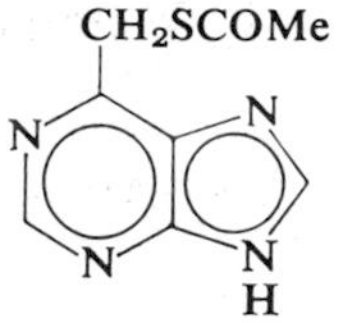   (12) 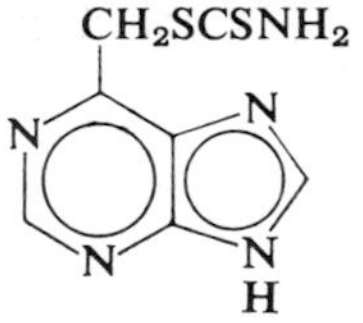   (13)

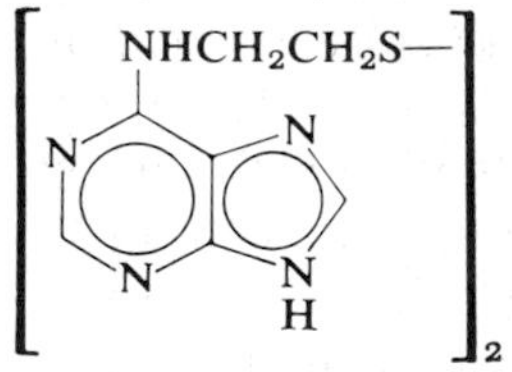

(14) 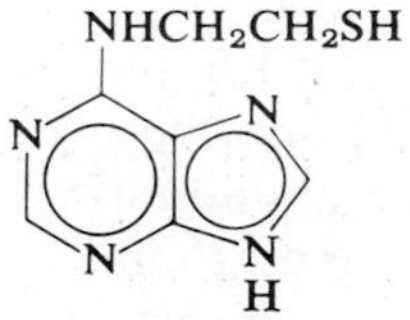   (15) 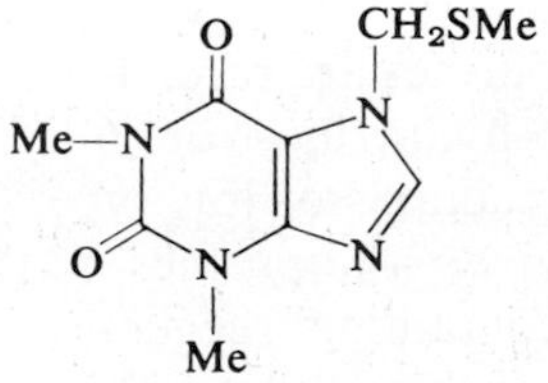   (16)

known.[30, 31] A novel alkylation procedure[32] that allows direct introduction of a methylthiomethyl group is demonstrated by the formation
of the 7-methylthiomethyl derivative (16) when theophylline in dimethylsulphoxide containing phosphorus pentoxide undergoes prolonged
heating at 75°.

C. Reactions of 2-, 6-, and 8-Thiopurines

Although there exists considerable evidence[33, 34] that both in the
solid state and in solution structures of thiopurines are best represented
in the cyclic thioamide forms, it must be emphasised that the "thiol"
tautomers, nevertheless, play an essential part in some reactions.

a. *Removal of Thio Groups*

Before the introduction of suitable nickel catalysts for replacement of
sulphur-containing groups by a hydrogen atom, the methods available
involved oxidation of the groups under acid conditions and hydrolytic
cleavage of the product. This latter procedure is uncertain in outcome as
replacement of the thio group by an oxo group is also possible.

Desulphurisation, sometimes referred to as "dethiation," with Raney
nickel is usually carried out in water, aqueous ammonia or alkali, or
alcohols under reflux conditions, the time varying from one hour
upwards. Complete solution of the purine is not a prerequisite as
suspensions have been successfully employed.

More than one thio group can be removed concurrently, purine itself
being obtained from either 1,6-dihydro-6-thiopurine or 1,2,3,6-tetra-
hydro-2,6-dithiopurine by heating with the reagent for two hours in
dilute ammonia solution.[35] Homologues likewise prepared include
2-methylpurine, from 1,6-dihydro-2-methyl-6-thiopurine,[36] 6-methyl-
purine[10] from both the 2-thio and 8-thio analogues, and likewise
8-methylpurine from either the 2,6-dithiopurine[37] or the 6-thiopurine.[38]
Desulphurisation in boiling water of 1,6-dihydro-1-methyl-6-thiopurine
(17) was used to prepare 1-methylpurine which can only be depicted in
the quinonoid form (18).[39] However, failure to convert either 3,6-
dihydro-3-methyl-6-thio- (19) or 1,2,3,6-tetrahydro-3-methyl-2,6-dithio-
purine to the isomeric 3-methylpurine is reported.[40] Likewise unsuccessful was the attempted dethiation of 1,6-dihydro-2-phenyl-6-thiopurine
which gave an inseparable mixture of products.[41] Sulphur removal is
usually facile if amino groups are present; adenine arises from 2-thio-
adenine.[42] Also resulting from deletion of a thio group at $C_{(2)}$ are

3-methyl- and 8-methyladenine.[37] Examples of 8-aminopurines are seen in the formation of the parent compound from 8-amino-1,6-dihydro-6-thiopurine[43] and 2,8-diaminopurine from the corresponding 6-thio analogue.[43] The oxothiopurines are conveniently desulphurised in dilute sodium hydroxide, with the products being recovered on acidification. Representative examples are the formation of 7,8-dihydro-8-oxo-,[44] 2,3-dihydro-8-methyl-2-oxo-,[37] 2,3-dihydro-3-methyl-2-oxo,[40] and 2,3,-7,8-tetrahydro-3-methyl-2,8-dioxopurine[40] from the appropriate 6-thiopurine. Removal of a 2-thio group has given 1,6,7,8-tetrahydro-6,8-dioxopurine,[44] 3-methylhypoxanthine,[15, 40] and other oxopurines.[37, 40] Both sulphur-containing groups were replaced with hydrogen in conversion of the 2,6-dithiopurine (20) into 7,8-dihydro-3-methyl-8-oxopurine (21),[40] although an attempted dethiation of the corresponding purine lacking an 8-oxo group failed.[40] Removal of a sulphur atom

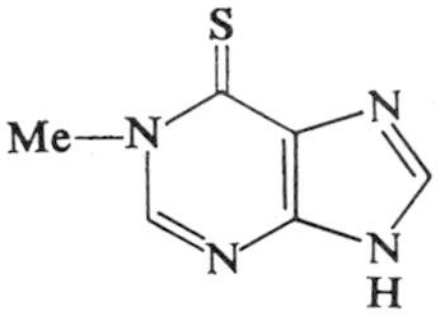

(17) 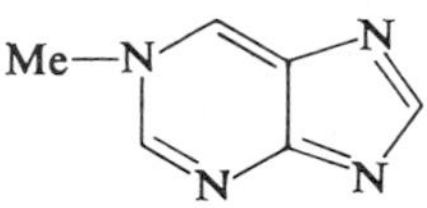   (18) 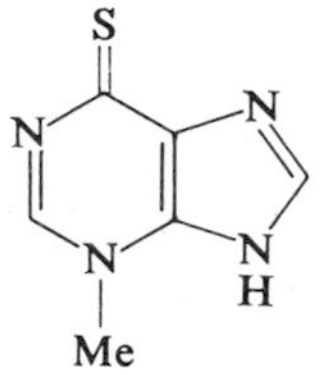   (19)

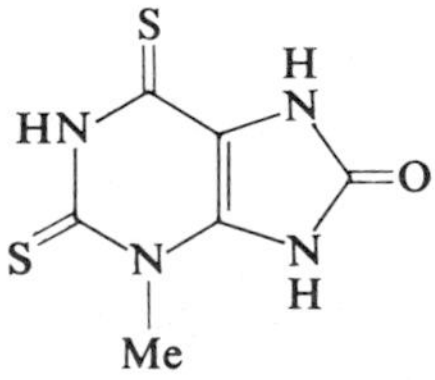

(20) 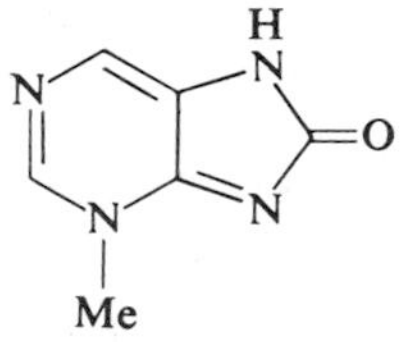   (21)

located at the 8-position is equally facile, being successful in the formation of 9-isobutylxanthine[45] and 1,3,9-trialkylxanthine derivatives.[46]

Desulphurisation of N-methylated thiopurines which, when shown by the classical formulae, contain fixed doubly bound sulphur atoms, can lead to dihydro analogues. Illustrative of this is the conversion of the 6-thio analogue (22) of theophylline to 1,2,3,6-tetrahydro-1,3-dimethyl-2-oxopurine (23) by heating for one hour in dilute ammonia solution with Raney nickel.[12] Effective desulphurisations were likewise carried out with the 2-thio analogue[12] of theophylline and 6-thio analogues of theobromine[47] and caffeine.[48] No correspondingly reduced* purines

---

* Biltz,[50] using nitric acid or iodine-sodium carbonate, had also obtained these desulphurised derivatives nearly 40 years earlier but considered them to be reduced purines, as is revealed by his naming them as "desoxyuric acids."

are obtained from 8-thio derivatives of 7,9-dimethyl-, 3,7,9-trimethyl-(24), and 1,3,7,9-tetramethylxanthine, which give either the corresponding betaine form (25) or the salt on heating their aqueous solutions with Raney nickel[49] or by the action of nitric acid–sodium nitrite mixtures.[49]

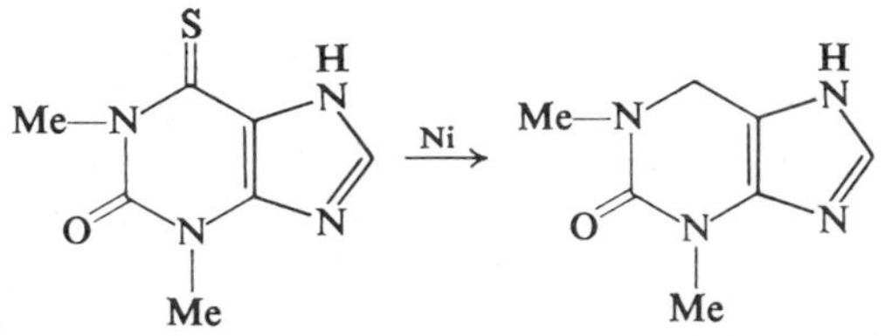

The earlier desulphurisation procedures using acid reagents have limited preparative value, any amino groups present being liable to suffer hydrolysis under the conditions used. In the main most of the purines undergoing this type of reaction are either 2-thio or, more commonly, 8-thio derivatives of oxopurines. The technique is simple: the thiopurine being dissolved in dilute hydrochloric or sulphuric acid, and aqueous sodium nitrite then added slowly. The reaction is completed by heating for a few minutes on a water bath but this step is sometimes unnecessary.

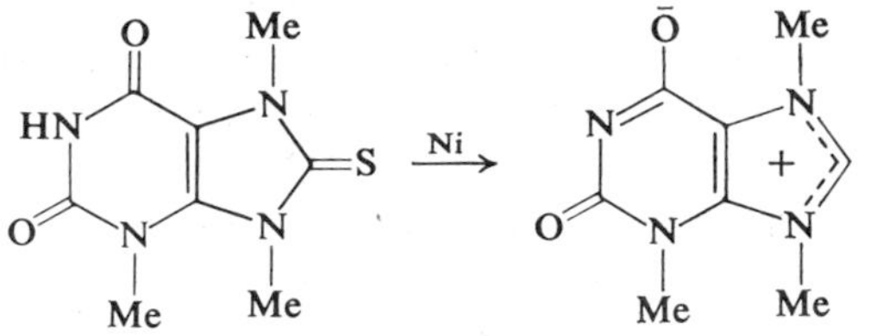

The literature is replete with examples of 8-thio group removal with nitrous acid, representative derivatives prepared in this way are 9-methyl-,[51] 9-ethyl-,[51] 1,3-dimethyl-,[52] 1,9-dimethyl-,[51] 3-methyl-9-phenyl-, 1,3,9-trimethyl-,[46, 51] 9-ethyl-1,3-dimethyl-,[46, 51] and 9-benzyl-1,3-dimethylxanthines.[46]

Although dilute nitric acid has had frequent use as a hydrolysing medium for converting thio groups to oxo groups, some application for the oxidative removal of sulphur groups is found. At water bath temperature, 2-thiohypoxanthine is transformed into hypoxanthine[53] and 3-methylhypoxanthine is also derived from the 2-thio analogue.[54] Hydrogen peroxide acidified with 20% sulphuric acid is an alternative oxidising medium effectively affording adenine from 2-thioadenine.[53]

Although nickel boride is an excellent desulphurising agent for certain thioheterocyclic systems, it failed to react with either 7,8-dihydro-8-thio- or 1,2,3,6-tetrahydro-2,6-dithiopurine.[55]

b. *S-Alkylation*

This operation, in almost every case, is carried out under standard conditions, which consist of treating a solution of the thiopurine, containing just over one equivalent of sodium or potassium hydroxide to each mercapto group present, with the alkyl halide. Any solubility difficulties at this stage can be overcome by the addition of an organic solvent, ethanol or dioxan being suitable. By keeping the working temperature below 40° the risk of contamination with *N*-alkylated products is minimised. For preparation of methylthio derivatives dimethyl sulphate can replace methyl iodide equally well, the desired thioether precipitating from the reaction mixture. The examples below illustrate the scope and general applicability of the procedure. Methylation to the appropriate mono-, di-, or trimethylthiopurine occurs with the following: 2-thio-,[1] 6-thio-,[2] 2-methyl-6-thio-,[13] 3-methyl-6-thio-,[40] 7-methyl-6-thio-,[56] 8-methyl-6-thio-,[38] 9-methyl-6-thio-,[57] 9-ethyl-6-thio-,[58] 8-phenyl-6-thio-,[59] 2,6-dithio-,[60] 7-methyl-2,6-dithio-,[61] 2,8-dithio-,[60, 62] 6,8-dithio-,[63] and 2,6,8-trithiopurines.[60] Aminothiopurines also react with methyl iodide or dimethyl sulphate under these conditions, exemplified in the preparation of 2-amino-6-methylthio-,[64, 65] 2-amino-3-methyl-6-methylthio-,[66] 2-amino-8-methyl-6-methylthio-,[67] 2-amino-6,8-dimethylthio-,[67] and 6-amino-2,8-dimethylthiopurine.[68] Many oxothiopurines *S*-methylate readily as, for example, do both 8-oxo-2-thio-[69] and 8-oxo-6-thiotetrahydropurine[44] but not the isomeric 2-thiohypoxanthine under the same conditions.[20] Dimethylthio derivatives arise with 8-oxo-2,6-dithio-[60, 70] and 6-oxo-2,8-dithio-1,2,3,6,7,8-hexahydropurines.[60]

Homologous alkylthiopurines are formed from the appropriate alkyl halides.[38, 67, 71–73] Benzyl chloride gives the appropriate benzylthio ethers of 2-thio-,[74] 6-thio-,[72] 3-benzyl-6-thio-,[75] 7-benzyl-6-thio-,[72] 9-benzyl-6-thio-,[72] 8-thio-,[74] and 9-ethyl-8-thiopurine,[74] the reactions being carried out in dimethylformamide containing potassium carbonate, gentle heat being required in some cases. The more usual medium of sodium or potassium hydroxide solution, generally at room temperature, is used to benzylate 6-thio-,[71] 1-methyl-6-thio-,[39] 7-methyl-6-thio-,[76] 9-benzyl-6-thio-,[76] 2,6-dithio-,[20, 23] and substituted thiopurines such as 2-amino-6-thio-,[20] 2-amino-3-methyl-6-thio-,[66] 2-amino-7(and 9)-methyl-,[77] 2-oxo-6-thio-,[20] 8-oxo-6-thiopurine,[44] and others.[67, 78]

The protons associated with ring nitrogen atoms appear to be available for enolisation of a doubly bound sulphur atom as both 6-thio analogues of theobromine[78a, 47] and theophylline[16] form the appropriate 6-methylthio derivative with methyl iodide in sodium hydroxide at room temperature as, for example, **26**. Similar S-alkylations occur with the 1-methyl[79] and 3-benzyl derivatives of 6-thiopurine in aqueous[40] or aprotic solvents.[75]

Di(purin-6-yl)alkanes of the type **27** ($n = 1$ or 4) follow from interaction of 1,6-dihydro-6-thiopurine, in excess, and an α,ω-dihalogenoalkane in alkali or dimethylformamide containing potassium carbonate.[17, 80] Corresponding dipurin-8-ylthioalkanes are formed from 8-thiotheobromine.[81] An exception arises from the action of 1,2-dibromoethane on the above 6-thiopurine, the product being 7,8-dihydrothiazolo[2,3-*i*]purine (**28**).[17] Somewhat similar end products are found when 8-thiopurines and chloroacetone are reacted in ethanol, the primary products, 8-acetonylthiopurines[52, 82–84] being capable of spontaneous cyclodehydration to thiazolo[2,3-*f*]purines. Illustrated is 1,2,3,4-tetrahydro-6-methyl-2,4-dioxothiazolo[2,3-*f*]purine (**30**) [RRI 2341] derived from 8-acetonylthioxanthine (**29**).[82]

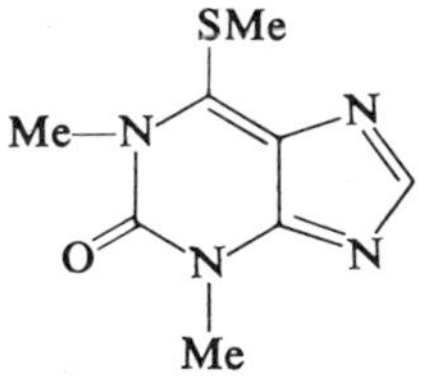

(26) 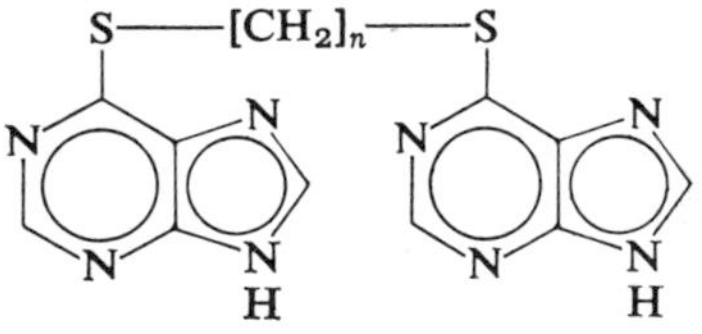   (27)

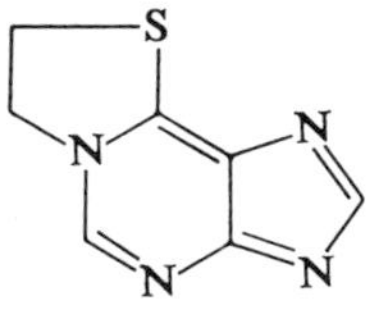

(28) 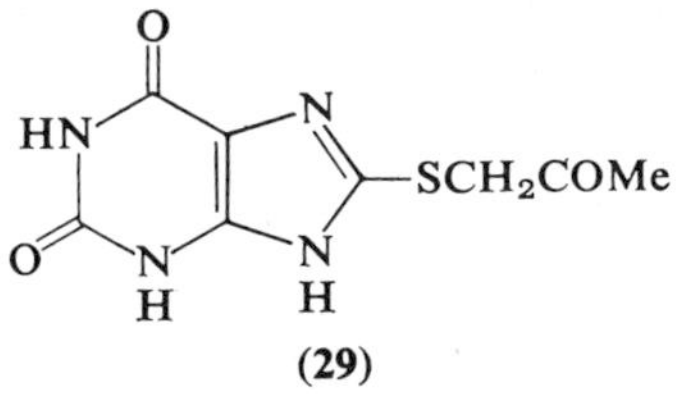   (29) 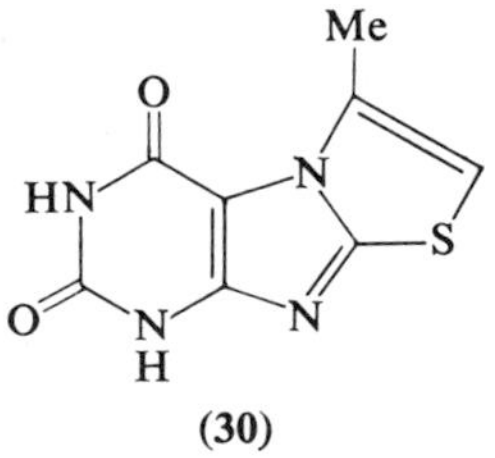   (30)

The interaction of thiopurines with chloroacetic acid and homologous acids is dealt with in Chapter IX, Section 9Ba. Miscellaneous reactions of alkylating character include transformation of 6-thio- to 6-aminoethylthiopurines by means of aziridines[85] and reaction of 2-, 6-, and 8-thiopurines with cyanogen bromide giving thiocyanatopurines. Derivatives of the latter type are discussed in Chapter IX, Section 7A.

c. *N-Alkylation*

Under forcing conditions both *S*- and *N*-alkylation is possible but in many cases the product is only the *N*-alkylated thiopurine. Studies made of this reaction[13] indicate that an *S*-alkylated purine is most likely formed initially but is converted back to a thiopurine during, or after, the ensuing *N*-alkylation. This is well illustrated by the formation of the 6-methylthioderivative (**32**) from the 6-thio analogue of theobromine (**31**) using methyl iodide in dilute sodium hydroxide at room temperature, whereas the 6-thio analogue of caffeine (**33**) results with dimethyl sulphate in sodium hydroxide at 40°.[78a] Comparable *S*-methyl[16] and *N*-methyl[78a] forms are obtained with the 6-thio analogue of theophylline. An interesting point is that under similar conditions the related 6,8-dithio analogue only suffers *S*-methylation at $C_{(8)}$.[86] Because of the initial *S*-methylation, the order of *N*-alkylation is generally found to be the same irrespective of whether a thio- or alkylthiopurine is the starting material. In aprotic solvents,which include dimethylformamide, dimethylacetamide, and dimethyl sulphoxide, both *S*- and *N*-alkylation can be effected although the higher reaction temperatures usually employed with them can be utilised to favour formation of only *N*-alkylated thiopurines. As found in other purines the site of primary *N*-alkylation is dependent on the presence or absence of a base, usually potassium carbonate, the products being either a mixture of 7- and 9-alkyl or only 3-alkyl derivatives, respectively. Benzyl chloride and 1,6-dihydro-6-thiopurine react (dimethylformamide/K_2CO_3/70°) forming 7-benzyl- and 9-benzyl-6-benzylthiopurine.[72] To overcome involvement of the sulphur function use of bromodiphenylmethane as a blocking reagent has been tried.[86a] Alkylation of 6-diphenylmethylthiopurine with substituted benzyl halides gives predominantly the 9-alkyl derivative which is unblocked to the 9-alkyl-6-thiopurine in a mixture of trifluoroacetic acid and phenol under reflux.[86a]

Formation of $N_{(3)}$-alkyl derivatives on omission of the base is seen in the isolation of 3-methyl-6-methylthiopurine from 1,6-dihydro-6-thiopurine in dimethylformamide containing methyl iodide.[87] The $C_{(6)}$-alkylthio group seems remarkably stable in this derivative as its preparation with methyl *p*-toluenesulphonate in dimethylacetamide at 170°[88] is reported.

Whether in the presence or absence of potassium carbonate 1-benzyl-1,6-dihydro-6-thiopurine gives a mixture of 1,7- and 1,9-dibenzyl-1,6-dihydro-6-thiopurine[89] but different ratios of products are found, the presence of carbonate favouring a preponderance of the 1,7-isomer whereas its absence reversed the effect.[89] This is surprising in view of the known fact that other 1-alkyl-1,6-dihydro-6-thiopurines give only

1,7-dialkyl derivatives in the absence of carbonate but 1,9-dialkyl isomers if it is present.[39, 75]

Substituted thiopurines can undergo concurrent methylation in the imidazole ring and at the sulphur atom, as with 2-thiohypoxanthine which at 170° forms the quaternary salt of 1,6-dihydro-7,9-dimethyl-2-methylthio-6-oxopurine from which the betaine form (34) follows reaction with cold concentrated ammonium hydroxide. Variations allowed by this procedure are seen in the formation of the toluene-*p*-sulphonate of 2-amino-7,9-dimethyl-6-methylthiopurine (35), from either 2-amino-6-thio-, 2-amino-7-methyl-6-thio-, or 2-amino-9-methyl-6-thio-1,6-dihydropurine.[23, 88] Other examples of combined *S*- and *N*-alkylation include 28, arising from the action of 1,2-dibromoethane on a 6-thiopurine,[68] noted previously, and related dihydrothiazolopurine formation from 2- and 8-thiopurines.[68, 90] In passing it should be noted that the same products arise from thermal cyclisation of

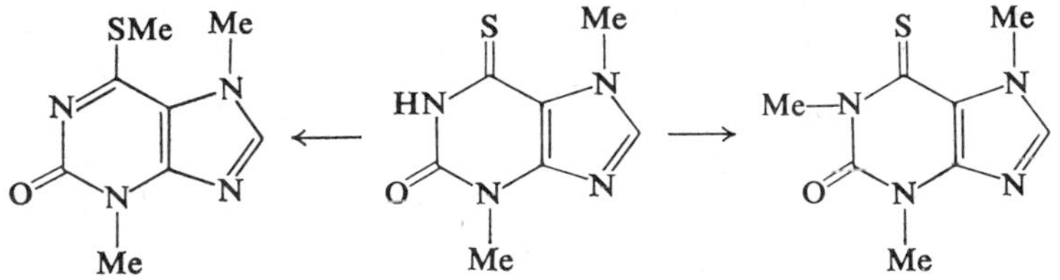

(32)   (31)   (33)

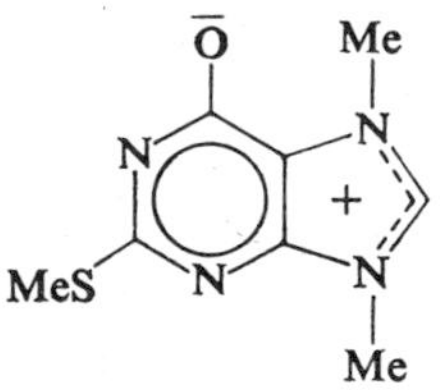

(34) 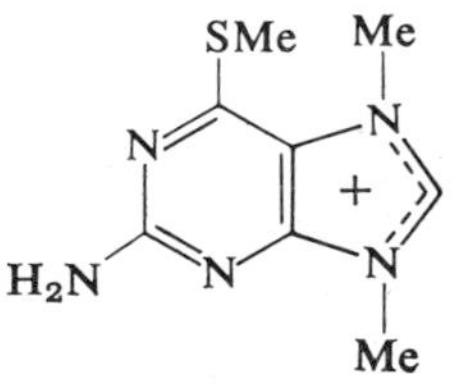   (35) 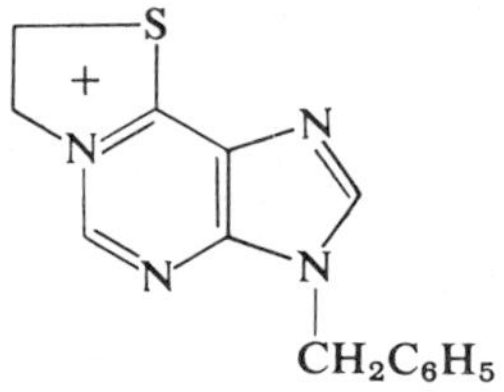   (36)

2-halogenoethylthiopurines[68] but if the ring nitrogen atom is unprotonated, as in 9-benzyl-6-(2-bromoethylthio)purine, a quaternary bromide such as 36 results.[17]

Mannich-type reactions provide a route to 9-aminomethyl derivatives as in the preparation of the 9-piperidinomethylpurine on treating 1,6-dihydro-6-thiopurine with formaldehyde and piperidine in ethanol. Morpholine forms the analogous 9-morpholinomethyl derivative.[91]

d. *Thio- to Oxopurines*

The various methods for effecting these transformations are fully described in Chapter VI, Section 1F.

e. *Thio- to Chloropurines*

Details of this facile and useful conversion of a thio group to a chlorine (also bromine and iodine) atom are given in Chapter V, Sections 1C, 1Eb, and 1F.

f. *Thio- to Aminopurines*

Few examples of direct replacement of a thio group by ammonia or amines exist. In contrast to the relative ease with which the corresponding thioether group is displaced, the thio group requires extremely vigorous conditions that can in some instances lead to cleavage of the molecule.*

Although refractory towards ammonia, 1,6-dihydro-6-thiopurine reacts to a detectable extent with an aqueous solution of ethylamine in a closed vessel at 140° for 15 h.[2] Concentrated ammonium hydroxide converts the 3-methyl homologue to 3-methyladenine[15] but alcoholic ammonia was successful with the 3-benzyl analogue.[93] Similar procedures with 1,2,3,6-tetrahydro-3-methyl-2,6-dithiopurine resulted in a 45% yield of 6-amino-2,3-dihydro-3-methyl-2-thiopurine.[15] Various rearrangements result from amination of thio groups. Although aqueous ammonia under pressure degrades 1,6-dihydro-1-methyl-6-thiopurine (37) to 4-amino-5-carbamoylimidazole, the use of alcoholic ammonia gives 6-methylaminopurine (38). Formation of the latter is readily explained by the known facility with which 1-alkylated adenine derivatives, under basic conditions, undergo the Dimroth rearrangement (see Ch. VIII, Sect. 1K) to 6-alkylaminopurines. This has been demonstrated by the fact that the expected product, 1-methyladenine, can be completely converted to 6-methylaminopurine in alkaline solution at 100°.[94] Under like conditions the 9-β-D-riboside of 1-benzyl-1,6-dihydro-6-thiopurine rearranges to that of 6-benzylaminopurine.[75] Another rearrangement that must also involve ring fission, bond rotation, and recyclisation is occasioned by prolonged heating in alcoholic ammonia (150°) of 1,2,3,6-tetrahydro-1-methyl-2,6-dithiopurine (39) from which 6-amino-2,3-dihydro-2-thiopurine (40) is isolated as the main product.[15] The precise mechanism of this interesting reaction, however, still awaits elucidation.

* Nevertheless, the statement by one author[92] that thio groups in purines cannot be replaced by amines is erroneous.

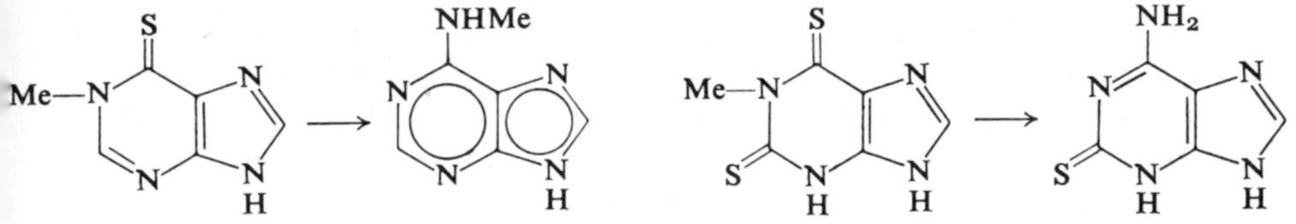

(37)   (38)   (39)   (40)

g. *Thio- to Nitropurines*

This reaction, which is specific for some 8-thiopurines, is dealt with in Chapter X, Section 1Aa.

h. *Oxidation*

Under the appropriate oxidising conditions, thiopurines can give rise to S-oxides (Sect. 8), sulphinic or sulphonic acids (Sect. 4), or disulphides (Sect. 3). The valuable oxidative hydrolysis of a thio group to an oxo group by means of chlorine is treated in Chapter VI, Section 1F.

i. *Acylation*

Depending on the particular thiopurine used, acylation may produce an S-acyl or N-acyl derivative. In some cases an NS-diacylated variant

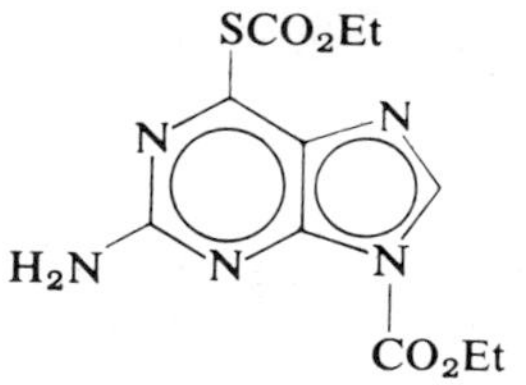

(41)

is obtained. If the imidazole proton is present in 6-[94a] or 8-thiopurines,[94b] reaction with acetic anhydride or ethyl chlorocarbonate yields the appropriate 9-acyl-6- or 8-thiopurine. These findings corrected earlier claims[95, 96] to have formed the 6- or 8-acylthiopurines by these procedures. With $N_{(9)}$-substituted analogues, as in the case of 1,6-dihydro-9-methyl-6-thiopurine, S-acylation is found to occur.[96] An example of $N_{(9)}$S-diacylation is the formation of **41** from 2-amino-1,6-dihydro-6-thiopurine (thioguanine) on treatment with ethyl chlorocarbonate in aqueous base.[94a] Acylating conditions vary with reagent; acetic and other acid anhydrides are used alone or in toluene under reflux whereas ethyl chlorocarbonate is employed in cold dimethylformamide containing either potassium carbonate or trimethylamine.[94b] The mechanism proposed[94a] for these acylations, which postulates an initial S-acylation followed by an $S \rightarrow N_{(9)}$-transacylation step, is given support by the observation that 9-acetyl-1,6-dihydro-6-thiopurine is the product when 6-chloropurine and potassium thiolacetate (MeCOSK) are reacted in dimethylformamide.[94a] Where blocking of the sulphur group by alkylation is effected, direct $N_{(9)}$-acylation is possible.

D. Reactions of Extranuclear Mercaptopurines

Extranuclear mercapto groups undergo most of the reactions associated with the nuclear-located sulphur groups. Exemplifying this are the desulphurisation of 6-mercaptomethylpurine to 6-methylpurine with Raney nickel[11] or conversion to the alkylthiomethyl derivative

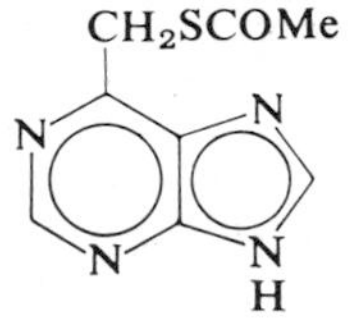

CH$_2$SCOMe

(42)

with an alkyl halide. The acetylated form (42) is produced on treatment of 6-mercaptomethylpurine with thiolacetic acid.[11] Reaction with hydrazine in ethanol under reflux gives the hydrazone of purine-6-aldehyde[97] (see Ch. IX, Sect. 5Bb).

2. The Thioethers: Alkyl(and Aryl)thiopurines

Although, strictly speaking, dipurinyl sulphides should also be included under this heading they are accorded separate treatment in Section 3.

A. Preparation of 2-, 6-, and 8-Alkylthiopurines

The major routes have been described already but a brief summary of available methods is given.

a. *By Direct Synthesis*

Traube syntheses performed with 2- or 6-alkylthio-4,5-diamino-pyrimidines afford 2- or 6-alkylthiopurines. By carrying out cyclisations of 4,5-diaminopyrimidines with NN'-dimethylthiourea under fusion conditions, an 8-methylthiopurine may result, as in the case of 8-methylthiotheophylline.[98] This route, however, is not specific as use of this substituted thiourea is known to give 8-thiopurines also. An unambiguous synthesis of 8-alkylthiopurines is found in the cyclisation of derivatives of 2-alkylthioimidazoles (Ch. III, Sects. 1Bg and h). A unique example of rearrangement of a purine-3-oxide in the presence of methionine to an 8-methylthiopurine is given in Chapter XI, Section 3C.

b. *By S-Alkylation of Thiopurines*

Due to the facile nature of this reaction and the standard conditions required it is the most important means of preparing purine thioethers and has been fully discussed in Section 1Cb.

c. *From Halogenopurines*

Reaction of halogenopurines, usually chloropurines, with alkyl mercaptans is an excellent route to 6-alkylthiopurines but is of less use in the case of 2- or 8-alkylthiopurines due to the general reluctance of halogen atoms at these positions in many purines to undergo nucleophilic displacement (see Ch. V, Sect. 5H).

B. Reactions of Alkylthiopurines

The relatively facile displacement of many methylthio groups by nucleophilic reagents makes this type of purine exceptionally valuable for purine transformation reactions and is, in this respect, second in importance only to a halogenopurine.

a. *Reductive Removal of Alkylthio Groups*

Raney nickel desulphurisation is normally carried out in methanol, ethanol, or propanol, and occasionally water, under reflux conditions, aqueous bases not being required as is usual with thiopurines. Removal of the thioether group is denoted by a pronounced odour of the appropriate alkyl mercaptan. The method is effective for removal of $C_{(2)}$-, $C_{(6)}$-, or $C_{(8)}$-alkylthio groups. Adenine,[99] 9-methyladenine,[100] and 6-dimethylaminopurine[101] arise from the respective 2-methylthio-6-aminopurines. Examples of 6-methylthio group removal occur in the formation of 3-methyl-[102] and 3-methyl-8-phenylpurine,[59] although difficulty was reported with the preparation of 2-phenylpurine[41] in this way. The 6-methylthiopurine (43) gives 2,3-dihydro-1,3-dimethyl-2-oxopurine (44).[47] Removal of sulphur at $C_{(8)}$ is illustrated by the conversion of the 8-methylthio derivative in boiling water to 1,3,9-trimethylxanthine.[46] Examples of concurrent displacement of two methylthio groups are seen in the formation of 9-methyladenine,[103] 6-dimethylaminopurine,[101] as well as the 7- and 9-methyl (and -ethyl) derivatives[104] from the respective 2,8-dimethylthiopurines. Only the 8-methylthio group is displaced from 6-amino-9-methyl-2,8-dimethylthiopurine on reducing the reflux period from four hours to one hour.[68]

A novel application is to the preparation of the four possible mono-N-ethyl derivatives of purine by heating the appropriate dihydrothiazolopurine with Raney nickel in propanol. The preparation of 3-ethylpurine (46) from 7,8-dihydrothiazolo[3,2-e]purine (45) exemplifies this route.[74]

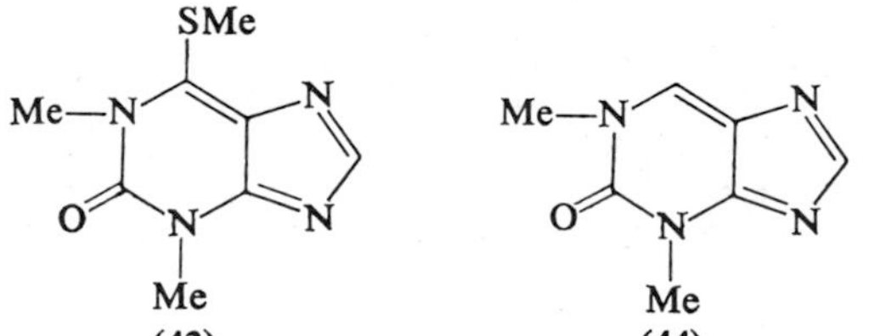

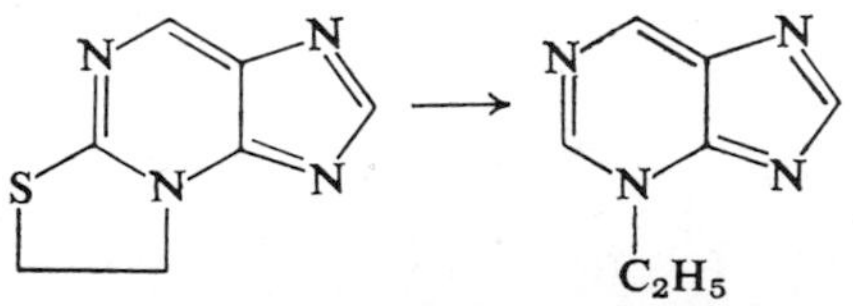

b. *Alkylthio- to Oxopurines*

The direct approach entails acid hydrolysis, usually with hydrochloric acid, under reflux conditions. This procedure, however, with

6-methylthiopurine and the 7- and 9-methyl homologues, leaves the methylthio group untouched but pyrimidine ring fission occurs giving the appropriate 4-amino-5-methylthiocarbonylimidazole even with dilute acid.[105] Successful conversions from alkylthio to oxo group under acid conditions are given in Chapter VI, Section 1F.

Alkylthio groups may also be oxidised to methylsulphonyl groups which are readily hydrolysed, by acid or alkali, to oxo groups. This aspect is dealt with later in the chapter (Sect. 7B).

c. *Alkylthio- to Aminopurines*

This reaction, surprisingly, was first reported over fifty years ago, by Johns who converted 2,3-dihydro-8-methylthio-2-oxopurine to the 8-methylamino analogue with aqueous methylamine at 100°.[106] Subsequently, the reaction was extended to include aminations at $C_{(2)}$ and $C_{(8)}$. The ease of replacement of a 2-alkylthio group depends largely on the nature of other substituent groups present. Both 2-methylthiopurine[1, 107] and 2-methylthioadenine[108] are unreactive or decompose to the 4,5-diaminopyrimidine on heating in aqueous alkylamine at 140° but 2-methylthiohypoxanthine is converted in fair yield to the 2-alkylamino analogue.[108] Although some activation of the methylthio group might be expected in the 3-methylpurine (47), no conversion to 3-methylguanine (48) occurs in aqueous or alcoholic ammonia, but in formamide at 190° some 48 is formed.[15]

As adenine derivatives are generally more accessible by direct synthesis, the conversion of 6-methylthio and 6-benzylthio groups to amino groups is not widely practised but 3-methyl-[88] and 3,7-dibenzyl-[22] adenine are formed from the respective 6-methylthio- and 6-benzylthio-purine with methanolic ammonia. The temperature (100°) and pressure conditions used in the former case seem unnecessary in view of the increased electrophilic nature of $C_{(6)}$ due to alkylation of $N_{(3)}$, this being borne out by the fact that in the latter case amination is possible at room temperature. A contrast is provided by the preparation of 9-methyladenine with ammonium hydroxide for which the usual vigorous conditions (140°, 18 h) are needed.[15] Examples of reactions between 6-alkylthiopurines and alkyl- and dialkylamines and arylamines are numerous. Amines most frequently used include methylamine, dimethylamine, diethylamine, butylamine, benzylamine, and aniline. These and others have been condensed with 6-methylthio-,[2, 71, 109–114] 3-methyl-6-methylthio-,[88] 3-methyl-6-methylthio-8-phenyl-,[59] 2-amino-6-methylthio-,[65, 115] 2-amino-3-methyl-6-methylthio-,[66] 6-methylthio-2-

oxo-,[116] 6-benzylthio-2-oxo-,[20] and 6-benzylthio-1-methylpurine.[117] Condensation of 2,6-dimethylthiopurine with aqueous methylamine, at 130° for 16 h, gives only the 6-methylamino derivative; the corresponding monoaminopurines are similarly obtained with dimethylamine and butylamine.[20] 2,6-Dibenzylthiopurine, likewise, only forms the 6-aminopurine.[20] The failure of alkylthio groups at C$_{(2)}$ to undergo replacement is not surprising in view of the known[108, 118] failure to react 6-amino-2-methylthiopurine with amines. An unusual rearrangement recorded in the early literature, involving a methylthio group so placed, is transformation of 1,6-dihydro-1-methyl-2-methylthio-6-thiopurine (49) to 6-amino-2-methylaminopurine (50) on heating with aqueous ammonia.[15]*

In contrast to 2-methylthiopurine the 8-methylthio analogue is stable in hot aqueous alkali; with ammonia solution in the presence of copper acetate and copper bronze at 160° prolonged heating gives 8-aminopurine.[1] Also produced, but without catalyst, were 8-methylamino- and 8-dimethylaminopurine.[1] Unlike the 2-methylthio derivative the 8-methylthio derivative of adenine reacts with aqueous ammonia at 160° to give 6,8-diaminopurine.[63] The preparation of 2,3-dihydro-8-methylamino-2-oxopurine from the 8-methylthio analogue[106] is described above. Kinetic studies on the aminolysis of alkylthiopurines indicate reactions to be of a first-order type.[107]

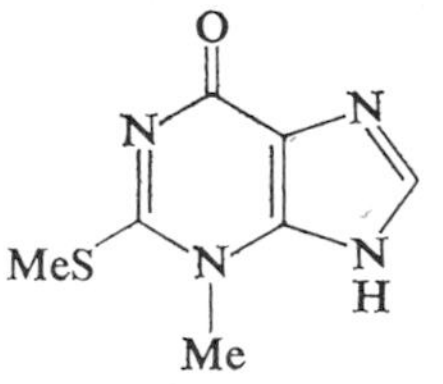

(47) 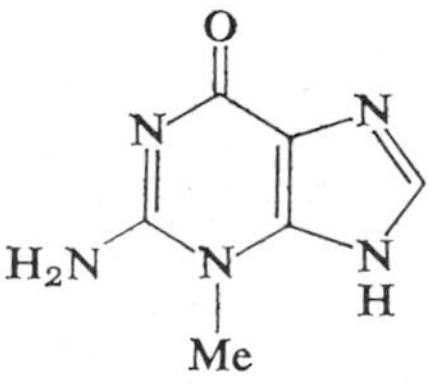 (48)

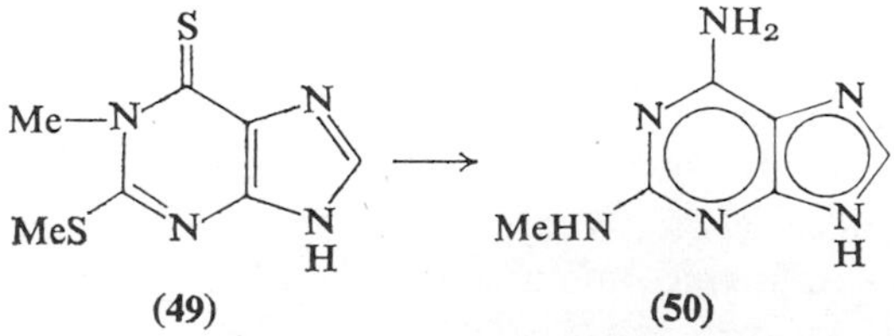

(49) → (50)

---

* The mechanism of this reaction still awaits elucidation, no known rearrangement in the purine series can be fitted to this case. In the light of the facts already given it seems unlikely that any simple exchange of methylthio group by an amino group takes place at the purine level.

d. *Alkylthio- to Thiopurines*

The methods available for carrying out this conversion are described in Section 1.Ae.

e. *Oxidation*

The various oxidation products resulting from oxidation of alkyl-thiopurines are described in Section 7A.

f. *Alkylthio- to Halogenopurines*

These transformations are discussed in Chapter V, Sections 1C and 1Eb.

g. *Alkylation of Alkylthiopurines*

The possibility of displacement of thioether groups discourages alkylation of alkylthiopurines in aqueous alkaline solutions. However, mixtures of 9- and 7-benzyl isomers with the former predominating result from the prolonged action of benzyl bromide on 6-methylthio-purine in dimethylformamide at 100° (18 h) in the presence of anhydrous potassium carbonate.[22] In the absence of the base, under reflux conditions, methyl iodide and 6-methylthiopurine give first the 3-methyl analogue which, in an excess of the reagent, is further converted to 3,7-dimethyl-6-methylthiopurinium bromide (**51**).[13] An alternative route to **51** is by a similar alkylation of 7-methyl-6-methylthiopurine.[13] Corresponding alkylations occur with 2-methyl- and 8-methyl-6-methyl-thiopurines.[12] Benzyl halides give mixtures of the 3-benzyl- and 3,7-dibenzyl-6-methylthiopurines[22] in which the halogen of the alkylating agent appears to control the ratio of the derivatives formed, the highest yield of dibenzylated purine coinciding with the highest atomic weight halogen (i.e., iodide) used.[22] Alkylation of 1-methyl-6-methylthiopurine affords the unstable 1,9-dimethyl quaternary salt (**52**).[13] Attempts to prepare **52** by methylation of 9-methy-6-methylthiopurine fail giving instead the 7,9-dimethyl-6-methylthiopurinium salt (**53**),[22] the structure of which was erroneously given by the same workers, in an earlier paper,[13] as being of the 3,9-dimethyl-6-methylthio derivative. The analogous 7,9-dibenzyl-6-benzylthiopurine arises with benzyl bromide[22] in acetonitrile. If vigorous heating is employed S-debenzylation to the thiopurine betaine **(54)** occurs.[22] Some cases of direct $N_{(9)}$-alkylation

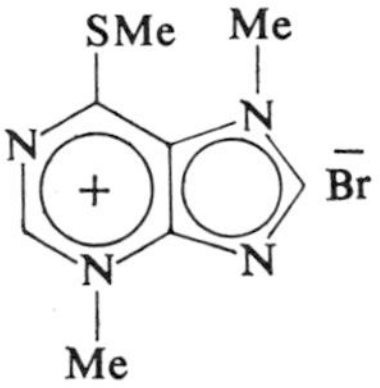

(51) 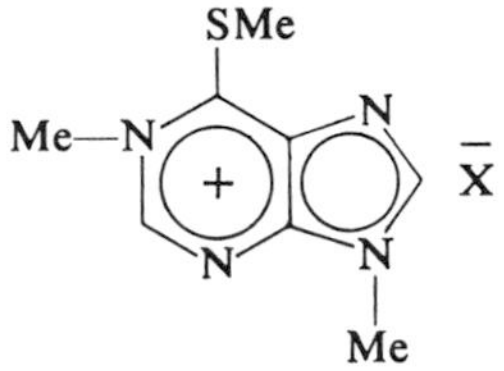   (52)

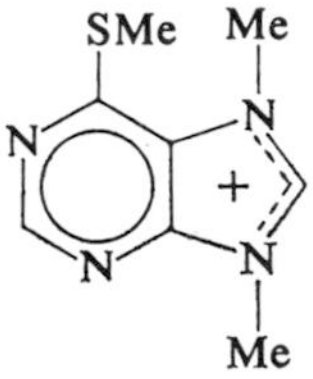

(53) 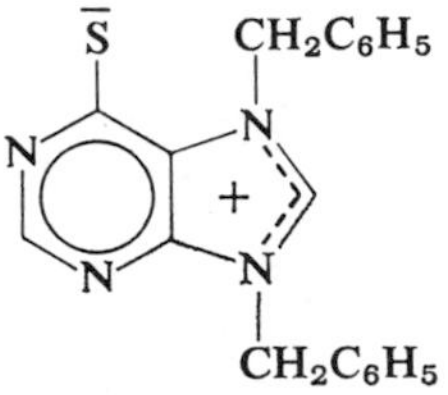   (54)

under aprotic conditions are known, such as the preparation of 2-amino-9-(2-hydroxyethyl)-6-methylthiopurine from interaction of 2-amino-6-methylthiopurine with ethylene bromohydrin in dimethylsulphoxide at room temperature in the presence of potassium carbonate.[119]

More widely exploited routes to 9-substituted alkylthiopurines are addition reactions of the Mannich and Michael type. After some hours

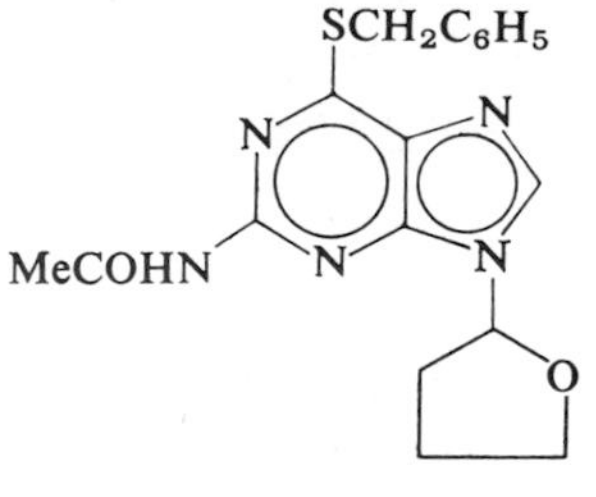

(55) 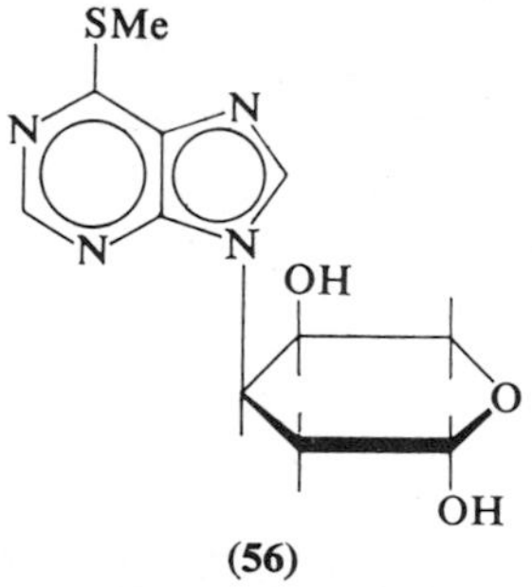   (56)

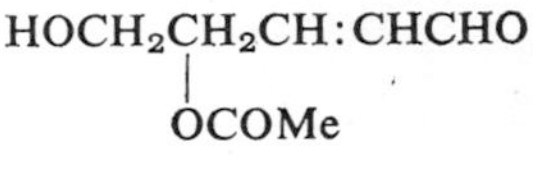

$HOCH_2CH_2CH:CHCHO$
|
$OCOMe$

(57) 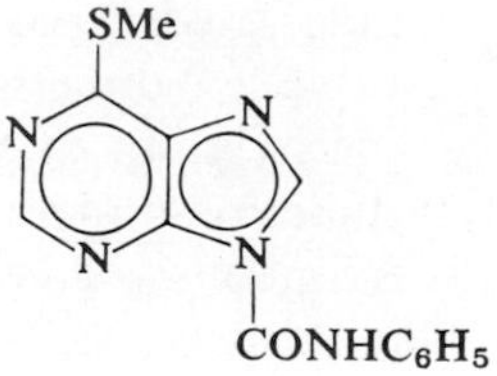   (58)

in a mixture of 37% aqueous formaldehyde and sodium carbonate solution, 6-alkylthiopurines afford the corresponding 9-hydroxymethyl derivatives.[91] With morpholine and formaldehyde 6-propylthiopurine is converted to 9-morpholinomethyl-6-propylthiopurine.[91] Through addition across a double bond 2-acetamido-6-benzylthiopurine is converted to the 9-(tetrahydrofur-2-yl) derivative (55) when treated with 2,3-dihydrofuran in ethyl acetate containing *p*-toluenesulphonic acid.[77]

Under neutral or alkaline conditions a mixture of *erythro* and *threo* forms of the 2,3-dideoxypentose (56) arise from 6-methylthiopurine and 4-acetoxy-5-hydroxypent-2-enaldehyde (57), the reaction taking place best in dimethylformamide containing triethylamine but water alone can be used.[120] In benzene containing a base *N*-alkyl and *N*-aryl isocyanates and 6-methylthiopurine react to form 9-alkyl(or aryl)carbamoyl-6-methylthiopurines of the type 58.[96]

3. Dipurinyl Disulphides

A. Preparation

Dipurinyl disulphides are known but uncommon. Of these dipurin-6-yl disulphide has been prepared by oxidation of 1,6-dihydro-6-thiopurine, the usual reagent being iodine in sodium iodide solution at room temperature under neutral or slightly alkaline conditions,[25, 121] but hydrogen peroxide in dimethylformamide (at 35°)[122] or butyl nitrite in aqueous methanol (reflux)[123] are other suitable oxidants. The disulphide form of 1,6-dihydro-6-thiopurine-3-oxide is formed with the latter reagent in ethanol.[124] Thioguanine gives di(2-aminopurin-6-yl)-disulphide with iodine solution[25]; corresponding di(purin-8-yl) disulphides being obtained from 8-thiocaffeine and -theobromine with the same reagent.[127]* 8-Thiotheobromine derivatives afford the disulphide analogues with hydrogen peroxide also.[128] The novel cyclic dipurin-8-yl disulphide (59) is obtained by iodine oxidation of 1,3-di(8-thiotheophyllin-7-yl)propane in ethanol.[129] Other derivatives of this type are known.[130]

Extranuclear disulphides are formed in various ways. Heating 8-methylsulphonylpurine with di(2-aminoethyl) disulphide in propanol

---

* This reaction should be compared with that carried out by earlier workers[125, 126] in which a series of 8-thioxanthines including 8-thiotheophylline, -theobromine, and -caffeine on treatment with iodine in aqueous sodium bicarbonate gave the corresponding 8-iodopurines.

under reflux (5 h) gives di[2-(purin-8-ylamino)ethyl] disulphide (**60**, R = H),[30] while the 6-amino analogue (**60**, R = NH$_2$) is similarly derived from 8-bromoadenine[131] or by atmospheric oxidation of the product obtained from sodium-liquid ammonia reduction of 6-amino-8-(2-benzylthioethylamino)purine.[31] A similar attempted debenzylation of 6-(2-benzylthioethylamino)purine gave mixed products containing some of the disulphide (**61**).[30] After 3 h in hot 0.1N-sodium hydroxide, 7,8-dihydrothiazolo[2,3-i]purine (**62**) is converted in 30% yield to di[2-(hypoxanthin-1-yl)ethyl] disulphide (**63**).[74] The attempted hydrolysis of the isothiouronium derivative (**64**) in aqueous sodium carbonate at room temperature failed to give 7-(2-mercaptoethyl)theophylline, the

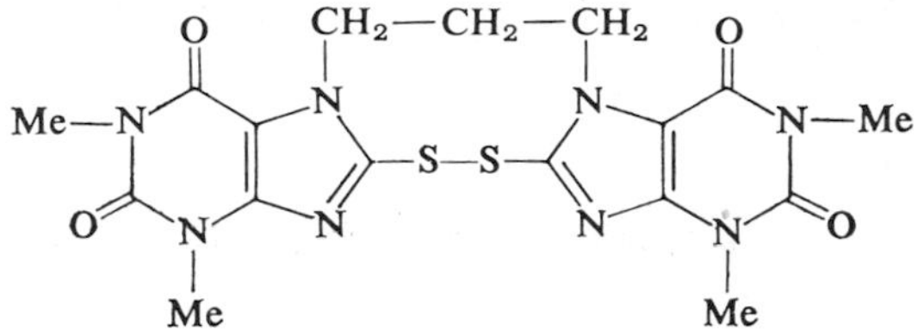

(**59**)

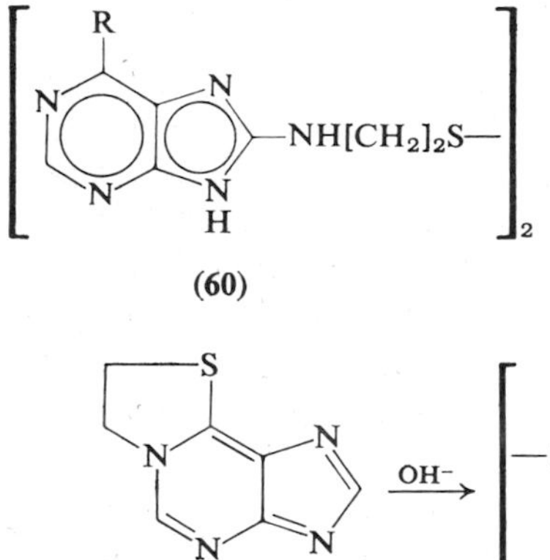

(**60**) 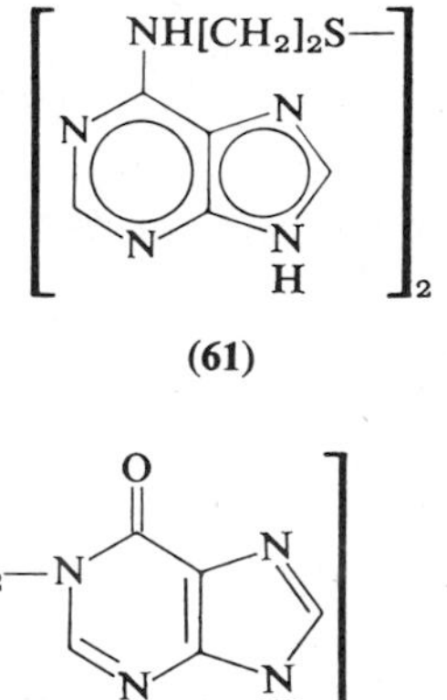   (**61**)

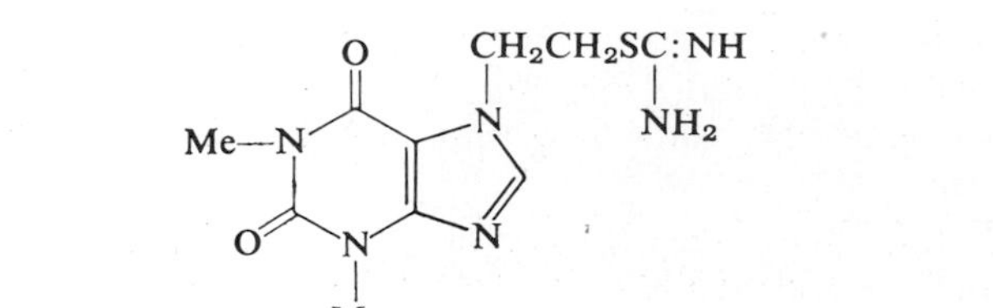

(**62**)   (**63**)

(**64**)

product being di[2-(theophyllin-7-yl)ethyl] disulphide.[132] Other preparations of this type are known.[133]

B. Reactions

The few reactions recorded are typical of those shown by heterocyclic disulphides in general. Oxygenation of a dilute alkaline solution of dipurin-6-yl disulphide (over 24 h) gives a 90% yield of purine-6-sulphinate.[25] In the absence of oxygen a mixture of 1,6-dihydro-6-thiopurine and purine-6-sulphinate is obtained. The disulphides are unstable to both acid and basic conditions. In N-hydrochloric acid at room temperature dipurin-6-yl disulphide is rapidly converted in good yield to 1,6-dihydro-6-thiopurine.[25] The above reactions are shown by di(2-aminopurin-6-yl) disulphide also.[25]

Reduction of extranuclear disulphides can lead to formation of the parent thiopurine. This occurs on hydrogenolysis of **65** to 6-(2-mercaptoethylamino)purine (**66**) using a palladium catalyst in sodium hydroxide solution[30] but, on reduction in dilute hydrochloric acid, rearrangement to 6-(2-aminoethylthio)purine (**67**) occurs.[30] Other examples of disulphide reductions are known.[30] Desulphurisation with Raney nickel of the disulphide linkage in the hypoxanthine derivative (**63**) gives 1-ethylhypoxanthine.[70]

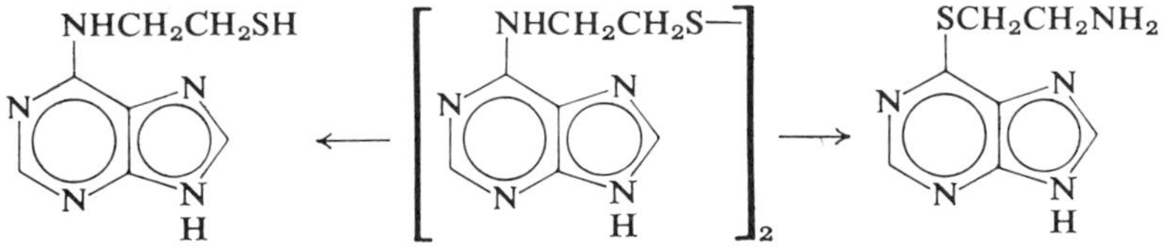

(66)   (65)   (67)

4. Purine Sulphinic and Sulphonic Acids

A. Preparation

Sulphinates, of which only one or two are known, are formed by mild oxidation of purine disulphides in alkaline solution. Purine-6-sulphinate (**68**, R = H) and the 2-amino analogue (**68**, R = NH$_2$) are produced on passage of oxygen (24 h) through dilute sodium hydroxide solutions of the disulphides.[25]

The higher oxidation state sulphonates show greater stability, being produced either by oxidation of thiopurines or from the action of sodium sulphite solutions on halogenopurines. Both 6-sulphopurine (purine-6-sulphonate) and 2-amino-6-sulphopurine result from alkaline potassium permanganate oxidation of the respective thiopurine.[25] The same reagent also transforms 6-sulphino- to sulphopurines.[25] Conversion of a thio group to a sulphonic acid group can also be effected with chlorine in methanolic solution at low temperatures.

Note: This versatile reaction under suitable conditions can lead to replacement of a thio group by a chlorine atom. Full treatment of this aspect is given in Chapter V, Section 1C, but the point can be illustrated by the two following examples. In a simple case 1,2,3,6,7,8-hexahydro-8-oxo-2,6-dithiopurine (69, R = O) is converted to the 2,6-disulpho analogue (70), the product being isolated as the diammonium salt, the ammonium ions resulting from partial degradation of the purine.[134] With the 2,6,8-trithio derivative (69, R = S), however, both sulphonation and chlorination occur together with hydrolysis of the group at $C_{(8)}$ giving 6-chloro-7,8-dihydro-8-oxo-2-sulphopurine (71).[134]

Heating a halogenopurine with aqueous sodium sulphite provides an alternative means of inserting sulphonic acid groups. Early preparations of 8-sulphocaffeine, from the 8-chloro [135, 136] and 8-bromo [137] analogues,

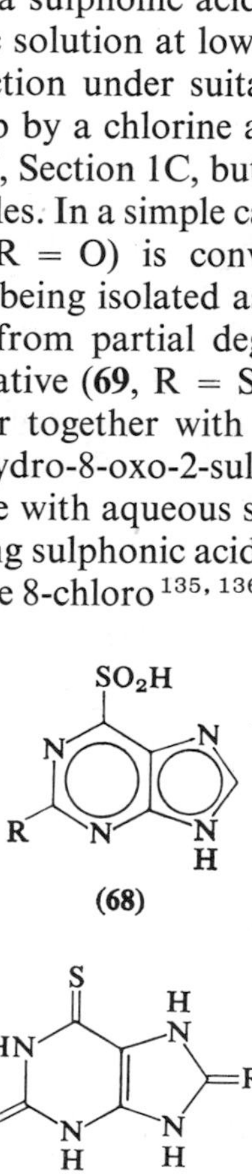

(68)

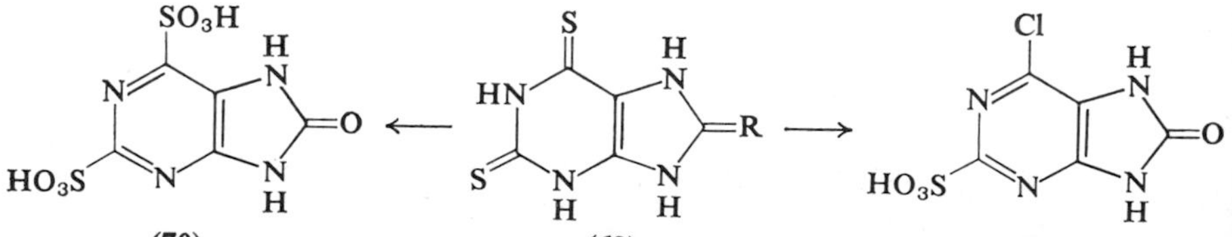

(70)   (69)   (71)

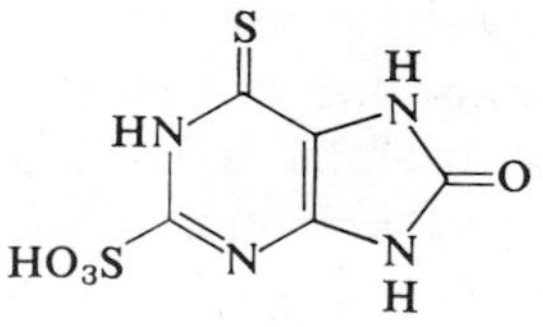

(72)

have been followed by those of 8-sulphotheophylline[138] and derivatives.[139] Examples of sulpho groups at $C_{(2)}$ and $C_{(6)}$ formed in this way are 6-sulpho-[25] and 2-amino-6-sulphopurine,[25] and the 2-sulphopurine (**72**).[70] The last purine (**72**) was also obtained from **71** with thiourea in ethanol.[134]

B. Reactions

Both sulphinic and sulphonic acid groups are readily transformed to an oxo group with acid. The more unstable 6-sulphinates of purine and 2-aminopurine are immediately converted to hypoxanthine and guanine, respectively, on treatment with 0.1N-hydrochloric acid.[25] The corresponding sulpho derivatives usually require a short period of reflux to effect hydrolysis, as in the preparation of uric acid from 6-chloro-7,8-dihydro-8-oxo-2-sulphopurine.[134] Formic acid on a steam bath removes the sulphinic acid group from purine-6-sulphinic acid (**73**, R = H) and the 2-amino analogue (**73**, R = NH$_2$) giving purine and 2-aminopurine[25] while thionyl chloride at room temperature for some hours gives the corresponding 6-chloropurines (**74**, R = H or NH$_2$).[25] The failure of sulpho groups in 1,6,7,8-tetrahydro-8-oxo-2-sulpho-6-thio- and 7,8-dihydro-8-oxo-2,6-disulphopurine to react with 2N-sodium hydroxide or other nucleophilic reagents can be attributed to the deactivating effect of an 8-oxo group.[134]

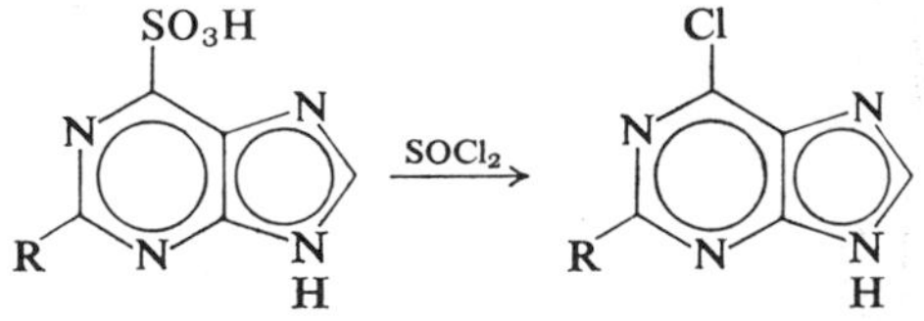

(73)   (74)

5. Halogenosulphonyl Purines

These highly reactive compounds are usually encountered only as transitory derivatives during the formation of sulphamoylpurines but some stable derivatives have been isolated.

A. Preparation

Although chlorosulphonylpurines are present as intermediates[140] during the oxidation of thio- to sulphopurines, when chlorine in methanolic hydrogen chloride is employed, such derivatives are probably too reactive to be isolated. The more stable sulphonyl bromides are represented only by those derived from 8-thiotheophylline derivatives.[45] Using a bromine–potassium bromide mixture in 0.1N-hydrochloric acid at 5°, containing ferric chloride in catalytic amounts as oxidising agent, purines of this type obtained include **75**, R = benzyl, phenylethyl, phenylpropyl, cyclohexylmethyl, phenyl, and hexyl.[140] The wider range of purine sulphonylfluoride derivatives available is a reflection of the lower activity of the sulphonyl fluoride group compared with that of the sulphonyl bromide or chloride.

Both 2- and 6-thiodihydropurine in methanol at 6° containing hydrofluoric acid and potassium fluoride give, on treating with chlorine, almost quantitative yields of the 2- (**76**) or 6-fluorosulphonylpurine.[141] An excess of potassium fluoride in this reaction converts both groups in 1,2,3,6-tetrahydro-2,6-dithiopurine to fluorosulphonyl groups but with restricted amounts monofluorosulphonation at $C_{(2)}$ is accompanied by chlorination at $C_{(6)}$ giving 6-chloro-2-fluorosulphonylpurine.[141] 2-Thioadenine and 2-thiohypoxanthine also give the respective 2-fluorosulphonylpurine by this procedure.[141] An extension of this route allows

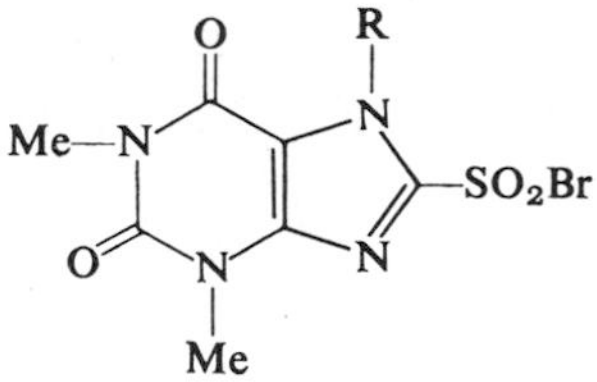

(75) 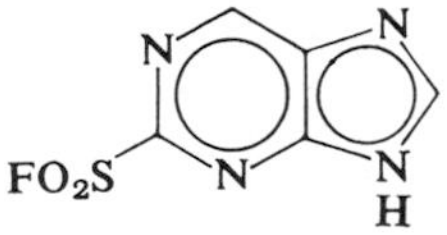   (76)

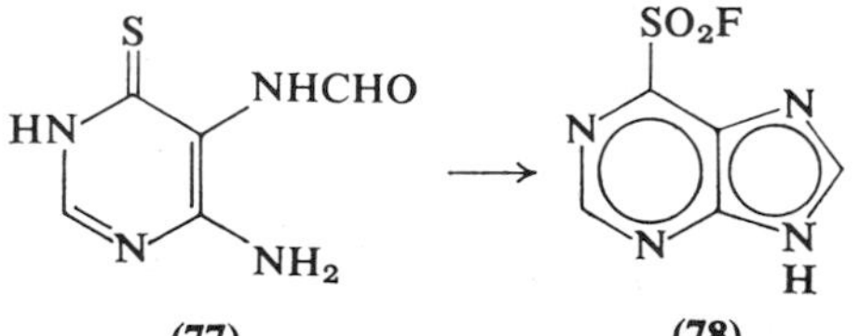

(77)   (78)

the direct formation of 6-fluorosulphonylpurine (**78**) from 4-amino-5-formamido-1,6-dihydro-6-thiopyrimidine (**77**) in one step.[141]

B. Reactions

The interactions of halogenosulphonylpurines with ammonia and amines to form sulphamoyl derivatives is discussed below (Sect. 6A).

Under sealed-tube conditions at 100° with aqueous ammonia, complete replacement of the fluorosulphonyl group occurs in 6-fluorosulphonylpurine giving adenine.[141] In this reaction the fluorosulphonyl group is more reactive than a halogen atom at this position; 6-chloropurine does not react under these conditions. The expected differences in reactivity toward nucleophilic replacement are found between 2- and 6-fluorosulphonyl groups. Whereas N-hydrochloric acid or N-sodium hydroxide at 100° converts 6-fluorosulphonylpurine to hypoxanthine in the case of 2-fluorosulphonylpurine these reagents afford mainly 2-sulphopurine.[141] Similarly, whereas 6-fluorosulphonylpurine is quickly transformed in boiling water to 6-sulphopurine, the 2-fluorosulphonyl derivative is unchanged.[141]

6. Sulphamoylpurines

A. Preparation

Amination of halogenosulphonylpurines constitutes the most important route; in some preparations the crude derivative resulting from chlorine oxidation of the thiopurine is used. In liquid ammonia 2- and 6-fluorosulphonylpurine give the respective 2- and 6-sulphamoylpurines.[141] Analogous sulphamoylpurines are produced from 6-amino-2-fluorosulphonyl- and 6-chloro-2-fluorosulphonylpurine and 2-fluorosulphonylhypoxanthine.[141] Under these conditions 2,6-difluorosulphonylpurine is converted to the ammonium salt of 2-sulphamoyl-6-sulphopurine (79).[141] Aqueous solutions of ammonia at ambient temperature convert fluorosulphonyl groups to sulphamoyl groups but reactive halogens present may be replaced, as in the formation of 6-amino-2-sulphamoylpurine from 6-chloro-2-fluorosulphonylpurine.[141] With bromosulphonylpurines dilute ammonia (25%) effectively forms the corresponding 8-sulphamoyl derivatives of 7-[140] and 9-benzyltheophylline[139] also 1-benzyltheobromine.[139] Aliphatic amines react under comparable conditions with 2-halogenosulphonyl-,[141] 6-halogenosulphonyl-,[141, 142] and 8-halogenosulphonylpurines[139] giving N-alkyl- or NN-dialkylsulphamoylpurines.

An oxidative procedure starting from the thio analogue involves double oxidation in which the intermediate sulphenamide form as, for example, 80 is involved. The conversion of 8-thiocaffeine to the 8-sulphamoyl derivative (81) in this way followed from treatment of 8-thiocaffeine first with potassium ferricyanide and then further

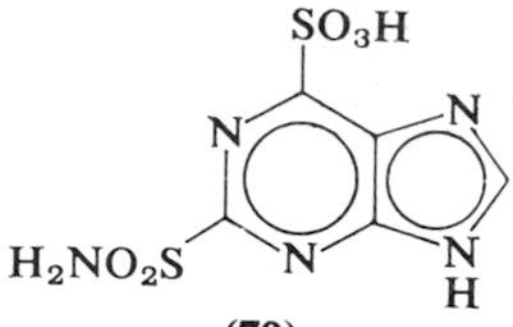

(79)

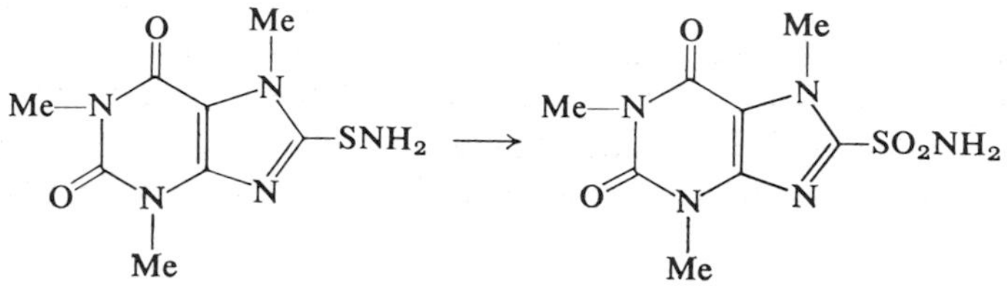

(80)   (81)

oxidation, at 0°, with ammoniacal potassium permanganate.[140]

B. Reactions

Few reactions are recorded; a sulphamoyl group at $C_{(2)}$[141] or $C_{(8)}$ seems to be inert toward acid or alkaline conditions, exemplified by loss of only the benzyl group from 7-benzyl-8-sulphamoyltheophylline when in 45% hydrobromic acid under reflux conditions.[140] Conversion of 6-sulphamoylpurine to hypoxanthine occurs at 100° in N-hydrochloric acid whereas in water at this temperature hydrolysis to 6-sulphopurine takes place.[141] Replacement of sulphamoyl group by thio group occurs to a limited extent on treating 6-chloro-2-sulphamoylpurine with sodium hydrosulphide. Although the major product is the 2-sulphamoyl-6-thio derivative, a small amount of 1,2,3,6-tetrahydro-2,6-dithiopurine has been isolated from the reaction mixture.[141]

7. Alkylsulphinyl- and Alkylsulphonylpurines

Alkylsulphinylpurines are readily oxidised to the corresponding alkylsulphonylpurines, the latter being valuable in purine metatheses by virtue of the generally facile replacement of the alkylsulphonyl group by nucleophilic reagents. In some examples the group shows activity paralleling that of a halogen atom located on the same carbon atom.

A. Preparation

a. Of Methylsulphinylpurines

The methylsulphinylpurines are best represented by their adenine derivatives, preparation of which consists of gentle oxidation of the methylthiopurine with restricted amounts of oxidising agent. 2-Methyl-thioadenine in a hydrogen peroxide–acetic acid mixture at room temperature affords 6-amino-2-methylsulphinylpurine (**82**, R = H)[143] on

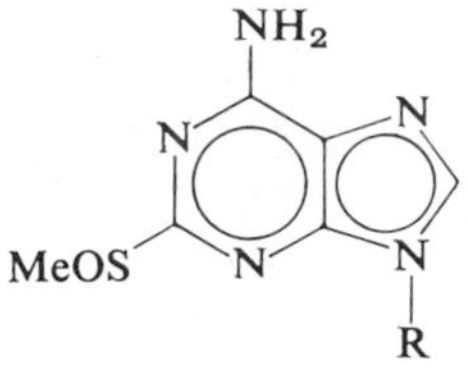

(**82**) 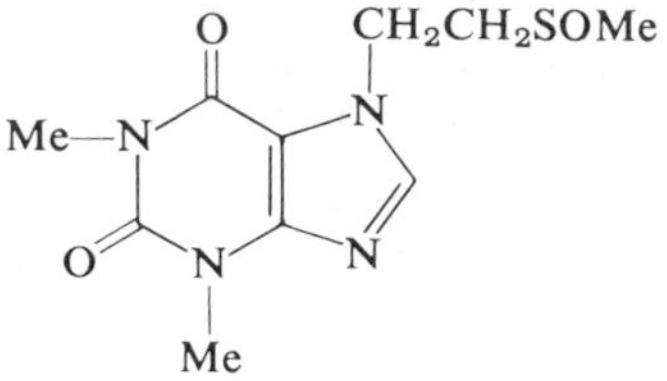   (**83**)

leaving for some hours. Oxidation of the 9-methyl analogue to **82** (R = Me) was effectively carried out with chlorine in a dioxan solvent.[118] Either acidified hydrogen peroxide or chlorine in methanol gives the 2-methylsulphinyl derivative from 6-amino-2-methylthiopurine-1-oxide.[143] An extranuclear methylsulphinyl group is formed on peroxide oxidation of 7-(2-methylthioethyl)theophylline to **83**.[132]

b. Of Methylsulphonylpurines

The above oxidising agents applied with more vigorous conditions or in larger amounts will convert methylthio groups to methylsulphonyl groups directly. Hydrogen peroxide or chlorine are most commonly employed but use of the latter agent requires controlled conditions owing to the danger of side reactions taking place. These include replacement of the methylsulphonyl group by a chlorine atom or hydrolysis to an oxo group. Illustrating the use of acidified hydrogen peroxide are preparations of the 8-methylsulphonyl derivatives of caffeine (**84**, R = Me),[144] 7-benzyltheophylline (**84**, R = $CH_2C_6H_5$),[139] and extranuclear methylsulphones of the type **85**.[32, 132, 133] Alkaline peroxide, at room temperature for 24 h, gives an 84% yield of 6-amino-2-methylsulphonylpurine[143] but, as a generalisation, yields are lower, with either this reagent or potassium permanganate, compared with those obtained by chlorine oxidation. The majority of methylsulphones are derived from passage of chlorine through an aqueous or aqueous methanolic solution of the methylthiopurine, cooling (below 10°) being required to reduce the risk of hydrolysis of the resulting methylsulphonyl group. It should be noted, however, that on using anhydrous methanol as solvent preferential replacement of the thioether group by a chlorine atom is favoured (see Ch. V, Sect. 1C). In place of chlorine some of the foregoing oxidations have been effected with an excess of N-chloro-succinimide in methanol–dimethylformamide mixtures (at 50°) with good results.[145]

From the appropriate methylthiopurines this procedure gives 2-methylsulphonyl-,[60] 6-methylsulphonyl-,[60, 145] 8-methylsulphonyl-,[33] 2,6-dimethylsulphonyl-,[60, 145] and 2,6,8-trimethylsulphonylpurine.[60] Homologous alkylsulphonylpurines are likewise prepared.[59] An exception occurs with 6,8-dimethylthiopurine which undergoes concomitant hydrolysis of the group at $C_{(8)}$ giving 7,8-dihydro-6-methylsulphonyl-8-oxopurine.[60] Methylthio groups of oxopurines are oxidised in methanol alone at higher temperatures. At 60°, 2-methylthiohypoxanthine,[60, 145] 2-methylthio-8-oxo-,[60] and 2,6-dimethylthio-8-oxo-,[60] but not 8-methyl-thio-2-oxodihydropurine,[62] afford the appropriate mono- or dimethyl-sulphonylpurines. These reaction conditions also give 2-amino-8-methylsulphonyl-,[62] 6-amino-2-methylsulphonyl-,[118, 143] and 6-amino-9-methyl-2-methylsulphonylpurine.[118]

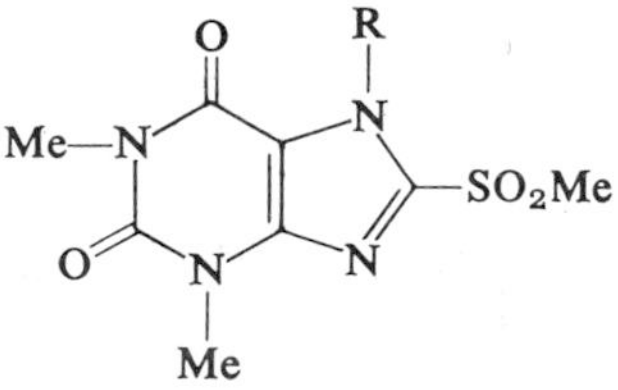

(84) 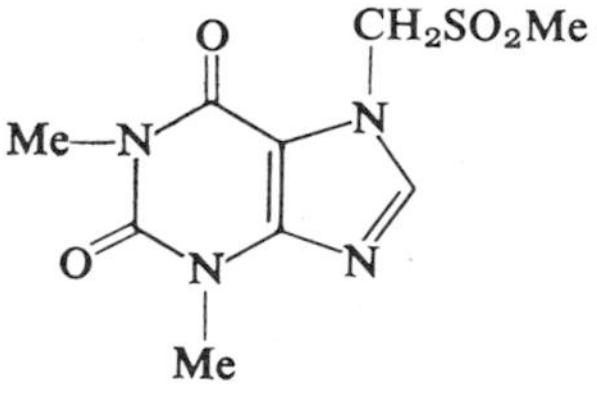   (85)

The N-chlorosuccinimide method has successfully oxidised methyl-thiopurine ribosides. Less well defined products result from using N-bromosuccinimide.[145]

B. Reactions

Due to the unstable nature of methylsulphinylpurines they have little application in purine transformation reactions. With hydrogen peroxide oxidation to the methylsulphonyl analogue readily occurs.[143] Heating

6-amino-2-methylsulphinylpurine with concentrated hydrochloric acid causes rapid reduction to 6-amino-2-methylthiopurine.[143] Reaction of 6-amino-2-methylsulphinylpurine-1-oxide with 2-hydroxyethylamine in dimethylformamide under reflux gives the corresponding 2-(2-hydroxy-ethylamino)purine but, surprisingly, no amination occurs with mor-pholine.[143] With N-sodium hydroxide at room temperature (40 h) a 77% yield of isoguanine-1-oxide is obtained.[143]

Methylsulphonylpurines are generally stable compounds, although 2,6,8-trimethylsulphonylpurine appears to be the exception.[60] Where more than one sulphone group is present the normal order of displace-ment is found on reaction with nucleophilic reagents. Heating with dilute acids or alkali leads to the appropriate oxopurine. Hypoxanthine, for example, results from a brief reflux period (5 min) in $0.1N$-hydro-chloric acid, or a longer one (60 min) in $0.1N$-sodium hydroxide, of 6-methylsulphonylpurine.[145] Isoguanine and the 9-methyl homologue are obtained on dilute alkaline hydrolysis of the corresponding 2-methylsulphonylpurine.[118] Under acid conditions the group at $C_{(2)}$ shows a typically inert character as illustrated by the action of boiling N-hydrochloric acid on 2,6,8-trimethylsulphonylpurine and 1,6-di-hydro-2,8-dimethylsulphonyl-6-oxopurine which gives the respective 2,6-dimethylsulphonyl-8-oxo- and 2-methylsulphonyl-6,8-dioxo-purines.[60]

Replacement of methylsulphonyl groups by ether groups occurs with hot solutions of the appropriate sodium alkoxide. Both the 2-benzyloxy and 2-ethoxy derivatives of 6-amino-9-methylpurine are prepared[118] under reflux in ethanol, but sealed-tube conditions (180°, 4 h) are required for converting 8-methylsulphonylpurine to 8-methoxypurine.[33]

Ammonia and amines give analogous products to those obtained with the corresponding halogenopurines. Thus, 2,6-dimethylsulphonyl-purine and methanolic ammonia (150°, 1 h) give 6-amino-2-methyl-sulphonylpurine.[145] Aqueous methylamine on the steam bath gives the 6-methylamino analogue whereas at 125° 2,6-bis-methylaminopurine results.[60] With 2,6,8-trimethylsulphonylpurine and the aqueous amine at 90° only replacement of the $C_{(6)}$ group occurs.[60] The conversion of 8-methylsulphonylpurine to an 8-substituted aminopurine is reported on prolonged heating with the amine in propanol.[30]

8. Thiopurine-*S*-Oxides

The yellow product, obtained in good yield (75%) from hydrogen peroxide oxidation, in a mixed solvent (EtOH/CHCl$_3$/NMe$_3$) at moderate temperature, of the 6-thio analogue of theophylline is accorded the *S*-oxide structure (**86**).[132, 146] Although no other purine-*S*-oxide has been isolated, the presence of these derivatives in solution is claimed to be demonstrated by a blue colouration obtained with ferric chloride, in the case of 6-thio-(DMSO) and 1-methyl-6-thio-1,6-dihydropurine (MeOH/CHCl₃/NEt₃). An oxidation in dimethylformamide gives the corresponding oxide of the 6-thio analogue of theobromine (1,2,3,6-tetrahydro-3,7-dimethyl-2-oxo-6-thiopurine) in solution.[122] Although the corresponding caffeine analogue does not oxidise under these conditions the oxide has been obtained by methylation of **86** with diazomethane.[146] The nature of the solvent system used and the pH of the medium appear to influence the course of oxidation; in the case of 1,6-dihydro-6-thiopurine reaction in dimethyl sulphoxide gives the oxide form whereas in dimethylformamide dipurin-6-yl disulphide is found to be the product.[122]

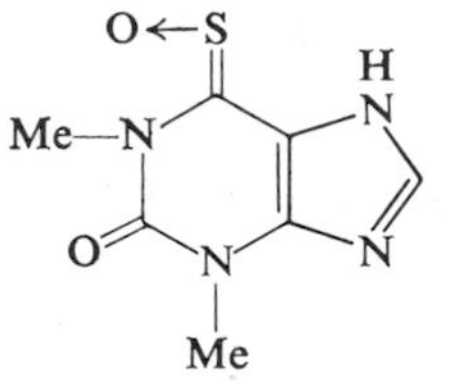

(86)

9. "6-Mercaptopurine" (1,6-dihydro-6-thiopurine) and Analogues

As a result of the universal acceptance of the trivial name "6-mercaptopurine" for 1,6-dihydro-6-thiopurine, it will be used in this section but elsewhere the substitutive name will be given.

Commercial preparation of 6-mercaptopurine[147, 148] (6MP, Purinthiol) utilises the same procedure as was employed in the original synthesis, namely, interaction of hypoxanthine with phosphorus pentasulphide in tetralin or pyridine.[141] The patent literature also details a synthesis from 4-amino-5-formamido-1,6-dihydro-6-oxopyrimidine under similar reaction conditions, thiation of the oxo group in this case occurring concomitantly with ring closure to the purine.[121, 147] A further variant employs cyclisation, with formic acid or formamide, of 4,5-diamino-1,6-dihydro-6-thiopyrimidine.[149, 150] Using a labelled form of the former reagent 6-mercaptopurine[14] C₍₈₎ results.[151] The alternative ring cyclisation procedure using 4-amino-5-thiocarbamoylimidazole is effected[152] in formamide at 200° but hypoxanthine is produced if formic acid is used instead.[153] Of less importance are replacement of the halogen in 6-chloropurine using alkali metal sulphides, this route leading to[151] 6-mercaptopurine ^{35}S, and rearrangement with alkali of 7-aminothiazolo[5,4-d]pyrimidine.[33, 154] Minor routes include the reduction of dipurin-6-yl disulphide[25] or heating 6-hydrazino- or 6-hydroxyaminopurine with thiolacetic acid.[10]

Of more restricted clinical value is 2-amino-1,6-dihydro-6-thiopurine (thioguanine) formed either by phosphorus pentasulphide treatment of guanine[67, 121, 155] or ring closure of 2,4-diamino-5-formamido-1,6-dihydro-6-thiopyrimidine in sodium hydroxide or formamide.[149]

6-Mercaptopurine monohydrate occurs as yellow prisms from water in which it has a 1% solubility at boiling point. In acetone and ether it is almost insoluble and only faintly soluble (1 part in 950) in ethanol. On heating the monohydrate to 140° the anhydrous form, melting point (with decomp.) 313–314°C, is obtained. Two value are shown[156] for the acid pK_a (7.8, 10.8) with the basic pK_a lying below 2.5.

Complex formation is observed with ions of divalent metals such as cobalt, nickel, lead, copper, and zinc[157, 158] with resulting structures containing a 2:1 ratio of purine to metal ion. A tris-purinyl complex is reported to be formed with cadmium in sodium acetate buffer.[157] Polarographic studies of these complexes have been made.[159] The derivative formed with methyl mercuric chloride has antibacterial action.[160]

Clinically 6-mercaptopurine has found wide applications in the treatment of myeloid and acute forms of leukaemia, especially those of children where startling remissions are sometimes found in the course of the disease. Unfortunately, the slow development of drug resistance renders the compound of only temporary, but nevertheless valuable, palliative value. One serious side effect is the suppression of the immune response mechanism in the subject under treatment. This property has been put to good use in heterografting procedures as, for example, in human kidney transplants. The chemotherapeutic application to neoplastic diseases arose from the observation that it induced regressions specifically in Sarcoma 180, a mouse tumour hitherto unresponsive to chemical agents.[161] Surprisingly, the results of tests made on a wide spectrum of other animal tumours were negative.[162]

Metabolic pathway studies indicate that drug action may occur at more than one enzyme level but specific interference with the *de novo* synthesis of nucleic acid at the inosinic acid (hypoxanthine ribotide) stage is found, the effect being to inhibit subsequent formation of adenylic and guanylic acids. An excellent recent review[163] covers all aspects of the chemotherapy of tumours by antimetabolites including 6-mercaptopurine and its derivatives. Like hypoxanthine 6-mercaptopurine is enzymically oxidised at $C_{(2)}$ and $C_{(8)}$ by normal metabolic processes to 1,2,3,6,7,8,hexahydro-2,8-dioxo-6-thiopurine, the so-called "6-thiouric acid."[70] Mercaptopurine was introduced in the addendum of 1960 to the *British Pharmacopae* (1958). It is marketed under the synonyms Puri-Nethol® and Leukerin®.

References

1. Albert and Brown, *J. Chem. Soc.*, **1954**, 2060.
2. Elion, Burgi, and Hitchings, *J. Amer. Chem. Soc.*, **74**, 411 (1952).
3. Johns and Baumann, *J. Biol. Chem.*, **15**, 515 (1913).
4. Isay, *Ber.*, **39**, 250 (1906).
5. Baker, Joseph, and Williams, *J. Org. Chem.*, **19**, 1793 (1954).
6. Traube, *Annalen*, **432**, 266 (1923).
7. Biltz, *Ber.*, **61**, 1409 (1928).
8. Cook and Smith, *J. Chem. Soc.*, **1949**, 2329.
9. Shaw and Butler, *J. Chem. Soc.*, **1959**, 4040.
10. Giner-Sorolla, Thom, and Bendich, *J. Org. Chem.*, **29**, 3209 (1964).
11. Giner-Sorolla and Bendich, *J. Medicin. Chem.*, **8**, 667 (1965).
12. Merz and Stähle, *Beitrage zur Biochemie und Physiologie von Naturstoffen*, Fischer-Verlag, Jena, 1965, p. 285.
13. Neiman and Bergmann, *Israel J. Chem.*, **3**, 161 (1965).
14. Boehringer, Ger. Pat., 128,117 (1902); *Chem. Ztbl.*, II, 314 (1902).
15. Elion, *J. Org. Chem.*, **27**, 2478 (1962).
16. Neiman and Bergmann, *Israel J. Chem.*, **3**, 85 (1965).
17. Montgomery, Balsiger, Fikes, and Johnston, *J. Org. Chem.*, **27**, 195 (1962).
18. Elion and Hitchings, *J. Amer. Chem. Soc.*, **69**, 2138 (1947).
19. Elion, Goodman, Lange, and Hitchings, *J. Amer. Chem. Soc.*, **81**, 1897 (1959).
20. Montgomery, Holum, and Johnston, *J. Amer. Chem. Soc.*, **81**, 3963 (1959).
21. Haggerty, Springer, and Cheng, *J. Medicin. Chem.*, **8**, 797 (1965).
22. Neiman and Bergmann, *Israel J. Chem.*, **5**, 243 (1967).
23. Bredereck, Schellenberg, Nast, Heise, and Christmann, *Annalen*, **99**, 944 (1966).
24. Saneyoshi and Chihara, *Chem. and Pharm. Bull.* (*Japan*), **15**, 909 (1967).
25. Doerr, Wempen, Clarke, and Fox, *J. Org. Chem.*, **26**, 3401 (1961).
26. Neiman, *Chem. Comm.*, **1968**, 200.
26a. Walentowski and Wanzlick, *Chem. Ber.*, **102**, 3000 (1969).
27. Jones and Robins, *J. Amer. Chem. Soc.*, **82**, 3773 (1960).
28. Chaman and Golovchinskaya, *Zhur. obshchei Khim.*, **33**, 3342 (1963).
29. Golovchinskaya and Chaman, *Zhur. obshchei Khim.*, **32**, 3245 (1962).
30. Johnston and Gallagher, *J. Org. Chem.*, **28**, 1305 (1963).
31. Chu and Mautner, *J. Org. Chem.*, **26**, 4498 (1961).
32. Onodera, Hirano, Kashimura, and Yajima, *Tetrahedron Letters*, **1965**, 4327.
33. Brown and Mason, *J. Chem. Soc.*, **1957**, 682.
34. Willets, Decius, Dille, and Christensen, *J. Amer. Chem. Soc.*, **77**, 2569 (1955).

35. Beaman, *J. Amer. Chem. Soc.*, **76**, 5633 (1954).
36. Prasad, Noell, and Robins, *J. Amer. Chem. Soc.*, **81**, 193 (1959).
37. Bergmann and Tamari, *J. Chem. Soc.*, **1961**, 4468.
38. Koppel and Robins, *J. Org. Chem.*, **23**, 1457 (1958).
39. Townsend and Robins, *J. Org. Chem.*, **27**, 990 (1962).
40. Bergmann, Levin, Kalmus, and Kwietny-Govrin, *J. Org. Chem.*, **26**, 1504 (1961).
41. Bergmann, Kalmus, Unger-Waron, and Kwietny-Govrin, *J. Chem. Soc.*, **1963**, 3729.
42. Bendich, Tinker, and Brown, *J. Amer. Chem. Soc.*, **70**, 3109 (1948).
43. Lewis, Beaman, and Robins, *Canad. J. Chem.*, **41**, 1807 (1963).
44. Bergmann and Kalmus, *J. Org. Chem.*, **26**, 1660 (1961).
45. Koppel and Robins, *J. Amer. Chem. Soc.*, **80**, 2751 (1958).
46. Blicke and Schaaf, *J. Amer. Chem. Soc.*, **78**, 5857 (1956).
47. Kalmus and Bergmann, *J. Chem. Soc.*, **1960**, 3679.
48. Seyden-Penne, Le Thi Minh, and Chabrier, *Bull. Soc. chim. France*, **1966**, 3934.
49. Bredereck, Küpsch, and Wieland, *Chem. Ber.*, **92**, 566 (1959).
50. Biltz, Bülow, and Heidrich, *Annalen*, **426**, 237 (1922).
51. Biltz and Strufe, *Annalen*, **423**, 200 (1921).
52. Ochiai, *Ber.*, **69**, 1650 (1936).
53. Traube, *Arch. Pharm.*, **244**, 11 (1906).
54. Traube and Winter, *Arch. Pharm.*, **244**, 11 (1906).
55. Clark, Grantham, and Lydiate, *J. Chem. Soc. (C)*, **1968**, 1122.
56. Fischer, *Ber.*, **31**, 431 (1898).
57. Robins and Lin, *J. Amer. Chem. Soc.*, **79**, 490 (1957).
58. Montgomery and Temple, *J. Amer. Chem. Soc.*, **79**, 5238 (1957).
59. Bergmann, Neiman, and Kleiner, *J. Chem. Soc. (C)*, **1966**, 10.
60. Noell and Robins, *J. Amer. Chem. Soc.*, **81**, 5997 (1959).
61. Prasad and Robins, *J. Amer. Chem. Soc.*, **79**, 6401 (1957).
62. Albert, *J. Chem. Soc. (B)*, **1966**, 438.
63. Robins, *J. Amer. Chem. Soc.*, **80**, 6671 (1958).
64. Montgomery and Holum, *J. Amer. Chem. Soc.*, **79**, 2185 (1957).
65. Leonard, Skinner, Lansford, and Shive, *J. Amer. Chem. Soc.*, **81**, 907 (1959).
66. Townsend and Robins, *J. Amer. Chem. Soc.*, **84**, 3008 (1962).
67. Daves, Noell, Robins, Koppel, and Beaman, *J. Amer. Chem. Soc.*, **82**, 2633 (1960).
68. Blackburn and Johnson, *J. Chem. Soc.*, **1960**, 4347.
69. Noell and Robins, *J. Org. Chem.*, **24**, 320 (1959).
70. Elion, Mueller, and Hitchings, *J. Amer. Chem. Soc.*, **81**, 3042 (1959).
71. Skinner, Shive, Ham, Fitzgerald, and Eakin, *J. Amer. Chem. Soc.*, **78**, 5097 (1956).
72. Johnston, Holum, and Montgomery, *J. Amer. Chem. Soc.*, **80**, 6265 (1958).
73. Koppel, O'Brien, and Robins, *J. Org. Chem.*, **24**, 259 (1959).
74. Balsiger, Fikes, Johnston, and Montgomery, *J. Org. Chem.*, **26**, 3446 (1961).
75. Montgomery and Thomas, *J. Org. Chem.*, **28**, 2304 (1963).
76. Robins, Godefroi, Lewis, and Jackson, *J. Amer. Chem. Soc.*, **83**, 2574 (1961).
77. Bowles, Schneider, Lewis, and Robins, *J. Medicin. Chem.*, **6**, 471 (1963).
78. Baker and Kozma, *J. Medicin. Chem.*, **10**, 682 (1967).
78a. Wooldridge and Slack, *J. Chem. Soc.*, **1962**, 1863.
79. Bergmann, Kleiner, Neiman, and Rashi, *Israel J. Chem.*, **2**, 185 (1964).
80. Yun-Feng and Yuen-Yin, *Acta Pharm. Sinica*, **10**, 298 (1963).
81. Gräfe, *Arzneim.-Forsch.*, **15**, 1206 (1965).
82. Todd and Bergel, *J. Chem. Soc.*, **1936**, 1559.
83. Elderfield and Prasad, *J. Org. Chem.*, **24**, 1410 (1959).

84. Gordon, *J. Amer. Chem. Soc.*, **73**, 984 (1951).
85. Johnston, McCaleb, and Montgomery, *J. Medicin. Chem.*, **6**, 669 (1963).
86. Dietz and Burgison, *J. Medicin. Chem.*, **9**, 160 (1966).
86a. Carroll and Philip, *J. Org. Chem.*, **33**, 3776 (1968).
87. Neiman and Bergmann, *Israel J. Chem.*, **1**, 477 (1963).
88. Jones and Robins, *J. Amer. Chem. Soc.*, **84**, 1914 (1962).
89. Montgomery, Hewson, Clayton, and Thomas, *J. Org. Chem.*, **31**, 2202 (1966).
90. Cacace and Masironi, *Ann. Chim.* (*Italy*), **46**, 806 (1956).
91. Bryant and Harman, *J. Medicin. Chem.*, **10**, 104 (1967).
92. Plaut, *Organic Sulfur Compounds*, Ed., Kharasch, Pergamon Press, Oxford, 1961, Vol. I, p. 512.
93. Montgomery and Thomas, *J. Heterocyclic Chem.*, **1**, 115 (1964).
94. Brookes and Lawley, *J. Chem. Soc.*, **1960**, 539.
94a. Dyer, Russell, Farris, Minnier, and Tokizawa, *J. Org. Chem.*, **34**, 973 (1969).
94b. Dyer and Minnier, *J. Heterocyclic Chem.*, **6**, 23 (1969).
95. Farris, *Diss. Abstracts*, **25**, 3265 (1964).
96. Dyer and Bender, *J. Medicin. Chem.*, **7**, 10 (1964).
97. Giner-Sorolla and Bendich, *J. Org. Chem.*, **31**, 4239 (1966).
98. Merz and Stähle, *Arzneim.-Forsch.*, **15**, 10 (1965).
99. Taylor, Vogel, and Cheng, *J. Amer. Chem. Soc.*, **81**, 2442 (1959).
100. Howard, Lythgoe, and Todd, *J. Chem. Soc.*, **1945**, 556.
101. Baker, Joseph, and Schaub, *J. Org. Chem.*, **19**, 631 (1954).
102. Townsend and Robins, *J. Heterocyclic Chem.*, **3**, 241 (1966).
103. Cook and Smith, *J. Chem. Soc.*, **1949**, 3001.
104. Baker, Schaub, and Joseph, *J. Org. Chem.*, **19**, 638 (1954).
105. Albert, *Chem. Comm.*, **1969**, 500.
106. Johns, *J. Biol. Chem.*, **21**, 319 (1915).
107. Brown, Ford, and Tratt, *J. Chem. Soc.* (*C*), **1967**, 1445.
108. Elion, Lange, and Hitchings, *J. Amer. Chem. Soc.*, **78**, 217 (1956).
109. Wellcome Foundation Ltd., *Brit. Pat.*, 713,259 (1954); through *Chem. Abstracts*, **49**, 13301 (1955).
110. Skinner and Shive, *J. Amer. Chem. Soc.*, **77**, 6692 (1955).
111. Miller, Skoog, Okumura, Von Saltza, and Strong, *J. Amer. Chem. Soc.*, **77**, 2662 (1955).
112. Ham, Eakin, Skinner, and Shive, *J. Amer. Chem. Soc.*, **78**, 2648 (1956).
113. Skinner, Gardner, and Shive, *J. Amer. Chem. Soc.*, **79**, 2843 (1957).
114. Okumura, Kotarn, Ariga, Masumura, and Kuraisui, *Bull. Chem. Soc. Japan*, **32**, 883 (1959).
115. Montgomery and Holum, *J. Amer. Chem. Soc.*, **80**, 404 (1958).
116. Leonard, Orme-Johnson, McMurtray, Skinner, and Shive, *Arch. Biochim. Biophys.*, **99**, 16 (1962).
117. Townsend, Robins, Loeppky, and Leonard, *J. Amer. Chem. Soc.*, **86**, 5320 (1964).
118. Andrews, Anand, Todd, and Topham, *J. Chem. Soc.*, **1949**, 2490.
119. O'Brien, Westover, Robins, and Cheng, *J. Medicin. Chem.*, **8**, 182 (1965).
120. Carbon, *J. Amer. Chem. Soc.*, **86**, 720 (1964).
121. Hitchings and Elion, *U.S. Pat.*, 2,697,709 (1954); through *Chem. Abstracts*, **50**, 1933 (1956).
122. Walter, Voss, and Curts, *Annalen*, **695**, 77 (1966).
123. Giner-Sorolla, personal communication.
124. Giner-Sorolla, Gryte, Bendich, and Brown, *J. Org. Chem.*, **34**, 2157 (1969).
125. Biltz and Bülow, *Annalen*, **426**, 306 (1921).

126. Biltz and Beck, *J. prakt. Chem.*, **118**, 1149 (1928).
127. Berdichevskii, Rachinskii, and Novoselova, *Zhur. obshchei Khim.*, **31**, 689 (1958).
128. Gräfe, *Arch. Pharm.*, **300**, 111 (1967).
129. Merz and Stähle, *Arch. Pharm.*, **293**, 801 (1960).
130. Merz and Gräfe, *Arch. Pharm.*, **297**, 146 (1964).
131. Lyashenko, Kolesova, Aleksandr, and Shermeteva, *Zhur. obshchei Khim.*, **34**, 2752 (1964).
132. Eckstein, Gorczyca, Kocwa, and Zejc, *Dissertationes Pharm.* (*Poland*), **10**, 239 (1958); through *Chem. Abstracts*, **53**, 18046 (1959).
133. Eckstein and Sulko, *Dissertationes Pharm.* (*Poland*), **13**, 97 (1961), through *Chem. Abstracts*, **55**, 23548 (1961).
134. Robins, *J. Org. Chem.*, **26**, 447 (1961).
135. Fritz, *Pharm. Post*, **28**, 130 (1895).
136. Balaban, *J. Chem. Soc.*, **1926**, 569.
137. Meister, Lucius, Brüning, and Höchst, Ger. Pat., 74045 (1893); *Frdl.*, **3**, 979 (1890–1894).
138. Giani and Molteni, *Farmaco, Ed. Sci.*, **12**, 1016 (1957); through *Chem. Abstracts*, **52**, 12874 (1958).
139. Dolman, Van der Goot, and Moed, *Rec. Trav. Chim.*, **84**, 193 (1965).
140. Dolman, Van der Goot, Mos, and Moed, *Rec. Trav. Chim.*, **83**, 1215 (1964).
141. Beaman and Robins, *J. Amer. Chem. Soc.*, **83**, 4038 (1961).
142. Lewis, Noell, Beaman, and Robins, *J. Medicin. Chem.*, **5**, 607 (1962).
143. Cresswell and Brown, *J. Org. Chem.*, **28**, 2560 (1963).
144. Long, *J. Amer. Chem. Soc.*, **69**, 2939 (1947).
145. Ikehara, Yamazaki, and Fujieda, *Chem. and Pharm. Bull.* (*Japan*), **10**, 1075 (1962).
146. Walter and Voss, *Annalen*, **698**, 113 (1966).
147. Wellcome Foundation Ltd., Brit. Pat., 713,286 (1954); through *Chem. Abstracts*, **49**, 12546 (1955).
148. Hitchings and Elion, U.S. Pat., 2,691,654 (1955); through *Chem. Abstracts*, **50**, 1933 (1956).
149. Elion, Lange, and Hitchings, *J. Amer. Chem. Soc.*, **78**, 2858 (1956).
150. Elion and Hitchings, U.S. Pat., 2,724,711 (1955); through *Chem. Abstracts*, **50**, 8748 (1956).
151. Elion and Hitchings, *J. Amer. Chem. Soc.*, **76**, 4027 (1954).
152. Hitchings and Elion, U.S. Pat., 2,756,228 (1956); through *Chem. Abstracts*, **51**, 2887 (1956).
153. Ikehara, Nakazawa, and Nakayama, *Chem. and Pharm. Bull.* (*Japan*), **10**, 660 (1962).
154. Elion and Hitchings, U.S. Pat., 2,721,866 (1955); through *Chem. Abstracts*, **50**, 8748 (1956).
155. Elion and Hitchings, *J. Amer. Chem. Soc.*, **77**, 1676 (1955).
156. Albert, *Heterocyclic Chemistry*, Athlone Press, London, 1968, p. 441.
157. Ghosh and Chatterjee, *J. Inorg. Nuclear Chem.*, **26**, 1459 (1964).
158. Cheney, Freiser, and Fernando, *J. Amer. Chem. Soc.*, **81**, 2611 (1959).
159. Christian and Purdy, *Biochim. Biophys. Acta*, **54**, 587 (1961).
160. Saka, Sasada, and Shiobara, Jap. Pat. 12,239 (1960); through *Chem. Abstracts*, **55**, 9799 (1961).
161. Clarke, Philips, Sternberg, Stock, Elion, and Hitchings, *Cancer Res.*, **13**, 593 (1959).
162. Elion, Hitchings, and Van der Werff, *J. Biol. Chem.*, **192**, 505 (1951).
163. Stock, *Experimental Chemotherapy*, Eds., Schnitzer and Hawking, Academic Press, New York, 1966, Vol. 4, p. 104.

CHAPTER VIII

The Amino (and Amino-oxo) Purines

Roughly one-half of the purines found in nature belong to this group, the remainder being oxopurine derivatives. For convenience guanine and related derivatives, in which both amino and oxo groups are present, are treated as aminopurines. This choice is a purely arbitrary one as their inclusion with the oxopurines would be equally valid. A survey of the naturally occurring aminopurines concludes this chapter but for more detailed treatment a topical review should be consulted.[1]

1. Preparation of 2-, 6-, and 8-Aminopurines

In most cases primary routes to aminopurines have been given in Chapters II and III, but for completeness these are outlined below together with the more specialised methods available.

A. By the Traube Synthesis

Cyclisation of 4,5-diaminopyrimidines, having amino- or substituted-amino groups at $C_{(2)}$ and $C_{(6)}$, affords the appropriate 2- or 6-aminopurine. In some cases the procedure lends itself to 8-aminopurine preparation by employing a cyclising agent such as guanidine or cyanogen bromide. Complete treatment of these aspects is given in Chapter II.

B. By Cyclisation of Imidazoles

Of more restricted value than the Traube synthesis this approach is nevertheless useful for the preparation of 6-aminopurines and occasionally for 2-amino derivatives. Ring closure of either 4-amino-5-cyano- or

4-amidino-5-aminoimidazole with reagents like formic acid, triethyl orthoformate, or formamide affords the appropriate adenine derivative whereas with urea or phosgene an isoguanine (6-amino-2,3-dihydro-2-oxopurine) derivative results. Some 2-substituted-amino-1,6-dihydro-6-oxopurines are derived by interaction of amines with 4-[3-(N-alkyl)thio-ureido]-5-ethoxycarbonylimidazoles in the presence of mercuric oxide. Although the foregoing methods do not lend themselves to direct formation of 8-aminopurines, which is a consequence of the difficulty in obtaining suitable 2-aminoimidazoles, they have some value for such preparations in that the readily available 2-methylthioimidazoles on ring closure give 8-methylthiopurines, which are readily aminated to the requisite 8-aminopurine. A full discussion of these routes is to be found in Chapter III.

C. From Acyclic Precursors

The concurrent formation of pyrimidine and imidazole ring in the same reaction mixture has been used to prepare both 2- and 6-amino-purines. Examples are the formation of 2,6-diaminopurine by condensation of isonitrosomalononitrile with guanidine in the presence of a reducing agent and production of various 2-alkyladenines by using amidines in place of the guanidine. These and related reactions are the subject of Chapter III, Sections 2A and 2B.

D. From Halogenopurines

The reaction of ammonia and amines with halogenopurines is fully elaborated in Chapter V, Sections 5Ca to g.

E. From Thio- and Alkylthiopurines

Thiopurines have limited application due to the forcing conditions generally needed to aminate the thio group. However, a few examples are to be found (Ch. VII, Sect. 1Cf). By contrast extensive replacements of alkylthio groups by amino groups have been effected, the major application being to the preparation of 6-aminopurines (Ch. VII, Sect. 2Bc). Carboxymethylthiopurines are also used for this purpose as in the formation of 6-(2-hydroxyethylamino)-, 6-anilino-, 6-piperidino-, and other 6-substituted-aminopurines from 6-carboxymethylthiopurine.[2]

Prolonged heating with ammonia solution (18 h) converts 6-carboxy-methylthio-7-methylpurine to 7-methyladenine.[3]

F. From Oxo- and Alkoxypurines

These approaches have had little attention in the past as more suitable precursors of aminopurines, such as chloro- or alkylthiopurines, are readily to hand. The direct conversion of oxopurines to dialkylamino-purines is a reaction of recent origin, involving heating the oxo deriva-tive with $NN'N''$-trialkylphosphoramides, details of which have been already given (Ch. VI, Sect. 6G). This section also includes reference to the apparent transformation of oxo to dialkylamino groups by means of phosphoryl chloride–trialkylamine mixtures.

Also of extremely restricted application in the purine series, in con-trast to that of the pyrimidines, is replacement of methoxy groups by amino groups. Available examples are noted in Chapter VI, Section 8C.

G. By Amination of Other Groups

Methylsulphonyl groups are replaceable by amines, but ease of substitution depends markedly on the position of the group. This is ably demonstrated in the formation of 2,6-bis-dimethylaminopurine (3) from 2,6-dimethylsulphonylpurine (1) in aqueous dimethylamine at 125° but only the monosubstituted form (2) is obtained if the reaction proceeds at a lower ($< 100°$) temperature.[4]

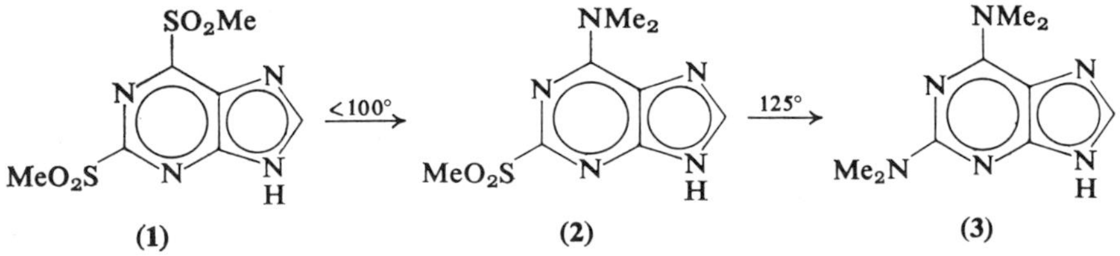

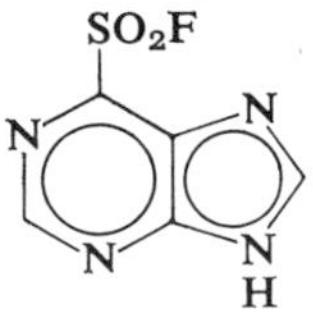

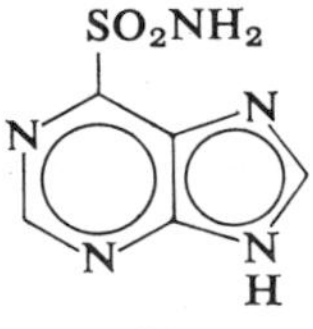

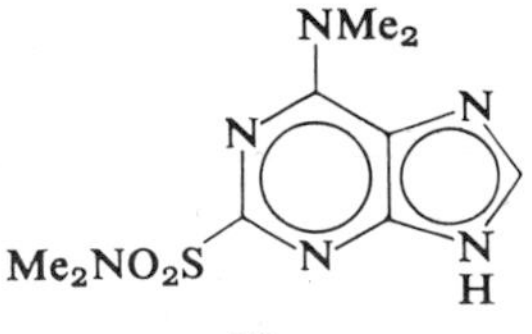

Fluorosulphonylpurines can be aminated but the route is limited to the formation of 6-aminopurines, adenine resulting from interaction of 6-fluorosulphonylpurine (**4**) with aqueous ammonia at 100°,[5] while with cold liquid ammonia only 6-sulphamoylpurine (**5**) is obtained. Cold aqueous dimethylamine converts 2,6-difluorosulphonylpurine to 6-dimethylamino-2-(*NN*-dimethylsulphamoyl)purine (**6**).[5]

H. By Reduction of Nitrogen-Containing Groups

Adenine results from heating 6-nitrosopurine with Raney nickel in aqueous solution.[6] As nitropurines, except for an isolated instance of a 2-nitropurine,[7] are found only as 8-nitro derivatives, their reduction affords only 8-aminopurines (Ch. X, Sect. 1B).

Azidopurines, arising from nitrous acid treatment of the corresponding hydrazinopurine, undergo facile hydrogenation, in ethanol over palladium catalyst, to the aminopurine. Good yields of 6-amino-2-chloro-, 2,6-diamino-, 6,8-diamino-, and 2,6,8-triamino-7-methylpurine are formed in this way.[8] Although 6-azidopurine was found not to give adenine under these conditions, a successful reduction over Raney nickel is reported.[9] Reductions of the above type are found to consume one mole of hydrogen for each azido group with an accompanying liberation of one mole of nitrogen.

Hydroxyaminopurines are reduced readily with hydrogen and a palladium catalyst. Unfortunately, relatively few hydroxamino derivatives are known but adenine has been obtained from 6-hydroxyaminopurine.[10] In boiling dilute ammonia solution containing Raney nickel 2,6-dihydroxyaminopurine is converted to 2,6-diaminopurine.[11] The 9-β-D-ribosides of the above hydroxyaminopurines have been reduced similarly.[12, 13]

On reduction of ethyleneimino(aziridinyl)purines the cyclic alkylamino group is opened to afford the corresponding ethylaminopurines. The usual reducing conditions are hydrogenation over palladium in ethanol at atmospheric pressure but by using sodium in liquid ammonia instead any *N*-benzyl group present may be concomitantly removed, as in the formation of 6-ethylaminopurine from 9-benzyl-6-ethyleneiminopurine.[13a]

I. By Direct Amination

A report,[14] without detail, assumes the formation of 8-amino derivatives from theophylline and caffeine by reaction with chloramine.

Supporting evidence for electrophilic substitution at $C_{(8)}$ by an amine is found in the isolation of 8-substituted-amino derivatives of guanosine when the riboside is treated with aromatic amines under neutral conditions.[15, 16]

J. By Other Routes

One use of hydrazinopurines, via the azido derivative, is noted above (Sect. H) but examples of direct transformation of hydrazino- to aminopurine are to be found. Thus, 8-aminocaffeine results (43%) when the 8-hydrazino analogue is heated for some hours in boiling dimethylformamide or phenol.[17] Presumably the isolation of 8-amino-guanosine from prolonged treatment (36 h) of 8-bromoguanosine in aqueous hydrazine, under reflux, is a further illustration of this mechanism.[18] Formation of a protected amino group is seen in the reaction between 8-bromocaffeine and potassium phthalimide in dimethylformamide, 8-aminocaffeine resulting from dilute acid hydrolysis of the product.[17] Other protected amino groups initially employed in aminopurine formation include benzylamino, converted to amino by hydrogenolysis over a palladium catalyst[19] or furfurylamino which, in the case of 6-furfurylaminopurine (kinetin), is cleaved in hot mineral acid producing adenine.[20] The isolation of 8-aminoxanthine following treatment of 3-hydroxyxanthine with a pyridine-*p*-toluenesulphonyl chloride mixture and alkaline hydrolysis of the resulting 8-pyridinium-xanthinyl chloride is more appropriately detailed elsewhere (Ch. XI, Sect. 3C).

K. Secondary and Tertiary Aminopurines Derived from Primary Aminopurines (The Dimroth Rearrangement)

From a practical standpoint alkylation under the usual conditions involves only ring nitrogen atoms, any amino groups present remaining inviolate.*

* Nevertheless, direct alkylation of an amino group does sometimes seem to have taken place. The formation of 6-furfurylaminopurine (kinetin) on heating a mixture of adenine, furfuryl chloride, and sodium bicarbonate is one example[21] while others involving 9-substituted adenines are known.[21a] As strong bases, which include sodium hydride and alkali metal alkoxides, are present the reactions are presumed to take place through anionic forms of the purines. Methylation of a 2-amino group is observed to occur to a limited extent when guanine is alkylated, one of the products being 1,6-dihydro-2-methylamino-6-oxopurine.[22] In none of the above examples can rearrangement be excluded, as no proof of direct amino group alkylation has been obtained.

The formation of 6-benzylimino-3,7-dibenzylpurine (**9**) on treatment of either 3-benzyl- (**7**) or 7-benzyladenine (**8**) with benzyl bromide in dimethylacetamide containing potassium carbonate[23] may occur through benzylation of an activated amino group but is more likely to have been due to a rearrangement. Numerous examples are available of $N_{(1)}$-alkyl-6-aminopurines rearranging to 6-alkylaminopurines, the simplest case being the conversion of 1-methyladenine (**10**) to 6-methylaminopurine (**11**) in hot aqueous ammonia solution.[24]

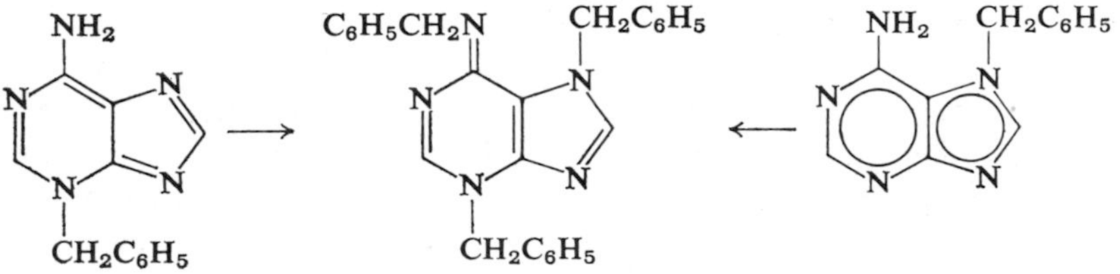

The alkali-induced Dimroth rearrangement is also shown by 1-(3-methylbut-2-enyl)-,[25, 26] 1,7-dimethyl-,[27] 1,9-dimethyl-,[28] 1,7-dibenzyl-,[29] 1,9-dibenzyl-,[29] and 1-butyl-7-methyladenine.[27] Migration of an $N_{(1)}$-hydroxyethyl group is reported with an adenosine derivative.[30] Although less apparent the formation of 6-carboxymethylaminopurine (**13**) on brief heating of an aqueous solution of 6-(*C*-aminoacetamido)purine (**12**, R = NH$_2$) is a rearrangement of this type and most likely proceeds through an initial internal alkylation at $N_{(1)}$, to 7,8-dihydro-8-oxo-imidazole[2,1-*i*]purine (**14**), followed by hydrolysis to **15** and then isomerisation to **13**. The feasibility of this route is demonstrated by the formation of **14** from 6-(*C*-chloroacetamido)purine (**12**, R = Cl) and its subsequent rearrangement to **13** in boiling water.[31] Mechanistically, the Dimroth reaction is envisaged as acting by prior nucleophilic attack at C$_{(2)}$ by hydroxyl ions producing fission of the $N_{(1)}$–C$_{(2)}$ bond, thus allowing rotation of the *N*-alkylamidino group formed (**16**), and a recyclisation, with the primary amino group of the amidine moiety, giving a 6-alkylaminopurine.[3] Appreciation of this reaction makes clear the obtaining of 6-methylaminopurine (**11**) from attempted amination

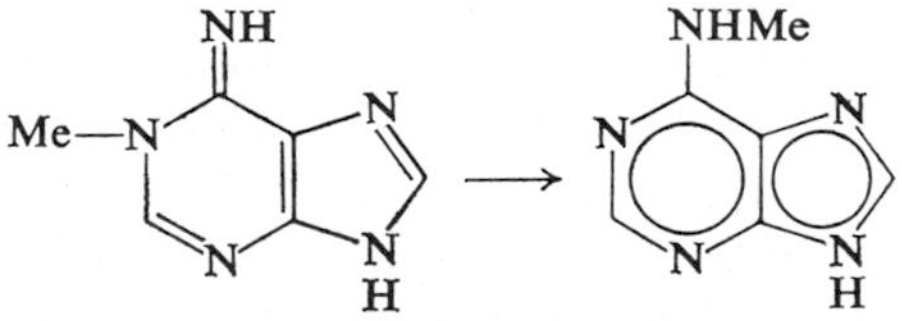

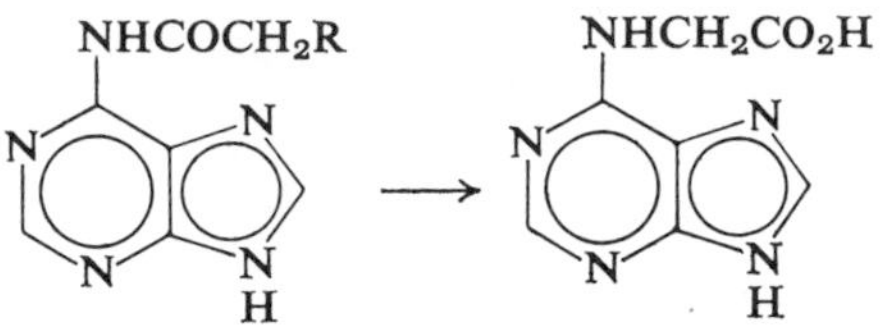

(12) → (13) 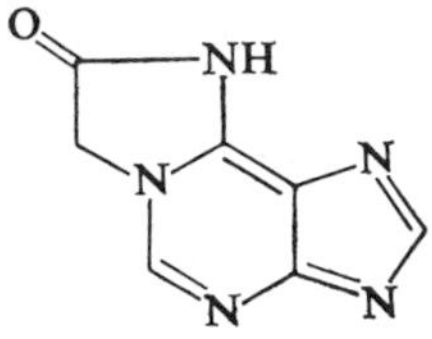   (14)

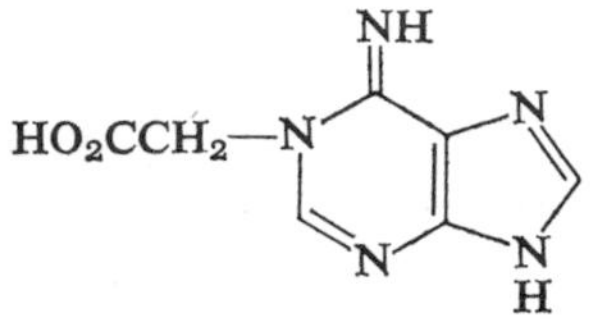

(15) 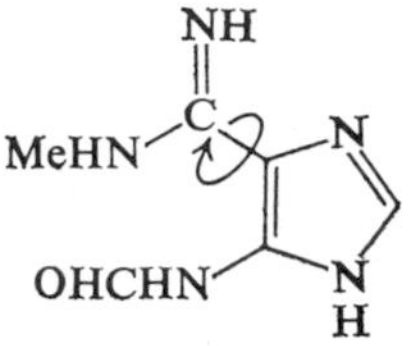   (16)

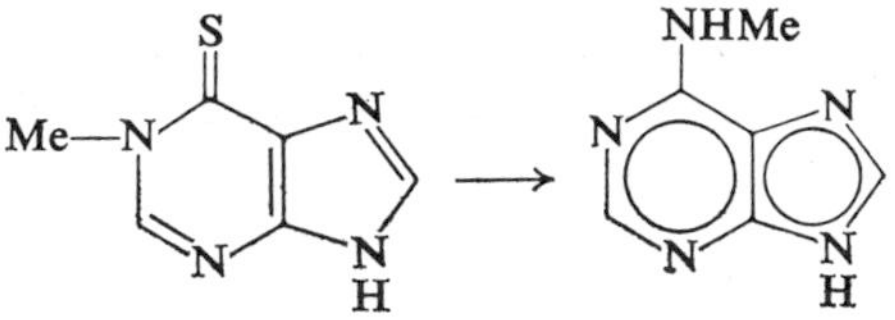

(17) → (11)

of 1,6-dihydro-1-methyl-6-thiopurine (17), the alkaline conditions employed isomerising the 1-methyladenine primarily formed.[3] The analogous conversion of the 9-β-D-riboside of 1-benzyl-1,6-dihydro-6-thiopurine to the 6-benzylamino derivative is known.[32] As yet not fully explained is the rearrangement of 1,6-dihydro-1-methyl-2-methylthio-6-thiopurine (18) to 6-amino-2-methylaminopurine (19) in ammonium hydroxide at 150°[3] but a related mechanism most likely obtains. Several conversions of 6-aminopurines to the 6-benzylamino analogues through heating with sodium benzyloxide in benzyl alcohol are documented, illustrated by the formation of 21 from 6-amino-9-methyl-2-methylsulphonylpurine (20).[33] The structure of the latter is confirmed by acid hydrolysis to 6-benzylamino-2,3-dihydro-9-methyl-2-oxo-purine.[33]

Many heterocyclic systems will exchange primary amino groups for secondary or tertiary groups. Metatheses of this type are possible with adenine, which on heating in a mixture of benzylamine and benzylamine hydrochloride (170°, 8 h) gives 6-benzylaminopurine in moderate (54%) yield.[34] 6-Anilino- and 6-furfurylaminopurine are formed in the same way[34] and successful replacements of this type have been extended to 3-alkyladenine derivatives.[35] Clarification of the mechanism of this

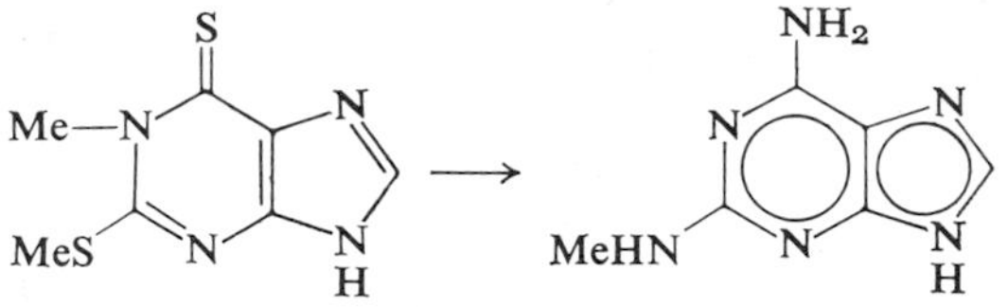

S

Me—N

MeS   N   N
          H

(18)

→

NH$_2$

N   N

MeHN   N   N
            H

(19)

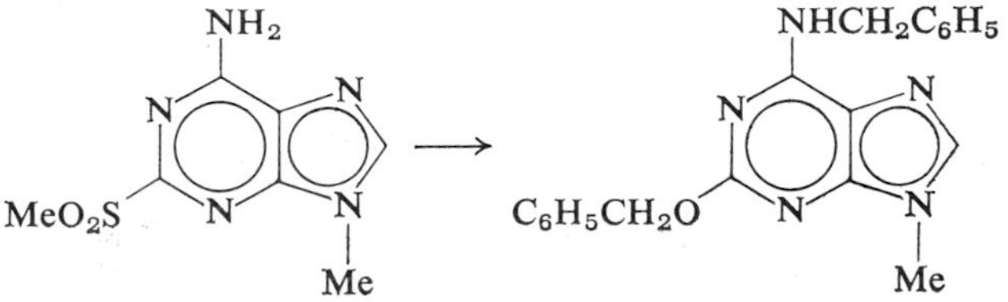

NH$_2$

N   N

MeO$_2$S   N   N
                |
               Me

(20)

→

NHCH$_2$C$_6$H$_5$

N   N

C$_6$H$_5$CH$_2$O   N   N
                      |
                     Me

(21)

acid-catalysed reaction is still awaited. A further route to alkylamino-purines involving acylation of the primary amino group followed by reduction of the resulting amide is given later (Sect. 5Dd).

2. Preparation of *N*-Aminopurines

Examples of 1-, 3-, 7-, and 9-aminopurines are known. In the majority of cases direct synthesis has been employed for their formation. Although direct amination procedures have so far provided only one example in the purine series (noted below), a number of purine nucleosides have been so converted to their *N*-amino analogues.

Aqueous hydrazine, at ambient temperatures, cyclises 5-cyano-4-ethoxymethyleneamino-1-methylimidazole to 1-amino-1,6-dihydro-6-imino-7-methylpurine (**22**).[27] A series of 3-aminoxanthines, including **23**, result on ring closure of 3-substituted-amino-4,5-diaminopyrimidines with triethyl orthoformate.[36] Both the 3-morpholino and 3-piperidino analogues of **23** are known.[36] The most numerous examples are provided by 9-aminopurines, for which alternative cyclisation routes exist. Ring closure of 5-amino-4-hydrazinopyrimidines is the more versatile in allowing a variety of substituted purines to be prepared. The parent member, 9-aminopurine, arises from formic acid ring closure of 4-amino-5-hydrazinopyrimidine[37] as does the 6-methyl homologue likewise.[38] 8-Oxo- and 8-thio-9-aminopurines result from cyclisations with phosgene and thiophosgene, respectively.[39] The very real chance that the alternative cyclisation, through involvement of the β-nitrogen of the hydrazino group, could take place giving a 1,2-dihydropyrimido-

[5,4-*e*]*as*-triazine (**24**) can be obviated by using the anil of the hydrazino-pyrimidine, an example being **25**.[37, 38]

Similar conditions and reagents are employed for the other route in which *N*-aminoimidazole derivatives are cyclised. Thus, 1,5-diamino-4-carbamoyl-2-methylimidazole in triethyl orthoformate–acetic anhydride gives 9-amino-8-methylhypoxanthine.[40] Similarly formed is 9-dimethyl-aminohypoxanthine.[41] Urea fusions afford corresponding 9-amino-xanthine analogues.[40] The unique case of *N*-amination mentioned previously is the transformation of 6-chloropurine-3-oxide into the 9-cyanoaminopurine (**26**) on short heating (85°, 30 min) in a cyanamide melt.[42] By a reaction akin to methylation, and to date reported only with nucleosides, the 9-β-ribosyl derivatives of hypoxanthine and guanine when interacted with hydroxylamine-*O*-sulphonic acid in cold *N*-sodium hydroxide (18–20 h) afford the appropriate 1-amino analogues. In the case of 6-amino-7,8-dihydro-8-oxopurine riboside the product is the 7-amino derivative.[42a]

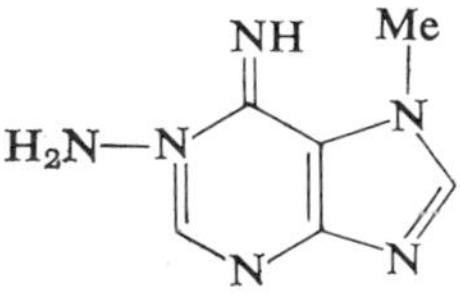

(22) 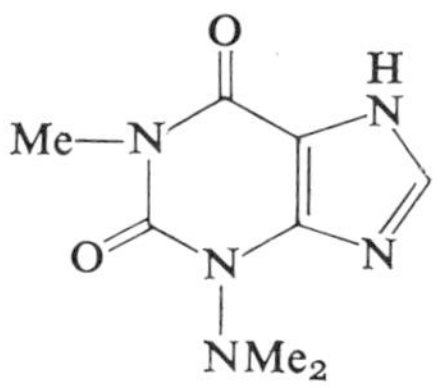 (23) 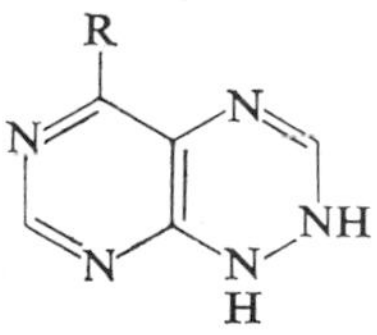 (24)

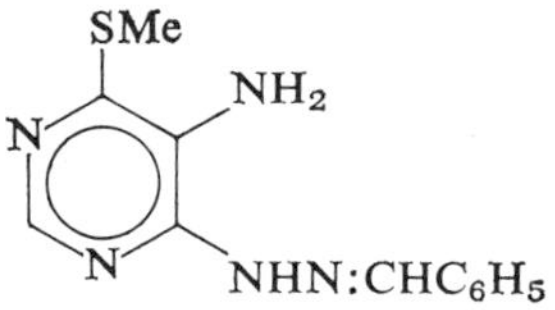

(25) 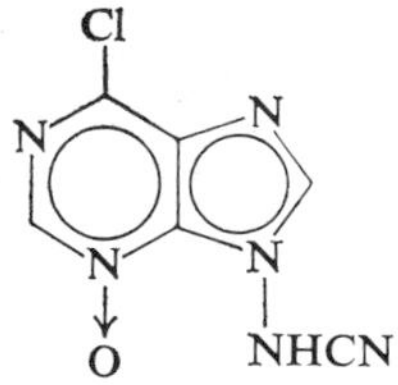 (26)

3. Preparation of Extranuclear Aminopurines

Most of the general methods of forming aliphatic amines can be utilised. The main importance of these derivatives lies in the pharmacological properties many of them possess, particularly those derived from theophylline, some examples of which are mentioned below.

The Mannich reaction, in which an amine and formaldehyde react with an acidic hydrogen atom to form an aminomethyl derivative, has been extensively applied to theophylline. Amines employed include diethylamine,[43, 44] morpholine,[43–46] piperidine,[43, 44] and N-methyl-piperazine[43] giving 7-aminomethylpurines of the type 27. With difunctional amines, for example, piperazine, bis derivatives e.g. 28 are possible.[43, 46] Theobromine forms the Mannich base at $N_{(1)}$ with morpholine and piperidine.[46] Hypoxanthine affords the $N_{(1)}N_{(9)}$-diaminomethylpurine while xanthine gives the 1,3,9-triaminomethyl derivative (29, R = morpholino). The structure (30) assigned to the diaminomethylated product arising with adenine is based on nmr spectral evidence.[47]

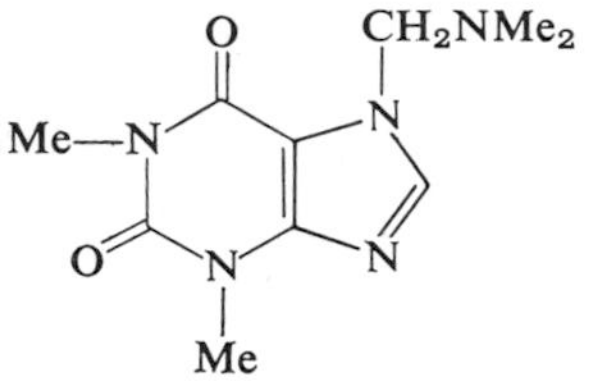

(27) 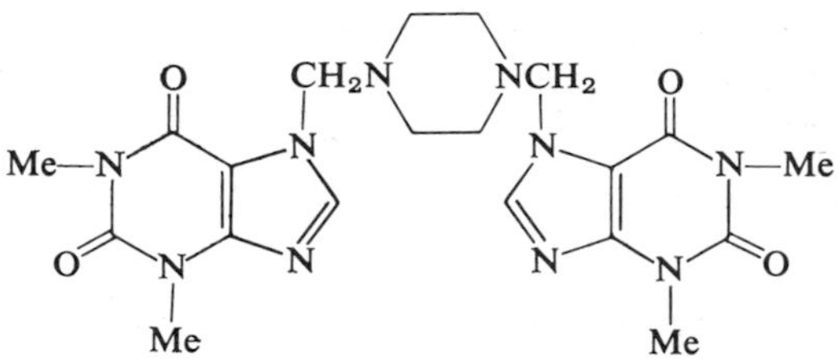   (28)

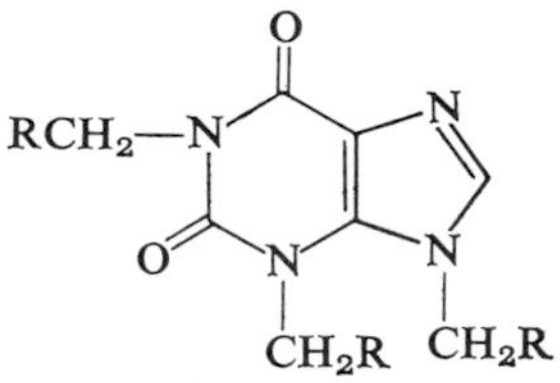

(29) 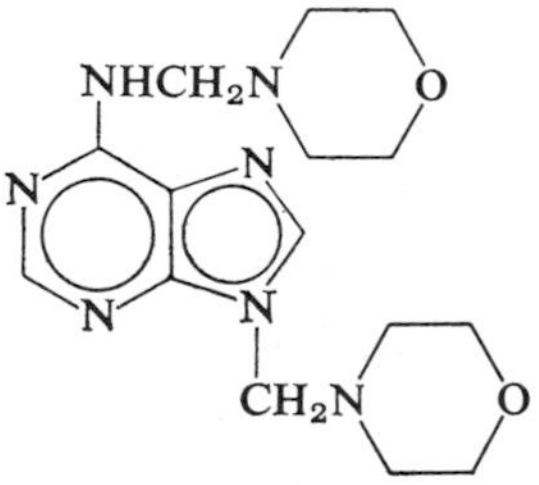   (30)

Base formation is also possible with reactive methylene groups outside the ring; an interesting case is found with 7-acetonyltheophylline (31) which with dimethylamine gives 32 but, if the amine hydrochloride is used, then through reaction with the terminal methyl group, 33 results.[48, 49]

A variety of 7-aminoalkyltheophyllines arise from direct alkylation procedures, exemplified by the 7-(2-diethylaminoethyl) derivative from reaction of sodium theophyllinate with 2-chloroethylamine.[50] Other similar alkylations are known.[51, 52] The parent 7-(2-aminoethyl)-theophylline is produced by heating theophylline in dimethylformamide (60°, some hours) with ethylene imine.[53] Substituted ethylene imines give the corresponding 7-(2-alkylaminoethyl) derivatives.[53] Traube-type syntheses have afforded purines with amino-alkyl groups attached to the pyrimidine moiety.[54] By using 5-amino-4-aminoalkylaminopyrimidines, ring closure gives 9-aminoalkylpurines.[54] A detailed account of the amination of extranuclear halogen atoms is found elsewhere (Ch. V, Sect. 6A). Exceptions to the usually vigorous conditions required are

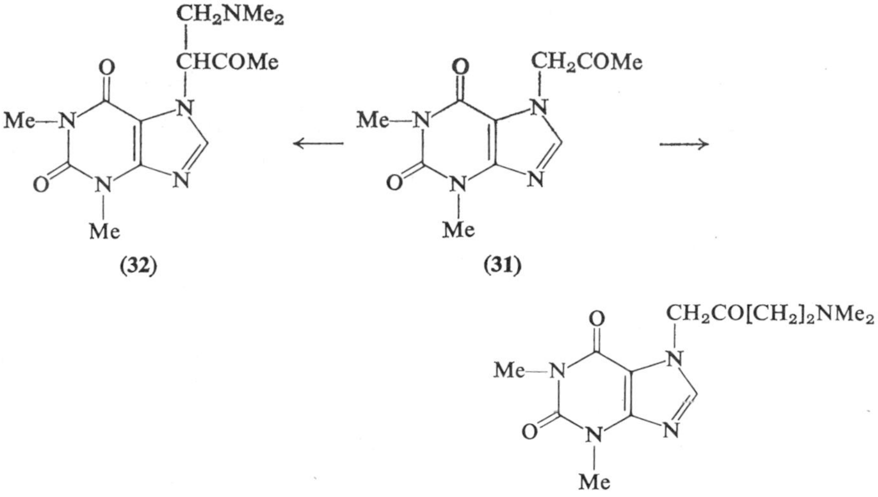

(32)   ←   (31)   ⟶

(33)

sometimes encountered, as in the formation of 6-(3-aminopropyl)purine from the 6-(3-chloropropyl) analogue on prolonged contact (3 days) at 80°, with ethanolic ammonia.[55] Potassium phthalimide in toluene under reflux converts 8-chloromethylcaffeine to the intermediate 8-phthalimidomethylpurine which requires further heating in ethanolic hydrazine to obtain 8-aminomethylcaffeine.[56] The Hofmann reaction has had some application to extranuclear amine formation; bromine in cold sodium hydroxide on 6-propionamidopurine gives a low yield of 6-(2-aminoethyl)purine.[55] 8-Aminomethylcaffeine has been prepared in this way.[56]

The remaining routes utilise reductive procedures. Hydrogenation of the anil (34) of caffeine-8-aldehyde in ethanol over Raney nickel gives 8-aminomethylcaffeine.[56]

Catalytic reduction of 6-cyanopurine affords the unstable 6-aminomethylpurine in good yield, using a palladium catalyst in methanol.[10] Similarly both 7-(3-aminopropyl)theophylline[49, 57] and 1-(3-aminopropyl)theobromine[57] arise from the respective cyanoethylpurine. The established conversion of a carbonyl group oxygen atom to an amino

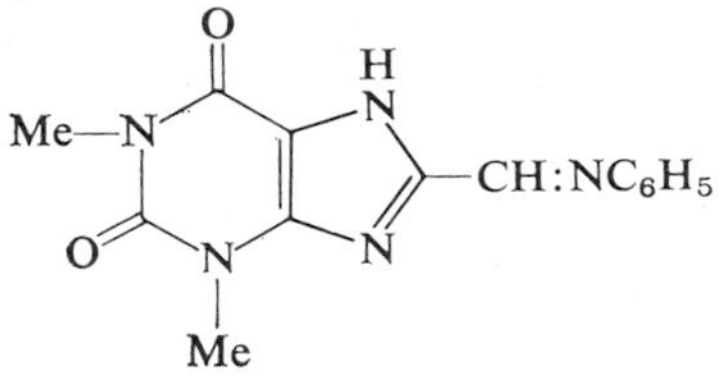

(34)

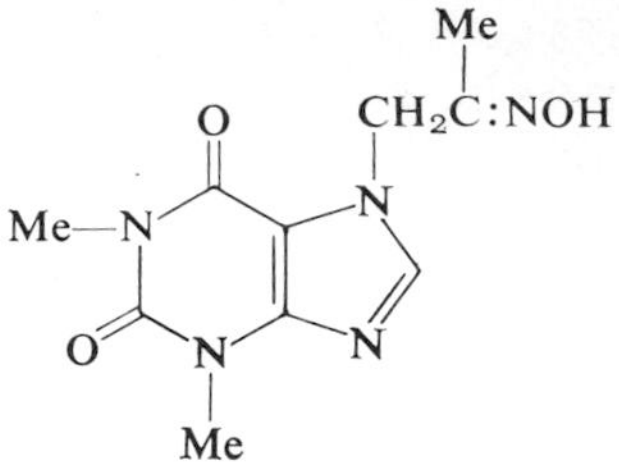

(35)

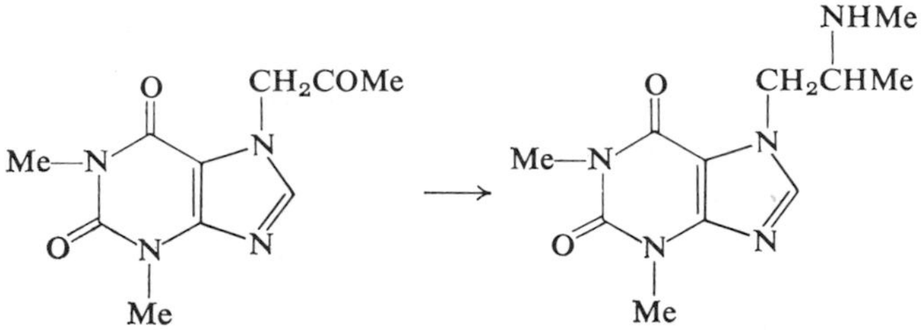

(36)

$\longrightarrow$

(37)

group via the oxime is exemplified by reduction of the oxime (35) to 7-(2-aminopropyl)theophylline.[58] The analogous 7-(2-aminobutyl), 7-(2-amino-3-phenylpropyl), and other derivatives followed likewise.[58] In some instances reduction of the ketone in the presence of an amine is successful. Under pressure conditions (80 atm/100°) 7-acetonyltheophylline (36) with methylamine in ethanol containing Raney nickel in a hydrogen atmosphere gives 7-(2-methylaminopropyl)theophylline (37).[58] When other amines are used, mixtures of the amino derivative and the alcohol formed by reduction of the ketone are obtained. A related reaction is the conversion of 7-(3-hydroxypropyl)theophylline to the 7-(3-aminopropyl) analogue when hydrogenated over platinum oxide in the presence of ammonia.[59] With amines in place of ammonia, 7-(3-alkylaminopropyl)purines are obtained.[59]

4. Properties of Aminopurines

The low solubility and generally high melting points, or, more accurately, decomposition points, shown by aminopurines (Table 7) are a reflection of the considerable degree of hydrogen bonding present in the crystal lattice. These effects are intensified when oxo groups are also present. Crystallographic examination of the hydrohalides of adenine,[60] guanine,[61, 62] and the 9-methyl derivatives[63, 64] by X-ray

TABLE 7.   Melting Points of Aminopurines.

Position of substitution	Amino	Methylamino	Dimethylamino
2-	278°	276°	223°*
6-	>350°	314°	245°*
8-	>350°	334°	292°
2,6-	302°	>350°	254°
2,8-	300°(H_2SO_4)	—	—
6,8-	>350°	305°(HCl·H_2O)	293°
2,6,8-	—	—	—

*But higher values are also reported, see Table 22.

techniques and of the neutral molecules by polarised absorption spectroscopy[65] has identified the sites of hydrogen bonding and protonation. These studies have also indicated that the amino groups in adenine and guanine, in the solid state, exist in the amino rather than imino form, a finding that is upheld by infrared examination[66] and nmr studies.[67, 68] Adenine and guanine derivatives in solution are shown to predominate as the amino tautomer by ultraviolet spectroscopy.[69] Although both 1- and 3-methyladenine can be drawn as either imino (**38a, 39a**) or amino (**38b, 39b**) tautomers, physical evidence indicates that only 1-methyladenine is composed mainly of the imino (**38a**) tautomer,[24,69a] the 3-methyl isomer existing largely in the amino (**39b**) configuration.[66,69a]

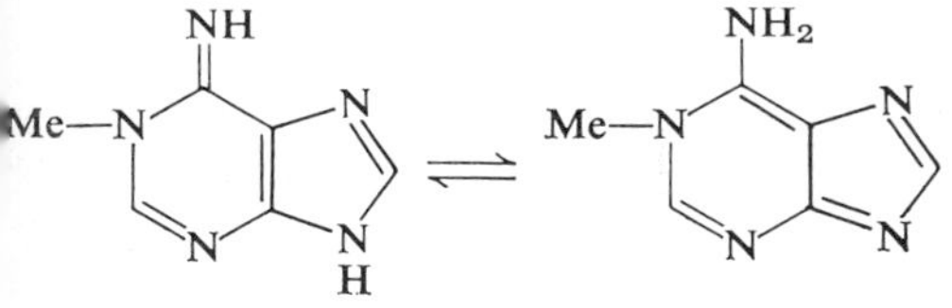

(38a)   (38b) 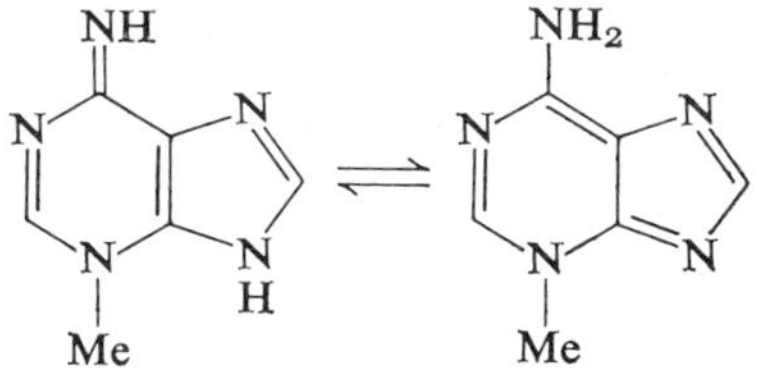   (39a)   (39b)

5. Reaction of Aminopurines

A. Replacements of Amino by Oxo Groups

The use of nitrous acid to hydrolyse amino groups is one of the earliest known purine metatheses, exemplified in the conversion of adenine to hypoxanthine[70] or guanine to xanthine.[71] In the latter example the ready conversion to xanthine is in marked contrast to the failure to obtain this purine from the isomeric isoguanine under the same conditions.[72, 73] To a lesser degree hydrochloric acid or sodium hydroxide are sometimes appropriate for hydrolysis purposes. Isoguanine can be converted to xanthine on prolonged heating with 25% hydrochloric acid.[72, 73] A comprehensive survey of this reaction is given in Chapter VI, Sections 1Da and b.

B. Replacement of Amino Groups by Halogens

Examples of the Schiemann reaction, a modified Sandmeyer reaction, occur in special circumstances with 2-, 6-, and 8-aminopurines. Adenine and its simple derivatives do not undergo the reaction[74]; adenosine, likewise, does not respond to nitrous acid in the presence of fluoboric acid but the 2',3',5'-tri-O-acetyl derivative is converted to 9-(2,3,5-tri-O-acetyl-β-D-ribofuranosyl)-6-fluoropurine.[75] The same reagent mixture transforms 2-aminopurine into 2-fluoropurine.[76] Low yields of 2-chloropurine result when the fluoboric acid is replaced by hydrochloric acid.[76, 77] Spectroscopic techniques indicate that 6,8-diaminopurine is converted to 6-amino-8-chloropurine in this way.[78] Fuller treatment of these aspects is given in Chapter V, Sections 1D and 1Gc.

C. Formation of Azomethine and Related Derivatives

At ambient temperatures the amino group in adenine derivatives reacts with dimethylformamide dimethylacetal forming the corresponding 6-dimethylaminomethyleneaminopurine (40). The $N_{(2)}$-analogue forms with guanine but due to the lower solubility of this purine in dimethylformamide higher reaction temperatures (80°, 5 h) are needed.[79] By contrast, stable anils or Schiff bases are not formed by interaction of 2-, 6-, or 8-aminopurines with aromatic aldehydes.[21] The formation of

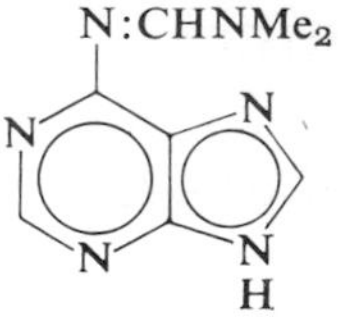

(40) 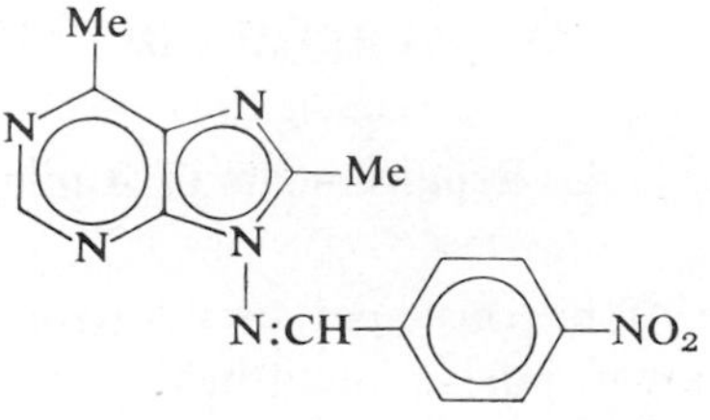   (41)

a Schiff base between adenine and hypoxanthine, reported[80] in the early literature, is now discounted.[81] Anils of the $N_{(9)}$-aminopurines are, however, known,[37, 38] being derived by cyclisation of a 5-amino-4-benzylideneaminopyrimidine, the 9-benzylideneaminopurine (41) being so derived.[38] Both hypoxanthine[37] and 6-methylthiopurine[38] analogues are known.

Schiff bases derived by the reverse condensations, that is, between purine aldehydes and amines, are described in Chapter IX, Section 5Bb.

D. Acylation of Aminopurines

Acylation in the general sense, taken to include formylation, acetylation, and benzoylation, is possible with amino groups at the 2-, 6-, and 8-positions. As aminopurines are frequently prepared by a cyclisation reaction, in which acylating agents are employed, the product is often the acylaminopurine. Under the milder conditions employed in the Schotten-Baumann procedure, only an imidazole ring nitrogen atom is acylated.

a. *Acetylation*

Kossel first demonstrated the formation of an acetylated derivative of adenine on heating it in acetic anhydride[82] but it was left to later workers to show that the product was 6-acetamidopurine.[83] More recent studies show that initially a diacetylated purine is produced in which the second, highly labile, acetyl group is located at either $N_{(7)}$ or $N_{(9)}$.[84–86] Removal of this group, forming 6-acetamidopurine, occurs readily on heating in water[84, 85] or ethanol[86] or on contact with cold dilute alkali.[49] Mixtures of mono- and diacetylated purines are best separated by utilising the much greater solubility of the diacetyl derivative in petroleum ether.[84] Active esters, such as *p*-nitrophenyl acetate or 2,4,5-trichlorophenyl acetate, with adenine in cold dimethylformamide containing triethylamine, give unspecified diacetyl derivatives from which 6-acetamidopurine is isolated on heating in dimethylacetamide.[86]

Diacetylation of 2-amino-6-chloropurine occurs on heating in acetic anhydride with a phosphoric acid catalyst[87] or with the anhydride in dimethylacetamide.[88] Deacetylation to 2-acetamido-6-chloropurine follows treatment with methanolic ammonia[87] or aqueous alkali.[88] Guanine forms the 2-acetamido analogue on prolonged heating with the anhydride.[89] Toluene was the solvent employed in acetylation of some 2-aminopurines.[88] Both amino groups in 2,6-diaminopurine are acetylated,[83] but the $N_{(9)}$-acetylated derivative (42) has been shown to be formed in the reaction.[90] The facility with which this acetyl moiety is lost during recrystallisation of the product from water explains the failure of the earlier workers[83] to recognise the triacetylated purine. Designation of $N_{(9)}$ as the site of nuclear acylation is based on the observed disappearance of the imidazole imino group absorption bands

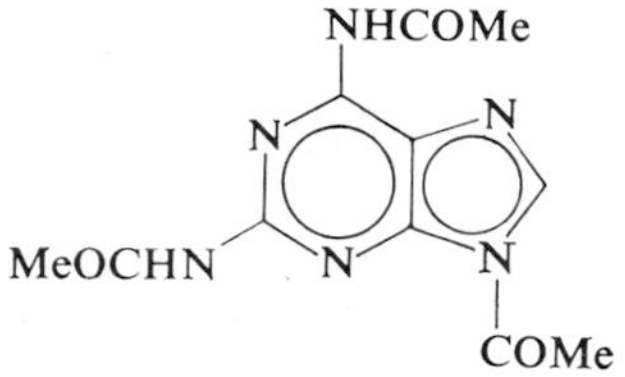

(42) 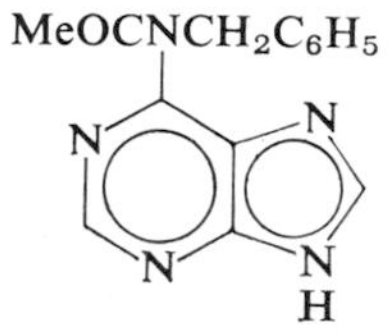   (43)

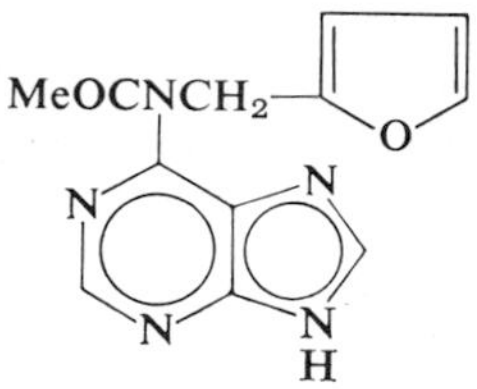

(44) 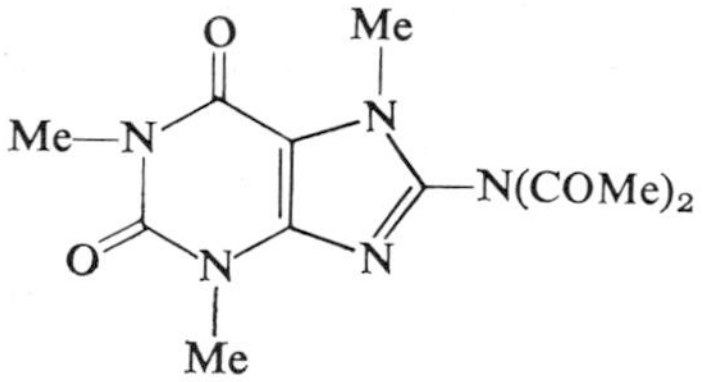   (45)

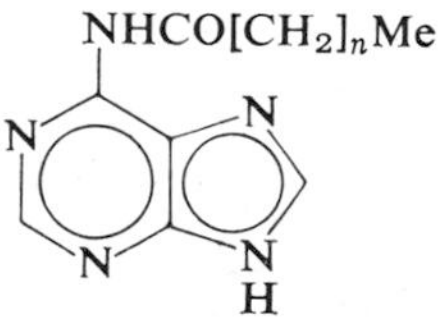

(46)

in the infrared.[90] Further support is provided by the fact that 9-methyladenine forms only a monoacetylated derivative.[91] 3-Alkyladenine likewise gives only the 6-acetamidopurine.[92] Although a secondary amino group at $C_{(6)}$ can be acylated, as seen in 6-(N-acetylbenzylamino)-purine (43),[93] the kinetin analogue (44) is not formed,[21]* the product being either an $N_{(7)}$- or $N_{(9)}$-acetyl derivative.[95] Diacetylation of an amino group, although rarely encountered, is reported[17] to occur in 8-aminopurines as, for example, 45 derived from 8-aminocaffeine. Extensive use of homologous anhydrides has been made with a variety of 6-aminopurines. Chloroacetic acid and adenine give 6-chloroacet-

---

* A claim [94] to have produced 44 under acetylating conditions has now been retracted.

amidopurine[96] in toluene solution. Heating with the anhydride alone or in pyridine affords the corresponding 6-acylaminopurines with propionic,[84, 85, 97] butyric,[84, 85, 91] valeric[85] and higher anhydrides.[85, 97a] Guanine forms the respective 2-acylamino derivative with octanoic and palmitic anhydrides in pyridine likewise.[97a] The use of acyl chlorides overcomes the problem of the inaccessibility of the anhydride for preparing high molecular weight acylamino groups. Condensation of the appropriate acyl chloride with adenine in cold $2N$-sodium hydroxide[98] or pyridine,[99] under reflux, gives derivatives of the type **46** ($n = 4$ to 16).

b. *Formylation*

Formamidopurines are rarely encountered in the literature, those described are usually the result of concomitant formylation occurring during formic acid cyclisation of triaminopyrimidines to aminopurines. These are usually derivatives of 2-formamidopurines.[100, 101]

c. *Benzoylation and Other Acylations*

Fusion of benzoic anhydride with adenine was the first preparation of 6-benzamidopurine,[82] this route being still favoured by later workers.[93, 102] This derivative also arises unexpectedly on benzoylation of 3-benzhydryladenine (**47**) through loss of the benzhydryl group during the fusion (190°, 15 min) reaction.[92]* Secondary aminopurines undergo benzoylation exemplified by formation of 6-(*N*-methyl)benzamidopurine (**48**), from 6-methylaminopurine (210°, 20 min), in good yield.[103] Both amino groups are benzoylated in 2,6-diaminopurine under these conditions.[83] Corresponding amides are formed by adenine with furoic, thienoic, toluic[102] and phthalic anhydrides[91] as, for example, **49** in the last case. Of related interest is 2-phthalonamidohypoxanthine (**50**) obtained on sodium metaperiodate oxidation of the adduct formed between guanine and ninhydrin.[104] With benzoic anhydride, even on prolonged heating at 100°, guanine affords only a poor yield of the 2-benzamido analogue.[89] As an alternative to fusion some of the above aroylations have been carried out in xylene under reflux.

Although Kossel failed to react benzoyl chloride with adenine,[82] later condensations performed in cold sodium hydroxide solution[98] or

---

* By contrast the benzhydryl groups remain *in situ* during the milder conditions used for acetylation.[92]

boiling pyridine[105] afforded a dibenzoylated derivative which, in boiling water, was converted to 6-benzamidopurine.[105] In pyridine at room temperature on long standing (96 h), adenine-9-β-D-xylofuranoside is reported to give the 6-dibenzamidopurine nucleoside,[106] but the structure is not positively established. More detailed studies of these reactions are needed in view of other claims that only 7- or 9-benzoyl-adenines are formed under Schotten-Baumann conditions.[107] Furoyl chloride gives 6-furamidopurine[98, 105] with adenine.

Carbobenzoxy chloride reacts with adenine giving a mixture of 7- and 9-carbobenzoxy derivatives[107, 108] but if $N_{(9)}$ is blocked, as in 9-benzyladenine, fission of the imidazole ring ensues producing the triaminopyrimidine (51).[107] Instability in the five-membered ring in this case seems most likely to be the result of the production of a quaternised intermediate.

Carboxylic acids or derivatives find little application in acylation but derivatives of 2-phenyloxazolone (52) in tetrahydrofuran have given

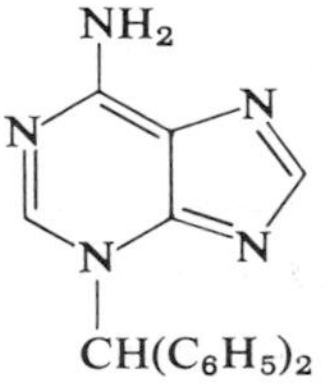

(47) 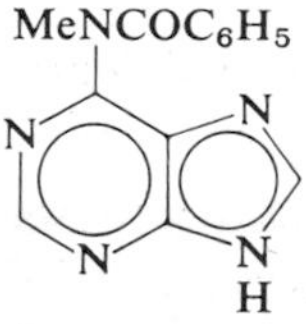　　　　　　　　(48) 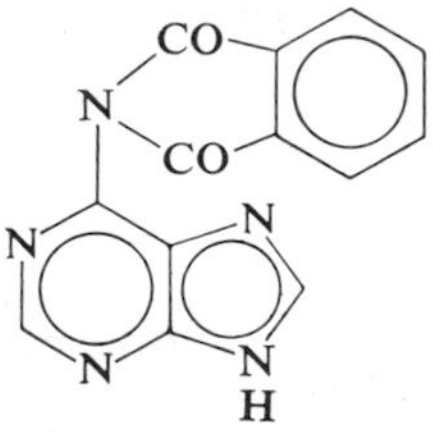　　　　　　　　(49)

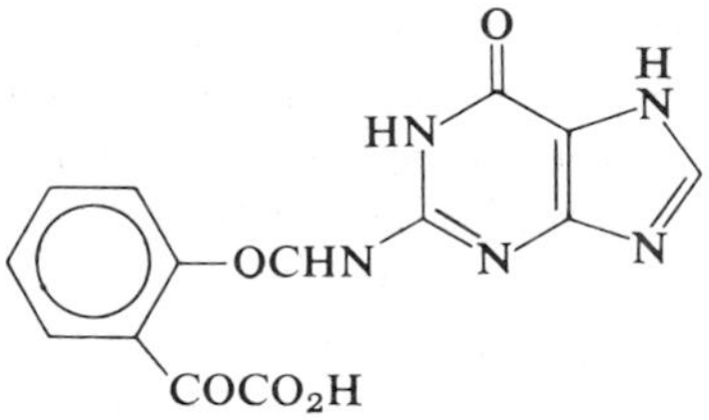

(50) 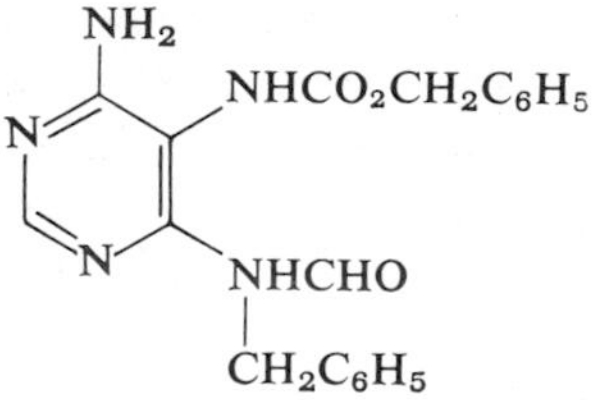　　　　　　　　　　　　　　　(51)

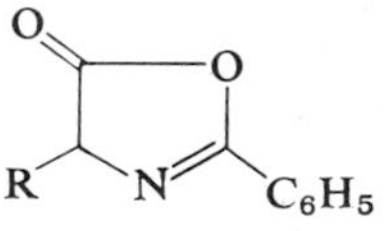

(52) 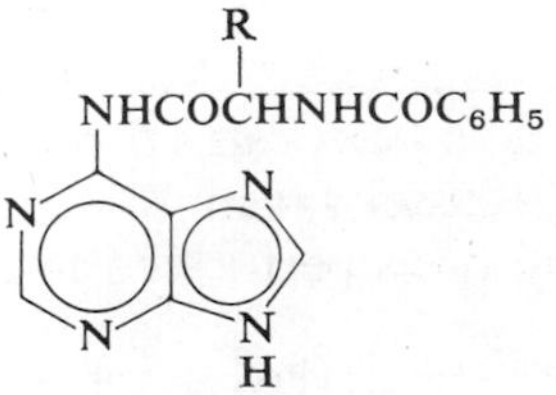　　　　　　　　　　　　　　　(53)

6-aminoacetamidopurines of the type **53** (R = H or Me),[109] long periods (60–70 h) under reflux being necessary.

d. *Reduction of Acylaminopurines*

In all cases the reducing agent is lithium aluminium hydride, the method being useful for converting a primary amino group to the secondary state. The majority of applications have been to preparations of 6-arylmethylaminopurines. The most widely used solvent, tetrahydrofuran, was employed in the reduction of 6-acetamidopurine to 6-ethylaminopurine in 88% yield.[102, 110, 111] 6-Benzylaminopurine, in over 90% yield, results from the 6-benzamido analogue[102, 111, 112] in this way but a much reduced yield is obtained using pyridine as solvent.[105] Kinetin, a plant growth factor, results from reduction of 6-furamidopurine.[102, 105] Other 6-arylmethylaminopurines have been similarly derived.[102] Both carbonyl groups in compounds of the type **53** (R = H or Me) are reduced under the above conditions.[109]

c. *Hydrolyses of Acylaminopurines*

Generalisations cannot be made for the conditions required for deacylation, great variations being found. Dilute alkali and fairly vigorous heating were required to convert 2-acetamidopurines to the 2-amino analogues.[88] Less successful was the attempt to obtain the parent adenine derivative from 6-benzimido-3,7-dibenzylpurine,[23] although sodium methoxide treatment effectively converts 6-benzamido-9-(3-methylbut-2-enyl)purine to 6-amino-9-(3-methylbut-2-enyl)purine.[113] Aqueous 2*N*-sodium hydrosulphide transformed a 2-acetamido-9-alkyl-6-chloropurine into a 2-amino-9-alkyl-1,6-dihydro-6-thiopurine.[88]

f. *Rearrangement of Acylaminopurines*

6-Benzamidopurine (**54**) and benzyl bromide in dimethylformamide at 80° give a mixture of 3-benzyl (**55**) and 9-benzyl (**56**) derivatives.[114] As the yield of the 9-benzyl isomer increases as the heating time is extended, benzyl group migration from $N_{(3)}$ to $N_{(9)}$ is postulated. The isolation of the 9-benzyl isomer (**56**), in 75% yield, when the hydrobromide of the 3-benzyl derivative (**55**) suffers prolonged heating

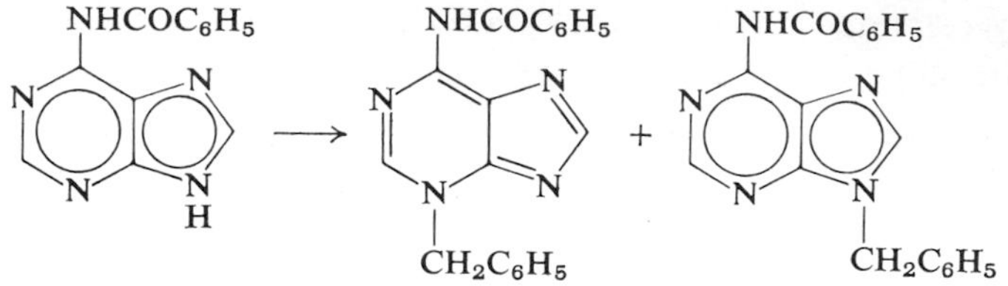

(120°, 70 h) in the same solvent, lends support to this.[114] Additional evidence is in the isolation of 6-benzamido-9-benzylpurine containing 52% of ^{14}C activity on heating together a mixture of 6-benzamido-purine [14]$C_{(8)}$ and 6-benzamido-3-benzylpurine hydrobromide.[113] Corresponding $N_{(3)}$ to $N_{(9)}$ rearrangements are shown by 3-allyl-6-benzamido-,[114] 6-benzamido-9-(3-methylbut-2-enyl)-,[113] and 6-acet-amido-9-(3-methylbut-2-enyl)purine.[113] As the reaction appears to be catalysed by bromide ion the hydrobromide salts are employed although in some instances, mercuric bromide and the base are found to give better yields.[113] In no case could analogous rearrangements be observed with the parent 3-alkyladenine derivatives.[114]

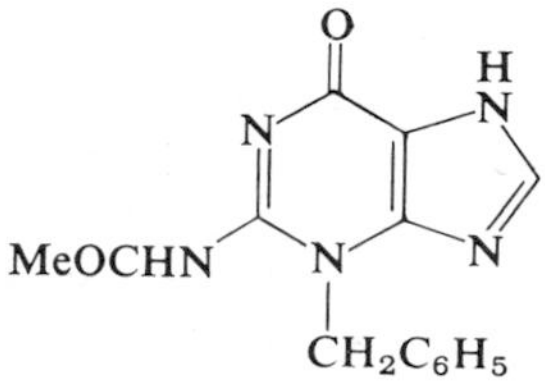

Of related interest is the formation of both 7- and 9-benzyl derivatives when $N_{(2)}$-acetylated guanine is heated (90°, 7 h) with benzyl bromide in dimethylacetamide. The isolation of trace amounts of the 3-benzyl isomer (57) from the product mixture may point to the reaction occur-ring by way of an $N_{(3)}$ to $N_{(7)}$ or $N_{(9)}$ migration of the benzyl group, this being supported by the observation that in dimethylacetamide (120°, 30 h) rearrangement of the 3-benzyl derivative (57), as the hydro-bromide, to a mixture of 7- and 9-benzyl isomers does take place.[115]

g. *Sulphanilylaminopurines*

The product obtained from interaction of adenine and *p*-nitro-benzenesulphonyl chloride, in pyridine at 100°, was initially thought to be 58 (R = NO$_2$),[116] but this was shown to be erroneous when an unambiguous synthesis of **58** (R = NH$_2$), from 6-chloropurine and sodium sulphanilamide, in dimethyl formamide at 100°, showed it to be different from the reduction product of the original derivative.[117] It seems, therefore, that the reaction product is either formed by sulphanilylation of adenine at $N_{(7)}$ or, more likely, at $N_{(9)}$ to give **59** as it reverts to adenine readily on hydrolysis, and is insoluble in sodium hydroxide solution. Other $N_{(7)}$- or $N_{(9)}$-substituted derivatives of this type have been prepared.[118]

A rare example of direct sulphonylation of an amino group is given by the formation of the 2-sulphanilylaminopurine (**60**) from 2-amino-9-methylpurine and p-acetamidobenzenesulphonyl chloride in pyridine followed by deacetylation in dilute mineral acid.[117] A wide range of examples embracing 2-, 6-, and 8-sulphanilylaminopurines has been prepared from the corresponding halogenopurine and the sodium or potassium salt of sulphanilamide either under fusion conditions or in a high boiling solvent. Both 7- and 9-methyl derivatives of 2,6-dichloro-purine are converted to the respective 2-chloro-6-sulphanilylaminopurine in dimethylformamide (90°, 3 h).[117] Hydrogenation removes the halogen atom of 2-chloro-9-methyl-6-sulphanilylaminopurine, the product being the same as that from the action of sodium sulphanilamide on 6-chloro-9-methylpurine.[117] Similar conditions convert 2,6,8-trichloro-9-methylpurine to 2,6-dichloro-9-methyl-8-sulphanilylaminopurine which on reduction gives 9-methyl-8-sulphanilylaminopurine. Another

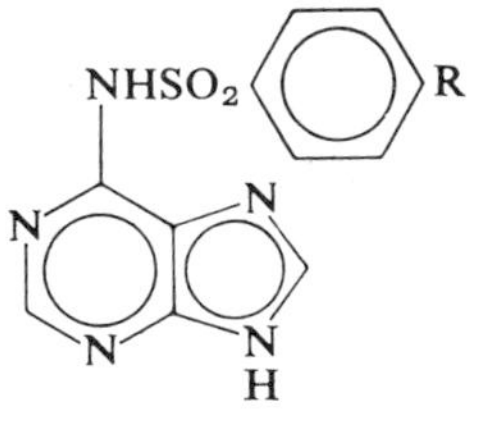

(58) 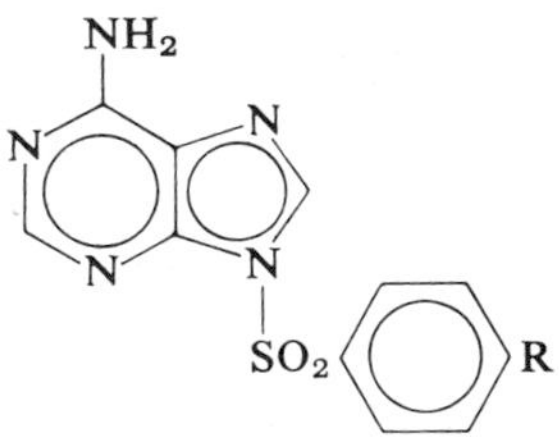   (59)

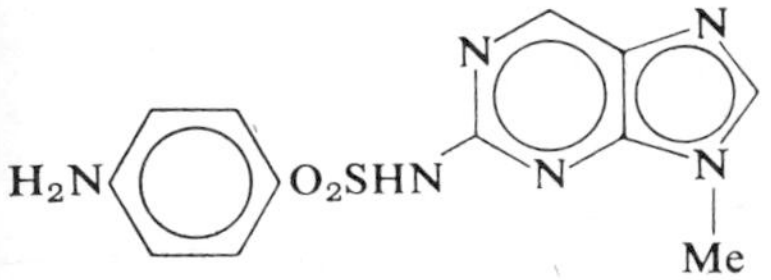

(60) 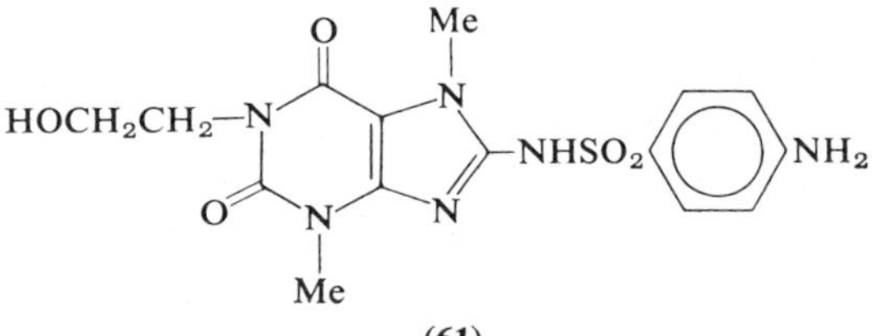   (61)

route to this last derivative is by sulphanilylamination of 8-chloro-9-methylpurine.[117] Other condensations have been carried out with 8-bromoxanthine derivatives giving products of the type 61.[119-121] 8-Sulphanilylaminomethylcaffeine is obtained likewise from 8-chloro-methylcaffeine.[121] A halogen at $C_{(2)}$ shows the expected low reactivity in this reaction, only poor yields of 9-methyl-2-sulphanilylaminopurine being obtained from the 2-chloro analogue, a better route to this derivative being from 2-amino-9-methylpurine, described above.[117]

E. With Formaldehyde

Kinetic studies carried out on adenine in aqueous formaldehyde at room temperature show that the unstable (reversible) and very soluble 6-hydroxymethylaminopurine (62) is the product[122] which can be

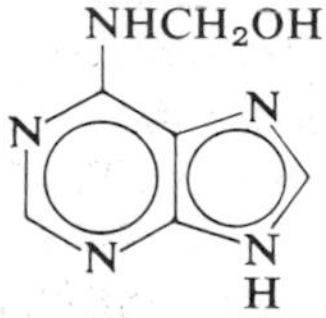

(62) 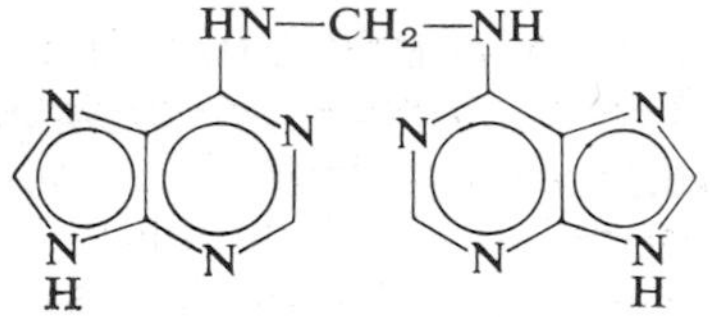   (63)

detected by ultraviolet spectroscopy.[123] The formation of an *NN*-dihydroxymethylaminopurine has been proposed[124] but seems unlikely. Further hydroxymethylation may, however, be possible at $N_{(9)}$ or $N_{(7)}$ in this and other purines,[122, 123] especially in alkaline solution. Evidence in favour of this is in the observation that uptake of formaldehyde by adenine is twice that found with a 9-alkyladenine.[123] Some additional support would appear to be given by Mannich-base formation in adenine derivatives (Sect. 3). Equimolar amounts of formaldehyde and adenine in aqueous solution on standing deposit crystals of the di-adeninyl derivative (63). The rate of formation is pH dependent, being much higher in an acid medium (3 days) than in a neutral one (30 days).[125]

Di(purin-6-ylamino)methane (63) melts in the range 336–340°, is insoluble in water, dilute alkali, and most organic solvents. In dilute hydrochloric or sulphuric acid a slow decomposition to adenine occurs.[125] The analogous dipurinyl derivative of adenosine is also known.[125] Guanine derivatives are reported[123] to react with form-aldehyde under the same conditions but the products are not specified.

F. Diazo- and *N*-Nitrosoaminopurines

a. *Preparation of Diazopurines*

A feature of many 8-amino derivatives of hydroxy- and amino-purines is the formation of fairly stable 8-diazopurines on treatment with nitrous acid. The first preparation[126] of this type was made from 8-aminocaffeine but the product structure was not established. Related compounds derived from 8-aminoxanthine and -theophylline were originally formulated[127] as being of the type **64**. More recently 8-di-

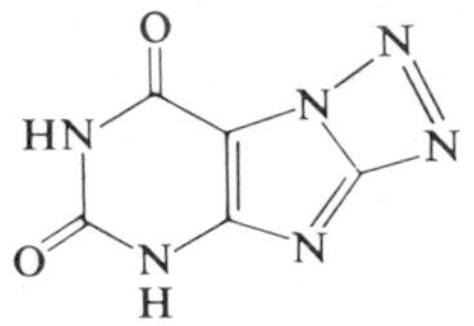

(64)

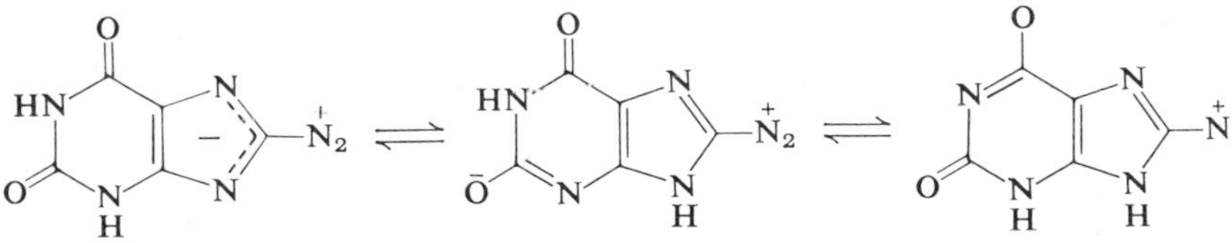

(65)   **(66)**   **(67)**

azoxanthine, shown in the betaine form **(65)**, has been prepared pure by addition of an aqueous solution of the potassium salt of 8-amino-xanthine and sodium nitrite to cold (5–10°) concentrated hydrochloric acid, the product precipitating fairly rapidly.[78] This procedure also affords the 8-diazo derivatives of hypoxanthine, 1,6-dihydro-2-methyl-amino-6-oxopurine, isoguanine, and theophylline.[78] Other canonical forms of these charged structures, with **65**, for example, both **66** and **67** are alternative representations, but of these **65** is preferred as no imidazole imino group bands can be assigned in the infrared, and from the fact that 8-amino-9-methylxanthine does not form a stable diazo derivative.[78] Analogous diazo derivatives do not arise from purines with amino groups at $C_{(2)}$ or $C_{(6)}$, 6,8-diaminopurine shows this in forming only 6-amino-8-diazopurine. With 2,8-diamino-1,6-dihydro-6-oxopurine nitrous acid alone gives 8-diazoguanine but in sodium nitrite and fluoboric acid the $C_{(2)}$-amino group is also hydrolysed giving **65**.[78]

b. *Reactions of Diazopurines*

The readily displaced diazo group provides a facile means of converting 8-aminopurines to a variety of 8-substituted purines. Hot dilute acid treatment gives the 8-oxo analogue, exemplified by the preparation of uric acid from 8-diazoxanthine or the conversion of 8-diazoguanine to 2-amino-1,6,7,8-tetrahydro-6,8-dioxopurine.[78] Other examples are known.[78] The corresponding 8-thiopurines are formed with hot potassium hydrosulphide; an example of this is found with 8-diazoguanine.[78]

Like the diazonium salts the diazo derivatives couple readily with amines and mercaptans in ethanol or acetone solution at room temperature. Fairly stable triazenopurines of the type **68** result with simple dialkylamines,[78, 128] these being formed with hypoxanthine, xanthine, guanine, and adenine diazo derivatives. 2-Mercaptoethanol and 8-diazotheophylline give 8-(2-hydroxyethylthioazo)theophylline while similar condensations with thioalkylcarboxylic acids[128] give, for example, **69**. Typical coupled products are obtained with phenols and benzenesulphonic acids.[126, 129] The addition product (**70**) is produced on leaving the diazopurine with cyanoacetic ester (3 h) in an ethanol–pyridine solution.[78]

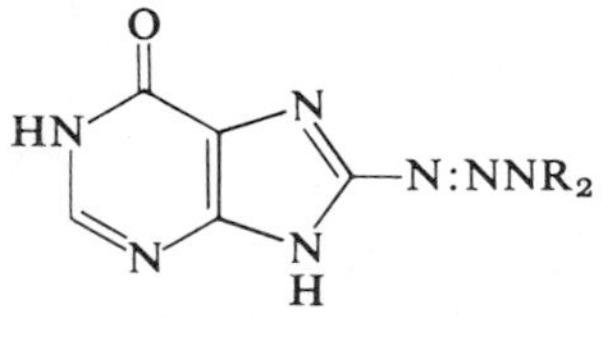

(68)

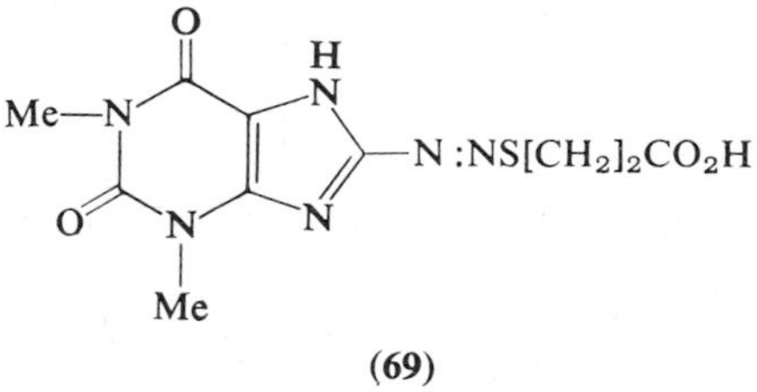

(69)

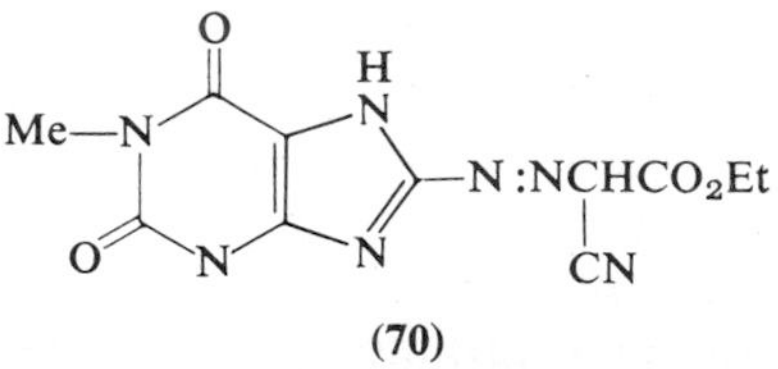

(70)

c. *N-Nitrosoaminopurines*

This system is poorly represented in the purines. The action of sodium nitrite in acetic acid on 6-benzylaminopurine gives 6-(*N*-nitrosobenzylamino)purine.[93] Similarly produced are 6-(*N*-nitrosomethylamino)purine and the 2-(*N*-nitroso) analogue of 2-ethylamino-1,6-dihydro-6-oxo-purine.[129a] The dipurinyl derivative (71) was likewise formed in hydrochloric acid from 1,2-di(6-methylaminopurin-9-yl)ethane.[130] A series of exocyclic nitrosoaminopurines follows from nitrosation of

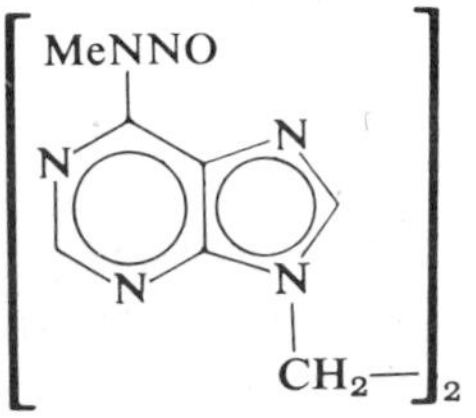

(71) 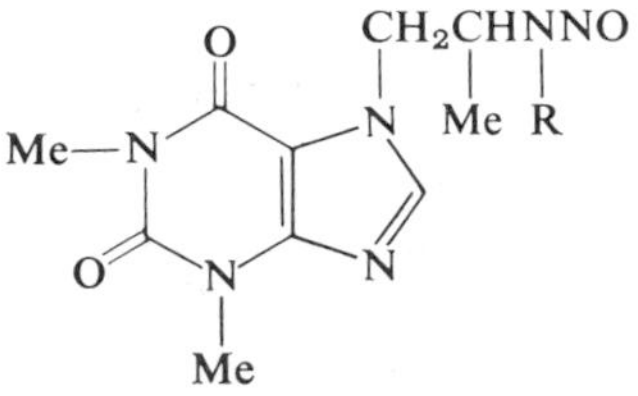   (72)

7-(2-alkylaminopropyl)theophyllines in acetic acid. Representative products include the methyl (72, R = Me), ethyl (72, R = Et), and benzyl (72, R = $CH_2C_6H_5$) analogues.[58] Other similar derivatives[131] have been reported.

G. Other Reactions of the Amino Groups

Adenosine and adenylic acid react under alkaline conditions (pH 10.5) with diazotised aromatic amines forming unstable derivatives formulated as being the 6-triazenopurines (73, R = β-D-ribofuranosyl and β-D-ribofuranosyl-5-phosphate).[132, 133] The analogous derivative is also reported from adenosine-1-oxide.[134] Guanine appears to form the 2-triazeno analogue but other products (unidentified) are simultaneously formed.[132] Evidence for the formation of the triazenopurines is based upon ultraviolet spectroscopic data. In acid media reversion to the parent aminopurine occurs rapidly. That these derivatives are not triazeno- but azopurines, arising through a coupling at $C_{(8)}$, seems unlikely, in view of the reversible nature of the reaction under acid conditions. Support for the postulated structures is given by the observation that on carrying out the reaction in the presence of formaldehyde, an avid reagent for amino groups, the above compounds are not formed.[132] The instability of these 2- and 6-triazenopurines is in contrast with that of the 8-triazenopurines previously described (Sect. 5Fb). Both adenine and guanine with phenyl isocyanate in hot pyridine or dimethylformamide afford the respective 6- or 2-(3-phenylureido)-purine.[134a]

6. Reactions of Extranuclear Amino Groups

Acylation requires conditions similar to those already described for nuclear aminopurines (Sects. 5Da, b, and c). Methylation of primary and secondary amino groups has been carried out by the Eschweiler synthesis. In this way on prolonged heating (100°) with formaldehyde and hydrochloric acid, 8-aminoethylcaffeine is converted to the 8-methylamino analogue.[56] A similar example, using formic acid in place of the mineral acid, is 7-(2-dimethylaminopropyl)theophylline from the 7-(2-methylaminopropyl) derivative.[58] Further alkylation to quaternary salts, as in the above case (74), occurs with an alkylating agent in ethanol–ether solution.[58] The nitrosation of alkylamino groups attached

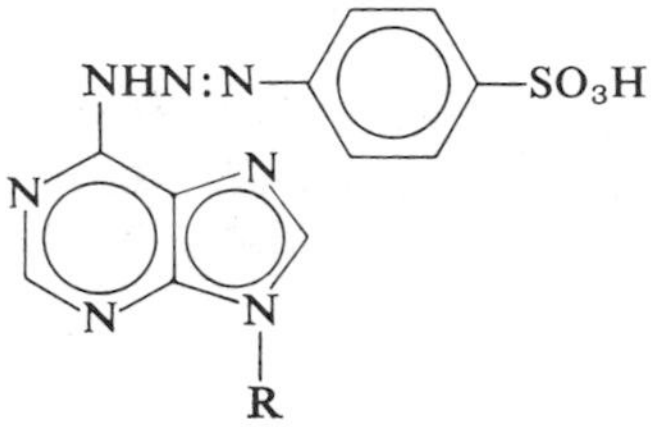

(73)

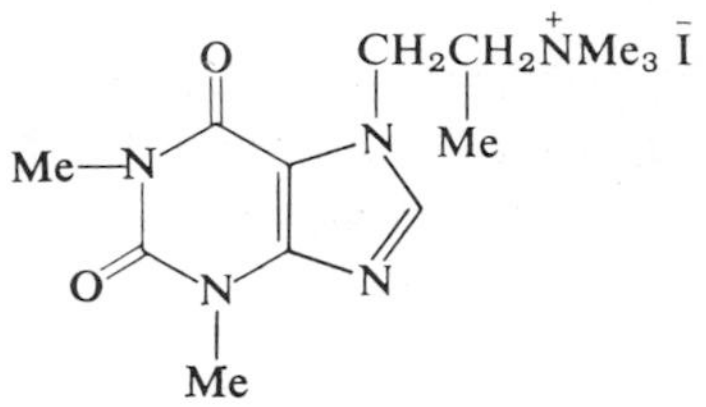

(74)

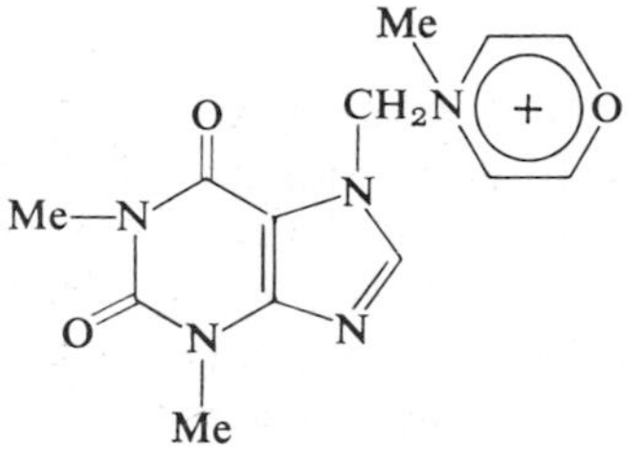

(75)

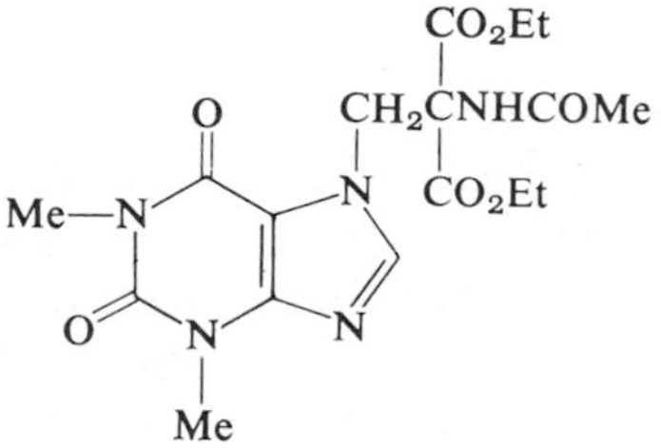

(76)

to side chains has been noted already (Sect. 5Fc). The product (76) occurs when the Mannich-base methiodide (75) and diethyl acetamidomalonate react together in ethanolic sodium ethoxide at room temperature.[135]

7. Other Substituted Aminopurines

Purines bearing modified forms of the amino group are described in these sections. Not included are simple alkyl- and N-nitrosoaminopurines which have been discussed already (Sect. 5Fc).

A. Hydrazinopurines

a. *Preparation*

Because hydrazine reacts readily with many cyclising reagents, the preparation of hydrazinopurines by ring closure of hydrazinopyrimidines or -imidazoles is not a practical route. Generally, replacement of halogen atom or methylthio group is made, the reaction being carried out under reflux conditions using either anhydrous or hydrated hydrazine. Cases where no external heating is required are known but for the majority prolonged heating (12–20 h) is necessary. Examples of hydrazino group insertion at $C_{(2)}$, $C_{(6)}$, and $C_{(8)}$ by these means include 2-hydrazino-(chloro),[136] 6-amino-2-hydrazino-(chloro),[136] 6-substituted-amino-2-hydrazino-(chloro),[137] 2-hydrazino-6-oxo-(chloro, 91%; methylthio, 30%),[136] 6-hydrazino-(chloro, 80%[136]; thio, 79%[138]; methylthio, 68%[139–141*]), 2-amino-6-hydrazino-(methylthio),[136] 6-hydrazino-7-methyl-(chloro),[142] 6-hydrazino-9-methyl-(chloro),[143] and 6-hydrazino-9-ethylpurine (chloro).[144] Chloropurines were also employed to obtain the 7- and 9-benzyl-6-hydrazinopurines[145] and a series of 8-hydrazino derivatives of *N*-methylated xanthines.[146, 147] With anhydrous hydrazine at room temperature or boiling aqueous hydrazine, both 2,6-dichloropurine[136] and the 7-methyl homologue form only the corresponding 2-chloro-6-hydrazinopurine.[136] Both chlorine atoms are replaced by heating for some hours (80°) in anhydrous hydrazine.[8, 136] Reflux conditions convert 6-amino-2,8-dichloropurine to the 2,8-dihydrazino analogue.[137] The strongly nucleophilic character of hydrazine is shown by the conversion of 2,6,8-trichloropurine to the 2,6,8-trihydrazino analogue under the above conditions.[137] In ethanolic hydrazine only the 6,8-dihydrazinopurine is obtained from 2,6,8-trichloro-7-methylpurine.[8] With stoicheometric amounts of hydrazine and halogenopurine a dipurinylhydrazide may result, vigorous treatment being required, the derivative (**77**) from 8-bromocaffeine being shown.[147] A further and older illustration is *NN*-di(2-chloro-7-methylpurin-6-yl)-hydrazine[148]† from 2,6-dichloro-7-methylpurine.

The presence of two hydrazine-reactive centres has led to the production of tricyclic derivatives such as **78** (R = Me) when 7-acetonyl-8-bromotheophylline is reacted in boiling ethanol.[149, 150] The analogous

---

* Through a misprint, the melting point (144–5°) given for 6-hydrazinopurine in *Chemical Abstracts*[138] is 100° too low.

† Fischer's structure shows this as the symmetrical *NN'*-dipurinylhydrazine:

derivative (**78**, R = C_6H_5) forms with 8-bromo-7-phenacyltheophylline.[149]

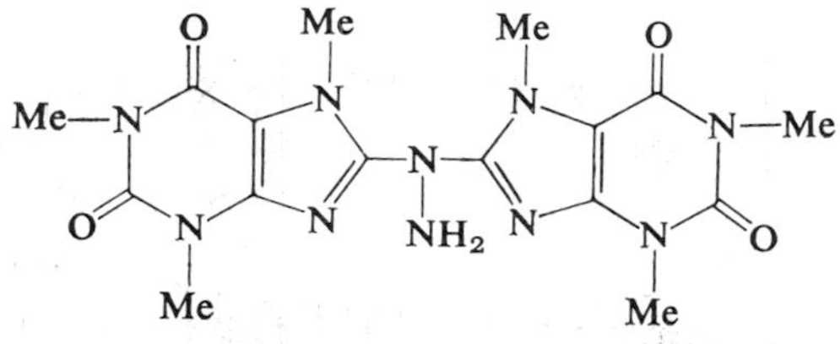

(77) 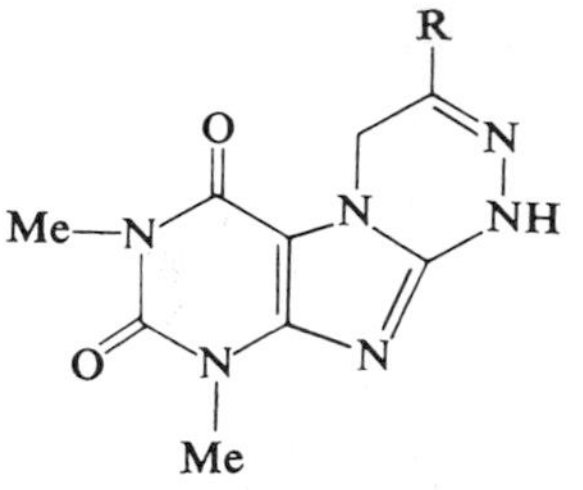   (78)

b. *Reaction with Nitrous Acid*

The azidopurines that result from this treatment are the subject of a later section (7Ca).

c. *Other Reactions*

Removal of hydrazino groups has been effected under various conditions. Aerial oxidation in alkaline solution of 6-hydrazinopurine gives purine,[151] also 7-benzylpurine from the 6-hydrazino derivative.[151] Caffeine is obtained when an aqueous solution of 8-hydrazinocaffeine sulphate is treated with ferric ammonium sulphate at room temperature.[146] Replacement of hydrazino group by chlorine occurs in low yield (6%) when 6-hydrazinopurine in dilute hydrochloric acid containing ferric chloride is left to stand for 30 mins.[10] This reagent also converts 8-hydrazino- to 8-chlorocaffeine.[146] Other hydrazino-group transformations are production of 1,6-dihydro-6-thiopurine (60%) on prolonged heating of 6-hydrazinopurine in thiolacetic acid[152] and degradation of 8-hydrazinocaffeine to 8-aminocaffeine in dimethylformamide or phenol under reflux.[17]

Typical addition reactions are also shown by the group. At room temperature 8-hydrazinocaffeine reacts with potassium cyanate and isothiocyanate to give the respective semicarbazide or thiosemicarbazide derivative.[153] Exocyclic ring formation through involvement of the hydrazino group is represented by a number of derivatives of 8-hydrazinocaffeine. Ethyl acetate in alkaline solution affords the 8-(pyrazol-3-on-2-yl) derivative (**79**)[146] while the 8-(1,2,4-triazol-1-yl) and 8-(1,2,3,6-tetrahydro-3,6-dioxopyridazin-1-yl) analogues arise from heating in formamide and reaction with maleic anhydride in hydrochloric acid, respectively.[153] 6-Hydrazinopurine with diethoxymethyl acetate undergoes annelation to s-triazolo[3,4-i]purine (**80**), which isomerises to a s-triazolo[5,1-i]purine (**81**) in formamide at 180°.[151] Both 7- and 9-benzyl-6-hydrazinopurine form analogous tricyclic derivatives.[151] In acidified dimethylformamide, 6-hydrazinopurine and benzaldehyde derivatives react to form anil-type compounds.[154]

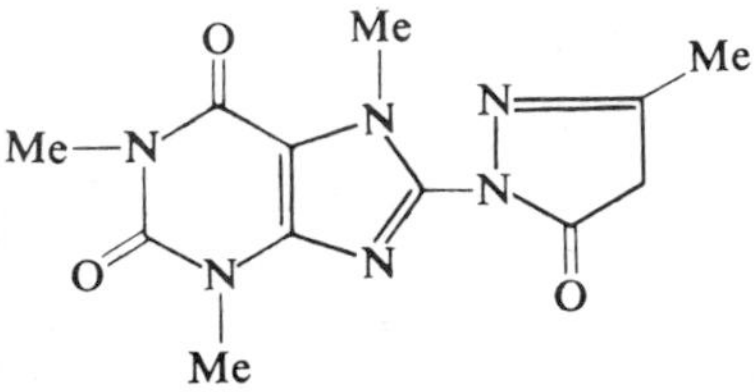

(**79**) 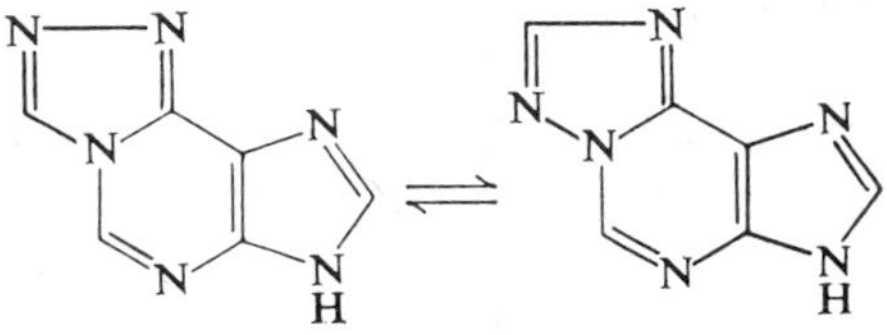   (**80**)   (**81**)

B. Hydroxyaminopurines

Of the small number of derivatives known the majority are 6-hydroxyaminopurines, the remainder being 2-hydroxyamino compounds. No purines with the group at $C_{(8)}$ are as yet known.

a. Preparation

Nucleophilic displacement of either halogen atom or alkylthio group in ethanol under reflux are the usual routes followed. The most readily formed are by substitution of a chlorine atom at $C_{(6)}$, this giving rise to 6-hydroxyamino-,[10] 2-amino-6-hydroxyamino-,[13] 2,3-dihydro-6-hydroxyamino-2-oxo-,[13] and 6-hydroxyamino-9-methylpurine.[13] Alternative preparations of some of the above have been carried out from analogous 6-methylthio- and 6-benzylthiopurines.[13] Formation of 2-hydroxyamino derivatives by replacement of chlorine has not been successful, as observed by the isolation of only 2-chloro-6-hydroxyaminopurine from 2,6-dichloropurine.[11] The more reactive 2-fluoropurines can be converted under reflux conditions as, for example, 2-fluorohypoxanthine.[77] This procedure also allows the preparation of 2,6-dihydroxyaminopurine from 6-chloro-2-fluoropurine.[11] 6-Chloropurine is transformed into 6-methoxyaminopurine in methanol containing methoxyamine.[155]

b. *Reactions*

Probably the most noteworthy reaction is oxidation of 6-hydroxy-aminopurine to 6-nitrosopurine[155] (Ch. X, Sect. 2A). Nitrous acid converts the same hydroxyaminopurine into 6-(*N*-nitrosohydroxy-amino)purine (**82**),[156] while with cold nitric acid hypoxanthine, as nitrate salt, is produced.[10] Similarly, in cold hydrogen peroxide in trifluoroacetic acid hydrolysis occurs both with 2-amino-6-hydroxy-amino- and 2,3-dihydro-6-hydroxyamino-2-oxopurine giving guanine (68%) and xanthine (91%), respectively.[13] In boiling 3*N*-hydrochloric acid alone the above conversion to xanthine is effected in a few minutes.[156] In 2*N*-sodium hydroxide in air 6-hydroxyaminopurine is rapidly transformed to the deep red 6,6'-azoxypurine (**83**)[9] while on heating in thiolacetic acid 1,6-dihydro-6-thiopurine is formed.[10] Hydrogenation of 6-hydroxyaminopurines over a palladium catalyst is a facile operation and has been used to prepare adenine[10] and other 6-amino-purines.[12, 13]

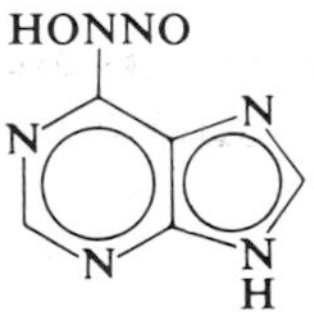

(82) 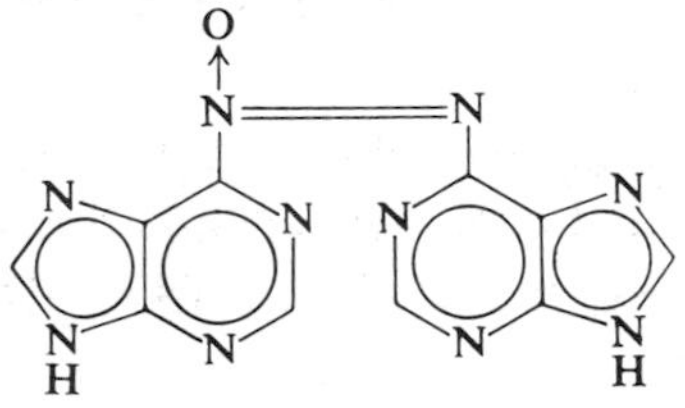   (83)

C. Azidopurines

Owing to the facility with which the azido group is able to annelate through an adjacent ring nitrogen atom to form the tetrazole tautomer, the precise formulation of azidopurines is difficult. Investigations carried out mainly by infrared and pmr studies[157, 158] show that in both solid state and solution an equilibrium mixture of azido- and tetrazolo-purines is usually present. As illustration, in the case of 2-azidopurine (**84**) the tetrazolo[1,5-*a*]purine (**84a**) and tetrazolo[5,1-*b*]purine (**84b**) may also be components of the reaction mixture.[157] While the solvent employed does play some not insignificant part in determining the extent of tautomer formation, as a generalisation acid conditions, involving some degree of protonation, favour a predominance of the azido form whereas in more neutral solution the tetrazole modification is more likely to be found. For convenience and ease of recognition only the azido form will be used both in the text and formulae.

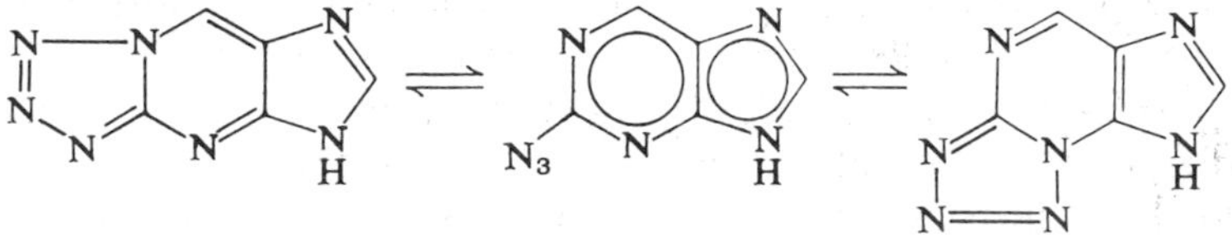

(84a)   (84)   (84b)

a. *Preparation*

Of the two routes that are available, the more usual one is through cold nitrous acid treatment of a hydrazinopurine. The earliest example appears to be of 8-azidocaffeine from the 8-hydrazino analogue[159] but others include 2-azido-,[160] 6-azido-,[9] 6-azido-7-, and 9-benzyl-,[158] 2,6-diazido-7-methyl-,[8] and 2,6,8-triazido-7-methylpurine.[8] The formation of azidopurines in the presence of other groups is illustrated by 6-amino-2-azido-,[158] 2-amino-6-azido-,[158] 6-azido-2-chloro-,[158] 6,8-diazido-2-chloro-7-methyl-,[8] and 6-azido-2-chloro-8-ethoxypurine.[8] This route also gives 8-azidotheobromine[8] and has been applied to the formation of a 2-azidopurine nucleoside.[161]

The alternative route, by nucleophilic displacement of a halogen atom (usually chlorine) by the azide ion, requires vigorous conditions. The appropriate chloropurine on heating with sodium azide in a variety of solvents, including ethanol, dimethylsulphoxide, and dimethylformamide, has given 6-azido-7-benzyl-,[158] 6-azido-9-benzyl-,[158] 2,6-diazido-,[158, 162] and 2,6,8-triazidopurine.[162] 2-Chlorohypoxanthine is converted to the 2-azido derivative in this way,[158] while a 75% yield of 8-azidoadenosine results from treatment of the 8-bromo analogue.[18]

b. *Reactions*

The chemistry of these derivatives is largely unexplored. As they undergo facile catalytic reduction to the corresponding aminopurines they serve as useful intermediates in the synthesis of the latter, especially as the azido group will replace a halogen atom at $C_{(2)}$, $C_{(6)}$, or $C_{(8)}$ under relatively mild conditions compared with those required with ammonia. Thus, a 2,6,8-triazidopurine gives a 58% yield of the triaminopurine on hydrogenation in ethanol over palladium.[8] Other reductions of this type are known.[18]

The azido group can function as a pseudo-halogen in that with morpholine or piperidine in aqueous solution 2,6-diazidopurine affords the corresponding 2-azido-6-morpholino- or piperidinopurine (reflux, 60 min).[162] The same reagents and conditions applied to 2,6,8-tri-azidopurines give the 2,8-diazido-6-aminopurines.[162] Acid or alkaline hydrolysis of 6-azidopurines affords the corresponding 6-oxopurine provided an electron-releasing group is present at $C_{(2)}$. In the absence of this, cleavage of the pyrimidine ring occurs forming a 4-formamido-5-tetrazol-5-yl)imidazole (**85**).[158]

D. Trimethylpurinylammonium Chlorides

Only 6-trimethylpurinylammonium chlorides have so far been described, the preparation of which by interaction of a 6-chloropurine with trimethylamine is described elsewhere (Ch. V, Sect. 5Ca). Few reactions of these quaternary compounds have been documented but these are sufficient to show the charged moiety to be a good leaving group in nucleophilic displacements. The most notable example of this is the formation of 6-fluoropurine under mild conditions (Ch. V, Sect. 1Gb). In aqueous alkali, conversion of **86** to hypoxanthine is effected on warming (60°, 60 min)[88] while on heating in acetonitrile or digol, at 90°, a Hofmann-Martius rearrangement gives 6-dimethyl-aminopurine (**87**).[163, 164] An attempted recrystallisation of **86** from methanol is reported to give some of the methoxy salt (**88**) also.[163]

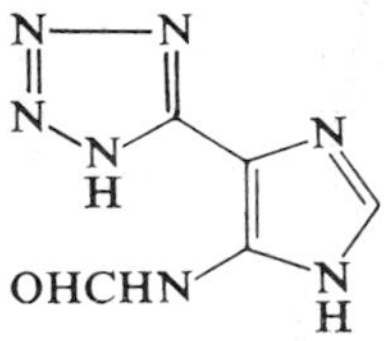

(85) 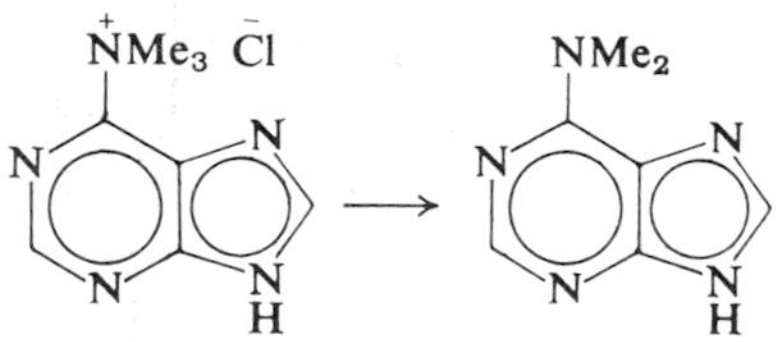   (86)   (87) 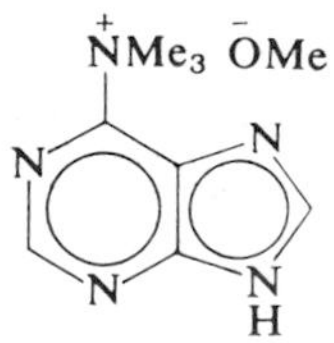   (88)

8. The Nuclear *N*-Alkylated Aminopurines

Almost all examples of these compounds are derived from adenine and guanine derivatives.

A. Preparation

Synthetic routes are available which provide unambiguous modes of formation. In contrast, although direct alkylation is often a valuable preparative procedure the risk of subsidiary by-products is present although this can often be minimised by using optimum conditions.

a. *By Direct Synthesis*

The Traube synthesis is applicable to the formation of 1-, 3-, 7-, and 9-alkylpurines by using the appropriate alkyl-4,5-diaminopyrimidines. Representative examples are 1-methylguanine,[165] 3-methyladenine,[166] 7-benzyladenine,[166] and 9-methyl-6-methylaminopurine[143] prepared in this way. An alternative and more general route is amination of existing *N*-alkylpurines, usually the chloro or alkylthio derivatives. Illustrating this are the formation of 1-methylguanine from the corresponding 2-methylthiopurine, with formamide as aminating agent,[3] and the use of methanolic ammonia on 3-methyl-6-methylthiopurine giving 3-methyladenine.[167] More numerous are replacements of the halogen atom in 6-chloro-7-alkyl-[142, 168] and 6-chloro-9-alkylpurines.[143, 144] An unusual application is chlorination with phosphoryl chloride, of the methylated hypoxanthine (89) which, without isolation of the 6-chloropurine intermediate (depicted as 90), is then aminated to the iminopurine (91, R = H or Me).[169]

Syntheses from imidazoles have found little application to *N*-alkyl derivatives. Some 7-[29, 142] and 9-alkyladenines[170] are derived by cyclisation of the appropriate 1(3)-alkyl-5(4)-cyano-4(5)-ethoxymethyl-ideneaminoimidazole. If alkylamines are introduced during the cyclisation procedure, 1-alkyladenine derivatives are formed[27, 29] (see Ch. III, Sect. 1Ea). Although of limited scope the foregoing route has some merit in permitting the formation of 1-alkyladenines which, due to their susceptibility to undergo the Dimroth rearrangement in alkali (Sect.

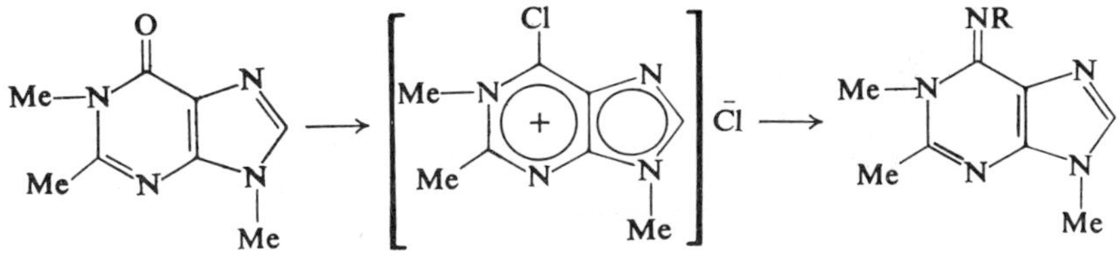

(89)   (90)   (91)

1K), cannot be formed by amination of either 1-alkyl-6-chloro- or 1-alkyl-6-methylthiopurines.[3] A rearrangement of a 2-chloro-6-methyl-aminopurine in alkali to a derivative of 1-methylguanine[171] is described in Chapter V, Section 5Fa.

b. *By Alkylation*

In adenine all ring nitrogen atoms are available for alkylation. The alkylation pathway is determined by the presence or absence of base and, to a lesser extent, by the type of solvent used. In distinctly acid media alkylation does not take place at all. Until recently[168] the view was held that alkylation under alkaline conditions, in aqueous or ethanolic media, gave either only the 9-alkyladenine or a mixture of the 7- and 9-alkyl isomers in which the 9-isomer predominated.[108, 172–178] That any significant amount of the 7-alkyl derivative is formed under these conditions is now known not to be the case, the other isomer obtained being the 3-alkyladenine. The long delay in recognising the true position, although appearing remarkable, is understandable if account is taken of three factors, two of which, on the face of it, give direct support to the product being the 7-alkyl isomer. These factors are, first, that the ultraviolet spectra of 3- and 7-alkyladenines show a remarkable similarity;[35, 168, 179, 180] second, that most simple 6-sub-stituted-purines on alkylation do give a mixture of both 7- and 9-alkyl derivatives and, last, that the unambiguous synthesis of 7-alkyl-purines has until recently generally presented some difficulty so that few derivatives have been available for comparison. An exception to the above alkylation mode is found when the sodium salt of adenine in ethanol is reacted with 3-methylbut-2-enyl bromide the main product being the 3-alkyl isomer (92) (the naturally occurring "triacanthine"), together with a small amount of the 9-alkyl isomer (93).[35] An un-

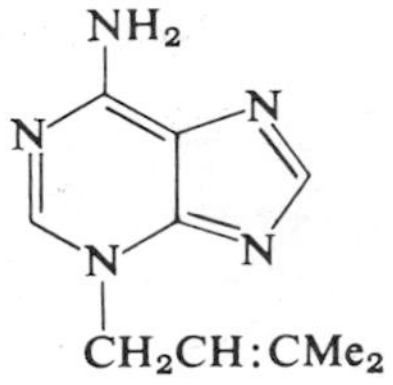

(92) 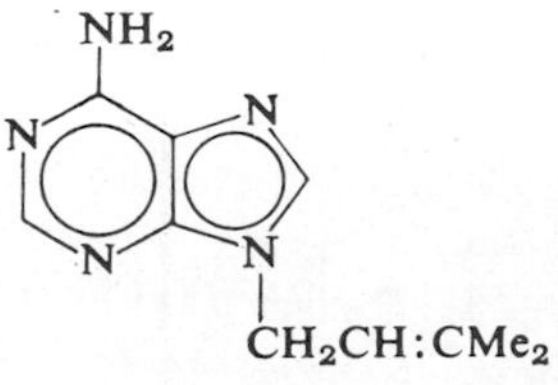   (93) 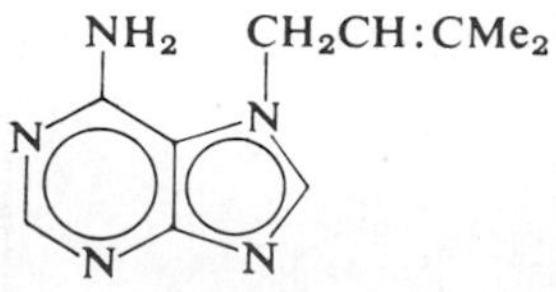   (94)

expected feature of this reaction is that with 3-methylbut-2-enyl chloride the 7-alkyl derivative (94) is obtained in low yield.[35]

In neutral media, which may be an aqueous solution or a dipolar aprotic solvent, such as dimethylformamide or dimethylacetamide, the major alkylation product is the 3-alkyladenine but some of the 9- and 1-alkyl isomers may also be present. The formation of 3-, 9-, and 1-methyladenine occurs with dimethyl sulphate in cold dimethylformamide.[179] An indication of the relative amount of the three isomers produced is given by reaction studies using alkyl bromides in dimethylacetamide in which yields are of the order 3-alkyl (55–66%), 9-alkyl (10–14%), and 1-alkyl (7–13%).[181] In aqueous solution (phosphate buffer, pH 7) ethyl methanesulphonate affords the corresponding three N-ethyl derivatives, but yields (25%, 9%, and 8%, respectively) are lower.[179] 6-Dimethylaminopurine gives analogous derivatives on methylation.[66] A mixture of triacanthine (92) and the $N_{(9)}$-isomer (93) can be formed in dimethylformamide (30%/30 h) in this way from 3-methylbut-2-enyl bromide.[113] These procedures have afforded a facile route to various 3-alkyladenines, no isolation of the other isomers being attempted; these include the 3-allyl[181, 182] (allyl bromide), 3-benzyl (benzyl bromide or chloride),[23, 181, 182] 3-methyl (methyl p-toluene-sulphonate),[167] 3-propyl (propyl bromide),[182] 3-(2-hydroxyethyl) (ethylene oxide),[166] 3-(3-hydroxypropyl),[183] and 3-benzyloxymethyl (benzyloxymethyl chloride)[92] derivatives. Conversion of 6-dimethylaminopurine to the 3-methyl analogue[184] and 6-benzylaminopurine to the 3-benzyl derivative are brought about with methyl iodide and benzyl bromide, respectively.

Although some direct influence on the site of alkylation is exerted by the $C_{(6)}$-amino group, surprisingly, no significant effect is noted following the conversion to an amide moiety. In either dimethylformamide[114] or dimethylacetamide[23] benzyl bromide and 6-benzamidopurine give the 3-benzylpurine. At higher temperatures (110°, 16 h) mixtures of 6-benzamido-3-benzyl- and 6-benzamido-9-benzylpurines are formed, the amount of the latter increasing as heating times are lengthened.[114] Alkyl group migrations from $N_{(3)}$ to $N_{(9)}$, which are discussed more fully in Section 5Df, may also be responsible for the formation of 3,9-dimethyl derivatives when isoguanine and 2,6-diaminopurine undergo high-temperature methylation (dimethyl sulphate/145°) in dimethylacetamide.[185]

Further alkylation of 3-alkyladenines occurs at $N_{(7)}$ as in 3,7-dimethyladenine (96) being obtained on leaving 3-methyladenine (95) for some hours in cold methanolic potassium hydroxide with methyl iodide.[28] Methylation of 7-methyladenine (97) in dimethylformamide also affords 96 but higher temperatures (~100°) were used.[28] Other examples of $N_{(7)}$-methylation of 3-alkyladenines, in either dimethyl-

(95)   (96)   (97)

formamide or acetone, are to be found.[181, 183] Likewise, benzylation of either 3- or 7-benzyladenine gives 3,7-dibenzyladenine.[23] In this reaction benzyl chloride was ineffective, the bromide being used.[23] Benzylation of 3-benzyladenine in the presence of sodium carbonate gives rise to the tribenzyl derivative (98), the structure of which is confirmed by the synthesis on benzylation of 6-benzylaminopurine.[23] This apparent example of direct $C_{(6)}$-amino group alkylation would be most readily explained in terms of a Dimroth rearrangement of an intermediate 1,3,7-tribenzyl quaternary salt taking place under the alkaline conditions employed (Sect. 1K).

As in aqueous alkaline solution, adenine in aprotic media containing a base forms the 9-alkyl derivative as primary product. Both 9-allyl-[186] and 9-benzyladenines[23] result from interaction with the appropriate alkylating agent in dimethylacetamide containing anhydrous potassium carbonate. Also successful was benzylation carried out in dimethylformamide containing sodium hydride.[29] Alternatively, heating a metal salt of adenine with the alkyl halide in xylene is effective.[187] Further alkylation of a 9-alkyl derivative introduces the second alkyl group at $N_{(1)}$. While vigorous conditions (125°, 2 h) were employed in converting 1-methyl- (99) to 1,9-dimethyladenine (100), using methyl p-toluene-sulphonate in dimethylacetamide,[28] the analogous formation of 1,9-dibenzyladenine in the same solvent was effected on long standing (70 h) at room temperature.[29] Other alkylations of this type are known.[186]

From the foregoing studies have emerged routes to derivatives obtained only with difficulty by other methods. The fact that 3-alkyl-adenines give 3,7-dialkyladenines and 9-alkyl derivatives the 1,9-dialkyl homologues allows the formation of 7-alkyl- and 1-alkyladenines by employing as starting material either 3-benzyl- or 9-benzyladenine and removing by hydrogenation the benzyl group after the second alkylation stage.[181] As an alternative to benzyl, the cyanoethyl group is removed at $N_{(9)}$ from a series of 1-alkyladenines on prolonged heating (reflux, 45 h) in methanolic potassium t-butoxide.[187a]

Where imino forms of the purine, as for example 100, are precluded by the presence of a dialkylamino group at $C_{(6)}$ alkylation, when it occurs it requires vigorous conditions and gives rise to an $N_{(3)}$-quaternised derivative. An illustration is **101** from prolonged heating (24 h) of an ethanolic solution of 6-diethylamino-9-methylpurine with methyl

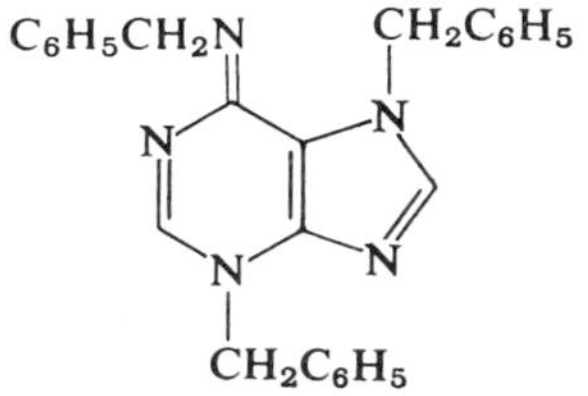

C₆H₅CH₂N   CH₂C₆H₅

CH₂C₆H₅
(98)

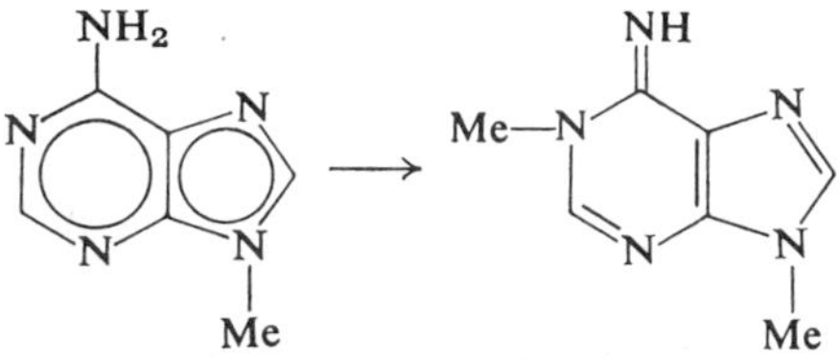

NH₂

Me
(99)

→

NH

Me—N

Me
(100)

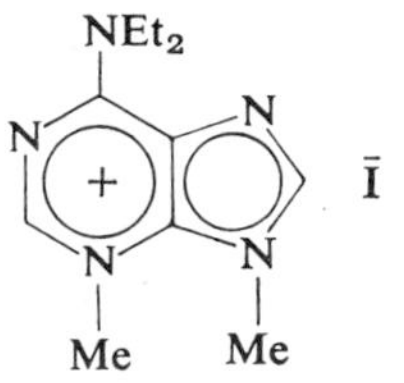

NEt₂

Me   Me
(101)

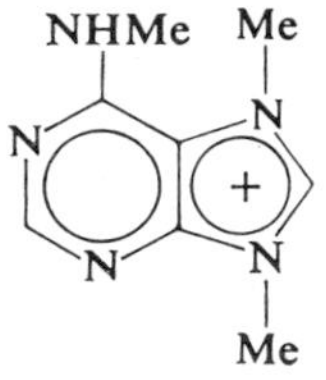

Ī

NHMe   Me

Me
(102)

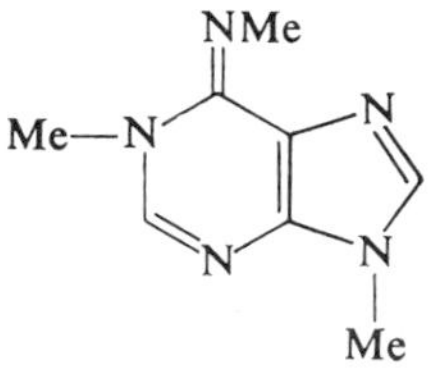

NMe

Me—N

Me
(103)

iodide.[188] Preparation of 3,9-dibenzyl-6-dimethylaminopurinium bromide necessitated extensive boiling (3 days) of the 9-benzylpurine with benzyl bromide in acetonitrile.[189] Quaternisation of the imidazole ring is claimed to produce **102** when 9-methyl-6-methylaminopurine and methyl iodide interact (140°, 60 min) in the absence of solvent.[190] However, as 6-aminopurines have not been found[167] to give 7,9-dialkyl derivatives under these alkylating conditions, it seems that the product is more likely to be 1,6-dihydro-1,9-dimethyl-6-methyliminopurine (**103**).

Intramolecular alkylation can occur through transformation of existing groups into ones with alkylating character, as for example, in obtaining 9-ethyl-7,8-dihydroimidazo[2,1-*i*]purine (**104**) on heating 6-(*N*-ethyl)hydroxyethylaminopurine with thionyl chloride.[191] An associated reaction is the conversion of 6-(3-methylbut-2-enyl)aminopurine (**105**), by means of trifluoroacetic or fluoboric acids, to 3,7,8,9-tetrahydro-7,7-dimethylpyrimido[2,1-*i*]purine (**106**). Isomerisation of the latter to the 9,9-dimethyl analogue (**107**) results on prolonged heating (85°, 36 h) in dilute ammonium hydroxide.[26] The isomeric triacanthine, as hydrochloride, undergoes similar ring closure forming pyrotriacanthine chloride (**108**) using high temperatures (280°) for a brief time.[35] In the last two examples adduct formation is most likely involved in which protonation of the double bond precedes cyclisation.

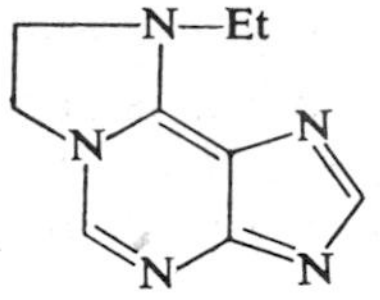

(104)

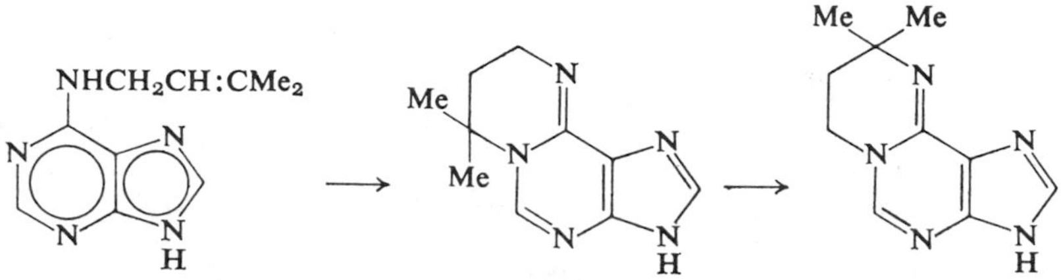

(105)   **(106)**   **(107)**

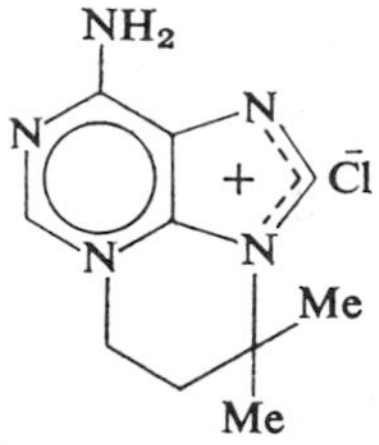

(108) 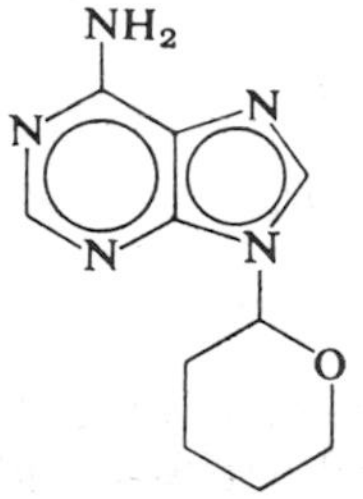   **(109)**

This mechanism is presumed to operate in forming 9-(tetrahydropyran-2-yl)adenine **(109)** on reaction between adenine and 2,3-dihydropyran in dimethyl sulphoxide containing hydrogen chloride.[192] Although use of ethyl acetate as solvent and p-toluenesulphonic acid as acid catalyst failed in the above reaction,[193] these conditions were successful in condensing 2,3-dihydrofuran with 2-acetamido-6-chloropurine.[88] It is noteworthy that the more basic 2-amino-6-chloropurine fails to give any 9-(tetrahydrofur-2-yl) derivative.[88]

Adducts also arise under alkaline conditions (Michael reaction); in ethanolic sodium methoxide adenine and ethyl acrylate under reflux give a near quantitative yield of 9-(2-ethoxycarbonylethyl)adenine. The 9-(2-cyanoethyl) analogue results with acrylonitrile.[178]

Alkylations in the solid state have been carried out by first evaporating an equimolecular mixture of adenine and a tetra-alkylammonium hydroxide in aqueous solution and subjecting the residue to sublimation (150–250°) *in vacuo*. Only 9-alkyl derivatives are isolated; utilisation of tetramethyl-, tetraethyl-, and tetrapropylammonium hydroxides gives the corresponding 9-methyl-, 9-ethyl-, and 9-propyladenines.[194] 6-Substituted-aminopurines, such as kinetin, are similarly $N_{(9)}$-alkylated in this way.[194]

Alkylation modes of guanine in almost every respect differ profoundly from those of adenine. A common feature, however, is the formation of a 9-alkyl derivative as major component when an alkaline solution is employed. This is surprising as until recently a widely held view was that the initial product from guanine was the 7-alkyl derivative.[179] Detailed examination of the mixture obtained from the action of methyl chloride on a sodium hydroxide solution of guanine at 70° shows that the products include 9-methyl- (33%), 7-methyl- (18%), 3-methyl- (11%), and 1-methylguanine (trace). In addition, some 1,6-dihydro-2-methylamino-6-oxopurine was also recovered.[22] This work represents a repeat of the conditions used by Traube,[165] who reported the isolation of 7-methyl- and 1,7-dimethylguanine only. The failure of the latter to isolate the 9-methyl derivative is ascribed to the solubility of this compound in water[22] but the true nature of the "1,7-dimethylguanine" remains to be established, the later workers finding only traces of dimethylated derivatives.[22] In passing, a specific route to 9-alkylguanines involves fusion (260°) of guanine with a tetra-alkylammonium hydroxide.[194]

With 1-alkylguanines further alkylation gives the 1,7-dialkyl analogue.[165, 195, 196] Numerous derivatives of 1-methylguanine so prepared include 7-ethyl, 7-propyl, and 7-(2-hydroxyethyl) analogues.[195] Under more drastic conditions (sealed tube, 60°, 6 h) in sodium hydroxide with methyl iodide, the quaternary iodide of 1,7,9-trimethylguanine is obtained.[196]

Because of the recent synthesis of 3-methylguanine[197] alkylation studies are still awaited. Secondary alkylation of 9-alkylguanines under aqueous alkaline conditions leads to the 7,9-dialkyl form, as for example, the 7,9-dimethyl betaine (110), which it has been suggested is the compound erroneously claimed to be 1,7-dimethylguanine by Traube, noted above.

More usually 7,9-dialkylguanines are prepared under acidic conditions, either by heating guanine with the alkylating agent alone or in an aprotic solvent, dimethylacetamide being favoured. Either 7- or 9-alkylguanines can also be used as starting material. Thus, 110, in excellent yield, results from treatment of guanine in the above solvent (90°) with dimethyl sulphate[167, 185] or from the action of methyl p-toluenesulphonate (130°,1 h) on 7-(111) or 9-methylguanine (112).[198] The tosylate salt formed in the last two cases is converted to the betaine (110) using methanolic ammonia.[198] Appropriate 7- or 9-alkylguanines

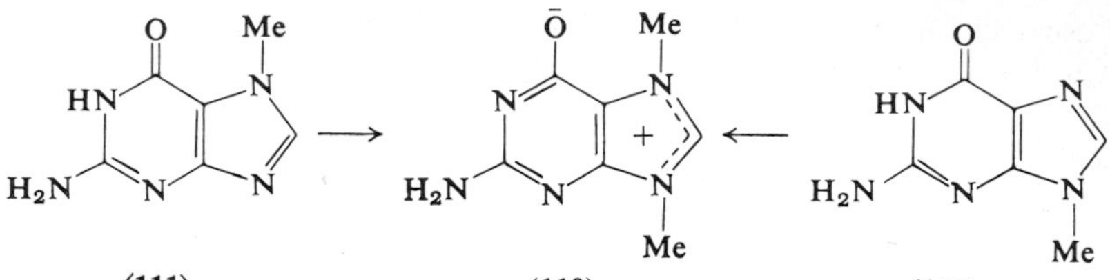

$$
\text{(111)} \longrightarrow \text{(110)} \longleftarrow \text{(112)}
$$

and alkylating agent afford 9-benzyl-7-methyl-,[199, 200] 7,9-diethyl-,[199] and 9-ethyl-7-methylguaninium betaines.[199] In glacial acetic acid either 7- or 9-(2-hydroxyethyl)guanine are transformed by ethylene oxide to the 7,9-di(2-hydroxyethyl)betaine.[201] In the absence of an ionisable proton, so precluding betaine formation, the appropriate 7,9-dialkyl quaternary salt results, the isolation of 1,7,9-trimethylguaninium p-toluenesulphonate from 1-methylguanine illustrates this.[196]

B. Reactions

These reactions can be divided into two categories; those involving the N-alkyl group itself and those in which a substituent group or atom exhibits unusual or enhanced activity because of the presence of an adjacent N-alkyl moiety.

The most significant reaction of the first type with $N_{(1)}$-alkyladenines is the alkali-induced isomerisation to 6-alkylaminopurines (Dimroth rearrangement) for which detailed treatment has been accorded previously (Sect. 1K). Migration of benzyl and most probably methyl groups from $N_{(3)}$ to $N_{(9)}$ has been found to occur thermally with acylated aminopurines (Sect. 5Df). Removal of groups attached to ring nitrogen atoms can be effected in a variety of ways. Hydrogenolysis over a palladium catalyst of $N_{(3)}$-benzyl groups from the appropriate purine has given 7-methyl-,[181] 7-(2-hydroxyethyl)-,[183] and 7-benzyladenine.[181] This procedure also cleaves $N_{(3)}$-benzhydryl groups from adenine derivatives.[23] Sodium in liquid ammonia successfully converts 6-amino-9-benzyl-2-fluoropurine to 2-fluoroadenine.[76] Oxidative conditions, provided by cold alkaline potassium permanganate solution, removes an $N_{(9)}$-propenyl group to give 1-methyladenine.[186] Loss of the 3-methylbut-2-enyl group at $N_{(9)}$ from triacanthine follows prolonged heating (water bath) with fairly concentrated hydrochloric acid.[35] Hydrobromic acid (48%) under reflux conditions provides an alternative means to hydrogenation of removing an $N_{(3)}$-benzyl group.[202] Adenine and 7-methyladenine result by such debenzylations. It is noteworthy that 3,7-dibenzyladenine, as under hydrogenolysis, loses only the $N_{(3)}$-benzyl group by this treatment.[202]

In reactions of the second type some good examples of nucleophilic displacement of amino groups are to be found in 3-methylpurines. Activation of groups at $C_{(2)}$ and $C_{(6)}$ are seen in the formation of 3-methylxanthine from 3-methylguanine[197] and the conversion of both 3-methyladenine and 6-dimethylamino-3-methylpurine into 3-methyl-hypoxanthine[66] in dilute sodium hydroxide under reflux. Only long

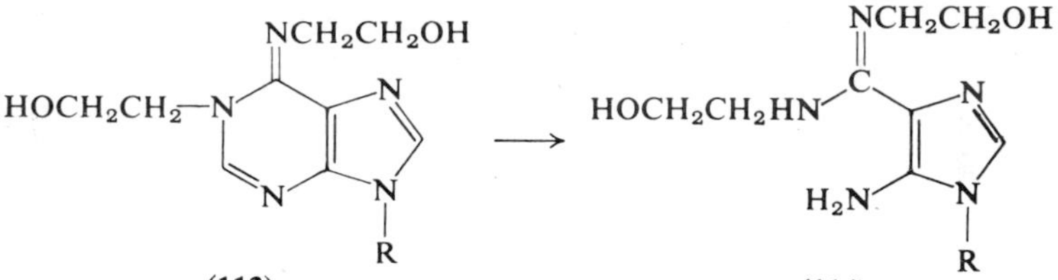

(113)   (114)

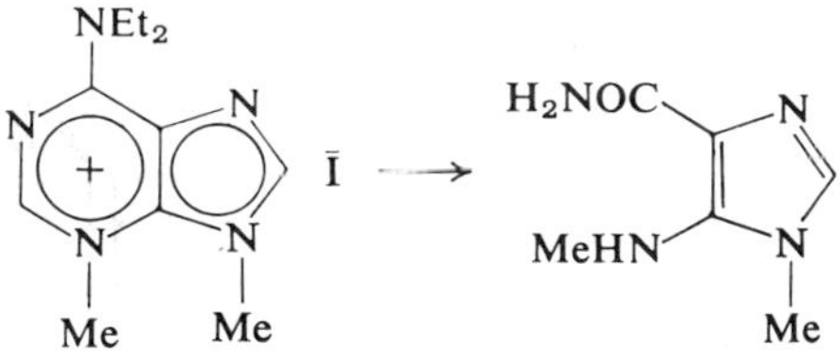

(115)   (116)

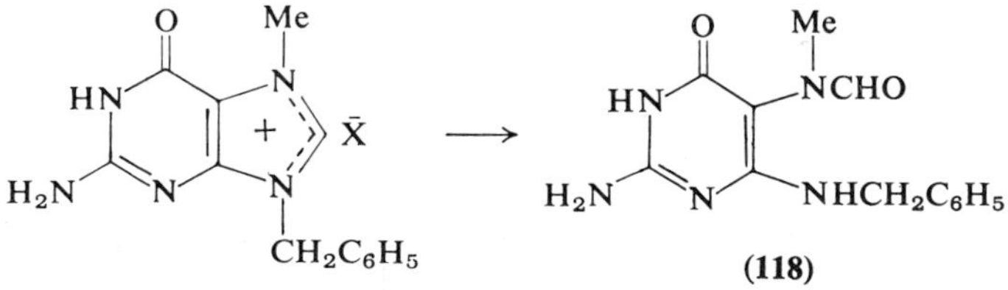

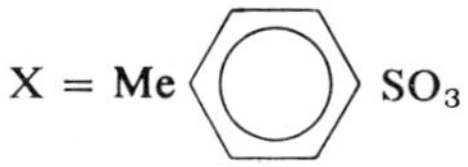

(117)   (118)

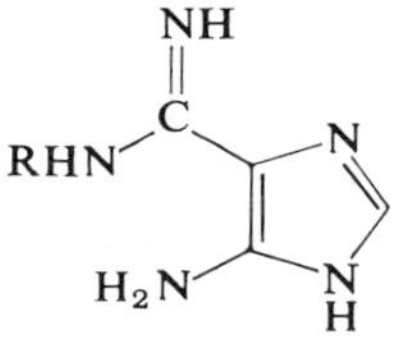

(119)

standing (4 days) in alkaline solution was needed to obtain 3,7-dimethyl-hypoxanthine from the 6-aminopurine.[28] To a lesser extent hydrolyses of 9-alkyladenines to the appropriate hypoxanthine occurs on prolonged heating in normal alkali.[203] The appearance of some 6-alkylamino-4,5-diaminopyrimidines in the above reactions is a reflection of the general alkali-unstable nature of purines which are unable to form anions. A variety of other alkali-induced degradations, involving either pyrimidine or imidazole ring, are known. In all cases the mechanism proceeds through a nucleophilic displacement at $C_{(2)}$ or $C_{(8)}$. In the case of 1-alkyl-6-alkyliminopurines of the type (113) for which valency requirements preclude the Dimroth rearrangement taking place, nucleophilic attack at $C_{(2)}$ produces the imidazole derivative (114).[30] Instability is found also among the quaternised derivatives, ring opening occurring in the ring that carries the formal charge. Examples are the breakdown of 6-diethylamino-3,9-dimethylpurinium iodide (115) to 4-carbamoyl-1-methyl-5-methylaminoimidazole (116)[188, 189] or the recovery of the pyrimidine (118) after heating the tosylate of 9-benzyl-7-methylguanine (117) in dilute alkali.[200] Similar degradations of this type are known.[201, 204] The rearrangement of 6-amino-2-chloro-7-methylpurine in alkali to 7-methylguanine, first noted by Fischer,[171] has been shown to involve ring opening to an imidazole intermediate[205] (Chapt. V, Sect. 5Fa).

Aminopurines are usually stable toward all but the most rigorous acid conditions. Exceptions to this are 1-benzyl-[206] and 1-methyl-adenine[24] which in boiling dilute acid afford the respective 4-amino-5-amidinoimidazole (119, R = $CH_2C_6H_5$ and Me). A significant feature is that hydrolytic degradation of the 1-benzyl derivative is faster, by a factor of ten, than the 1-methyl analogue.[206]

9. Naturally Occurring Aminopurines

In the succeeding sections detailed sketches of adenine, guanine, and isoguanine together with notes on the occurrence, discovery, and isolation of some related aminopurines are given.

A. Adenine

Adenine, which is present in all living matter, was first described in 1885 by Kossel who had extracted it from cattle pancreas and spleen[207] and who, in the same year, demonstrated the conversion to hypoxanthine on treatment with nitrous acid.[208] Complete structural elucidation followed as a result of the elegant syntheses of Fischer, starting from uric acid, in which the final stage was hydriodic acid reduction of 6-amino-2,8-dichloropurine,[148, 209] and from the total synthesis, by Traube,[210] which involved an initial condensation of malononitrile and thiourea and, after a multistep transformation, desulphurisation of the resulting 2-thioadenine by means of hydrogen peroxide.

From the many routes to adenine to be found in the literature probably the best for quantity preparation utilises a modified Traube cyclisation with formamide.[211, 212]

Adenine is stable in hot alkali[213] and toward boiling dilute hydrochloric acid but in 6N-acid at 150°, 4(5)-amino-5(4)-carbamoylimidazole is formed.[214] Complete degradation to glycine, ammonia, carbon dioxide, and carbon monoxide occurs in concentrated hydrochloric acid at 180°[214] or in sulphuric acid at high temperature.[215] The transformation to hypoxanthine with nitrous acid is noted above. Oxidation to urea and glycine is reported following the use of hot acid permanganate solution,[82, 216] but a repeat of this preparation did not produce any glycine.[214]

Reduction with zinc and hydrochloric acid leads to the formation of an aminoimidazole.[82] Adenine is unchanged on hydrogenation over a platinium catalyst but is reduced at a dropping mercury electrode in dilute perchloric acid.[217, 218] Exposure of a dilute (0.1%) aqueous solution to ionising radiation gives a mixture of products from which hypoxanthine, 6-amino-7,8-dihydro-8-oxopurine, and 4,6-diamino-5-formamidopyrimidine have been isolated.[219] In the dry state reaction with bromine gives 8-bromoadenine.[220–222] Coupling with diazotised p-aminobenzene derivatives takes place at $C_{(8)}$[127, 223–225] but the products are unstable.

Adenine, being an ampholyte (pK_a 4.2, 9.8), forms salts with acids and bases. Mineral acids usually give hydrated crystalline derivatives which include the sulphate (as hemi- or monohydrate),[70, 82, 209] hydrochloride (hemihydrate),[70] nitrate (hemihydrate),[70] and metaphosphate.[89] Organic acid salts are exemplified by the oxalate (monohydrate)[70] and picrate (monohydrate).[226] The latter salt finds application in the spectrophotometric and gravimetric determination of adenine.[226]

With monovalent metal ions simple salts arise through replacement of the imidazole ring proton. Ammoniacal silver nitrate precipitates the silver salt under basic conditions,[70, 83, 220] while silver sulphate or the oxide can be used in dilute sulphuric acid,[227, 228] near-quantitative precipitation being given by this latter procedure.[229] Boiling with cuprous oxide in a citrate buffer (pH 5) affords the cuprous derivative,[230] another route to which is to heat the purine with aqueous cupric sulphate under reducing conditions provided by the addition of sodium bisulphite.[231]

The sodium salt of adenine with a molecular proportion of mercuric chloride gives the chloromercuri derivative (120)[83, 172, 232] as a crystalline precipitate. The covalent attachment of the metal residue to $N_{(9)}$ has been demonstrated.[233]

Divalent metal ions show little evidence of reaction under neutral conditions,[234] but in acid media complexes containing two molecules of purine to each metal ion[235] arise. Structures of the type **121** are sug-

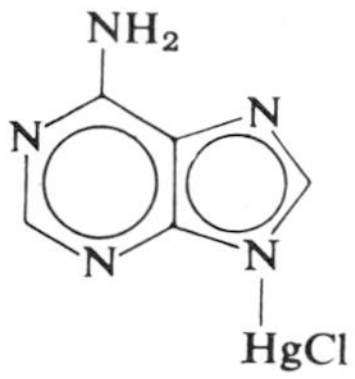

(120) 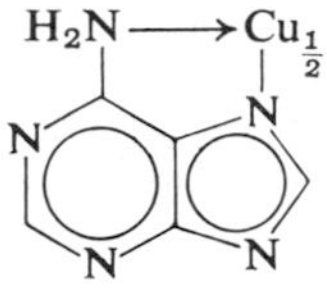   (121)

gested for the complexes formed with Cu^{2+}, Co^{2+}, and Ni^{2+} ions,[236] but $N_{(3)}$ and $N_{(9)}$ have been recently proposed as alternative binding sites.[237] The complexes form well-defined, highly coloured crystals, the trihydrated form of **121** being a deep violet.[238] Complex formation in acetic acid with mercuric acetate has been studied.[239]

Adenine does not melt below 350°.[209] On slow heating to 220° a sublimate of fine featherlike needles appears, which decompose on heating to 250°. Crystallisation from a concentrated aqueous solution gives anhydrous adenine in a microcrystalline form but, with dilute solutions, the trihydrate emerges on standing; this can be converted to the anhydrous state on heating to 110°. Solubility in water varies from 40 parts in the cold to nearly 2000 parts at 100°. It is soluble in glacial acetic acid, dilute mineral acids, and alkali, including ammonium hydroxide, from which it can be precipitated by passing carbon dioxide through the solution.[222] In sodium carbonate solution, surprisingly, solubility is poor.[70, 172, 208] It dissolves moderately well in hot ethanol but is insoluble in chloroform and ether.

Numerous nucleosides having adenine or a related derivative as aglycone have been isolated; the more important of these are summarised below together with some purine bases derived from natural sources.

Adenine, in addition to its ubiquitous distribution as the 9-β-D-ribofuranoside, adenosine, is found combined with various other sugar residues. The range of such compounds is exemplified by the occurrence of a thiosugar moiety in "adenine thiomethylpentoside"[240] and extraction of the 3'-deoxyadenosine, cordycepine,[241, 242] from the mould *Cordyceps militaris L.* From the latter source is also obtained the 3'-amino-3'-deoxyadenosine (**122**, R = H)[243] which is found present in the mould *Helminothosporium sp.*[244] also. Two related antibiotics, homocitrullylaminoadenosine [**122**, R = $H_2NOCNH(CH_2)_4CH(NH_2)$-CO-] and lysylaminoadenosine [**122**, R = $H_2N(CH_2)_4CH(NH_2)CO$-] are found associated with **122** (R = H).[243] Psicofuranine, isolated from *Streptomyces hygroscopicus decoyinine*[245] has been shown by synthesis[246, 247] to be the 9-D-psicofuranosyl derivative and to be identical with the antibiotic angustmycin C[248] present in a different *Streptomyces* culture. A further adenine nucleoside, angustmycin A, extracted from the latter mould, is identical with decoyinine, from *S. hygroscopicus decoyinine*. The original structure proposed[249] has been revised[250] and confirmed by conversion[250a] of psicofuranine to angustmycin A. Nucleocidin, an antibiotic found in *S. calvus*[251] is of interest in being the first naturally occurring purine nucleoside found to contain fluorine. This halogen is located on $C_{(5)}$ of the ribosyl group and carried on the same carbon atom is a sulphamic acid residue esterified through the hydroxymethyl group.[251a] Another fungal product, eritadenine, known also as lentinacin or lentysine, from *lentinus edodes*,[251b] is shown by various syntheses[251c] to be the 9-(3-carboxy-D-*erythro*-2,3-dihydroxy-propyl) derivative of adenine.

Pseudoadenosine, in which the ribosyl group is attached through $N_{(7)}$ arises on degradation of pseudovitamin B_{12}.[252] A pseudonucleoside containing 2-methyladenine is known,[253] the base itself having been recovered from nucleic acid derived from various sources.[254] Another unusual adenine nucleoside, found in the antibiotic septacidin,[255] is thought to have an aminohexose residue linked through the amino group of the purine. In view of the facility for rearrangement, isolation of 1-methyladenine from the siliceous giant sponge (*Geodia gigas*) is noteworthy, the product, before identification, having been named spongopurine.[256] Triacanthine, first isolated from the leaves of *Gleditsia triacanthos L.*,[257] in 1954, is proved by synthesis to be 3-(3-methylbut-2-enyl)adenine.[35, 258] The same purine has been extracted from other plant material and reported under the names togholamine[259] and chidlovine.[260]

A number of 6-substituted aminopurines are known to be present in the plant and animal kingdoms, their presence being associated with growth control. Kinetin, 6-furfurylaminopurine was first isolated from nucleic acid[261] and synthesised three years later.[94] More recently, the cell division factor, zeatin, isolated from sweet corn (*Zea mays*) is identified as the *trans* form of 6-(4-hydroxy-3-methylbut-2-enyl)amino-purine (**123**)[262] and the structure established by synthesis.[263, 264] A related nucleoside (**124**) is found in soluble RNA,[265, 266] the base being

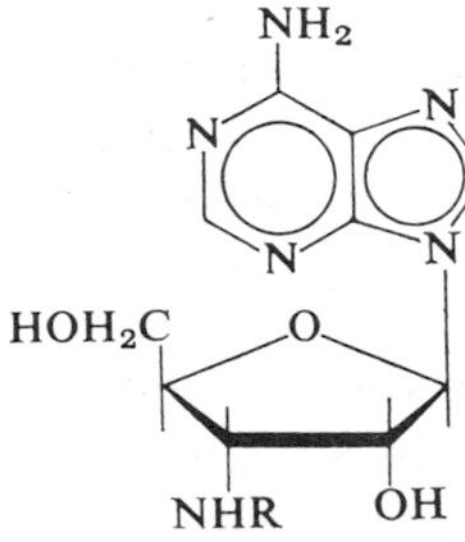

NH$_2$

N

N

N

N

HOH$_2$C O

NHR   OH

(**122**)

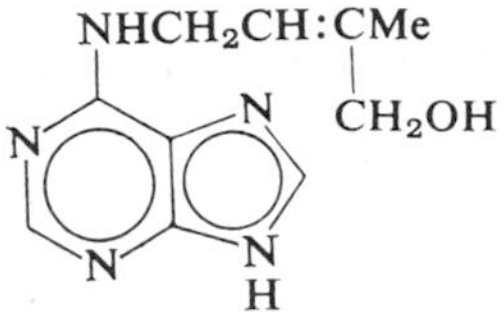

NHCH$_2$CH:CMe

CH$_2$OH

N

N

N

N
H

(**123**)

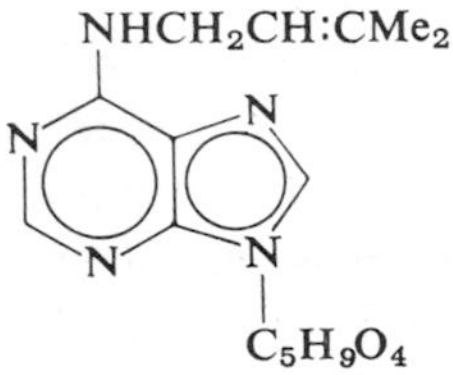

NHCH$_2$CH:CMe$_2$

N

N

N

N

C$_5$H$_9$O$_4$

(**124**)

CO$_2$H

NHCHCH$_2$CO$_2$H

N

N

N

N

R

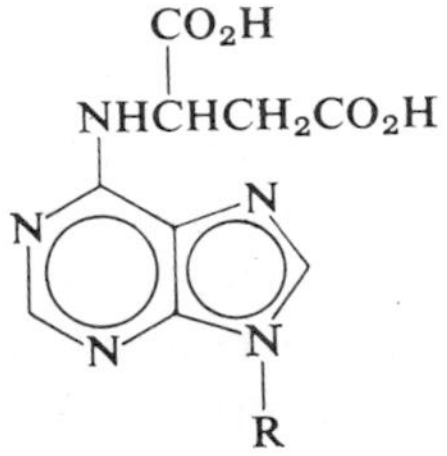

(**125**)

an isomer of triacanthine. Syntheses of both base and nucleoside have been reported.[266, 267] Elucidation of the structure of the complex amino-nucleoside, puromycin, present in cultures of *Streptomyces alboniger*,[268] followed from a series of synthetic studies.[269] The aglycone of this antibiotic, 6-dimethylaminopurine, is also present in ribonucleic acid from various sources.[254] A probable intermediate in the *in vivo* conversion of inosinic acid to adenylic acid is the ribotide of 6-(1,2-di-carboxyethylamino)purine (**125**, R = β-D-ribofuranosyl-5-phosphate). Since the first isolation from liver,[270] in 1956, syntheses of the nucleoside form, the so-called "succinyl adenosine" (**125**, R = β-D-ribofurano-syl)[271] and the purine base (**125**, R = H)[272] have been accomplished. The presence of 6-methylaminopurine in both ribonucleic and deoxy-ribonucleic acids of bacterial origin has been demonstrated by various workers.[273–275]

B. Guanine

This purine is present in nature to the same extent as adenine and together they comprise the majority of naturally occurring purines.

The first isolation, reported in 1844, was from guano, the sea bird excreta used as a source of fertilizer, from which the name was derived.[276, 277] Some fifty years were to elapse before the structure was established by Fischer, who demonstrated the conversion of uric acid to guanine.[209] The first total synthesis, due to Traube,[278] still provides possibly the best route for large-scale preparation.[196]

The chemistry of guanine is concisely reviewed in a recent monograph.[279]

Guanine is stable in boiling dilute potassium hydroxide but prolonged heating with dilute mineral acids transforms it to xanthine.[280, 281] With stronger acid ($3.4N$) at high temperatures (160°) fission of the pyrimidine ring to give 4(5)-amino-5(4)-carbamoyl imidazole occurs.[282, 283] Complete degradation to glycine, ammonia, carbon dioxide, and carbon monoxide follows the use of concentrated acid at 180°.[89] Other studies of this type have been made.[215] The hydrolysis to xanthine with nitrous acid[71, 284, 285] was employed by Strecker as an aid to structure determination.[71, 284] Oxidation occurs more readily than with adenine; after leaving for some days in hydrochloric acid containing potassium chlorate, guanidine is the main product.[284] No significant amount of parabanic acid was found in this preparation—this fact being in contrast with earlier findings.[71] On exposure of aqueous solutions of guanine containing dyestuffs to visible light, photodynamic degradation occurs. In the presence of lumichrome at room temperature the products identified were guanidine, parabanic acid, and carbon dioxide.[286] With alkaline permanganate oxidation products obtained are urea, oxalic acid, ammonia, and carbon dioxide.[216] No reduction occurs on heating guanine with zinc and hydrochloric acid,[82] nor is it reduced at a dropping mercury cathode,[217] but it is electrolytically reduced to 2-amino-1,6-dihydropurine (deoxyguanine) in 60% sulphuric acid at a lead cathode.[287] With bromine an 8-bromo analogue is formed.[288] Coupling with diazonium salts gives the corresponding 8-azoguanines.[127, 289, 290]

The usual hydrated salts are obtained with mineral acids such as sulphate (dihydrate),[291] hydrochloride (monohydrate but on heating to 100° the anhydrous salt results),[291] hydrobromide ($2\frac{1}{2}H_2O$), nitrate (sesquihydrate), and metaphosphate (polyhydrate). The weak basic character of this purine (pK_a 3.3 and 9.2) shows in the ease of hydrolysis of the above salts[291] and in the failure to form salts with monobasic organic acids. A well-defined picrate (monohydrate) can be obtained.[88, 292] Metal salts such as barium, disodium, and ammonium form readily. Complexes of the base-acid type are formed with oxides of Ag and Cu^{2+} with a general formula $C_5H_5N_5O \cdot M_2O$. Other complexes are derived with silver nitrate,[71] zinc chloride,[293] mercuric chloride,[293] mercuric nitrate,[239] platinic chloride,[277] and other metal halides.

Guanine does not melt below 350°, the presence of oxo and amino groups in the molecule promotes a high degree of intermolecular hydrogen bonding with a resulting high crystal lattice energy. The free base is virtually insoluble in water (1 part in 200,000) but readily soluble in dilute acids and bases, including ammonium hydroxide, although in concentrated solutions of the latter it is insoluble. Precipitation, by neutralisation of an acid or alkaline solution, gives an amorphous product but crystalline guanine results from spontaneous evaporation of an ammonia solution,[294] or by passing carbon dioxide through an alkaline solution containing 30% alcohol,[295] or on neutralisation of the alcoholic solution with acetic acid.[296] Solubility is poor in ethanol and most organic solvents.

Apart from guanosine, the 9-β-D-ribofuranoside derivatives of guanine are by no means as well represented naturally as those of adenine. The nucleotide "neoguanylic acid" isolated from brewers yeast was first formulated[297] with the glycoside linkage through $N_{(1)}$ but this was later shown[298] to be on the amino group at $C_{(2)}$. Subsequent studies indicate that this unusual nucleoside may not exist naturally but arises during acid treatment of RNA.[299] Cultures of a *Fusarium* species[300, 301] and *Eremothecium ashbyii*[302] both produce 2-(1-carboxy-ethylamino)-1,6-dihydro-6-oxopurine (126), the structure of which was

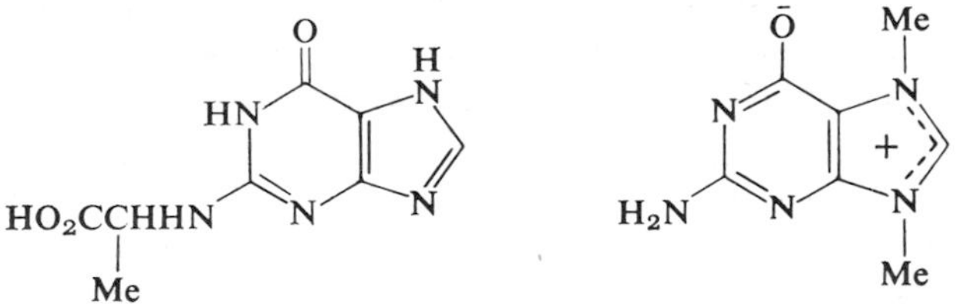

(126)   (127)

confirmed by synthesis.[303] Guanine analogues with a $C_{(2)}$-methylamino or -dimethylamino group occur in nucleic acid hydrolysates[274, 304] and human urine.[305] Methylated guanines include the 1-methyl and 7-methyl derivatives in urine[305] and the 7,9-dimethyl betaine form (127), herbipoline, extracted from the giant sponge (*Geodia gigas*).[306] *In vivo* hydroxylation would account for the presence of the 8-oxo derivatives of 7-methylguanine in human urine[305] and of guanine in the sea squirt (*Microcosmus polymorphus*).[307]

C. Isoguanine (6-Amino-2,3-dihydro-2-oxopurine)

Although synthesised by Fischer in 1897 by hydriodic acid reduction of 6-amino-8-chloro-2-ethoxypurine,[209] the occurrence of the base in nature was not appreciated until the isolation, as the 9-β-D-ribofuranoside, from croton beans (*Croton tiglium L.*) was made in 1932.[308] It was later found to be a component of butterfly wing pigments,[309] having been originally wrongly identified as a pteridine derivative under the name "guanopterin".[310]

Cyclisation in formamide of 4,5,6-triamino-2,3-dihydro-2-oxopurine gives a 93% yield of isoguanine monohydrate.[73] Alternative routes are nitrous acid treatment of 2,6-diaminopurine, in which only the $C_{(2)}$-amino group is hydrolysed,[311, 312] or cyclisation of 4(5)-amino-5(4)-carbamoylimidazole with urea or phosgene,[214] or by oxidative hydrolysis of 2-methylthioadenine.[33, 313] Photolysis of an aqueous solution of adenine-1-oxide results in an isomerisation to isoguanine, but the product also contains appreciable amounts of adenine.[314]

Isoguanine, because anion formation is possible, shows the expected stability toward alkali. Dilute hydrochloric acid hydrolyses convert it fairly rapidly to xanthine,[72, 150, 309] but no reaction occurs with nitrous acid.[72, 209, 308, 311] Reduction by zinc in hydrochloric acid at 100° results in ring fission and formation of an aminoimidazole.[315] Unlike adenine, exposure of an aqueous solution to ultraviolet light causes a rapid decrease in absorption.[316] The 8-diazo derivative is formed with diazonium salts.[289, 308]

No melting point is apparent, slow decomposition ensuing on heating over 250°. Solubility in water at 20° is 1 part in 16,000 rising to 1 part in 400 at 100°.[317] A more basic nature (pK_a 4.5 and 8.99) compared with the isomeric guanine is found. Salts formed with mineral acids include the sulphate (monohydrate), hydrochloride (anhydrous), hydrobromide (anhydrous), and nitrate. The hydrohalide salts are decomposed by boiling water. The picrate liberates the free base on treatment with dilute hydrochloric acid or metaphosphoric acid.[308] Crystallisation of isoguanine is best carried out by acidification of a solution in dilute sodium hydroxide with acetic acid.[72, 317] The monohydrate thus obtained retains the water of crystallisation on heating *in vacuo* at 130°.[308]

Synthesis of the naturally occurring isoguanine nucleoside crotonoside, by Davoll[312] in 1952, confirmed the structure proposed by Cherbuliez[308] and Spies[72] as being 6-amino-2,3-dihydro-2-oxo-9-β-Dribofuranosylpurine. A recent but impractical preparation, in 20% yield, results from ultraviolet irradiation of a solution of adenosine-1-oxide.[318]

The O-methylated analogue of crotonoside (2-methoxyadenosine) has been isolated[319] from a Caribbean sponge (*Cryptotethia crypta*), the structure being established by various syntheses.[161, 320]

References

1. Fox, Watanabe, and Bloch, *Progr. Nucleic Acid Res.*, **5**, 252 (1966).
2. Huber, *Angew. Chem.*, **68**, 706 (1956).
3. Elion, *J. Org. Chem.*, **27**, 2478 (1962).
4. Noell and Robins, *J. Amer. Chem. Soc.*, **81**, 5997 (1959).
5. Beaman and Robins, *J. Amer. Chem. Soc.*, **83**, 4038 (1961).
6. Giner-Sorolla, personal communication.
7. Shapiro, *J. Amer. Chem. Soc.*, **86**, 2948 (1964).
8. Itai and Ito, *Chem. and Pharm. Bull.* (*Japan*), **10**, 1141 (1962).
9. Bendich, Giner-Sorolla, and Fox, *Chemistry and Biology of Purines*, Eds., Wolstenholme and O'Connor, J. and A. Churchill Ltd., London, 1957, p. 3.
10. Giner-Sorolla and Bendich, *J. Amer. Chem. Soc.*, **80**, 3932 (1958).
11. Giner-Sorolla, Nanos, Burchenal, Dollinger, and Bendich, *J. Medicin. Chem.*, **11**, 52 (1968).
12. Giner-Sorolla, Medrek, and Bendich, *J. Medicin. Chem.*, **9**, 143 (1966).
13. Giner-Sorolla, O'Bryant, Burchenal, and Bendich. *Biochemistry*, **5**, 3057 (1966).
13a. Montgomery, Hewson, and Temple *J. Medicin. Chem.*, **5**, 15 (1962).
14. Brooks and Rudner, *J. Amer. Chem. Soc.*, **78**, 2339 (1956).
15. Kriek, *Biochem. Biophys. Res. Comm.*, **20**, 793 (1965).
16. Kriek, Miller, Juhe, and Miller, *Biochemistry*, **6**, 177 (1967).
17. Zimmer and Mettalia, *J. Org. Chem.*, **24**, 1813 (1959).
18. Holmes and Robins, *J. Amer. Chem. Soc.*, **87** 1772 (1965).
19. Borowitz, Bloom, Rothschild, and Sprinson, *Biochemistry*, **4**, 650 (1965).
20. Miller, Skoog, Okumura, Von Saltza, and Strong, *J. Amer. Chem. Soc.*, **77**, 2662 (1955).
21. Miller, Skoog, Okumura, Von Saltza, and Strong, *J. Amer. Chem. Soc.*, **78**, 1375 (1956).
21a. Young and Letham, *Phytochemistry*, **8**, 1199 (1969).
22. Litwack and Weissman, *Biochemistry*, **5**, 3007 (1966).
23. Montgomery and Thomas, *J. Heterocyclic Chem.*, **1**, 115 (1964).
24. Brookes and Lawley, *J. Chem. Soc.*, **1960**, 539.
25. Grimm and Leonard, *Biochemistry*, **6**, 3625 (1967).
26. Martin and Reese, *J. Chem. Soc.* (*C*), **1968**, 1731.
27. Taylor and Loeffler, *J. Amer. Chem. Soc.*, **82**, 3147 (1960).
28. Broom, Townsend, Jones, and Robins, *Biochemistry*, **3**, 494 (1964).
29. Leonard, Carraway, and Helgeson, *J. Heterocyclic Chem.*, **2**, 291 (1965).
30. Windmuller and Kaplan, *J. Biol. Chem.*, **236**, 2716 (1961).
31. Chheda and Hall, *Biochemistry*, **5**, 2082 (1966).
32. Montgomery and Thomas, *J. Org. Chem.*, **28**, 2304 (1963).

33. Andrews, Anand, Todd, and Topham, *J. Chem. Soc.*, **1949**, 2490.
34. Whitehead and Traverso, *J. Amer. Chem. Soc.*, **82**, 3971 (1960).
35. Leonard and Deyrup, *J. Amer. Chem. Soc.*, **84**, 2148 (1962).
36. Kazmirowski, Dietz, and Carstens, *J. prakt. Chem.*, **19**, 162 (1963).
37. Montgomery and Temple, *J. Amer. Chem. Soc.*, **82**, 4592 (1960).
38. Taylor, Barton, and Paudler, *J. Org. Chem.*, **26**, 4961 (1961).
39. Krackov and Christensen, *J. Org. Chem.*, **28**, 2677 (1963).
40. Naylor, Shaw, Wilson, and Butler, *J. Chem. Soc.*, **1961**, 4845.
41. Leese and Timmis, *J. Chem. Soc.*, **1961**, 3818.
42. Giner-Sorolla, Gryte, Bendich, and Brown, *J. Org. Chem.*, **34**, 2157 (1969).
42a. Broom and Robins, *J. Org. Chem.*, **34**, 1025 (1969).
43. Burckhalter and Dill, *J. Org. Chem.*, **24**, 562 (1959).
44. Okuda, *J. Pharm. Soc. Japan*, **80**, 205 (1960).
45. Mauvernay, Brit. Pat., 881,827 (1960); through *Chem. Abstracts*, **56**, 8725 (1962).
46. Roth and Brandes, *Arch. Pharm.*, **298**, 765 (1965).
47. Brandes and Roth, *Arch. Pharm.*, **300**, 1000 (1967).
48. Doebel and Spiegelberger, *Helv. Chim. Acta*, **39**, 283 (1956).
49. Doebel and Spiegelberger, U.S. Pat., 2,761,861 (1956); through *Chem. Abstracts*, **51**, 3676 (1957).
50. Quevauviller, Chabrier, and Morin, *Bull. Soc. Chim. biol.*, **31**, 352 (1949).
51. Richter, Rubinstein, and Elming, *J. Medicin. Chem.*, **6**, 192 (1963).
52. Damiens and Delaby, *Bull. Soc. chim. France*, **1955**, 888.
53. Stieglitz and Stamm, Ger. Pat., 1,122,534 (1962); through *Chem. Abstracts*, **56**, 14306 (1962).
54. Lister and Timmis, *J. Chem. Soc.*, **1960**, 327.
55. Lettré and Woenckhaus, *Annalen*, **649**, 131 (1961).
56. Golovchinskaya and Chaman, *Zhur. obshchei Khim.*, **22**, 535 (1952).
57. Zelnik and Pesson, *Bull. Soc. chim. France*, **1959**, 1667.
58. Zelnik, Pesson, and Polonovski, *Bull. Soc. chim. France*, **1956**, 1773.
59. Konz and Zeile, Ger. Pat., 1,018,869 (1957); through *Chem. Abstracts*, **54**, 1570 (1960).
60. Cochran, *Acta Cryst.*, **4**, 81 (1951).
61. Broomhead, *Acta Cryst.*, **4**, 92 (1951).
62. Iball and Wilson, *Nature*, **198**, 1193 (1963).
63. Bryan and Tomita, *Acta Cryst.*, **15**, 1179 (1962).
64. Sobell and Tomita, *Acta Cryst.*, **17**, 126 (1964).
65. Stewart and Davidson, *J. Chem. Phys.*, **39**, 255 (1963).
66. Pal and Horton, *J. Chem. Soc.*, **1964**, 400.
67. Kokko, Goldstein, and Mandell, *J. Amer. Chem. Soc.*, **83**, 2909 (1961).
68. Miles, Howard, and Frazier, *Science*, **142**, 1458 (1963).
69. Mason, *J. Chem. Soc.*, **1954**, 2071.
69a. Pullman, Berthod, and Dreyfus, *Theor. Chim. Acta*, **15**, 265 (1969).
70. Kossel, *Z. physiol. Chem.*, **10**, 248 (1886).
71. Strecker, *Annalen*, **118**, 151 (1861).
72. Spies, *J. Amer. Chem. Soc.*, **61**, 350 (1939).
73. Bendich, Tinker, and Brown, *J. Amer. Chem. Soc.*, **70**, 3109 (1948).
74. Giner-Sorolla and Bendich, *J. Amer. Chem. Soc.*, **80**, 5744 (1958).
75. Montgomery and Hewson, *J. Org. Chem.*, **34**, 1396 (1969).
76. Montgomery and Hewson, *J. Amer. Chem. Soc.*, **82**, 463 (1960).
77. Gerster and Robins, *J. Org. Chem.*, **31**, 3258 (1966).

78. Jones and Robins, *J. Amer. Chem. Soc.*, **82**, 3773 (1960).
79. Zemlicka and Holý, *Coll. Czech. Comm.*, **32**, 3159 (1967).
80. Bowne, *Experientia*, **17**, 175 (1961).
81. Ludowieg and Benmamam, *Experientia*, **23**, 36 (1967).
82. Kossel, *Z. physiol. Chem.*, **12**, 241 (1888).
83. Davoll and Lowy, *J. Amer. Chem. Soc.*, **73**, 1650 (1951).
84. Birkofer, *Ber.*, **76**, 769 (1943).
85. Schein, *J. Medicin. Chem.*, **5**, 302 (1962).
86. Mehrotra, Jain, and Anand, *Indian J. Chem.*, **4**, 146 (1960).
87. Iwamoto, Acton, and Goodman, *Nature*, **198**, 285 (1963).
88. Bowles, Schneider, Lewis, and Robins, *J. Medicin. Chem.*, **6**, 471 (1963).
89. Wulff, *Annalen*, **17**, 468 (1893).
90. Baker and Hewson, *J. Org. Chem.*, **22**, 959 (1957).
91. Farris, *Diss. Abstracts*, **25**, 3265 (1964).
92. Montgomery and Thomas, *J. Amer. Chem. Soc.*, **87**, 5442 (1965).
93. Overbeek, U.S. Pat., 3,013,885 (1960); through *Chem. Abstracts*, **56**, 14306 (1962).
94. Hull, *J. Chem. Soc.*, **1958**, 2746.
95. Hull, personal communication.
96. Craveri and Zoni, *Chimica (Italy)*, **34**, 407 (1958); through *Chem. Abstracts*, **53**, 13163 (1959).
97. Dyer, Reitz, and Farris, *J. Medicin. Chem.*, **6**, 289 (1963).
97a. Furukawa and Honjo, *Chem. and Pharm. Bull. (Japan)*, **16**, 1076 (1968).
98. Okumura, Jap. Pat., 8526 (1959); through *Chem. Abstracts*, **54**, 6769 (1960).
99. Ueda, Kato, and Toyoshima, Jap. Pat., 9773 (1957); through *Chem. Abstracts*, **52**, 15599 (1958).
100. Lister and Timmis, *J. Chem. Soc.*, **1961**, 1113.
101. Clark and Lister, *J. Chem. Soc.*, **1961**, 5048.
102. Baizer, Clark, Dub, and Loter, *J. Org. Chem.*, **21**, 1276 (1956).
103. Walton, Holly, Boxer, Nutt, and Jenkins, *J. Medicin. Chem.*, **8**, 659 (1965).
104. Shapiro and Hachmann, *Biochemistry*, **5**, 2799 (1966).
105. Bullock, Hand, and Stokstad, *J. Org. Chem.*, **22**, 568 (1957).
106. Benitez, Crews, Goodman, and Baker, *J. Org. Chem.*, **25**, 1946 (1960).
107. Altman and Ben-Ishai, *Bull. Res. Council. Israel*, **11A**, 4 (1962).
108. Blackburn and Johnson, *J. Chem. Soc.*, **1960**, 4347.
109. Lettré and Ballweg, *Annalen*, **649**, 124 (1961).
110. Lettré and Ballweg, *Angew. Chem.*, **69**, 507 (1957).
111. Lettré and Ballweg, *Chem. Ber.*, **91**, 345 (1958).
112. Fillippi and Guern, *Bull. Soc. chim. France*, **1966**, 2617.
113. Miyaki, Iwasi, and Shimizu, *Chem. and Pharm. Bull. (Japan)*, **14**, 87 (1966).
114. Shimizu and Miyaki, *Tetrahedron Letters*, **1965**, 2059.
115. Shimizu and Miyaki, *Chem. and Pharm. Bull. (Japan)*, **15**, 1066 (1967).
116. Berlin and Sjögren, *Svensk Kem. Tidskr.*, **53**, 457 (1941).
117. Beaman, Tautz, Duschinsky, and Grundberg, *J. Medicin. Chem.*, **9**, 373 (1966).
118. Jensen, Falkenberg, Thorsteinsson, and Lauridsen, *Dansk. Tidsskr. Farm.*, **16**, 141 (1942).
119. Pon, Belg. Pat., 451,086 (1943); through *Chem. Abstracts*, **42**, 215 (1948).
120. Satoda, Fukui, Matsuo, and Okumura, *Yakugaku Kenkyu*, **28**, 621 (1956).
121. Morishita, Nakano, Satoda, Yoshida, Fukui, Matsuo, and Okumura, Jap. Pat., 535 (1959); through *Chem. Abstracts*, **54**, 6767 (1960).
122. Lewin, *J. Chem. Soc.*, **1964**, 792.

123. Fel'dman, *Biochem. (U.S.S.R.)*, **25**, 432 (1960).
124. Alderson, *Nature*, **187**, 485 (1960).
125. Fel'dman, *Biochem. (U.S.S.R.)*, **27**, 378 (1962).
126. Gomberg, *Amer. J. Chem.*, **23**, 51 (1901).
127. Fischer, *Z. physiol. Chem.*, **60**, 69 (1909).
128. Usbeck, Jones, and Robins, *J. Amer. Chem. Soc.*, **83**, 1113 (1961).
129. Burgison, Abstracts 133rd A.C.S. Meeting, San Francisco, 1958, p. 14M.
129a. Shapiro and Shiuey. *Biochim. Biophys. Acta*, **174**, 403 (1969)
130. Lister, *J. Chem. Soc.*, **1963**, 2228.
131. Johnston, McCaleb, and Montgomery, *J. Medicin. Chem.*, **6**, 669 (1963).
132. Kössel, *Z. physiol. Chem.*, **340**, 210 (1965).
133. Kössel, *Angew. Chem.*, **76**, 689 (1964).
134. Kössel and Doehring, *Biochim. Biophys. Acta*, **95**, 663 (1965).
134a. Jones and Warren, *Tetrahedron*, **26**, 791 (1970).
135. Okuda, *J. Pharm. Soc. Japan*, **80**, 208 (1960).
136. Montgomery and Holum, *J. Amer. Chem. Soc.*, **79**, 2185 (1957).
137. Breshears, Wang, Bechtolt, and Christensen, *J. Amer. Chem. Soc.*, **81**, 3789 (1959).
138. Huber, Ger. Pat., 1,085,532 (1956); through *Chem. Abstracts*, **56**, 3494 (1962).
139. Elion, Burgi, and Hitchings, *J. Amer. Chem. Soc.*, **74**, 411 (1952).
140. Wellcome Foundation Ltd., Brit. Pat., 713,259 (1954); through *Chem. Abstracts*, **49**, 13301 (1955).
141. Hitchings and Elion, U.S. Pat., 2,691,654 (1954); through *Chem. Abstracts*, **50**, 1933 (1956).
142. Prasad and Robins, *J. Amer. Chem. Soc.*, **79**, 6401 (1957).
143. Robins and Lin, *J. Amer. Chem. Soc.*, **79**, 490 (1957).
144. Montgomery and Temple, *J. Amer. Chem. Soc.*, **79**, 5238 (1957).
145. Montgomery and Temple, *J. Amer. Chem. Soc.*, **83**, 630 (1961).
146. Priewe and Poljack, *Chem. Ber.*, **88**, 1932 (1955).
147. Libermann and Rouaix, *Bull. Soc. chim. France*, **1959**, 1793.
148. Fischer, *Ber.*, **31**, 104 (1898).
149. Zelnik, Pesson, and Polonovski, *Bull. Soc. chim. France*, **1956**, 888.
150. Polonovski, Pesson, and Zelnik, *Compt. rend.*, **236**, 2519 (1953).
151. Temple, Kussner, and Montgomery, *J. Org. Chem.*, **30**, 3601 (1965).
152. Giner-Sorolla, Thom, and Bendich, *J. Org. Chem.*, **29**, 3209 (1964).
153. Klosa, *J. prakt. Chem.*, **23**, 34 (1964).
154. Elderfield, Prasad, and Liao, *J. Org. Chem.*, **27**, 573 (1962).
155. Giner-Sorolla, *Galenica Acta*, **19**, 97 (1966).
156. Parham, Fissekis, and Brown, *J. Org. Chem.*, **32**, 1151 (1967).
157. Temple, Thorpe, Coburn, and Montgomery, *J. Org. Chem.*, **31**, 935 (1966).
158. Temple, Kussner, and Montgomery, *J. Org. Chem.*, **31**, 2210 (1966).
159. Cramer, *Ber.*, **27**, 3089 (1894).
160. Johnson, Thomas, and Schaeffer, *J. Amer. Chem. Soc.*, **80**, 699 (1958).
161. Schaeffer and Thomas, *J. Amer. Chem. Soc.*, **80**, 3738 (1958).
162. Smirnova and Postovskii, *Zhur. Vsesoyuz Khim. obshch. im D. I. Mendeleeva*, **9** 711 (1965); through *Chem. Abstracts*, **62**, 9130 (1964).
163. Reist, Benitez, Goodman, Baker, and Lee, *J. Org. Chem.*, **27**, 3274 (1962).
164. Kiburis and Lister, *Chem. Comm.*, **1969**, 381.
165. Traube and Dudley, *Ber.*, **46**, 3839 (1913).
166. Denayer, *Bull. Soc. chim. France*, **1962**, 1358.
167. Jones and Robins, *J. Amer. Chem. Soc.*, **84**, 1914 (1962).

168. Townsend, Robins, Loeppky, and Leonard, *J. Amer. Chem. Soc.*, **86**, 5320 (1964).
169. Ovcharova and Golovchinskaya, *Zhur. obshchei Khim.*, **34**, 2472 (1964).
170. Shaw and Butler, *J. Chem. Soc.*, **1959**, 4040.
171. Fischer, *Ber.*, **31**, 542 (1898).
172. Krüger, *Z. physiol. Chem.*, **18**, 422 (1894).
173. Kruger, *Ber.*, **26**, 1914 (1893).
174. Baddiley, Lythgoe, and Todd, *J. Chem. Soc.*, **1944**, 318.
175. Baker, Joseph, Schaub, and Williams, *J. Org. Chem.*, **19**, 1780 (1954).
176. Baker, Schaub, and Joseph, *J. Org. Chem.*, **19**, 638 (1954).
177. Reiner and Zamenhof, *J. Biol. Chem.*, **228**, 475 (1957).
178. Lira and Huffman, *J. Org. Chem.*, **31**, 2188 (1966).
179. Pal, *Biochemistry*, **1**, 558 (1962).
180. Leonard and Deyrup, *J. Amer. Chem. Soc.*, **82**, 6202 (1960).
181. Leonard and Fujii, *J. Amer. Chem. Soc.*, **85**, 3719 (1963).
182. Abshire and Berlinguet, *Canad. J. Chem.*, **42**, 1599 (1964).
183. Schaeffer and Vince, *J. Medicin. Chem.*, **8**, 710 (1965).
184. Neiman and Bergmann, *Israel J. Chem.*, **3**, 161 (1965).
185. Okano, Goya, and Kaizu, *J. Pharm. Soc. Japan*, **87**, 469 (1967).
186. Montgomery and Thomas, *J. Org. Chem.*, **30**, 3235 (1965).
187. Parikh and Burger, *J. Amer. Chem. Soc.*, **77**, 2386 (1955).
187a. Lira, *J. Heterocyclic Chem.*, **5**, 863 (1968).
188. Marsico and Goldman, *J. Org. Chem.*, **30**, 3597 (1965).
189. Montgomery, Hewson, Clayton, and Thomas, *J. Org. Chem.*, **31**, 2202 (1966).
190. Brown and Jacobsen, *J. Chem. Soc.*, **1965**, 3770.
191. Johnston, Fikes, and Montgomery, *J. Org. Chem.*, **27**, 973 (1962).
192. Nagasawa, Kumashiro, and Takenishi, *J. Org. Chem.*, **31**, 2685 (1966).
193. Robins, Godefroi, Taylor, Lewis, and Jackson, *J. Amer. Chem. Soc.*, **83**, 2574 (1961).
194. Myers and Zeleznick, *J. Org. Chem.*, **28**, 2087 (1963).
195. Mann and Porter, *J. Chem. Soc.*, **1945**, 751.
196. Pfleiderer, *Annalen*, **647**, 167 (1961).
197. Townsend and Robins, *J. Amer. Chem. Soc.*, **84**, 3008 (1962).
198. Bredereck, Christmann, and Koser, *Chem. Ber.*, **93**, 1206 (1960).
199. Bredereck, Christmann, Koser, Schellenberger, and Nast, *Chem. Ber.*, **95**, 1812 (1962).
200. Pfleiderer and Sagi, *Annalen*, **673**, 78 (1964).
201. Brookes and Lawley, *J. Chem. Soc.*, **1961**, 3923.
202. Neiman and Bergmann, *Israel J. Chem.*, **6**, 9 (1968).
203. Mian and Walker, *J. Chem. Soc. (C)*, **1968**, 2577.
204. Lawley and Brookes, *Nature*, **192**, 1081 (1961).
205. Shaw, *J. Org. Chem.*, **27**, 883 (1962).
206. Brookes, Dipple, and Lawley, *J. Chem. Soc. (C)*, **1968**, 2026.
207. Kossel, *Ber.*, **18**, 79 (1885).
208. Kossel, *Ber.*, **18**, 1928 (1885).
209. Fischer, *Ber.*, **30**, 2226 (1897).
210. Traube and co-workers, *Annalen*, **331**, 64 (1904).
211. Robins, Dille, Willets, and Christensen, *J. Amer. Chem. Soc.*, **75**, 263 (1953).
212. Fujimoto and Ono, *J. Pharm. Soc. Japan*, **86**, 364 (1966).
213. Jones, Mian, and Walker, *J. Chem. Soc. (C)*, **1966**, 692.
214. Cavalieri, Tinker, and Brown, *J. Amer. Chem. Soc.*, **71**, 3973 (1949).

215. Marsh, *J. Biol. Chem.*, **190**, 633 (1951).
216. Jolles, *J. prakt. Chem.*, **62**, 67 (1900).
217. Heath, *Nature*, **158**, 23 (1946).
218. Smith and Elving, *J. Amer. Chem. Soc.*, **84**, 1412 (1962).
219. Ponnamperuma, Lemmon, and Calvin, *Radiation Res.*, **18**, 540 (1963).
220. Bruhns, *Ber.*, **23**, 225 (1890).
221. Bruhns and Kossel, *Z. physiol. Chem.*, **16**, 1 (1892).
222. Krüger, *Z. physiol. Chem.*, **16**, 329 (1892).
223. Burian, *Z. physiol. Chem.*, **51**, 425 (1907).
224. Burian, *Ber.*, **37**, 696 (1904).
225. Fargher and Pyman, *J. Chem. Soc.*, **115**, 217 (1919).
226. Hitchings and Fiske, *J. Biol. Chem.*, **141**, 827 (1941).
227. Schmidt and Levene, *J. Biol. Chem.*, **126**, 423 (1938).
228. Kerr, Seraidarian, and Wargon, *J. Biol. Chem.*, **181**, 761 (1949).
229. Loring, Fairley, Bortner, and Seagran, *J. Biol. Chem.*, **197**, 809 (1952).
230. Graff and Maculla, *J. Biol. Chem.*, **110**, 71 (1935).
231. Hitchings and Fiske, *J. Biol. Chem.*, **140**, 491 (1941).
232. Bruhns, *Z. physiol. Chem.*, **4**, 533 (1890).
233. Thomas and Montgomery, *J. Org. Chem.*, **31**, 1413 (1966).
234. Albert, *Biochem. J.*, **54**, 646 (1953).
235. Albert, *Biochem. J.*, **76**, 621 (1960).
236. Harkins and Freiser, *J. Amer. Chem. Soc.*, **80**, 1132 (1958).
237. Sletten, *Chem. Comm.*, **1967**, 1119.
238. Weiss and Venner, *Z. physiol. Chem.*, **333**, 169 (1963).
239. Bayer, Posgay, and Majlat, *Pharm. Zentralhalle*, **101**, 476 (1962).
240. Mandel and Dunham, *J. Biol. Chem.*, **11**, 85 (1912).
241. Cunningham, Hutchinson, Manson, and Spring, *J. Chem. Soc.*, **1951**, 2299.
242. Frederiksen, Malling, and Klenow, *Biochim. Biophys. Acta*, **95**, 189 (1965).
243. Guarino and Kredich, *Biochim. Biophys. Acta*, **68**, 317 (1963); *Fed. Proc.* **23**, 371 (1964).
244. Gerber and Lechevalier, *J. Org. Chem.*, **27**, 1731 (1962).
245. Eble, Hoeksema, Boyack, and Savage, *Antibiotics and Chemotherapy*, **9**, 419 (1959).
246. Schroeder and Hoeksema, *J. Amer. Chem. Soc.*, **81**, 1767 (1959).
247. Farkaš and Šorm, *Coll. Czech. Chem. Comm.*, **28**, 882 (1963).
248. Yünsten, *J. Antibiotics (Japan)*, *Ser. A*, **11**, 244 (1958).
249. Yünsten, *J. Antibiotics (Japan)*, *Ser. A*, **11**, 79 (1958).
250. Hoeksema, Slomp, and Van Tamelen, *Tetrahedron Letters*, **1964**, 1787.
250a. McCarthy, Robins, and Robins, *J. Amer. Chem. Soc.*, **90**, 4993 (1968).
251. Waller, Patrick, Fulmor, and Meyer, *J. Amer. Chem. Soc.*, **79**, 1011 (1957).
251a. Morton, Lancaster, Van Lear, Fulmor, and Meyer, *J. Amer. Chem. Soc.*, **91**, 1535 (1969).
251b. Chibata, Okumura, Takeyama, and Kotera, *Experientia*, **25**, 1237 (1969).
251c. Kamiya, Saito, Hashimoto, and Seki, *Tetrahedron Letters*, **1969**, 4729; Okumura and co-workers, *Chem. Comm.*, **1970**, 1045, 1047.
252. Friedrich and Bernhauer, *Chem. Ber.*, **89**, 2507 (1956).
253. Friedrich and Bernhauer, *Chem. Ber.*, **90**, 465 (1957).
254. Littlefield and Dunn, *Biochem. J.*, **70**, 642 (1958).
255. Dutcher, Van Saltza, and Pansy, *Antimicrobiol. Agents and Chemotherapy*, **1963**, 83.
256. Ackermann and List, *Z. physiol. Chem.*, **323**, 192 (1961).
257. Belikov, Bankowsky, and Tsarev, *Zhur. obshchei Khim.*, **24**, 919 (1954).

258. Denayer, Cave, and Goutarel, *Compt. rend.*, **253**, 2994 (1961).
259. Janot, Cave, and Goutarel, *Bull. Soc. chim. France*, **1959**, 896.
260. Monseur and Adriaens, *J. Pharm. Belg.*, **1960**, 279.
261. Miller, Skoog, Von Saltza, and Strong, *J. Amer. Chem. Soc.*, **77**, 1392 (1955).
262. Letham, Shannon, and McDonald, *Proc. Chem. Soc.*, **1964**, 230; Letham, Mitchell, Cebalo, and Stanton, *Austral. J. Chem.*, **22**, 205 (1969).
263. Shaw and Wilson, *Proc. Chem. Soc.*, **1964**, 231.
264. Ceballo and Latham, *Nature*, **213**, 96 (1967).
265. Biemann and Tsunakawa, *Angew. Chem. Internat. Edn.*, **5**, 590 (1966).
266. Hall, Robins, Stasink, and Thedford, *J. Amer. Chem. Soc.*, **88**, 2614 (1966).
267. Leonard and Fujii, *Proc. Nat. Acad. Sci. U.S.A.*, **51**, 73 (1964).
268. Porter and co-workers, *Antibiotics and Chemotherapy*, **2**, 409 (1952).
269. Baker, Schaub, Joseph, and Williams, *J. Amer. Chem. Soc.*, **77**, 12 (1955).
270. Joklik, *Biochim. Biophys. Acta*, **22**, 211 (1956).
271. Hampton, *J. Amer. Chem. Soc.*, **79**, 503 (1957).
272. Baddiley, Buchanan, Hawker, and Stephenson, *J. Chem. Soc.*, **1956**, 4659.
273. Dunn and Smith, *Biochem. J.*, **68**, 627 (1958).
274. Adler, Weissmann, and Gutman, *J. Biol. Chem.*, **230**, 717 (1958).
275. Vanyushin, Kokurina, and Belozerskii, *Doklady Akad. Nauk, S.S.S.R.*, **161**, 1451 (1965).
276. Magnus, *Annalen*, **51**, 395 (1844).
277. Unger, *Annalen*, **58**, 18; **59**, 58 (1846).
278. Traube, *Ber.*, **33**, 1371, 3035 (1900).
279. Shapiro, *Progr. Nucleic Acid Res.*, **8**, 73 (1968).
280. Fischer, *Ber.*, **43**, 805 (1910).
281. Bang, *Biochem. Z.*, **26**, 293 (1910).
282. Hunter, *Nature*, **137**, 405 (1936).
283. Hunter, *Biochem. Z.*, **30**, 1183 (1936).
284. Strecker, *Annalen*, **108**, 129, 141 (1858).
285. Plentl and Schoenheimer, *J. Biol. Chem.*, **153**, 203 (1944).
286. Sussenbach and Berends, *Biochim. Biophys. Acta*, **95**, 184 (1965).
287. Tafel and Ach, *Ber.*, **34**, 1165 (1901).
288. Fischer and Reese, *Annalen*, **221**, 336 (1883).
289. Spies and Harris, *J. Amer. Chem. Soc.*, **61**, 351 (1939).
290. Burian, *Ber.*, **37**, 696 (1904).
291. Unger, *Ann. Physik.*, **65**, 225 (1845).
292. Capricanica, *Z. physiol. Chem.*, **4**, 233 (1880).
293. Neubauer and Kerner, *Annalen*, **101**, 318 (1857).
294. Lippmann, *Ber.*, **29**, 2645 (1896).
295. Dreschel, *J. prakt. Chem.*, **24**, 45 (1840–1844).
296. Horbaczewski, *Z. physiol. Chem.*, **23**, 226 (1897).
297. Hemmens, *Biochim. Biophys. Acta*, **68**, 284 (1963).
298. Shapiro and Gordon, *Biochem. Biophys. Res. Comm.*, **17**, 160 (1964).
299. Hemmens, *Biochim. Biophys. Acta*, **91**, 332 (1964).
300. Ballio, Delfini, and Russi, *Nature*, **186**, 968 (1960).
301. Ballio, Delfini, and Russi, *Gazzetta*, **96**, 337 (1966).
302. Al-Khalidi and Greenberg, *J. Biol. Chem.*, **236**, 189 (1961).
303. Al-Khalidi and Greenberg, *J. Biol. Chem.*, **236**, 192 (1961).
304. Smith and Dunn, *Biochem. J.*, **72**, 294 (1959).
305. Weismann, Bromberg, and Gutman, *J. Biol. Chem.*, **224**, 407 (1957).

306. Ackermann and List, *Z. physiol. Chem.*, **309**, 286 (1957).
307. Karrer, Manunta, and Schwyzer, *Helv. Chim. Acta*, **31**, 1214 (1948).
308. Cherbuliez and Bernhard, *Helv. Chim. Acta*, **15**, 464 (1932).
309. Purrmann, *Annalen*, **544**, 182 (1940).
310. Schöpf and Becker, *Annalen*, **524**, 49 (1936).
311. Trattner, Elion, Hitchings, and Sharefkin, *J. Org. Chem.*, **29**, 2674 (1964).
312. Davoll, *J. Amer. Chem. Soc.*, **73**, 3174 (1951).
313. Taylor, Vogl, and Cheng, *J. Amer. Chem. Soc.*, **81**, 2442 (1959).
314. Brown, Levin, and Murphy, *Biochemistry*, **3**, 880 (1964).
315. Friedman and Gots, *Arch. Biochem. Biophys.*, **39**, 254 (1952).
316. Stimson, *J. Amer. Chem. Soc.*, **64**, 1604 (1942).
317. Albert and Brown, *J. Chem. Soc.*, **1954**, 2060.
318. Cramer and Schlingoff, *Tetrahedron Letters*, **1964**, 3201.
319. Bergmann and Burke, *J. Org. Chem.*, **21**, 226 (1956).
320. Bergmann and Stempien, *J. Org. Chem.*, **22**, 1575 (1957).

CHAPTER IX

The Purine Carboxylic Acids and Related Derivatives

Apart from the title purines, this chapter embraces derivatives with ester, amide, and nitrile functions and, where applicable, the thio analogues. Aldehydo- and ketopurines are also included under this heading.

Simple purinecarboxylic acids do not seem to play a significant part, if any at all, in biological processes. This is in direct contrast with the purinylamino acids, which have an amino group interposed between purine and carboxyl moieties, representatives of which are key intermediates in metabolic processes.

1. The Carboxypurines

Of the simplest members, while 6- and 8-carboxypurine are known, the synthesis of the 2-carboxy isomer is still awaited. No examples of either di- or tricarboxypurines exist. Considerable attention has been paid to the study of the carboxy derivatives of N-methylated oxopurines, mainly those of caffeine, and to a lesser degree those of theophylline and theobromine, also.

A. Preparation of Carboxypurines

Procedures are based mainly on hydrolytic or oxidative modification of existing groups, the direct introduction of a carboxyl group is limited to the formation of 8-carboxypurines.

a. *By Hydrolysis*

Purine-6-carboxylic acid (**1**) can be obtained by alkaline hydrolysis of 6-cyano-,[1, 2] 6-carbamoyl-,[1] or 6-thiocarbamoylpurine[1] using $2N$-

sodium hydroxide under reflux conditions. Less alkaline media, such as aqueous sodium bicarbonate or sodium acetate, are suitable ones in which to prepare the acid from 6-trichloromethylpurine[3] (2), while only boiling water[1] was required to convert the formamidino derivative (3). The α-hydroxy acid (5), obtained by the prolonged action of cold sodium carbonate solution on 6-(2-hydroxy-2-trichloromethylethyl) purine (4), is converted to the acrylic acid derivative (6) on standing in sodium ethoxide.[4] Catalytic reduction over a palladium catalyst converts this to 6-carboxyethylpurine (7).

CO_2H (1) CCl_3 (2) $H_2NC:NH$ (3) $CH_2CHOHCCl_3$ (4)

$CH_2CHOHCO_2H$ (5) $CH:CHCO_2H$ (6) $CH_2CH_2CO_2H$ (7)

Caffeine-8-carboxylic acid in 90% yield results when nitrogen trioxide is bubbled through a cold solution of the 8-carbamoyl derivative in 50% sulphuric acid.[5, 6] Heating the amide to 80° in the acid alone gives a poor yield of the 8-carboxypurine and at higher temperatures the product, through decarboxylation, is caffeine.[5] The same 8-carboxy derivative[7] arises through heating 8-trichloromethylcaffeine in water. The 3-methylxanthine analogue can be likewise hydrolysed to the 8-carboxy compound,[7] but under these conditions the 8-trifluoromethyl groups in the isocaffeine[8] and theobromine[9] derivatives are removed completely giving the parent 8-unsubstituted purines. By using ethanolic instead of aqueous solutions the corresponding 8-ethoxycarbonyl derivatives are produced.[7, 8] Acid hydrolysis of the ester is required in those cases when an ester group, already present in the intermediate, is not converted to the acid during purine formation.[10, 11] Acid conditions are used to form carboxymethylpurines from the corresponding malonic ester derivatives. Reactions of this type, involving hydrolysis in aqueous media, have been widely employed to introduce carboxymethyl groups into N-alkylxanthines. An early example was the condensation of

8-bromocaffeine with the sodium derivative of diethyl malonate followed by hydrolysis of the resulting diester (8), with 18% hydrochloric acid, to 8-carboxymethylcaffeine (9).[12] The latter has been obtained from 8-chlorocaffeine similarly and also by alkaline hydrolysis of the 8-carboxymethyl amide.[11] The range of 8-carboxyalkylpurines can be extended by alkylation of the malonic ester derivative (8). Hydrolysis of the product affords the appropriate 8-(α-alkyl)carboxymethylcaffeine (10, R = Me, Et, Pr, and Bu).[11, 13] 8-Carboxymethylisocaffeine is likewise prepared from 8-chloroisocaffeine[14] but the less reactive halogen in 8-chloromethylisocaffeine requires more forcing conditions, condensation with the malonic ester sodium salt being effected in boiling toluene.[15] Dilute hydrochloric acid readily converts the product to the 8-(2-carboxyethyl)purine which also is obtained on dilute acid hydrolysis of the 8-carbamoylethyl analogue.[15] Other examples prepared by the

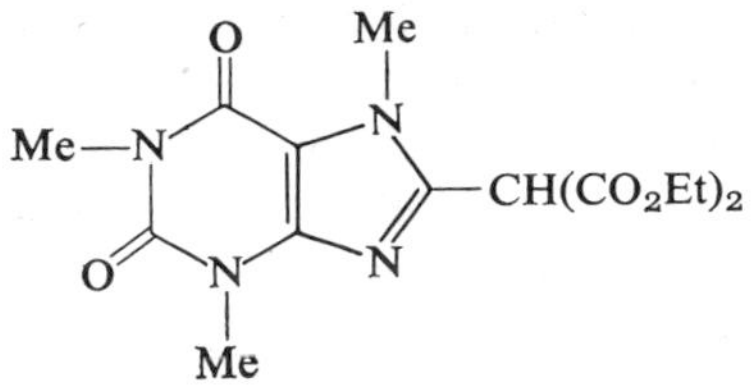

(8)

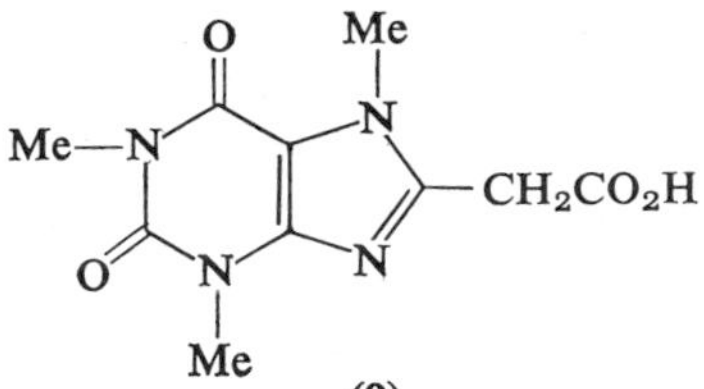

(9)

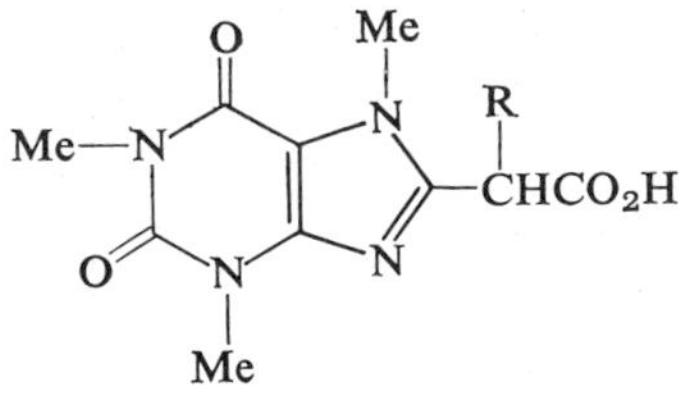

(10)

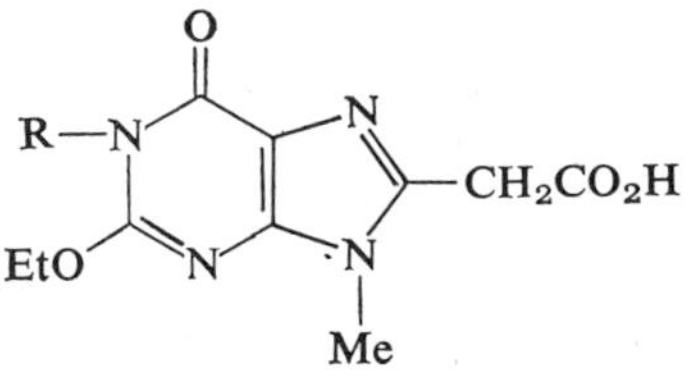

(11)

above routes are 1-benzyl-8-carboxymethyltheobromine, readily reduced to 8-carboxymethyltheobromine,[16] and the 8-carboxymethyl derivatives (11, R = H or Me) which, under mild hydrolysis conditions, retain the 2-ethoxy group intact but lose it with more vigorous treatment.[17] With 8-chloro-2,6-diethoxy-9-methylpurine, after malonic ester treatment, both ethoxy groups were hydrolysed giving 8-carboxymethyl-9-methylxanthine.[17] Purines with chlorine atoms at either $C_{(2)}$ or $C_{(6)}$ also react with malonic ester but the products readily lose both carboxyl groups on acid treatment with formation of the respective 2-methyl[18] or 6-methyl derivatives.[19, 20] This aspect is discussed elsewhere (Ch. IV, Sect. 2Ad).

b. *By Oxidation*

Oxidation of methyl groups is a poor route to carboxypurines, as seen in the 8% yield of 6-carboxypurine obtained from permanganate oxidation of 6-methylpurine.[21] Modified methyl groups, on the other hand, are more readily converted. With purine-6-pyridinium methiodide (**12**) a good yield (77%) of 6-carboxypurine (**1**)[22] results on stirring with aqueous permanganate at room temperature for 2 hours. Other routes to this purine include oxidation of 6-formylpurine[22] (82%) or the 6-thioformyl analogue[23] with permanganate. A poor yield results, by aerial oxidation, on simply boiling[22] an aqueous solution of the aldehyde. Both purine-6-acrylic acid and 6-carboxyethylpurine are reported to be oxidised to purine-6-carboxylic acid.[4] Other 6-substituted-purines used include the 6-hydrazide with sodium periodate as oxidising agent[22] and 6-benzoylpurine (**13**), which is oxidised on standing for some hours at room temperature in alkaline hydrogen peroxide.[24] Hydroxymethyl groups undergo ready conversion to carboxy groups in aqueous permanganate. In this way both 6-[21] and 8-carboxypurine[25] are obtained from the respective 6- or 8-hydroxymethylpurine. Others likewise formed include the 8-carboxy derivatives of theophylline,[26] isocaffeine,[27] and caffeine.[28] The same oxidising agent successfully converts 8-formylpurine to the corresponding acid form, being used to prepare theophylline-,[26] theobromine-,[26] and caffeine-8-carboxylic acids.[28] Hydroxymethyl groups frequently result from hydrolysis of chloromethyl groups which can themselves be oxidised to carboxyl groups. The permanganate oxidation of 2-chloro-6-chloromethyl-9-methylpurine to 6-carboxy-2-chloro-9-methylpurine illustrates the reaction.[19] This reagent, in acetone at 25°, also converts 8-styryladenine to the 8-carboxy analogue.[28a]

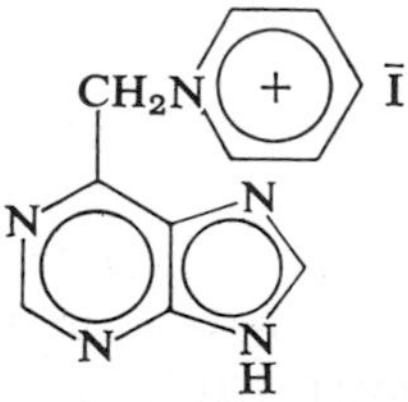

(12) 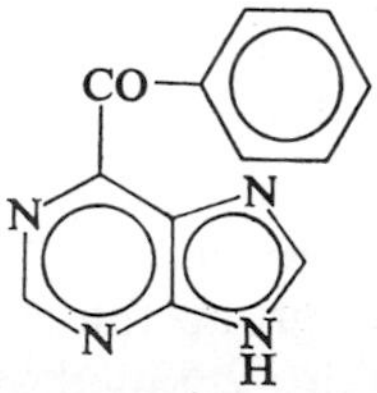   (13)

c. *By Other Means*

Although Traube syntheses can be utilised for direct insertion of carboxyl groups, they are restricted to the formation of purines with carboxyl groups attached, either directly or indirectly, to the 8-position. In the general procedure one of the carboxyl group of a dicarboxylic acid is condensed with a 4,5-diaminopyrimidine. Oxalic acid, the simplest dicarboxylic acid, gives anomalous results. With 4,5-diamino-1,2,3,6-tetrahydro-3-methyl-2,6-dioxopyrimidine the expected 8-carboxy-3-methylxanthine results. Similarly formed is 8-carboxytheophylline.[29, 30] By contrast other 4,5-diaminopyrimidines under these conditions undergo the Isay reaction with the production of 6,7-dihydroxy-pteridines.[31]

The range of the reactions can be extended by use of acid precursors such as cyanoacetic acid which, with the above pyrimidine, undergoes hydrolysis of the cyano group during ring closure to give 8-carboxymethyl-3-methylxanthine.[29, 30] Likewise derived is 8-carboxymethyl-theophylline.[29] Succinic acid with the appropriate diaminopyrimidine affords 8-carboxymethyl-3-methylxanthine and 8-carboxymethyl-guanine.[29, 30] A more detailed account of the use of acids in Traube synthesis is given in Chapter II, Section 1D.

By using 4,5-diamino-6-ethoxycarbonylpyrimidines a series of ethyl esters of 6-carboxypurines has been obtained by Traube-type cyclisations[10, 32, 33]; these are noted in a succeeding ester section. Unsaturated acid derivatives of theophylline[26] and caffeine[34] result from condensations of the purine-8-aldehydes with malonic acid or ester. 8-Formyl-theophylline reacts with malonic acid in boiling pyridine containing catalytic amounts of pyridine giving 8-carboxyvinyltheophylline (**14**, R = H) directly[26] but, if malonic ester is used, the 8-diethoxycarbonyl-vinyl derivative (**15**) results. Similar compounds form with caffeine-8-

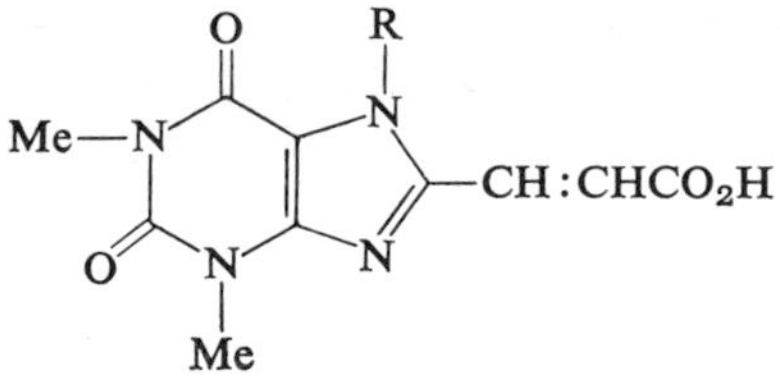

(14) 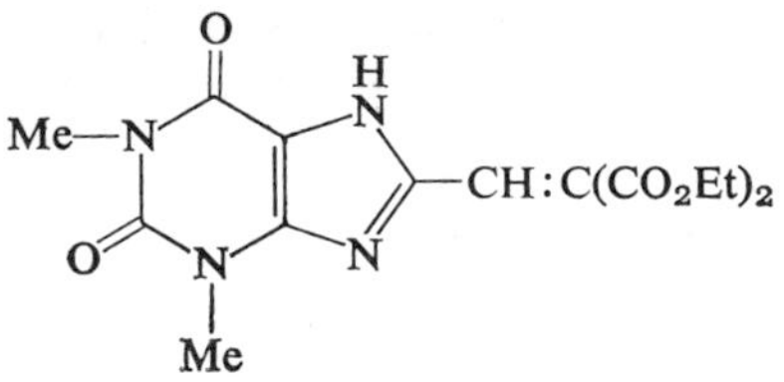 (15)

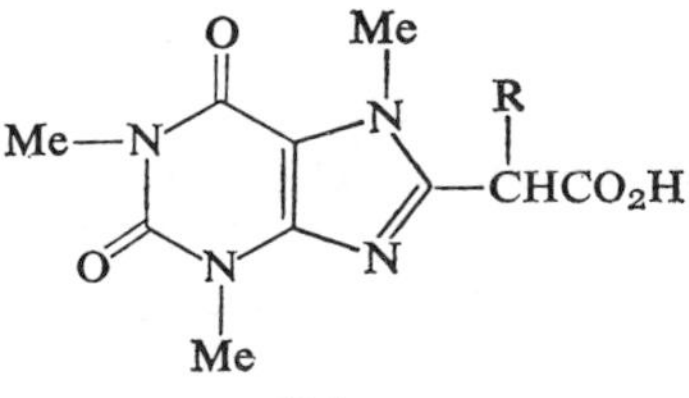

(16)

aldehyde,[34] condensation in acetic acid gives the dicarboxyvinyl compound, which undergoes decarboxylation to **14** (R = Me) in boiling quinoline.[34]

B. Reactions of Carboxypurines

Whether attached to the nucleus directly or through alkyl groups the carboxyl groups show characteristic reactions.

a. *Decarboxylation*

This follows from the usual procedures, the acid being heated to above the fusion point or in a high boiling solvent. Purine itself is derived by fusing 6-carboxy-[1] or 8-carboxypurine,[25] the carboxyl group of the latter being the more easily removed. 9-Methyl-2-methylaminopurine follows from decarboxylation of the 6-carboxy analogue in petroleum jelly at 160°.[19] A variety of conditions have been applied to carboxyl group removal in methylated xanthines. With 8-carboxy-3-methylxanthine fusion at 160°[7, 29] was employed while with the caffeine[5, 28] and isocaffeine analogues[27] decarboxylation was effected in boiling water. Derivatives with 8-carboxymethyl groups give the corresponding 8-methyl analogues, as in formation of 8,9-dimethyl- and 1,8,9-trimethylxanthine[17] in boiling dimethylformamide. Conversions of 8-carboxymethylcaffeine (**16**, R = H) to the 8-methyl derivative follows hydrolysis in dilute acid.[11, 12] Homologous 8-alkylcaffeines arise on decarboxylation of appropriate 8-carboxyalkyl derivatives (**16**, R = Et, Pr, Bu).[13] The use of boiling quinoline[34] or dilute hydrochloric acid to partially or completely decarboxylate malonic ester derivatives has been noted in the previous section.

b. *Esterification*

The majority of examples result from the Fischer-Speier technique, the methyl or ethyl ester being obtained on heating the acid in methanol or ethanol through which dry hydrogen chloride is passing. Examples include 6-ethoxycarbonylethylpurine,[4] 6-ethoxycarbonylvinylpurine,[4] also the 8-methoxy- or 8-ethoxycarbonyl derivatives of 3-methylxanthine,[7, 29] 9-methylxanthine,[17] 1,9-dimethylxanthine,[17] and theophylline.[26, 30] The corresponding 8-alkoxycarbonylmethyl analogues of

3-methylxanthine,[29] theobromine,[16] caffeine[11, 13] and isocaffeine,[14] and the 8-methoxycarbonylethyl derivatives of 3-methylxanthine[29] and guanine[30] are known. Sulphuric acid was used in place of hydrogen chloride to esterify an ethanolic solution of 8-(2,2-dicarboxyethyl)-caffeine,[34] also 8-carboxyethylisocaffeine.[15] Preparation of the methyl and ethyl esters of 8-carboxycaffeine by interaction of the silver salt of the acid and methyl or ethyl iodide in ethanol is recorded in the early literature.[5]

c. *Acid Chloride Formation*

This class of compound is poorly defined and invariably represents an intermediate stage in the amide or ester synthesis. Due to the reactive nature purification is difficult and the crude reaction product is generally employed with advantage. 6-Carboxypurine is converted to the acid chloride with hot phosphorus pentachloride[35] but thionyl chloride, either alone or in a solvent such as benzene, is the usual reagent employed. This gives 8-chlorocarbonyltheophylline[26] and the 8-chloro-carbonylethyl derivatives of caffeine[34, 36] and isocaffeine.[15]

d. *Other Reactions*

Unsaturated acid moieties are readily reduced at the double bond, 6-carboxyvinylpurine (**17**) being hydrogenated in ammonium hydroxide over a palladium catalyst to the propionic acid derivative (**18**).[4] Other

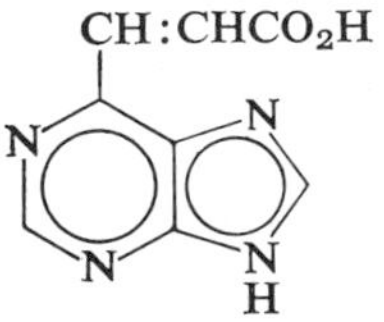

(17) 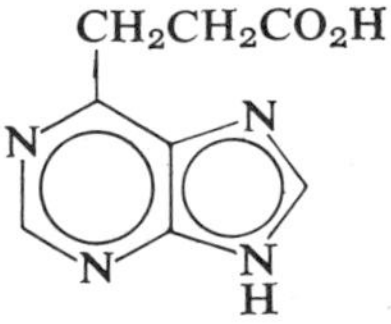   (18) 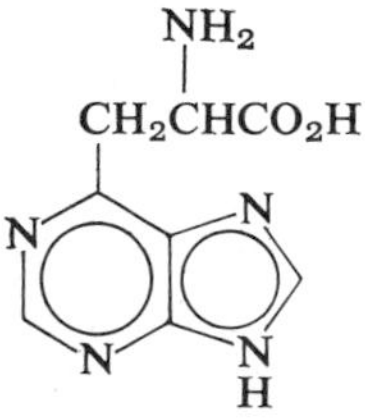   (19)

reductions include use of Raney nickel in sodium hydroxide for transforming 8-carboxyvinylcaffeine and the dicarboxy analogue to their respective reduced forms.[34]

An unusual reductive amination is observed when the vinyl acid (**17**) undergoes prolonged heating under reflux with hexamethyldisilazane

[HN(SiMe$_3$)$_2$] the product, after dilute mineral acid treatment, is 6-(2-amino-2-carboxyethyl) purine (19).[37]

2. Alkoxycarbonylpurines (Purine Esters)

Of little importance by themselves they are useful in the formation of amides or for reduction to the alcohol derivative.

A. Preparation

a. *From Pyrimidine Esters*

Under Traube synthesis conditions 4,5-diamino-6-ethoxy (and -methoxy)-carbonylpyrimidines form the appropriate 6-alkoxycarbonyl-purines with various cyclising reagents. With formic acid[33] or dimethyl-formamide–phosphoryl chloride mixture[10, 32] 8-unsubstituted-purine esters arise, whereas acetic anhydride gives 8-methyl derivatives.[33] Carbon disulphide in pyridine leads to the 8-thiopurine[33] analogues, while urea in the same solvent gives the corresponding 8-oxo-6-ethoxy-carbonylpurines. If the more usual fusion conditions employed for urea cyclisations are adopted, conversion of ester groups to amides occurs concomitantly.[33]

b. *By Other Methods*

Esters, in addition to direct esterification of the acid (Sect. 1Bb), also result by acid alcoholysis of trichloromethyl groups. In methanol under reflux 6-methoxycarbonylpurine is formed[3] via hydrolysis of the intermediate 6-trimethoxy derivative (20). This route was also used for the 8-methoxy (and -ethoxy)-carbonyl derivatives of theobromine[7] and isocaffeine.[8] Thiocarboxy esters result from interaction of acid chlorides with thioalcohols, thus 8-benzylthiocarboxy- (21)[38] and 8-benzylthio-carboxyethylcaffeine[36] are obtained from the appropriate 8-carbonyl chloride and benzylmercaptan. Formation of the unsaturated acid ester (22) provides an example of the Wittig reaction and results from interaction of 8-formyltheophylline with ethoxycarbonylmethylenetri-phenylphosphorane (23) in boiling dioxane for some hours.[26]

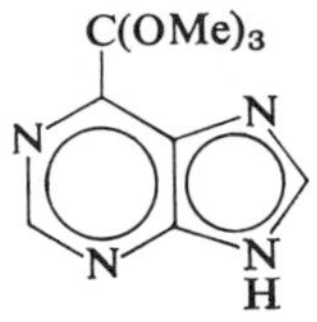

(20)

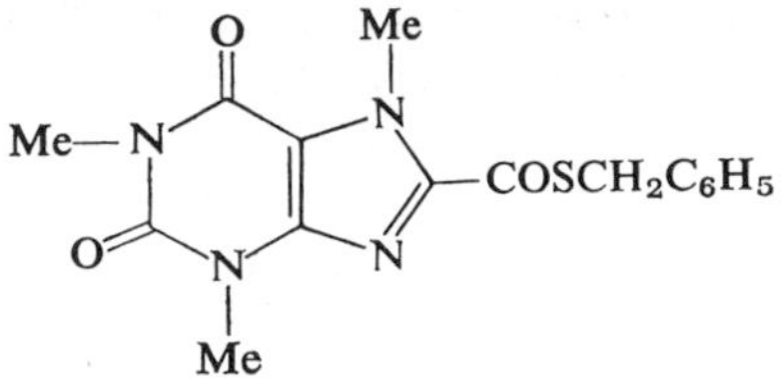

(21)

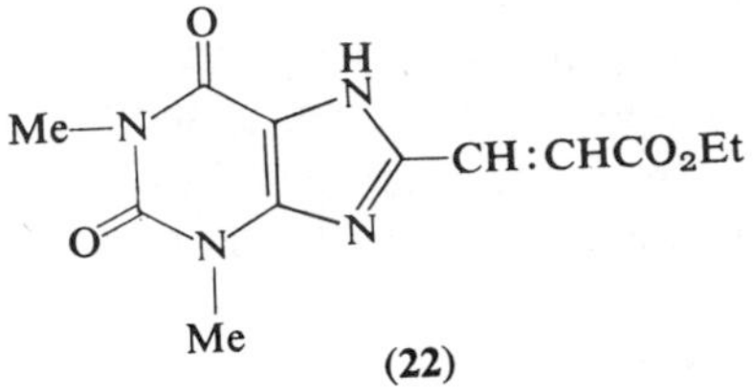

(22)

$(C_6H_5)_3P:CHCO_2Et$

(23)

B. Reactions

a. *Amide Formation*

Esters are readily transformed to their amide analogues on gentle treatment with ammonia; appropriate examples are the formation of 8-carbamoylmethyl derivatives of 9-methyl-[17] and 1,9-dimethyl-xanthine,[17] also the corresponding caffeine,[11] isocaffeine,[14] and related derivatives[13, 29] from the 8-methyl esters,[17] likewise, 6-carbamoyl-ethylpurine, the reaction in this case being carried out in a sealed tube.[4]

b. *Reduction*

The reagent of choice is lithium aluminium hydride, reduction taking place in tetrahydrofuran. This procedure readily converts 6-ethoxy-carbonylethylpurine to 6-hydroxypropylpurine.[4] Examples involving $C_{(8)}$-ester groups include conversion of 8-ethoxycarbonylmethyltheobromine and the corresponding 1-benzyl homologue to the respective 8-hydroxyethylpurines.[16] By reductive debenzylation, with sodium in ammonia,[16] the latter product is converted to the former one. Because of their ease of reduction, 8-benzylthiocarboxyalkylcaffeines have been utilised for preparing 8-hydroxyalkylcaffeine derivatives. In dry dioxan containing Raney nickel a smooth conversion of the benzylthio ester (21) to 8-hydroxymethylcaffeine can be brought about on leaving at room temperature for some hours.[38] The 8-hydroxypropyl analogue from 8-benzylthiocarboxyethylcaffeine[36] is a further example.

c. *Other Reactions*

The general application of malonic ester to the formation of purine-carboxylic acids has been made (Sect. 1Aa). The initial product arising from the condensation between a 6-chloropurine and sodium derivative of malonic ester will react with halogenating reagents, such as thionyl chloride or bromine in chloroform, to form a halogenoester of the type **24**, which undergoes hydrolytic decarboxylation in ethanolic hydrogen chloride to afford the 6-chloromethylpurine (**25**).[19] Similarly derived is 2-chloromethyl-1,9-dimethylhypoxanthine from the corresponding 2-chloropurine.[18] The 8-chloromalonic ester derivatives of caffeine (**26**) and isocaffeine with dilute hydrochloric acid undergo further hydrolysis at the 8-chloromethyl stage, giving finally the 8-hydroxymethylpurine.[27] The bromomalonic derivatives are more stable than the chloro derivatives. With the bromo analogue of **26** some 8-bromomethylcaffeine can be recovered after acid hydrolysis.[27]

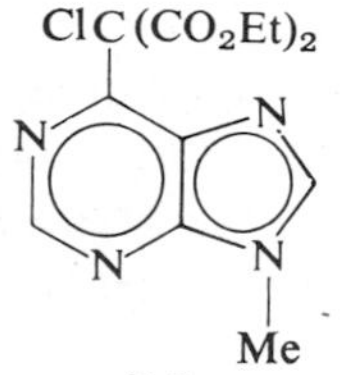

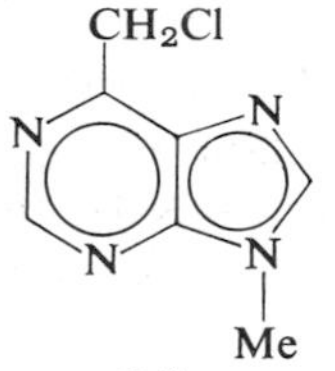

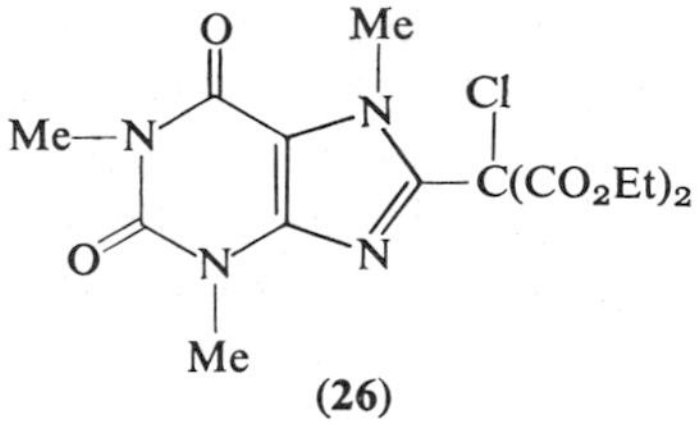

3. Carbamoylpurines (Purine Amides) and Related Compounds (Thio Analogues, Hydrazides, and Azides)

Like the acids and esters the amides of purines are of no great importance either synthetically or biologically, and only a few are known. Some amides and related derivatives undergo the Hofmann and Curtius reactions and so provide a means of transforming a carboxy group to an alkylamino group. Dehydration of the amide to the nitrile can sometimes be made and this, through subsequent reduction, is a means of introducing aminomethyl groups.

A. Preparation

a. *From Esters and Acid Chlorides*

The ammonolysis of esters has been described (Sect. 2Ba), including one example of inadvertent amide formation resulting from the ammonia liberated during a urea fusion.[33] In some instances acid chlorides may replace esters, thus 6-chlorocarbonylpurine in ethanolic ammonia for 1 hour at room temperature gives a 60% yield of amide.[35] Methylamine or dimethylamine in ethanol under these conditions afford the respective 6-N-methyl- and 6-NN-dimethylcarbamoylpurine.[35] The acid chlorides of 8-carboxytheophylline,[26] 8-carboxyethylcaffeine,[34] and -isocaffeine[15] give the appropriate amide with cold ammonium hydroxide. N-Alkylamides can be prepared similarly from alkylamines.[34] N-Arylamides are formed when 6-trichloromethylpurine and aniline derivatives such as p-nitroaniline, p-aminobenzoic acid, and sulphanilic acid, react.[39] Aniline itself behaves anomalously in giving the NN'-diphenylamidine derivative (27). Amino acids have also been condensed in this way to the purine amide.[39]

b. *By Other Means*

Other routes to amides involve regulated hydrolysis of cyano derivatives; 1 hour's hydrolysis of 6-cyanopurine with an equivalent of 2N-sodium hydroxide gives 6-carbamoylpurine (28).[1, 2] In preparing 2-amino-6-carbamoylpurine from the 6-cyanopurine[40] hydrogen peroxide was added to the alkaline medium. Heating in water for some minutes was sufficient to convert 8-cyanocaffeine to 8-carbamoyl-caffeine.[5]* Dilute ammonium hydroxide treatment converts 6-thio-carbamoyl- (29) to 6-carbamoylpurine,[1] which also results from allowing a solution of 6-trichloromethylpurine in ammonium hydroxide to stand for some hours,[3] or from the action of alkaline potassium ferricyanide solution on 6-hydrazidopurine (30).[22, 35]

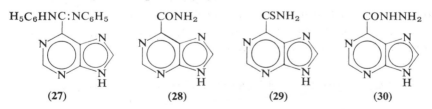

$H_5C_6HNC:NC_6H_5$ $CONH_2$ $CSNH_2$ $CONHNH_2$

(27)   (28)   (29)   (30)

* This compound was claimed in an early paper[6] to be formed directly on prolonged heating of 8-chlorocaffeine with an alcoholic solution of potassium cyanide.

Thioamides are conveniently prepared from cyanopurines by passing hydrogen sulphide through their solutions in cold ethanolic ammonia. Examples resulting from this procedure are 6-thiocarbamoylpurine (29)[1, 2, 41] and 2-amino-6-thiocarbamoylpurine.[41]

B. Reactions of Amides

The hydrolysis of amides to carboxypurines has been described previously (Sect. 1Aa). Two other general reactions of amide groups encountered in carbamoylpurines are dehydration to nitriles and the Hofmann rearrangement.

Heating 8-carbamoylcaffeine with phosphorus pentoxide for some hours at 250° gives 8-cyanocaffeine[5] but current practice, favouring milder conditions, employs phosphoryl chloride under reflux as dehydrating agent.[42]. Other examples are 8-cyanomethylcaffeine[11] and -isocaffeine.[43]

With alkaline hypobromite the appropriate aminoalkyl purine results from a Hofmann rearrangement. 6-Aminoethylpurine is derived in this fashion from the 6-carbamoylethyl analogue.[4] This procedure when carried out with 8-carbamoylethylcaffeine gave the expected 8-amino-ethylpurine,[44] but the isocaffeine analogue (31) suffered fission of the

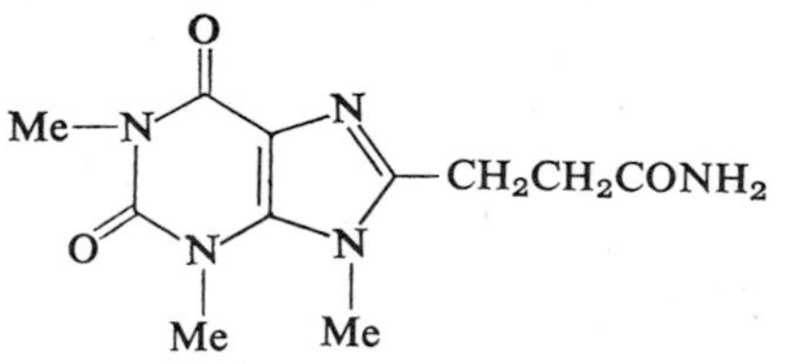

(31) 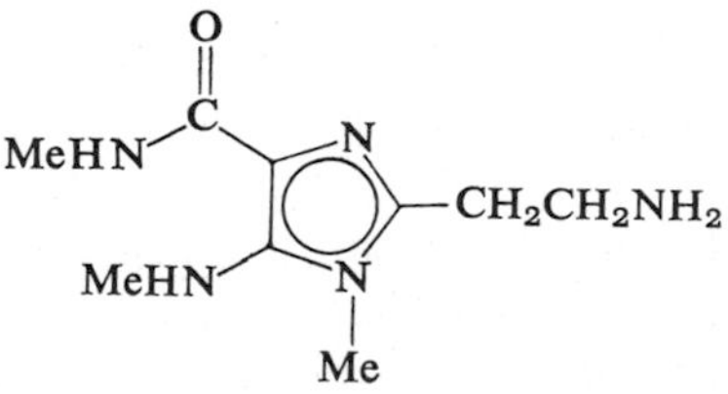   (32)

pyrimidine ring forming the imidazole derivative (32).[43] This result can be attributed to the demonstrated instability of the expected 8-amino-ethylisocaffeine towards alkali.[43]

Reactions of amides with hydrazine lead to formation of hydrazido derivatives as with 6-hydrazidopurine (30)[35] and the 9-methyl analogue[40] from the appropriate amide, only short periods under reflux with aqueous hydrazine being necessary.

C. Preparation and Reactions of Hydrazides and Azides

In addition to the above reaction, between amide and hydrazine, 6-hydrazidopurine is obtained in poorer yields from hydrazine interaction with either 6-chlorocarbonyl-[35] or 6-carboxypurine[35] or by alkaline hydrolysis of 6-cyanopurine in the presence of hydrazine.[22] The hydrazide of 8-carboxytheophylline follows from hydrazine treatment of the methyl ester.[26]

With acetone the hydrazides give isopropylidene derivatives of the type **33**[40] and are sulphonated with benzenesulphonyl chloride giving **34**.[22] Conversion of the hydrazide to the azide by nitrous acid treatment (Curtius reaction) was successful with 6-hydrazidopurine (**30**), the 6-azidocarbonylpurine (**35**) forming in near-quantitative yield,[35, 40] but failed with the 9-methyl homologue.[40] Azides react under reflux con-

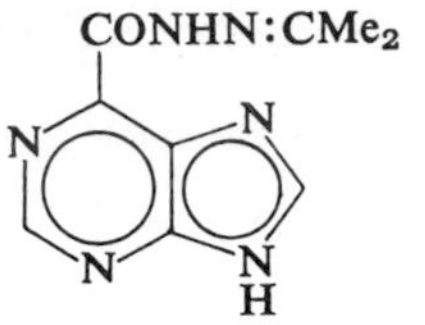

CONHN:CMe₂ — (33) CONHNHSO₂C₆H₅ — (34) CON₃ — (35)

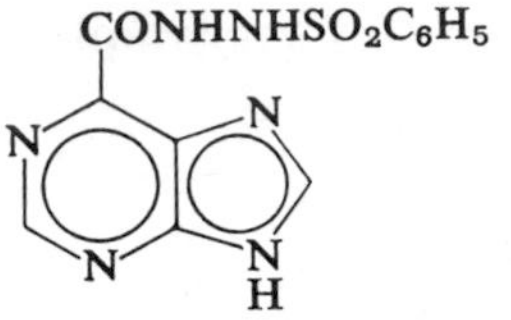

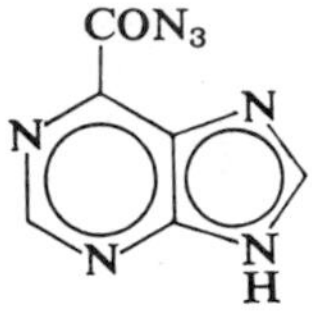

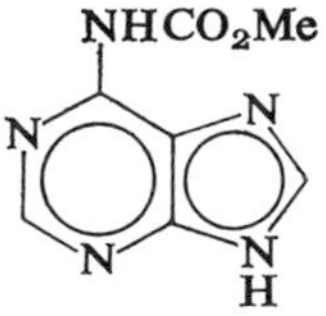

NHCO₂Me — (36) NHCONH₂ — (37)

ditions with methanol,[35] ethanol,[35] and butanol,[40] urethanes of the type **36** being formed which are capable of reaction with ammonia to give the corresponding ureidopurine (**37**).[35]

4. Purine Nitriles (Cyanopurines)

Few examples of this versatile class of compounds exist but sufficient have been examined to show that typical reactions are shown by the cyano groups attached directly to a purine nucleus. As with the related acids, esters, and amides the chemistry is limited to purines with cyano groups at the 6- and 8-positions.

A. Preparation

Two main methods are available (*a*) replacement of a halogen atom, (*b*) modification of an existing amide or aldoxime group.

a. *By Replacement of Halogen Atoms*

Heating 6-iodopurine in pyridine containing cuprous cyanide for 2 hours produces 6-cyanopurine, but the reaction fails if 6-chloropurine is used instead.[1, 2, 41] These conditions also produce 6-cyano-2-methyl-,[40] 6-cyano-9-methyl-,[40] and 2-amino-6-cyanopurine[41] from the appropriate 6-iodopurine. The reluctance of a chlorine atom to suffer displacement by a nitrile group is also noted with 8-chlorocaffeine. With potassium cyanide even fusion at 200° for 50 h gives a mixture containing only 60% of 8-cyanocaffeine.[5] Exocyclic nitrile groups are introduced with less vigour, for example, 8-chloromethylcaffeine with

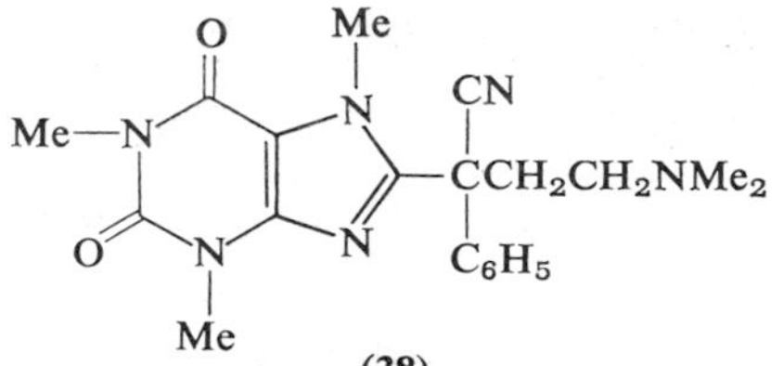

(38) 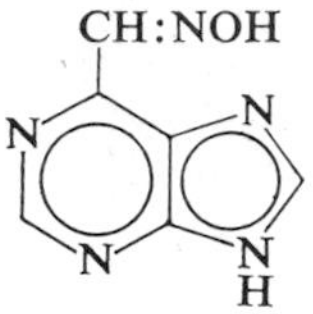   (39)

sodium cyanide in ethanol forms the (possibly) dimerised form of 8-cyanomethylcaffeine[11] in good yield. Heating 8-chlorocaffeine with 1-benzyl-3-dimethylaminopropyl cyanide in alcohol in the presence of sodamide[45] affords the nitrile **38**.

b. *By Modification of Amide or Aldoxime Groups*

Treatment of amide groups with dehydrating agents has already been dealt with (Sect. 3B). Aldehyde groups, in the form of their oxime derivative, undergo thermal rearrangement in the presence of acetic anhydride to nitriles. The oxime (**39**) of 6-formylpurine affords 6-cyano purine on prolonged heating under reflux (5 h) in acetic anhydride.[21] In some instances dry heating of the acetylated oxime at more elevated temperatures ($\sim 200°$) is adopted. This route provides the 8-cyano derivatives of theophylline,[26] theobromine,[26] and 6-diethylamino-2-methylpurine.[46]

B. Reactions of Cyanopurines

A number of the more important examples have been described in preceding sections, these include alkaline hydrolysis to the acid (Sect.

1Aa) and amide (Sect. 3Ab), also ammoniacal hydrogen sulphide treatment giving the thioamide (Sect. 3Ab).

a. *General Reactions*

In addition to the three noted in the introduction, a variety of other reactions are known. Ethanol alone converts 6-cyanopurine to the ethylimino ester (**40**).[1, 2] Ethanolic ammonia under pressure gives the formamidine (**41**),[1, 2] while hydrazine[22, 35] and phenylhydrazine[22] give the respective 6-N-aminoamidinopurines (**42**, R = H and C_6H_5).[22, 35] The former purine (**42**, R = H) undergoes the Curtius reaction with nitrous acid producing the azide (**43**).[22] The 6N-hydroxyamidinopurine (**44**) is obtained with hydroxylamine.[22, 35] On the water bath in 2N-sulphuric acid 6-cyanopurine is hydrolysed to hypoxanthine. Ethanolic hydrogen chloride at 4° produces the amide (**45**), which reverts to the amidine (**41**) with ethanolic ammonia under pressure.[2] With Grignard reagents ketones result. 8-Acetyl- (**46**, R = Me) and 8-propionylcaffeine (**46**, R = Et) are obtained from the respective reaction of methyl

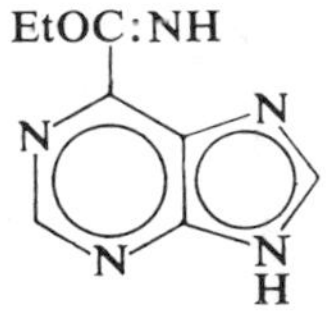

EtOC:NH (40) H₂NC:NH (41)

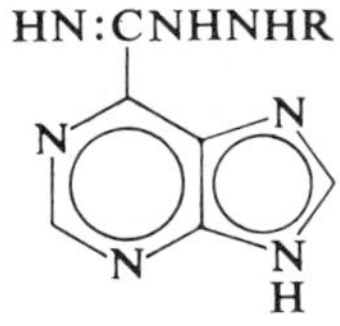

HN:CNHNHR (42) HN:CN₃ (43) HN:CNHOH (44)

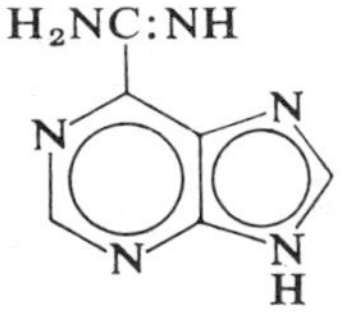

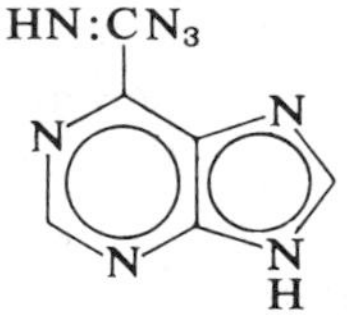

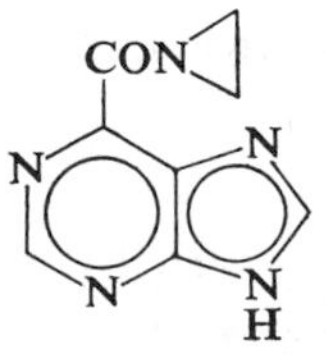

CON (45) 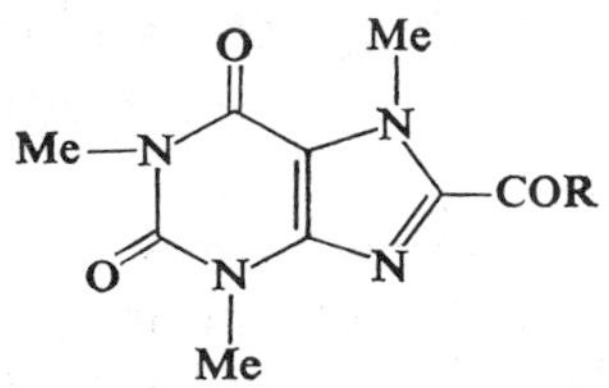 (46)

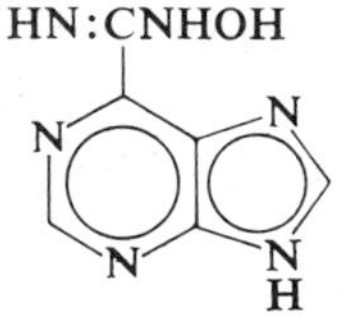

magnesium iodide or ethyl magnesium bromide on 8-cyanocaffeine.[42] Removal of an exocyclic cyano group from a purine, presumably by a decarboxylation process, is reported to take place in hot 70% sulphuric acid.[45]

b. *Reduction*

Conversion of a cyano group to an aminomethyl group is possible by means of sodium in alcohol, as in the formation of 8-aminomethyl-caffeine.[5] More usually, catalytic hydrogenation is employed as, for example, in the preparation of 6-aminomethylpurine (47)[35] in methanol using a palladium–charcoal catalyst. 8-Aminomethylisocaffeine results from reduction over Raney nickel.[43] In the latter acetic anhydride was the solvent from which the product was isolated as the acetylated amine; hot dilute hydrochloric acid treatment is required for deacetylation. When 6-cyanopurine is subject to controlled hydrogenation over Raney nickel in aqueous methanol containing semicarbazide, a low yield of the semicarbazone (48) is obtained.[22] This reaction could occur

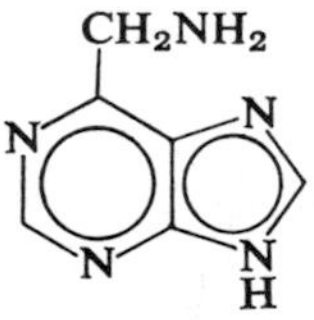

CH_2NH_2 $CH:NNHCONH_2$ $CH:NH$

(47) 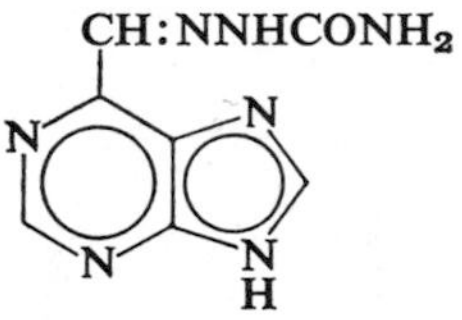   (48) 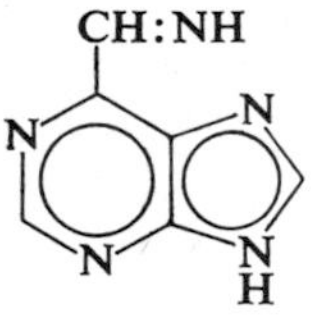   (49)

by replacement of the imino function of the intermediate aldehyde imine (49) by the semicarbazone group as other transformations of this type have been described.[47] Using hydrazine in place of semicarbazide only a low yield (30%) of the hydrazone resulted, the main product being the 6-aminoamidinopurine (42, R = H).[47]

5. Purine Aldehydes (*C*-Formylpurines) and Thio Analogues

Relatively few *C*-formylpurines are known, those described are simple derivatives of 6- or 8-formylpurines, the latter including *N*-methylated forms of xanthine. No examples of purine with an aldehyde group at $C_{(2)}$ appear to exist.

A. Preparation

The routes available require either hydrolytic or oxidative treatment of a substituent group. Direct syntheses from aldehyde intermediates or by C-formylation have not been made.

a. By Hydrolysis

6-Formylpurine (50), in low yield, is indirectly prepared from 6-methylpurine by converting this first to the pyridinium salt (51) and then, by reaction with a p-nitrosodialkylaniline, to the nitrone derivative (52).[22, 48] Warm dilute mineral acid readily converts the latter to the aldehyde (50).[22] 2-Amino-6-formylpurine is similarly derived.[21] Cold acid was sufficient to hydrolyse the $C_{(8)}$-nitrone derivatives of theophylline and theobromine to the respective 8-formylpurines.[49]

Dichloromethyl derivatives are valuable precursors of aldehydes which they give on hydrolysis. The 8-formyl derivatives of caffeine,[28]* isocaffeine,[50] and theobromine[8] followed from some hours' heating of the appropriate 8-dichloromethyl analogue. This approach forms a useful, practical route to 8-formylpurines as the starting materials are the readily accessible 8-methylpurines which are easily converted to the dichloromethyl derivatives.[8, 50] Aldehyde derivatives are occasionally used as a source of the parent compound, thus the hydrazone of 6-formylpurine gives a good yield of the aldehyde on treatment with nitrous acid,[21, 22] while on heating with thiolacetic acid conversion to 6-thioformylpurine (53) occurs[23].

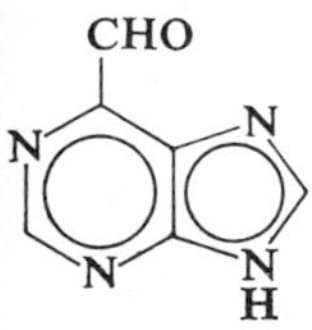

(50) 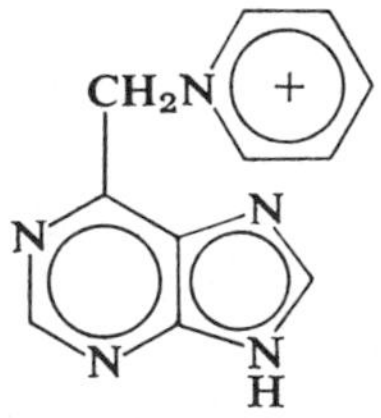   (51) 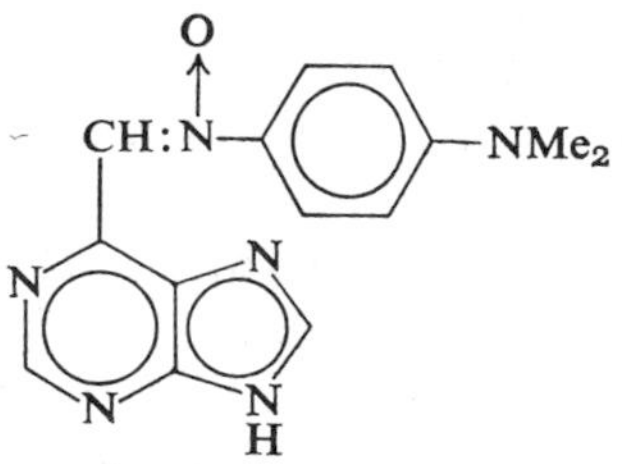   (52) 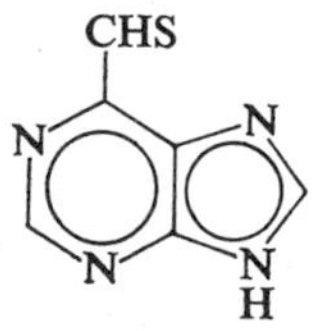   (53)

b. By Oxidation

Direct oxidation of a methyl group is not a practical means of forming an aldehyde group. More successful results are obtained using the

---

* This 8-formylpurine appears to be the first purine aldehyde prepared and predates by some ten years a recent claim [22] to this distinction.

corresponding hydroxymethyl derivative instead. Oxidations with sodium dichromate in acetic acid have given 8-formyltheophylline,[49] -theobromine,[49] and -caffeine,[49] and the analogous derivatives of 3-methyl-[49] and 1,3-dimethyl-7-phenacylxanthine,[49] and 6-diethyl-amino-2-methylpurine[46] from the corresponding 8-hydroxymethyl-purines. Caffeine-8-aldehyde has also resulted from a selenium dioxide oxidation.[51] Homologues of hydroxymethyl groups react likewise with oxidising reagents. Examples of 8-formylpurines resulting from oxidative fission of $C_{(8)}$-polyhydroxyalkyl chains are known, the oxidant being sodium metaperiodate.[46]

B. Reactions of Purine Aldehydes

a. *Oxidation and Reduction*

Examples of both these types of reaction are included in the sections devoted to the resulting products, i.e., carboxylic acids (Sect. 1Ab) and hydroxymethyl derivatives (Ch. VI, Sect. 3), respectively.

b. *Formation of the Usual Aldehyde Derivatives*

Characteristic derivatives are formed by the aldehyde groups at the 6- and 8-positions with the usual aldehyde-characterising agents. Some examples of these are cited below together with special reactions undergone by such derivatives.

Oximes have been recorded for purine-6-aldehyde[22, 23, 52] and the 8-formyl derivatives of theobromine,[8, 26] theophylline,[26] caffeine,[28] and 6-diethylamino-2-methylpurine.[46] The action of acetic anhydride on aldoximes giving nitriles has already been outlined (Sect. 4Ab). The hydrazone, which is formed by 6-formyl- (50)[22, 23, 52] or 6-thioformyl-purine (53), is readily converted back to the thioaldehyde (53) by means of thiolacetic acid[23] or to the aldehyde (50) with cold nitrous acid.[22] With ethyl nitrite in hot acetic acid the bis-aldazine (54) is most likely formed through reaction between unchanged hydrazone and some of the aldehyde formed.[22] With an excess of hydrazine under reflux 6-methylpurine is the product.[23] Various phenylhydrazones are given by 6-formylpurine,[22, 23] and the 8-formyl derivatives of theophylline,[26] theobromine,[8] and 6-diethylamino-2-methylpurine.[46] Semicarbazones of 6-formylpurine[22] and 8-formyltheophylline are known[26]; more use has been made of thiosemicarbazone derivatives which include those of 6-formylpurine[22, 23] (6-thioformyl),[23] and 8-formyltheophylline,[26] -theobromine,[26] and -3-methylxanthine.[26] Among the more unusual are hydrazones of the type **55** prepared from theophylline, theobromine, and caffeine aldehydes and NN-di(2-hydroxyethyl)hydrazine[53] and the thiocarboxy ester (**56**) from methylthiocarbonylhydrazine.[46] Isonicotinic acid hydrazide gives a derivative with 8-formyltheophylline.[26]

Schiff bases, formed with aniline, are given by 8-formyltheophylline, -theobromine,[8] -caffeine,[28, 44, 49] and -isocaffeine.[50] The last purine affords the azine ester (**57**) with glycine.[50] Anils reduce readily, that of caffeine gives the 8-phenylaminomethyl analogue on catalytic (Raney nickel) hydrogenation.[44]

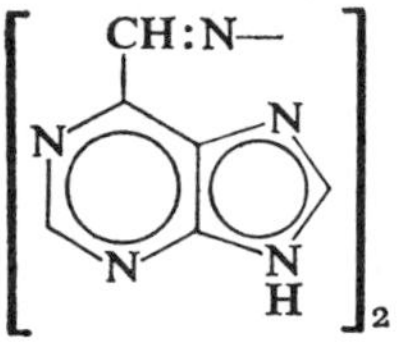

(54)

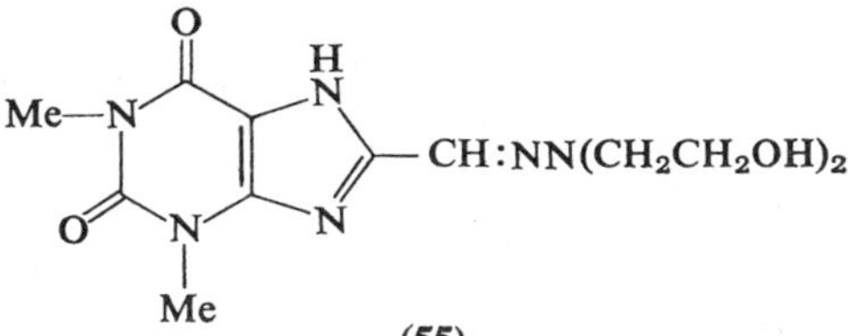

(55)

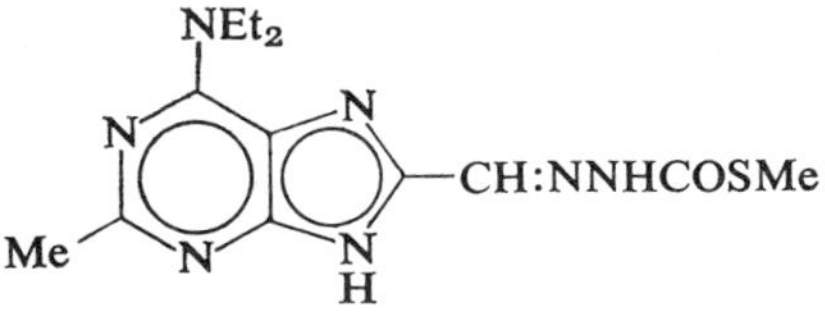

(56)

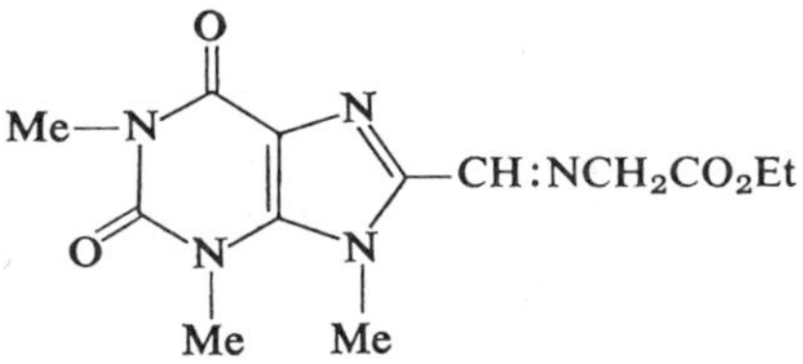

(57)

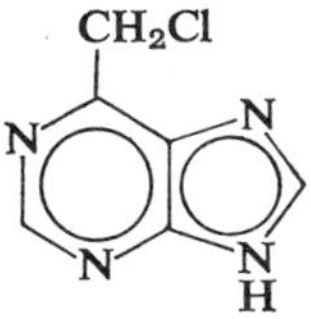

(58)

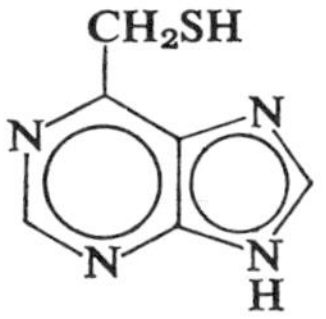

(59)

Some of the above derivatives can be derived using 6- or 8-nitrones, e.g. **52**, instead of the free aldehyde.[22, 26] An unusual procedure, at present restricted to forming 6-formylpurine derivatives, affords the oxime (**39**), for example, on prolonged heating in an ethanolic solution of hydroxylamine of 6-chloromethyl- (**58**), 6-bromomethyl- or even 6-mercaptomethylpurine (**59**).[52] In aqueous hydrazine the hydrazone results.[52] A mechanism for this reaction has been advanced.[52] The oxime (39), which also results from treating 6-methylpurine with cold nitrous acid, is converted to the hydrazone on prolonged heating in aqueous hydrazine.[20]

c. Other Reactions

Examples of reaction between purine-8-aldehydes and an active methylene group have been quoted (Sect. 1Ac); in these malonic acid (or the ester) is condensed with 8-formyltheophylline[26] and -caffeine.[34] With 1,3-diketones, for example acetylacetone or benzoylacetone, the respective 8-(2,2-diacetylvinyl)- (60, R = R' = Me) and 8-(2-acetyl-2-benzoylvinyl)theophylline (60, R = Me, R' = C_6H_5) result using pyridine, which contains catalytic amounts of piperidine, as solvent.[26] Styryl derivatives arise when active methyl groups are present.

In acetic anhydride, or acetic acid under reflux conditions, alkiodides of 2- and 4-methylpyridine, and other heterocycles react with 8-formyl-theobromine, for example, forming products of the type 61.[54] Caffeine[55] and isocaffeine[56] aldehydes react analogously.

Examples of the Cannizzaro rearrangement are not a feature of this class of aldehydes but benzoin-type condensations are known. 8-Formyltheophylline, -theobromine, and -caffeine in dimethylformamide or ethanol in the presence of potassium cyanide give the bis-purinyl forms, the example (62) being depicted as the internally hydrogen bonded tautomer rather than with the classical benzoin link.[26] A

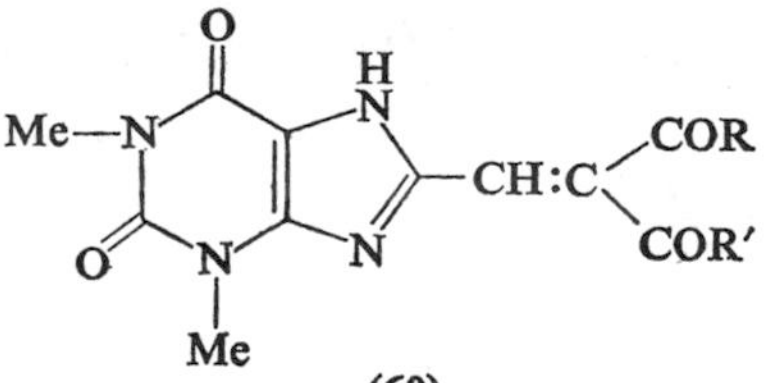

(60) 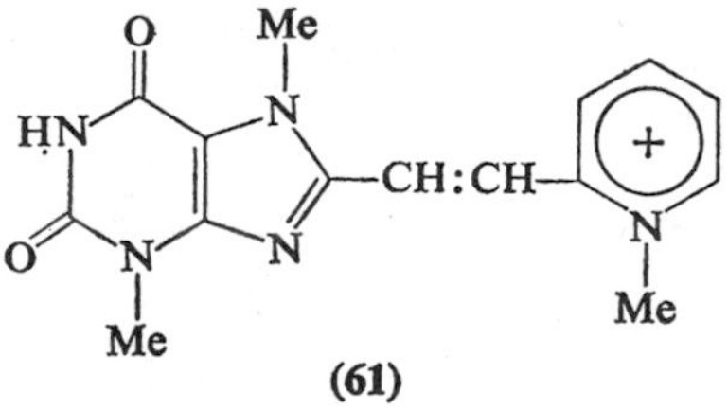 (61) 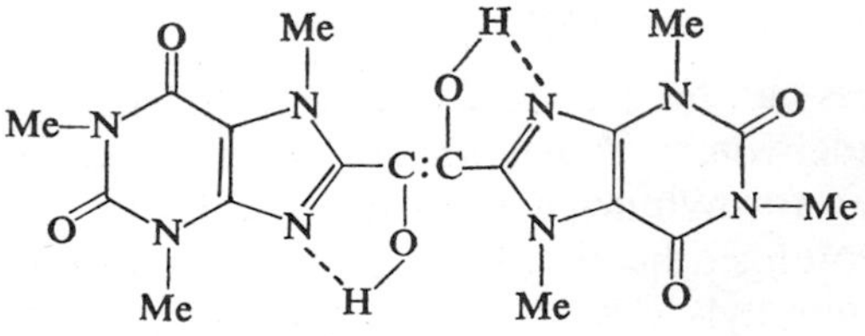 (62) 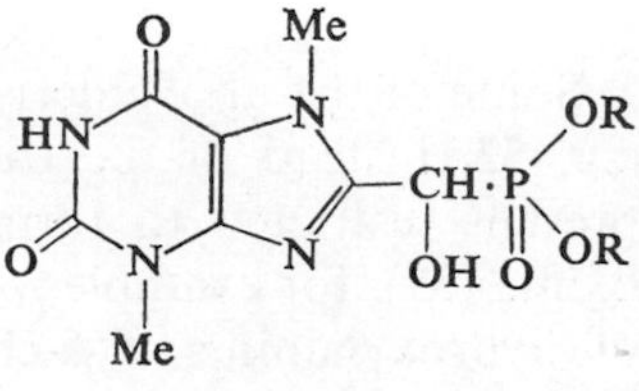 (63)

parallel can be drawn between this condensation and that which occurs between aldehydes and dialkyl hydrogen phosphonates. 8-Formyltheobromine gives the 8-(α-dialkylphosphonyl-α-hydroxy)methyl derivative[57] (63). The analogous derivative arises from caffeine-8-aldehyde.[51] In both series the methyl-, ethyl-, propyl-, and butylhydrogenphosphonates were used. The Wittig reaction undergone by theophylline-8-aldehyde with a triphenylphosphorane derivative, which gives rise to an unsaturated acid ester derivative, has been outlined already (Sect. 2Ab).

Conversion of an aldehyde group to a dichloromethyl group is accomplished with thionyl chloride in chloroform; 8-dichloromethylisocaffeine is obtained by this procedure and is converted back to the aldehyde in boiling water.[50]

6. Purine Ketones and Derivatives

This area of purine chemistry is largely unexplored and as such purine ketones are rare compounds, the syntheses of the first examples are of fairly recent date.

A. Preparation

When 6-trichloromethylpurine and sodium phenoxide react in the presence of sodium methoxide, the resulting dimethylketal (64) is converted to 6-p-hydroxybenzoylpurine (65) on heating in dilute hydrochloric acid.[24] The sodium salt of p-cresol gives 6-(2-hydroxy-5-methylbenzoyl)purine likewise.[24] In these examples a Reimer-Tiemann type C-acylation occurs at either the *ortho* or *para* positions of the phenol

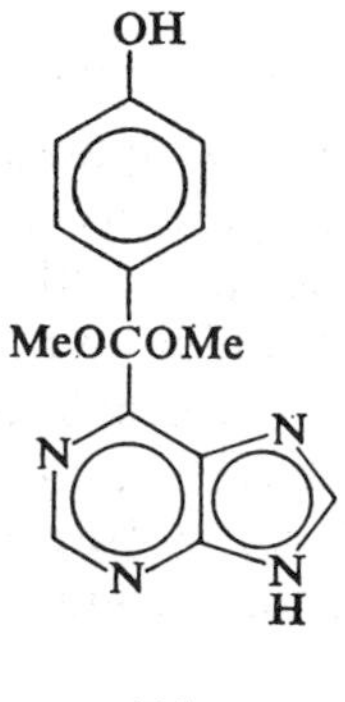

(64) 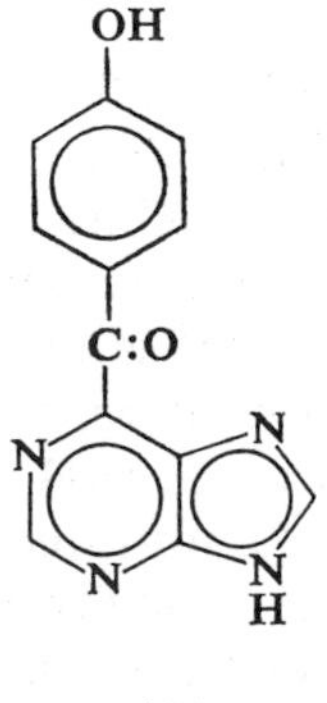   (65) 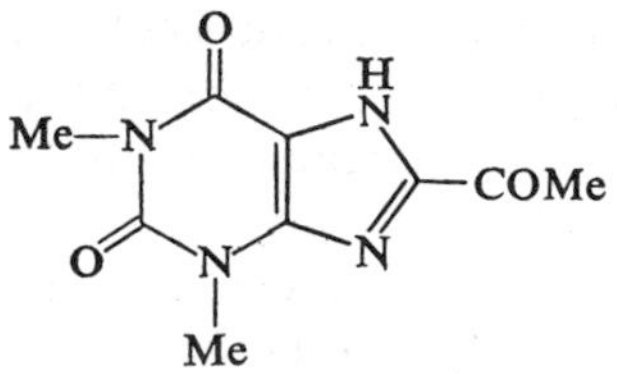   (66)

through the trichloromethyl group of the purine. The formation of aliphatic ketone moieties at $C_{(8)}$ from the action of Grignard reagents on 8-cyanocaffeine has been noted previously (Sect. 4Ba). 8-Acetyltheophylline (66) results from chromium trioxide oxidation of 8-(1-hydroxyethyl)theophylline in acetic acid at 100°.[42] Conversion of 66 to the known 8-acetylcaffeine was by methylation with alkaline dimethyl sulphate.

Exocyclic ketones are more common, usually being N-ketonylpurines derived by alkylation of the parent purine with a halogenoketone.[58] Other routes to such derivatives include Dakin-West reactions on purinylcarboxylic acid (Sects. 8A and 9Bb), and a conversion of 7-chlorocarbonylmethyltheophylline to the 7-benzoylmethyl analogue by a Friedel-Crafts reaction in benzene is reported.[58]

B. Reactions

a. *Reduction*

Reducing agents, both catalytic and chemical, have been successfully used. With palladium on charcoal catalyst 6-p-hydroxybenzoylpurine is hydrogenated to the corresponding carbinol, and other examples are known.[24] Aluminium isopropoxide reduces 8-acetylcaffeine to the 8-(1-hydroxyethyl) analogue[42] and 8-bromoacetyl- and 8-(α-bromopropionyl)-caffeine to the respective 8-bromoalkanol derivative.[42]

b. *Other Reactions*

Both 6-p-hydroxybenzoyl- and 6-(2-hydroxy-5-methylbenzoyl)purine form the hydrazone on heating with excess of anhydrous hydrazine. Oxidation of the former ketone with alkaline hydrogen peroxide gives 6-carboxypurine whereas with the same reagent made acid with trifluoracetic or acetic acid hypoxanthine[24] is formed. Bromine in acetic acid gives the 3,5-dibromobenzoyl derivative.[24] Dioxan dibromide converts 8-acetylcaffeine to the 8-bromoacetyl derivative,[42] and the 8-propionyl analogue to the 8-(α-bromopropionyl) form.[42] 8-Bromoacetylcaffeine undergoes reaction with phenylmagnesium bromide to give 8-(α-bromomethyl-α-hydroxy)benzylcaffeine.[42]

7. Purine Thiocyanates (Thiocyanatopurines)

Only a few examples of these compounds, which are of recent introduction, are known. The related cyanates, isocyanates, and isothiocyanates do not seem to have been made.

A. Preparation and Reactions

Under reflux conditions 6-chloropurine in methanol reacts with
potassium thiocyanate* to give 6-thiocyanatopurine (**67**).[59]
This approach appears to be limited in application due to the inactivity of halogen atoms at $C_{(2)}$ and $C_{(8)}$ towards such replacements.
A general route affording 2-, 6-, and 8-thiocyanatopurines is by treating
the appropriate thiopurine in dilute sodium hydroxide at 0° with
cyanogen bromide. Examples so formed include 6-thiocyanato-, 2,6-
dithiocyanato-, 2-amino-6-thiocyanato-, and 6-amino-8-thiocyanato-
purine, also 2-thiocyanatohypoxanthine.[60] The homologous 6-thio-
cyanatomethylpurine (**68**) is obtained[61] from 6-bromomethyl- or, in
better yield, from 6-chloromethylpurine[61] with potassium thiocyanate
in methanol. Other exocyclic thiocyanatopurines have been reported.[62]
Few reactions of thiocyanatopurines have been described; they are
stable in acid or neutral media but with cold 0.1N-sodium hydroxide
rapid formation of the appropriate thiopurine occurs.[59, 60] Warming a
6-thiocyanatopurine with methanolic sodium methoxide gives a mixture
of the 6-thio and 6-methylthiopurine[60] but prolonged heating gives the
thiourethane derivative (**69**), which dilute acid hydrolysis converts to
the 6-alkoxycarbonylpurine (**70**).[63] Unlike the thiocyanatopyrimidines

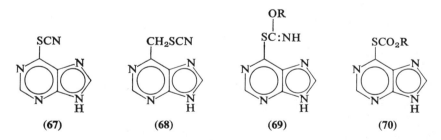

SCN	CH₂SCN	OR | SC:NH	SCO₂R
(**67**)	(**68**)	(**69**)	(**70**)

* It is interesting to note that when the more reactive iodopurine was treated with
potassium or silver cyanate in boiling pyridine or tetrahydrofuran, no reaction was
observed.[40]

which isomerise to the isothiocyanato analogue on treatment with alcohol,[59] the thiocyanatopurines are unchanged when likewise treated. Replacement of thiocyanato groups at $C_{(2)}$, $C_{(6)}$, and $C_{(8)}$ by hydrogen is effected in aqueous solution by heating with Raney nickel.[60]

8. The Purine-N-Carboxylic Acids and Derivatives

In their general properties and reactions they resemble the purine-C-carboxylic acids. The most important derivatives of this class are those of theophylline.

A. Preparation and Reactions

Alkylation of an existing purine is the usual procedure adopted, the acid or ester function being introduced directly by using a halogeno-carboxylic acid or ester or, alternatively, some group is inserted which may be transformed fairly readily to carboxyl group at a later stage of the synthesis. Examples of the former type are shown by adenine[40] which with ethyl chloroformate or ethyl chlorothioformate, in one equivalent of aqueous alkali at room temperature, gives the respective 9-ethoxycarbonyladenine (**71**, R = O or S). The fact that both 1,6-dihydro-6-thio-[63a] and 7,8-dihydro-8-thiopurine[63b] also give 9-ethoxycarbonyl derivatives, in dimethylformamide containing anhydrous carbonate, is noteworthy as the products were originally formulated as the isomeric 6- and 8-S-acylated purines (see Sect. 9Ba). Analogous $N_{(9)}$-acids have been formed similarly from 6- and 8-methylthiopurine.[64] Benzyl chloroformate with 6-amino-2-methylthiopurine gives 6-amino-9-benzylcarbonyl-2-methylthiopurine.[64] Theophylline, as the sodium salt, in benzene, affords the 7-ethoxycarbonyl derivative.[65] These ester groups, being directly attached to ring nitrogen atoms, are rapidly cleaved in alkali and the parent purine is regenerated. More stable homologues are obtained with chloroacetic esters, the 9-ethoxycarbonyl-methyl derivative, from 6-chloropurine, can be saponified with barium hydroxide solution at room temperature to 6-chloro-9-carboxymethyl-purine.[66] An ethoxycarbonylethyl group can be directly inserted at $C_{(9)}$ in adenine by a Michael reaction using ethyl acrylate in benzene.[67] Alkylation of theophylline with chloroacetic acid and related halogeno acids[68, 69, 70] and halogeno esters afford the corresponding 7-carboxy-alkyl- or 7-alkoxycarbonylalkyltheophyllines. Esters used include those of α-bromo-[71] and β-bromopropionic acid[65] and bromomalonic acid.[72] Analogous 7-carboxyalkyl derivatives of 8-bromotheophylline[68, 72] and 3-methylxanthine[69] are known. With 8-thiotheophylline and ethyl chloroacetate, two ester groups are introduced through concurrent $N_{(7)}$- and S-alkylation occurring.[73] A Michael addition between theophylline and ethyl acrylate provides an alternative route to 7-ethoxy-carbonylethyltheophylline (**72**, R = Et), the reaction taking place in dioxan containing benzyltrimethylammonium hydroxide as basic catalyst.[74]

Examples of alkylation with compounds having potential acid function are the formation of 7-carboxyethyltheophylline (**72**, R = H) from theophylline either by alkylation with 3-chloropropanol followed by oxidation of the 7-(3-hydroxypropyl)purine with acidified potassium dichromate[65] or by a Michael addition using acrylonitrile and subsequent acid hydrolysis of the product (**73**).[74] Other reactions with

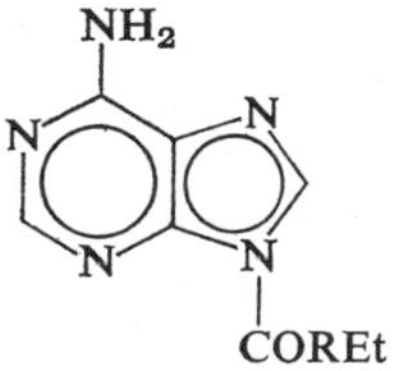

(71) 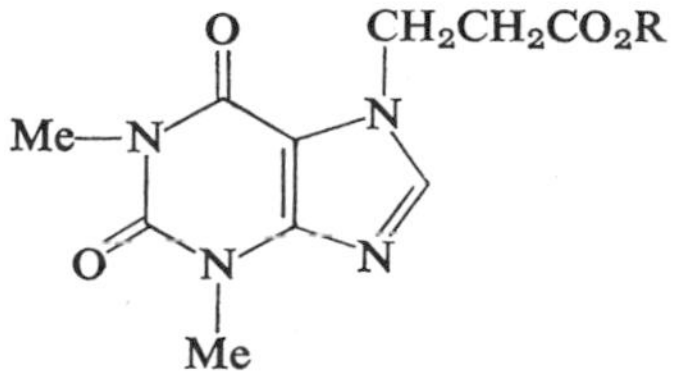   (72)

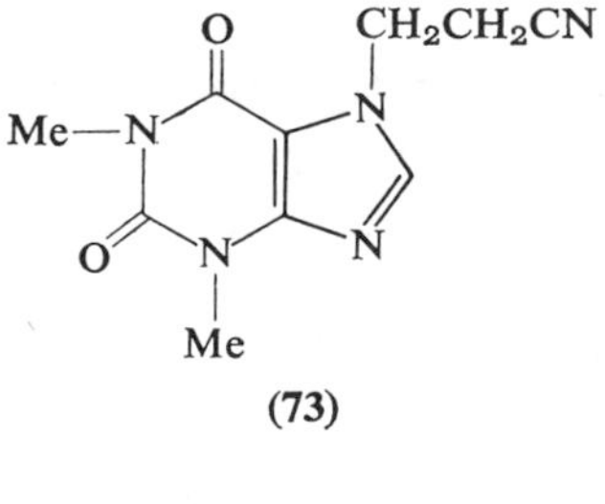

(73) 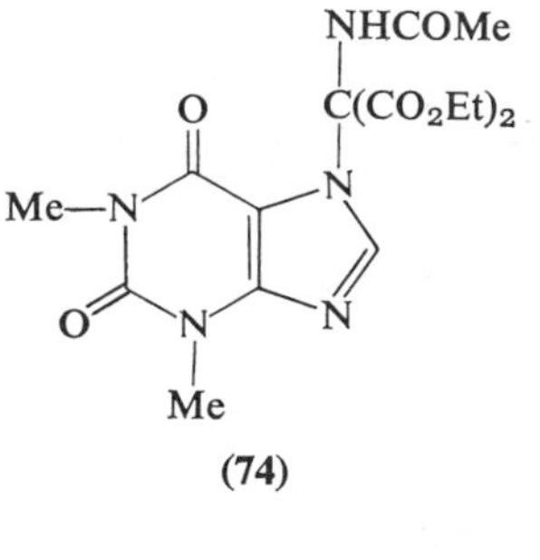   (74)

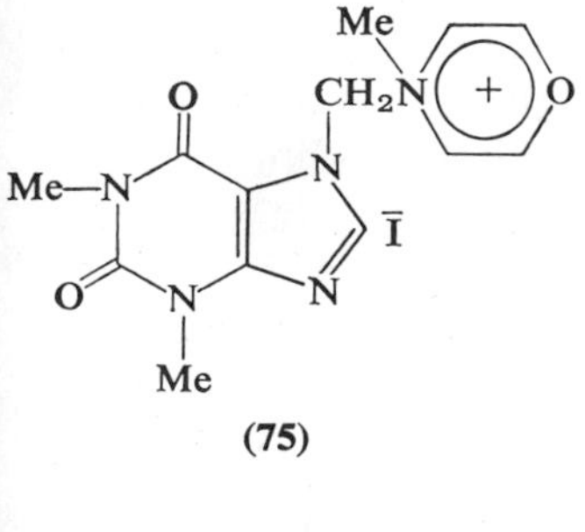

(75) 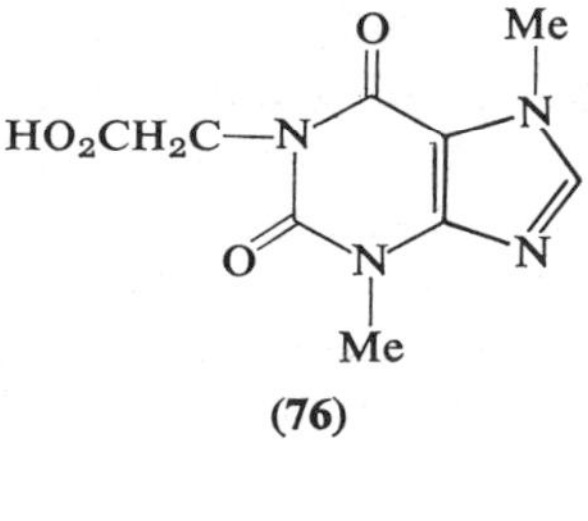   (76) 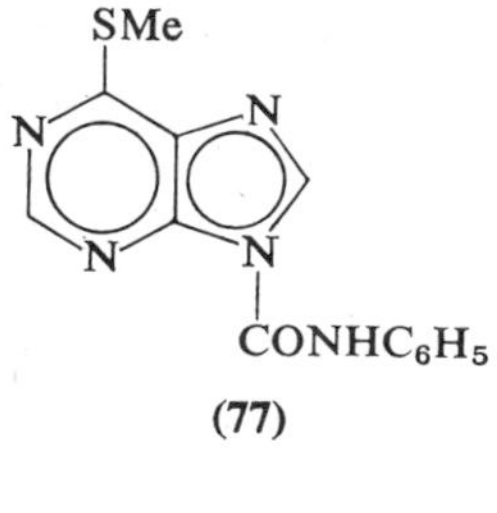   (77)

acrylonitrile leading to 9-carboxyethyl-6-substituted-purines are shown by adenine[67] and 6-chloropurine.[75]

The $N_{(7)}$-dicarboxylic ester (74) is formed on treating the Mannich base, 7-morpholinomethyltheophylline methiodide (75), with acetamidomalonic ester in ethanolic sodium methoxide at room temperature.[76]

Theobromine alkylates at $N_{(1)}$, forming 1-carboxymethyltheobromine (76) with chloroacetic acid.[69, 77] With acrylonitrile a Michael condensation affords the 1-cyanoethyl derivative but the analogous condensation with ethyl acrylate does not seem to take place.[74]

The above acids and esters show reactions typical of C-carboxylic acids, examples of conversion to acid chlorides with thionyl chloride,[65] esterification with ethanolic hydrogen chloride,[65, 74] and formation of amides[65, 66, 72] and hydrazides[66] from the ester or acid chloride are known.[77, 78] On heating with acetic anhydride, under reflux, 7-carboxymethyltheophylline undergoes the Dakin-West reaction affording the 7-acetonyl analogue.[79] No such transformation appears to occur with 1-carboxymethyltheobromine.[79] Both 1-carbamoyltheobromine[77] and 7-carbamoyltheophyllines can be derived also by direct alkylation of the purine with the appropriate halogenoamide.[80]

A route to 9-carbamoylpurines of the type 77 is by condensation of 6-methylthiopurine with phenyl isocyanate in benzene containing triethylamine.[63] It should be noted that if amino groups are present, e.g., as in adenine or guanine, these will interact with the reagent also (Ch. VIII, Sect. 5G).

9. The Purinylamino and Purinylthio Acids

A. The Purinylamino Acids (and Esters)

a. Preparation and Reactions

Purines of this type are almost invariably derived by condensation of a halogenopurine with an amino acid at controlled pH. Few examples with the amino acid moiety in the 2-position are to be found. 2-Chlorohypoxanthine has reacted with alanine under vigorous conditions[81] but milder treatment is effective if derivatives of 2-fluoropurine are used.[81] A wide range of amino acids react with 6-chloropurine[82–85] and its derivatives.[86, 87] A novel route, possibly of more general application, for formation of 6-carboxyaminopurines, involves an intramolecular conversion of 6-aminoacetylaminopurine (78) to 6-carboxymethylaminopurine (79) brought about on boiling an aqueous solution for a short time.[88] The methyl and ethyl esters (81, R = Me and Et) result through rearrangement of 6-azidocarbonylpurine (80) on heating

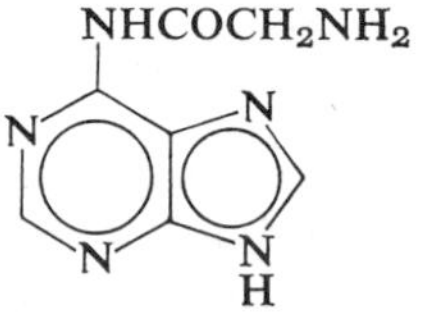

(78) 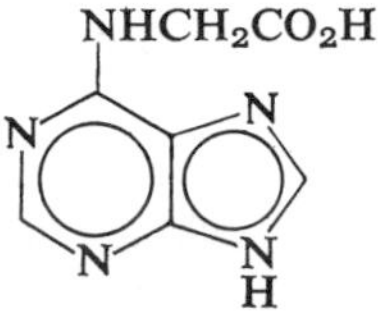   (79) 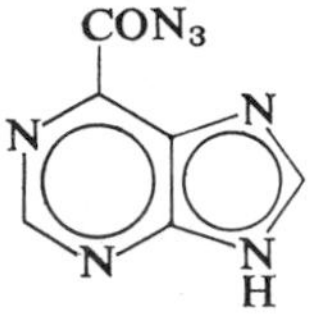   (80)

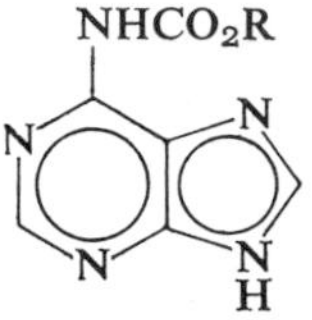

(81) 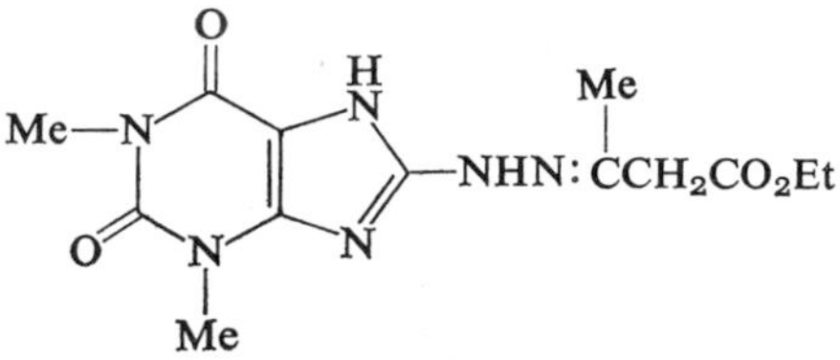   (82)

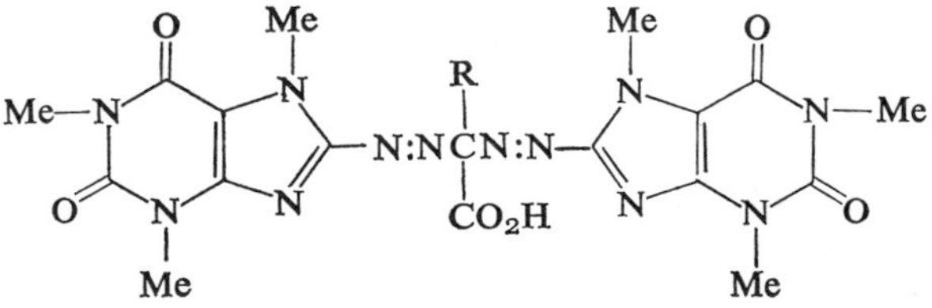

(83)

briefly with methanol or ethanol.[35] Unsaturated acids of the type 82 are rapidly formed on condensation of ethyl acetoacetate with the 8-hydra-zino derivatives of theophylline, theobromine, caffeine, and 1,7-dimethyl-xanthine in boiling ethanol.[89] The somewhat related bis-purinyl acids (83, R = Pr or benzyl) are produced on treating diazotised 8-amino-caffeine with the potassium salt of the appropriate α-alkyl derivative of ethyl acetoacetate.[90]

B. The Purinylthio Acids (and Esters)

a. Preparation

Unlike an amino group a thio group will react fairly readily with halogenocarboxylic acids or esters although with the simplest exam-ples of the latter, i.e., alkyl chloroformates, anomalous behaviour is encountered. Thus, both 1,6-dihydro-6-thio- and 7,8-dihydro-8-thio-purine with an alkyl chloroformate give $N_{(9)}$-alkoxycarbonylthiopurines (see Sect. 8) rather than the corresponding 6- or 8-S-carboxy analogue as was originally thought.[63, 91] In purines lacking an ionisable proton in the imidazole ring, however, S-carboxylation by these reagents does occur, seen in the formation of **84** from 1,6-dihydro-9-methyl-6-thiopurine.[63] These results are explicable if in all the above cases an S-carboxy derivative is assumed to be formed initially which then undergoes transacylation from sulphur to a ring nitrogen if the latter carries an acidic proton.[63a] Two examples supporting this are the formation of 9-ethoxycarbonyl-1,6-dihydro-6-thiopurine (**85**) either by treating 6-chloropurine with potassium thiolacetate[63a] or by heating 6-thiocyanatopurine with sodium ethoxide followed by prolonged hydrolysis (48 h) in dilute mineral acid.[63*]

Chloro- or bromoacetic acids will condense with thiopurines in water but aqueous sodium hydroxide or sodium carbonate are better reaction media. This route provides carboxymethylthio derivatives of 2-,[92] 6-,[59, 93, 94] and 8-thiodihydropurines,[92] also of 1,2,3,6-tetrahydro-6-oxo-2-thio-,[95] 2-amino-1,6-dihydro-6-thio-[96] (thioguanine), and 1,6-dihydro-7-(and 9-)methyl-6-thiopurines.[97] The 8-thio analogues of xanthine, 3-phenylxanthine,[98] 9-alkylxanthines,[99] theophylline,[100] theobromine,[100] caffeine,[100] among others,[101, 102] form the corresponding acids. Both sulphur groups in 1,6,7,8-tetrahydro-6,8-dithiopurine[103] and in 2,8-dithiohypoxanthine[95] can be converted to carboxymethylthio groups. In place of chloroacetic acid chloroacetonitrile can be used[104] to form carboxymethylthiopurines directly. Homologues of chloroacetic acid are frequently employed in the form of the ethyl ester, the 6-alkoxycarbonylalkylpurine produced is hydrolysed to the acid form with aqueous potassium hydroxide. 6-Carboxyethylthio-, 6-carboxypropylthio-, 6-carboxybutylthiopurines[105] and others[106] are formed by this procedure. Appropriate carboxybutylthiopurines are obtained from 2-amino-6-thio-,[96, 107] 2-oxo-6-thio-,[107] 2-methyl-6-thio-,[107] 8-methyl-6-thio-,[107] 9-alkyl-6-thio-,[108] and 6-oxo-2-thiopurine derivatives.[107]

It should be noted that instances are known where N-alkylation in addition to S-alkylation has occurred giving rise to diester derivatives.[63a, 73]

The halogenoalkylamides[96, 107] can be used to prepare the amides directly as an alternative to ammonia or amine treatment of the 6-alkoxycarbonylalkylthiopurine.

* In the original paper [63] the product is incorrectly given as 6-ethoxycarbonylthio-purine.

The other synthesis of carboxyalkylthiopurines, by reaction between halogenopurine and mercapto acid, has been less exploited and generally yields are inferior. 6-Chloropurine and mercaptosuccinic acid react in aqueous dimethylformamide, at 25° containing potassium hydrogen carbonate but the product (86) is better obtained using bromosuccinic ester and 1,6-dihydro-6-thiopurine.[109] The formation of 6-o-carboxy-phenylthiopurine (87), from 6-chloropurine and o-thiobenzoic acid in dimethyl formamide at 50° in the presence of potassium carbonate[110]

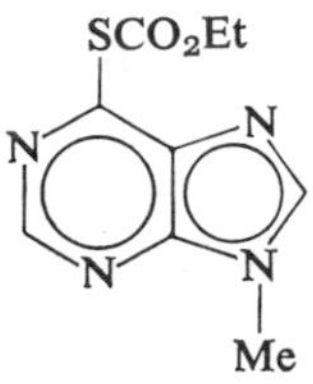

(84) 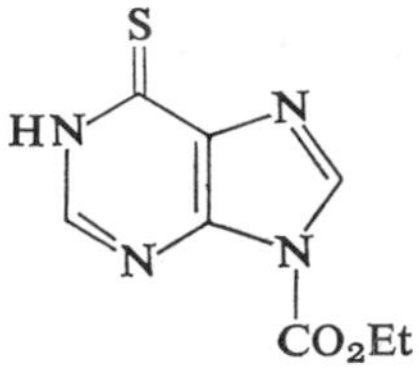 (85) 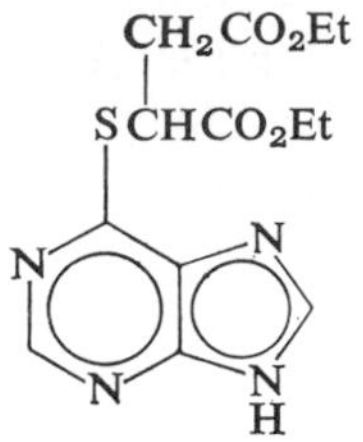 (86) 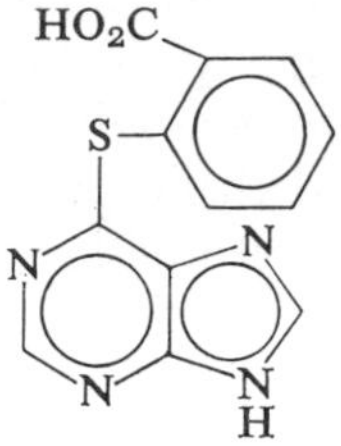 (87)

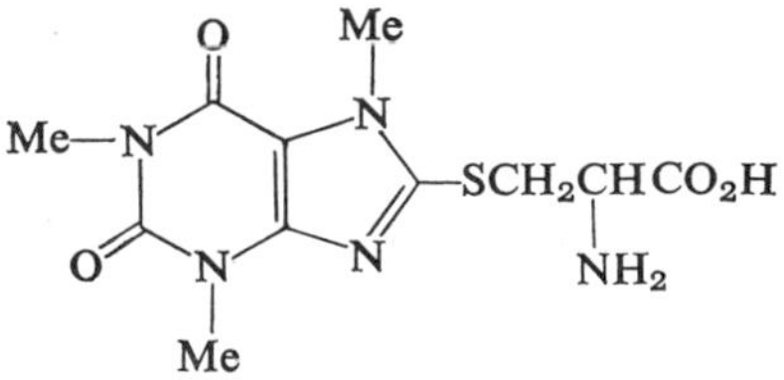

(88) 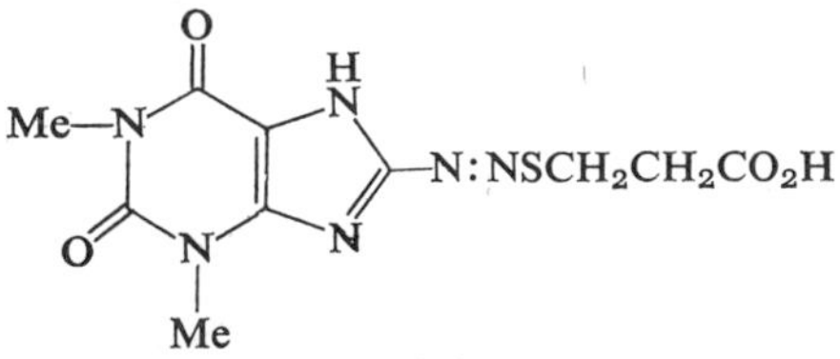 (89)

or of 8-(2-amino-2-carboxyethylthio)caffeine (88) from 8-chlorocaffeine and cysteine in aqueous sodium hydroxide are other examples.[111] The unusual acid stability shown by 6-amino-2-carboxymethylthiopurine enables its synthesis to be made by ring closure of 4,5,6-triamino-2-carboxymethylthiopyrimidine with formic acid.[112]

The 8-carboxyethylthioazopurine (89) results by interaction of the diazotisation product of 8-aminotheophylline with 3-mercaptopropionic acid in acetone at room temperature.[113]

b. Reactions

The acid function shows many reactions typical of simple C-carboxylic acids. Examples of the formation of esters,[93, 101] acid chlorides,[107] and amides[107] are known. With many nucleophilic reagents the carboxyalkylthio group can be displaced under conditions similar to those used to replace a methylthio group. Thus, 6-carboxymethylthiopurine reacts with aliphatic or aromatic amines under reflux giving the appropriate 6-substituted-aminopurine.[93] Ammonium hydroxide in a sealed tube at 140° effectively converts 6-carboxymethylthio-7-methylpurine to 7-methyladenine.[108] Replacement of an 8-carboxymethylthio group in this way has given only a poor yield of the 8-aminopurine.[92] Although mineral acid frequently hydrolyses a carboxymethylthio group to an oxo function,[95, 100] cases of resistance to this are known.[92, 98] In acetic anhydride, under reflux conditions, the 8-carboxymethylthiopurine (90) undergoes cyclodehydration forming the dihydrothiazolo[2,3-*f*]purine derivative[101] (91) [RRI 2341]. Under parallel conditions 6-(1-carboxyethylthio)purine (92) undergoes the Dakin-West reaction affording the 6-ketonylthiopurine (93).[114]

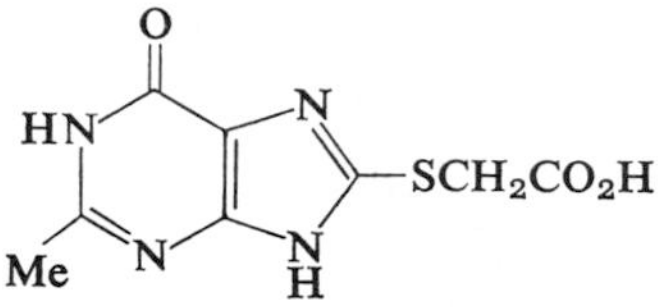

(90) 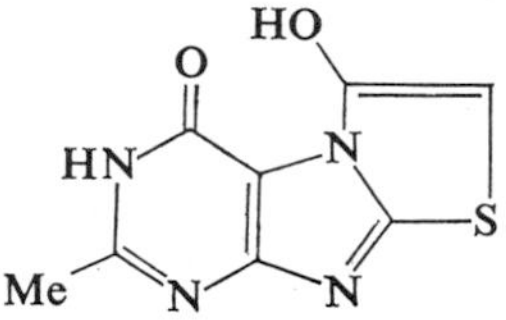   (91)

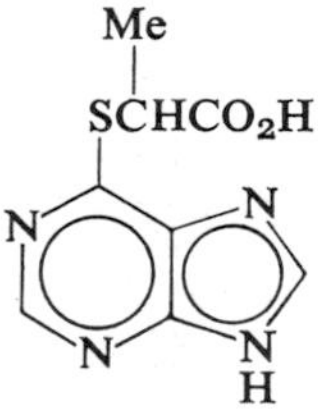

(92) 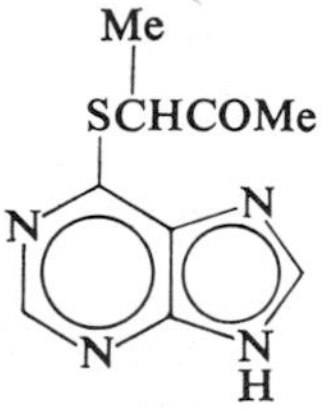   (93)

References

1. Mackay and Hitchings, *J. Amer. Chem. Soc.*, **78**, 3511 (1956).
2. Hitchings, Elion, and Mackay, U.S. Pat., 3,098,074 (1963); through *Chem. Abstracts*, **60**, 1771 (1964).
3. Cohen, Thom, and Bendich, *J. Org. Chem.*, **27**, 3545 (1962).
4. Lettré and Woenckhaus, *Annalen*, **649**, 131 (1961).
5. Gomberg, *Amer. J. Chem.*, **17**, 403 (1895).
6. Gomberg, *Amer. J. Chem.*, **14**, 611 (1892).
7. Boehringer, Ger. Pat., 153,121 (1902); *Frdl.*, **7**, 674 (1902–1904).
8. Chaman, Cherkasova, and Golovchinskaya, *Zhur. obshchei Khim.*, **30**, 1878 (1960).
9. Golovchinskaya, Fedosova, and Cherkasova, *Zhur. priklad. Khim.*, **31**, 1241 (1958).

10. Clark and Ramage, *J. Chem. Soc.*, **1958**, 2821.
11. Golovchinskaya, *Sbornik Statei obshchei Khim.*, *Akad. Nauk. S.S.S.R.*, **1**, 692 (1953).
12. Bargioni, *Boll. Chim.-Farm.*, **64**, 869 (1935); through *Chem. Abstracts*, **30**, 2320 (1935).
13. Golovchinskaya, *Sbornik. Statei obshchei Khim.*, *Akad. Nauk. S.S.S.R.*, **1**, 702 (1953).
14. Golovchinskaya and Chaman, *Zhur. obshchei Khim.*, **30**, 3628 (1960).
15. Golovchinskaya and Chaman, *Zhur. obshchei Khim.*, **32**, 3245 (1962).
16. Ebed, Chaman, and Golovchinskaya, *Zhur. obshchei Khim.*, **36**, 816 (1966).
17. Nikolaeva and Golovchinskaya, *Zhur. obshchei Khim.*, **34**, 1137 (1964).
18. Ovcharova and Golovchinskaya, *Zhur. obshchei Khim.*, **34**, 3254 (1964).
19. Chaman and Golovchinskaya, *Zhur. obshchei Khim.*, **33**, 3342 (1963).
20. Lettré, Ballweg, Maurer, and Rehberger, *Naturwiss.*, **50**, 224 (1963).
21. Giner-Sorolla, *Chem. Ber.*, **101**, 611 (1968).
22. Giner-Sorolla, Zimmerman, and Bendich, *J. Amer. Chem. Soc.*, **81**, 2515 (1959).
23. Giner-Sorolla, Thom, and Bendich, *J. Org. Chem.*, **29**, 3209 (1964).
24. Cohen, Thom, and Bendich, *J. Org. Chem.*, **28**, 1379 (1963).
25. Albert, *J. Chem. Soc.*, **1960**, 4705.
26. Bredereck and Föhlisch, *Chem. Ber.*, **95**, 414 (1962).
27. Chaman and Golovchinskaya, *Zhur. obshchei Khim.*, **31**, 2645 (1961).
28. Golovchinskaya, *Zhur. obshchei Khim.*, **18**, 2129 (1948).
28a. Giner-Sorolla and Brown, *J. Chem. Soc. (C)*, **1971**, 126
29. Traube, Schottländer, Goslich, Peter, Meyer, Schlüter, Steinbach, and Bredow, *Annalen*, **432**, 266 (1923).
30. Bayer, Ger. Pat., 213,711 (1908); *Frdl.*, **9**, 1010 (1908–1910).
31. Albert, *Quart. Rev.*, **6**, 225 (1952).
32. Clark and Lister, *J. Chem. Soc.*, **1961**, 5048.
33. Clark, Kernick, and Layton, *J. Chem. Soc.*, **1964**, 3221.
34. Golovchinskaya and Chaman, *Zhur. obshchei Khim.*, **22**, 528 (1952).
35. Giner-Sorolla and Bendich, *J. Amer. Chem. Soc.*, **80**, 3932 (1958).
36. Golovchinskaya and Chaman, *Zhur. obshchei Khim.*, **22**, 2220 (1952).
37. Woenckhaus and Stock, *Z. Naturforsch.*, **20B**, 400 (1965).
38. Golovchinskaya and Chaman, *Zhur. obshchei Khim.*, **22**, 2225 (1952).
39. Cohen, Thom, and Bendich, *Biochemistry*, **2**, 176 (1963).
40. Dyer, Reitz, and Farris, *J. Medicin. Chem.*, **6**, 289 (1963).
41. Hitchings, Elion, and Mackay, U.S. Pat., 3,128,274 (1964); through *Chem. Abstracts*, **60**, 14523 (1964).
42. Ehrhart and Hennig, *Arch. Pharm.*, **289**, 453 (1956).
43. Golovchinskaya, Kolganova, Nickolaeva, and Chaman, *Zhur. obshchei Khim.*, **33**, 1650 (1963).
44. Golovchinskaya and Chaman, *Zhur. obshchei Khim.*, **22**, 535 (1952).
45. Ehrhart, *Arch. Pharm.*, **290**, 16 (1957).
46. Hull, *J. Chem. Soc.*, **1958**, 4069.
47. Plieninger and Werst, *Chem. Ber.*, **88**, 1956 (1955).
48. Schulze and Willitzer, *J. prakt. Chem.*, **33**, 50 (1966).
49. Bredereck, Siegel, and Föhlisch, *Chem. Ber.*, **95**, 403 (1962).
50. Chaman and Golovchinskaya, *Zhur. obshchei Khim.*, **32**, 2015 (1962).
51. Lugovkin, *Zhur. obshchei Khim.*, **30**, 2427 (1960).
52. Giner-Sorolla and Bendich, *J. Org. Chem.*, **31**, 4239 (1966).

53. Koppel, Springer, Robins, Schneider, and Cheng, *J. Org. Chem.*, **27**, 2173 (1962).
54. Lugovkin, *Zhur. obshchei Khim.*, **33**, 3205 (1963).
55. Lugovkin, *Zhur. obshchei Khim.*, **32**, 452 (1962).
56. Lugovkin, *Zhur. obshchei Khim.*, **33**, 2942 (1963).
57. Lugovkin, *Zhur. obshchei Khim.*, **31**, 3406 (1961).
58. Klosa, *Arch. Pharm.*, **288**, 301 (1955).
59. Elion, Goodman, Lange, and Hitchings, *J. Amer. Chem. Soc.*, **81**, 1898 (1959).
60. Saneyoshi and Chihari, *Chem. and Pharm. Bull. (Japan)*, **15**, 909 (1967).
61. Giner-Sorolla and Bendich, *J. Medicin. Chem.*, **8**, 667 (1965).
62. Eckstein and Sulko, *Dissertationes Pharm. (Poland)*, **13**, 97 (1961); through *Chem. Abstracts*, **55**, 23548 (1961).
63. Dyer and Bender, *J. Medicin. Chem.*, **7**, 10 (1964).
63a. Dyer, Farris, Minnier, and Tokizawa, *J. Org. Chem.*, **34**, 973 (1969).
63b. Dyer and Minnier, *J. Heterocyclic Chem.*, **6**, 23 (1969).
64. Blackburn and Johnson, *J. Chem. Soc.*, **1960**, 4347.
65. Cacace, Fabrizi, and Zifferero, *Ann. Chim. (Italy)*, **45**, 983 (1955).
66. Montgomery and Temple, *J. Amer. Chem. Soc.*, **83**, 630 (1961).
67. Lira and Huffmann, *J. Org. Chem.*, **31**, 2188 (1966).
68. Cacace, Criserà, and Zifferero, *Ann. Chim. (Italy)*, **46**, 99 (1956).
69. Merck, Wolfes, and Kornick, Ger. Pat., 352,980; through *Chem. Abstracts*, **17**, 1306 (1923).
70. Baisse, *Bull. Soc. chim. France*, **1949**, 769.
71. Morishita, Nakano, Satoda, Yoshida, and Fukuda, Jap. Pat., 6473 (1958), through *Chem. Abstracts*, **54**, 1571 (1960).
72. Aliprandi, Cacace, and Montifinale, *Farmaco, Ed. Sci.*, **12**, 751 (1957).
73. Cacace and Masironi, *Ann. Chim. (Italy)*, **46**, 806 (1956).
74. Polonovski, Pesson, and Zelnick, *Compt. rend.*, **241**, 215 (1955).
75. Baker and Tanna, *J. Org. Chem.*, **30**, 2857 (1965).
76. Okuda, *J. Pharm. Soc. Japan*, **80**, 205 (1960).
77. Cacace and Zifferero, *Ann. Chim. (Italy)*, **45**, 1026 (1955).
78. Klosa, *J. prakt. Chem.*, **12**, 212 (1961).
79. McMillan and Wuest, *J. Amer. Chem. Soc.*, **75**, 1998 (1953).
80. Weissenberger, *Arch. Pharm.*, **288**, 532 (1955).
81. Gerster and Robins, *J. Amer. Chem. Soc.*, **87**, 3752 (1965).
82. Clark, *J. Biol. Chem.*, **223**, 139 (1956).
83. Lettré and Ballweg, *Annalen*, **633**, 171 (1960).
84. Ward, Wade, Walborg, and Osdene, *J. Org. Chem.*, **26**, 5000 (1961).
85. Tretyakova, Kapran, and Nedelkina, *Khim. geterotsikl Soedinenii*, **1967**, 170.
86. Ballio and DiVittorio, *Gazzetta*, **90**, 501 (1960).
87. Baddiley, Buchanan, and Stephenson, *Arch. Biochem. Biophys.*, **83**, 54 (1959).
88. Chheda and Hall, *Biochemistry*, **5**, 2082 (1966).
89. Priewe and Poljack, *Chem. Ber.*, **88**, 1932 (1955).
90. Gomberg, *Amer. J. Chem.*, **23**, 60 (1901).
91. Farris, *Diss. Abstracts*, **25**, 3265 (1964).
92. Albert and Brown, *J. Chem. Soc.*, **1954**, 2060.
93. Huber, *Angew. Chem.*, **68**, 706 (1956).
94. Skinner, Claybrook, Ross, and Shive, *J. Org. Chem.*, **23**, 1223 (1958).
95. Johns and Hogan, *J. Biol. Chem.*, **14**, 299 (1913).
96. Daves, Noell, Robins, Koppel, and Beaman, *J. Amer. Chem. Soc.*, **82**, 2633 (1960).
97. Elion, *J. Org. Chem.*, **27**, 2478 (1962).

98. Kishikawa and Yuki, *Chem. and Pharm. Bull.* (*Japan*), **14**, 1365 (1966).
99. Biltz and Sauer, *Ber.*, **64**, 752 (1931).
100. Johns and Baumann, *J. Biol. Chem.*, **15**, 515 (1913).
101. Elderfield and Prasad, *J. Org. Chem.*, **24**, 1410 (1959).
102. Blicke and Schaaf, *J. Amer. Chem. Soc.*, **78**, 5857 (1956).
103. Ishidate and Yuki, *Chem. and Pharm. Bull.* (*Japan*), **5**, 244 (1957).
104. Gordon, *J. Amer. Chem. Soc.*, **73**, 984 (1951).
105. Sermonský, Černý, and Jelinek, *Coll. Czech. Chem. Comm.*, **25**, 1091 (1960); **33**, 3823 (1968).
106. Skinner, Shive, Ham, Fitzgerald, and Eakin, *J. Amer. Chem. Soc.*, **78**, 5097 (1956).
107. Černý, Sermonský, and Jelinek, *Coll. Czech. Chem. Comm.*, **27**, 57 (1962).
108. Kotva, Černý, Semonský, Vachek, and Jelinek, *Coll. Czech. Chem. Comm.*, **34**, 2114 (1969).
109. Hampton, *J. Biol. Chem.*, **237**, 529 (1962).
110. Johnston, Holum, and Montgomery, *J. Amer. Chem. Soc.*, **80**, 6265 (1958).
111. Long, *J. Amer. Chem. Soc.*, **69**, 2939 (1947).
112. Bendich, Tinker, and Brown, *J. Amer. Chem. Soc.*, **70**, 3109 (1948).
113. Usbeck, Jones, and Robins, *J. Amer. Chem. Soc.*, **83**, 1113 (1961).
114. Dyer and Minnier, *J. Org. Chem.*, **33**, 880 (1968).

CHAPTER X

Nitro-, Nitroso-, and Arylazopurines

Although preparations of 8-nitro- and 8-arylazopurines were reported over 60 years ago, neither class of derivative has been extensively studied. Both on reduction give 8-aminopurines but such compounds are more usually obtained by other routes. The presence of more than one strong electron-releasing group is necessary before direct introduction of either a nitro or an arylazo group can be made; the nitro group requiring the stronger nucleophilic 8-carbon atom of the two. Although at the time of writing nitrosopurines are represented by a single example, an erroneous claim to the synthesis of a number of 8-nitroso derivatives occurs in the early literature.[1]

1. The Nitropurines

Until very recently only 8-nitropurines were known but a stable 2-nitropurine derivative has now been claimed. The effect of the 8-nitro group on the reactivity of any other substituents present is, due to the paucity of experimental evidence, not fully realised. However, a rough analogy may be drawn with 8-azapurine* derivatives, which show only slight differences in group reactivity toward nucleophilic reagents when compared with the corresponding purines. The versatility of some nitro groups is shown by their replacement by halogen atom or an oxo group or in being reduced to an amino group.

A. Preparation of Nitropurines

Direct nitration is successful only with polysubstituted purines containing two or more strong electron-releasing groups. In the absence of

* In these purine analogues, the nitrogen atom, which replaces the methine group at the 8-position, is considered to exert a similar electron-demanding effect on the ring system as a nitro group located at $C_{(8)}$.

these, an 8-nitropurine can sometimes arise through conversion of the 8-diazopurine. No route using an intermediate containing a nitro group appears to be recorded.

a. *By Nitration*

The earliest preparations utilised direct action of nitric acid on the purine, without solvent; the 8-nitro derivatives of caffeine (**1**, R = R' = Me)[2] and theobromine (**1**, R = H, R' = Me)[3, 4] are obtained on evaporation of the reaction[5] mixture. Although theophylline nitrates[6] under these conditions the reaction, in most cases, is best carried out in acetic acid.[7] Modifications of this procedure have been made.[5, 8] One approach to 8-nitrotheophylline (**1**, R = Me, R' = H)[9]

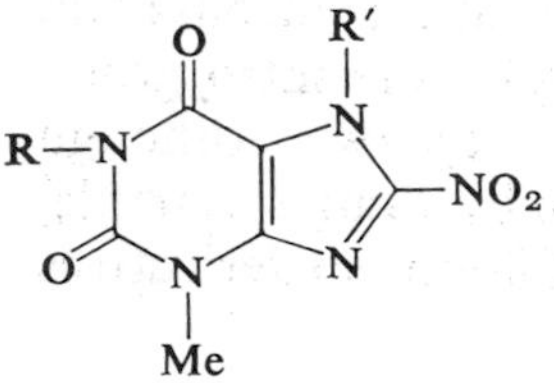

(1) 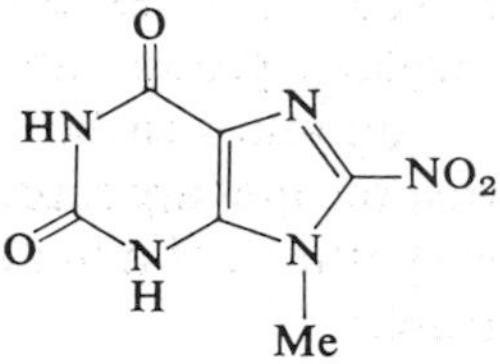   (2) 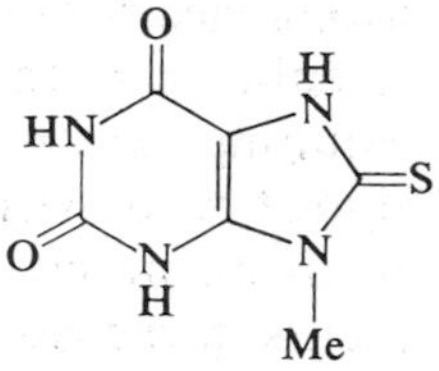   (3)

uses fuming nitric acid in aqueous ethanol as reaction medium, this also being used for the corresponding 7-benzyl (**1**, R = Me, R' = CH$_2$C$_6$H$_5$)[9] and 7-*p*-cyanobenzyl (**1**, R = Me, R' = CH$_2$·C$_6$H$_4$·CN)[9] derivatives. Although the 8-carbon atom in xanthine is not sufficiently nucleophilic to react with a nitronium ion, the 9-methyl homologue is converted to 9-methyl-8-nitroxanthine (**2**) on heating in 50% nitric acid.[10, 11] The same nitropurine can also be derived by sulphur replacement in the 8-thio analogue (**3**). For this a mixture of nitric acid and sodium nitrite[1, 11] was originally employed as nitrating reagent but later 50% nitric acid was found more suitable.[10] The course of this reaction appears to proceed by an initial conversion of the sulphur group to a higher oxidation state, thus facilitating its removal, this being followed by nitration of the now unsubstituted 8-position. The oxidative removal of thio groups from purines by means of nitric acid is well known[12, 13] (Ch. VII, Sect. 1Ca), and the occurrence of some 9-methyl-xanthine along with the above 8-nitropurine (**2**) points towards this mechanism operating. Corresponding 8-nitropurines have been obtained from a variety of 9-alkyl-8-thiopurines.[1]*

---

* In the original paper[1] the products were thought to be 8-nitrosopurines, this assumption being based largely on the analytical data. A recent reexamination[10] of

A novel route giving 8-nitroxanthine is by rearrangement of 3-acetoxyxanthine in aqueous sodium nitrite at ambient temperature.[13a] This and other related rearrangements are discussed in Chapter XI, Section 3C.

b. From Aminopurines

This approach provides a means of forming nitropurines which, due to the absence of strong electron-releasing groups, cannot be derived by direct nitration procedures. Diazotisation of the aminopurine is followed by treatment with neutral aqueous sodium nitrite solution causing displacement of the diazonium group by nitrite ion. In this way 8-nitro-(**4**) is obtained from 8-amino-hypoxanthine (**5**)[10] as are also the 8-nitro derivatives of xanthine,[10] guanine,[10] and theophylline.[10] Although 6,8-diaminopurine underwent diazotisation to 8-diazoadenine, the conversion to 8-nitroadenine is not reported.[10]

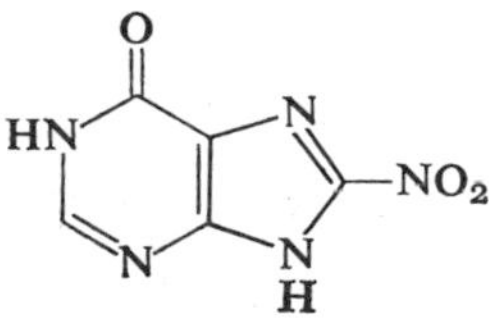

$$R = \beta\text{-D-ribofuranosyl}$$

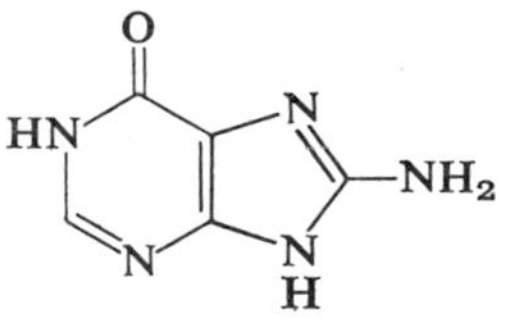

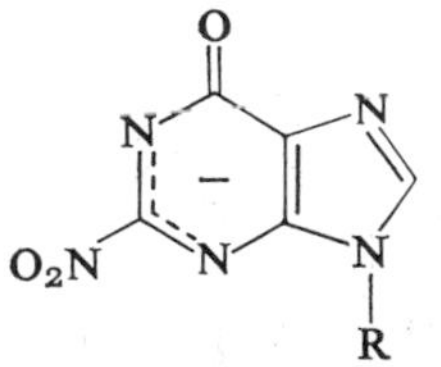

(**4**)   (**5**)   (**6**)

A unique example of the formation of a 2-nitropurine is claimed[14] in 5% yield on treating guanosine with an excess of sodium nitrite solution in acetate buffer at 0°, the product (**6**) being isolated as the ammonium salt.

B. Reduction of Nitropurines

Nitropurines are readily converted to the analogous aminopurine by most of the usual reducing agents. Stannous chloride with 8-nitro-theophylline[15] gives a 70% yield of the 8-amino derivative (**7**), other

---

these preparations, however, has shown that the products, in fact, were 8-nitropurines but which, due to contamination with 8-unsubstituted-purines arising through oxidative desulphurisation of the 8-thiopurines, gave analysis figures approximating those of the 8-nitroso analogues.

suitable reagents are sodium hydrosulphide[10] and hydriodic acid.[5] The isomeric 8-aminotheobromine follows reduction of nitrotheobromine with ammonium sulphide.[3] Sodium hydrosulphite reductions are the most suitable for alkali-soluble derivatives, as in the derivation of the respective 8-aminopurine from 8-nitrohypoxanthine and 9-methyl-8-nitroxanthine.[10] Although 8-aminocaffeine has been reported to be formed by a sodium amalgam reduction of the 8-nitro analogue,[3] later workers suggest that under these conditions a reductive hydrolysis occurs giving 1,3,7-trimethyluric acid and ammonia as products.[16] Sodium hydrosulphite effectively converts the ammonium salt of 2-nitroinosine (6) to guanosine.[14]

C. Other Reactions of Nitropurines

Only a limited range of reactions has been studied. An 8-nitro group in methylated xanthines is readily replaced by nucleophilic reagents, 8-nitrotheophylline (8) on heating with strong hydrochloric acid gives a

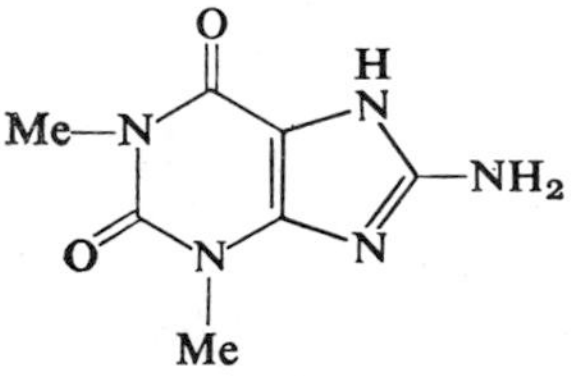

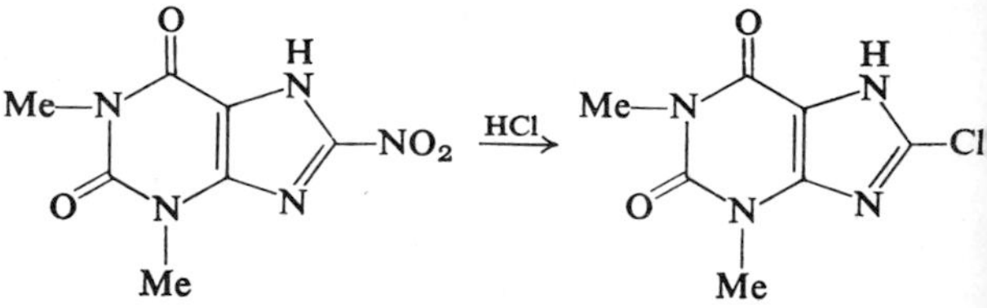

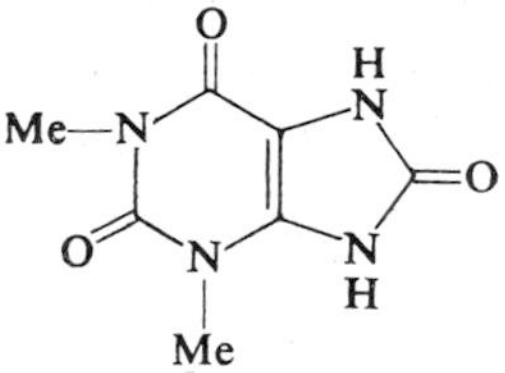

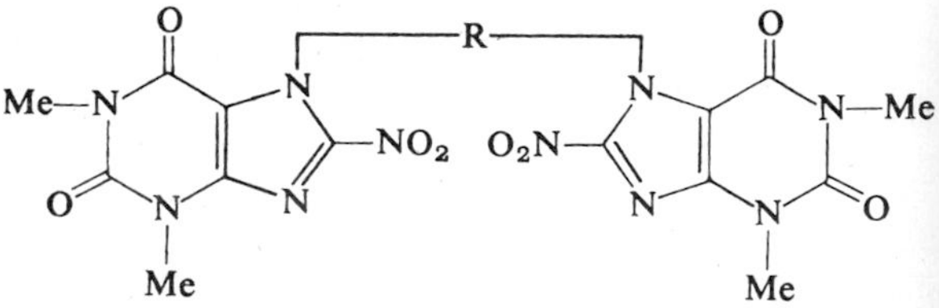

good yield of 8-chlorotheophylline (9),[5] while with 48% hydrobromic acid the 8-bromo derivative results. In boiling dilute sulphuric acid hydrolysis is reported to give 1,3-dimethyluric acid (10)[5] as product.

8-Nitrotheophylline readily forms an alkyl derivative in boiling aqueous sodium bicarbonate, the respective $N_{(7)}$-acetonyl[17] and -benzyl[9] derivatives being produced with chloroacetone and benzyl chloride, respectively. The 1,3-di(theophyllin-7-yl)propanol (**11**) arises with α,γ-dichlorohydrin likewise.[18]

2. The Nitrosopurines

A. Preparation, Properties, and Reactions

Oxidation of 6-hydroxyaminopurine (**12**) in neutral aqueous solution with activated manganese dioxide[19] or in dilute sulphuric acid by potassium dichromate[20] affords 6-nitrosopurine (**13**) as an orange-red crystalline solid, melting explosively at 195°. Reduction to adenine is readily achieved.[19] Examples of *N*-nitrosopurines are given in Chapter VIII, Section 5Fc.

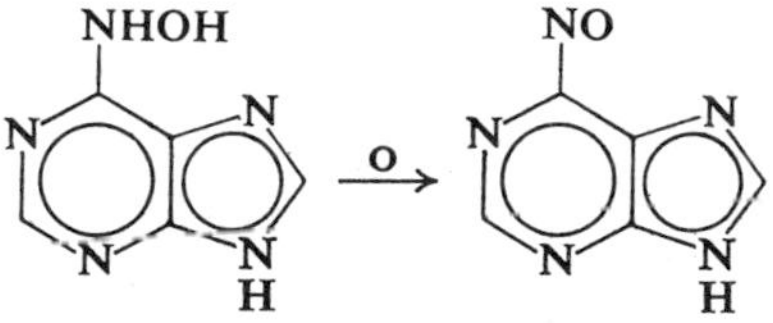

(12)   (13)

3. The Arylazopurines

These derivatives serve the same purpose as their nitro analogues in providing a facile means of inserting an amino group at the 8-position.

A. Preparation of Arylazopurines

The 8-arylazopurines generally result from coupling an 8-unsub-stituted purine with a diazotised aniline derivative in alkaline solution. A few examples of the reverse procedure exist in which an 8-amino-purine is diazotised and coupled with an amine or phenol. Because a weaker nucleophilic 8-position is required than for nitration, less highly substituted purines can be used, such as xanthine, which couples readily but does not nitrate.

a. *By Coupling Reactions*

Adenine[21, 22] and 2-aminopurine[21] do not seem to couple success-fully with diazotised aromatic amines* while the situation with hypo-xanthine is dubious. An early claim[23] to have obtained a product with diazotised *p*-aminobenzenesulphonic acid, although supported by correct analytical data, has been discounted by later workers.[24, 25]

With disubstituted purines containing oxo or amino groups, well-defined 8-arylazo derivatives are obtained exemplified in the coupling of xanthine with the diazonium chlorides of *p*-aminobenzenesulphonic acid,[23] *p*-chloroaniline,[10] and 2,4-dichloroaniline[24, 26] to give, respec-tively, **14**, R = H, R′ = SO$_3$H; **14**, R = H, R′ = Cl; and **14**, R = R′ = Cl. The $N_{(1)}N_{(3)}$-dimethylated homologue, theophylline, reacts likewise with the diazonium chlorides of benzenesulphonic acid,[22] dichloroaniline,[26, 27] and *p*-nitroaniline.[15] With purines *N*-alkylated in the imidazole ring, for example, theobromine or caffeine,[22] coupling does not occur. Although steric hindrance may play some part in inhibiting this reaction a more likely explanation, in the light of a recent theory,[28] is that unlike theophylline, both these purines are prevented from forming an anion because of replacement of the acidic imidazole proton by a methyl group. The moderately electrophilic character shown at C$_{(8)}$ is, therefore, retained in these derivatives but is weakened by the presence of charged forms in the case of theophylline.

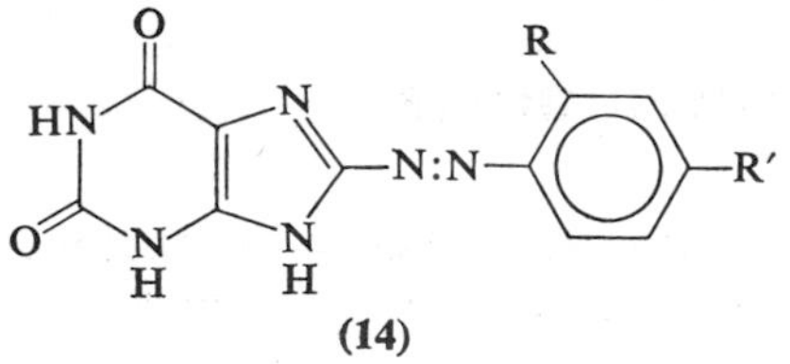

(14) 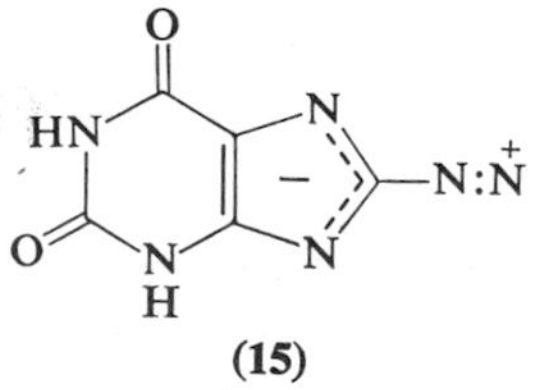   (15)

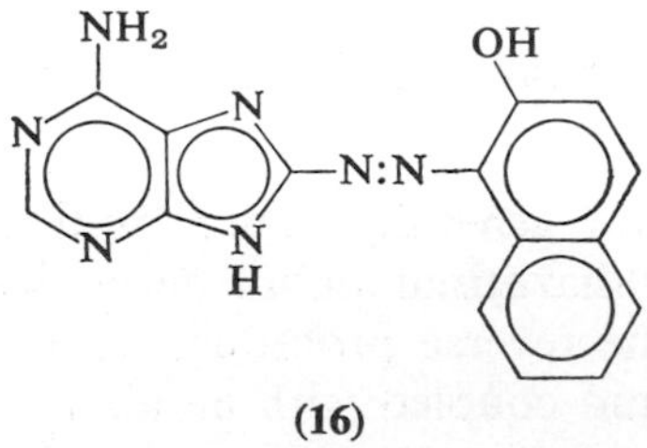

(16)

* Products have been isolated from attempts to couple with adenine. Burian's[22] compound was highly sensitive to alkali, being decomposed on contact with ammonium hydroxide. That obtained by Cavalieri and Bendich[24] was reported to show a spectrum akin to that of 2,6,8-triaminopurine.

Guanine readily forms 8-arylazo derivatives with diazotised forms of p-aminobenzenesulphonic acid[23] and dichloroaniline.[24, 26] The latter reagent also affords the isomeric derivative with isoguanine.[24, 29, 30] In spite of a report to the contrary[21] 2,6-diaminopurine does form 8-arylazo derivatives, which is illustrated by the formation of 2,4-diamino-8-(2,4-dichlorophenylazo)purine.[24] The above procedures form the basis for a commercial production of 8-aminopurines.[27]

b. *Other Methods*

Nitrous acid treatment of 8-aminoxanthine affords the stable 8-diazopurine (**15**), this purine betaine couples with N-phenyldiethanol-amine in cold methanol giving the 8-phenylazoxanthine (**14**, R = H, R′ = N[C$_2$H$_4$OH]$_2$).[31] With 6,8-diaminopurine only the amino group at C$_{(8)}$ is diazotised, the product coupling with β-naphthol forming the 8-naphthylazoadenine (**16**).[32] These examples are of little preparative significance as they represent a reverse of the usual practice whereby 8-arylazopurines are used as intermediates for the preparation of 8-aminopurines.

B. Reduction of Arylazopurines

The alkali-soluble natures of most 8-arylazopurines lend themselves to reduction with sodium hydrosulphite. 8-Aminoxanthine results in this way from either 8-(p-chlorophenylazo)-[10] or 8-(2,4-dichlorophenyl-azo)xanthine.[24, 26] A variety of 8-arylazotheophyllines[15, 26, 27] have been reduced to 8-aminotheophylline. Similarly derived are 6,8-diamino-2,3-dihydro-2-oxopurine from 8-(2,4-dichlorophenylazo)isoguanine[24, 29, 30] and 8-aminoguanine,[24, 26] 8-amino-1,6-dihydro-2-methylamino-6-oxopurine,[10] and 2,6,8-triaminopurine[24] from the appropriate 8-phenyl-azopurine.

References

1. Biltz and Sauer, *Ber.*, **64**, 752 (1931).
2. Schultzen, *Z. Chem.*, **10**, 614 (1867).
3. Brunner and Leins, *Ber.*, **30**, 2584 (1897).
4. Knoll and Co., Ger. Pat., 399,903 (1922); *Frdl.*, **14**, 1322 (1921–1925).
5. Cacace and Masironi, *Ann. Chim. (Italy)*, **47**, 366 (1957).

6. Marquardt and Müller-Erbeling, Ger. Pat., 859,470 (1952); through *Chem. Abstracts*, **47**, 11237 (1953).

7. Duesel, Bermann, and Schachter, *J. Amer. Pharm. Assoc. Sci. Edn.*, **43**, 619 (1954).

8. Morozowich and Bope, *J. Amer. Pharm. Assoc. Sci. Edn.*, **47**, 173 (1958).

9. Serchi, Sancio, and Bichi, *Farmaco, Ed. Sci.*, **10**, 733 (1955).

10. Jones and Robins, *J. Amer. Chem. Soc.*, **82**, 3773 (1960).

11. Biltz and Strüfe, *Annalen*, **423**, 200 (1921).

12. Traube, *Annalen*, **331**, 64 (1904).

13. Traube and Winter, *Arch. Pharm.*, **244**, 11 (1906).

13a. Wölke, Birdsall, and Brown, *Tetrahedron Letters*, **1969**, 785.

14. Shapiro, *J. Amer. Chem. Soc.*, **86**, 2948 (1964).

15. Cacace and Masironi, *Ann. Chim. (Italy)*, **47**, 362 (1957).

16. Biltz and Nachtwey, *J. prakt. Chem.*, **145**, 84 (1936).

17. Serchi and Bichi, *Farmaco, Ed. Sci.*, **11**, 501 (1956).

18. Serchi and Bichi, *Farmaco, Ed. Sci.*, **12**, 594 (1957).

19. Giner-Sorolla, *J. Heterocyclic Chem.*, **7**, 75 (1970)

20. Giner-Sorolla, *Galenica Acta*, **19**, 97 (1966).

21. Albert and Brown, *J. Chem. Soc.*, **1954**, 2060.

22. Burian, *Z. physiol. Chem.*, **51**, 425 (1907).

23. Burian, *Ber.*, **37**, 696 (1904).

24. Cavalieri and Bendich, *J. Amer. Chem. Soc.*, **72**, 2587 (1960).

25. Robins, *J. Amer. Chem. Soc.*, **80**, 6671 (1958).

26. Fischer, *Z. physiol. Chem.*, **60**, 69 (1909).

27. Kalle and Co., Ger. Pat., 230,401 (1909); through *Chem. Abstracts*, **5**, 2733 (1911).

28. Sutcliffe and Robins, *J. Org. Chem.*, **28**, 1662 (1963).

29. Spies and Harris, *J. Amer. Chem. Soc.*, **61**, 351 (1939).

30. Cherbuliez and Bernhard, *Helv. Chim. Acta*, **15**, 464 (1932).

31. Usbeck, Jones, and Robins, *J. Amer. Chem. Soc.*, **83**, 1113 (1961).

32. Burgison, Abstracts of 133rd A.C.S. Meeting, San Francisco, Calif. 1958, p. 14M.

CHAPTER XI

Purine-*N*-oxides

Owing to the comparatively recent commencement of studies (*ca* 1955) of this interesting class of compounds large areas of the chemistry still await investigation. Already a remarkable number of rearrangements have been found to take place and, in addition, an enhanced reactivity towards nucleophilic substitution is often shown by a group which is inert in the parent purine. This latter feature, because of the ready reduction of the oxide to the purine, can be usefully applied in some purine transformations.

Three diagrammatic representations of purine-*N*-oxides are current. Purine-1-oxide, for example, is usually shown either as the donor bond form (**1**) or with the dipolar bond (**2**) but the formation of *N*-alkoxy

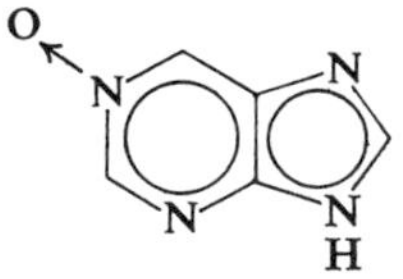

(**1**) 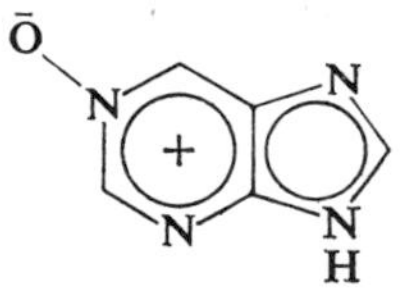   (**2**) 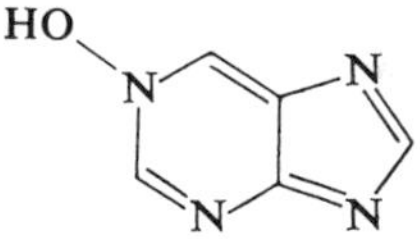   (**3**)

purines shows that the *N*-hydroxy form (**3**) is also a contributing tautomer. In some oxides of oxopurines tautomeric considerations dictate that the *N*-hydroxy rather than the *N*-oxide tautomer is the dominant species present. In this text the group will be depicted as either type **1** or **3**.

Oxide formation involving $N_{(1)}$, $N_{(3)}$, and $N_{(7)}$ has been reported but to date no example of a purine-9-oxide is known.

1. Preparation of Purine-*N*-oxides

Direct oxidation is of value for forming certain 1- and 3-oxides but is limited by the fact that any thio groups or halogen atoms present may be hydrolysed by the acid conditions used. Various unambiguous syntheses are available for the three types of N-oxides.

A. Purine-1-oxides

The majority arise by direct oxidation on leaving the purine for some days in a mixture of acetic acid and 30% hydrogen peroxide at room temperature. Perbenzoic acid and perphthalic acid can serve as alternative oxidising agents. Purine itself is converted to the oxide (2 weeks) with perbenzoic acid[1] but peroxide–acetic acid mixtures were employed to oxidise 6-methylpurine (12 h at 80°),[1] adenine ($2\frac{1}{2}$ days),[2, 3, 4] and related 2,6-diamino- (3 days),[2] 6-amino-7,8-dihydro-8-oxo- (7 days),[5] 6-amino-2-methyl- (5 days),[6] 6-amino-9-methyl-, 6-amino-9-ethyl-, and 6-amino-9-benzylpurine.[7] Examples of similar 1-oxide formation are 7,8-dihydro-8-oxo- (8 days)[1] and 7,8-dihydro-6-methyl-8-oxopurine (5 days).[1] Adenosine[2, 8] and the mono-,[8, 9, 10] di-,[9, 10] and triphosphate[10] and related nucleosides[11] give the respective oxides under these conditions. Perphthalic acid in ethereal solution is another reagent for such oxidations.[12, 13, 14]

Peracetic acid treatment of hypoxanthine is not a productive route to the oxide[2] which is best obtained by controlled nitrous acid hydrolysis of the amino group of adenine-1-oxide.[15] Similar conversions of adenosine-1-oxide (**4**, R = β-D-ribofuranosyl)[15, 16] and adenylic acid-1-oxide (**4**, R = β-D-ribofuranosyl-5′-phosphate)[8, 16] to the appropriate 1,6-dihydro-1-hydroxy-6-oxopurine riboside (**5**) have been made with nitrous acid[8] or nitrosyl chloride.[16] An unambiguous route to hypoxanthine-1-oxide (**5**, R = H) consists of brief heating of 4-amino-5-N-hydroxycarbamoylimidazole (**6**) with triethyl orthoformate.[17] A more sophisticated synthesis of xanthine-1-oxide (**9**, R = H)[18] and the 7-methyl (**9**, R = Me)[19] and 7-benzyl (**9**, R = $CH_2C_6H_5$)[19] analogues involves a Lossen-type rearrangement which takes place when the 4,5-di(N-hydroxycarbamoyl)imidazoles (**7**, R = H, Me, and $CH_2C_6H_5$) in tetrahydrofuran at 20° are treated with benzenesulphonyl chloride. The appropriate xanthine oxide is obtained by alkaline hydrolysis of the resulting 1-N-benzenesulphonyloxyxanthine (**8**, R = H or Me) intermediate. Likewise obtained are the 7-benzyl and 7-methyl derivatives of 8-methylthioxanthine-1-oxide.[19] 6-Hydroxymethylpurine-1-oxide arises from oxidative hydrolysis of 6-acetoxymethylpurine (see Sect. 3C). Selenium dioxide in dimethylformamide (21 h at 25°) converts 6-methylpurine-1-oxide to the 6-formyl analogue.[19a]

An alternative preparation of adenine-1-oxide to direct oxidation is ring closure of 4-amino-5-N-hydroxyformamidinoimidazole (**10**) with triethyl orthoformate in dimethylformamide.[20] With carbon disulphide in pyridine-methanol **10** gives the light-sensitive 2-thioadenine-1-oxide

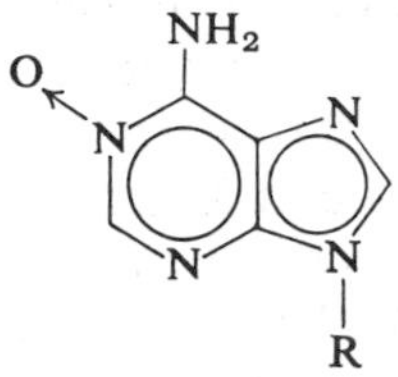

(4) 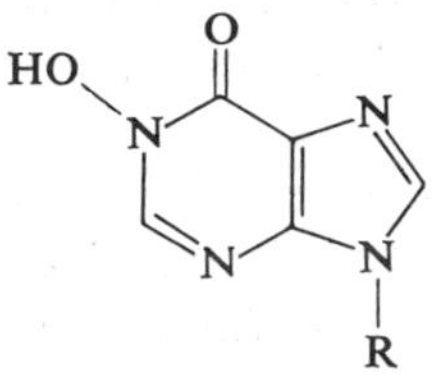   (5) 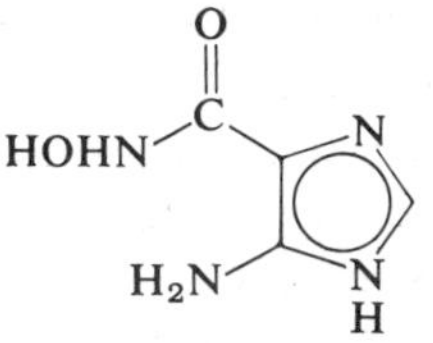   (6)

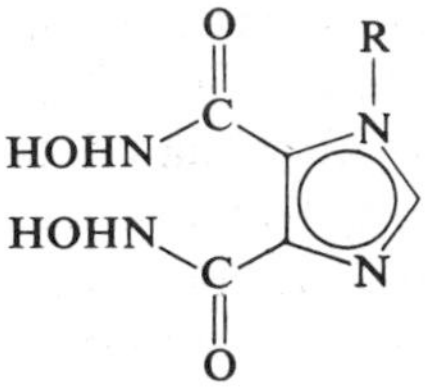

(7) 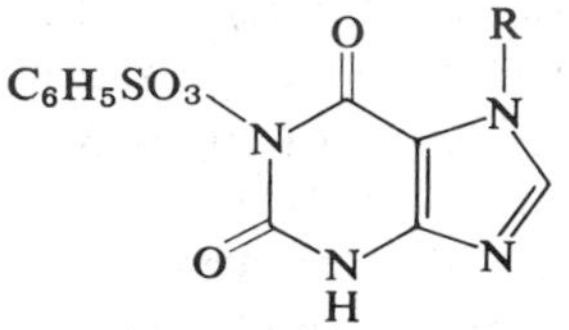   (8) 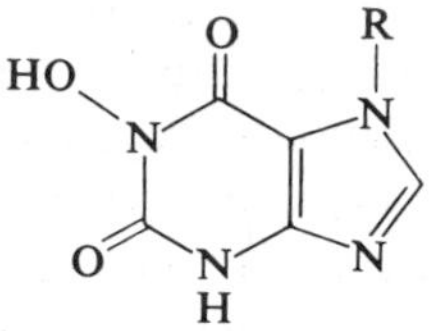   (9)

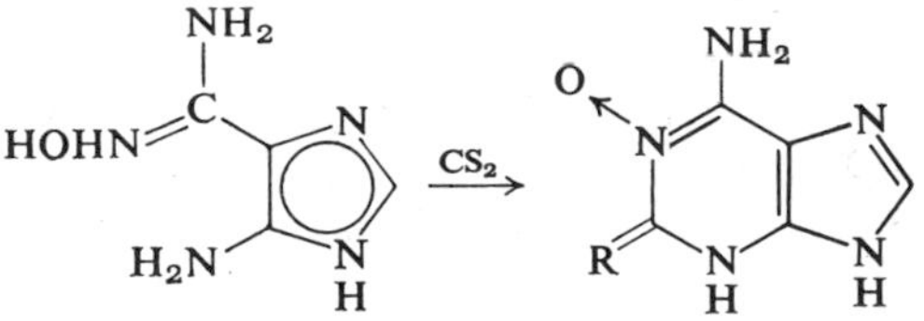

(10)   (11)

(**11**, R = S) which, on treatment with alkaline peroxide followed by cold 3*N*-hydrochloric acid hydrolysis (room temperature, 40 days), is converted to isoguanine-1-oxide (**11**, R = O).[21]

B. Purine-3-oxides

Guanine is oxidised directly to the 3-oxide (**12**) by hydrogen peroxide in trifluoroacetic acid, the latter, unlike acetic acid, being a good solvent for guanine. Conversion of **12** to xanthine-3-oxide (**13**, R = H) is effected by prolonged boiling in 6*N*-hydrochloric acid. Although the above derivatives were originally formulated as 7-oxides,[22, 23] the 3-oxide structures have now been established.[24, 25] Oxidation in ethanol solution with perphthalic acid or by means of an acetic acid–hydrogen peroxide mixture of 6-chloro-,[26] 6-methoxy-,[26, 26a] and 6-ethoxypurine[26]

at ambient temperature affords the corresponding 3-oxides. The latter
reagent, under more vigorous conditions (12 h at 80°), oxidises 6-
cyanopurine in 50% yield.[19a] A low yield (11%) of 6-chloropurine-3-oxide
is obtained using *m*-chloroperbenzoic acid in ether, but this procedure
is unsuccessful with either 6-bromo- or 6-iodopurine.[27] Purine itself,
under these conditions, gives almost exclusively the 3-oxide whereas
6-methylpurine affords about equal amounts of the 1- and 3-oxide.[19a]

Nonoxidative approaches are varied and include the formation of
13 (R = H) by cyclisation of 6-amino-5-formamido-1-hydroxyuracil
(**14**, R = H) in hexamethyldisilazane under reflux.[25] It should be noted
that an earlier cyclisation of the above pyrimidine in a mixture of formic
acid and acetic anhydride was reported[28] to give **13** (R = H) but the
product due to rearrangement is actually uric acid (see Sect. 3C). The
analogous cyclisation product of **14** (R = Me) originally formulated as
3-hydroxy-1-methylxanthine (**13**, R = Me), is, therefore, 1-methyluric
acid.[29] Prolonged heating (20 h) in ethanolic formamidine acetate converts 5-cyano-4-hydroxyamino-1-methylimidazole (**15**) to 6-amino-7-
methylpurine-3-oxide (**16**).[30]

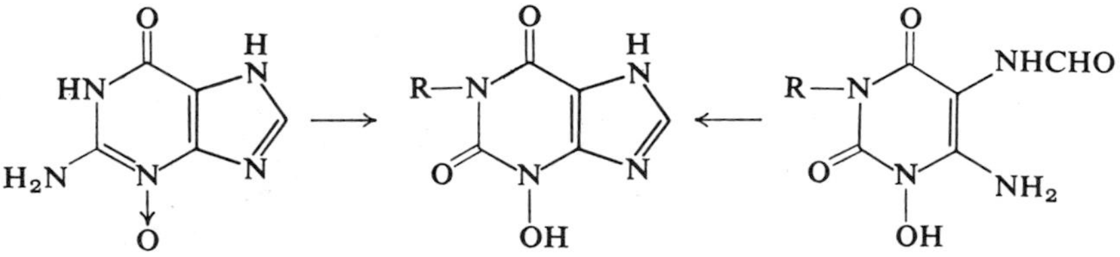

(12)   (13)   (14)

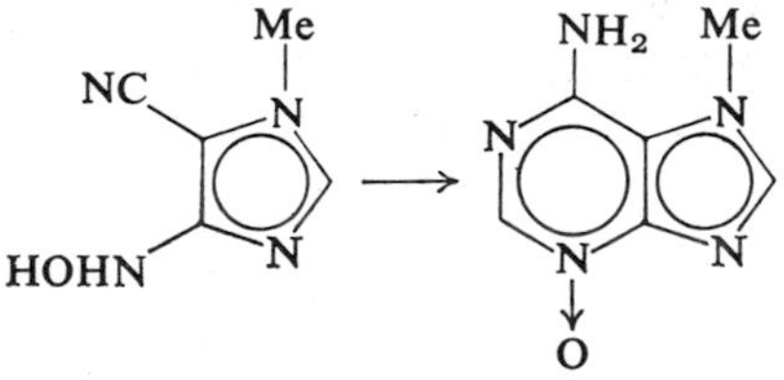

(15)   (16)

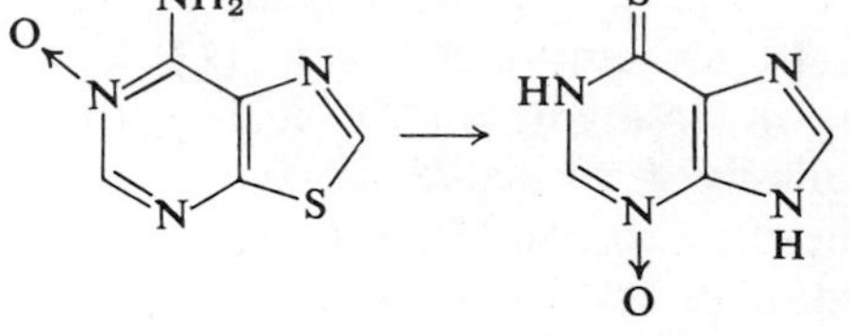

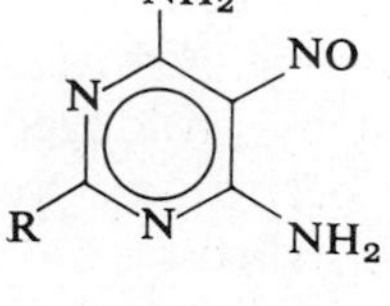

(17)   (18)   (19)

The oxide of 1,6-dihydro-6-thiopurine (**18**) arises from rearrangement of 7-aminothiazolo[5,4-*d*]pyrimidine-1-oxide (**17**) in hot alkali.[31]

Other purine-3-oxides not accessible by direct oxidation or by the other methods given above can be obtained by interconversion reactions, an example of which, the transformation of guanine-3-oxide to the xanthine analogue, has been given already. The variety of nucleophilic displacements possible with purine-3-oxides is a reflection of their pronounced stability. Thus, 6-chloropurine-3-oxide in *N*-sodium hydroxide under reflux affords hypoxanthine-3-oxide,[26] the 6-methoxy derivative follows from sodium methoxide treatment.[26, 27] The same halogenopurine-oxide is converted to 6-bromo-,[27] 6-iodo-,[27] and 6-sulphopurine-3-oxide[27] using bromine-hydrobromic acid, hydriodic acid, and aqueous sodium sulphite, respectively. Typical reactions of 1,6-dihydro-6-thiopurine are shown by the 3-oxide also. These include oxidation to dipurin-6-yl-3-oxide disulphide,[27] 6-sulphino-[19a, 26a] and 6-sulphopurine-3-oxides,[26a] conversion to 6-chloropurine-3-oxide by means of chlorine in acid medium,[27] and to 6-methylthiopurine-3-oxide through methyl iodide.[27, 31] Oxidation of the latter purine to 6-methylsulphonylpurine-3-oxide can be carried out with *N*-chlorosuccinimide[26] or trifluoroacetic acid-hydrogen peroxide mixtures.[26a] A successful route to 6-hydroxyaminopurine-3-oxide was to allow the 6-sulpho analogue to stand for a considerable time (25 days) in ethanolic hydroxylamine.[31a] The same starting material with ammonia solution and vigorous conditions (18h at 100°) provides the first preparation of adenine-3-oxide.[26a] Two unambiguous routes to purine-3-oxide are thermal decarboxylation of 6-carboxypurine-3-oxide, obtained by alkaline hydrolysis of the 6-cyano analogue,[19a] and desulphonation in formic acid (30 min at 80°) of the 6-sulpho derivative.[19a]

C. Purine-7-oxides

Direct oxidation has not so far provided a route to purine-7-oxides. Attention, however, must be drawn to the remarks in the preceding section concerning the oxidation product of guanine which was initially formulated[23] as the 7-oxide but later amended to the 3-oxide.[25] Various synthetic approaches lead to 7-oxides of which the earliest, reported without detail,[32] involved interaction between benzaldehyde anils and derivatives of 4,6-diamino-5-nitrosopyrimidines (**19**) giving 6-amino-purine-7-oxides of the type **20** (R = NH$_2$ or SMe). Further studies of this kind with 4-amino-1,2,3,6-tetrahydro-1,3-dimethyl-5-nitroso-2,6-dioxopyrimidine (**21**) and benzaldehyde anil show that 8-phenyltheophylline-7-oxide (**22**) is produced on heating in acetic acid under reflux for 3 hours.[33] Although the earlier workers[32] reported no interaction when benzaldehyde rather than the anil was used on heating **21** and the aldehyde in dimethylformamide, mainly 8-phenyltheophylline results although this is admixed with an appreciable quantity of the 7-oxide (**22**).[34]

Extensive synthetic studies by one school[35] have also utilised nitrosopyrimidine intermediates but under oxidative conditions. 4-Alkylaminouracil derivatives of the type **23** are directly converted in an

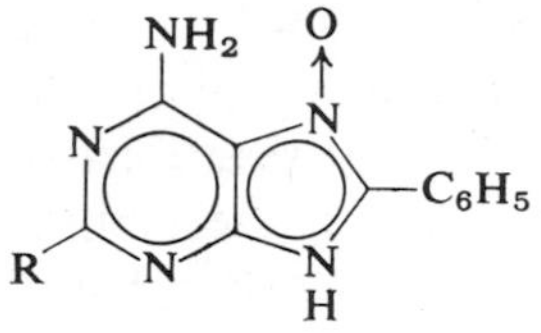

(20) 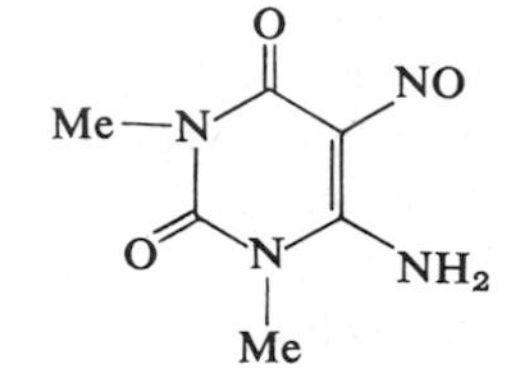   (21) 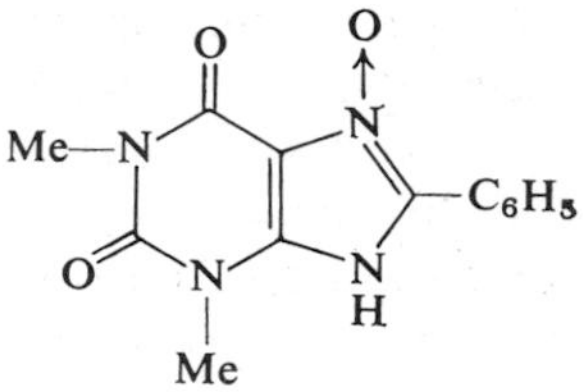   (22)

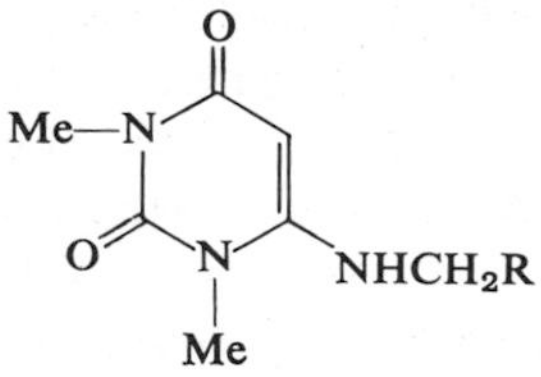

(23) 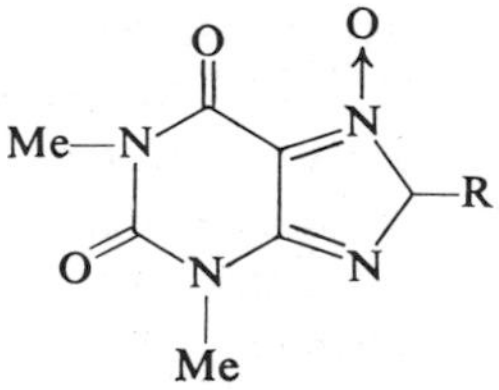   (24) 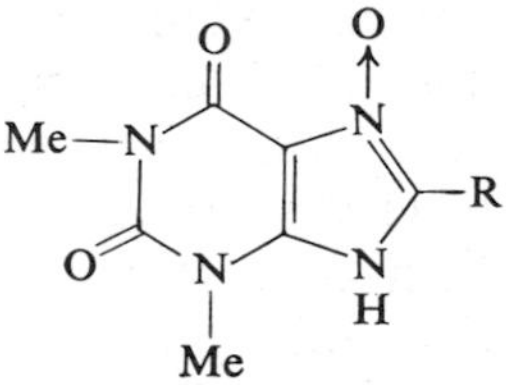   (25)

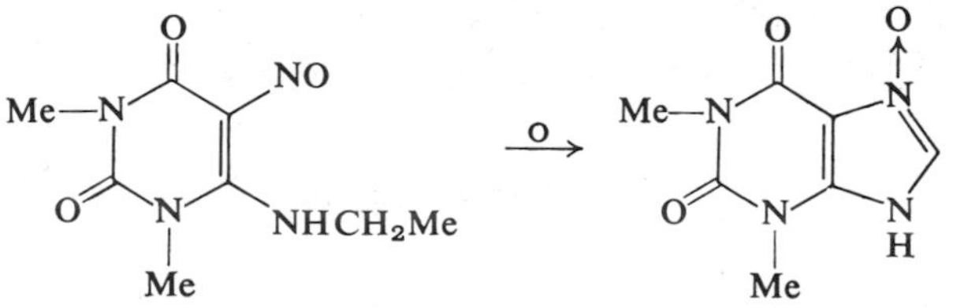

(26)   (27)

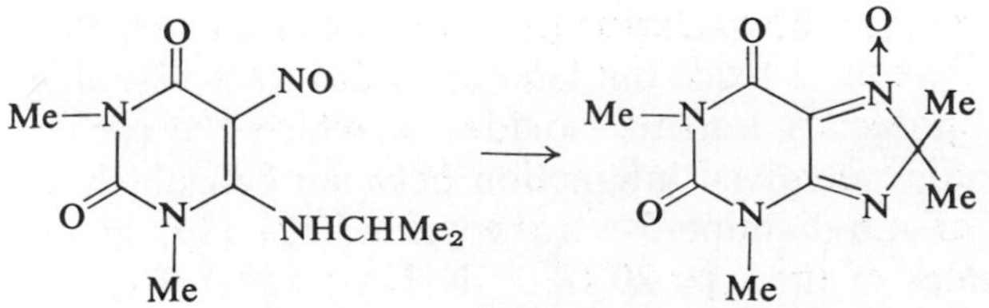

(28)   (29)

excess of nitrous acid to the thermolabile 8H-xanthine-7-oxides (**24**), which rearrange in hot ethanol or butanol to the more usual 9H-xanthine-7-oxides (**25**). In practice nitrosation and cyclisation are carried out concurrently using an excess of isoamyl nitrite in warm ethanol containing a trace of hydrochloric acid.[33] The parent member of the series, theophylline-7-oxide (**25**, R = H),[36] also the 8-methyl (**25**, R = Me),[33] 8-ethyl (**25**, R = Et),[33] 8-propyl (**25**, R = Pr and isoPr),[33] and 8-benzyl (**25**, R = $CH_2C_6H_5$)[33] analogues arise in this way. Purines of the type **25** may be formed by the action of isoamyl nitrite on 4-alkylamino-5-nitrosopyrimidines,[36, 37] exemplified by the conversion of the appropriate 4-diethylamino-5-nitrosopyrimidine to 8-methyltheophylline-7-oxide (**25**, R = Me) on standing three days at-room temperature.[37] Oxidising agents, such as aqueous solutions of nitric acid or acidified potassium permanganate have been employed but their use may lead to oxidative removal of the 8-alkyl group. The formation of theophylline-7-oxide (**27**) occurs in this way when the nitrosopyrimidine (**26**) is used.[36] With other oxidising agents, for example potassium dichromate or acidified hydrogen peroxide, fission of the five-membered ring usually occurs.[37] An exceptional case is 4-isopropylamino-5-nitrosopyrimidine (**28**) which, with acidified potassium dichromate in aqueous ethanol at room temperature (2 h), gives a good yield (73%) of 1,3,8,8-tetramethylxanthine-7-oxide (**29**).[38] Other derivatives in which this fixed "8H configuration" exists are known.[37, 39]

D. Other N-oxides

Extranuclear N-oxides of the type **30**,[40] also N-methyl derivatives of xanthine (**31**)[41] with an azamethine-N-oxide group at the 8-position are described elsewhere (Ch. IX, Sect. 5Aa).

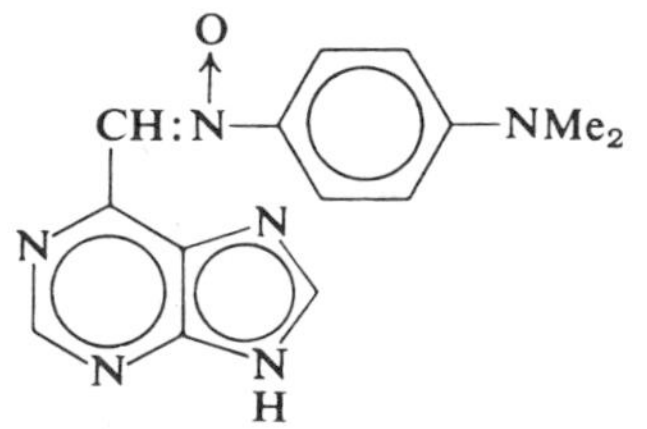

(30)

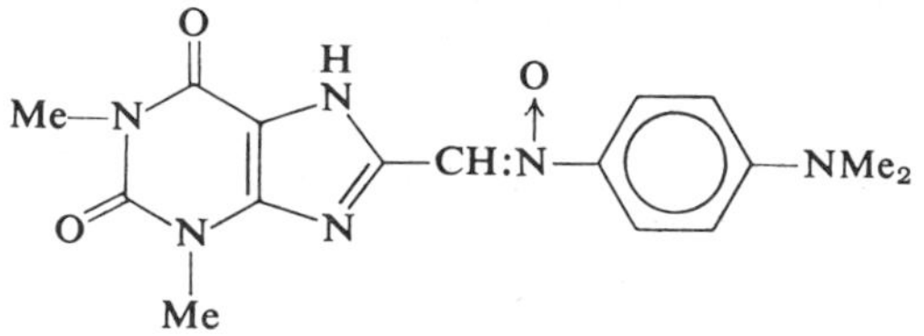

(31)

2. Properties of Purine-*N*-oxides

These compounds are usually colourless, crystalline solids, the melting points of which are of the same order as the parent purine. They give a positive Hantzsch test[42] for hydroxylamine derivatives, a blue or purple colour developing with ferric chloride. The existence of the oxide in the hydroxy form shows in the readiness with which the *O*-alkyl derivative results under mild alkylation conditions. Although moderately stable in dilute alkali and organic acids, in aqueous solutions of mineral acid degradation products result. In ultraviolet light removal of the oxygen and concurrent oxo formation at an adjacent carbon atom occurs. Adenine-1-oxide forms well-defined complexes with many divalent metal ions. Structures proposed are based on results of potentiometric studies.[43]

3. Reactions of Purine-*N*-oxides

Two aspects are considered, the former deals with reactions peculiar to the oxide molecule as a whole, whereas the other covers reactions of individual atoms or groups that are different from those of the parent purine.

A. Reduction

Catalytic hydrogenation in dilute ammonia with Raney nickel gives adenine in 94% yield from the 1-oxide,[2] less facile reductions have been carried out with phosphorus pentasulphide in pyridine under reflux[28] or in phosphorus trichloride at room temperature.[28] Adenine is also obtained on heating an aqueous solution of 6-hydroxyaminopurine-3-oxide with Raney nickel.[31a] Hydrogenation of the appropriate *N*-oxide is also a route to 6-amino-2-methyl- (Ni, $MeCO_2H$),[6] 6-amino-7,8-dihydro-6-oxo- (Ni, $MeCO_2H$),[5] 6-amino-7-methyl- (Ni, aq. NH_4OH),[2] and 6-methylpurine (Ni, H_2O).[1] Guanine (Ni, aq. NaOH) arises likewise.[22] Oxides of oxopurines may show indifferent behaviour towards reduction. Hypoxanthine-1-oxide is not reduced over a nickel catalyst and only slowly with platinic oxide.[17] The 3-oxide is hydrogenated over Raney nickel in ammoniacal solution but prolonged treatment (22 h) is necessary.[26] Although hydrogenation over Raney nickel is without effect on xanthine-1-oxide,[18, 21] reduction of both 1- and 3-oxides of 7,9-dimethylxanthine occurs readily by this means.[25] A slow hydrogen absorption was noted with xanthine-7-oxide (Ni, aq. NaOH)[22] and 7-benzylxanthine-1-oxide[19] whereas 8-methyltheophylline resulted from a facile reduction of the 7-oxide (Ni, EtOH).[33] The corresponding 8-isopropyl and 8-phenyl derivatives were reduced by heating the oxide in chloroform containing phosphorus trichloride.[33] Examples of deoxygenation resulting from thermal treatment alone are known. In dimethylformamide, under reflux conditions, 8-phenyltheophylline-7-oxide readily affords 8-phenyltheophylline.[34] A similar reduction occurs on amination of 6-methylthiopurine-3-oxide with morpholine giving 6-morpholinopurine.[44] Likewise reduced and converted to the appropriate parent 6-aminopurine is 6-chloropurine-3-oxide when treated with ammonia, hydrazine, hydroxylamine, and morpholine.[27]

In addition to removal of the oxide function other reducible groups present may also be lost. In the case of 6,6'-azoxypurine-3,3'-dioxide (32) boiling a solution of ammonium salt with Raney nickel produces *NN'*-dipurin-6-ylhydrazine (33).[31a]

With 1,6-dihydro-6-thiopurine-3-oxide reduction to purine itself occurs rapidly on adding Raney nickel to an ammoniacal solution at room temperature.[44] In contrast, only the thio group is displaced when 6-amino-2,3-dihydro-2-thiopurine-1-oxide is heated with nickel in *N*-sodium hydroxide giving adenine-1-oxide.[20] After 60 minutes in boiling dilute ammonia solution containing Raney nickel the 3-oxides of 6-bromo-, 6-iodo-, 6-methoxy-, and 6-sulphopurine are converted to the parent purines.[27] Hypoxanthine is similarly obtained but the comparable reduction of 6-chloropurine-3-oxide requires extensive (72 h) treatment.[27] Purine arises from the 3-oxide on catalytic hydrogenation in water whereas 6-methylpurine results when an aqueous solution of the oxide is heated with Raney nickel.[19a]

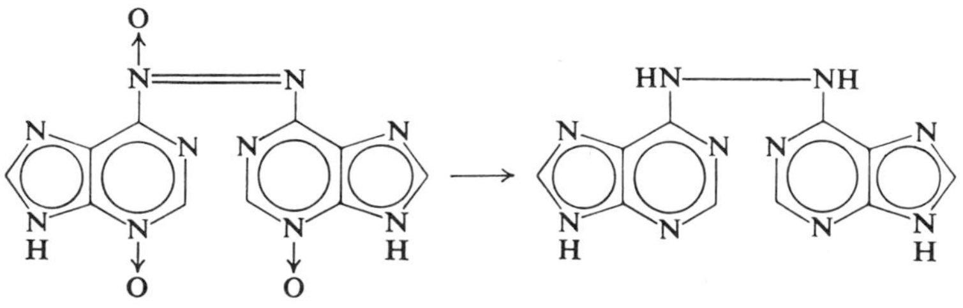

(32)   (33)

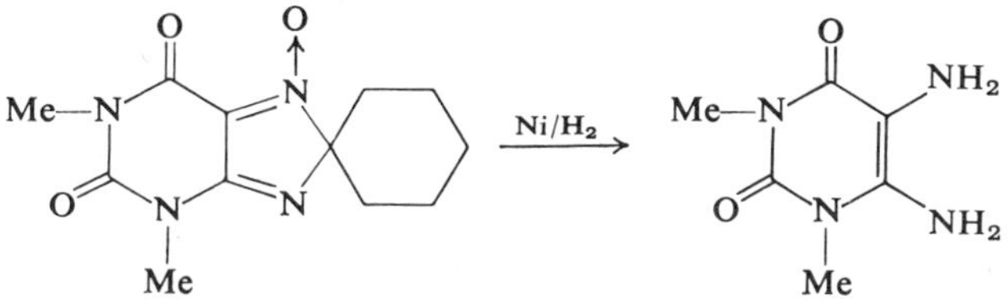

(34)   (35)

Hydrogenation of 8*H*-theophylline-7-oxides, for example the 8,8-pentamethylene derivative (**34**), over Raney nickel in ethanol causes imidazole ring fission and production of the corresponding 4,5-di-aminopyrimidine (**35**).[38]

B. Alkylation and Acylation

The 1-oxides of adenine[45] and the 9-methyl, 9-ethyl, and 9-benzyl homologues[7] (**36**, R = H, Me, Et, and $CH_2C_6H_5$) form the corresponding 1-methoxy derivatives (**37**) when treated with methyl iodide in dimethylacetamide at room temperature.[7, 45] The 1-ethoxy and 1-benzyloxy analogues are similarly derived with ethyl iodide and benzyl bromide, respectively.[45]

Only methylation of both of the imidazole nitrogen atoms occurs with either the 1- or 3-oxide of xanthine, using dimethyl sulphate in dimethylformamide at 40°. The resulting methosulphates are converted to the respective 1-hydroxy- (**38**) or 3-hydroxy-7,9-dimethylxanthinium betaine form by means of Amberlite I.R.45 resin.[25] Various methylating agents, such as methyl iodide in hot acetone containing potassium carbonate,[36] dimethyl sulphate in aqueous alkali at room temperature,[36] and ethereal diazomethane in methanol[36] afford the appropriate 7-methoxypurines (**39**, R = H,[29] Me,[36] isoPr,[33, 35, 36] and C_6H_5[33]) from the theophylline-7-oxides. Methylation of a thiopurine-oxide has given only the corresponding methylthiopurine-oxide.[20, 27, 31] Like a number

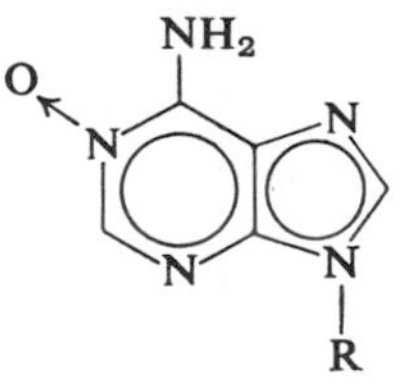

(36) 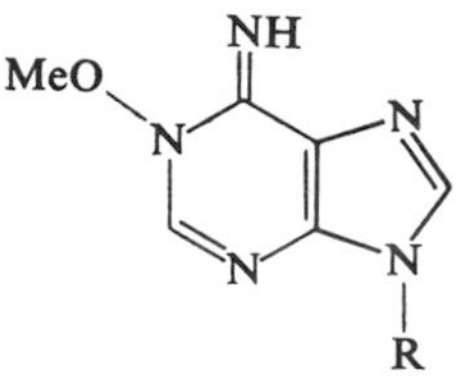   (37) 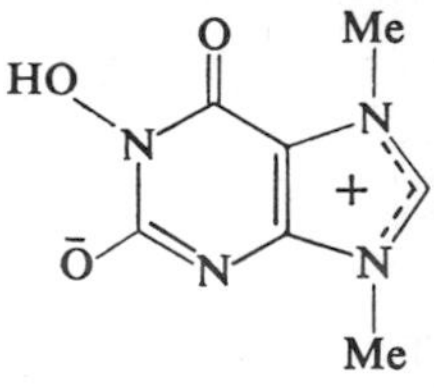   (38)

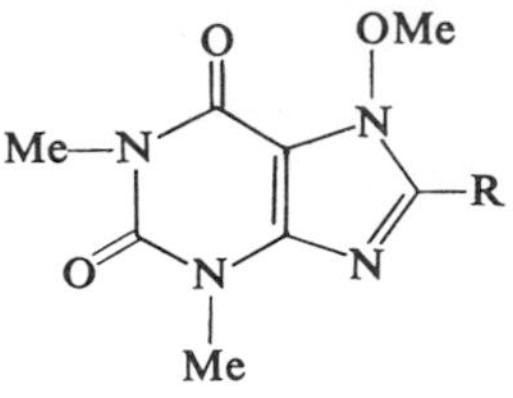

(39) 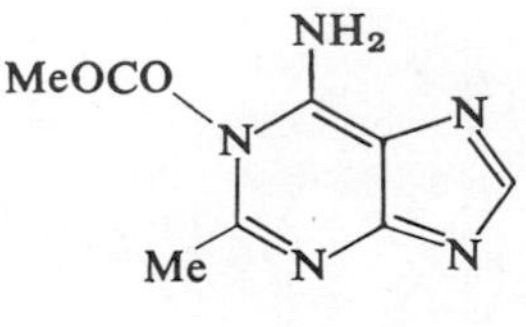   (40) 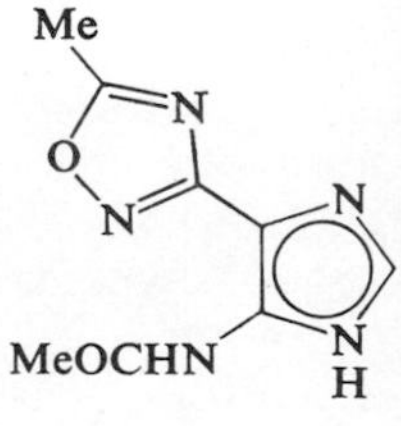   (41)

of other heterocyclic *N*-oxides the action of acetic anhydride on purine oxides may induce rearrangement of the oxide function (see succeeding Section C) but cases of simple *O*-acylation occurring under mild conditions are known, an example being 1-acetoxy-6-amino-2-methylpurine (**40**).[6] An unstable diacetylated derivative of adenine-1-oxide is formed under like conditions which reverts to the purine oxide in the presence of warm acid or base. Under reflux conditions it behaves like 2,6-diaminopurine-1-oxide in that cleavage of the pyrimidine ring occurs giving 5-methyl-3-(5-acetamidoimidazol-4-yl)-1,2,4-oxadiazole (**41**).[6]

C. Rearrangement

This can be effected in a number of ways, the most frequently used being heating with acetic anhydride but other reagents and means can induce isomerisation. That a common mechanism is not followed is seen in the different products obtained from the same purine oxide with different treatments. The majority of rearrangements involve loss of the *N*-oxide function and formation of an oxo group at a vicinal carbon atom. However, examples of oxygen migration to more distant carbon atoms are known.

With acetic anhydride 7,8-dihydro-8-oxopurine-1-oxide (**42**) is converted to a mixture of 2,8-dioxo- (**43**) and 6,8-dioxodihydropurine (**44**).[1] Although a similar rearrangement might be expected in the case of 6-methylpurine-1-oxide in practice the methyl group is the site of attack, the product being 6-acetoxymethylpurine (**45**). The latter in peracetic acid undergoes oxidative hydrolysis to 6-hydroxymethylpurine-1-oxide (**46**).[1] Isomerisations with either the above reagent or trifluoroacetic anhydride, or a mixture of both, of purine-3-oxides may produce either the analogous 2-oxo- or 8-oxopurine depending on the particular purine-oxide used. The 3-oxides of adenine, hypoxanthine, and 6-methoxypurine give isoguanine,[46] xanthine,[26] and 2,3-dihydro-6-methoxy-2-oxopurine,[46] respectively, whereas the corresponding oxides of xanthine, 7,9-dimethylxanthine, and guanine are converted to uric acid, 7,9-dimethyluric acid, and 6-amino-2,3,7,8-tetrahydro-2,8-dioxopurine by the same procedure.[46] A mechanism for oxygen migration from $N_{(3)}$ to $C_{(8)}$ has been postulated.[46] An extension to this rearrangement uses 3-acetoxyxanthine which is converted to 8-nitro-, 8-methylthio-, and 8-chloroxanthine on treatment with aqueous solutions of sodium nitrite, methionine, and sodium chloride, respectively, at ambient temperature.[46a] The preparation of 8-chloroguanine on heating guanine-3-oxide with acid chlorides in dimethylformamide is also of this type.[46a]

Rearrangement of a 7-oxide on acetylation is exemplified by that of theophylline which gives 8-acetoxytheophylline (**47**).[36] A novel preparation of guanine results from heating (60 min) in N-sodium hydroxide of 2-(pyridin-1-yl)hypoxanthine acetate (**48**), formed when hypoxanthine-3-oxide and pyridine are left in acetic acid for some hours.[26] A further example is the formation of 8-(pyridin-1-yl)xanthine (**49**) from xanthine-3-oxide and the subsequent hydrolysis to 8-aminoxanthine,[46a] The obtaining of 1,6,7,8-tetrahydro-6,8-dioxopurine, in near theoretical yield, on heating hypoxanthine-3-oxide in formic or acetic acid is in contrast to the isolation of xanthine on using acetic anhydride.[26]

Although thiolacetic acid can act as an acetylating agent, with hypo-

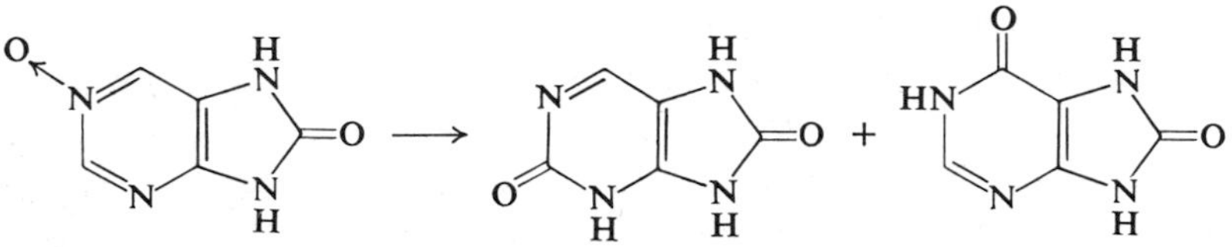

(42)   (43)   (44)

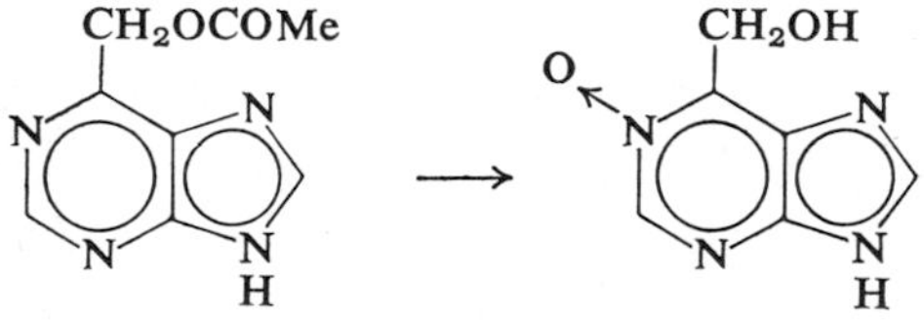

(45)   (46)

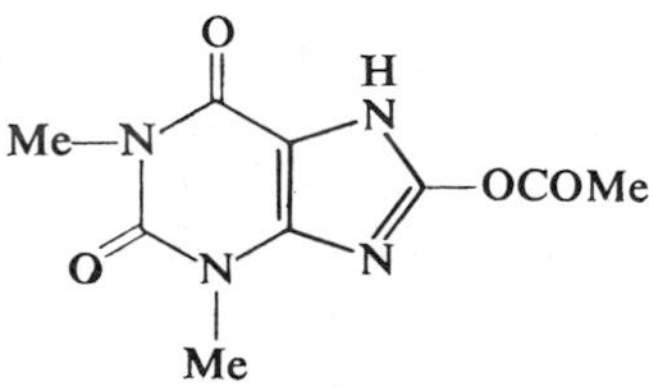

(47) 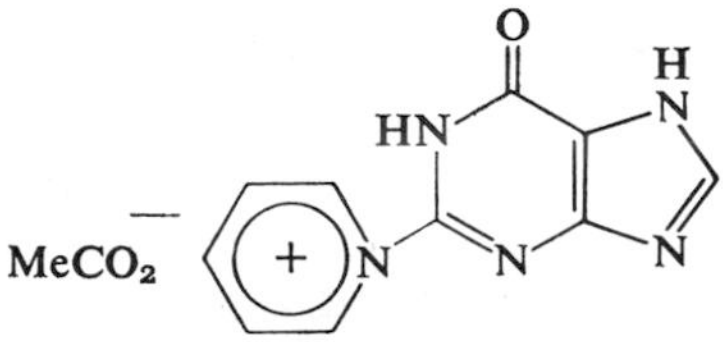   (48)

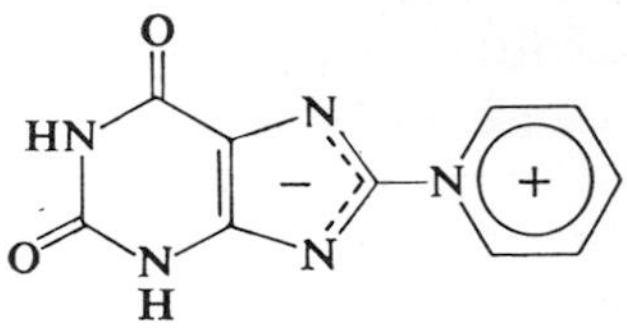

(49) 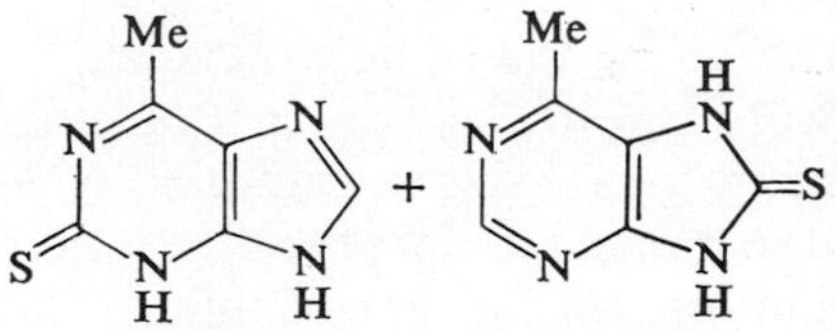   (50)   (51)

xanthine-3-oxide, after 4 hours under reflux, 8-thiohypoxanthine is the main product (88%) containing a minor amount (10%) of the 2-thio isomer.[26] The same reflux time with 6-methylpurine-3-oxide gives, conversely, 2,3-dihydro-6-methyl-2-thiopurine (50) as major component but, if heating is prolonged (12 h), the 8-thiopurine (51) now predominates.[47] In the case of 6-chloropurine-3-oxide this reagent gives a mixture of 2,6-dithio- and 6,8-dithiotetrahydropurines, the former derivative being in greater yield.[26] From the same treatment of 6-methoxypurine-3-oxide only 2-thiohypoxanthine is isolated.[26]

At least three possible mechanisms can be advanced to explain the formation of 2,6-dichloropurine when either hypoxanthine-3-oxide[26, 48] or 6-chloropurine-3-oxide[26] are heated with phosphoryl chloride or other chlorinating agent. In both reactions 6,8-dichloropurine occurs as a minor constituent.[26] 6-Alkoxypurine-3-oxides are converted to 6-alkoxy-2-chloropurines on this treatment.[26]

As noted previously (Sect. B) xanthine-3-oxide in dimethylformamide at 40° is methylated to the 7,9-dimethyl analogue. At higher temperature (80°) isomerisation to 7,9-dimethyluric acid takes place,[46] the reacting species appears to be a 3-methoxyxanthine, as thermal rearrangement

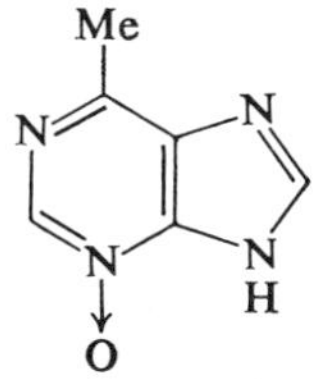

(52) 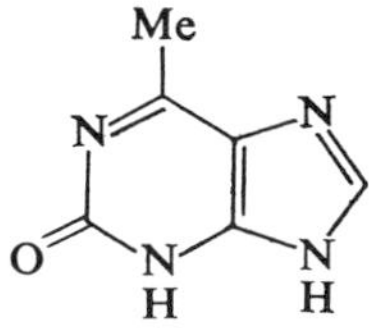   (53)

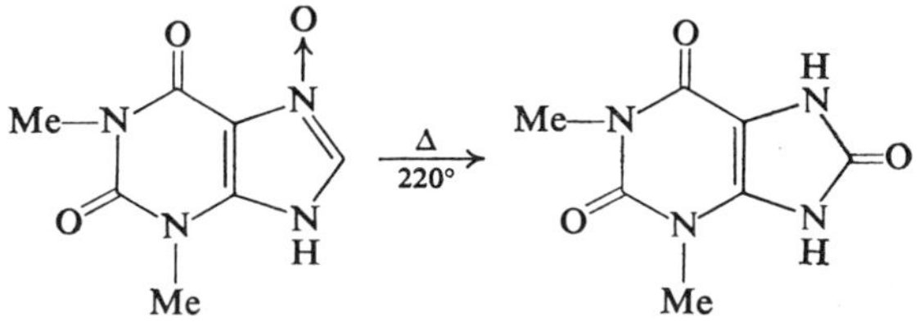

(54)   (55)

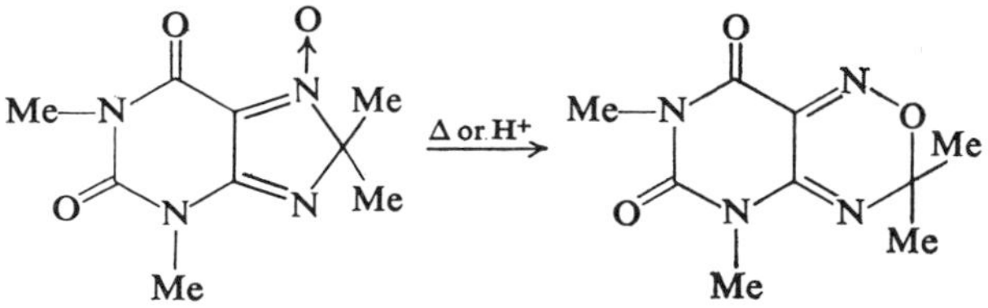

(56)   (57)

of 7,9-dimethylxanthine-3-oxide cannot be effected in dimethylform-amide.[46] An unambiguous synthesis has confirmed the identity of the rearranged product.[49]

Transformations can also be brought about by exposure to ultra-violet light (253.7 mµ). Adenine-1-oxide in buffered solution gives adenine and isoguanine as major products.[50] The corresponding ribo-sides result from photolysis of adenosine-1-oxide.[50, 51] With either the 1-oxide or 3-oxide (52) of 6-methylpurine mainly 2,3-dihydro-6-methyl-2-oxopurine (53)[50] results whereas the main product from 1,6-dihydro-6-thiopurine-3-oxide is the parent 6-thiopurine contaminated with a small amount of the 2-oxo analogue.[44] Kinetic studies of re-arrangements of this type have been made.[52]

Thermal rearrangements of theophylline-7-oxide (54) give 1,3-di-methyluric acid (55) directly.[35, 36] Similar treatment of the 8-methyl and 8-benzyl analogues gives theophylline as main product through con-current deoxygenation and loss of the 8-alkyl group.[50]

Under fusion conditions or hot mineral acid treatment 8,8-dialkyl-theophylline-7-oxides of the type 56 are transformed to the appropriate 3,3-dialkyl-5,6,7,8-tetrahydro-5,7-dimethyl-6,8-dioxo-$3H$-pyrimido[5,4-c]-1,2,5-oxadiazole (57).[39]

D. Group Reactivity

Oxide formation can result in activation of groups towards a nucleo-philic displacement which in the parent purine are usually inert. As an example can be cited the conversion of 6-amino-2-methylthiopurine-1-oxide to 6-amino-2-chloropurine-1-oxide in methanolic solution with

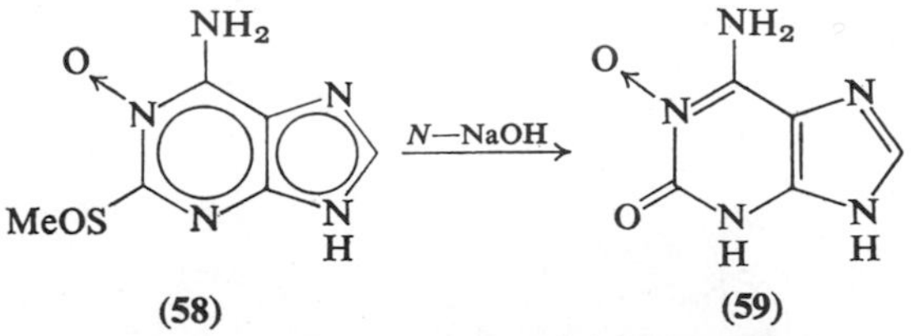

(58)   (59)

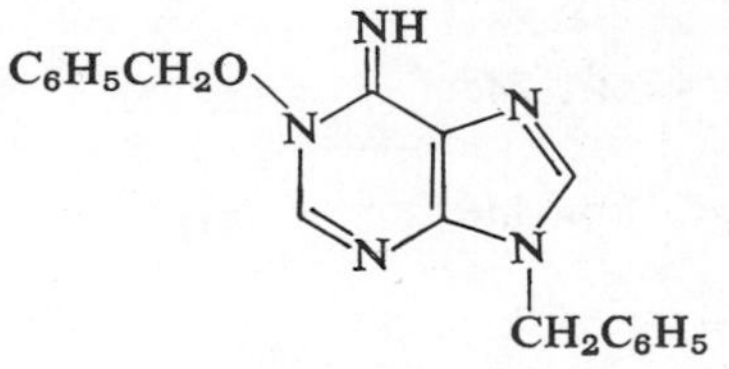

(60)

chlorine.[20] Under the same conditions 6-amino-2-methylthiopurine does not afford 6-amino-2-chloropurine.[53] With 6-amino-2-methylsulphinyl-purine-1-oxide (**58**), arising from controlled oxidation of the 2-methyl-thiopurine, a facile conversion to isoguanine-1-oxide (**59**) occurs in *N*-sodium hydroxide at room temperature or in hot dilute hydrochloric acid.[20] On subjecting 6-amino-2-methylsulphinylpurine to these procedures no hydrolysis to isoguanine is found.[20] Likewise, condensation of the oxide with β-hydroxymethylamine in boiling dimethylformamide to give 6-amino-2-(2-hydroxyethylamino)purine-1-oxide occurs within 10 minutes but prolonged heating of the parent purine with the amine is without effect.[20] Unlike 6-hydroxyaminopurine, which forms 6,6′-azoxypurine only slowly (some days) in dilute ammonia solution, the 3-oxide derivative gives 6,6′-azoxypurine-3,3′-dioxide (**32**) almost immediately.[31a]

4. Purine-*N*-alkoxides

A. Preparation

Preparation to date has been by the alkylation of the purine-oxides, described previously (Sect. 3B). Like the parent oxides they undergo a number of rearrangements.

B. Reactions

Reductive removal of alkoxide groups is readily effected by hydrogenolysis over Raney nickel catalyst. Adenine is obtained from the 1-methoxy, 1-ethoxy, or 1-benzyloxy derivatives (**62**, R = Me, Et, and $CH_2C_6H_5$, R′ = H),[45] also 9-methyl-, 9-ethyl-, and 9-benzyladenine from the appropriate 1-alkoxide.[45] Under similar conditions, using a palladium catalyst, 6-imino-9-benzyl-1-benzyloxypurine (**60**) is partially reduced to 6-amino-9-benzylpurine-1-oxide (**61**).[7] Removal of a 7-methoxy group with Raney nickel was used in a preparation of 8-phenyltheophylline.[34]

In hot aqueous solution (100°) the 9-alkyl-1-alkoxy-6-iminopurines (**62**, R = R′ = Me and Et) undergo an isomerisation[54] akin to the Dimroth rearrangement of 1-alkyladenines, the products being 6-alkoxyamino-9-alkylpurines (**63**, R = Me and Et).

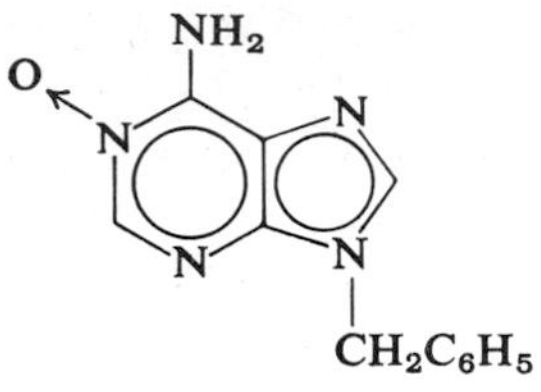

(61) 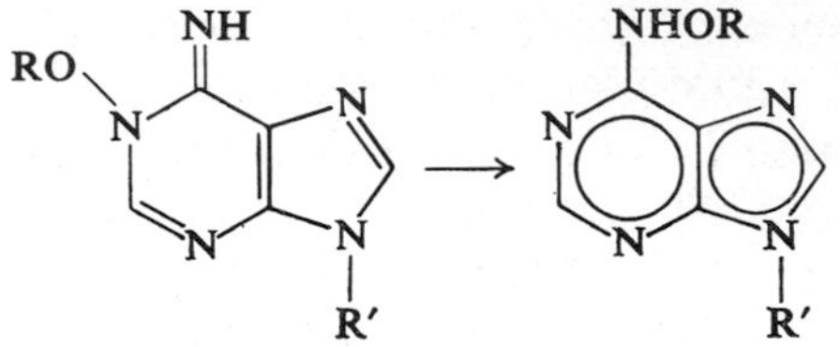   (62)   (63)

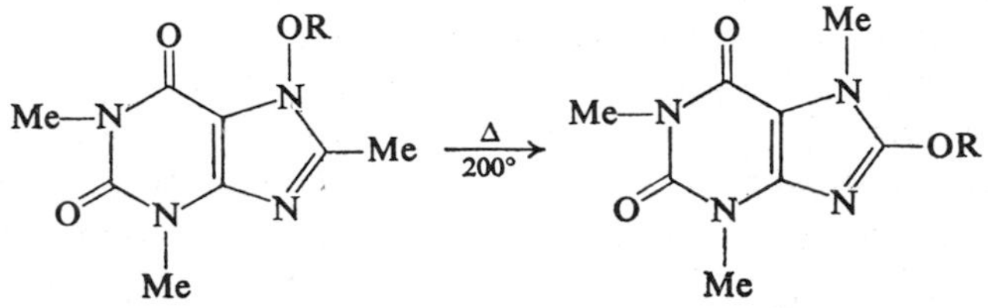

(64)   (65)

The products of thermal rearrangement of 7-alkoxypurines depend on the nature of the substituents and the conditions employed. When the 7-alkoxy-8-methyltheophyllines (64, R = Me, Et, Pr) are fused (200°, 30 min) or heated in dimethylformamide under reflux (10 min), an apparent exchange of groups between $N_{(7)}$ and $C_{(8)}$ occurs giving 8-alkoxy derivatives (65) of caffeine.[36] With less stringent heating only relocation of the alkoxy group is observed, the formation of 8-methoxy-8-methyl-8H-theophylline (67) when 7-methoxy-8-methyltheophylline (66) is heated to the m.p. (188°) illustrates this.[36] The product (67) is readily transformed into 8-methyltheophylline (68) on hydrogenation

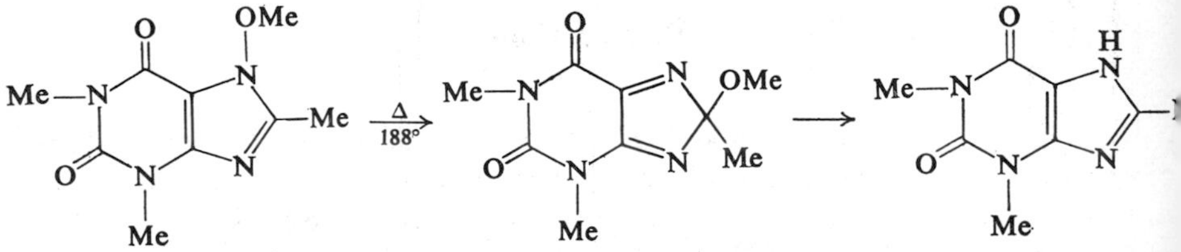

(66)   (67)   (68)

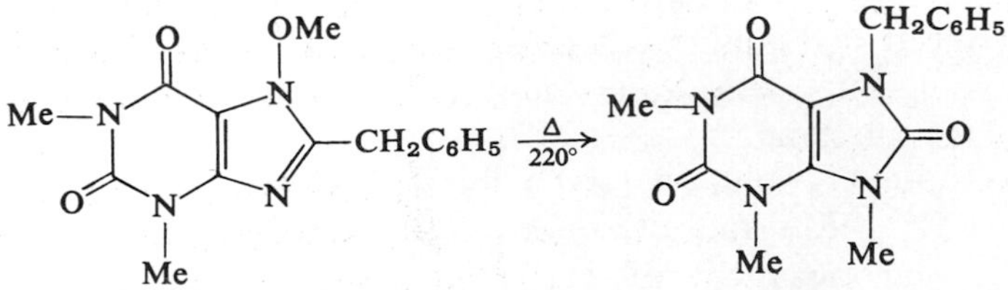

(69)   (70)

of an aqueous solution over Raney nickel.[36] Exceptions are found in the cases of 8-benzyl-7-methoxytheophylline (69), which at 220° gives 7-benzyl-1,3,9-trimethyluric acid (70),[36] and 7-methoxy-8-phenyltheophylline, which rearranges to 7-phenyl-1,3,9-trimethyluric acid.[36]

Complex rearrangements occur on alkylation of 1-alkoxyadenines as illustrated by the reaction of ethyl iodide with 1-benzyloxyadenine in dimethylacetamide at room temperature. At least six purines have been isolated from the product mixture which, in addition to the expected 1-benzyloxy-9-ethyl-, includes 9-benzyl-1-benzyloxy- and 1-ethoxy-9-ethyladenine.[55]

5. Biological Activity

Pronounced oncogenic (tumour producing) properties are reported[56] for the 3-oxides of guanine and xanthine. At higher dose levels 6-thiopurine-3-oxide shows similar activity but xanthine-1-oxide is inactive, as are the parent purines, xanthine and guanine.

References

1. Stevens, Giner-Sorolla, Smith, and Brown, *J. Org. Chem.*, **27**, 567 (1962).
2. Stevens, Magrath, Smith, and Brown, *J. Amer. Chem. Soc.*, **80**, 2755 (1958).
3. Stevens and Brown, *J. Amer. Chem. Soc.*, **80**, 2759 (1958).
4. Von Euler and Hasselquist, *Arkiv. Kemi.*, **13**, 185 (1958).
5. Brown, Stevens, and Smith, *J. Biol. Chem.*, **233**, 1513 (1958).
6. Stevens, Smith, and Brown, *J. Amer. Chem. Soc.*, **82**, 1148 (1960).
7. Fujii, Wu, Itaya, and Yamada, *Chem. and Ind.*, **1966**, 1598.
8. McCormick, *Biochemistry*, **5**, 746 (1966).
9. Stevens, Smith, and Brown, *J. Amer. Chem. Soc.*, **81**, 1734 (1959).
10. Cramer and Randerath, *Angew. Chem.*, **70**, 571 (1958).
11. Reist, Calkins, and Goodman, *J. Medicin. Chem.*, **10**, 130 (1967).
12. Cramer, Randerath, and Schafer, *Biochim. Biophys. Acta*, **72**, 150 (1963).
13. Klenow and Frederiksen, *Biochim. Biophys. Acta*, **52**, 384 (1961).
14. Frederiksen, *Biochim. Biophys. Acta*, **76**, 366 (1963).
15. Parham, Fissekis, and Brown, *J. Org. Chem.*, **31**, 966 (1966).
16. Sigel and Britzinger, *Helv. Chim. Acta*, **48**, 433 (1965).
17. Taylor, Cheng, and Vogl, *J. Org. Chem.*, **24**, 2019 (1959).
18. Bauer and Dhawan, *J. Heterocyclic Chem.*, **2**, 220 (1965).
19. Bauer, Nambury, and Dhawan, *J. Heterocyclic Chem.*, **1**, 275 (1964).
19a. Giner-Sorolla, Gryte, Cox, and Parham, *J. Org. Chem.*, (in press).
20. Cresswell and Brown, *J. Org. Chem.*, **28**, 2560 (1963).
21. Parham, Fissekis, and Brown, *J. Org. Chem.*, **32**, 1151 (1967).
22. Brown, Suguira, and Cresswell, *Cancer Res.*, **25**, 986 (1965).

23. Delia and Brown, *J. Org. Chem.*, **31**, 178 (1966).
24. Brown, *Prog. Nucleic Acid Res., and Mol. Biol.*, **8**, 209 (1968).
25. Wölke and Brown, *J. Org. Chem.*, **34**, 978 (1969).
26. Kawashima and Kumashiro, *Bull. Chem. Soc. Japan*, **42**, 750 (1969).
26a. Scheinfeld, Parham, Murphy, and Brown, *J. Org. Chem.*, **34**, 2153 (1969).
27. Giner-Sorolla, Gryte, Bendich, and Brown, *J. Org. Chem.*, **34**, 2157 (1969).
28. Cresswell, Maurer, Strauss, and Brown, *J. Org. Chem.*, **30**, 408 (1965).
29. McNaught and Brown, *J. Org. Chem.*, **32**, 3689 (1967).
30. Taylor and Loeffler, *J. Org. Chem.*, **24**, 2035 (1959).
31. Levin and Brown, *J. Medicin. Chem.*, **6**, 825 (1963).
31a. Giner-Sorolla, *J. Medicin. Chem.*, **12**, 717 (1969).
32. Timmis, Cooke, and Spickett, *Chemistry and Biology of Purines*, Eds., Wolstenholme and O'Connor, Churchill, London, 1957, p. 134.
33. Goldner, Dietz, and Carstens, *Annalen*, **691**, 142 (1965).
34. Taylor and Garcia, *J. Amer. Chem. Soc.*, **86**, 4721 (1964).
35. Goldner, Dietz, and Carstens, *Z. Chem.*, **4**, 454 (1964).
36. Goldner, Dietz, and Carstens, *Annalen*, **693**, 233 (1966).
37. Goldner, Dietz, and Carstens, *Annalen*, **699**, 145 (1966).
38. Goldner, Dietz, and Carstens, *Annalen*, **692**, 134 (1965).
39. Goldner, Dietz, and Carstens, *Tetrahedron Letters*, **1965**, 2701.
40. Giner-Sorolla, Zimmerman, and Bendich, *J. Amer. Chem. Soc.*, **81**, 2515 (1959).
41. Bredereck, Siegel, and Föhlisch, *Chem. Ber.*, **95**, 403 (1962).
42. Hantzsch and Besch, *Annalen*, **323**, 23 (1902).
43. Perrin, *J. Amer. Chem. Soc.*, **82**, 5642 (1960).
44. Brown, Levin, Murphy, Sele, Reilly, Tarnowski, Schmid, Teller, and Stock, *J. Medicin. Chem.*, **8**, 190 (1965).
45. Fujii, Itaya, and Yamada, *Chem. and Pharm. Bull.* (*Japan*), **13**, 1017 (1965).
46. Wölke, Pfleiderer, Delia, and Brown, *J. Org. Chem.*, **34**, 981 (1969).
46a. Wölke, Birdsall, and Brown, *Tetrahedron Letters*, **1969**, 785.
47. Giner-Sorolla, Thom, and Bendich, *J. Org. Chem.*, **29**, 3209 (1964).
48. Kawashima and Kumashiro, *Bull. Chem. Soc. Japan*, **40**, 639 (1967).
49. Brown, Pfleiderer, and Delia, *J. Org. Chem.* (in press).
50. Brown, Levin, and Murphy, *Biochemistry*, **3**, 880 (1964).
51. Cramer and Schlingloff, *Tetrahedron Letters*, **1964**, 3201.
52. Levin, Setlow, and Brown, *Biochemistry*, **3**, 883 (1964).
53. Noell and Robins, *J. Amer. Chem. Soc.*, **81**, 5997 (1959).
54. Fujii, Itaya, Wu, and Yamada, *Chem. and Ind.*, **1966**, 1967.
55. Fujii, Itaya, and Yamada, *Chem. and Pharm. Bull.* (*Japan*), **14**, 1452 (1966).
56. Suguira and Brown, *Cancer Res.*, **27**, 925 (1967).

CHAPTER XII

The Reduced Purines

Although some reduced forms of purines were described over seventy years ago, they are not generally well known derivatives. Purines reduced in either pyrimidine or imidazole ring have been prepared but examples of the latter are restricted to a few special cases. While the position of dihydropurines is well authenticated, many such compounds having been prepared, that for tetrahydropurines still needs further investigation and has been hindered by the apparently unstable nature of the one or two derivatives that have been produced so far.

Little help is provided by the formal nomenclature for the reader wanting to differentiate at a glance between a true reduced purine and an *N*-methylated oxo-, amino-, or thiopurine. This is illustrated by the case of 1-methylxanthine (**1**) which systematically named is 1,2,3,6-tetrahydro-1-methyl-2,6-dioxopurine although, in fact, it possesses con-

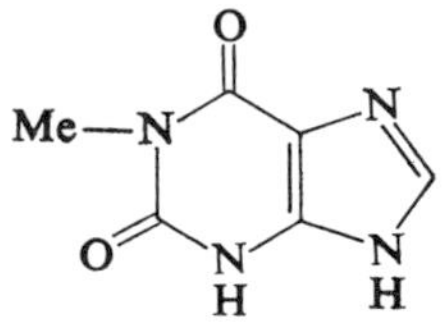

siderable associated aromatic character which can be represented by the canonical form (**2**). If, however, the oxo function at $C_{(6)}$ is replaced by two hydrogen atoms, the resulting 1,2,3,6-tetrahydro-1-methyl-2-oxopurine (**3**) now has an authentic reduced state which can only be restored to full aromaticity by oxidation.

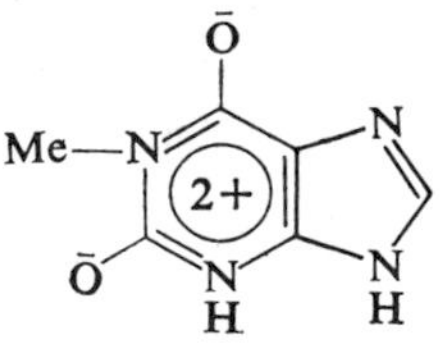

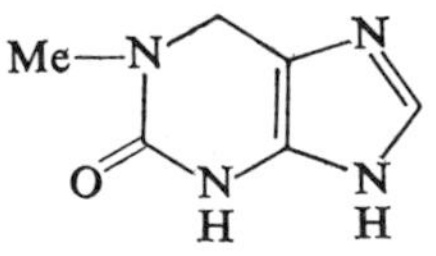

1. Purines Reduced in the Pyrimidine Ring

This class embraces nearly all the known hydropurines, the majority of which are dihydropurines.

427

A. Preparation of Hydropurines

Reduction of an existing purine is usually employed, which may involve electrochemical, catalytic, or chemical means. A unique example of direct synthesis is known.

a. *By Electrochemical Reduction*

Studies of this form of reduction have been both quantitative and qualitative. Purine itself, on macro-scale reduction[1] at a mercury cathode in acetic or in hydrochloric acid, is reduced to 1,6-dihydropurine (4), isolated as a complex with sodium tetraphenylborate. Further reduction leads to an unstable tetrahydro derivative, the presence of which has been shown polarographically. Much attention has been devoted to the reduction of oxopurines. The first such reduction was carried out by Tafel[2] in 1899. Using a lead cathode and a dilute sulphuric acid electrolyte, he converted caffeine (5) into the 1,6-dihydro derivative (6) in 70% yield. The same reduced purine, originally termed "desoxycaffeine," results when 8-chloro-7-chloromethyltheophylline (7)

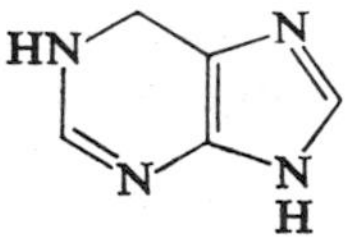

(4)

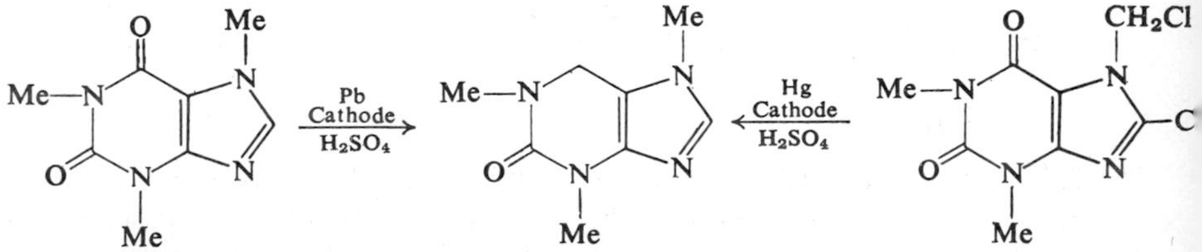

(5)   (6)   (7)

is electrolysed in sulphuric acid at a mercury cathode.[3] If a lead cathode is employed reduction is incomplete, producing a mixture of caffeine and 8-chlorocaffeine.

Tafel's original experiment has been repeated recently[4] and loss of oxygen at the 6-position was confirmed by nmr measurements. Subsequently, reduction was extended to various mono-, di-, and trioxopurines and their methylated derivatives. Hypoxanthine appears to be exceptional in undergoing reduction of the $C_{(2)}-N_{(3)}$ double bond as on degradation the product obtained is a carbamoylimidazole.[1] Xanthine and other N-methylated homologues, in addition to caffeine noted above, undergo replacement of the oxygen function at the 6-position forming 6-deoxyxanthines. Electrolyte acid strengths can be varied to suit the solubility of the xanthine derivative, xanthine itself being electrolysed to deoxyxanthine (8) in good yield at a lead cathode in 75% sulphuric acid.[5] Under the same conditions 1-methyl-,[6] 3-methyl-,[7] and 7-methylxanthine[7] afford the appropriate 6-deoxy derivative. Dimethyl-ated xanthine derivatives are likewise reduced: theophylline, in 30%

(8)   (9)   (10)

sulphuric acid,[8] gives the 6-deoxytheophylline (9), which can also be obtained by reduction of 8-chlorotheophylline in acid at a mercury cathode.[3] Polarographic studies in alkaline solution of 8-chlorotheo-phylline[9, 10] have demonstrated the presence of not only the 6-deoxy-purine (9) but also the intermediate 1,6-dihydro-6-hydroxytheophylline (10). Theobromine[11] and 1,7-dimethylxanthine[8] give the respective 6-deoxy form at a lead cathode in sulphuric acid.

The results obtained with uric acid depend mainly on the conditions adopted. At temperatures below 8° the derivative obtained arises through hydrogenolysis of the oxygen atom at $C_{(6)}$ with an accompany-ing reduction of the $C_{(4)}-C_{(5)}$ double bond.[12] Present knowledge suggests structure 11 for this compound which Tafel had originally named "purone." By carrying out the electrolysis at a slightly higher temperature (12–15°),[12] or by warming purone with alkali or acidified ethanol, the isomeric "isopurone" (12)[13] results through cleavage of the imidazole ring and aromatisation of the $C_{(4)}-C_{(5)}$ bond. With longer electrolysis times at temperatures above 20° a further product obtains to which the misleading name "tetrahydrouric acid" was originally given. Subsequent work[14] indicates the probability of this being the 5-ureido derivative (13) of 1,6-dihydrouracil. Methylated forms of uric acid by contrast reduce in an uncomplicated fashion, the analogous 1,4,5,6-tetrahydropurines being obtained from 3-methyl-, 1,3-dimethyl-3,7-dimethyl-, 7,9-dimethyl-, 1,3,7-trimethyl-, and 1,3,7,9-tetramethyl-uric acid (14).[15] Guanine, like xanthine, loses the oxygen atom under

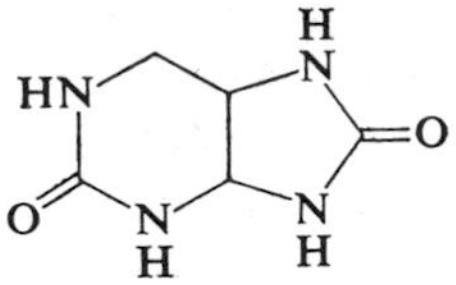

(11) 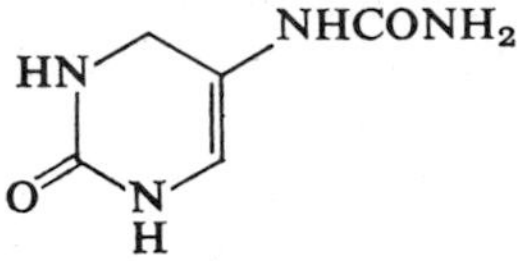   **(12)** 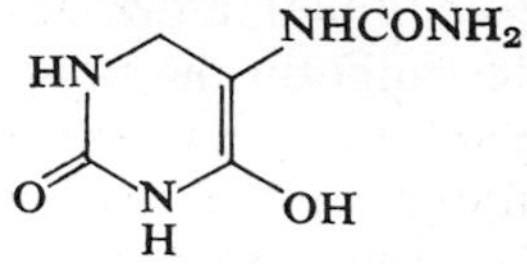   **(13)**

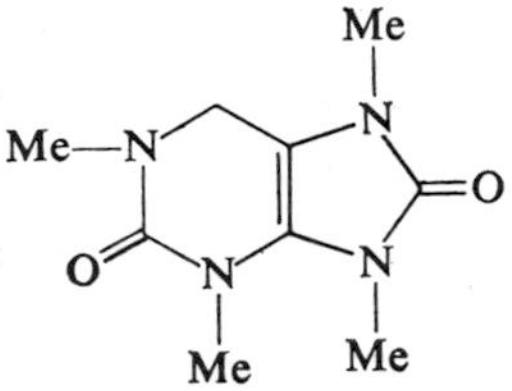

(14) 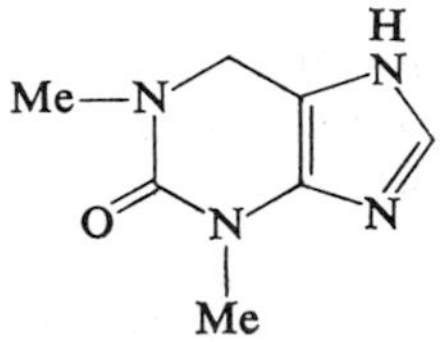   **(9)**

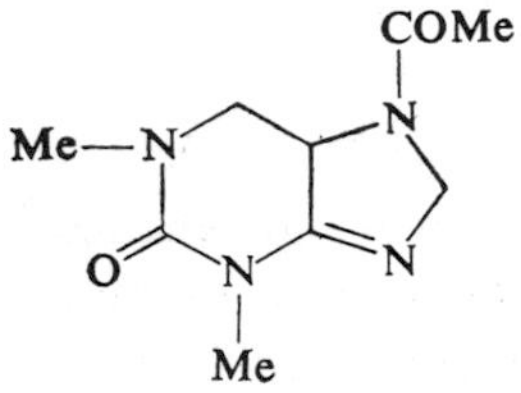

(15a) 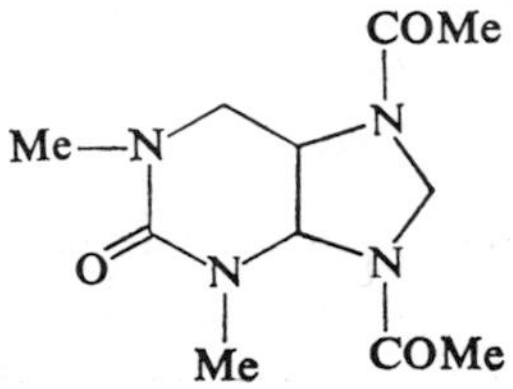   **(15b)** 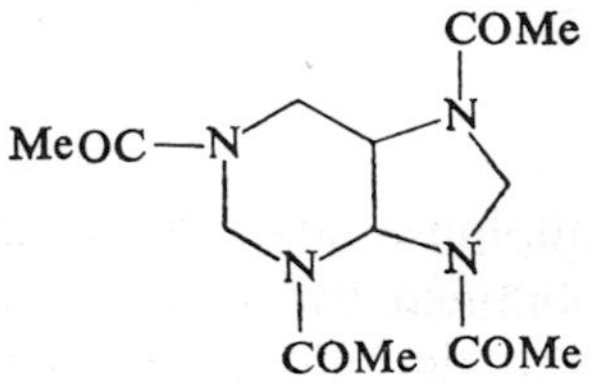   **(15c)**

these conditions forming 2-amino-1,6-dihydropurine in 75% yield.[16] No reduction, surprisingly, can be effected with this purine under polarographic conditions[1, 17, 18] although other aminopurines, for example, adenine[1, 17, 18, 19] and 2,6-diaminopurine,[18] show reduction waves under this treatment.

b. *By Catalytic Hydrogenation*

In alkaline or neutral solution purine is not reduced but with aqueous solutions of purine hydrochloride an uptake of one mole of hydrogen obtains. The product, presumed to be 1,6-dihydropurine, has not been isolated,[20, 21] but the 1,9-diacetyl derivative arises on hydrogenation of purine in acetic anhydride with either a palladium or platinum catalyst.[21a] Both 2-chloro- and 2-chloro-9-methyl-purine are reduced in this way to 1,9-diacetyl-2-chloro- and 1-acetyl-2-chloro-9-methyl-1,6-dihydropurine.[21a] It is noteworthy that under these conditions the chlorine atoms remain intact whereas in the reduction of 2,6,8-trichloropurine in acetic acid over palladium under pressure all halogen atoms are removed and the unstable product obtained is claimed to be a tetrahydropurine dihydrochloride.[22] As the nature of the solvent often profoundly affects the course of hydrogenation, it is not surprising that further reduction of deoxytheophylline (9) in a mixture of acetic acid and the anhydride gives 7-acetyl-1,2,3,5,6,8-hexahydro-1,3-dimethyl-2-oxopurine (15a) using a palladium catalyst while in the anhydride alone with a platinum catalyst 7,9-diacetyl-1,2,3,4,5,6,8,9-octahydro-1,3-dimethyl-2-oxopurine (15b) is formed.[21a] A stable form of fully reduced purine itself is claimed, as the 1,3,7,9-tetra-acetyl derivative (15c), by the same procedures.[21a] Attempted reductions of 6-chloropurine, adenine, hypoxanthine, xanthine, and N-methylated xanthines in acetic anhydride with palladium or platinum catalyst have failed.[21a]

In neutral solution 2,3-dihydro-2-oxopurine is slowly reduced (palladium) absorbing 1 mole of hydrogen from which the reduced purine derivative can be isolated.[20] Under similar conditions neither hypoxanthine[20] nor adenine are reduced and hydrogenation of xanthine and the 3-methyl derivative have also failed using pressure conditions and a platinum, nickel or copper chromite catalyst.[23]

c. *By Chemical Reduction*

Initially, reductions with hydrogen produced *in situ* by amalgams or metals in acid solution were favoured. Purine and simple substituted derivatives with these reagents undergo reduction and concomitant rupture of the pyrimidine ring giving rise to aminoimidazole derivatives, the presence of these being demonstrated by positive reactions with Bratton-Marshall reagent, *i.e.* N-(1-naphthyl)ethylenediamine hydrochloride, which is specific for diazotisable amino groups. Examples of reductive degradation of this kind are found with purine,[24] using a zinc amalgam, and adenine[18] or hypoxanthine,[18] with a sodium amalgam, in dilute acid solution. When zinc dust replaces the amalgams[25, 26, 27] these purines again undergo cleavage as do xanthine[27] and isoguanine[27] also. However, no diazotisable amino groups could be detected in the products which arose when guanine,[18, 25, 27] 2,6-diaminopurine,[27] or uric acid[27] were similarly treated. By contrast, successful reductions with caffeine and theobromine have given the stable 6-deoxy analogues.[28]

Reduction of thio groups affords routes to 1,6- and 2,3-dihydro derivatives of N-methylated purines. Thus the 6-thio analogue (16) of theobromine on heating under reflux in dilute ammonium hydroxide with Raney nickel (1 h) gives 6-deoxytheobromine (17).[29] The 2- and 6-thio analogues of theophylline are converted in this way to 1,2,3,6-

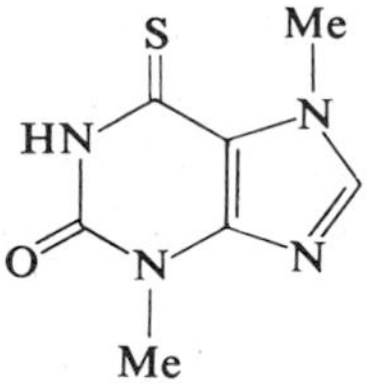

(16) 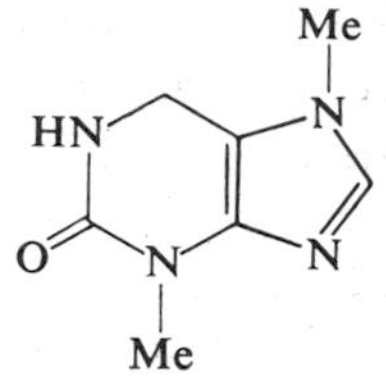   (17) 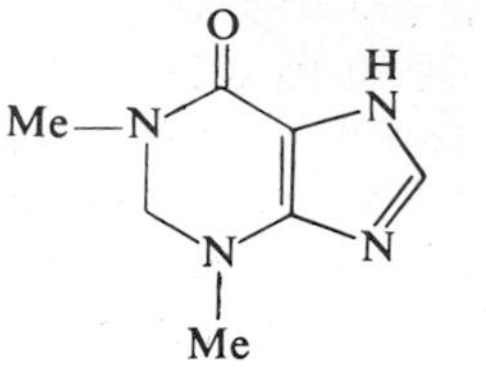   (18)

tetrahydro-1,3-dimethyl-6-oxopurine (2-deoxytheophylline) (18)[30] and 1,2,3,6-tetrahydro-1,3-dimethyl-2-oxopurine   (6-deoxytheophylline).[30] Similarly, the 6-thio modification of caffeine affords 6-deoxycaffeine.[31] However, although desulphurisation of both 2- and 6-thio groups is generally possible the simultaneous removal of both sulphur atoms in 2,6-dithio-N-methylpurines has not so far been achieved.[30, 32]

Sodium borohydride, in methanol ($\leqslant 0°$), successfully reduces cationic forms of some N-methylated oxopurines to dihydro derivatives. Either the six- or five-membered ring (See Section 2A) may suffer reduction; the ring involved being that associated with the positive charge. Thus, 1,3-dimethyl-8-phenylhypoxanthinium iodide is reported[32a] to give the 2,3-dihydro analogue.

d. *By Direct Synthesis*

When the S-methylated derivative (19) of 2,4-dibromo-5-thioureidomethylimidazole is cyclised by boiling in ethanol containing pyridine

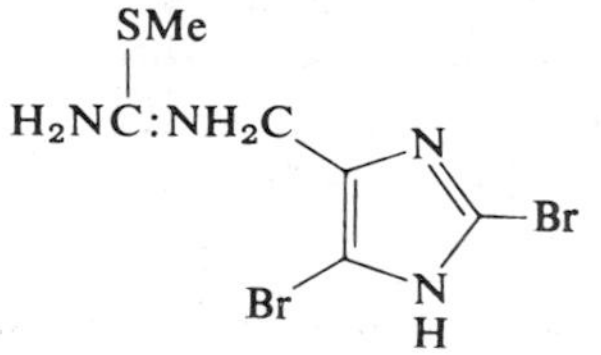

(19) 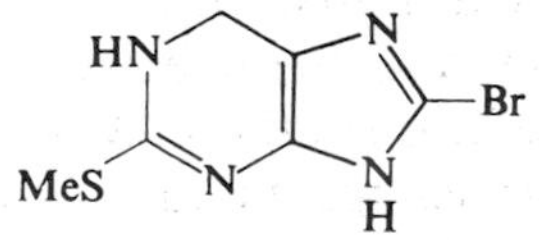   (20)

for some hours, 8-bromo-1,6-dihydro-2-methylthiopurine (20) is obtained.[33] A preparation, along similar lines, of derivatives of 1,6-dihydro-7-methylpurine, by cyclisation of 4-amino-5-carboxymethyl-amino-1-methylimidazole, has proved unsuccessful.[33a] The reported direct synthesis of tetrahydrouric acid from 1,2-diaminopropionic acid by successive reaction with potassium cyanate and hydrochloric acid[34] is erroneous as the product is the same as Tafel's[14] "tetrahydrouric acid," that is, the dihydrouracil derivative (13) (see Sect. IAa). With

1,2-dibenzylaminopropionic acid the dibenzylated uracil analogue of **13** results.[35]

Reduction through adduct formation is reported following ultra-violet irradiation of solutions of purine in methanol, ethanol, or propanol.[35a] The product with *n*-propanol, for example, is 1,6-dihydro-6-(1-hydroxyethyl)purine.[35a] This and other aspects of this reaction are discussed elsewhere (Ch. I, Sect. 4Dd).

B. Properties and Reactions

Reduced simple substituted purines are hydrolytically labile, the reduction products from purine, 2,3-dihydro-2-oxopurine, and adenine all being degraded to the same derivative, namely, 4-amino-5-amino-methylimidazole (**21**), when subjected to mild acid conditions.[1, 20, 36] The production of this imidazole from reduced adenine is attributed to an initial removal of the amino group, as ammonia, giving rise to the

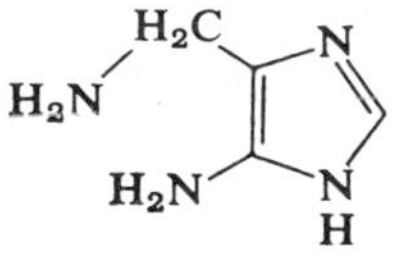

(21) 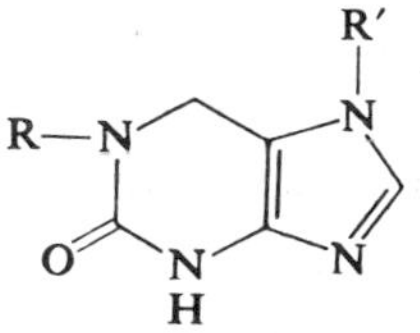   (22) 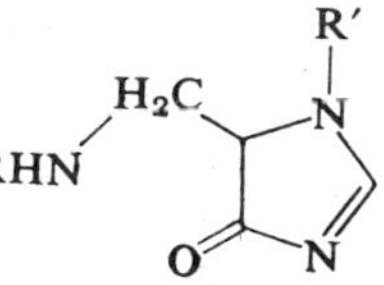   (23)

reduced form of purine.[1] With more highly substituted derivatives stability is dependent on the nature and location of the substitutents. Thus purone (**11**) is converted to the isomeric isopurone (**12**), on warming with alcoholic sulphuric acid[12] while among the *N*-alkylated deoxyxanthine derivatives the most acid labile compounds are those unsubstituted at $N_{(3)}$, e.g., **22**; with the above treatment these give imidazolones of the type **23**, the reaction being accompanied by evolution of ammonia and carbon dioxide.[37] Deoxyxanthine itself is likewise degraded. Alkylation at $N_{(3)}$ produces highly acid-stable derivatives which are broken down to complex products only on prolonged vigorous treatment.[37] A generally high degree of stability is found towards alkali irrespective of the substituents present.

Oxidation is usually facile; 1,6-dihydropurine in solution, for example, is slowly converted to purine on exposure to air.[1] Other satisfactory oxidising agents are lead peroxide and silver acetate, the latter being used to convert deoxyguanine (**24**) to 2-aminopurine (**25**)[16]; the yield is improved if bromine in acetic acid is employed instead.

Deoxyxanthine does not give recognisable products on oxidation[5] but the N-alkylated analogues can be oxidised with bromine in acetic acid. Derivatives prepared in this way include the 3-methyl,[7] 7-methyl,[7] 3,7-dimethyl,[11] and 1,3,7-trimethyl[38] homologues. In the case of deoxycaffeine, deoxytheophylline, deoxytheobromine, and the deoxy form of 1,7-dimethylxanthine the corresponding 1,6-dihydro-6-bromo-purine intermediate, e.g., 26, can be isolated.[7, 8, 11, 38] Treatment of the intermediate with a hydroxylic solvent may cause hydrolysis to the corresponding 1,6-dihydro-6-hydroxypurine, e.g., 27,[11] or take the process to completion through loss of the elements of water and give

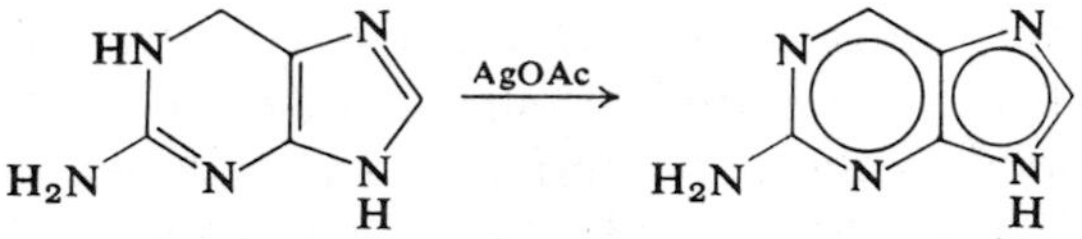

(24)   (25)

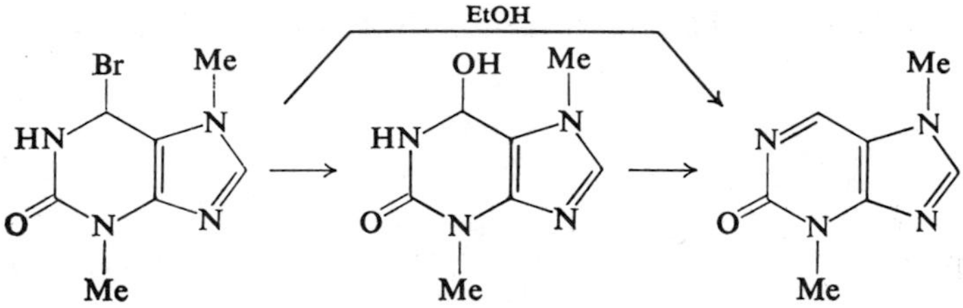

(26)   (27)   (28)

rise to the aromatic form (28). Direct conversion of the bromohydro-purine (26) to the purine (28), by removal of hydrogen bromide, is effected in boiling ethanol.[11, 27]

The deoxyxanthines show increasing basicity as the number of alkyl groups present increases. Thus the 1-methyl derivative[6] has no pro-nounced basic character and the 1,3-dimethyl derivative[8] is weakly basic but a strongly basic character is shown by the 1,3,7-trimethylated purine which forms well-defined salts with mineral acids.[38]

2. Purines Reduced in the Imidazole Ring

Not many examples of this type of purine are known, these being obtained by chemical reduction or by direct synthesis. Hydrogenation or electrochemical reduction does not seem to be applicable, such pro-cedures being successful with the pyrimidine ring only.

A. Preparation of 7,8- or 8,9-Dihydropurines

Replacing the thio group at $C_{(8)}$ by hydrogen has afforded a route to a series of dihydro-7,9-dialkylxanthine derivatives.[39] This is illustrated by the formation of 8,9-dihydro-7,9-dimethylxanthine (30) from the 8-thiopurine (29, R = R' = Me).[40] Corresponding reduced derivatives are likewise formed from 29 (R = Me, R' = Et),[41] 29 (R = Et, R' = Me),[42] and 29, (R = R' = Et)[43]; also from 1,7,9-trimethyl-[44] and 1,3,7,9-tetramethyl-8-thioxanthine.[43] As an alternative to the use of nitrous acid, removal of 8-thio groups in the above compounds can be effected with iodine in aqueous sodium bicarbonate.[39]* An alternative route to 8,9-dihydro-1,3,7,9-tetramethylxanthine is reduction, in aqueous solution of the 1,3,7,9-tetramethylxanthinium cation (31) by sodium or potassium borohydride.[47, 48] The reaction is complicated by formation of the pyrimidine (32) through concomitant degradation of the imidazolone ring in part of the product. Sodium borohydride, in methanol, has been used to convert 7,9-dimethylhypoxanthinium iodide to the 7,8(8,9)-dihydro analogue[32a] (cf. Section 1Ac).

In aqueous solution 4,5-diaminopyrimidine reacts with glyoxal giving 8,8'-bisdihydropurinyl (33).[49] This reaction is restricted to the use of glyoxal as diketones always give rise to pteridines. A detailed study of the condensation resulted in the preparation of a number of 8,8'-bis-

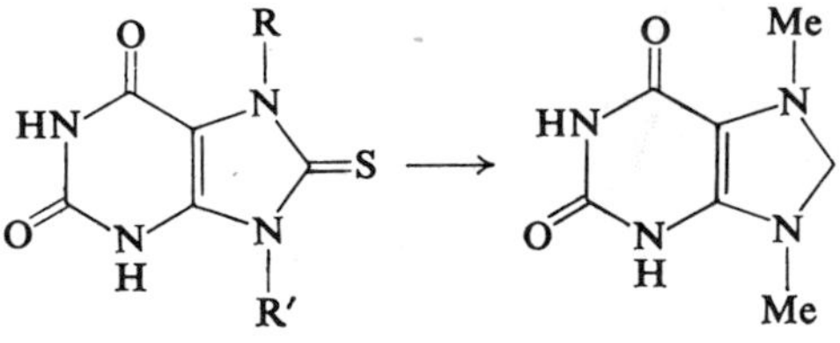

(29)   (30) 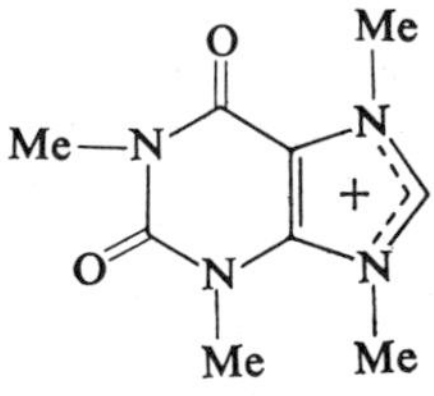   (31)

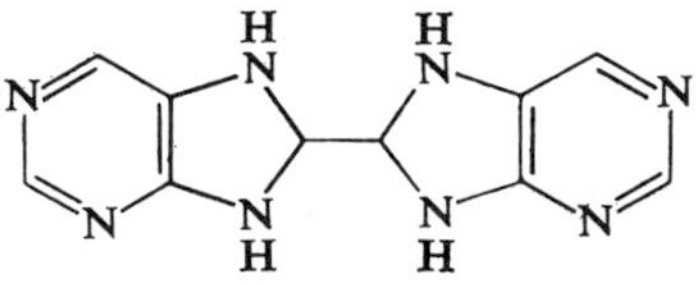

(33) 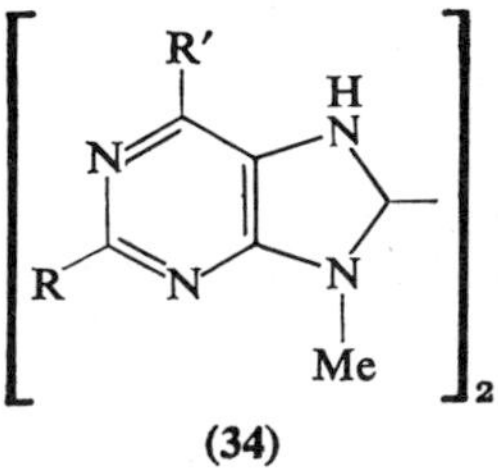   (34) 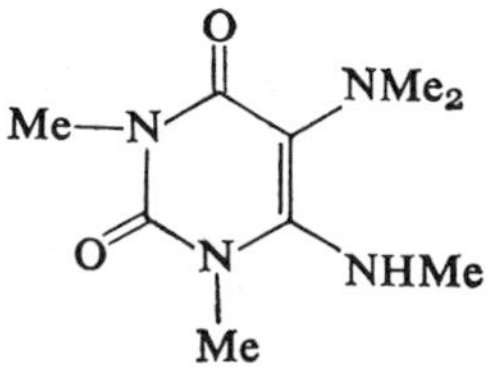   (32)

* This reaction should be compared with the effect of the same reagent and conditions on certain 9-alkyl-8-thiopurines,[45, 46] the products in this case being 9-alkyl-8-iodopurines.

7,8-dihydro-9-methylpurinyls of the general type (**34**).[50] Steric rather than electron density factors are proposed to explain the preferential formation of five- rather than six-membered rings in these cases.[51]

The bisdihydropurinyls are fairly high melting solids; no reactions of this system have been described.

3. The Reduced Uric Acids

Uric acid and many of the *N*-alkylated homologues form derivatives having the $C_{(4)}-C_{(5)}$ double bond reduced. Such compounds are best considered as arising through addition of water, alcohols, or hydrogen halides across the double bond, they cannot be formed by direct reduction methods, e.g., hydrogenation, and no example having a simple 4,5-dihydro bond is known. As their chemistry is peculiar to uric acid it is therefore elaborated elsewhere (Ch. VI, Sects. 10A to 10E).

References

1. Smith and Elving, *J. Amer. Chem. Soc.*, **84**, 1412 (1962).
2. Baillie and Tafel, *Ber.*, **32**, 68 (1899).
3. Yoshitomi, *J. Pharm. Soc. Japan*, **512**, 839 (1924).
4. Albert and Hoskinson, unpublished results.
5. Tafel and Ach, *Ber.*, **34**, 1165 (1901).
6. Tafel and Herterich, *Ber.*, **44**, 1033 (1911).
7. Tafel and Weinschenk, *Ber.*, **33**, 3369 (1900).
8. Tafel and Dodt, *Ber.*, **40**, 3752 (1907).
9. Urabe and Yasukochi, *J. Electrochem. Soc. Japan*, **22**, 469 (1954).
10. Urabe and Yasukochi, *J. Electrochem. Soc. Japan*, **22**, 525 (1954).
11. Tafel, *Ber.*, **32**, 3194 (1899).
12. Tafel, *Ber.*, **34**, 258 (1901).
13. Tafel and Houseman, *Ber.*, **40**, 3743 (1907).
14. Tafel, *Ber.*, **34**, 1181 (1901).
15. Tafel, *Ber.*, **34**, 279 (1901).
16. Tafel and Ach, *Ber.*, **34**, 1170 (1901).
17. Heath, *Nature*, **158**, 23 (1946).
18. Hamer, Waldron, and Woodhouse, *Arch. Biochem. Biophys.*, **47**, 272 (1953).
19. Luthy and Lamb, *J. Pharm. Pharmacol.*, **8**, 410 (1956).
20. Bendich, Russell, and Fox, *J. Amer. Chem. Soc.*, **76**, 6073 (1954).
21. Nakajima and Pullman, *J. Amer. Chem. Soc.*, **81**, 3876 (1959).
21a. Betula, Abstracts, Second Internat. Congress Heterocyclic Chem., 1969, p. 195.; *Annalen*, **729**, 73 (1969).
22. Breshears, Wang, Bechtolt, and Christensen, *J. Amer. Chem. Soc.*, **81**, 3789 (1959).
23. Johnson and Ambelang, *Science*, **90**, 68 (1939).

24. Loo and Michael, *J. Biol. Chem.*, **232**, 99 (1958).
25. Glazko and Wolf, *Arch. Biochem. Biophys.*, **21**, 241 (1949).
26. Woodhouse, *Arch. Biochem. Biophys.*, **25**, 347 (1950).
27. Friedman and Gots, *Arch. Biochem. Biophys.*, **39**, 254 (1952).
28. Fichter and Kern, *Helv. Chim. Acta*, **9**, 380 (1926).
29. Kalmus and Bergman, *J. Chem. Soc.*, **1960**, 3697.
30. Merz and Stahl, *Beitrage zur Biochemie und Physiologie von Naturstoffen*, Gustav Fischer Verlag, Jena, 1965, p. 285.
31. Seyden-Penne, Le Thi Minh, and Chabrier, *Bull. Soc. chim. France*, **1966**, 3934.
32. Bergmann, Levin, Kalmus, and Kwietny-Govrin, *J. Org. Chem.*, **26**, 1504 (1961).
32a. Neiman, *J. Chem. Soc. (C)*, **1970**, 91.
33. Mitter and Chatterjee, *J. Indian Chem. Soc.*, **11**, 867 (1934).
33a. Hoskinson, *Austral. J. Chem.*, **21**, 1913 (1968).
34. Frankland, *J. Chem. Soc.*, **97**, 1686 (1910).
35. Frankland, *J. Chem. Soc.*, **97**, 1316 (1910).
35a. Connolly and Linschitz, *Photochem. and Photobiol.*, **7**, 791 (1968).
36. Bendich, *Chemistry and Biology of Purines*, Eds., Wolstenholme and O'Connor, J. and A. Churchill, Ltd., London, 1957, p. 308.
37. Biltz and Meyer, *Ber.*, **41**, 2546 (1908).
38. Baillie and Tafel, *Ber.*, **32**, 3206 (1899).
39. Biltz, *Annalen*, **426**, 237 (1921).
40. Biltz and Bulow, *Annalen*, **426**, 246 (1921).
41. Biltz and Bulow, *Annalen*, **426**, 264 (1921).
42. Biltz and Heidrich, *Annalen*, **426**, 269 (1921).
43. Biltz and Bulow, *Annalen*, **426**, 299 (1921).
44. Biltz and Heidrich, *Annalen*, **426**, 290 (1921).
45. Biltz and Bulow, *Annalen*, **426**, 306 (1921).
46. Biltz and Beck, *J. prakt. Chem.*, **118**, 149 (1928).
47. El'tsov and Muravich-Alexander, *Tetrahedron Letters*, **1968**, 739.
48. El'tsov, Muravich-Alexander, and Tsereteli, *Zhur. org. Khim.*, **4**, 110 (1968).
49. Mautner, *J. Org. Chem.*, **26**, 1914 (1961).
50. Fidler and Woods, *J. Chem. Soc.*, **1956**, 3311.
51. Fidler and Woods, *J. Chem. Soc.*, **1957**, 3980.

CHAPTER XIII

The Spectra of Purines

This last chapter is a composite work and includes contributed essays, by two other authors, covering ultraviolet and infrared spectra. Additional annotations dealing with nmr and mass spectrometry follows these.

1. Ionisation and Ultraviolet Spectra*†

A. Introduction

For purines, the ionisation constants and ultraviolet absorption spectra are important criteria of identity and homogeneity, frequently the most useful. The estimation of purines, more especially in small quantities, e.g., in chromatography, also uses ultraviolet absorption as the principal technique. The maximal molar extinction coefficients are around 10^4, so that in a conventional 10 mm absorption cell 0.1 μmole or about 20 μg of a purine would give an optical density of about 0.3.

Absorption spectra are generally measured for aqueous solutions and a knowledge of pK_a values is almost always required in order to specify pH values at which the compounds exist virtually as a single species. For example, a compound will exist practically wholly ($>99\%$) as the cationic form at pH values numerically less than 2 units below the basic pK_a, as the neutral molecule more than 2 units above this pH, and less

---

* By P. D. Lawley, Chester Beatty Research Institute, Institute of Cancer Research, London, England.

† In this particular essay adoption of the "hydroxy" rather than "oxo" nomenclature for oxopurines has been preferred. Correspondingly, "mercapto" rather than "thio" is the favoured terminology for sulphur derivatives. However, it should be noted that in the remainder of the chapter and throughout this volume as a whole the "oxo" and "thio" forms are used.

439

than 2 units below the acidic pK_a, and as the anion 2 units above the latter.

A particularly useful criterion of purity is often provided by the existence of isosbestic points. These are wavelengths at which molecular extinction coefficients of two species are identical; the two species may be in rapidly established equilibrium, or one may be in process of conversion at a finite rate into the other. It will be clear that any mixture of the two species will have the same extinction at this wavelength, e.g., for a base and its cation the spectral curves for different pH's over the whole titration range will cross at the isosbestic points; or, in a slow reaction, the spectra at all times will cross at such points. Failure to obtain isosbestic points is characteristic of a system with more than two ultraviolet absorbing components and can therefore be taken as an indication of the presence of impurities, or of side reactions.

The standard deviations in determinations of absorption spectra by different observers have been estimated from a cooperative test[1] involving 80 observers to be less than 1 mμ in wavelength of absorption maxima and 1.5% in ϵ_{max}. For purines, variations of this order and sometimes rather more may be noted between published results of different observers, although differences in wavelengths are generally not greater than 1 mμ. Reported values of pK_a for a given purine sometimes differ by up to 0.5 units of pH. Precautions necessary for accurate measurements of optical density have been discussed by Harding and others.[2] For an account of experimental procedures for determinations of ionisation constants by the spectroscopic and other methods see Albert and Serjeant.[3] In the present work, the designation pK_a should not be taken to imply that any correction for activity coefficients has been made, and for the condition of solvent and temperature the original references should be consulted, as given in Tables 10 to 19 at the end of this chapter. Where more than one set of results is available for a given compound, the choice for inclusion in the tables has generally been made so as to give reference to the most comprehensive data.

B. Use of Spectra for Structural Assignments

The value of measurements of ultraviolet spectra in this field may be illustrated principally by the determination of positions of substituents in the purine ring system. Thus Gulland and co-workers showed that the glycosyl linkage between purines and the sugar-phosphate backbone in nucleic acids was at $N_{(9)}$ by comparison of the absorption spectra of adenosine[4] and guanosine[5] with those of the respective 9-methyl-purines.

It should be noted, however, that although Falconer, Gulland, and Story[6] found that the spectra of 9-methylisoguanine were almost identical with those of crotonoside, the ribosyl derivative of isoguanine, their values for the latter do not agree with those found by Davoll,[7] whose wavelengths for the longer wavelength peaks are consistently 10 mμ higher. The possibility might be considered that a partial rearrangement had occurred during the synthesis of 9-methylisoguanine by Falconer, Gulland, and Story, possibly analogous to that reported by Fischer[8] in the attempted synthesis of 7-methylisoguanine by a similar method. In this way 9-methylguanine instead of 9-methyliso-guanine could have been obtained. Whereas, however, Davoll's results show, as expected, analogy between the spectra of crotonoside and of isoguanine, those of the earlier workers resemble neither those for isoguanine nor those for 9-methylguanine. Thus, although the data of Falconer, Gulland, and Story for 9-methylisoguanine are included in the tables, they must be regarded as almost certainly incorrect.

It should also be noted that neoguanylic acid, a minor guanine nucleotide obtained by Hemmens from commercial guanylic acid, was shown by its spectra to differ from the normal nucleotide in that the sugar moiety is not attached to the base at $N_{(9)}$. Shapiro and Gordon[9] assigned the structure $N_{(2)}$-ribosyl to neoguanosine by the analogy between its spectra and pK_a values and those of $N_{(2)}$-methylguanine. The earlier formulation of neoguanylic acid by Hemmens as a 1-ribosylguanine phosphate was shown to be incorrect.

The elucidation of the products of chemical alkylation of nucleic acids *in vitro* and *in vivo*,[10] and of biomethylation of purines in nucleic acids,[11] also rest largely on comparisons of the ultraviolet spectra of the products with those of authentic alkylpurines.

The naturally occurring base herbipoline, isolated from the sponge *Geodia gigas*, was shown[12] by ultraviolet spectra and pK_a determination to be 7,9-dimethylguanine. A further product, spongopurine, was similarly identified[13] with 1-methyladenine.

Zeatin, a cell division inducing factor from kernels of sweet corn, was shown by ultraviolet spectra and pK_a determinations to be a 6-alkyl-aminopurine derivative,[14] the structure of the side chain then being elucidated by pmr and mass spectroscopic determinations and by synthesis.[15]

Triacanthine, from leaves of *Gleditsia triacanthos*, was shown to be 6-amino-3(3,3-dimethylallyl)purine, the positions of substitution of the side-chain being assigned by comparison of the ultraviolet spectra with those of known *N*-alkyladenines.[16, 17]

The structure of a number of glycosyl derivatives of 6-dimethyl-aminopurine, of interest in the puromycin field, and previously assigned[18] the structures of 7-glycosyl derivatives, were shown[19] by ultraviolet spectral comparison to be 3-glycosyl derivatives, and the use of spectra to distinguish between N-disubstituted adenines has been thoroughly reviewed,[20] reflecting its importance in this and related fields.

C. Use of Spectra for Investigating Tautomerism in Neutral Molecules

Structures of predominant tautomeric forms of hydroxy, mercapto and aminopurines have often been deduced from comparisons of their ultraviolet absorption spectra and ionisation constants with those of corresponding O-, S-, and N-alkylated derivatives.

For example, Mason[21] concluded that 2-, 6-, and 8-amino-substituted purines exist predominantly in the amino- and not the imino forms. An exception is provided by 1-methyladenine (1), which is indicated from ultraviolet spectral and pK_a data[22] to exist as an imino-base ($pK_a = 7.2$). On the other hand, 3-methyladenine, unlike the 1-methyl isomer, shows no acidic pK_a in the region of pH 11 and is, therefore, indicated[23, 24, 25] to exist as the amino form (2).

With hydroxypurines* the evidence from ultraviolet spectra alone is sometimes ambiguous; for example, 6-methoxypurine, hypoxanthine, and 1-methylhypoxanthine have similar ultraviolet absorption spectra, and assignments of the predominance of the oxo as opposed to the hydroxy form rests on infrared spectral evidence.[26] It may be noted that hypoxanthine resembles other hydroxy-substituted heterocycles in that the hydrogen atom prefers the α-nitrogen atom of the ring to that in the γ-position to the substituent.[26, 27]

For the case of 8-hydroxypurine the ultraviolet spectra of the neutral molecules of the parent base and its 7- and 9-methyl derivatives are similar and significantly different from that of 8-methoxypurine. 2-Hydroxypurine is of some interest since its 9-methyl derivative differs significantly in its absorption spectrum from that of the parent base, and the structure 3 was therefore considered more probable for the latter,[26, 27] and 4 for the 9-methyl isomer.

From the near identity of the ultraviolet spectra of theophylline (1,3-dimethylxanthine) and caffeine (1,3,7-trimethylxanthine), the

---

* The use of terms "hydroxypurine" and "mercaptopurine" is retained for convenience irrespective of whether predominance of oxo and thio forms of neutral molecules of such purines has been established.

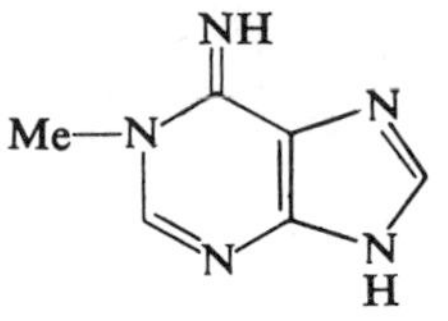

(1) 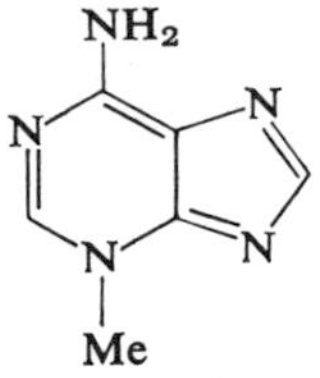   (2) 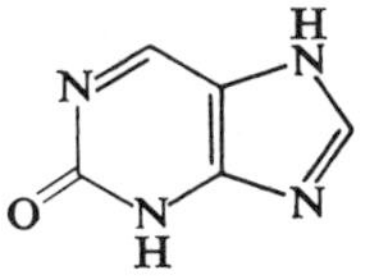   (3)

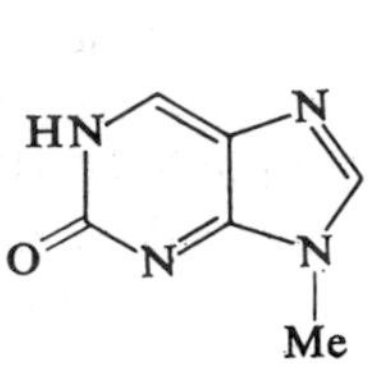

(4) 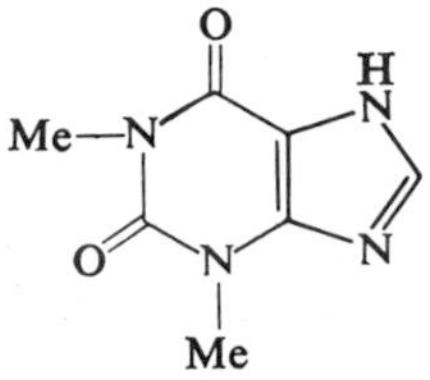   (5)

structure **5** was deduced for the former. The spectra of xanthine and its 1- and 1,7-dimethyl derivatives are also closely similar to those of caffeine and predominance of the oxo forms of these neutral molecules is therefore indicated.[28]

Predominance of thione, as opposed to thiol, forms for 2-, 6-, and 8-mercaptopurines is supported by the marked differences between their ultraviolet absorption spectra and those of the corresponding methylthio derivatives.[26] An exception is provided by 3-methyl-6-mercaptopurine; its structure was deduced by Elion[29] to be probably analogous to that of 3-methyladenine, i.e., **2** with —SH replacing —NH$_2$, since, as with the latter, methylation at N$_{(3)}$ in contrast to N$_{(1)}$ produces a relatively large bathochromic shift from the absorption peak of the parent base.

Montgomery and Hewson[30] found that 2-fluoro-6-mercaptopurine and its ribosyl derivative presented interesting features. These appear to be cases where transformation of the thione to thiol forms can be demonstrated. The ultraviolet spectra of the bases in ethanol resemble those of 2-fluoro-6-methylthiopurine. When the anions existing in aqueous alkali were acidified, the absorption peaks shifted immediately to positions characteristic of the corresponding thiones, but then a further shift at measurable rate to the peak associated with the thiol was discerned. The interpretation was that the powerful inductive effect of the 2-fluoro substituent withdraws electrons from the N$_{(1)}$ atom, so that the thione base becomes the metastable form; thus while the initial site of protonation of the anion is at N$_{(1)}$, slow rearrangement to the thiol form ensues (Scheme 1).

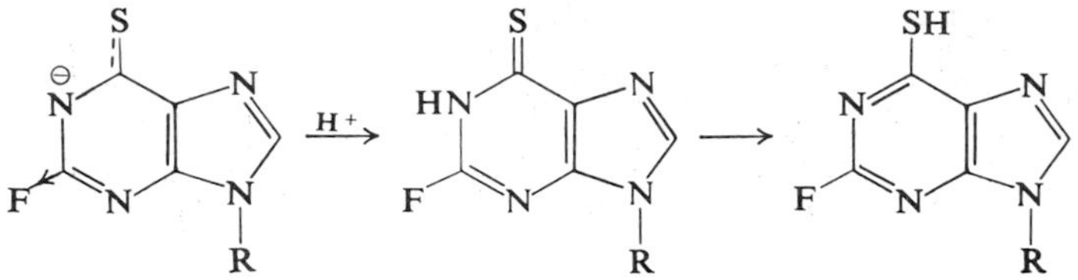

SCHEME 1

The aminohydroxypurines, like the hydroxypurines, appear to exist generally in the oxo forms; the possible existence of a zwitterionic structure for guanosine was ruled out[31] by comparison of its ultraviolet spectrum as a neutral molecule with that of a 7,9-dialkylguanine which must be a zwitterion; evidence from nmr and infrared spectroscopy enabled conclusive assignment of the structure to be made for guanosine.

D. Assignments of Structure of Ionic Forms

The position of protonation* of purine appears not to have been established unequivocally. It may be noted that protonation of the base causes a small hypsochromic shift (3 mμ); the similarity in spectra for the neutral and cationic forms of purine and 9-methylpurine, respectively, suggests that both protonate at the same position,[32] which on the general grounds that the pyrimidine ring tends to withdraw π-electrons from the imidazole ring, would be expected to be in the former moiety. The much more pronounced shift due to protonation of 7-methylpurine suggested a site for its protonation different from that of the 9-methyl isomer.[32] Studies of di(N-alkyl)purines, as yet unavailable, might clarify these questions; in this connection it may be noted that attempts to methylate purine under a variety of conditions failed.

The question of the sites of protonation of adenine in aqueous acid solution also remains incompletely resolved. On general grounds, Mason[21] considered that heterocyclic bases with an amino group α- or γ- to a ring nitrogen atom were unlikely to protonate at the extranuclear amino group. In agreement with this concept, the ring nitrogen atom at $N_{(1)}$ in adenine hydrochloride was indicated to be the site of protonation by a crystallographic study,[34] but alternative tautomers of protonated adenine might exist in aqueous solution. For adenosine, in acid D_2O, a

---

* "Position of protonation" is used here to indicate the site of attachment of a proton to the purine ring, but does not imply that the positive charge on the resulting cation is there located. Correspondingly "ionisation" of a proton from a ring —NH— group does not imply location of the negative charge at that position.

proton magnetic resonance study[35] indicated $N_{(1)}$ as the principal site of protonation.

Lewin[36] showed that when adenine in aqueous solution at a pH near its basic pK_a value reacts with formaldehyde, a release of acid occurs. This result was interpreted as showing that the positive charge in cationic adenine is located on the amino group, but does not distinguish between the possible tautomers **6a, 6b** and **6c, 6d.**

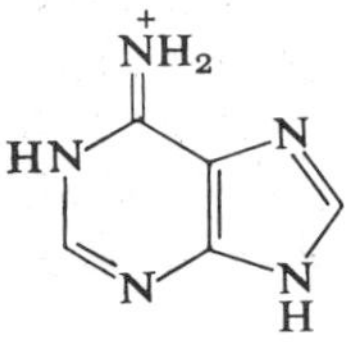

(6a) 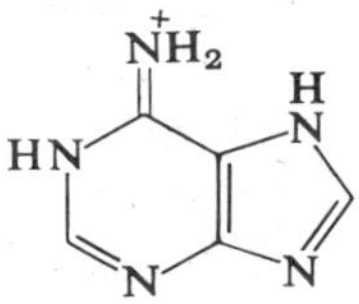   (6b) 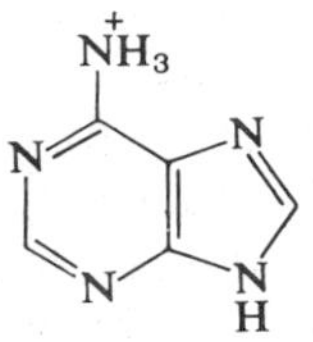   (6c)

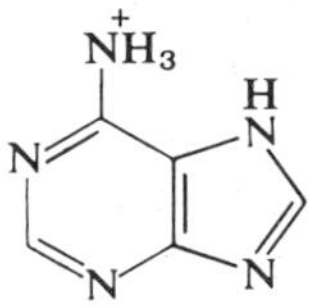

(6d) 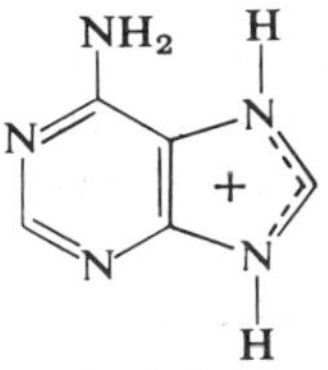   (6e)

The ultraviolet absorption spectrum of adenine in acid solution as the singly protonated form (λ_{max} 263 mμ) differs significantly from that of 1-methyladenine (λ_{max} 259 mμ); a doubly protonated form (λ_{max} 261 mμ) can also be found at pH values below zero. On the reasonable assumption that methylation and protonation would have closely similar effects on the adenine spectrum, it is then evident that $N_{(1)}$ is unlikely to be the sole site of protonation.

Also working on the assumption that effects of methylation and protonation should be similar, Börresen[36a] has compared the fluorescence and excitation spectra of adenine (monocation), 7-methyladenine, and 1-methyladenosine, at pH 2; he also found that 1-methyladenine, 3-methyladenine, and 3,7-dimethyladenine did not fluoresce. These comparisons indicated that protonation of adenine at $N_{(1)}$ could not account for the observed fluorescence of the monocation derived from adenine, but could account for the absorption spectrum of the nonfluorescent component. The fluorescence spectrum of adenine resembled that of 7-methyladenine, and the fluorescent tautomer was therefore indicated to be **6d** or **6e**; the latter was considered less likely on theoretical grounds. The observed fluorescence of adenosine at pH 2, which resembles that of 1-methyladenosine, could be ascribed to a tautomer of the type **6a** with a ribosyl substituent at $N_{(9)}$; the non-fluorescence of 1-methyladenine might then be ascribed to predominance of the tautomer of type **6b** with a methyl substituent at $N_{(1)}$.

2-Hydroxypurine is exceptionally unstable to acid,[33] the imidazole ring being opened. Protonation of this ring has been suggested as a possible cause, with $C_{(8)}$ being proposed as a possible site of attack.

For guanosine protonation of the imidazole ring is well established,[31] the predominant structure of the cation being **7**. The crystal structure of 9-methylguanine hydrobomide confirms[37] that here also protonation occurs at $N_{(7)}$. Comparisons of the ultraviolet spectra of the cations of the 9-alkylguanines show resemblance to those of 7,9-dialkylguanines, but the evidence from comparisons of their infrared spectra in D_2O was more conclusive.[31]

Acidic dissociation of guanosine[31] and hypoxanthine is indicated to yield anions with the negative charge principally on the oxygen atom as shown in **8**. The assignment in the former case followed from infrared evidence, in the latter from metal chelation studies[38, 39] indicating the structure **9** and from observed similarity between the ultraviolet absorption spectrum with that of the neutral molecule of adenine. In **10** the symbol M^+ represents a metal ion; a hydrogen-bonded proton could also be envisaged to occupy this position as in the anion **9**.

(7)   (8)   (9)   (10)

The ultraviolet spectrum of the anion of 8-hydroxypurine resembles that of the anion of its 7-, rather than of its 9-methyl derivative, and the first ionisation therefore probably occurs[27] by dissociation of the proton from $N_{(9)}$.

A detailed study of the ultraviolet spectra of xanthine and substituted xanthines led Cavalieri and co-workers[28] to assign structures for the predominant anionic forms of these compounds. For theophylline the structure **11** was considered most likely for the monoanion with a minor contribution from **12**.

The variation with pH of the spectra of 3,7-dimethylxanthine and of 1,3-dimethylxanthine are similar, and the same pattern was observed for the first dissociation of 3-methylxanthine, which was therefore concluded to give **13** as a monoanion and **14** as a dianion.

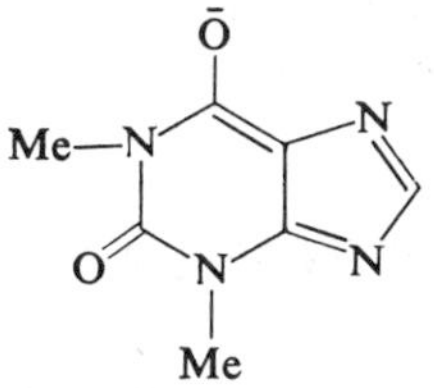

(11) 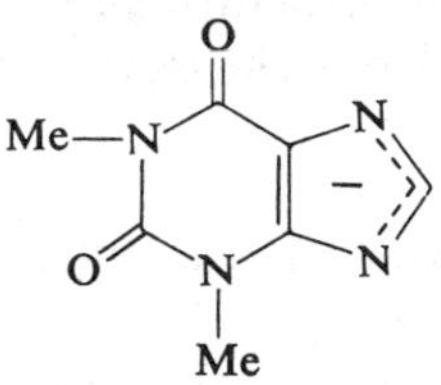   (12) 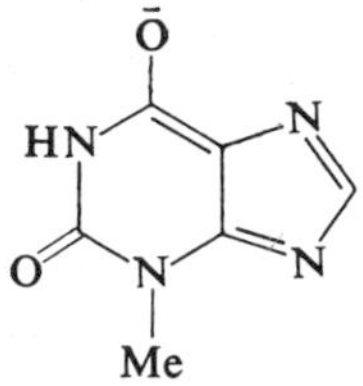   (13)

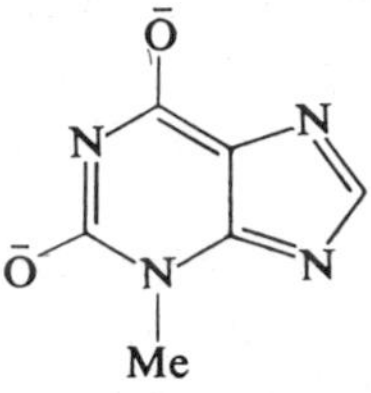

(14) 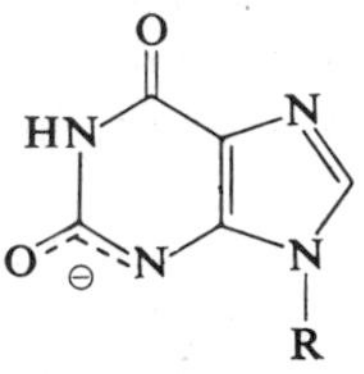   (15)

The changes in spectra with pH for xanthine, 1-methylxanthine, 1,7-dimethylxanthine, and xanthosine resemble each other but fall into a different category from the former group, and it was therefore deduced that the first ionisation involves $N_{(3)}$, e.g., xanthosine gives **15** (R = β-D-ribofuranosyl) and the second, $N_{(1)}$.

Elion[29] found it difficult to assign the ionising groups of hypoxanthine or its 1-, 3-, 7-, or $O_{(6)}$-methyl derivatives, and suggested that more than a single group might contribute in each case. As previously noted, for hypoxanthine Albert[38] considered ionisation of a proton from $N_{(1)}$ to be most likely.

For 6-mercaptopurine derivatives Elion pointed out that the observed marked increase in acidic pK_a due to methyl substitution at $N_{(1)}$ or at the extranuclear sulphur atom was consistent with the predominance of ionisation at $N_{(1)}$ in other monomethylated derivatives.

2-Mercapto- and 2-methylthiopurine are stronger bases than the corresponding 6-substituted isomers, and it was therefore deduced[39] that here $N_{(1)}$ is the most basic nitrogen, since "resonance" stabilisation of a cation formed by protonation at $N_{(3)}$ would be expected to be greater for the 6-substituted derivatives.

Bergmann and Dikstein[40] studied the effects of ionisation on ultraviolet absorption for uric acid and its methylated derivatives. They concluded that $N_{(9)}$ was the first site of ionisation in uric acid, then $N_{(3)}$. For xanthine the first dissociation was attributed to $N_{(3)}$ (i.e., in agreement with Cavalieri and co-workers[28]), but the second to $N_{(7)}$ or $N_{(9)}$, rather than to $N_{(1)}$.

E. Influence of Structure and Environment on
Ionisation—Theoretical Studies

Albert and Brown[33] reviewed the ionisation of 34 simple purines, their principal conclusions being as follows.

The acidic strength of purine (pK_a 8.9) is greater than that of glyoxaline (pK_a 13) or of benzimidazole (pK_a 12.3), and the basic strength (pK_a 2.4) weaker (glyoxaline, pK_a 7.0; benzimidazole, pK_a 5.5), although greater than that of pyrimidine (pK_a 1.3). These values reflect the withdrawal of electrons from the imidazole ring by the pyrimidine ring in purine.

As expected, methyl substituents show feeble acid-weakening and base-strengthening properties, and 8-phenylpurine (pK_a 8.1) is a somewhat stronger acid than purine; 7- and 9-methylpurine, of course, lack acidic properties since the ionisable proton is absent. The 2-fluoro substituent causes a slight increase in acidity of purine, but a much more marked effect in decreasing the basicity of adenine[30] (no basic pK_a could be detected for 2-fluoro-6-aminopurine in 50% ethanol). The trifluoromethyl group in 2-, 6-, or 8-positions causes larger increases in acidity of purine and amino- or hydroxypurines, and correspondingly marked decreases in basicity.[41] 6-Chloropurine was noted[32] to be a somewhat stronger acid than purine, and a weaker base, although hydrolysis to hypoxanthine prevented detailed study of its protonation.

The 2-, 6-, and 8-hydroxypurines possess two anionic dissociations, the first of which for the 2- and 8-derivatives occurs at slightly lower pH than for purine, while the pK_a values for 6-hydroxypurine (hypoxanthine) and its 9-methyl derivative are the same as for the parent heterocycle.

A nitro substituent at $C_{(2)}$ in inosine (9-β-D-ribofuranosylhypoxanthine) causes a marked increase in acidity, a single pK_a value of 3.3 being found.[42]

The basic strength of 2- and 6-hydroxypurines is less than, and that of 8-hydroxypurine slightly greater than, that of purine.

As expected, amino substituents exert a generally base-strengthening action, rather greater than the base-weakening effect of the 2- and 6-hydroxy substituents, confirming the stabilisation of cations by the presence of guanidino groups. Methylation of amino substituents decreases acid strength to an extent greater than the corresponding increase in basic strength; in one case, 6-methylaminopurine, pK_a 3.9, compared with adenine, pK_a 4.2, basic strength is actually decreased.

Mercaptopurines are more acidic than hydroxypurines, e.g., 6-mercaptopurine, pK_a 7.8, 10.8 compared with hypoxanthine, pK_a 8.9, 12.1; they are correspondingly weaker bases. Methylthio substituents have less effect on acidic pK_a values; 8-methylthiopurine is notable in being a weaker acid and stronger base than purine, and is thus in line with the general tendency for substituents at the 8-position to have a weaker acid-strengthening property, but a stronger base-strengthening effect, than at the 2- or 6-positions. The methylthio substituent in the 6-position may also be noted as weakening base strength significantly more than as a 2-substituent; as already noted this has been deduced to correlate with $N_{(1)}$ as the site of protonation.[38]

Cavalieri and co-workers[28] have discussed the relationship between structure and ionisation constants for xanthine and its N-substituted derivatives. The pK_a of 1,3-dimethylxanthine (8.8) is greater than that of 3-methylxanthine (8.3) and that of 1,7-dimethylxanthine (8.7) is similarly greater than that of 7-methylxanthine (8.3); these comparisons reflect the increased electron density in the vicinity of the ionisable groups (deduced to be $N_{(7)}$ and $N_{(1)}$, respectively) in the disubstituted purines.

The pK_a of 7-methylxanthine (8.3) is higher than that of xanthine (7.5), which in turn is higher than that of 9-methylxanthine (6.3). The explanation suggested was that xanthine could be regarded as a composite of two tautomers each with zwitterionic resonance forms (Scheme 2).

In 7-methylxanthine only forms of the type 16a, 16b can contribute; the anion being derived by ionisation at $N_{(3)}$, and because of the

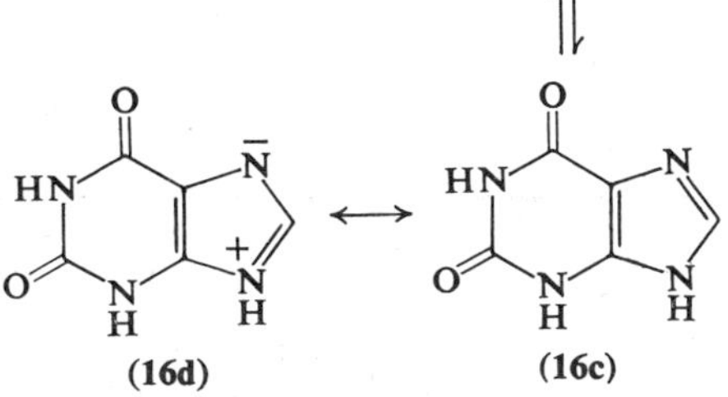

SCHEME 2

proximity of the negative charge at $N_{(9)}$ in **16a**, ionisation will occur less readily than for xanthine. On the other hand, the positive charge at $N_{(9)}$ in 9-methylxanthine, confined to structures **16c** and **16d**, will exert an opposite effect.

Theoretical studies of the ionisation of purines have been made principally by A. and B. Pullman and their co-workers. Nakajima and B. Pullman[43] pointed out that the determining factor for basicity of heterocyclic bases should be not the electron density at the ring nitrogen atoms, but rather the ionisation potential of the lone pair of electrons on these atoms. The lower the value of this parameter, given by $I_d =$ constant $+ \sum Q_p(dd/pp)$, where Q_p is the net charge of an atom p, and (dd/pp) the Coulomb integral between the lone pair electrons of an N-atom and of the π-electrons of all the atoms p in the ring system, the greater the basicity. The positions of highest basicity could thus also be predicted. More recently the calculations of Nakajima and Pullman were refined by Veillard and Pullman[44] by the use of the self-consistent field method of Pariser and Parr, and Pople.[45] The results of these later calculations are shown in Fig. 1.

It will be noted that a good correlation between $-I_d$ and basic pK_a values was obtained. The most basic nitrogen atom of purine and of adenine was indicated to be $N_{(1)}$, the latter in agreement with experi-

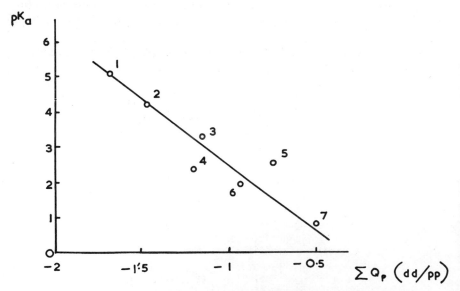

FIG. 1. Comparison of experimental and calculated basicities of purines (from data of Veillard and Pullman[44]). 1. 2,6-diaminopurine, $N_{(3)}$; 2. adenine, $N_{(1)}$; 3. guanine, $N_{(7)}$; 4. purine, $N_{(1)}$; 5. 8-hydroxypurine, $N_{(1)}$; 6. hypoxanthine, $N_{(7)}$; 7. xanthine, $N_{(7)}$.

mental assignments as discussed in Section 1D, while for 2,6-diamino-purine $N_{(3)}$ is taken to be the most basic site.

For hypoxanthine, xanthine, and guanine $N_{(7)}$ is the predicted site of protonation, and for 8-hydroxypurine, $N_{(1)}$. It may be noted that experimental evidence favours $N_{(7)}$ as the site of protonation of guanosine and of 9-methylguanine (*cf.* Sect. 1D), which may be considered analogous to guanine if the mobile H atom is predominantly attached to $N_{(9)}$.

Where definitive studies of sites of protonation are lacking, support for the assignments of basic ring N atoms may be sought from studies of alkylation of neutral molecules. This follows from the concept that the basic and nucleophilic centres of heterocyclic bases are identical. It should be noted that this theoretically sound concept has also received general experimental support, but there are interesting differences in certain cases. Whereas the alkylation sites of neutral molecules are without exception ring N atoms, their relative reactivities are not always the same as their relative basicities. Thus, adenine alkylates at $N_{(3)}$[22, 23] but the susceptibility of the adenine nucleus to perturbation by methyl substitution is illustrated by the predominant alkylations of 3-methyl-adenine at $N_{(7)}$, of 7-methyladenine at $N_{(3)}$, and of 9-methyladenine at $N_{(1)}$.[46]

Minor sites of alkylation react simultaneously with the principal sites, illustrating the small differences in basicity of the various ring N atoms; in adenine, for example, $N_{(1)}$ and $N_{(9)}$ are the less reactive sites.

With regard to these observed differences between sites of alkylation and protonation, and also to the admittedly minor discrepancies between the semiempirical theoretical and empirical assignments of basic sites in the adenine nucleus, Kasha[47] has discussed some problems presented by the former.

Apart from the lack of sufficient experimental data with which to determine resonance integrals and heteroatom parameters with precision, there are difficulties in allowing for the important factor of solvation, which are not generally considered in the theoretical studies. At present, therefore, discrepancies between theory and experiment in this field are not unexpected.

Extensive studies[49] of alkylation of hydroxypurines showed that hypoxanthine and xanthine react at $N_{(7)}$ and $N_{(9)}$, and their 7- and 9-substituted derivatives at $N_{(9)}$ and $N_{(7)}$, respectively; this predominance of imidazole ring alkylation agrees with the theoretical studies. It will be recalled that Albert and Brown[33] made the sole proposal for protonation at a carbon atom, $C_{(8)}$ of 2-hydroxypurine; although theory indicates[49] that electrophilic attack would occur at this site for guanine, and this agrees with experimental results, no study of 2-hydroxypurine appears to have been made; the general lack of correspondence between nucleophilic centres and receptor sites of electrophilic attack in purines may however be noted.

6-Mercaptopurine alkylates[48] at S and at $N_{(3)}$, but it has been deduced[39] to protonate at $N_{(1)}$; the analogy with adenine, as in other respects (*cf.* Sects. 1C and 1D), will again be noted. No theoretical study of mercaptopurines appears to have been reported yet.

For acidity of ring N atoms, the corresponding calculations by Nakajima and Pullman[43] took as their basis the relationship between the energy change due to liberation of a proton from a ring —NH— group and the net charge on that N atom, the conclusion being that acidity should increase as the value of $\Sigma\,Q_p(N_HN_H/pp)$ increased. Figure 2 shows the results of Veillard and Pullman's refinement[44] of this type of calculation.

It will be noted that for purine, adenine, guanine, hypoxanthine, and uric acid, ionisation at $N_{(9)}$ is predicted, while for xanthine $N_{(7)}$ or $N_{(3)}$ is indicated. The assignment for hypoxanthine is not that deduced by Albert,[38, 39] who gave $N_{(1)}$ as the acidic site. That for xanthine agrees with Cavalieri's conclusion[28] that $N_{(3)}$ is the primary ionising atom,

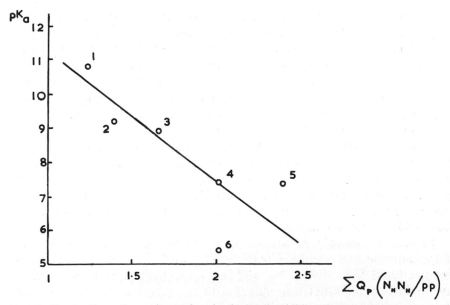

FIG. 2. Comparison of experimental and calculated acidities of purines (from data of Veillard and Pullman[44]). 1. 2,6-diaminopurine, $N_{(9)}$; 2. guanine, $N_{(9)}$; 3. hypoxanthine, $N_{(9)}$; 4. xanthine (H on $N_{(7)}$), $N_{(3)}$; 5. xanthine (H on $N_{(9)}$), $N_{(9)}$; 6. uric acid, $N_{(9)}$.

provided that $N_{(7)}$ is the predominant site of attachment of the remaining mobile hydrogen of the imidazole ring. For uric acid, the theoretical and empirical[40] assignments are in agreement.

F. Influence of Structure and Environment on Spectra—
Empirical Correlations and Theoretical Studies

The ultraviolet absorption spectra of purines may be classified[21] broadly into three groups of bands. At the longer wavelengths (around 300 mμ and above), the presence of weak ($\epsilon < 10^3$) $n \to \pi^*$ bands can be shown in certain cases. For this type of transition interaction of polar solvents with the ground state of the molecule would be expected to be greater than for the excited state[50, 51]; thus bathochromic shifts in passing from polar to nonpolar solvents are generally diagnostic for such transitions. They also characteristically possess low extinction coefficients, i.e., the transition is "forbidden," with low transition moment, since it involves removal of a nonbonding electron from an sp^2 hybrid atomic orbital "in plane" with the heterocyclic ring to a π^* molecular orbital extending above and below the plane of this ring. These transitions are polarised at right angles to this plane.

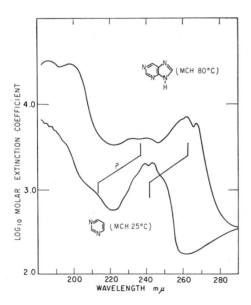

Fig. 3. Absorption spectra of pyrimidine and purine in methylcyclohexane; the lines indicate a possible correlation of bands (From Clark and Tinoco,[52] by permission of the American Chemical Society.)

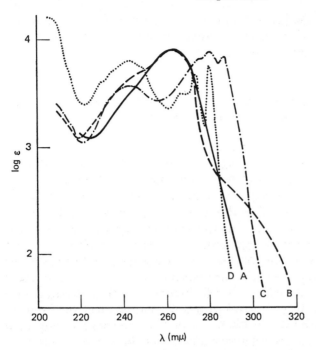

FIG. 4. Absorption spectra of: A. 9-methylpurine, neutral molecule, pH 5.1; B. 9-methylpurine, in cyclohexane; C. 1:3:4-triazaindene, in cyclohexane; D. benzimidazole, in cyclohexane. (From Mason,[21] by permission of the Chemical Society.)

Also lying within the longer wavelengths in the ultraviolet (230–300 mμ) are found the second group of transitions, the $\pi \rightarrow \pi^*$ transitions of lower energy. Here the excited state is more polar than the ground state and interacts more with solvents of high dielectric constant; hypsochromic shifts in passing to less polar solvents are therefore expected, and extinction coefficients are relatively high (2×10^3 to 2×10^4). The direction of polarisation is in the plane of the purine ring. This group was designated x bands by Mason,[21] who further discerned at least two bands in this category.

The third group, Mason's y bands, being the $\pi \rightarrow \pi^*$ transitions of higher energy, occur below 230 mμ and the maxima require care for their accurate determination, since even the highest wavelength bands in this group may lie below the limit of about 210 mμ generally accessible with conventional spectrophotometers. Extinction coefficients are higher again than for the second group, generally exceeding 2×10^4.

Empirical correlations of the ultraviolet spectra of purine with those of simpler ring systems take as their starting point pyrimidine (Fig. 3),[52] related to benzene, or benzimidazole and 1,3,4-triazaindene, related to naphthalene and styrene (Fig. 4).[21] For 2- or 6-substituted purines, analogies with the spectra of the corresponding substituted 4,5-diamino-pyrimidines have been discerned (Fig. 5).[21] For purine in methylcyclo-hexane, and for 9-methylhypoxanthine in the vapour phase,[53] analogies with benzimidazole and with acetophenone vapour spectra were considered significant (Fig. 6).

From a comparison of the spectra of 9-methylpurine in cyclohexane, and less obviously in neutral aqueous solution, with those of 1,3,4-triazaindene and benzimidazole, Mason[21] deduced that the x band was divisible into two components, principally the longitudinally polarised x_1 transition, with a minor contribution from a transverse polarisation, x_2. This assignment was supported by study of the effects of substituents on the distribution of intensities within these bands, and on their wavelengths.

Polar substituents cause the most marked effects in the 2-position, while 8- and 6-substituents have, progressively, less effect. The 6-sub-stituents, being transversely disposed with respect to the longer axis of the purine molecule, were considered to lower the transition energy of only the weaker component of the x system, and generally to raise that of the stronger component. This could result therefore in an envelope

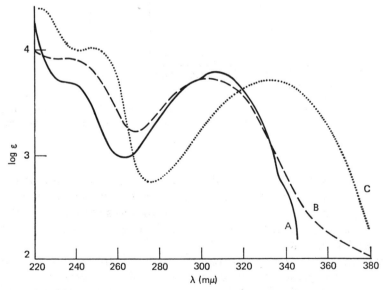

FIG. 5. Absorption spectra of: A. 2-aminopurine, neutral molecule; B. 2,4,5-triamino-pyrimidine, neutral molecule; C. 2-dimethylaminopurine, neutral molecule. (From Mason,[21] by permission of the Chemical Society.)

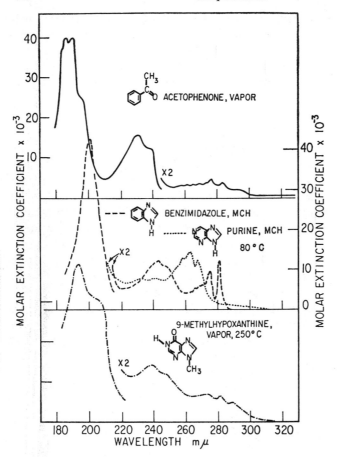

Fig. 6. Absorption spectra of 9-methylhypoxanthine, purine, benzimidazole, and acetophenone. (From Clark and Tinoco,[52] by permission of the American Chemical Society.)

showing a single peak. The gradual merging of the two components, according to Clark and Tinoco,[52] is seen as the purine system is increasingly perturbed by chloro, methyl, and amino substituents at the 6-position (Fig. 7), the shoulder at about 240 mμ (x_2 band?) being least obvious for adenine.

It should be noted, however, that the detailed assignment of two bands in this series presents difficulties in view of the overlapping and lack of defined peaks; for example, whether the hidden x_2 band lies to the longer or shorter wavelength side of the adenine maximum appears difficult to decide.

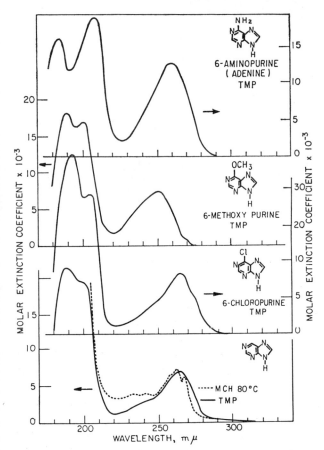

FIG. 7. Absorption spectra of purine and some 6-substituted purines, in trimethyl-phosphate solution (TMP) and methylcyclohexane (MCH). (From Clark and Tinoco,[52] by permission of the American Chemical Society.)

The 2-substituents were also considered by Mason[21] to cause more marked shifts in the x_2 band than in the x_1, whereas the 8-substituents could shift the x_1 more specifically, to enable its separation from the x_2 component. Thus 8-hydroxy and 8-aminopurine show two well-defined x-bands at 235, 277 mμ and at 241, 283 mμ, respectively. 2-Hydroxy-purine also shows two x bands, at 238 and 315 mμ better defined for the anion, at 271 and 313 mμ (Fig. 8); for 2-aminopurine, however, the x_2 band remains as a shoulder at 236 mμ, the main peak being at 305 mμ.

For the di- and trisubstituted purines, simple additivity of effect of substituents is not observed, and purines are distinguished from

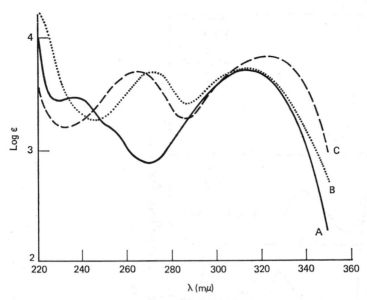

Fig. 8. Absorption spectra of 2-hydroxypurine: A. neutral molecule; B. anion; C. cation. (From Mason,[21] by permission of the Chemical Society.)

pyrimidines in this respect. An amino or hydroxy group introduced into the 6-position of an 8- or 2-substituted purine causes a greater hypsochromic shift than it does as a single substituent in purine itself; in the 2-position such a group additional to an existing 6- or 8-substituent conversely causes a smaller bathochromic shift than it does as a single substituent. An 8-hydroxy substituent in 2- or 6-substituted purines gives little effect. Conjugation between the substituents is thus clearly indicated; detailed analyses of the observed shifts in wavelength and extinction are given by Mason.[21]

Clark and Tinoco[52] drew conclusions similar to those of Mason regarding the composite nature of the x band system, but their nomenclature differs in deriving from that of the benzene spectrum. The 280 mμ band of 9-methylhypoxanthine (x_1 band of Mason) was assigned, by analogy with that of benzophenone at the same wavelength, as correlating with the $A_{1_g} \rightarrow B_{2_u}$ (260 mμ) band of benzene. The x_2 band (240 mμ) was correlated with the 230 mμ band of benzophenone (Fig. 6), probably analogous to the $A_{1_g} \rightarrow B_{1_u}$ transition of benzene (203 mμ). The y bands (180–210 mμ) were related to the $A_{1_g} \rightarrow E_{1_u}$ transition of benzene (180 mμ). See Table 8.

These assignments derived further support from comparison of the spectra of 9-methylhypoxanthine and 9-ethylguanine with those of

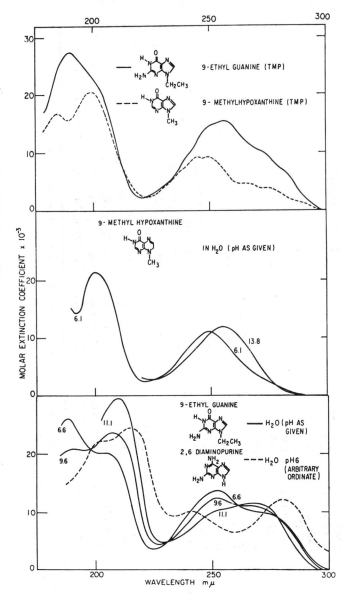

FIG. 9. Absorption spectra of 9-ethylguanine, 9-methylhypoxanthine, and 2,6-diamino-purine, in trimethyl phosphate and aqueous solution. (From Clark and Tinoco,[52] by permission of the American Chemical Society.)

TABLE 8. Spectral Data for Purines and Related Compounds and Suggested Assignment of Bands, λ mμ $(10^{-3}\epsilon)$.†

	Solvent[a]	B_{2u}	B_{1u}	E_{1u}	Other
Purine	TMP	265 (6.9)	240[i] (3.0)	200 (18.1), 188 (21.1)	290 (0.6), N($n\pi*$)
Pyrimidine	MCH	242 (2.0)	210[i] (1.0)	190 (6.0)	310 (0.35), N($n\pi*$)
Benzimidazole	MCH	275 (4.7)	243 (6.2)	201 (42.8)	
Acetophenone	vapour	276 (2.2)	230 (15.6)	190 (40.0)	320 (0.15), O($n\pi*$)
6-Chloropurine	TMP	265 (8.2)	245[i] (3.9)	205 (19.0), 193 (24.2)	
Adenine	vapour	252		207	
	TMP	260 (12.6)		208 (18.7), 185 (15.8)	
	H$_2$O, pH 7	260		207	
9-Methyladenine	vapour	249			
	MCH	258			
	acetonitrile	259		212	
	TMP	260		210	
	H$_2$O, pH 7	260			
Hypoxanthine	acetonitrile	269[i]	[249, 244][b]		
	TMP	280[i]	250	196	
	H$_2$O, pH 7	251 (7.3)	249		
6-Methoxypurine	TMP	[290, 281, 273][b]		200 (16.7), 190 (18.0)	
9-Methylhypoxanthine	vapour	278 (3.8)	239	205[i], 193	
	TMP	260–270[i]	247 (9.3)	198 (20.4), 185 (16.8)	
	H$_2$O, pH 6	255 (11.8)	249 (11.1)	200 (21.5), 183	
	H$_2$O, pH 13.8				
2,6-Diaminopurine	H$_2$O, pH 6	280	242	215, 202	
9-Ethylguanine	TMP	275[i] (9.4)	256 (15.4)	203[i] (20.0), 190 (27.4)	
	H$_2$O, pH 6.6	275[i] (9.6)	252 (13.6)	205 (20.0), 188 (26.0)	
	H$_2$O, pH 9.6	270[i] (10.8)	252 (12.2)	208 (23.7), 190 (21.0)	
	H$_2$O, pH 11.1	269 (11.5)	252[i] (10.2)	210 (29.6), 200 (16.7), 190 (18.0)	

† From Clark and Tinoco,[52] and Clark, Peschel, and Tinoco,[53] by permission of the American Chemical Society.
[a] TMP, trimethyl phosphate; MCH, methylcyclohexane.
[b] Vibrational structure.
[i] Inflexion or shoulder.

2,6-diaminopurine (Fig. 9). The 6-hydroxy substituent shifts the x_1 band hypsochromically, so that it lies beneath the long wavelength tail of the x_2 component. For the anion of the former, cf. **9**, the spectrum resembles that of adenine. However, in this case the intensity of the longer wavelength band should be increased, reflecting the increased transition moment, and support for this view came, therefore, from the observation that in the spectrum of the anion of 9-ethylguanine, where the two components are well separated, the longer wavelength components of both the x and y systems become of higher intensity than those of the shorter wavelength components. The overall spectrum resembles that of 2,6-diaminopurine (Fig. 9).

Solvent effects on the absorption spectra of purine,[54] 2-aminopurine, and 2,6-diaminopurine [55] have been studied in some detail. The results were considered to amplify, and to some extent modify, Mason's conclusions regarding the classification of the absorption bands of purine and its 2- and 6-substituted derivatives.

Purine[54] in methylcyclohexane showed a long wavelength tail more prominently than in water, i.e., the existence of a hidden $n \rightarrow \pi^*$ band could be deduced (Fig. 10). Vibrational structure of the 263 mμ band was resolved in the organic solvent and the 242 mμ band also appeared

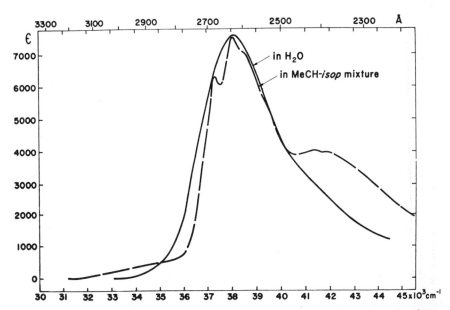

FIG. 10. Absorption spectra of purine in a methylcyclohexane-isopropanol mixture (1500:1, v/v), and in water; concentration, 10^{-4} M, 1 cm cell. (From Drobnik and Augenstein,[54] by permission of Pergamon Press Ltd.)

as a distinct peak (Fig. 11). For a series of solvents of increasing dielectric constant, the vibrational structure, separation of x_1 and x_2 bands, and long wavelength tail, all decreased. These changes were considered to depend on the ability of the solvents of high dielectric constant to form hydrogen bonds with the nonbonding electrons on the ring nitrogen atoms.

With 2,6-bis(diethylamino)purine,[55] the long wavelength tail is, contrary to the predictions concerning $n \rightarrow \pi^*$ transitions, most prominent for water as solvent; the 245 mμ band is the predominant x band, and both this and the 295 mμ bands have shoulders, the nature of which was considered uncertain (Fig. 12). In water the 295 mμ peak was shifted hypsochromically relative to the position in nonpolar solvents while that at 245 mμ shifted bathochromically.

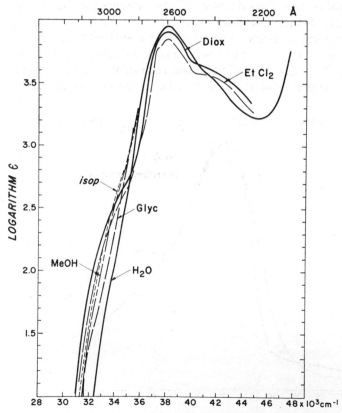

FIG. 11. Absorption spectra of purine in various solvents. Only the long wavelength part is given in isopropanol, methanol, and glycerol. The ordinate is $\log_{10}\epsilon$. (From Drobnik and Augenstein,[54] by permission of Pergamon Press Ltd.)

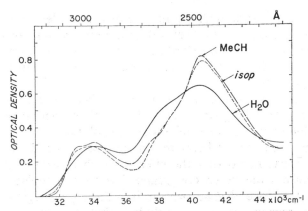

FIG. 12. Absorption spectra of 2,6-di-(diethylamino)purine in various solvents. 20 μl of a saturated solution in dioxane diluted by 3 ml of either methylcyclohexane, isopropanol, or water; 1 cm cell. (From Drobnik and Augenstein,[55] by permission of Pergamon Press Ltd.)

For 2-aminopurine in dioxane, the 242 mμ band is a shoulder to the y band; in isopropanol or dioxane a bathochromic shift and resolution of another inflexion 1000 cm^{-1} to the red occur; the longer wavelength band lies at 306 mμ in isopropanol and at 302 mμ in water or dioxane.

Evidently the simple correlations between solvent polarity, as indicated by dielectric constant, and spectral shifts are inadequate to explain these effects in detail for purine. Ability of solvents to form hydrogen bonds with the —NH— group at $N_{(9)}$, and with the ring nitrogen atoms according to their electronegativity, decreasing in the order $N_{(7)} > N_{(3)} > N_{(1)}$, was therefore suggested[54] to play a significant role. In this connexion it will be recalled (Sect. 1E) that the question of the possible significance of solvation as a determinant of ionisation properties of purines was raised. Drobnik and Augenstein[55] further concluded that the nature and position of substituents influenced not only the energy of the transitions but also their direction of polarisation.

In further studies of 6- and 9-substituted purines, Kleinwächter, Drobnik, and Augenstein[55a] were of the opinion that a useful correlation between the x_1 and x_2 bands could be obtained by consideration of solvent effects. The x_2 band was characterised by its much more marked sensitivity to solvent shifts. Three groups of purines were distinguished with respect to their absorption patterns: (a) purine and 9-ribosylpurine, for which the excited states in order of increasing energy are $n\pi^*$, $\pi\pi^*$ reached by the x_1 transition, and $\pi\pi^*$, x_2 transition; (b) adenine, neutral and anionic, anionic hypoxanthine, and 6-methoxypurine; no low-lying $n\pi^*$ state; x_1 and x_2 transitions difficult to distinguish; and (c) protonated adenine, neutral or protonated adenosine, 9-methyladenine or hypoxanthine, and neutral or anionic inosine and 9-methylhypoxanthine; no low-lying $n\pi^*$ state, x_2 transition of lower energy than x_1, but blue shifted in passing from the vapour phase through nonpolar to polar solvents.

Effects of thio and seleno substituents in purine have been studied in some detail by Mautner and Bergson.[56] The general phenomenon of bathochromic shifts due to replacement of oxygen by sulphur, or more so by selenium, is observed; this reflects the increasing polarisability of the larger atoms. It was noted that the effect of such replacement at the 6-position in purines was relatively greater than that of replacement at the 2-position and extended absorption into the visible region. For 2,6-disubstituted purines it was concluded that the marked bathochromic shift, causing yellow colour, is associated with the

—NH—CS—$\overset{|}{\text{C}}$= group including the substituent at $C_{(6)}$, rather than with the —NH—CS—NH— group including the substituent at $C_{(2)}$.

An interesting effect of steric hindrance on purine spectra was found with hydroxyphenyl purin-6-yl ketones by Cohen, Thom, and Bendich.[57] 4-Hydroxyphenylpurin-6-yl ketone (17) shows a long wavelength band at 310.5 mμ compared with 267 mμ for purine-6-carboxaldehyde. The analogue with a hydroxy substituent *ortho* to the purinoyl substituent (18) has $\lambda_{max} = 276$ mμ in water, but peaks with maxima at 312 mμ and 277 mμ were found in chloroform. Evidently conjugation of the ring systems as in 17 is inhibited in 18 in aqueous solution, an effect which can be attributed to steric hindrance to the assumption of the coplanar configuration because of the presence of the *ortho* hydroxy substituent in 18. In chloroform intramolecular hydrogen bonding of this group to the carbonyl group was suggested, to account for the similarity between the spectrum of 18 and that of salicylaldehyde.

The introduction of electrophilic substituents into the 8-position of purines causes considerable bathochromic shifts in spectra. Mason[21] noted the marked effect of a phenyl group at $C_{(8)}$, second only to that of a thio substituent. Jones and Robins,[58] from studies of 8-diazo- and 8-nitropurines concluded that the imidazole ring tended to acquire a stabilising negative charge and, therefore, proposed structure 19 for 8-diazotheophylline. As a consequence of the enhanced mobility of the π-electrons, absorption extends into the visible region. It was noted[58] that 8-nitroxanthine in basic solution has the longest λ_{max} recorded for a purine, viz. 430 mμ, giving a deep red-orange colour. Substitution of the nitro group at an electron-deficient position, $C_{(2)}$ in the purine ring, was reported by Shapiro,[42] who isolated 2-nitroinosine from the reaction of guanosine with nitrous acid; the bathochromic effect of the nitro substituent is again very marked, giving rise to yellow colour.

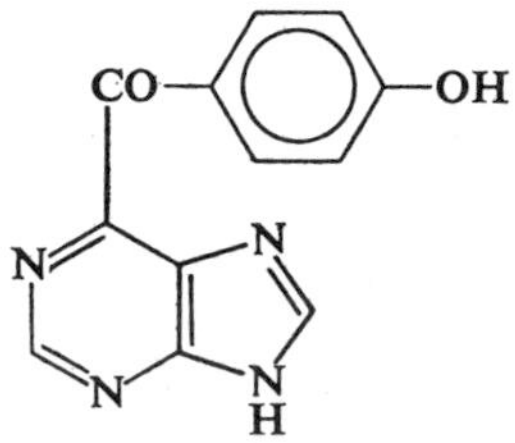

(17)

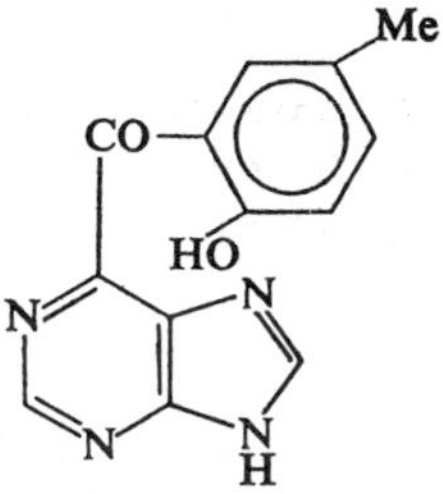

(18)

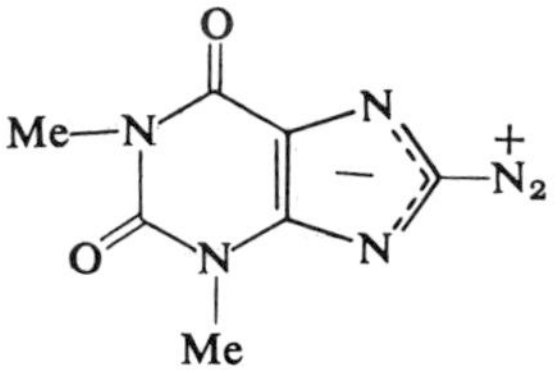

(19)

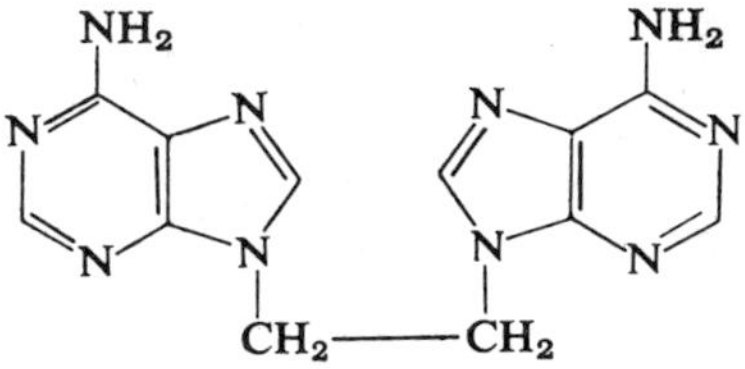

(20)

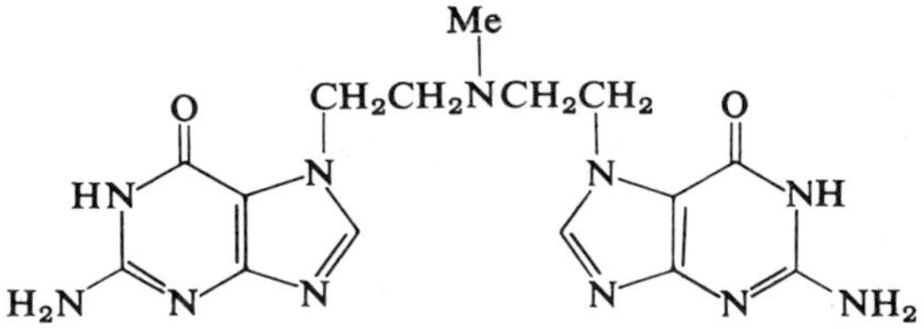

(21)

The direction of polarisation of electronic spectra can be determined from study of the absorption of polarised ultraviolet light by oriented arrays of molecules; e.g., by single crystals, about 10^{-5} cm thick, or by solutions of polymers in which asymmetric molecules are oriented by flow.

The principal studies for crystalline purines so far are those by Stewart and co-workers[59, 60] for 9-methyladenine; and by Callis, Rosa, and Simpson[61] for 9-ethylguanine.

The absorption spectra of 9-methyladenine in three environments are shown in Fig. 13; the hypochromic and bathochromic shifts in passing from solution to crystalline 9-methyladenine or crystalline "dimer" are evident. In Fig. 14 the absorption spectra of the crystal are shown for light polarised in the plane of the molecules, with the direction of polarisation parallel to either the long or short axis of the molecule.

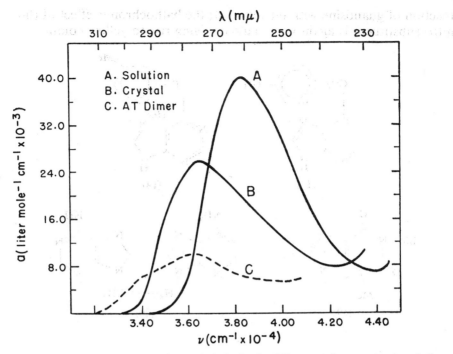

Fig. 13. Absorption spectra of 9-methyladenine in different environments: A. solution; B. crystal; C. AT dimer. (From Stewart and Davidson,[59] by permission of the American Institute of Physics.)

From these results it was concluded that the more intense band ($\lambda_{max} = 275$ mμ) is polarised approximately paralled to the $C_{(4)}$–$C_{(5)}$ bond, i.e., the short axis of the molecule. The observed absorption polarised parallel to the long axis of the molecule was divided into a component of the 275 mμ band (the dichroic ratio being constant at 5.1 ± 0.5 down to 280 mμ), and at shorter wavelengths a weak component polarised parallel to the long molecular axis, and peaked at 255 mμ, as represented by the broken lines in Fig. 14.

In order to reconcile this result with Mason's[21] assignment of the directions of polarisation of the x bands of 9-methyladenine, it was therefore necessary to postulate a bathochromic shift of the short-axis polarised band from its position in the solution spectrum as part of the main band at 261 mμ, to 275 mμ; the weaker long-axis polarised component at 255 mμ in the crystal is possibly hypsochromically shifted if it is identified with the 267 mμ shoulder of the solution spectrum.

The results of Callis, Rosa, and Simpson[61] for crystalline 9-ethyl-guanine showed that here also the principal longer wavelength bands

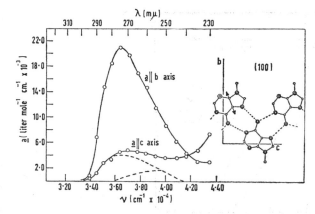

FIG. 14. Absorption spectra of crystalline 9-methyladenine in (100) plane. The molecular arrangement in the unit cell is shown for molecules near a|4 (*cf.* Stewart and Jensen[60]). The analysis of the a‖c band into a component of the stronger a‖b band and a weak band with its main polarization direction along the c-axis is shown in dashed lines. Data were obtained mainly from a 0.16 μ and a 0.61 μ section. The estimated error in absorptivity (a) is about 15%. (A revised drawing corresponding to Fig. 2 of Stewart and Davidson,[59] kindly supplied by Dr. R. F. Stewart.)

are polarised at right angles to each other, but no experimental assignment of which band is longitudinally polarised has yet been made.

Vapour phase spectra of adenine, guanine, 9-methyladenine, and 9-methylhypoxanthine have been studied by Clark, Peschel, and Tinoco.[53] These spectra are similar in form to those for aqueous solutions but show the expected hypsochromic shift of about 10 mμ from the latter, characteristic of $\pi \to \pi^*$ transitions, reflecting the greater attraction shown by London forces of the solvent for the excited states than for the ground states of the molecules.[62] No evidence of bands which shift bathochromically in passing from solution to vapour phase was found; i.e., no $n \to \pi^*$ transition, for which interaction of the polar solvent would be greater for the ground state, was detected. The $n \to \pi^*$ transitions were concluded to be either hidden under the $\pi \to \pi^*$ bands, or to be less than 0.05 of the intensity of the latter.

Vibrational fine structure is apparent in the vapour phase spectrum of 9-methylhypoxanthine in the region 280–300 mμ (Fig. 15),[53] and was also observed for trimethyl phosphate solutions (Fig. 9), but not for aqueous solutions (Fig. 15). The regularity of bathochromic shifts with increasing solvent polarity is evident from the results presented in Table 8.

Some hypochromicity in passing from solution to vapour phase spectra was observed.[53] When the function log (product of band area

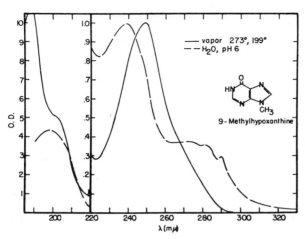

FIG. 15. Spectra of 9-methylhypoxanthine in the vapour phase, and in neutral solution in water. (From Clark, Peschel, and Tinoco,[53] by permission of the American Chemical Society.)

of vapour phase spectrum and absolute temperature) was plotted *versus* reciprocal of temperature, straight lines with negative slopes giving values for heats of vaporisation of the purines, in agreement with those from measurements of variation of vapour pressure with temperature, were obtained. These values, ranging from 20 kcal/mole for 9-methylhypoxanthine to 27 kcal/mole for 9-methyladenine, are much greater than the heats of solution, 6.9 and 8 kcal/mole, respectively, showing that a considerable negative enthalpy change (about 20 kcal/mole) would accompany the solution of isolated molecules of these purines in water. This gives a direct indication of the importance of their solvation. For guanine vapour spectra were obtained only with concomitant decomposition of the purine; fine structure with maxima at 284 and 293 mμ was observed, but quantitative data on oscillator strength could not be obtained.

Much attention has been devoted to effects on ultraviolet absorption associated with plane-to-plane interaction between purine ring moieties of synthetic and naturally occurring polynucleotides.

The salient feature of such effects is that the extinction coefficients (ϵ_P) per base residue are less for the polymers than for their constituent monomers. This hypochromicity, although dependent to some extent on the degree of polymerisation, has been detected for oligonucleotides and even for dimers. Hypochromicity for neutral molecules of non-nucleotide purines containing two rings has also been observed. Thus the examples may be quoted of 1,2-di(purin-6-ylamino)ethane (**20**)[63] (ϵ_{max}, pH 13 = 21,000 compared with 11,900 for 9-methyladenine); 1,6-di(purin-9-yl)hexane,[63] ϵ_{max} = 14,500, *cf.* 9-methylpurine, 7,900; 1,2-di(6-methylaminopurin-9-yl)ethane,[63] ϵ_{max} 27,800, *cf.* 9-(2-hydroxyethyl)-6-dimethylaminopurine, 14,700; and di(2-guanin-7'-ylethyl)-methylamine, (**21**),[64] ϵ_{max} at pH 7, 12,500, *cf.* 7-ethylguanine, 7,700.

Hypochromicity as observed particularly for polynucleotides is not necessarily uniform over the whole range of wavelengths,[65] and it was evident that hydrogen bonding was most unlikely to be the major cause of this phenomenon, which was therefore proposed to be associated with the plane-to-plane "stacking" of the heterocyclic rings.

Theoretical studies attempting to account for the phenomenon have been made by application of exciton theory,[66-68] but these were criticised[69] on the grounds that there are no marked changes in the

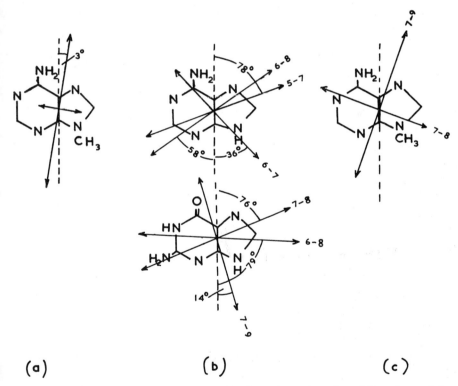

(a)   (b)   (c)

FIG. 16. Directions of polarization of transitions in purines (transitions denoted by orbital jumps): (a) observed for crystalline 9-methyladenine (Stewart and Davidson[59]); (b) theoretical for adenine and guanine (Berthod, Giessner-Prettre, and A. Pullman[73]); (c) theoretical for 9-methyladenine (Nagata, Imamura, Tagashira, and Kodama[75]).

TABLE 9. Results of SCF–MO Calculations of Electronic Energy Levels in Purine.

Adenine	Nesbet[68,a]	281 (0.06)	260 (0.42
	Nagata et al. (PP)[75,b]	(6 → 7) 262 (0.39)	(6 → 8) 232 (0.47
	(MN)	(6 → 7) 253 (0.47)	(6 → 8) 238 (0.38
	Berthod et al.[73]	(6 → 7) 239 (0.4)	(6 → 8) 226 (0.31
	Experimental[c]	249 (vap.)(0.24)	
		263 (H_2O)	
9-Methyladenine	Nagata et al.[d] (PP)[b]	(7 → 8) 263 (0.38)	(7 → 9) 231 (0.48
	(MN)	(7 → 8) 254 (0.48)	(7 → 9) 238 (0.39
	Experimental[c]	252 (vap.)(0.16)	
		261 (H_2O)(0.26)	
Guanine	Nesbet	325 (0.46)	273 (0.36
	Nagata et al. (PP)[b]	(7 → 8) 299 (0.4)	(7 → 9) 239 (0.58
	(MN)	(7 → 8) 325 (0.36)	(7 → 9) 268 (0.42
	Berthod et al.	(7 → 8) 282 (0.4)	(7 → 9) 230 (0.6)
	Experimental[c]	293, 284[e] (vap.)	
		276 (H_2O)	

[a] Nesbet gave no assignment of orbital jumps.
[b] PP—Pariser-Parr[45]; MN—Mataga-Nishimoto.[74]
[c] Clark, Peschel, and Tinoco.[53]
[d] Hyperconjugation by CH_3 is included[75] giving 7 as the highest filled orbital.
[e] Possibly vibrational structure.

shape of the absorption curves associated with hypochromicity. A semiclassical treatment of the problem,[69, 70] in which the dissipation of energy absorbed by an oscillator is reduced by regeneration of part of this energy by neighbouring oscillators in a favourable configuration, seems to account for the observed facts more satisfactorily.

Another question at issue is that raised by the "sum rule of oscillator strengths"[71]; according to earlier theories the loss of absorbancy in the 260 mμ band might have been expected to be accompanied by hyperchromicity in the 200 mμ region; but this expectation was contradicted by experiment. Weiss[70] considered that his theory would predict hyperchromism for transitions with about three times the energy of those involved in the hypochromic effect, and would thus be inaccessible to spectroscopy by current techniques.

Theoretical studies of purine spectra have been concerned mainly with the constituent bases of nucleic acids and their analogues.

The Hückel molecular orbital calculations of Ladik and Hoffman[72] led to a value of 396 mμ for the 6 → 7 (or B_{2u}) transition* for adenine, and of 353 mμ for the 7 → 8 transition of guanine. These values are

* The symbolism $i \rightarrow j$ denotes the orbital jump, the highest filled orbital in purine being 5; i.e., the filled orbitals accommodate 10 π electrons.

Transitions Given as Orbital Jump, Wavelength in mμ, and Oscillator Strength (in parentheses)).

224 (0.45)	208 (0.30)	201 (0.51)	184 (0.21)
5 → 7) 178 (0.18)			
5 → 7) 199 (0.16)			
5 → 7) 204 (0.5)	(6 → 9) 197 (0.01)	(4 → 7) 193 (0.2)	(5 → 8) 180 (0.2)
207 (H_2O)	190 (H_2O)		
→ 10) 199 (0.19)	(6 → 8) 180 (0.21)	(6 → 9) 178 (0.45)	(6 → 10) 164 (0.36)
→ 10) 204 (0.19)	(6 → 8) 200 (0.17)	(6 → 9) 184 (0.45)	(6 → 10) 172 (0.41)
206 (H_2O)			
229 (0.25)	198 (0.45)	192 (0.08)	
	(6 → 8) 173 (0.15)		
	(6 → 8) 194 (0.15)		
→ 10) 207 (0.1)	(6 → 8) 196 (0.1)	(5 → 8) 187 (0.2)	(6 → 9) 182 (0.1)

clearly too high in comparison with the observed 252 mμ for adenine (vapour phase) and 284, 293 mμ for guanine (vapour phase).

These early calculations were criticised (cf. Berthod, Giessner-Prettre, and A. Pullman[73]) on the grounds that the choice of the value of the resonance integral was unrealistic. Furthermore, the approximations due to Pariser and Parr, and Pople,[45] and to Mataga and Nishimoto,[74] have generally been admitted to yield better values for electronic energy levels by the semiempirical self-consistent field method,[68, 73, 75] and these methods have been employed by the later workers. Comparisons of some of the results are shown in Table 9. In addition, Berthod, Giessner-Prettre, and A. Pullman[76] have given results for purine, B_{2u} band, 253 mμ; experimental, 263 mμ; B_{1u}, 255 mμ (calc.), 240 mμ (exptl.); E_{1u}, 197 and 184 mμ (calc.), 203 and 187 mμ (exptl.), and also for hypoxanthine and xanthine. In the latter case it was noted that the calculated values for the tautomer with the —NH— group of the imidazole ring at $N_{(7)}$ gave better agreement with experimental data than those with the group at $N_{(9)}$.

The predicted directions of polarisation as given by Berthod, Giessner-Prettre, and A. Pullman,[73] and by Nagata and co-workers,[75] are compared with the observed value[59] for 9-methyladenine in Fig. 16. It will be noted that the lowest energy $\pi \rightarrow \pi^*$ transition of adenine is predicted to be polarised transversely at an angle of $-36°$ to the $C_{(4)}-C_{(5)}$ axis, away from $N_{(7)}$, by the former group. This can be compared with the observed polarisation for the 9-methyladenine crystal at $+3°$ to this bond. On the other hand, Mason,[21] on empirical grounds, and

Nagata and co-workers' theoretical treatment,[75] assign longitudinal polarisation to this transition. The correlation between the calculated values, presumably referring to the vapour phase, and the crystal and solution spectra, remains therefore unclear.

G. Fluorescence and Phosphorescence in Purines

Emission spectra are quite often observable with purines, and thus, in common with aromatic and other heterocyclic systems, stability of excited states for times appreciably longer than those required for absorption and emission of light energy (about 10^{-15} sec) is encountered in certain cases.

As is well known, the emission spectra are generally at longer wavelengths than the excitation spectra. The energy differences reflect loss of absorbed energy through intersystem crossing of excited states (radiationless transitions), or by thermal losses of vibrational and rotational energy. Fluorescence, of relatively short lifetime (about 10^{-8} sec), can generally be assumed to occur by return of a molecule from a $\pi\pi^*$ singlet excited state to the ground state. Furthermore, if an $n\pi^*$ state lies below the lowest $\pi\pi^*$ singlet, efficient intersystem crossing enabling radiationless transition to such a state will reduce the probability of fluorescence emission; but intersystem crossing to a $\pi\pi^*$ triplet state, which can be assumed more likely to result if an $n\pi^*$ state of intermediate energy exists, can result in phosphorescence, i.e., a longer-lived emission with half-life of the order of seconds or more. For general theoretical reviews of this subject the reader is referred to Kasha[77] and to McGlynn, Smith, and Cilento.[78] For analytical applications, see Udenfriend,[79] and Williams and Bridges.[80]

As a very general rule, reduction of the extent of delocalisation of π-electrons in a heteroaromatic system, e.g., by introduction of a nitrogen atom, will tend to reduce fluorescence. The presence of —N̄H— or —N̄— groups in the ring may enhance the probability of such emission, but the situation is complex, since the latter substitutions may also introduce $n\pi$ states of lower energy than the lowest $\pi\pi$ state. For this reason the relationship between structure, solvation, and ionisation of purines and their emission properties have merited close study. It should be noted that quantum efficiency of emission (F) must be of the order of 10^{-3} or more to enable its quantititive study.

Börresen[81] found that purine itself exhibits a low fluorescence intensity as a neutral molecule, but this increases markedly on either acid or alkaline titration. The plot of quantum efficiency *versus* pH thus takes the form of two titration curves, and the apparent pK_a values derived from these curves are in agreement with those for the cationic and anionic dissociations derived titrimetrically (Fig. 17). For the monocation emission at 400 mμ with $F = 0.008$ was observed, while at lower pH values another quenching process, possibly involving dication formation, was observed; for the anion emission at 370 mμ $F = 0.045$ was found.

Appreciable fluorescence at room temperature can be observed for the cation of adenine (380 mμ, 0.003), while similarly guanine in neutral solution is nonfluorescent but emission becomes appreciable on protonation ($F = 0.005$) or on acidic dissociation ($F = 0.024$). However, with an increase in pH beyond 11, with an apparent pK_a at 12.7, decreased fluorescence is observed.

A case where the neutral molecule is more fluorescent than either the cation (nonfluorescent) or the anion is presented by 2,6-diaminopurine.

Interesting alterations in the fluorescence spectra of guanine and its derivatives were found[82] at low temperature. Between $-90°$ and $-143°$ blue shifts occurred for guanine in acid solution from 384 mμ to 327 mμ with a smaller shift at alkaline pH. These effects were associated with the change in viscosity of the medium by application of the Franck-Condon principle; the equilibrium distances between the solute and the surrounding solvent molecules would not be the same in the excited state as in the ground state; high solvent viscosity would prolong the time necessary for the excited state to reach equilibrium. At higher temperatures and lower viscosity the relaxation time of the excited state is relatively short (10^{-11} sec) compared with the lifetime of the excited

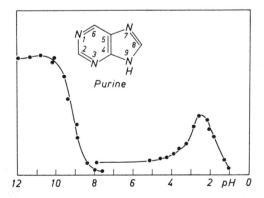

FIG. 17. Fluorescence intensity versus pH for purine in aqueous solution. (From Börresen,[81] by permission of the Editors of *Acta Chemica Scandinavica*.)

state (10^{-8} sec), and loss of vibrational and rotational energy to yield an equilibrium excited state can occur; some difference between the vibrational levels of the ground states reached at the two temperatures may also accompany this effect. At lower temperatures, when viscosity is high enough to prevent occurrence of these energy losses before emission takes place, the fluorescence will correspond with the whole of the absorbed quanta for the lowest excited state, less only the vibrational energy differences between the electronic ground state and the higher vibrational levels of this state reached on return. The shortest wavelength emission could thus in the latter case coincide with the longest wavelength absorption. It was confirmed that the increase in emission energy occurred at temperatures where the viscosity would be enhanced sufficiently to increase the relaxation time of the excited state to about 10^{-8} sec.

At pH 10.5–11.5, where only guanine fluoresces, the excitation spectrum is not congruent with the absorption spectrum, particularly at room temperature. It was suggested that two tautomeric forms of the guanine ring with a single negative charge exist, only one of which fluoresces at room temperature.

Solvent effects on the fluorescence of purine have been studied in some detail.[54] In methylcyclohexane or dioxane no fluorescence (greater than $F = 10^{-4}$) could be detected; isopropanol as solvent gave the highest yield ($F = 0.004$) and shortest emission wavelength (362 mμ), but quenching was observed as the concentration of purine increased from 0.04 to 0.1 mM. Fluorescence was also found with solvents of high dielectric constant such as methanol and water.

Börresen[81] has discussed his results in terms of the general theory that the presence of a lowest excited state of the $n\pi^*$ type would preclude fluorescence and possibly permit phosphorescence. This theory cannot be applied strictly to the cases of either purine or pyrimidine, which fluoresce, albeit weakly, as neutral molecules, but possess low-lying $n\pi^*$ states. However, the observed enhancement of fluorescence by ionisation is consistent with this view. Protonation, by binding a lone pair of electrons, increases the energy of the lowest $n\pi^*$ state leaving a $\pi\pi^*$ state as the lowest singlet level. In order to explain the observation that lowering at pH well below that of the first basic pK_a in turn decreases fluorescence, Börresen argues that a second protonation of the excited state occurs stabilising $n\pi^*$ states involving lone pairs on atoms other than those involved in protonated.

Dissociation of purine to yield an anion conversely liberates a second lone pair and might therefore be thought to yield a new low-lying $n\pi^*$ state; however, $n \rightarrow \pi^*$ transitions involving other nitrogen atoms might be opposed by the delocalisation of the negative charge. The balance of these effects will be determined by the extent to which the negative charge does in fact remain associated with the acidic nitrogen. This is expected to be lower the more acidic the heterocycle, i.e., the lower its pK_a. It was therefore considered significant[81] that purine (pK_a 8.9) and guanine (pK_a 9.2) yield fluorescent anions, while adenine (pK_a 9.8) and 2,6-diaminopurine (pK_a 10.7) do not. The nonfluorescence of the dianion of guanine (second pK_a 12.7) is also explicable.

For neutral molecules Börresen points out that the more strongly the π orbitals attract electrons,[81] and thus the lower the acidic pK_a, the more will the charge displacements involved in $n \rightarrow \pi^*$ transitions be favoured. Fluorescence of a neutral molecule is thus expected to be favoured by a high acidic pK_a, as with 2,6-diaminopurine, although it does not explain the nonfluorescence of adenine; the latter according to Drobnik and Augenstein[55] should be attributed to thermal quenching at room temperature.

From their studies of solvent effects on the fluorescence of purine, Drobnik and Augenstein[54] concluded that in solvents of low dielectric constant, such as methylcyclohexane, the existence of low-lying $n\pi^*$ states, shown by the absorption "tail," correlates with the lack of fluorescence. The effectiveness of dioxane in destroying vibrational structure and shifting the x_2 band in the absorption spectrum (cf. Sect. F) was correlated with the lack of fluorescence in this solvent, and attributed to hydrogen bonding between dioxane and the —NH— group of the imidazole ring of purine at $N_{(9)}$. The observation that for purine in isopropanol maximal fluorescence accompanies an appreciable $n \rightarrow \pi^*$ "tail" raises some difficulties, and the possibility was suggested that solvation involving the lone pair electrons on some or all of the ring nitrogen atoms could yield a variety of types of solvated purine molecules, some of which possess low-lying $n\pi^*$ states, others even lower-lying $\pi\pi^*$ states. The nature of the particular sites solvated could thus be significant.

Marked effects of substituents in the 2- and 6-positions of purine on emission spectra may be expected[55]; by analogy with pyrimidines, the negative charge of the promoted electron in an $n \rightarrow \pi^*$ transition will have a distribution giving increased electron density at $C_{(6)}$, but decreased electron density at $C_{(2)}$. Thus an electron-donating substituent at $C_{(6)}$ would be expected to favour a low-lying $n\pi^*$ state; whereas such substitution at $C_{(2)}$ would have the opposite effect. Thus the observed relatively high fluorescence of 2-aminopurine, in contrast to the lack of fluorescence of neutral adenine, might thus be explained; however, as will be noted later, adenine does fluoresce at low temperatures. The fluorescence from 2,6-bis(diethylamino)purine was found to be stronger in methylcyclohexane at room temperature than in either isopropanol or water, thus confirming in this case the expected lack of solvent effects associated with $\pi\pi^* \rightarrow n\pi^*$ conversions observed with purine itself.

The effect of temperature on fluorescence was small for 2-aminopurine, but adenine and guanine, while nonfluorescent at 20°, fluoresce strongly at −123°. It was considered[55] that thermal quenching was a more likely explanation of this effect than the postulate that distribution of different molecular species varied with temperature.

Phosphorescence of purine is readily observed at low temperatures and detailed emission spectra in glasses of appropriate composition to retain purine in solution at −196° have been recorded by Drobnik and Augenstein[54] (*cf.* Fig. 18). These show vibrational spacing falling into two progressions; the average lifetimes were of the order of 1–2 sec. Emission from powders of purine at this temperature was also recorded.

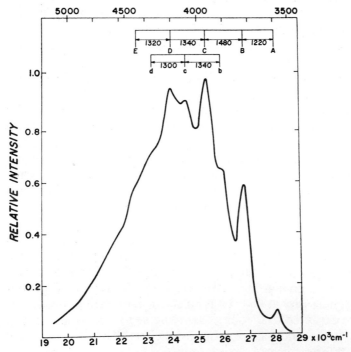

FIG. 18. The emission spectrum from purine (10^{-4} M) in a glass of ethyl ether : isopentane : ethyl alcohol (5:5:2) at −196°; excitation at 265 mμ. The scale at the top designates the vibrational bands. (From Drobnik and Augenstein,[54] by permission of Pergamon Press Ltd.)

This phosphorescence from purine was considered[54] from comparisons of its wavelength, vibrational spacing, and decay time, to resemble that of quinoline,[83] rather than that of pyrimidine.[84] It was therefore ascribed to a $\pi\pi^*$ triplet state, rather than to an $n\pi^*$ triplet. Solvent effects on distribution of emission between peaks, and on their wavelengths, were found, but the spacing of the peaks was not affected.

With the substituted purines examined[55] 2-aminopurine showed no phosphorescence, as neutral molecule, anion, or cation, at $-196°$. 2,6-Diaminopurine, which gave less fluorescence than 2-aminopurine, showed, correspondingly, appreciable phosphorescence. With 2,6-di-(diethylamino)purine in methylcyclohexane-isopentane glass, structure similar to the BCDE progression of purine phosphorescence (Fig. 18) was observed. Guanine, with lower fluorescence than 2-aminopurine, was also phosphorescent.

Cohen and Goodman[85] have studied phosphorescence and polarised phosphorescence excitation spectra, for purine, purine anion, and adenine and some of its derivatives, at $-196°$ in various solvent systems. In ether:isopentane:ethanol (5:5:2, by vol.) the phosphorescence maxima were 400 mμ for purine, 407 mμ for adenine, 402 and 403 mμ for 9-n-butyl- and 9-methyladenine, respectively, and 440 mμ for purine anion. The corresponding singlet-triplet intervals were 13,200, 13,600–13,900, and 14,200 cm^{-1}, and lifetimes were about 1 sec. Negative polarisation of phosphorescence was found for purine anion and for adenine at all wavelengths of excitation from 240–310 mμ; but for purine as neutral molecule positive polarisation was found above 280 mμ, negative at lower wavelengths.

These observations therefore permitted the conclusions that the singlet-triplet transition moment is predominantly perpendicular to the molecular plane, i.e., as expected, perpendicular to that of the $\pi \rightarrow \pi^*$ absorption bands, but parallel to that of the $n \rightarrow \pi^*$ tail of purine absorption. The singlet-triplet separation of 1.7 eV observed for the first triplet state of adenine is in reasonable agreement with the value 1.5 eV calculated by Berthod, Giessner-Prettre, and Pullman.[73]

In summary, the existence of low-lying $n\pi^*$ states of purine is thus indicated by both absorption and emission studies. The latter also show the existence of $\pi\pi^*$ triplet states. Substitution of an amino group at $C_{(2)}$ changes the relationship between the $n\pi^*$ and $\pi\pi^*$ states, the former presumably being relatively increased in energy, leaving a $\pi\pi^*$ state as the lowest singlet; fluorescence rather than phosphorescence then predominates in emission. Hydroxy or amino substituents at $C_{(6)}$ partially reverse the effect of the 2-amino substituent; the lowest-lying singlet remains $\pi\pi^*$ in character, but intersystem crossing to a $\pi\pi^*$ triplet is enhanced especially by the 6-hydroxy substituent.

TABLE 10.   U.V. Data for Purine, Alkyl- and Arylpurines.

Substituents	pH or solvent	$\lambda_{max}(10^{-3}\epsilon)$	pK_a	References
None	0.3	260 (6.2)	2.4	21, 32, 86[a]
	5.7	263 (7.9)	8.9	
	11	219, 271 (8.3, 7.6)		
1-Methyl	1	268 (6.6)		87
	11	275 (6.1)		
2-Methyl	1	266 (5.4)		88
	11	275 (8.0)		
3-Methyl	1	275 (9.8)		89
	MeOH	277 (7.0)		
	11	276 (7.0)		
6-Methyl	0	265 (7.6)	2.6	21, 33
	5.9	261 (8.3)	9.0	
	11.5	271 (8.5)		
7-Methyl	0.2	258 (6.7)	2.3	32
	9.2	267 (8.1)		
8-Methyl	0	264 (8.3)	2.9	21, 33
	5.9	266 (10.2)	9.4	
	12	274 (8.3)		
9-Methyl	0	262 (5.5)	2.4	21, 32, 33
	5.1	264 (7.8)		
2,6-Dimethyl	1	270 (7.0)		88
	11	275 (9.2)		
2,8-Dimethyl	1	269 (7.2)		88
	11	278 (9.6)		
2,6,8-Trimethyl	1	273 (5.0)		88
	11	278 (6.3)		
6,8-Dimethyl	1	269 (7.7)		88
	11	278 (8.9)		
8-Phenyl	0	237, 304 (13.5, 25.7)	2.7	33, 21
	5.4	231, 298 (11.5, 26.3)	8.1	
	10.3	233, 304 (17.4, 26.3)		
6-Methyl-8-phenyl	0	232, 293, 300 (14.2, 25.4, 25.0)		90
	6	234, 294 (19.6, 21.4)		
	13	238, 303 (20.4, 22.0)		
1,6-Di-(9-purinyl)-hexane[b]	1	264 (11.3)		63
	13	265 (14.5)		

[a] Detailed study of acidic pK_a at 20–50°.
[b] Full name of compound, not substituent.

TABLE 11.   U.V. Data for Halogen-Substituted Purines, Including Trifluoro-methylpurines.

Substituents	pH or solvent	$\lambda_{max}(10^{-3}\epsilon)$	pK_a	References
2-Fluoro	0	264 (8.3)	8.2[a]	30
	7	268 (8.4)		
	13	273 (8.8)		
2-Chloro	1	271 (8.1)		91
	7	272 (8.0)		
	13	279 (8.2)		

[a] In 50% (v/v) EtOH—H$_2$O.

TABLE 11 (*continued*).

Substituents	pH or solvent	$\lambda_{max}(10^{-3}\epsilon)$	pK_a	References
6-Chloro	5.2	265 (9.1)	7.7	32
	13	274 (8.8)		
6-Bromo	1	266 (9.7)		92
	11	275 (8.6)		
6-Iodo	1	276 (10.4)		92
	11	280 (9.3)		
2,6-Dichloro	1	250^i, 275 (3.9, 8.9)		92
	11	280 (8.5)		
2,6,8-Trichloro	6	281 (11.2)		93
2,6-Dibromo	1	247, 279 (3.9, 8.0)		92
	11	283 (7.8)		
6-Chloro-2-fluoro	1	270 (9.7)		30
	7	272 (9.5)		
	13	275 (9.0)		
2-Fluoro-6-methyl	1	262 (9.9)		30
	7	266 (9.8)		
	13	272 (10.4)		
2-Fluoro-9-methyl	1	265 (8.7)		94
	11	267 (8.6)		
8-Fluoro-9-methyl	1	261 (6.4)		94
	11	280, 287^i (9.3, 8.8)		
2-Chloro-7-methyl	1	273 (7.4)		94
	11	275 (6.8)		
2-Chloro-9-methyl	1	271 (8.3)		94
	11	272 (8.0)		
8-Chloro-7-methyl	1	283 (11.8)		94
	11	225, 286 (4.4, 11.5)		
8-Chloro-9-methyl	1	269 (8.3)		94
	11	267 (11.0)		
6-Chloro-7,9-dimethyl-8-oxo	1	283 (15.1)		94
	11	283 (15.1)		
2,6-Difluoro-7-methyl	1	254 (7.7)		94
	11	267, 272^i (10.2, 9.4)		
2,6-Difluoro-9-methyl	1	255 (7.2)		94
	11	268, 273^i (15.0, 13.6)		
2,6-Dichloro-7-methyl	1	280 (7.7)		94
	11	277 (7.5)		
2,6-Dichloro-9-methyl	1	276 (10.6)		94
	11	275 (11.2)		
2,6-Difluoro-7-methyl-8-oxo	1	275 (7.5)		94
	11	285 (11.0)		
6-Amino-2-fluoro	0	266 (11.8)	9.6^a	30
	7	262 (12.6)		
	13	269 (12.7)		
2-Fluoro-6-hydroxy	7	257 (11.0)		30
	13	262 (11.0)		
2-Fluoro-6-mercapto	1^b	328 (14.2)		30
	7	316 (14.6)		
	13	314 (16.2)		
2-Fluoro-6-methylthio	1	296 (18.6)		30
	7	295 (18.8)		
	13	296 (16.3)		

^{b} Immediately after acidification.

^{i} Inflexion

continued

TABLE 11 (*continued*).

Substituents	pH or solvent	$\lambda_{max}(10^{-3}\epsilon)$	pK_a	References
2-Fluoro-6-mercapto-9-β-D-ribofuranosyl	1[b]	326 (13.7)		30
	7	315 (14.7)		
	13	315 (16.0)		
	ethanol	294		
	1[c]	296		
6-Amino-2-chloro	1	265 (12.5)		95
	7	266 (12.0)		96
	13	272 (12.5)		
6-Amino-2-chloro-9-β-D-ribofuranosyl	1	265 (13.9)		30
	13	265 (15.0)		
2-Chloro-6-methylamino	1	273 (14.4)		97
	7	271 (15.0)		
	13	226, 272 (5.3, 15.3)		
2-Chloro-6-dimethylamino	1	285 (13.6)		97
	7	277 (18.3)		
	13	284 (17.1)		
6-n-Butylamino-2-chloro	1	276 (15.7)		97
	7	272 (17.2)		
	13	278 (16.5)		
2-Amino-6-chloro-8-(3-chlorophenyl)	1	250[i], 342 (14.6, 19.3)		98
	11	240, 335 (20.0, 18.6)		
8-Chloro-2,6-diethoxy	6	234, 269 (5.6, 11.2)		93
2,8-Dichloro-6-succinylamino	1	273 (20.0)		99
	12	280 (20.9)		
2,6-Dichloro-8-hydroxy	1	248, 288 (5.7, 12.1)		100
	6	247, 287 (5.0, 11.2)		
	11	295 (14.4)		
6-Trifluoromethyl	−0.3	268 (8.4)	0	41
	3.2	270 (8.1)	7.4	
	10.3	275 (7.5)		
8-Trifluoromethyl	−0.3	268 (7.0)	1.0	41
	3.0	264 (7.7)	5.1	
	8.0	271 (8.8)		
2-Amino-6-trifluoromethyl	0.2	323 (6.0)	1.9	41
	5.1	323 (5.9)	8.9	
	12.6	283, 323 (4.8, 5.0)		
2-Amino-6,8-bis-trifluoro-methyl	−0.3	334 (3.4)	0.3	41
	2.1	328 (7.3)	5.0	
	7.7	278, 327 (4.8, 7.7)		
2,6-Diamino-8-trifluoro-methyl	1.4	249, 287 (8.4, 7.9)	3.7	41
	5.9	284 (10.8)	7.6	
	14	245, 288 (2.6, 7.4)		
6-Hydroxy-2-trifluoromethyl	0	247 (9.1)	1.1	41
	3.2	253 (8.2)	5.1	
	7	258 (9.1)	11.2	
	14	263 (9.3)		
6-Hydroxy-8-trifluoromethyl	8.6	261 (12.2)	5	41
	13	268 (12.1)	10.9	

[c] 5 minutes after acidification.
[i] Inflexion.

TABLE 12.   U.V. Data for Oxo- and Alkoxypurines.

Substituents	pH or solvent	$\lambda_{max}(10^{-3}\epsilon)$	pK_a	References
2-Hydroxy	−0.7	264, 322 (4.7, 6.5)	1.7	27, 33
	6.0	238, 315 (2.9, 4.9)	8.4	
	10.2	271, 313 (4.8, 4.8)	11.9	
	13.0	219, 265, 312 (19.5, 4.0, 6.8)		
2-Methoxy	0	284 (6.8)	2.4	27, 33
	6	246, 283 (2.6, 8.1)	9.2	
	11.4	283 (7.6)		
6-Hydroxy[a]	−0.7	248 (10.5)	2.0	27, 33, 143
	5.2	249 (10.5)	8.9	
	10.4	258 (11.2)	12.1	
	13.0	262 (11.0)		
6-Methoxy	0.2	254 (10.2)	2.2	27, 33
	5.6	253 (9.8)	9.2	
	11.3	261 (9.8)		
8-Hydroxy	0	280 (10.5)	2.6	27, 33
	5.4	235, 277 (3.2, 11.2)	8.2	
	10.1	285 (12.9)		
8-Methoxy	1.0	271 (11.2)	3.1	27
	5.4	271 (10.7)	7.7	
	10.0	279 (9.5)		
2,6-Dihydroxy[b]	−0.8	231, 260 (6.4, 9.2)	0.8	33, 143
	5	267 (10.3)	7.4	
	10	241, 278 (8.9, 9.3)	11.1	
	14	284 (9.4)		
2,8-Dihydroxy	5.1	230[i], 310 (7.9, 5.0)	7.5	21, 33
	10	262, 306 (9.6, 7.4)		
	13	220, 310 (17.0, 9.6)		
6,8-Dihydroxy	5.1	257, 280 (12.0, 5.81)	7.7	21, 33
	8.7	265 (11.0)	9.9	
	12.0	271 (13.8)		
2,6,8-Trihydroxy[c]	2.3	231, 283 (8.5, 11.5)	5.8	40, 101, 172
	7	235, 292 (9.9, 12.4)	10.3	
	12	294.5 (13.5)		
2-Hydroxy-6-methyl	1	260, 318 (2.7, 7.4)		102
2-Methyl-6-hydroxy-8-phenyl	1	233, 288 (12.3, 21.6)		120
	6	236, 295 (17.0, 21.6)		
	13	237, 304 (19.9, 20.7)		
2,6-Dihydroxy 3-N-oxide	1.2	230, 284 (16.2, 10.3)		103
	12.5	216, 295 (22.0, 12.0)		

[a] Hypoxanthine.
[b] Xanthine.
[c] Uric acid.
[i] Inflexion.

TABLE 13. U.V. Data for *N*-Alkylhydroxypurines.

Substituents	pH or solvent	$\lambda_{max}(10^{-3}\epsilon)$	pK_a	References
2-Hydroxy-9-methyl	−2.5	246, 317 (2.0, 4.1)	9.2	27
	6.5	218, 246, 259, 316 (30.2, 2.6, 2.4, 4.8)		
	11.2	240[i], 308 (4.5, 8.1)		
6-Hydroxy-1-methyl	0	249 (9.4)	8.9	29
	5.1	251 (9.4)		
	13	260 (9.9)		
6-Hydroxy-3-methyl	0	253 (11.0)	2.6,	29
	5.1	264 (14.0)	8.4	
	13	224, 264 (7.3, 10.3)		
	14	226, 264 (6.6, 11.1)		
6-Hydroxy-7-methyl	0	250 (10.2)	2.2,	29
	5.1	255.5 (9.5)	9.0	
	13	262 (10.7)		
6-Hydroxy-9-methyl	0	250 (10.6)	1.9,	29
	5.5	250 (12.0)	9.3	
	12	254 (12.9)		
6-Hydroxy-9-β-D-ribofuranosyl[a]	−1	251 (10.9)	1.2,	143
	6	249 (12.25)	8.7	
	12	253 (13.1)		
6-Methoxy-9-methyl	1	254 (9.9)		104
	11	254 (10.6)		
6-Hydroxy-1,7-dimethyl	0	250 (10.0)	2.2	27
	7	255 (8.7)		
6-Hydroxy-1-methyl-3-benzyl	1	254, 280[i] (10.0, 3.3)		106
	7	256[i], 295[i] (6.7, 1.7)		
1-Benzyl-6-hydroxy	h	251 (9.2)		106
	j	261 (9.8)		
3-Benzyl-6-hydroxy	h	265 (13.8)		106
	j	264, 276 (10.5, 9.6)		
7-Benzyl-6-hydroxy	h	257 (9.5)		106
	j	264 (10.0)		
6-Hydroxy-1-methyl-2′-deoxy-β-D-ribofuranosyl	1	250 (10.0)		46
	MeOH	251 (10.0)		
	11	250 (10.6)		
6-Hydroxy-7-methyl-9-β-D-ribofuranosyl	1	252 (10.8)		107
	6	261 (8.4)		
	11	256 (6.6)		
1,3-Dibenzyl-6-hydroxy	1	254, 280[i] (10.2, 3.9)		105
	7	245[i], 304[i] (9.4, 0.5)		
1,7-Dibenzyl-6-hydroxy	1	257 (8.3)		105
	7	257 (7.6)		
	13	256 (7.8)		
1,9-Dibenzyl-6-hydroxy	1	252 (11.3)		105
	7	253 (11.2)		
	13	252 (11.1)		

[a] Inosine.
[h] Neutral molecule, pH not stated.
[i] Inflexion.
[j] Anion, pH not stated.

TABLE 13 (*continued*).

Substituents	pH or solvent	$\lambda_{max}(10^{-3}\epsilon)$	pK_a	References
3,7-Dibenzyl-6-hydroxy	*h*	266 (11.8)		106
7,9-Dibenzyl-6-hydroxy	1	256 (10.8)		105
	7	268 (9.2)		
1,7,9-Tribenzyl-6-hydroxy	1	257 (9.7)		105
	7	257 (9.2)		
	13	262, 290^i (6.6, 4.1)		
8-Hydroxy-7-methyl	0.3	284 (7.9)	2.7,	27
	5.5	240, 278 (4.0, 9.3)	8.2	
	12.0	186 (11.5)		
8-Hydroxy-9-methyl	0.3	281 (11.8)	2.8,	27
	5.5	235^i, 277 (2.8, 10.2)	9.1	
	12.0	257^i, 289 (4.3, 9.3)		
8-Hydroxy-7,9-dimethyl	0.3	285 (10.7)	2.8	27
	7.0	241, 279 (3.6, 10.2)		
2,6-Dihydroxy-1-methyl	5	223, 267 (3.3, 11.0)	7.9,	108
	10	242, 276 (8.5, 9.5)	12.2	
	14	242^i, 283 (4.9, 10.0)		
2,6-Dihydroxy-3-methyl	6	270 (11.2)	8.5,	108
	10	274 (12.6)	11.9	
	14	232^i, 274 (8.9, 13.1)		
2,6-Dihydroxy-7-methyl	6	268 (10.2)	8.4	108
	11	231, 287 (4.7, 8.5)	13	
2,6-Dihydroxy-9-methyl	3	234, 265 (8.3, 10.2)	6	108
	10	245, 277 (9.5, 9.3)	13	
2,6-Dihydroxy-9-β-D-ribo-furanosyl^b	2	235, 263 (8.4, 9.0)	0	143
	8	249, 278 (10.2, 8.9)	5.5	
	14	252, 276 (8.6, 9.3)	13.0	
2,6-Dihydroxy-3-methyl-9-β-D-ribofuranosyl	2	240^i, 267		109
	6	239, 267		
	13	238, 278		
2,6-Dihydroxy-7-methyl-9-β-D-ribofuranosyl	1	262 (9.8)		107
	6	253, 280 (9.1, 8.8)		
	11	269 (15.4)		
2,6-Dihydroxy-1,3-dimethyl^c	7	271 (10.2)	8.7	108
	11	274 (12.0)		
2,6-Dihydroxy-1,7-dimethyl^d	6	268 (10.2)	8.7	108
	11	233, 288 (5.0, 8.7)		
2,6-Dihydroxy-3,7-dimethyl^e	7	271 (10.5)	10.0	108
	13	234, 273 (7.1, 10.2)		
2,6-Dihydroxy-3,9-dimethyl	7	235, 268 (8.1, 9.8)	10.1	108
	13	240^i, 269 (6.6, 10.2)		
2,6-Dihydroxy-7,9-dimethyl	1	239, 262 (5.9, 9.5)	3.2,	110
	7^z	250, 284 (8.3, 8.3)	12.1	
2,6-Dihydroxy-8,9-dimethyl	3.0	238, 265 (8.5, 10.5)	6.2	108
	10.0	245, 278 (10.7, 10.0)		

continued

^{b} Xanthosine.
^{c} Theophylline.
^{d} Paraxanthine.
^{e} Theobromine.
^{h} Neutral molecule, pH not stated.
^{z} Zwitterion.

TABLE 13 (*continued*).

Substituents	pH or solvent	$\lambda_{max}(10^{-3}\epsilon)$	pK_a	References
8-Bromo-2,6-dihydroxy-9-methyl	3	241, 267 (10.5, 12.3)	5.5	110
	10	250, 281 (12.0, 11.0)		
2,6-Dihydroxy-1,3,7-tri-methyl[f]	6	272 (10.5)		108
2,6-Dihydroxy-1,3,9-tri-methyl[g]	6	237, 268 (9.8, 10.0)		108
2,6-Dihydroxy-1,7,9-trimethyl	0	232, 262 (4.4, 8.3)	3.2	110
	5[z]	251, 285 (6.5, 7.6)		
2,6-Dihydroxy-3,7,9-trimethyl	5	242, 266 (6.0, 8.7)	7.0	110
	9.8[z]	227, 270 (10.2, 9.1)		
2,6-Hydroxy-7,8,9-trimethyl	1	237, 262 (5.8, 8.7)	3.1	110
	7[z]	250, 284 (7.4, 7.9)	12.3	
2,6-Dihydroxy-1,3,7,9-tetra-methyl[k]	5	243, 265 (5.5, 8.7)		110
2,6-Dihydroxy-1,3,8,9-tetra-methyl	6	240, 273 (7.8, 8.7)		110
2,6,8-Trihydroxy-1-methyl	3	231, 284 (7.9, 11.4)	5.8	40, 134
	12	218, 293 (21.4, 12.3)	10.6	
2,6,8-Trihydroxy-3-methyl	3	232, 287 (7.9, 11.0)	5.8	40, 101
	12	214, 293 (18.4, 14.9)		
2,6,8-Trihydroxy-7-methyl	3	234, 286 (8.7, 11.4)		101
	8	237, 293 (9.7, 12.2)		
	12	222, 297 (23.7, 13.1)		
2,6,8-Trihydroxy-1,3-dimethyl	3	234, 286 (8.8, 11.6)	5.8	40, 101
	12	215, 294 (19.0, 16.9)		
2,6,8-Trihydroxy-1,7-dimethyl	3	234, 286 (8.4, 11.0)		134
	12	222, 296 (26.0, 12.7)		
2,6,8-Trihydroxy-9-β-D-ribo-furanosyl[l]	1	237, 288 (9.6, 11.3)		111
	11	244, 298 (12.0, 9.6)		
2,6,8-Trihydroxy-1,3,7,9-tetramethyl	6	294 (10.5)		93

[f] Caffeine.
[g] Isocaffeine.
[k] Cation.
[l] Uric acid riboside.
[z] Zwitterion.

TABLE 14.   U.V. Data for Mercapto- and Alkylthiopurines; Selenopurines.

Substituents	pH or solvent	$\lambda_{max}(10^{-3}\epsilon)$	pK_a	References
2-Mercapto	−1.2	230, 287, 382 (7.4, 18.7, 1.8)	0.5	27
	5	241, 286, 346 (12.3, 17.4, 1.5)	7.2	
	8.8	235, 263, 328 (13.2, 15.5, 3.1)	10.4	
2-Methylthio	0	242, 250[i], 314 (13.5, 12.0, 4.4)	1.9	27
	6	232, 250, 305 (16.6, 8.5, 6.0)	8.9	
	11.6	240, 301 (19.1, 6.2)		

TABLE 14   (*continued*).

Substituents	pH or solvent	$\lambda_{max}(10^{-3}\epsilon)$	pK_a	References
6-Mercapto	5.1	225, 325 (7.4, 18.7)	2.5	27
	9.3	228, 312 (9.6, 14.5)	7.8	
	12.9	231, 310 (13.9, 20.3)	10.8	
6-Methylthio	−3.5	222, 313 (11.7, 25.7)	0	27
	5.8	255^i, 290 (4.6, 22.4)	8.8	
	11.1	222, 290 (18.2, 20.0)		
8-Mercapto	−3.5	238, 280^i, 331 (14.8, 4.1, 18.2)	2.5	27
	4.5	231, 310 (10.3, 28.9)	6.6	
	8.9	228, 313 (13.5, 23.4)	11.2	
	13.0	230, 315 (15.2, 20.4)		
8-Methylthio	0	232, 305 (11.0, 20.9)	3.0	27
	5.1	246, 290 (3.9, 20.0)	7.7	
	9.9	220, 296 (17.0, 18.7)		
6-Ethylthio	1	225, 298 (10.8, 16.2)		112^a
	11	226, 292 (10.1, 15.3)		
6-Furfurylthio	1	324 (18.7)		113
	6	214, 291 (13.7, 17.9)		
	13	292 (16.8)		
6-Benzylthio	1	295 (16.6)		113
	6	292 (17.4)		
	11	292 (16.0)		
6-Mercapto-1-methyl	1	229, 321 (11.2, 18.7)	3	29
	5.1	234, 320 (9.4, 20.7)	8.6	
	11.1	237, 321 (10.7, 24.0)		
6-Mercapto-3-methyl	0	244, 334 (5.5, 29.2)	1.7	29
	5.1	245, 338 (9.2, 32.2)	7.5	
	11.1	245, 332 (11.3, 29.9)		
3-Methyl-6-methylthio	1	235, 274, 317 (9.2, 4.9, 25.2)		48
	11	311 (17.2)		
6-Mercapto-7-ethyl	0	328 (17.7)	7.9	29
	5.1	329 (20.4)		
	11.1	232, 316 (8.8, 17.5)		
6-Mercapto-9-methyl	0	326 (18.5)	1.4	29
	5.1	229, 321 (12.7, 26.1)	8.0	
	11.1	234, 309 (13.0, 21.4)		
1,7-Dibenzyl-6-mercapto	b	244, 274 (10.0, 11.9)		106
1,9-Dibenzyl-6-mercapto	b	323 (23.4)		106
7-Benzyl-6-mercapto-1-methyl	1	324 (17.4)		87
	11	238, 324 (10.2, 18.7)		
7-*n*-Butyl-1-methyl-6-mercapto	1	323 (18.0)		87
	11	238, 324 (9.5, 19.5)		
6-Benzylthio-1-methyl	1	312 (15.9)		87
6-Hydroxy-7,9-dimethyl-2-methylthio	1	269 (17.2)		48
	11	241, 275 (17.9, 10.9)		
2-Amino-7,9-dimethyl-6-methylthio	1	248, 320 (10.1, 11.1)		48
2,6-Dimercapto	10.4	253, 280^i, 345 (21.0, 15.9, 14.1)		114

^{a} Contains data for thirteen other 6-alkylthiopurines.
^{b} Neutral molecule, pH not stated.
^{i} Inflexion.

continued

TABLE 14 (*continued*).

Substituents	pH or solvent	$\lambda_{max}(10^{-3}\epsilon)$	pK_a	References
	11	241, 260, 333 (12.6, 14.7, 31.4)		
2,6-Dimercapto-8-phenyl	0	220, 265, 333, 352 (23.4, 6.1, 26.3, 17.1)		90
6,8-Dimercapto	1	270, 357 (17.7, 32.0)		115
	13	249[i], 344 (12.3, 17.3)		
6-Amino-2-mercapto	6.4	230, 259[i], 285 (9.6, 7.3, 12.6)		102, 116
	6.8	229, 282 (13.7, 13.9)		
2-Amino-6-mercapto	1	258, 347 (8.1, 20.9)		117
	11	242, 270, 322 (8.7, 7.2, 16.0)		
2-Amino-6-methylthio	1	242, 273, 318 (7.5, 10.2, 12.7)		115[c]
	11	312 (10.9)		
6-Amino-2-methylthio	1	221, 246, 284 (9.9, 13.3, 11.2)		102
2-Methylthio-6-amino-9-β-ribofuranosyl	1	270 (16.0)		95
	13	235, 277 (21.2, 14.7)		
2-Amino-3-methyl-6-mercapto	1	237, 256, 340 (4.4, 6.3, 33.6)		118
	11	235, 279, 330 (8.6, 12.0, 33.6)		
2-Amino-6-methylthio-3-methyl	1	275, 318 (14.2, 15.9)		118[d]
	11	232, 264, 319 (10.6, 11.4, 16.3)		
9-Methyl-6-methyl-amino-2-methylthio	0.8	211, 253, 274, 282[i] (16.2, 12.9, 14.1, 13.8)	3.0	152
	5.5	214, 241, 281 (13.2, 20.9, 15.5)		
2-Amino-6-mercapto-8-methyl	1	256, 350 (8.0, 20.0)		120[e]
	11	323 (17.2)		
2-Amino-6-benzylthio-8-methyl	1	242, 270, 322 (9.2, 7.3, 14.2)		120
	11	318 (11.7)		
2-Amino-6-mercapto-8-(4-chlorophenyl)	1	268, 370 (18.0, 12.1)		121
	11	250, 350 (20.0, 16.2)		
6-Mercapto-2-methyl-amino	1	261, 350 (11.5, 17.1)		122
	11	245, 275, 325 (11.1, 10.2, 13.9)		
2-Ethylamino-6-mercapto	1	263, 350 (11.5, 17.1)		122
	11	250, 276, 325 (11.2, 10.1, 12.6)		
2-Dimethylamino-6-mercapto	1	268, 358 (12.7, 17.2)		122
	11	253, 283, 322 (11.8, 11.8, 11.2)		
2-Anilino-6-mercapto	1	278, 352 (17.6, 17.9)		122
	11	283, 328 (28.8, 11.7)		
2-Piperidino-6-mercapto	1	272, 359 (14.5, 14.4)		122
	11	257, 282, 328 (13.1, 11.1, 9.4)		
2-Amino-6,8-dimercapto	1	272, 372 (25.4, 35.0)		115
	11	251, 273, 347 (16.8, 14.6, 25.0)		
2-Amino-8-hydroxy-6-mercapto	1	250, 350 (8.6, 22.0)		115
	11	240, 325 (15.4, 17.0)		
6-Amino-2-hydroxy-8-mercapto	1	269, 303 (16.4, 15.6)		115
	11	237, 300 (16.8, 15.3)		
6-Amino-2-hydroxy-8-methylthio	1	263, 291 (12.1, 15.5)		115
	11	265, 290[i] (9.9, 12.7)		
2-Hydroxy-6-mercapto	10.4	250, 340 (7.4, 21.0)		114

[c] Contains data for eleven 6-alkyl- and 6-arylalkylthiopurines, five 6-arylthiopurines, and thirteen 2-amino-6-alkylthiopurines.

[d] Contains data for five other 2-amino-6-alkylthio-3-methylpurines.

[e] Contains data for thirty-six 2-amino-6-alkyl (or aryl) thiopurines.

[i] Inflexion.

TABLE 14   (*continued*).

Substituents	pH or solvent	$\lambda_{max}(10^{-3}\epsilon)$	pK_a	References
6-Hydroxy-2-mercapto	1	285 (21.6)		122
	11	278 (16.8)		
8-Hydroxy-6-mercapto	1	237, 290^i, 333 (13.2, 6.3, 19.5)		115
	11	237, 310 (17.8, 22.9)		
6-Hydroxy-8-mercapto	1	233, 288 (7.9, 22.0)		115
	11	233, 289 (26.7, 21.8)		
2,8-Dihydroxy-6-mercapto	1	260, 355 (8.1, 28.7)		100
	11	237, 344 (17.0, 22.2)		
8-Hydroxy-2,6-dimercapto	1	263, 298, 367 (9.4, 22.8, 19.0)		100
	11	250, 253 (30.0, 16.4)		
6-Hydroxy-2,8-dimercapto	1	243, 310 (11.2, 20.4)		100
	11	243, 295 (12.8, 17.2)		
6-Hydroxy-2,8-dimethylthio	1	283 (19.6)		100
	11	289 (15.2)		
2-Chloro-8-hydroxy-6-mercapto	1	240, 302, 341 (11.8, 7.7, 17.2)		100
	11	247, 320 (16.3, 23.9)		
6-Seleno	^{b}	225, 325 (7.4, 18.6)		56
2,6-Dimercapto-9-phenyl	1	330 (27.3)		124
	11	344 (28.6)		
2-Amino-6-mercapto-9-methyl	1	277, 250, 320 (10.1, 7.2, 11.5)		124
	11	260, 350 (6.0, 13.2)		
2-Amino-6-mercapto-9-phenyl	1	337 (30.7)		124
	11	340 (26.6)		
2,8-Diamino-6-mercapto	1	252, 355 (10.5, 25.3)		58
	11	260, 327 (9.6, 15.7)		
2,8-Diamino-6-methylthio	1	234, 333 (17.2, 11.8)		58
	11	234, 325 (19.4, 12.4)		

^{b} Neutral molecule, pH not stated.

TABLE 15.   U.V. Data for Amino- and Alkylaminopurines.

Substituents	pH or solvent	$\lambda_{max}(10^{-3}\epsilon)$	pK_a	References
2-Amino	−3.5	235^i, 325 (6.5, 4.2)	−0.3	21, 33
	1.8	237^i, 314 (4.2, 4.0)	3.8	
	7	236^i, 305 (5.0, 6.0)	9.9	
	12	276^i, 303 (4.1, 5.8)		
2-Methylamino	1.8	223, 244^i, 327 (35.5, 6.5, 3.7)	0.3	119
	7	219, 240, 319 (26.3, 7.6, 5.5)	4.0	
	12.5	226, 272, 316 (23.4, 5.2)		
2-Dimethylamino	1.7	228, 248^i, 340 (33.1, 9.3, 3.0)	4.0	21, 33
	7	223, 248, 332 (25.7, 10.5, 5.1)	10.2	
	12.7	232, 327 (24.6, 4.7)		
2-Hydrazino	1	297 (5.0)		96
	7	309 (4.7)		

^{i}Inflexion

continued

TABLE 15   (*continued*).

Substituents	pH or solvent	$\lambda_{max}(10^{-3}\epsilon)$	pK_a	References
6-Amino[a]	2	263 (13.2)	0	125, 143
	7	261 (13.4)	4.1	
	12	269 (12.3)	9.8	
6-Methylamino	2	267 (15.2)	4.2	21, 33
	7	266 (16.2)	10.0	
	12	273 (15.9)		
6-Ethylamino	1	270 (16.3)		121[b]
	11	273 (17.0)		
6-Dimethylamino	1.7	276 (15.5)	3.9	21, 33
	7	275 (17.8)	10.5	
	13	221, 281, (16.2, 17.8)		
6-Furfurylamino[c]	1	274 (16.9)		113[d]
	EtOH	212, 268 (26.1, 18.7)		
	13	273 (17.4)		
6-Cyclohexylamino	2	272 (17.1)	4.2	126
	7	270 (18.2)	10.2	
	12	275 (18.2)		
6-Succinamino	1	276 (17.6)		99, 127
	13	275 (17.0)		
6-(2′Imidazolinyl)	1	286, 296, 338, 353[i] (12.7, 11.9, 6.4, 5.3)		128
	11	245[i], 299, 311, 342[i] (2.5, 14.7, 13.8, 4.4)		
6-Azido	5	250, 258, 286 (4.6, 5.1, 7.4)		129
	10.3	233, 305 (10.7, 7.3)		
6-N-Hydroxylamino	1.2	271 (13.3)	3.8	130
	6.7	268 (11.8)	9.8	
			12	
6-Amino-1-N-oxide	1	258 (11.5)	2.6	124
	7	231, 263 (41.5, 8.1)	9.0	
	13	233, 275 (46.2, 7.4)	13	
6-Dibenzylamino	1[e]	287 (22.0)		20
	7[e]	278 (23.9)		
	13[e]	284, 292[i] (21.5, 18.6)		
6-Phenethylamino	2	274 (16.2)	4.2	126
	7	269 (17.5)	10.1	
	12	275 (17.7)		
8-Amino	2.4	288 (15.9)	4.7	21, 33
	7.1	241, 283 (3.2, 14.5)	9.4	
	12.0	230[i], 290 (9.6, 11.7)		
8-Methylamino	2.7	230[i], 296 (7.9, 17.4)	4.8	21, 33
	7.2	245, 290 (3.2, 16.6)	9.6	
	12.0	230[i], 298 (10.0, 14.5)		
8-Dimethylamino	2.7	230[i], 305 (11.0, 19.5)	4.8	21, 33
	7.3	250, 296 (3.0, 18.7)	9.7	
	12.0	230, 306 (8.1, 16.6)		

[a] Adenine; ref. 125 gives data for pK_a at 20–40°.
[b] Contains data for eight 6-substituted aminopurines.
[c] Kinetin.
[d] Contains data on thirteen related 6-alkylaminopurines.
[e] 95% ethanol.
[i] Inflexion.

TABLE 15   (*continued*).

Substituents	pH or solvent	$\lambda_{max}(10^{-3}\epsilon)$	pK_a	References
2-Amino-6-methyl	1	219, 313 (39.1, 4.5)		102
2-Amino-8-phenyl	1	257, 332 (24.0, 12.0)	4.0	21, 33
	6.5	238, 329 (16.6, 19.1)	9.2	
	11.4	239, 330 (20.4, 19.1)		
6-Amino-2-methyl	1	266 (12.9)		131^f, 136
	13	271 (10.7)		
6-Amino-8-phenyl	1	232, 293, 300 (14.2, 25.4, 25.0)		90
	6	234, 294 (19.6, 21.4)		
	13	238, 303 (20.4, 22.0)		
2,6-Diamino	3	241, 282 (9.1, 10.5)	1	21, 33
	7.5	246, 279 (7.1, 8.9)	5.1	
	13	243^i, 284 (4.7, 9.3)	10.8	
2-Amino-6-hydrazino	1	238^i, 285 (8.3, 7.6)		96
	7	240^i, 283 (7.5, 7.1)		
6-Amino-2-hydrazino	1	268 (10.2)		96
	7	263 (9.9)		
2,6-Dihydrazino	1	276 (7.9)		96
	7	273 (6.5)		
2,6-Diamino-1-N-oxide	1	248, 290 (8.6, 8.7)	1.0	124
	5.5	230, 290 (31.3, 7.0)	3.7	
	13	228, 295 (26.8, 7.9)	9.7	
			12	
2,6,8-Triamino	0.3	248, 305 (13.2, 12.9)	2.4	21, 33
	4.3	221, 250^i, 299 (20.4, 4.9, 17.4)	6.2	
	8.5	249, 293 (6.3, 12.0)	10.8	
	13	226, 261, 295 (20.4, 4.1, 12.0)		
2,6-Diamino-8-	1	303 (17.4)		132^g
(2-chlorophenyl)	11	305 (12.5)		
1,2-Di(6-purinylamino)-	1	278 (27.6)		133
ethane^h	13	276 (28.6)		

^{f} Contains data for eleven 2-substituted adenines.
^{g} Contains data on sixteen 2,6-diamino-8-arylpurines.
^{h} Full name of compound, not substituent.
^{i} Inflexion.

TABLE 16.   U.V. Data for Amino-ring-N-alkylpurines.

Substituents	pH or solvent	$\lambda_{max}(10^{-3}\epsilon)$	pK_a	References
6-Amino-1-methyl	4	259 (11.7)	7.2	22
	13	270 (14.4)	11.0	
1-Methyl-6-methyl-	1	261 (12.9)		46
amino	MeOH	276 (12.2)		
	11	274 (12.7)		
6-Dimethylamino-1-	1	293 (12.2)		19
methyl	7	298 (12.6)		
	12	301 (13.6)		

continued

TABLE 16   (*continued*).

Substituents	pH or solvent	$\gamma_{max}(10^{-3}\epsilon)$	pK_a	References
6-Amino-1,9-dimethyl	1	259 (10.6)		46
	11	259 (10.4)		
6-Amino-1-methyl-9-β-D-ribofuranosyl	1	257 (13.7)		107
	6	257 (14.6)		
	11	257 (14.6)		
1-Methyl-6-methyl-amino-9-β-D-ribofuranosyl	1	261 (14.2)		46
	MeOH	261 (15.9)		
	14	262 (14.9)		
6-Amino-1-methyl-9-β-D-2'-deoxyribofuranosyl	1	257 (15.1)		107
	6	258 (15.1)		
	11	258 (15.1)		
6-Amino-1-(2'-hydroxyethyl)-9-β-D-ribofuranosyl	1	258 (12.8)	8.3	134
	11	259 (12.9)		
1-Benzyl-6-benzylamino	1[a]	267 (12.8)		20
	95%EtOH	233, 280 (18.6, 12.8)		
	13[a]	275 (15.3)		
6-Amino-3-methyl	1	274 (15.9)	5.7	17, 25
	12	272 (12.8)		
6-Amino-3-ethyl	1	275 (14.9)	5.4	17[b]
	12	273 (10.5)		
3-Methyl-6-methylamino	1	281 (19.6)		52
	11	287 (14.5)		
6-Dimethylamino-3-methyl	1	290 (20.4)	5.8	19, 25
	7	222, 292 (11.3, 16.6)		
	12	293 (16.4)		
6-Amino-3-β-D-ribofuranosyl[c]	1	275 (18.2)	5.5	135[d]
	7	215, 277 (16.7, 12.9)		
	13	278 (13.1)		
6-Dimethylamino-3-β-D-ribofuranosyl	1	291 (22.7)		19
	MeOH	226, 298 (11.7, 15.1)		
	12	298 (17.5)		
6-Dimethylamino-3-β-D-glucopyranosyl	1	292 (19.6)		19
	7	222, 298 (11.7, 15.1)		
	12	299 (15.0)		
6-Amino-3-(3',3'-diethylallyl)[e]	1	274 (17.5)	5.4	17
	7	273 (13.8)		
3-Benzyl-6-benzylamino	1[a]	288 (24.8)		20
	95%EtOH	218, 293 (22.6, 17.3)		
	13[a]	294 (17.4)		
2,6-Diamino-3-methyl	1	243, 279 (11.4, 13.8)		118
	11	243, 283 (6.0, 16.7)		
2-Amino-6-n-butyl-amino-3-methyl	1	224, 249, 280 (14.7, 12.7, 16.5)		118
	11	228[i], 245[i], 284 (22.0, 8.2, 18.0)		

[a] 95% ethanol.
[b] Contains data for six other 3-alkyladenines.
[c] Isoadenosine.
[d] Contains data for five isoadenosine derivatives substituted in the ribosyl moiety.
[e] Triacanthine.
[i] Inflexion or shoulder.

TABLE 16   (*continued*).

Substituents	pH or solvent	$\lambda_{max}(10^{-3}\epsilon)$	pK_a	References
6-Amino-7-methyl	1	272 (13.8)	3.5[f]	17[g]
	12	270 (10.5)		
6-Amino-7-(3',3'-	1	276 (14.6)		17
dimethylallyl)	7	272 (9.8)		
6-Dimethylamino-7-	1	293 (18.1)		19
methyl	7	291 (14.4)		
6-Amino-2,7-dimethyl	1.3	273 (16.2)		136
	12	277 (13.2)		
6-Amino-9-methyl	1	260 (13.7)		29
	11	261 (14.0)		
6-Amino-9-β-D-ribo-	1.3	257 (14.6)	3.4	143
furanosyl[h]	6.4	260 (14.9)		
6-Methylamino-9-β-D-	1	261 (16.3)		107
ribofuranosyl	6	265 (16.3)		
	11	265 (15.0)		
6-Dimethylamino-9-β-	1	268 (18.4)		19
D-ribofuranosyl	7	215, 275 (15.6, 18.8)		
	12	275 (19.2)		
6-Amino-9-cyclohexyl	2	261 (14.6)	4.2	126
	7	262 (14.7)		
9-Cyclohexyl-6-cyclo-	2	267 (17.9)	4.4	126
hexylamino	7	271 (17.1)		
6-Amino-9-3'-dimethyl-	2	259 (13.7)		137[j]
aminopropyl	7	261 (13.8)		
6-Amino-9-(3',3'-	1	260 (14.0)	3.3[f]	16[k]
dimethylallyl)	7	260 (13.8)		
9-Methyl-6-methylamino	1	265 (15.3)		104
	11	268 (14.0)		
6-Dimethylamino-9-	1	270 (17.5)		104
methyl	7	276 (18.1)		
	14	277 (18.1)		
6-Hydrazino-9-methyl	1	263 (16.1)		104
9-Methyl-6-methyl-	1	267 (16.0)		104
hydrazino	11	277 (13.1)		
9-Benzyl-6-benzylamino	1[a]	266 (20.3)		20
	95%EtOH	271 (19.0)		
	13[a]	271 (19.1)		
1,6-Di-(6-amino-9-	1	258 (24.6)		63
purinyl)ethane[l]	13	258 (21.0)		
1,2-Di-(6-dimethylamino-	1	267 (29.2)		63
9-purinyl)ethane[l]	13	272 (27.8)		
6-Dimethylamino-9-(2'-	2	268 (14.2)		137
hydroxyethyl)	7	277 (14.7)		

[f] In *N,N*-dimethylformamide : water (3:2, by vol.).
[g] Contains data for eight 7-alkyladenines.
[h] Adenosine.
[j] Contains data for three 9-aminoalkyl-, three 9-hydroxyalkyl-, and two 9-chloroalkyl-adenines.
[k] Contains data for 7- and 9-(2',2'-diethoxyethyl) adenines.
[l] Full name of compound, not substituent.

TABLE 16   (*continued*).

Substituents	pH or solvent	$\lambda_{max}(10^{-3}\epsilon)$	pK_a	References
6-Amino-9-β-D-ribo-	1	258 (11.9)	2.1	124
furanosyl-1-N-oxide	5.3	233, 260 (40.8, 9.2)	12.5	
	13	230, 273, 312 (23.7, 8.4, 4.5)		
6,8-Diamino-9-β-D-	1	270 (13.5)		111[m]
ribofuranosyl	11	273 (16.4)		
6-Amino-2,9-dimethyl	1.3	264 (12.7)		136
	12	264 (16.2)		
6-Amino-9-d-xylosido-2-	1.3	261 (12.5)		136
methyl	12	263 (14.7)		
6-Amino-9-methyl-8-	1	238, 297 (15.5, 23.1)		98
phenyl	11	243, 313 (20.8, 20.4)		
6-(2'-Hydroxyethyl-	1	263 (17.9)	3.1	134
amino)-9-β-D-ribo-	13	267 (17.9)		
furanosyl				
9-(3-Deoxy-3-p-methoxy-	1	268 (19.5)		138
L-phenylalanylamino-	7	275 (20.3)		
D-ribofuranosyl)-6-	14	275 (20.3)		
dimethylamino[n]				
6-Amino-1,7-dibenzyl	1[a]	277 (8.4)		20
	95%EtOH	266, 275[i] (9.7, 8.7)		
	13[a]	266, 275[i] (11.0, 9.6)		
6-Amino-1,9-dibenzyl	1[a]	262 (14.5)		20
	95%EtOH	262, 269[i] (13.6, 11.2)		
	13[a]	262 (14.4)		
6-Amino-3,7-dimethyl	4	276 (15.7)	11	22[p], 48
	13	279 (14.1)		
6-Amino-3,7-dibenzyl	1[a]	224[i], 281 (14.9, 16.9)		20
	95%EtOH	281 (15.9)		
	13[a]	281 (13.6)		
6-Amino-3,9-dibenzyl	1	274[i], 287 (18.5)		138
	7	274[i], 286 (19.0)		
7,9-Dimethyl-6-methyl-	7	270[i], 280 (12.6, 14.1)	11	116
amino				

[m] Contains data for 8-methoxy-, 8-hydrazino, 8-diazo-, and 8-hydroxyadenosines.
[n] Puromycin.
[p] The compound for which the structure 1,3-dimethyladenine was suggested[22] was subsequently shown to be 3,7-dimethyladenine.[48]

TABLE 17.   U.V. Data for Amino- and Alkylaminohydroxypurines.

Substituents	pH or solvent	$\lambda_{max}(10^{-3}\epsilon)$	pK_a	References
6-Amino-2-hydroxy[a]	2	230[i], 284 (4.9, 10.7)	4.5	21, 33
	7	240, 286 (7.8, 7.9)	9.0	
	11.1	235[i], 284 (5.3, 12.3)		

[a] Isoguanine.
[i] Inflexion.

TABLE 17   (*continued*).

Substituents	pH or solvent	$\lambda_{max}(10^{-3}\epsilon)$	pK_a	References
6,8-Diamino-2-hydroxy	2.3	302 (13.9)		139
	6.5	298 (13.0)		
	9.2	295 (12.5)		
2-Amino-6-hydroxy[b]	1	249, 276 (11.4, 7.4)	0	143
	7	246, 276 (10.7, 8.2)	3.2	
	10.9	246, 274 (6.3, 8.0)	9.6	
	14	274 (9.9)	12.5	
6-Amino-8-hydroxy	2.3	272 (10.9)		139
	6.5	270 (12.8)		
	9.2	270 (13.0)		
2,6-Diamino-8-hydroxy	2.3	250, 305 (8.7, 9.6)		139
	6.5	246, 287 (8.1, 9.6)		
	9.2	245^i, 289 (6.7, 9.0)		
2,8-Diamino-6-hydroxy	2.3	247, 287 (12.8, 8.5)		139
	6.5	247, 291 (7.1, 10.0)		
	9.2	247, 290 (6.8, 9.7)		
6-Hydroxy-2-methyl-amino	1	250, 280 (13.9, 6.3)	3.3	9, 123
	11	244, 278 (9.5, 7.2)	8.9	
			12.8	
2-Dimethylamino-6-hydroxy	1	256, 288 (19.0, 6.5)		123
	11	245^i, 282 (12.9, 7.5)		
2-Ethylamino-6-hydroxy	1	253, 280^i (14.8, 8.1)		122
	11	245, 275 (9.5, 9.3)		
N-(2-Amino-6-hydroxy-purin-2-yl)-L-alanine[c]	1	249, 277 (15.9, 7.2)		123
	11	244, 277 (9.7, 7.2)		
6-Hydroxy-2-ribosyl-amino[d]	1.3	249, 273^i	3.3	9
	7	248, 272	8.8	
	11.2	249, 271	12.6	
	14	255^i, 274		
2-Anilino-6-hydroxy	1	270 (20.4)		122
	11	238, 274 (15.3, 20.4)		
2-p-Chloroanilino-6-hydroxy	1	274 (20.2)		122
	11	240, 280 (14.6, 21.1)		
6-Hydroxy-2-piperidino	1	260, 290^i (19.8, 6.2)		122
	11	252, 280^i (12.7, 8.1)		
2-Hydrazino-6-hydroxy	1	248 (10.6)		122
	7	248, 271^i (10.0, 6.5)		
2-Amino-6-hydroxy-8-phenyl	1	238, 268, 305 (15.4, 11.9, 17.4)		132
	11	238, 312 (19.1, 17.5)		
2-Amino-6-hydroxy-8-(p-toluyl)	1	254, 308 (15.8, 19.6)		90[e]
	6	231, 254, 307 (11.8, 13.4, 18.0)		
	13	241, 317 (17.5, 17.4)		
2-Amino-6-methoxy	1	286 (11.2)		144[f]
	7	240, 280 (7.9, 7.9)		
	13	246, 284 (4.5, 7.9)		

[b] Guanine.
[c] Full name of compound, not substituent.
[d] Neoguanosine.
[e] Contains data for 8-o,m- and p-chlorophenyl-, and 8-m-nitrophenyl-guanines.
[f] Contains data for O^6-ethyl-, -n-propyl-, -isopropyl-, and -n-butyl-guanines.

TABLE 18. U.V. Data for Amino-hydroxy-ring-N-alkylpurines.

Substituents	pH or solvent	$\lambda_{max}(10^{-3}\epsilon)$	pK_a	References
2-Amino-6-hydroxy-1-methyl	0	250, 270[i] (10.7, 7.1)	3.1	12
	6	249, 273 (10.2, 8.1)	10.5	
	13	262[i], 277 (7.9, 8.1)		
2-Amino-6-methoxy-3-methyl	1	235, 284 (5.0, 11.8)		47
	11	287 (14.0)		
2-Amino-6-hydroxy-3-methyl	1	244, 264 (8.3, 11.2)		47
	11	273 (13.7)		
2-Amino-6-hydroxy-7-methyl	0	250, 270[i] (10.6, 6.9)	3.5	12, 140
	6	248, 283 (5.7, 7.4)	10.0	
	13	281 (7.3)		
2-Amino-7-ethyl-6-hydroxy	1	250, 274 (11.1, 7.0)		64[a]
	7	245, 284 (5.9, 7.7)		
	12	280 (7.4)		
1,4-Di-(guanin-7-yl)butane-2,3-diol[b]	1	253, 274[i] (21.2, 14.3)		64
	7	250, 284 (13.9, 14.7)		
	12	281 (14.6)		
Di-(2-guanin-7'-ylethyl)methyl-amine[b]	−0.4	252 (20.7)		64
	7	284 (12.5)		
	12	281 (14.2)		
6-Amino-2-hydroxy-9-methyl	1.3	230[i], 270 (4.5, 10.0)		6[c]
	6	235, 285 (8.7, 9.2)		
	12.7	240, 275 (4.0, 7.5)		
6-Amino-2-hydroxy-9-β-D-ribofuranosyl[d]	1.3	235, 283 (6.1, 12.7)		7
	6	247, 293 (8.9, 11.1)		
	12.7	251, 285 (6.9, 10.6)		
2-Amino-6-hydroxy-9-methyl	0	251, 276 (12.0, 7.6)	2.8	12
	6	252, 270[i] (12.6, 9.3)	9.8	
	13	258[i], 268 (10.2, 11.2)		
2-Amino-9-ethyl-6-hydroxy	1	252, 280 (12.4, 8.1)		124
	11	270 (10.8)		
2-Amino-6-hydroxy-9-isopropyl	1	252, 280 (11.6, 7.9)		124
	11	270 (10.4)		
2-Amino-6-hydroxy-9-β-D-ribofuranosyl[e]	1	257 (12.2)	2.2	143
	6	253 (13.7)	9.5	
	11.3	258–266 (11.3)		
2-Amino-6-hydroxy-9-β-D-2'-deoxyribofuranosyl[f]	2	255 (12.3)		141
	12	260 (9.2)		
2,8-Diamino-6-hydroxy-9-β-D-ribofuranosyl	1	250, 289 (16.9, 9.6)		111
	11	258, 271[i] (13.5, 11.6)		
2-Amino-6-hydroxy-1-methyl-9-β-D-ribofuranosyl	1	258 (9.4)		46
	MeOH	256 (10.8)		
	11	254 (10.4)		

[a] Contains data for 7-(2-hydroxyethyl)guanine, 7-(4-hydroxy-n-butyl)guanine, and 7-(2,3,4-trihydroxy-n-butyl)guanine.
[b] Full name of compound, not substituent.
[c] See text, Section B, for discussion of these data, which may be inaccurate.
[d] Crotonoside.
[e] Guanosine.
[f] Deoxyguanosine.
[i] Inflexion or shoulder.

TABLE 18 (*continued*).

Substituents	pH or solvent	$\lambda_{max}(10^{-3}\epsilon)$	pK_a	References
2-Amino-6-hydroxy-1-methyl-9-β-D-2′-deoxyribofuranosyl	1	257 (12.1)		46
	MeOH	256 (14.3)		
	11	254 (13.6)		
6-Hydroxy-2-methylamino-9-β-D-ribofuranosyl	1	258, 281[i] (14.3, 7.9)		123[g]
	11	254, 270[i] (14.8, 11.4)		
2-Dimethylamino-6-hydroxy-9-β-D-ribofuranosyl	1	264, 293[i] (12.8, 5.9)		123
	11	262, 273[i] (12.2, 10.6)		
2-Amino-6-hydroxy-1,7-dimethyl	0	252, 273[i] (10.2, 6.8)	3.4	12
	6	250, 283 (5.6, 7.4)		
2-Amino-6-hydroxy-1,9-dimethyl	0	254, 279 (11.2, 7.6)	3.3	12
	6	255, 269[i] (12.3, 10.0)		
2-Amino-6-hydroxy-7,9-dimethyl[h]	4	253, 279 (11.8, 7.6)	7.2	12
	9.5[z]	252, 282 (5.9, 8.3)		
2-Amino-6-hydroxy-7-methyl-9-β-D-ribofuranosyl	3	258 (10.7)	7.2	24
	9[z]	259, 282 (5.8, 8.0)		
2-Amino-7-ethyl-6-hydroxy-9-β-D-ribofuranosyl	3	258 (12.0)	7.2	24
	9[z]	259, 282 (5.8, 7.8)		
2-Amino-6-hydroxy-7,9-di-(2-hydroxyethyl)	3	254, 281 (10.9, 7.1)	7.3	24
	9[z]	252, 283 (5.6, 7.8)		
2-Amino-6-hydroxy-1,7,9-trimethyl[k]	5	254, 280 (11.2, 7.8)		12
2-Amino-6-hydroxy-1,7-dimethyl-β-D-ribofuranosyl	1	259 (11.3)		46
	MeOH	261 (15.9)		

[g] Contains data for fifteen N^2-substituted guanosines.
[h] Herbipoline.
[i] Inflexion.
[k] Cation.
[z] Zwitterion.

TABLE 19. Miscellaneous U.V. Data: Cyano-, Aldehydo-, Keto-, Carboxy-, Nitro-, and Diazopurines.

Substituents	pH or solvent	$\lambda_{max}(10^{-3}\epsilon)$	pK_a	References
6-Cyano	1	289 (7.5)		129
	11	292 (6.5)		
Purine-6-carboxaldehyde[a]	1.6	264 (7.5)	2.4	142[b]
	7.2	268 (8.6)	8.8	
	11.9	272 (6.3)		
Purin-6-yl 4-hydroxyphenyl ketone[a]	0	248, 285[i], 325 (10.5, 10.5, 11.1)		108[c]
	6.2	240, 310 (7.5, 13.4)		
	14	250[i], 275, 323 (9.0, 9.5, 22.9)		

[a] Full name of compound, not substituent.
[b] Contains data for derivatives of purine-6-carboxaldehyde: semicarbazone, phenylhydrazone, thiosemicarbazone, and hydrazone.
[c] Contains data for corresponding carbinol, dimethylcarbinol, and hydrazone.

continued

TABLE 19   (*continued*).

Substituents	pH or solvent	$\lambda_{max}(10^{-3}\epsilon)$	pK_a	References
Purin-6-yl-2-hydroxy-5-	0	279 (11.9)		108[c]
methylphenyl ketone[a]	6.2	276 (11.8)		
	14	242[i], 276 (9.6, 10.8)		
Purine-6-carboxylic acid[a,d]	1	280 (7.7)		129
	11	279 (7.7)		
Purine-6-carboxamide[a]	1	240, 279 (4.5, 7.9)		129
	11	292 (6.7)		
Purine-6-thiocarboxamide[a]	1	285, 335[i] (6.0, 8.8)		129
	11	294 (8.7)		
Purine-6-carboxamidine[a]	1	294 (8.1)		129
	11	300 (7.2)		
6-Hydroxy-2-nitro-9-β-D-	1	222, 335 (12.0, 4.2)	3.3	42
ribofuranosyl	7	233, 343 (14.4, 3.8)		
2,6-Dihydroxy-8-nitro	1	360 (10.6)		58
	11	262, 430 (8.3, 12.8)		
6-Hydroxy-8-nitro	1	236, 334 (21.1, 14.0)		58
	11	240, 382 (15.8, 12.9)		
2,6-Dihydroxy-1,3-	1	245, 370 (8.8, 9.1)		58
dimethyl-8-nitro	11	240, 390 (8.5, 11.1)		
2,6-Dihydroxy-1,3,7-	1	244, 370 (9.1, 8.1)		58
trimethyl-8-nitro	11	244, 370 (9.8, 7.9)		
6-Amino-8-diazo	1	234, 358 (9.3, 11.6)		58
	11	288 (12.0)		
8-Diazo-6-hydroxy	1	234, 282, 364 (7.5, 4.2, 12.0)		58[e]
	11	243, 290, 342 (6.8, 8.2, 5.8)		

[d] Purinoic acid.

[e] Contains data for 8-diazo derivatives of guanine, $N_{(2)}$-methylguanine, xanthine, and theophylline; and for products from coupling of 8-diazohypoxanthine, 8-diazoguanine, and 8-diazoxanthine with di-*n*-butylamine.

[i] Inflexion.

2. Infrared Spectra*

A. Sampling Methods

A systematic study of the infrared spectra of a series of organic compounds is preferably carried out in the solid or liquid, solution, and vapour states, especially where strong interactions due to hydrogen bonding occur in the condensed phase. Frequency shifts are then observed in the absorption bands arising from groups involved in

* By R. Lumley Jones, Chester Beatty Research Institute, Institute of Cancer Research, London, England.

hydrogen bonding, e.g., O—H, N—H, and C=O, on going to the monomeric condition. These shifts are invaluable in making vibrational assignments. Where the hydrogen atom is labile, deuteration is useful in assigning absorptions arising from vibrational modes which primarily involve movement of the hydrogen atom.

Oxo- and aminopurines have low volatilities and low solubilities in the nonpolar solvents, such as carbon tetrachloride and carbon disulphide, normally used in infrared spectroscopic work.[145] Attempts have been made to overcome this problem by using molten antimony trichloride as a solvent.[146] Under the latter conditions, the positions of bands arising from NH stretching modes show little or no change on dilution, indicating the absence of intermolecular hydrogen bonding.

Alkyl-substituted purines, because of their higher solubilities, have been examined in chloroform solution by Brown and Mason[27] and Mason.[26] Similarly, the hydrogen-bonding properties in chloroform or deuterochloroform of analogues of the purine and pyrimidine bases found in DNA and RNA have been examined by infrared spectroscopic techniques both qualitatively and quantitatively.[147-152] Hydrogen bonding in deuterochloroform solution between 9-ethyladenine and both barbiturate[153] and riboflavin[154] derivatives has been observed. Some nucleosides with alkyl and aryl substituents on the sugar moiety have been found to be sufficiently soluble in carbon tetrachloride to enable the hydrogen-bonding properties of the constituent bases to be examined.[155] Hydrogen bonding between an adenosine derivative and $N_{(4)}$-hydroxycytosine together with $N_{(4)}$-methoxy cytosine has been studied in the same way.[156]

Aqueous solutions appear to have been used so far only in the examination of the Raman spectra of purines,[157-160] although the infrared spectrum of purine itself has been recorded under these conditions.[160] Vapour phase studies have been confined to the ultraviolet region of the electromagnetic spectrum.[53]

Solid films of purines, transparent to visible light, have been prepared by vacuum sublimation of the compounds on to rock salt plates.[161-165] The sublimed films give spectra of better quality than powdered samples, but the influence of hydration on molecular orientation in the sublimates may give rise to splitting and changes in the relative intensities of the absorption bands. This can be exemplified by the spectrum of guanine, in which the splitting of a band between 800 and 900 cm^{-1} is the major difference between the spectra of a sublimate exposed to environments of low and high humidity. X-ray diffraction studies show a parallel change on going from the anhydrous amorphous to the hydrated crystalline state.[165] A related point of interest is that hydration of the purine bases in the sodium salt of DNA has been postulated on the basis of changes in the infrared spectrum with variation in relative humidity of the surroundings.[166]

Interactions between aromatic hydrocarbons and caffeine in chloroform solution and in the solid state have been observed. Shifts in the infrared absorption bands arising from caffeine were interpreted in terms of mutual polarisation of the molecules and the formation of stable complexes.[167] Cook and Regnier have shown that theobromine and caffeine form weak complexes with acetic acid in the solid state, but stronger complexes, involving protonation on the purine $N_{(9)}$ atom in the imidazole ring are formed with salicyclic acid.[168] Protonation at the same position has been shown to occur in several salts of caffeine,[169] theobromine,[170] and theophylline.[168] Base-pairing in the solid state between 1-methylthymine and 9-methyladenine has been studied by infrared methods.[171]

Although the spectra of solids are usually recorded with the samples in the form of mulls or pastes in nujol[172] or perfluorokerosene,[173] the alternative potassium bromide disc technique[174–176] is advantageous in view of the high transparency of thin discs of the material down to 400 cm^{-1}. This latter method, which has been used extensively in the spectroscopic study of purine derivatives,[91, 177–180] has the further advantage of adaptation to quantitative studies, as suggested by Friedlander and DiPietro, and facilitates the examination of mixtures[177] as well as the determination of molecular structure.[178]

An interesting alternative method for spectroscopic study of solid samples is the recording of their infrared emission spectrum. Such a spectrum has been reported for uric acid.[181] The Raman spectrum of a material either in the solid state or in solution provides a useful complement to the infrared spectrum especially in the assignment of vibrational frequencies.[159]

B. Interpretation of Spectra in Terms of Structure

a. *Qualitative Aspects*

The infrared spectrum of an organic compound is its most characteristic physical property. The infrared absorption bands arise from the normal modes of vibration of a molecule, these modes being dependent upon: (*a*) the molecule's constituent atomic masses, (*b*) the strength of the bonds joining these atoms together, and (*c*) the geometry of the molecule as a whole. The infrared spectrum, therefore, has a distinctive character, especially in the region below 1500 cm^{-1} where absorptions occur due to skeletal modes of vibration involving the heavier atoms joined together by bonds of comparable strength. In these circumstances, considerable coupling between some vibrational modes can take place, and the absorption frequencies will be particularly sensitive to minor changes in molecular structure, leading to a change in the pattern of the observed infrared absorption bands.

The infrared spectrum below 1500 cm^{-1} is therefore useful in establishing the identity, or otherwise, of two given compounds. In this context the infrared spectra of purine derivatives[162, 178] are valuable since their melting points are high, and their ultraviolet spectra may be very similar. The infrared spectra of adenine and guanine show some similarities near 1600 cm^{-1} as do those of xanthine and hypoxanthine, but below this region there are few absorptions common to all four except in the range 957 to 935 cm^{-1}. The N-methylated purines, theophylline, theobromine, and caffeine show similarities to other purine spectra near 3300 cm^{-1} and between 1725 and 1220 cm^{-1}, except for the absence of the band at 1706 cm^{-1} in the spectrum of theobromine. Theophylline shows an anomalous strong band at 2618 cm^{-1}. All three compounds show a band near 738 cm^{-1}, but between 1180 and 770 cm^{-1} there are few similarities, and the spectra in this region are useful for identification and differentiation purposes.[162]

Willits and co-workers[163] have pointed out the variability of purine spectra below 1000 cm^{-1}, but distinctive features are a band in the range 975 to 925 cm^{-1}, which is sometimes resolved into several components, and a doublet near 800 cm^{-1} where a higher frequency band is always the more intense.[162, 163]

Friedlander and DiPietro have examined purine and a number of its derivatives in the region 800–650 cm^{-1} using the potassium bromide disc technique. Purine itself shows no absorption bands below 791 cm^{-1}, whereas the monosubstituted purines, adenine and hypoxanthine, show one absorption below this value. On the other hand, the disubstituted purines examined, guanine and xanthine, show three distinct bands. The authors discuss the application of these different spectral features in the identification of the compounds, either separately or as components of mixtures.[177]

b. General Structural Considerations

The usefulness of infrared spectra in structural diagnosis is illustrated by their use in the course of the synthesis by Montgomery and co-workers of purine derivatives as potential carcinostatic agents. Infrared spectra have been used, for example, to detect acetyl,[91, 182–184] amide,[182, 185–188] amino,[97, 187, 189] azide,[184, 188, 190, 191] phenyl,[182, 185–187, 192, 193] phosphate,[193, 194] and sugar[182, 186, 192–195] groups and residues, as well as the purine ring system itself,[193] in various compounds. The formation and loss of the labile 9-acetyl group in purine derivatives[182] have been confirmed by infrared spectroscopy.

Hydrazino derivatives give spectra similar to those of amino purines. Both show broad NH stretching [$\nu_{(NH)}$] bands between 2900 and 2400 cm^{-1}, and the 1750 to 1500 cm^{-1} region can be used in either case to differentiate between 2-monosubstituted, 6-monosubstituted, and 2,6-disubstituted purines.[96]

9-Ethylpurines[189] show spectra very similar to the corresponding purines but with some additional bands due to the 9-ethyl substituent. The spectra of 9-alkyl-6-substituted purines are very similar to the 9-ethyl analogues but with more absorption in the 3000 to 2800 cm^{-1} region due to the additional aliphatic CH stretching modes.[196]

The infrared spectra of 2-fluoropurines are very similar to those of the parent purines, but small shifts to higher frequencies are observed in the bands occurring in the 1800 to 1500 cm^{-1} region.[30] Similar shifts have been observed by Lister and Kiburis[197] in the spectra of 6-fluoropurines.

Infrared spectra have been used to differentiate between azides and tetrazoles,[190] the azido compounds showing strong bands near 2135 cm^{-1}, but the spectra are very similar to those of purines in the 1650 to 1500 cm^{-1} region. On the other hand, tetrazolopurines do not show an azide band near 2130 cm^{-1} but give rise to a band typical of this type of compound near 1045 cm^{-1}.

Balsiger and co-workers[198] illustrate the use of infrared spectra in distinguishing between 2-chloroethylthiopurines and the corresponding dihydrothiazolopurines. Whereas the spectra of the former resemble those of the corresponding alkylthiopurines, especially in the double-bond region 1650 to 1500 cm^{-1}, the spectra of the latter compounds are quite different. The latter do not show the 2800 to 2400 cm^{-1} bands typical of the spectra of the former compounds, attributed by the authors to a stretching mode of the "acidic" ($N_{(9)}$–H) bond.

The spectra of various series of purine derivatives have proved useful for structural diagnosis. Horák and Gut,[199] assuming the oxo tautomer for N-methyl derivatives of oxopurines in the solid state, have demonstrated the analytical value of successive shifts in the carbonyl stretching frequency. These shifts can be used as a means of distinguishing between oxopurines with various N-substituents.

The infrared spectra beyond 667 cm^{-1} have been used by Läufer[178] to differentiate between two main classes of purine derivatives, i.e., those in which the classical aromaticity of the purine nucleus is maintained, as in adenine for example, and N-methylated oxopurines related to caffeine which can adopt differing degrees of lactam structure.

In general, compounds of the first type show intense bands between 1640 and 1560 cm^{-1} due to the ring valence vibrations. In derivatives of the second type bands typical of the lactam group are observed between 1724 and 1640 cm^{-1}, although these are frequently split. Introduction of substituents such as phenyl or amino groups can cause confusion in the 1650 to 1500 cm^{-1} region and the use of the potassium bromide region, where the two types show very different spectra, is advocated. A compound of the first type will show its strongest band near 625 cm^{-1}, whereas a compound of the second type will show its strongest band between 526 and 455 cm^{-1}. Some complications arise on halogen substitution, but the pattern is not affected by exocyclic carbonyl groups. The bands observed for caffeine, which are similar to those for uric acid and xanthine between 526 and 455 cm^{-1}, confirm the lactam structure for these compounds. The N-methylated oxopurines show no bands between 667 and 625 cm^{-1}; this, together with the presence of strong bands at lower wave numbers, is indicative of the lactam structures for these molecules.

c. *Tautomerism*

In purine chemistry much speculation has centred around the nature of the predominating tautomeric forms of oxo- and aminopurines. Blout and Fields[162] suggested that the occurrence of partial enolisation would account for an anomaly in the spectrum of theobromine in which a band found at 1706 cm^{-1} in the spectra of theophylline and caffeine is absent. The authors have suggested that the omission is explicable if an enolic OH group in one molecule of theobromine is associated with a carbonyl group of a neighbouring molecule.

Brown and Mason suggest that 2- and 6-oxopurines exist in the keto form in view of the absence of OH stretching bands [$\nu_{(OH)}$] near 3400 cm^{-1} in their solid-state spectra, but they point out the difficulty of differentiating between $\nu_{(OH)}$ and $\nu_{(NH)}$ bands in the condensed phase spectra. The presence of a carbonyl absorption near 1670 cm^{-1} supports the oxo structure, and similarly the spectrum of the 8-isomer shows a strong band at 1740 cm^{-1}.[27] Deuteration studies of solid guanine in the 3300 cm^{-1} region support the oxo-amino structure for this compound.[200]

In the solution state, using molten antimony trichloride Lacher[146] found little or no absorption between 3700 cm^{-1} and 3500 cm^{-1} in the spectra of hydroxypurines, and concluded that these compounds exist primarily in the oxo form. Similarly, the spectrum of a guanine analogue[148] in deuterochloroform was consistent with the oxo-amino structure.

Mason[26] and Brown and Mason[27] examined 7- and 9-methyl derivatives of hydroxypurines in chloroform solution but found no $\nu_{(OH)}$ bands. Further studies involving comparison of the solution spectra of various oxopurines with the corresponding N-methylated derivatives in which the possibility of classical tautomerism was either absent or unlikely, suggested the absence of any appreciable contribution from the hydroxy forms.[26, 27]

Application of these observations to 6-oxopurine derivatives suggests that the tautomeric hydrogen atom is associated mainly with $N_{(1)}$. In general when the "hydroxy" group is located α- or γ- to a ring nitrogen atom in a heterocyclic system the proton usually tautomerises onto the α ring-nitrogen atom. Both 7- and 9-methyl analogues show an NH stretching band at 3390 cm^{-1} in the range of $\nu_{(NH)}$ for cyclic conjugated amides of the type **22**. Infrared evidence from $\nu_{(NH)}$ and $\nu_{(CO)}$ suggests that 6-oxopurines have the structure **23a** or **23b**, with **23b** being preferred. It is assumed that the 7- and 9-positions would be more or less equivalent in either the solid state or in aqueous solution.[27]

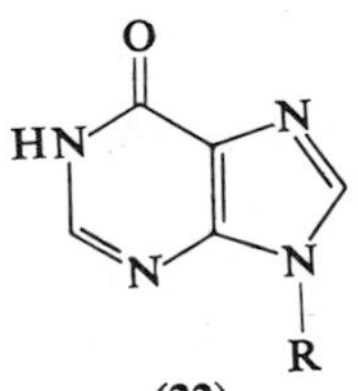

(22) 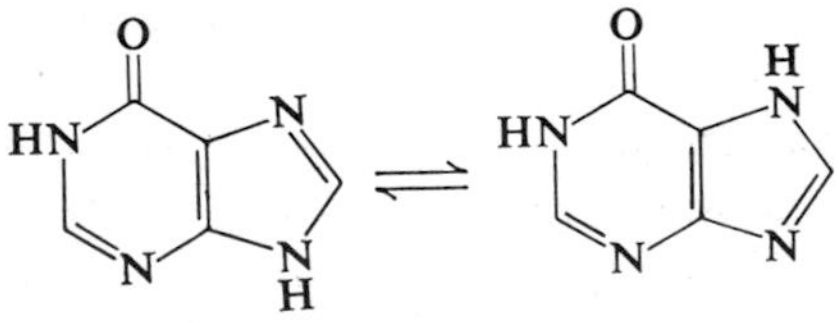   (23a)   (23b)

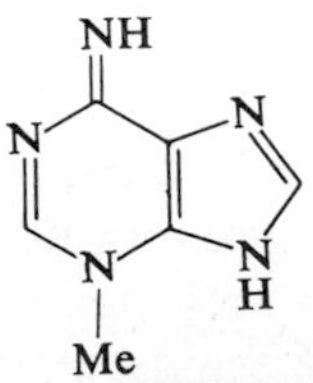

(24a) 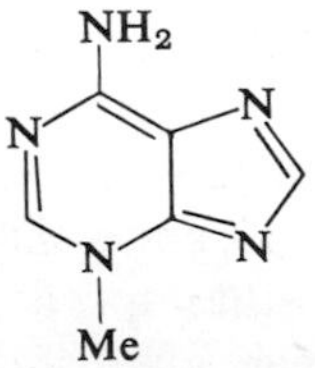   (24b)

Willits and colleagues compare the absence of strong absorptions in the solid-state spectrum of purine above 3000 cm^{-1} with the appearance of new bands between 3400 and 3100 cm^{-1} on substitution of an amino group into the 2- or 6-position. In addition, a new band appears at about 1670 cm^{-1}.[163] This band, located at 1672 cm^{-1} in the spectrum of adenine by Angell, can be assigned without doubt to the in-plane deformation mode of the amino group [$\delta_{(NH_2)}$].[200] 9-Ethyl-6-aminopurine in chloroform solution shows two $\nu_{(NH)}$ bands separated by about 100 cm^{-1}, which is characteristic of primary amines. This compound, therefore, and probably 6-aminopurine itself, is predominantly in the amino form. Mason suggests that aminopurines conform to this generalisation within the limits of available evidence.[26] The spectra of adenine in antimony trichloride[146] and of adenine analogues in chloroform or carbon tetrachloride[147-156] solution are consistent with the large majority of molecules being in the amino form.

Thus the greater weight of evidence from the infrared spectra of adenine and some of its derivatives, both in the solid state and in solution, shows that these compounds are predominantly in the amino form.

An imino structure (24a) was originally proposed for 3-methyladenine[201] but the infrared spectrum recorded by Pal and Horton[25] appears to be more consistent with the amino form (24b). An absorption band at 1672 cm^{-1} which arises from the in-plane deformation mode of the NH$_2$ group [$\delta_{(NH_2)}$] is found in the spectrum of 3-methyladenine as well as in the spectra of several adenine derivatives containing exocyclic amino groups. It does not appear, however, in the spectrum of the deuterated compound nor in that of 6-dimethylamino-3-methylpurine or of several analogues. The results obtained in the $\nu_{(NH)}$ region near 3300 to 3100 cm^{-1} and the changes observed on deuteration were equivocal. They did not give a clear indication of appropriate assignments in this region.

Infrared spectroscopy has also been used to examine the possibility of tautomerism in thiopurines. Willits and co-workers[163] suggest that the presence of a strong band at 1323 cm^{-1} and the absence of a band at 2500 cm^{-1} in the spectra of these molecules in the solid state favour the thione structure. But conclusions based on the absence of $\nu_{(SH)}$ bands, usually found near 2600 cm^{-1}, where many purines show weak absorptions, are not unequivocal due to the $\nu_{(SH)}$ bands' low intrinsic intensity and possible overlap from these other absorptions.[26, 27] A specific assignment to $\nu_{(SH)}$ was not facilitated by deuteration experiments.[27]

Brown and Mason found well-defined absorptions due to $\nu_{(NH)}$ in the chloroform solution spectra of 7,8-dihydro-9-methyl-8-thiopurine and 8,9-dihydro-7-methyl-6-thiopurine, and concluded that the parent compounds also exist partly, if not predominantly, in the thione form under these conditions.[26, 27]

C. Assignment of Absorption Bands

a. *From* 3600 *to* 2000 cm^{-1}

The fundamental stretching vibrations of the OH, NH, CH, and SH groups give rise to absorption bands in this region of the spectrum. The band positions for stretching modes of amino groups and their variation with the point of attachment to the purine nucleus have been tabulated by Katritsky and Ambler.[202]

From their spectra of sublimed films, Blout and Fields[162] assign bands at 3333 and 3125 cm^{-1} in the spectra of adenine and guanine to $\nu_{(NH)}$ and $\nu_{(OH)}$ modes, respectively. Since xanthine and hypoxanthine show no bands at frequencies higher than 3125 cm^{-1}, these authors assign the 3333 cm^{-1} band in adenine and guanine to $\nu_{(NH_2)}$. Further, the spectrum of a sublimed sample of purine itself shows no absorption above 3000 cm^{-1}, but on substitution of an amino group at the 2- or 6-position, new peaks appear between 3400 and 3100 cm^{-1}.[163] Two strong absorption bands near 3300 and 3100 cm^{-1} in the solid-state spectrum of adenine have been observed by Angell[200] and assigned by him to $\nu_{(NH_2)}$. These bands shift on deuteration, which lends support to the above assignment.

Pal and Horton[25] point to the unexpected appearance of bands near 3300 and 3100 cm^{-1} in the spectra of some dialkylamino and N-deuterated purine derivatives. These authors suggest that the appearance of these two bands can be taken as evidence for the presence of amino groups in aminopurines only if this is supported by the presence of other bands, e.g., the one at 1672 cm^{-1}, attributable with confidence to $\delta_{(NH_2)}$.

Lacher and co-workers[146] assign bands near 3450 and 3300 cm^{-1} in the spectra of a number of heterocyclic compounds dissolved in antimony trichloride to the antisymmetric and symmetric stretching modes of the NH$_2$ group, respectively. The band near 3333 cm^{-1} is attributed by these authors to the ring imide $\nu_{(NH)}$ but they also suggest the possibility of contributions from overtones of the strong fundamental bands near 1600 cm^{-1}. In chloroform solution, 9-methyl-6-amino purine shows two bands at 3530 and 3412 cm^{-1}, characteristic of a primary amino group.[26]

Absorption bands between 2941 and 2703 cm^{-1} in the spectra of sublimates have been tentatively assigned to aromatic $\nu_{(CH)}$.[162] Montgomery[91] assigns weak bands between 3115 and 3080 cm^{-1} to the same vibrational mode of N-acetylpurines, and three bands of medium intensity in purines between 3113 and 2920 cm^{-1} to this same mode. In N-acetylpurines, bands in the region 2960 to 2913 cm^{-1} are assigned to aliphatic $\nu_{(CH)}$ modes.[91] In antimony trichloride solution spectra of purines and pyrimidines, bands between 3125 and 3030 cm^{-1} and also between 2941 and 2778 cm^{-1} were assigned by Lacher[146] to aromatic and aliphatic $\nu_{(CH)}$, respectively.

Willits and co-workers[163] point out that in all the compounds which they studied a broad band appeared between 2700 and 2300 cm^{-1}, which was assigned to an associated $\nu_{(N-H)}$ mode, intermolecularly hydrogen bonded in the form $N—H \cdots N$. The breadth of the band was attributed to the formation of hydrogen bonds of varying strength. The same authors comment on a band at 2700 cm^{-1} in the spectra of theobromine and caffeine,[162] which has no "tail" at lower wavenumber values, and this unusually sharp band was attributed to the formation of an exceptionally strong hydrogen-bond at the $N_{(7)}$ position.[163] Angell, on the other hand, attributes a broad band between 2900 and 1500 cm^{-1} in the spectrum of adenine to $\nu_{(N_{(9)}-H)}$ since the band is absent in the spectrum of 9-methyladenine. Similar sets of bands between 2750 and 2540 cm^{-1} in the spectrum of purine disappear on methylation at the 9-position.[200] The same group of bands, observed between 2900 and 2400 cm^{-1} is attributed by Montgomery to the presence of an "acidic" $N_{(9)}$ hydrogen atom and disappears on $N_{(9)}$-acetylation.[91] The broad band between 2900 and 2400 cm^{-1} is also observed in hydrazino purines,[96] which show infrared spectra generally very similar to those of aminopurines.

b. *From* 2000 *to* 1500 cm^{-1}

Absorption in this region of the spectrum would be expected to arise from skeletal vibrations of the purine nucleus together with, for example, deformation modes of exocyclic amino groups and stretching modes of carbonyl groups. The strong bands appearing at 1622 and 1571 cm^{-1}, respectively, in the spectrum of purine are attributed to ring skeletal vibrations.[163] Analogous absorptions are observed at 1632 and 1568 cm^{-1} in the spectra of 2-diethylaminopurines.[163] Angell[200] notes the lack of similarity between adenine and purine spectra except for bands at 1610 and 1575 cm^{-1}, which are typical of purine and pyrimidine nuclei. The pyrimidine absorptions are usually found in the range 1610 $\pm$ 10 and 1575 $\pm$ 5 cm^{-1}, and the two bands can be assigned specifically therefore to the pyrimidine ring system. In the spectrum of adenine the 1575 cm^{-1} band appears as an unresolved shoulder but is clearly defined in the spectrum of deuterated adenine and 9-methyladenine.[200]

Some authors[146, 162] have attempted to assign absorption bands in this region to specific bonds (C=C, C=N) in the heterocyclic nucleus, but it is doubtful, in view of the strong coupling which would occur between vibrations involving neighbouring bonds in the heterocyclic system, whether such specific assignments can be justified. Montgomery assigns three bands of moderate intensity between 1609 and 1545 cm^{-1} in purines to the aromatic purine nucleus.[91] In a later publication[97] the range is given as 1630 to 1495 cm^{-1}.

Addition of an amino group to the purine nucleus gives rise to a new band at 1670 cm^{-1} mainly due to the in-plane deformation mode of this group [$\delta_{(NH_2)}$]. However, there is the possibility of a contribution due to coupling between these "external" modes and the skeletal modes of the purine nucleus.[163] Variations in the positions of bands arising from deformation modes of amino groups attached to different heterocycles have been tabulated.[66] Montgomery and Holum have assigned bands between 1670 and 1650 cm^{-1} to deformation of an amino group attached to the purine nucleus at $C_{(6)}$. Amino groups at $C_{(2)}$ absorb between 1650 and 1620 cm^{-1}. 2-Substituted- and 6-substituted-hydrazinopurines can be differentiated in a similar manner.[97] Bands in the range 1672 $\pm$ 8 cm^{-1} in several 2- and 6-aminopurines are assigned by Pal and Horton[25] to $\delta_{(NH_2)}$, an assignment confirmed by their behaviour on deuteration and methylation. Similarly, the disappearance of the band at 1672 cm^{-1} in the spectrum of adenine on deuteration leads to the assignment of this band to $\delta_{(NH_2)}$ while the bands appearing at 1510 and 1285 cm^{-1} are attributed to a shifted skeletal mode and $\delta_{(ND_2)}$, respectively.[200] The assignment is also supported by changes in the spectrum on protonation of adenine. Adenine hydrochloride shows an absorption at 1712 cm^{-1}, indicating that the proton is attached at $N_{(1)}$, affecting $\delta_{(NH_2)}$, but without giving rise to the deformation mode of a zwitterion $\delta_{(NH_3^+)}$.[200] However, Pal and Horton[25] found shoulders here in the spectra of neutral compounds and suggest that the appearance of a band in this region may be due to uncompensated water absorptions.

Addition of a hydroxy group to the purine nucleus would be expected to give rise to a new absorption around 1700 cm^{-1} due to the carbonyl stretching mode of the oxo tautomer.[163] In the case of guanine the band appears at 1692 cm^{-1} in antimony trichloride solution and at 1701 cm^{-1} in the solid state with the corresponding $\delta_{(NH_2)}$ bands at 1618 and 1681 cm^{-1}, respectively.[146, 200]

The 1555 cm^{-1} component of the 1567/1555 cm^{-1} doublet in the solid-state spectrum of guanine is assigned to $\delta_{(N_{(1)}-H)}$ since it disappears on deuteration.[200]

c. *From* 1500 *to* 1000 cm^{-1}

Angell[200] has pointed out the presence of a series of five fairly strong absorptions between 1500 and 1300 cm^{-1} in the spectra of adenine derivatives and also in the spectrum of 9-methylpurine. The latter has a spectrum much more similar to that of adenine than to the spectrum of purine itself. Adenine derivatives have a characteristic strong absorption at 1305 cm^{-1} which, like the other five bands, is not affected by deuteration.[200] Bands at 1477 and 1375 cm^{-1} in guanine derivatives have been assigned[200] to ring vibrations. In a series of purine derivatives Montgomery assigns bands in the range 1609 to 1549 cm^{-1} to "aromatic" purine bands. In chloropurines a weak band near 1450 cm^{-1} is assigned to a deformation mode of the CH group.[91] The assignment of a strong band at 1323 cm^{-1} in the spectra of thiopurines to $\nu_{(C=S)}$ has been made.[163]

d. *Below* 1000 cm^{-1}

Absorption bands in the ranges 980 to 900 cm^{-1} [162, 163, 187, 193] and 890 to 860 cm^{-1} [187, 193] have been attributed to the purine ring system. Angell[200] attributed a broad band at 870 cm^{-1} in the spectrum of adenine to the out-of-plane NH deformation mode $\gamma_{(N_{(9)}-H)}$. This band disappears on $N_{(9)}$-methylation and on deuteration, but Pal and Horton[25] claim that deuteration studies are inconclusive on this point. A similar broad band has been reported in the spectrum of purine but is absent in that of 9-methylpurine.[200] A strong band at 785–775 cm^{-1} in the spectrum of guanine is attributed by the same author to $\gamma_{(C_{(8)}-H)}$.

In the antimony trichloride solution spectrum of purines, Lacher assigns a band between 1229 cm^{-1} and 1215 cm^{-1} to $\gamma_{(CH)}$ since it is absent from the spectrum of uric acid. It is also absent from the antimony trichloride solution spectra of tetrasubstituted pyrimidines.[146]

The far infrared spectrum down to 33 cm^{-1} has been recorded recently for purine and some of its deuterated analogues, and some tentative assignments have been made.[160]

3. Magnetic Resonance Spectra

Initial applications of these techniques were with purine itself, the purpose of which was an attempt to correlate theoretical calculations of charge densities and related physical properties of the ring atoms with the order and position of the resonance peaks. Subsequent utilisations have been directed more towards resolution of structures, notably with some naturally occurring purines, as for example, zeatin,[203] or for ascertaining the tautomeric forms predominating under the prevailing conditions.

Although proton (pmr), carbon-13 satellite (nmr) and electron spin resonance (esr) spectra of purines have been studied the majority of the following section is devoted to pmr results. Carbon resonance (Sect. E) and electron spin resonance (Sect. F) spectra are, however, treated in outline.

A. Proton Resonance Spectra—Procedures

These studies necessitate having available samples of purine in which the protons attached to the $C_{(2)}$, $C_{(6)}$, and $C_{(8)}$ atoms are replaced by deuterium. Various routes to these derivatives have been followed. On brief heating of purine and substituted purines in deuterium oxide (10 min., 100°) replacement at $C_{(8)}$ occurs.[204, 205] On prolonged heating (72 h) purine gives a significant yield of the 6,8-dideuterated analogue.[205] Catalytic dehalogenation in the presence of deuterium is a more specific means of isotopic introduction being adopted in forming the 2-deutero (2-chloropurine, palladium catalyst)[206] and 6-deutero (6-chloropurine, palladium),[206, 207] (6-iodopurine, Adams catalyst)[204] derivatives. A less favourable group-replacement procedure involves sulphur removal in 6- and 8-thiopurines by means of Raney nickel containing occluded deuterium.[208] It should be noted that the rather vigorous conditions required in the latter method could induce proton exchange at other sites also. Applications of the Traube synthesis have given purines labelled at either $C_{(2)}$, $C_{(6)}$, or $C_{(8)}$. In the first two cases[209] the 2- or 6-deuterated-4,5-diaminopyrimidine was cyclised with formic acid while in the last case, using an unlabelled pyrimidine, closure was effected with dideuterated formic acid.[207]

Due to the poor solubility of purines generally in organic solvents* determinations are best carried out in deuterium oxide or dimethyl sulphoxide for the neutral molecule, while the spectra of anionic or

---

* The chemical shifts reported in an early publication,[210] for purine in chloroform solution, should be treated with caution as the concentration figure of purine reported (25%) is greater, by a factor of ten at least, than can be achieved in practice in this solvent.[204]

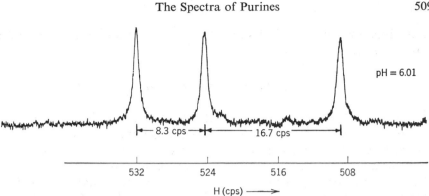

Fig. 19. Pmr spectrum of 0.42 M solution of purine (neutral molecule) in water at 38° (shown against tetramethylsilane standard). (From Coburn, Thorpe, Montgomery, and Hewson,[206] by permission of the American Chemical Society.)

cationic species are suitably resolved in sodium deuteroxide and trifluoroacetic acid, respectively. A variety of standards have been employed, either internally or externally, the most widely favoured being tetramethylsilane. Others used include hexamethyldisilyloxane, benzene, chloroform, and acetonitrile, while sodium 3-trimethylsilylpropane-1-sulphonate is useful as an internal standard in aqueous media.[206]

B. Assignment of Proton Resonance Peaks

a. *With Purine Itself*

In deuterium oxide a simple spectrum comprising three peaks corresponding to the protons located at $C_{(2)}$, $C_{(6)}$, and $C_{(8)}$ is shown, the noticeable broadening observed being ascribed to proton interaction with nitrogen nuclei of adjacent molecules (Fig. 19). The contribution due to the acidic imidazole ($N_{(7)}/N_{(9)}$) proton shows as a characteristic broad, low-intensity band at lower fields which is absent, due to proton exchange, if traces of moisture are present. From observations of peak disappearances in the spectra of 2-, 6-, and 8-deuteropurines the correct peak assignments in order of increasing field are found to be H-6, then H-2 with H-8 occurring at highest field.[204, 207, 209]

In nonaqueous solvents the same order of peak appearance is followed but in all cases with changes in chemical shift to lower field.[211] The above results invalidate earlier assignments in which the signals ascribed to H-6 and H-2 were interchanged.[35, 210] The various factors that influence the chemical shifts of the three main resonance bands are discussed later.

b. *With Substituted Purines*

The neutral molecules of 2-, 6-, and 8-methylpurine in either deuterium oxide or nonaqueous solvents show similar spectral patterns.[209, 210, 212] In each case the methyl group resonance band appears at very high field whereas the ring proton signals occur at much lower fields. In both 2- and 6-methylpurine the H-8 signal occurs at higher field than that due to the remaining proton.[209, 210, 212] With 8-methylpurine it is H-6 rather than H-2 which has the higher field position.[209] Results covering a number of 2- and 6-monosubstituted-[207, 213] and 2,6-disubstituted purines,[206, 213] obtained in dimethyl sulphoxide, show that in the monosubstituted derivatives the effect of a 2- or 6-substituent does not affect the chemical shift of H-2 significantly but can disturb profoundly that due to H-8. In the case of a 6-substituted purine the effects of the $C_{(6)}$-group are transmitted to H-2 purely by means of ring resonance whereas a combination of induction and enhanced resonance is responsible for the effect at H-8.[213] While in general the deshielding effect on H-8 is not sufficiently large enough to alter the relative field order of H-8 and H-2 exceptions to this are found, two notable examples being adenine and 6-iodopurine which show crossover of the two proton peaks, that due to H-2 appearing at the higher field in each case.[213] This phenomenon appears to be restricted to special cases in which a common feature is the presence of one or more strong electron-donating groups. The move to higher field of H-8 of monosubstituted purines can be intensified by further substitution at the remaining free carbon atom.[213]

The amino group in 1,9-dimethylguanine and other guanine derivatives gives rise to a peak upfield of the H-8 resonance thereby confirming that it is present as amino rather than the imino form. If the latter was the case two nonequivalent proton bands would be given, one due to the imino proton and the other from the protonated adjacent ring nitrogen atom.[31]

C. Factors Influencing pmr Spectra

a. *Ionisation*

As with the spectrum of the neutral molecule the anionic spectrum of purine shows as three, near equivalent, peaks but with the chemical shifts occurring at higher field.[207, 209] This increased shielding is largely a reflection of the distribution of the excess electron density of the anion, by means of the π-electron system, throughout the molecule.[211] In 6-substituted derivatives the same peak order of H-2 and H-8 is observed, the latter being at higher field.[207]

In the cation spectrum the resonance peaks occur in the same order as is found in that of the neutral molecule (and in the anion) but are moved downfield. Studies with purine hydrochloride in hydrochloric acid [212] show shift positions are pH dependent, the lowest fields being attained in the strongest acid solution. At high acid concentrations (but not at low concentrations) the largest shift change is demonstrated by H-6 and the least by H-8. The large absolute downfield shifts of the cation spectrum suggests that protonation is not confined to $N_{(1)}$, as has been generally thought, but may also take place at $N_{(3)}$ and $N_{(7)}$ with equal facility,[212] a view supported by data derived from ^{13}C-resonance spectra.[214]

Spectra of purine recorded in trifluoroacetic acid (TFAA) solutions (Fig. 20) show considerable broadening of the three resonance peaks as pH decreases. Below pH 2.5 splitting of the H-6 and H-2 peaks is ascribed to spin-spin coupling between H-2 and H-6 protons.[206, 214] In addition, a small splitting contribution affecting the H-6 peak may be provided by spin-spin coupling of H-6 with H-8.[206]

With monosubstituted purines the spectrum may be complicated by effects due to the substituent group present. Furthermore, crossover of the resonance peaks due to H-2 and H-8 has been reported,[207] the latter now being at the lower field. This result can be compared with that from purine itself in which no crossover of the H-8 resonance is found on changing from alkaline to acid solution. In the case of adenine no

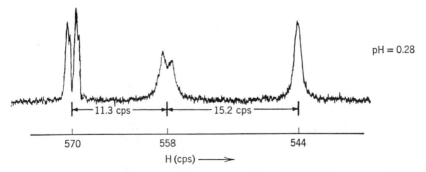

FIG. 20. Pmr spectrum of 0.42 M solution of purine (cation) in water at 38° (shown against tetramethylsilane standard). (From Coburn, Thorpe, Montgomery, and Hewson,[206] by permission of the Americal Chemical Society.)

evidence for protonation of the amino group has been found[213] using nmr techniques although with 6-dimethylamino-3-methylpurine location of the proton at the amino group has been demonstrated in acid solution.[215] Splitting of the *N*-methyl group signal results from spin-spin coupling between the proton and the alkyl group protons. A common feature of 2- or 6-substituted and 2,6-disubstituted purines is a pronounced downfield displacement of H-8 in acidic media.[206, 213] Deshielding of this type appears to be a consequence of ring protonation, the site of which remains to be assigned.[213] Unless, however, this is accompanied by extensive delocalisation of the positive charge the simple protonation postulation does not fully explain all the changes in chemical shift observed.[207]

b. *Solvent Interaction*

Comparison of the spectra of the neutral molecule of purine obtained in various solvents reveals that while the chemical shifts of H-2 and H-6 remain approximately the same a considerable variation is shown in the values for H-8.[211] This effect is a consequence of hydrogen bonding between the $C_{(8)}$ proton and solvent molecules, the degree of which varies according to the strength of the proton acceptor groups present in the solvent molecules. Table 20 showing the change in chemical shift with solvent demonstrates this, the largest downfield shifts being given by solvents possessing the strongest acceptor character. Of the most

TABLE 20. Chemical Shifts of Purine[a] Protons in Organic Solvents.

Solvent	Chemical Shift[b]		
	H-6	H-2	H-8
1,4-Dioxane	545.0	531.9	491.1
CDCl$_3$	554.0	543.3	496.6
Acetonitrile	543.5	534.4	497.0
Nitromethane	543.6	534.0	501.3
Acetone	543.9	533.4	508.0
Methanol	545.4	536.1	512.5
DMSO	547.9	535.6	516.4
DMF	548.5	536.3	520.5

[a] Concentration, 0.05*M*.
[b] In cps relative to internal tetramethylsilane.

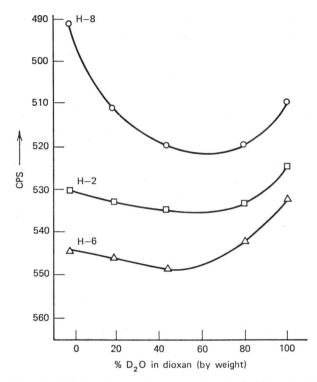

FIG. 21. Relative changes in proton chemical shifts of a 0.5 M solution of purine in dioxan on increasing dilution with deuterium oxide (against tetramethylsilane standard). (From Hruska, Bell, Victor, and Danyluk,[211] by permission of the American Chemical Society.)

commonly employed solvents the order followed for increasing low-field shift values[211] is dioxan, nitromethane, acetone, methanol, dimethylsulphoxide, dimethylformamide. Although deuterium oxide is far superior as a proton acceptor than any of the above organic solvents, the situation is anomalous in that in it all three proton resonances are shifted upfield. This effect, however, is the resultant of two separate effects and while a downfield shift of H-8 is in fact present, this is swamped by the much larger upfield shifts, affecting all three proton resonances, which result from the formation of vertical, face-to-face, stacks of purine molecules. In such complexes, for which an interface distance of 3 or 4 Å is found,[212] the enhanced magnetic anisotropy experienced by a particular purine in the stack, induced by ring current effects of adjacent molecules, gives rise to the considerable shielding observed.[211, 212] Figure 21 illustrates this effect in showing chemical shift changes of a dioxan solution of purine with increasing dilution with deuterium oxide.[211]

c. *Solute Concentration*

In nonaqueous solution the concentration of the purine has little effect on the pmr spectrum, indeed, a twelvefold increase of purine in dimethylsulphoxide only displaces the three resonance peaks by an average of 0.6 ppm downfield.[206] Although the resonance due to the acidic proton at $N_{(7)}/N_{(9)}$ is profoundly shifted downfield (5 ppm) under these conditions, due to hydrogen bonding with the unprotonated imidazole nitrogen atom of an adjacent molecule, the diffuse character of the absorption band renders it of little diagnostic importance.

By contrast, aqueous solution studies show that a near linear relationship exists between increase in solute concentration and the displacement to higher field of each of the three peaks. This phenomenon arises from the enhanced magnetic anisotropy in the purine produced by ring current induction effects of proximate molecules.[212] The subject of molecular stacking in solution which produces shielding of this type is elaborated at the end of the preceding section.

d. *Temperature*

The occurrence further upfield of each of the three resonance peaks of purine in aqueous solution is found when the solution temperature is raised. An exothermic molecular association is postulated to explain these results.[212] In contrast, in nonaqueous media little change in chemical shifts for H-2 and H-6 are found with rise in temperature although the upfield shifts of the H-8 resonance observed are a result of weakened hydrogen bonding between this proton and the solvent molecule.[211]

e. *Metal Complex Formation*

In dimethylsulphoxide solutions complexes of the type $[M(purine)]^{2+}$, formed with either zinc chloride or cupric chloride, show spectra in which the three main resonance peaks of purine are displaced downfield. As the greatest of these shifts is undergone by the H-8 signal the site of binding is deduced to be at $N_{(7)}$. Further evidence supporting this assignment is found in the spectrum of the copper complex in which the broadening of the H-6 and H-8 peaks is attributed to a relaxing effect of the paramagnetic metal ion on the carbon protons nearest to the binding site. The fact that no similar broadening of the H-2 signal is observed points to this being the case.[216]

D. Theoretical Aspects of pmr Spectra

Three main intrinsic properties of the molecule are responsible for the main resonance peaks of purine. These can be summarised as (a) the effect of the excess charge densities at each ring atom due to the π-electron system, (b) a contribution from the ring current of the molecule, and (c) magnetic anisotropic effects of ring nitrogen atoms on the carbon atoms. From the results of theoretical calculations[217] it is seen that the major contributions in predicting the chemical shift values arise from excess π-electron charge densities and ring current effects while magnetic anisotropy provides but little influence. The effect of substituent groups at the moment cannot be fully rationalised in terms of theoretical calculations of their inductive, or otherwise, effects.[206, 213] In addition, sufficient experimental data is not yet to hand to enable equations to be derived which will predict the effect of extrinsic factors, mainly those due to solute-solute interactions (i.e., stacking effects) and solvent interactions, on chemical shift values.

E. Carbon-13 Resonance Spectra

So far only purine itself has been investigated by this technique.[205, 214] As with pmr determinations deuterated analogues are made use of for resonance assignments, the deuterium-substituted carbon atom showing no resonance peak. In the absence of decoupling the carbon resonances occur as doublets but these are resolved into singlets when a decoupler is employed. The latter procedure allows unequivocal assignments for the ring carbon atoms to be made from which, surprisingly, of C-2, C-6, and C-8 the highest field resonance of the three is shown by C-6 and the lowest by C-2.[205] In this respect it is interesting to compare this result with that obtained by pmr in which the order, in increasing field, is H-6, H-2, then H-8. By adjustment of the coupling throughout the sweep of spectrum all five-ring carbon atom resonances are obtained, those due to C-4 and C-5 appearing at the lowest and highest fields, respectively, of the spectrum.[205]

low ⟶ high field (⟶TMS)

H-6   H-2   H-8

C-4   C-2   C-8   C-6   C-5

Spectral parameters from both anion and cation reveal a pH dependence suggesting that simple acid-base equilibrium exists. These findings have been used as a basis for suggesting that in the cation protonation is not confined to $N_{(1)}$ but occurs to some considerable degree also at $N_{(3)}$ and at one of the imidazole ring nitrogen atoms.[214]

F. Electron Spin Resonance (Electron Paramagnetic Resonance) Spectra

Although much useful structural data can be obtained from either proton or nuclear magnetic spectra the application of electron spin resonance spectra to this end has only limited utility. As this type of spectrum involves radical forms of the bases the main use to date in the purine field has been to investigate the fate of these derivatives following exposure to high-energy radiation.

a. *Procedures*

The purine is dispersed by freeze drying a solution of it on a quartz wool support, this being then inserted into the resonance cavity of the instrument. Radical formation occurs immediately on bombardment of the sample, *in vacuo*, at ambient temperature, with low velocity hydrogen or deuterium atoms. The resulting radicals are stable for some days *in vacuo* or a hydrogen atmosphere but are degraded rapidly in oxygen or moist air. Purine radicals of the same or related type have also been formed upon irradiation of the powdered derivative with gamma or X-rays,[218] in this case ambient or lower temperatures are adopted. The marker radical used is usually diphenylpicrylhydrazide (DPPH), *g* value = 2.0036. An innovation has been to examine ionic forms of purines produced by electrolysis, at a tungsten cathode, of solutions in dimethylformamide containing tetra-*n*-butylammonium iodide as electrolyte.[218a]

b. *Nature of the Spectra*

In addition to purine,[218a, 219] the spectra of adenine,[219, 220] guanine,[219, 220] hypoxanthine,[218] and xanthine[219, 220] have been reported. The derivative spectrum obtained in all cases closely resembles that of the parent compound (Fig. 22) which shows three lines only.[221] No effects due to the presence of substituent groups at $C_{(2)}$ or $C_{(6)}$ are

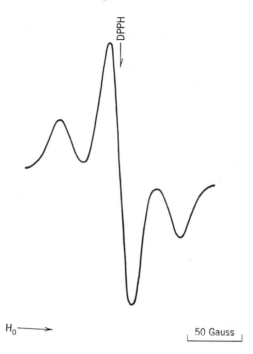

FIG. 22. Esr spectrum of purine.

detectable.[220] With the esr spectra derived from X-irradiated purines a similar picture is obtained but the resolution is less defined, some bands disappearing on cooling.[218]

c. *Theoretical Aspects*

Comparison of the spectrum of the purine obtained with hydrogen bombardment with that following deuterium bombardment shows that in all examples the latter treatment leads to loss of hyperfine structure. This suggests that the purine radicals formed are hydrogen atom adducts, the hydrogen from the beam being added at $C_{(8)}$. With adenine, for example, the structure **25a** is preferred to the alternative radical form **25b** in which hydrogen addition occurs at $C_{(2)}$. The $C_{(8)}$-proton coupling constants, which arise principally from hyperconjugative coupling of the two $C_{(8)}$-protons with the associated spin densities of the adjacent imidazole nitrogen atoms, are virtually the same (39 ± 1 gauss) for all purines investigated. This fact together with data adduced from molecular orbital calculations supports the $C_{(8)}$-adduct formation although the presence of larger central peaks in the purine triplet than would be expected if two equivalent protons were situated at $C_{(8)}$ leads to speculation that other radical forms may also be present. Evidence afforded by the spectra[218a] of purine and 6-cyanopurine indicates that in both cases the 9H rather than the 7H tautomer predominates.

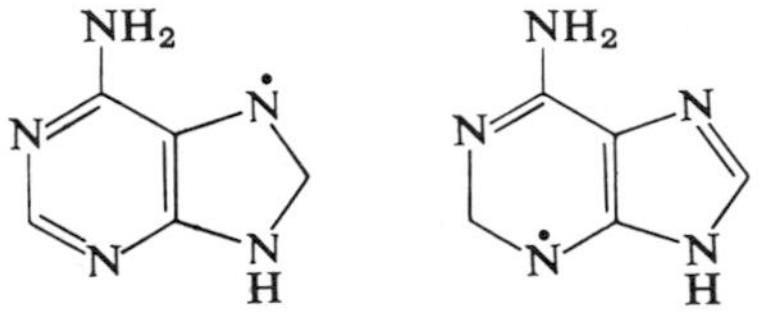

(25a)   (25b)

4. Mass Spectra

Although few direct studies on purine bases have been reported, fairly wide applications of this technique have been made in resolving structural and conformational problems arising with nucleosides, nucleotides, and purine-containing antibiotics.

A. Techniques and Method

In spite of the poorly volatile nature of some purines, for example, guanine, the use of moderately high temperatures (150–250°) and high vacuum conditions enables successful spectra to be taken. However, arising from gas chromatographic examination of purines, for which suitably volatile derivatives are mandatory, increasing use of the O- or N-trimethylsilylated analogues has been made. The advantages of this procedure have been demonstrated in a mass spectral study of guanine, but other suitable candidates for investigation are now available as preparations of the trimethylsilyl derivatives of purine,[222] adenine,[223] hypoxanthine,[222, 224] xanthine,[222, 224] and uric acid[222] have been reported.

For identification of fragments the use of deuterium labelled derivatives is often made, the preparation of which have been described earlier (Sect. 3A).

With derivatives that form stable molecular ions electron beams in the potential range 50–80 eV are employed but where it is necessary to suppress the formation of high-energy fragmentation pathways and

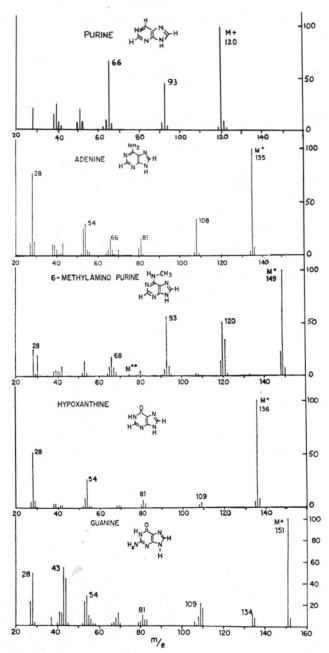

FIG. 23. Mass spectra of purine, adenine, 6-methylaminopurine, hypoxanthine, and guanine.

intensify the weak metastable ion peaks, potentials below 20 eV are adopted.

B. Nature of the Spectra

The inherent stability of the ring system of purines is demonstrated by the fact that in all cases the highest intensity peak observed is that due to the molecular ion (*cf.* Fig. 23). Initial degradation takes place mainly in the pyrimidine ring but the site of succeeding disruption is governed largely by the nature of the substituent groups present. The comparatively lower stability of monosubstituted derivatives is shown in the numerous low-intensity fragment ion peaks obtained. Spectra of this type are usually complex as molecular breakdown proceeds along multiple pathways with no particular one predominating. By contrast a more ordered fragmentation pattern is apparent with disubstituted derivatives which give high-intensity fragment ion peaks from which the major reaction pathway can be inferred.

a. *Purine and C-Alkyl Derivatives*

Purine itself undergoes initial fragmentation liberating hydrogen cyanide, derived by scission of the $C_{(2)}-N_{(3)}$ and $C_{(6)}-N_{(1)}$ bonds (Fig. 23a). With the 2-methyl and 6-methyl homologues, correspondingly, mixtures of hydrogen cyanide and acetonitrile are formed.[224] Parallel studies with deuterated purines give evidence for the proton in hydrogen cyanide being derived either from that at $C_{(2)}$ or $C_{(6)}$ rather than the one on a ring nitrogen atom. Although 8-methylpurine gives hydrogen cyanide and acetonitrile, the fragmentation scheme is not fully established.[224]

b. *Adenine and 6-Alkylaminopurines*

The three main fragmentation ion peaks given by adenine (Fig. 23b) correspond to a successive liberation of three molecules of hydrogen cyanide. Investigations with variously deuterated adenines reveal that this can occur by different pathways involving disruption at different sites.[225] While loss of the amino group does not play a significant part in the above reaction, in the case of 6-methylaminopurine (Fig. 23c) the initial fragmentation of the molecular ion involves the methylamino group following which expulsion of three molecules of hydrogen cyanide occurs.[225] The naturally occurring 6-alkylaminopurine, zeatin, shows likewise an initial degradation at the amino group.[203]

c. *Hypoxanthine, Xanthine and Uric Acid*

The first gives a spectrum notable for the low intensities of the fragment ion peaks (Fig. 23d). Examination of the ion products leads to the conclusion that one of the breakdown pathways concerned may be analogous to that followed by guanine (see d below) in which evolution of hydrogen cyanide, derived mainly from the $N_{(1)}-C_{(2)}$ bond,[225] is followed by decarbonylation. The existence of multiple fragmentation pathways together with the low ion peak intensities is taken to indicate that in the vapour phase hypoxanthine exists as the "oxo" rather than "hydroxy" tautomer. Xanthine initially loses the elements of hydrogen cyanate but the breakdown pathway subsequently follows that of hypoxanthine.[226] Uric acid behaves similarly but two molecules of hydrogen cyanate are eliminated during the early stages of degradation.[226]

d. *Guanine and N-Alkyl Derivatives*

The most characteristic and predominant fragmentation mode of molecular ions of guanine derivatives (Fig. 23e) is the expulsion of a neutral molecule of cyanamide, or cyanamide derivative, with subsequent fragmentations following a typical pattern.[225] That the cyanamide moiety is derived from $N_{(1)}-C_{(2)}$ and the amino group follows from the isolation of cyanamide from guanine, 3-methylguanine, and 7-methylguanine whereas methylcyanamide results with 1-methylguanine.[225] In the last derivative the methyl group at $N_{(1)}$ leads to a stabilisation of ion charge at this site and increases the tendency to fission of the "single" bonds associated with this atom. A consequence of the decreased stability is loss of carbonyl and methylamine fragments. This result may be contrasted with those from guanine and the other *N*-alkyl homologues in which decarbonylation does not appear to be a significant occurrence.[225] With the exception of 1-methylguanine, the similar mass spectra afforded by guanine and the other *N*-methyl derivatives can be related to the restricted fragmentation reactions possible due to the combined stabilising effects of the carbonyl and amino groups present.[225] However, the overall low intensity of the fragment ion peaks, which is also a feature of the spectrum of 1-methylguanine, provides evidence for the existence of the 6-oxo-1(H)-purine, as in the case of hypoxanthine, in the vapour phase.

e. *The N-Alkylated Xanthines*

Caffeine, theobromine, and theophylline show very high-intensity molecular ion peaks characteristic of highly stable polysubstituted purine derivatives.[227] Spectral evidence points to the initial disruption occurring at $N_{(1)}$ in all cases and liberation of corresponding imidazole fragment ions. The next most intense peak in all three spectra is attributed to formation of seven-membered heteroaromatic ions, derived directly from the molecular ion by rearrangement of the initial degradation product.[227] A corresponding highly stabilised hetero-aromatic ion of this type has been proposed during degradation of guanine derivatives.[225] Further studies with deuterated derivatives of caffeine has shed further light on the fragmentation patterns of this derivative and the associated theophylline and theobromine.[228]

f. *Thiopurines*

In addition to 1,6-dihydro-6-thiopurine a number of *C*- and *N*-alkyl analogues have been examined.[229] The molecular ion, as in other types of purines, is the most abundant but with the 7-methylated thioderivatives a considerable quantity of the $(M-1)^+$ ion is also obtained. This in the case of the 3,7-dimethyl analogue approaches near to that of the molecular ion in abundance. Stability of the $(M-1)^+$ ion is attributed to the formation of an internally bonded structure between the sulphur atom and the deprotonated 7-methyl group.[229]

C. Ionisation Studies

The application of mass spectrometry for determining ionisation potentials of a number of purines, including the parent member, adenine, hypoxanthine, xanthine, and 1,3,7,9-tetramethyluric acid,[230] has been made. These determinations, which are part of a study of the charge-transfer complexes formed by many purines with other organic molecules, are effected by comparing the purine molecular ion peak at 50 eV with an ion peak of the same intensity obtained with a rare gas (argon, krypton, or xenon) the ionisation potential of which is known. The ionisation potentials are computed from the known value of the standard used and the difference between the appearance potential of the purine and the standard gas.[230] Values obtained in this way show good agreement with those derived by means of quantum mechanical calculations.

References

1. Gridgeman, *Photoelectric Spectrometry Group Bulletin*, **4**, 67 (1951).
2. Harding, *Photoelectric Spectrometry Group Bulletin*, **4**, 79 (1951); *cf.* Scott, in *Physical Techniques in Biological Research*, Vol. I, Eds., Oster and Pollister, Academic Press, New York, 1956, p. 131.
3. Albert and Serjeant, *Ionization Constants of Acids and Bases*, Methuen, London, 1962.
4. Falconer and Gulland, *J. Chem. Soc.*, **1937**, 1912.
5. Gulland and Story, *J, Chem. Soc.*, **1938**, 692.
6. Falconer, Gulland, and Story, *J. Chem. Soc.*, **1939**, 784.
7. Davoll, *J. Amer. Chem. Soc.*, **73**, 3174 (1951).
8. Fischer, *Ber.*, **31**, 542 (1898).
9. Shapiro and Gordon, *Biochem. Biophys. Res. Commun.*, **17**, 160, (1964); *cf.* Hemmens, *Biochim. Biophys. Acta*, **68**, 284 (1963).
10. Lawley, *Progr. Nucleic Acid Res.*, **5**, 89 (1966).
11. Brown, *Progr. Nucleic Acid Res.*, **2**, 260 (1963).
12. Pfleiderer, *Annalen*, **647**, 167 (1961).
13. Ackerman and List, *Z. physiol. Chem.*, **323**, 192 (1961).
14. Letham, Shannon, and McDonald, *Proc. Chem. Soc.*, **1964**, 230.
15. Shaw and Wilson, *Proc. Chem. Soc.*, **1964**, 231.
16. Leonard and Deyrup, *J. Amer. Chem. Soc.*, **84**, 2148 (1962).
17. Denayer, *Bull. Soc. chim. France*, **1962**, 1358.
18. Baker, Joseph, Schaub, and Williams, *J. Org. Chem.*, **19**, 1780 (1954).
19. Townsend, Robins, Loeppky, and Leonard, *J. Amer. Chem. Soc.*, **86**, 5320 (1964).
20. Leonard, Carraway, and Helgeson, *J. Heterocyclic Chem.*, **2**, 291 (1965).
21. Mason, *J. Chem. Soc.*, **1954**, 2071.
22. Brookes and Lawley, *J. Chem. Soc.*, **1960**, 539.
23. Pal, *Biochemistry*, **4**, 558 (1962).
24. Lawley and Brookes, *Biochem. J.*, **89**, 127 (1963).
25. Pal and Horton, *J. Chem. Soc.*, **1964**, 400.
26. Mason, *Chemistry and Biology of Purines*, Eds. Wolstenholme and O'Connor, J. and A. Churchill, Ltd., London, 1957, p. 60.
27. Brown and Mason, *J. Chem. Soc.*, **1957**, 682.
28. Cavalieri, Fox, Stone, and Chang, *J. Amer. Chem. Soc.*, **76**, 1119 (1954).
29. Elion, *J. Org. Chem.*, **27**, 2478 (1962).
30. Montgomery and Hewson, *J. Amer. Chem. Soc.*, **82**, 463 (1960).
31. Miles, Howard, and Frazier, *Science*, **152**, 1458 (1963).
32. Bendich, Russell, and Fox, *J. Amer. Chem. Soc.*, **76**, 6073 (1954).
33. Albert and Brown, *J. Chem. Soc.*, **1954**, 2060.
34. Cochran, *Acta Cryst.*, **4**, 81 (1951).

35. Jardetzky and Jardetzky, *J. Amer. Chem. Soc.*, **82**, 222 (1960).
36. Lewin, *J. Chem. Soc.*, **1964**, 792.
36a. Börresen, *Acta Chem. Scand.*, **21**, 2463 (1967).
37. Sobell and Tomita, *Acta Cryst.*, **17**, 126 (1964).
38. Albert, *Biochem. J.*, **54**, 646 (1953).
39. Albert in *Physical Methods in Heterocyclic Chemistry*, Ed., Katritzky, Academic Press, New York, 1963, Vol. I, Chap. 1.
40. Bergmann and Dikstein, *J. Amer. Chem. Soc.*, **77**, 691 (1955).
41. Giner-Sorolla and Bendich, *J. Amer. Chem. Soc.*, **80**, 574 (1958).
42. Shapiro, *J. Amer. Chem. Soc.*, **86**, 2948 (1964).
43. Nakajima and Pullman, *Bull. Soc. chim. France*, **1958**, 1502.
44. Veillard and Pullman, *J. Theor. Biol.*, **4**, 37 (1963).
45. Pariser and Parr, *J. Chem. Phys.*, **21**, 466 and 767 (1953); Pople, *Trans. Faraday Soc.*, **49**, 1375 (1953).
46. Broom, Townsend, Jones, and Robins, *Biochemistry*, **3**, 494 (1963).
47. Kasha, *J. Chim. phys.*, **58**, 914 (1961).
48. Jones and Robins, *J. Amer. Chem. Soc.*, **84**, 1914 (1962).
49. Pullman and Pullman, *Biochim. Biophys. Acta*, **36**, 343 (1959).
50. Kasha, *Radiation Res. Suppl.*, **2**, 243 (1960).
51. Mason, *Quart. Rev.*, **3**, 287 (1961).
52. Clark and Tinoco, *J. Amer. Chem. Soc.*, **87**, 11 (1965).
53. Clark, Peschel, and Tinoco, *J. Phys. Chem.*, **69**, 3615 (1965).
54. Drobnik and Augenstein, *Photochem. and Photobiol.*, **5**, 13 (1966).
55. Drobnik and Augenstein, *Photochem. and Photobiol.*, **5**, 83 (1966).
55a. Kleinwächter, Drobnik, and Augenstein, *Photochem. and Photobiol.*, **6**, 133 (1967).
56. Mautner and Bergson, *Acta Chem. Scand.*, **17**, 1694 (1963).
57. Cohen, Thom, and Bendich, *J. Org. Chem.*, **28**, 1379 (1963).
58. Jones and Robins, *J. Amer. Chem. Soc.*, **82**, 3773 (1960).
59. Stewart and Davidson, *J. Chem. Phys.*, **39**, 255 (1963).
60. Stewart and Jensen, *J. Chem. Phys.*, **40**, 2071 (1964).
61. Callis, Rosa, and Simpson, *J. Amer. Chem. Soc.*, **86**, 2292 (1964).
62. Longuet-Higgins and Pople, *J. Chem. Phys.*, **27**, 192 (1957).
63. Lister, *J. Chem. Soc.*, **1960**, 3394.
64. Brookes and Lawley, *J. Chem. Soc.*, **1961**, 3923.
65. Michelson, *The Chemistry of Nucleosides and Nucleotides*, Academic Press, London, 1963, p. 445.
66. DeVoe and Tinoco, *J. Mol. Biol.*, **4**, 518 (1962).
67. Rhodes, *J. Amer. Chem. Soc.*, **83**, 3609 (1961).
68. Nesbet, *Biopolymers Symp.*, **1**, 129 (1964).
69. Bolton and Weiss, *Nature*, **195**, 666 (1962).
70. Weiss, *Nature*, **197**, 1296 (1963).
71. DeVoe, *Nature*, **197**, 1295 (1963).
72. Ladik and Hoffman, *Biopolymers Symp.*, **1**, 117 (1963).
73. Berthod, Giessner-Prettre, and Pullman, *Theor. Chim. Acta*, **5**, 53 (1966).
74. Mataga and Nishimoto, *Z. physik. Chem.*, **13**, 140 (1957).
75. Nagata, Imamura, Tagashira, and Kodama, *Bull. Chem. Soc. Japan*, **38**, 1638 (1965).
76. Berthod, Giessner-Prettre, and Pullman, *Compt. rend.*, **262**, 2657 (1966).
77. Kasha, *Discuss. Faraday Soc.*, **9**, 14 (1950).
78. McGlynn, Smith, and Cilento, *Photochem. and Photobiol.*, **3**, 269 (1964).

79. Udenfriend, *Fluorescence Assay in Biology and Medicine*, Academic Press, New York, 1962.
80. Williams and Bridges, *J. Clin. Path.*, **17**, 371 (1964).
81. Börresen, *Acta Chem. Scand.*, **17**, 921 (1963).
82. Börresen, *Acta Chem. Scand.*, **19**, 2100 (1965).
83. El Sayed, *J. Chem. Phys.*, **38**, 2835 (1963).
84. Krishna and Goodman, *J. Amer. Chem. Soc.*, **83**, 2042 (1963).
85. Cohen and Goodman, *J. Amer. Chem. Soc.*, **87**, 5487 (1965).
86. Lewin and Barnes, *J. Chem. Soc.*, **1966**, 478.
87. Townsend and Robins, *J. Org. Chem.*, **27**, 990 (1962).
88. Prasad, Noell, and Robins, *J. Amer. Chem. Soc.*, **81**, 194 (1959).
89. Townsend and Robins, *J. Heterocyclic Chem.*, **3**, 241 (1966).
90. Fu, Chinoporos, and Terzian, *J. Org. Chem.*, **30**, 1916 (1965).
91. Montgomery, *J. Amer. Chem. Soc.*, **78**, 1928 (1956).
92. Elion and Hitchings, *J. Amer. Chem. Soc.*, **78**, 3508 (1956).
93. Fromherz and Hertmann, *Ber.*, **69**, 2420 (1936).
94. Beaman and Robins, *J. Org. Chem.*, **28**, 2310 (1963).
95. Davoll and Lowy, *J. Amer. Chem. Soc.*, **74**, 1563 (1953).
96. Montgomery and Holum, *J. Amer. Chem. Soc.*, **79**, 2185 (1957).
97. Montgomery and Holum, *J. Amer. Chem. Soc.*, **80**, 404 (1958).
98. Falco, Elion, Burgi, and Hitchings, *J. Amer. Chem. Soc.*, **74**, 4897 (1952).
99. Baddiley, Buchanan, Hawker, and Stephenson, *J. Chem. Soc.*, **1956**, 4659.
100. Elion, Mueller, and Hitchings, *J. Amer. Chem. Soc.*, **81**, 3042 (1959).
101. Johnson, *Biochem. J.*, **51**, 133 (1952).
102. Robins, Dille, Willits, and Christensen, *J. Amer. Chem. Soc.*, **75**, 263 (1953).
103. Cresswell, Maurer, Strauss, and Brown, *J. Org. Chem.*, **30**, 408 (1965).
104. Robins and Lin, *J. Amer. Chem. Soc.*, **74**, 2420 (1952).
105. Montgomery, Hewson, Clayton, and Thomas, *J. Org. Chem.*, **31**, 2202 (1966).
106. Montgomery and Thomas, *J. Org. Chem.*, **31**, 1411 (1966).
107. Jones and Robins, *J. Amer. Chem. Soc.*, **85**, 193 (1963).
108. Pfleiderer and Nübel, *Annalen*, **647**, 155 (1961).
109. Adler and Gutman, *Nature*, **205**, 681 (1964).
110. Pfleiderer, *Annalen*, **647**, 161 (1961).
111. Holmes and Robins, *J. Amer. Chem. Soc.*, **87**, 1772 (1965).
112. Koppel, O'Brien, and Robins, *J. Org. Chem.*, **24**, 259 (1959).
113. Bullock, Hand, and Stokstad, *J. Amer. Chem. Soc.*, **78**, 3693 (1956).
114. Beaman, *J. Amer. Chem. Soc.*, **76**, 5633 (1954).
115. Elion, Goodman, Lange, and Hitchings, *J. Amer. Chem. Soc.*, **81**, 1898 (1959).
116. Bendich, Tinker, and Brown, *J. Amer. Chem. Soc.*, **70**, 3109 (1948).
117. Elion and Hitchings, *J. Amer. Chem. Soc.*, **77**, 1676 (1955).
118. Townsend and Robins, *J. Amer. Chem. Soc.*, **84**, 3008 (1962).
119. Brown and Jacobsen, *J. Chem. Soc.*, **1965**, 3770.
120. Daves, Noell, Robins, Koppel, and Beaman, *J. Amer. Chem. Soc.*, **82**, 2633 (1960).
121. Elion, Burgi, and Hitchings, *J. Amer. Chem. Soc.*, **74**, 411 (1952).
122. Elion, Lange, and Hitchings, *J. Amer. Chem. Soc.*, **78**, 217 (1956).
123. Gerster and Robins, *J. Amer. Chem. Soc.*, **87**, 3752 (1965).
124. Stevens, Magrath, Smith, and Brown, *J. Amer. Chem. Soc.*, **80**, 2755 (1958).
125. Lewin and Tann, *J. Chem. Soc.*, **1962**, 1466.
126. Leese and Timmis, *J. Chem. Soc.*, **1958**, 4107.
127. Carter, *J. Biol. Chem.*, **223**, 139 (1956).

128. Mackay and Hitchings, *J. Amer. Chem. Soc.*, **78**, 3511 (1956).
129. Bendich, Giner-Sorolla, and Fox, in *The Chemistry and Biology of Purines*, Eds., Wolstenholme and O'Connor, J. and A. Churchill, Ltd., London, 1957, p. 3.
130. Giner-Sorolla and Bendich, *J. Amer. Chem. Soc.*, **80**, 3932 (1958).
131. Taylor, Vogel, and Cheng, *J. Amer. Chem. Soc.*, **81**, 2441 (1959).
132. Elion, Burgi, and Hitchings, *J. Amer. Chem. Soc.*, **73**, 5235 (1951).
133. Lister, *J. Chem. Soc.*, **1960**, 3682.
134. Windmueller and Kaplan, *J. Biol. Chem.*, **236**, 2716 (1961).
135. Leonard and Laursen, *Biochemistry*, **4**, 356 (1965).
136. Baddiley, Lythgoe, and Todd, *J. Chem. Soc.*, **1944**, 318.
137. Lister and Timmis, *J. Chem. Soc.*, **1960**, 327.
138. Baker, Schaub, and Joseph, *J. Org. Chem.*, **19**, 638 (1954).
139. Cavalieri and Bendich, *J. Amer. Chem. Soc.*, **72**, 2587 (1950).
140. Lawley, *Proc. Chem. Soc.*, **1957**, 290.
141. Hotchkiss, *J. Biol. Chem.*, **175**, 315 (1948).
142. Giner-Sorolla, Zimmerman, and Bendich, *J. Amer. Chem. Soc.*, **81**, 2515 (1959).
143. Beaven, Holiday, and Johnson, in *The Nucleic Acids*, Vol. I, Eds. Chargaff and Davidson, Academic Press, New York, 1955, p. 493.
144. Balsiger and Montgomery, *J. Org. Chem.*, **25**, 1573 (1960).
145. Beaven, Holiday, and Johnson, in *The Nucleic Acids*, Vol. I, Eds. Chargaff and Davidson, Academic Press, New York, 1955, p. 545.
146. Lacher, Bitner, Emery, Seffl, and Park, *J. Phys. Chem.*, **59**, 615 (1955).
147. Hamlin, Lord, and Rich, *Science*, **148**, 1734 (1965).
148. Kyogoku, Lord, and Rich, *Science*, **154**, 518 (1966).
149. Kyogoku, Lord, and Rich, *J. Amer. Chem. Soc.*, **89**, 496 (1967).
150. Kyogoku, Lord, and Rich, *Proc. Nat. Acad. Sci. U.S.A.*, **57**, 250 (1967).
151. Miller and Sobell, *J. Mol. Biol.*, **24**, 345 (1967).
152. Pitha, Jones, and Pithova, *Canad. J. Chem.*, **44**, 1045 (1966).
153. Kyogoku, Lord, and Rich, *Nature*, **218**, 69 (1968).
154. Kyogoku and Yu, *Bull. Chem. Soc., Japan*, **41**, 1742 (1968).
155. Küchler and Derkosch, *Z. Naturforsch.*, **21b**, 209 (1966).
156. Brown and Hewlins, *Nature*, **221**, 656 (1969).
157. Malt, *Biochim. Biophys. Acta*, **120**, 461 (1966).
158. Lord and Thomas, *Biochim. Biophys. Acta*, **142**, 1 (1967).
159. Lord and Thomas, *Spectrochim. Acta*, **23A**, 2551 (1967).
160. Lautié and Novak, *J. Chim. phys.*, **65**, 1359 (1968).
161. Blout and Fields, *Science*, **107**, 252 (1948).
162. Blout and Fields, *J. Amer. Chem. Soc.*, **72**, 479 (1950).
163. Willits, Decius, Dille, and Christensen, *J. Amer. Chem. Soc.*, **77**, 2569 (1955).
164. Komatsu and Hachisu, *Nippon Kagaku Zasshi*, **81**, (1960); through *Chem. Abstracts*, **56**, 118d (1962).
165. Scott, Sinsheimer, and Loofbourow, *J. Amer. Chem. Soc.*, **74**, 275 (1952).
166. Falk, Hartman, and Lord, *J. Amer. Chem. Soc.*, **85**, 387 (1963).
167. Booth, Boyland, and Orr, *J. Chem. Soc.*, **1954**, 598.
168. Cook and Regnier, *Canad. J. Chem.*, **46**, 3055 (1968).
169. Cook and Regnier, *Canad. J. Chem.*, **45**, 2895 (1967).
170. Cook and Regnier, *Canad. J. Chem.*, **45**, 2899 (1967).
171. Kyogoku, Higuchi, and Tsuboi, *Spectrochim. Acta*, **23A**, 969 (1967).
172. Bradley and Potts, *Appl. Spectroscopy*, **12**, 77 (1958).
173. Blout and Mellors, *Science*, **110**, 137 (1949).

174. Stimson and O'Donnell, *J. Amer. Chem. Soc.*, **74**, 1805 (1952).
175. Schiedt and Reinwein, *Z. Naturforsch.*, **7b**, 270 (1952).
176. Schiedt, *Z. Naturforsch.*, **8b**, 66 (1953).
177. Friedlander and DiPietro, *Bull. Res. Council Israel Sect. A*, **11**, 140 (1962).
178. Laufer, *Z. analyt. Chem.*, **181**, 481 (1961).
179. Hayden and Sammul, *J. Amer. Pharm. Assoc., Sci. Edn.*, **49**, 489 (1960).
180. Sammul, Brannon, and Hayden, *J. Assoc. Offic. Agric. Chemists*, **47**, 918 (1964).
181. Low, Abrams, and Coleman, *Chem. Comm.*, **1965**, 389.
182. Baker and Hewson, *J. Org. Chem.*, **22**, 959 (1957).
183. Montgomery and Temple, *J. Org. Chem.*, **25**, 395 (1960).
184. Montgomery and Hewson, *J. Org. Chem.*, **33**, 432 (1968).
185. Baker and Hewson, *J. Org. Chem.*, **22**, 966 (1957).
186. Schaeffer and Thomas, *J. Amer. Chem. Soc.*, **80**, 4896 (1958).
187. Montgomery and Temple, *J. Amer. Chem. Soc.*, **82**, 4592 (1960).
188. Montgomery and Temple, *J. Amer. Chem. Soc.*, **83**, 630 (1961).
189. Montgomery and Temple, *J. Amer. Chem. Soc.*, **79**, 5238 (1957).
190. Johnson, Thomas, and Schaeffer, *J. Amer. Chem. Soc.*, **80**, 699 (1958).
191. Schaeffer and Thomas, *J. Amer. Chem. Soc.*, **80**, 3738 (1958).
192. Thomas and Montgomery, *J. Org. Chem.*, **31**, 1413 (1966).
193. Montgomery, Thomas, and Schaeffer, *J. Org. Chem.*, **26**, 1929 (1961).
194. Montgomery and Thomas, *J. Org. Chem.*, **26**, 1926 (1961).
195. Baker, Hewson, Thomas, and Johnson, *J. Org. Chem.*, **22**, 954 (1957).
196. Montgomery and Temple, *J. Amer. Chem. Soc.*, **80**, 409 (1958).
197. Lister and Kiburis, unpublished results.
198. Balsiger, Fikes, Johnston, and Montgomery, *J. Org. Chem.*, **26**, 3446 (1961).
199. Horák, and Gut, *Coll. Czech. Chem. Comm.*, **26**, 1680 (1961).
200. Angell, *J. Chem. Soc.*, **1961**, 504.
201. Elion, in *The Chemistry and Biology of Purines*, Eds., Wolstenholme and O'Connor, J. and A. Churchill, Ltd., London, 1957, p. 39.
202. Katritzky and Ambler in *Physical Methods in Heterocyclic Chemistry*, Vol. II, Ed., Katritzky, Academic Press, New York and London, 1963, p. 325.
203. Letham, Shannon, and McDonald, *Chem. Comm.*, **1964**, 230.
204. Schweizer, Chan, Helmkamp, and T'so, *J. Amer. Chem. Soc.*, **86**, 696 (1964).
205. Pugmire, Grant, Robins, and Rhodes, *J. Amer. Chem. Soc.*, **87**, 2225 (1965).
206. Coburn, Thorpe, Montgomery, and Hewson, *J. Org. Chem.*, **30**, 1110 (1965).
207. Bullock and Jardetsky, *J. Org. Chem.*, **29**, 1988 (1964).
208. Bonner, *J. Amer. Chem. Soc.*, **76**, 6450 (1954).
209. Matsuura and Goto, *J. Chem. Soc.*, **1965**, 623.
210. Reddy, Mandell, and Goldstein, *J. Chem. Soc.*, **1963**, 1414.
211. Hruska, Bell, Victor, and Danyluk, *Biochemistry*, **7**, 3721 (1968).
212. Chan, Schweizer, T'so, and Helmkamp, *J. Amer. Chem. Soc.*, **86**, 4182 (1964).
213. Coburn, Thorpe, Montgomery, and Hewson, *J. Org. Chem.*, **30**, 1114 (1965).
214. Read and Goldstein, *J. Amer. Chem. Soc.*, **87**, 3440 (1965).
215. Neiman and Bergmann, *Chem. Comm.*, **1968**, 1002.
216. Wang and Li, *J. Amer. Chem. Soc.*, **88**, 4592 (1966).
217. Giessner-Prettre and Pullman, *Compt. rend.*, **261**, 2521 (1965).
218. Kohnlein and Müller, *Internat. J. Radiation Biol.*, **8**, 141 (1964).
219. Herak and Gordy, *Proc. Nat. Acad. Sci. U.S.A.*, **54**, 1287 (1965).
220. Holmes, Ingalls, and Myers, *Internat. J. Radiation Biol.*, **13**, 225 (1967).
221. Holmes, Ingalls, and Myers, *Internat. J. Radiation Biol.*, **12**, 415 (1967).

222. Sasaki and Hashizume, *Analyt. Biochem.*, **16**, 1 (1966).
223. McCloskey, Lawson, Tsuboyama, Krueger, and Stillwell, *J. Amer. Chem. Soc.*, **90**, 4182 (1968).
224. Tatematsu, Goto, and Matsuura, *Nippon Kagaku Zasshi*, **87**, 71 (1966).
225. Rice and Dudek, *J. Amer. Chem. Soc.*, **89**, 2719 (1967).
226. Spiteller and Spiteller-Friedmann, *Monatsh.*, **93**, 632 (1962).
227. Votický, Kovácik, Rybár, and Antos, *Coll. Czech. Chem. Comm.*, **34**, 1657 (1969).
228. Lifschitz, Bergmann, and Pullman, *Tetrahedron Letters*, **1967**, 4583.

Systematic Tables of Simple Purines

Introduction

Tables 21 to 59 containing nearly 3,000 derivatives represent the fruits of a comprehensive survey of papers and patents covering the purine literature from the origin up to December 1969, with partial coverage extending to late 1970. As the emphasis of this text is on the practical aspect of purine chemistry, the references given in the accompanying general list have been selected so as to provide the best possible preparative details and data for the particular purine. The occurrence of multiple references to a compound usually indicates the availability of more than one route so that a choice of conditions and starting materials is possible. One factor in deciding the exclusion of papers is where insignificant yields or mixtures of products arise. Such preparations, however, are adequately covered in the text to which reference should be made. The omission of a number of references from the early literature will also be noted, but in these cases more recent and effective preparations are available which supersede those described earlier.

Purines Excluded from the Tables

For overall simplicity some types of purines have been ignored: these belong mainly to the following categories.

Those reduced in either pyrimidine or imidazole rings (this, however, does not exclude oxo- or thiopurines in which the ring nitrogens are alkylated as, for example, the pyrimidine ring in theophylline, or the imidazole ring in 7,9-dimethyluric acid).

Those containing heterocyclic substituents, e.g., morpholino and pyridino, with the exception of furfuryl derivatives, which are retained because of their biological significance.

Those fused with other ring systems.

Those containing substituents with more than six carbon atoms (with the exception of benzoyl and benzyl).

Those containing substituted phenyl groups.

Those with difunctional groups (except for carboxyalkylthio and carboxyalkylamino groups).

Those in which the derivative is formed by the linking together of two or more purine residues in any way.

Terms Used in Tables

Inclusive headings are used to keep the tables down to a workable number. Under these are grouped together derivatives with substituents having related functions.

Alkyl includes alkyl, aryl, cycloalkyl and, for convenience, furfuryl also.

Amino includes amino, alkylamino, dialkylamino, cycloalkylamino (e.g., aziridinyl), imino, anilino, hydrazino, hydroxyamino, and acyl (and aroyl)amino (e.g., acetamido, benzamido).

Carboxy includes carboxy (carboxylic acids), alkoxycarbonyl (esters), carbamoyl (amides), hydrazinocarbonyl (hydrazides), chlorocarbonyl (acid chlorides), azidocarbonyl, cyano (nitriles), formyl, acyl, and thiocyanato groups. These groups may be linked, directly or indirectly, with the nucleus through ring C- or N-atoms. Separate treatment is given to carboxyalkylthio and carboxyalkylamino groups.

Halogeno includes only the four halogen atoms.

Nitro includes nitro, nitroso, *N*-nitroso, diazo, and azido groups.

N-Oxide includes *N*-oxido, *N*-alkoxy, *N*-aryloxy, and *N*-acyloxy groups.

Oxo includes oxo, hydroxyalkyl, alkoxy, aryloxy, acyloxy, and *C*-oxido (betaines) groups.

Sulphonyl includes alkylsulphonyl (sulphones), sulpho (sulphonic acids), sulphino (sulphinic acids), sulphamoyl (sulphonamides), alkoxy-sulphonyl (sulphonic esters), and halogenosulphonyl (sulphonyl halides) groups.

Thio includes thio, alkylthio (thioethers), acylthio, and mercapto-alkyl groups. Selenium analogues are also incorporated.

In the column detailing melting points the various abbreviations used are as follows:

amm. (ammonium salt), bet. (betaine form), brom. (purinium bromide), chlor. (purinium chloride), expl. (explosive decomposition or melting), HBr, HCl, HI (hydrobromide, hydrochloride, and hydriodide, respectively), hy. (hydrate), iod. (purinium iodide), nit. (purinium nitrate), perch. (perchlorate), pic. (picrate), pot. (potassium salt), sod. (sodium salt), sul. (sulphate), tos. (tosylate).

Boiling points, where applicable, are given with the appropriate pressure. Many of the derivatives have melting points greater than 300°; therefore, the symbol > 300° can indicate either that the derivative has no definite melting point or that it melts over a wide range, or more likely, that a slow decomposition is observed over this temperature. Where the symbol for "greater than" occurs with a lower temperature figure, an indeterminate melting range, starting around this point, is indicated. In cases where the literature produces a spread of melting points for a particular compound, the limits of the range are given, e.g., 121 to 136°.

General Notes on Use of Tables

Each table, unless otherwise specified, incorporates both C- and N-alkyl derivatives of the particular class of purine. The four basic divisions adopted are for purines having either amino, oxo, or thio groups or halogen atoms. Compounds, in which any two of the above substituents are combined, make up six additional tables, which are complemented by subdivision of some of these into derivatives containing only C-alkyl groups and derivatives containing both C- and N-alkyl groups. While a table is devoted to amino-oxo-thiopurines, other purines containing three different substituents are classified under amino-oxo-, amino-thio-, or oxo-thiopurines containing a minor substituent. Thus 8-bromoguanine, for example, is found under "amino-oxo with a minor substituent." The remaining tables comprise purines containing specialist groups which may be in addition to any of the groups listed above. Alphabetical sequence is followed using systematic nomenclature for the compounds. For ease of reference, however, departures from substitutive nomenclature occur in certain tables as follows:

In the oxopurine tables 26 and 27, C- and N-alkyl derivatives of hypoxanthine, xanthine, and uric acid are listed under these trivial names, thus theobromine, for example, appears as xanthine/3,7-dimethyl-. It should be noted that hydroxyalkyl groups are included with alkyl groups in this treatment. Respective derivatives should therefore be sought under "h," "u," or "x" in the alphabetic sequence.

In the amino-oxopurine tables 34 and 35, the alkylated forms of guanine and isoguanine are correspondingly treated, the derivatives being listed under "g" and "i," respectively. This classification has also been extended to the tables of carboxy-oxo (39) and oxo-sulphonyl-purines (46). In the former, for example, 8-carboxycaffeine is to be found under "x" as xanthine/8-carboxy-1,3,7-trimethyl-. Similarly, *N*-oxide derivatives (58) of hypoxanthine, xanthine, and guanine (and the alkylated forms) are located under "h," "x," and "g," respectively.

A reference marked * gives an incorrect structural assignment to the compound quoted.

TABLE 21.   Alkyl and Aryl Purines.

Purine	M.p.	References
Unsubstituted	210–217°	6, 64, 98, 181, 191, 203, 365, 367, 368, 370, 374
	HCl 206–208°	377
1-Benzyl-	206–209°	826
3-Benzyl-	123–124°	826
7-Benzyl-	145–146°	67, 400, 826
9-Benzyl-	100–101°	400, 707, 826
2-Butyl-	144°	166
9-Butyl-	*ca* 30°	265
9-Cyclohexyl-	95–96°	265, 266
9-Cyclopentyl-	80–81°	265
2,6-Dimethyl-	271°	243
2,7-Dimethyl-	120–121°	387
2,8-Dimethyl-	217°	243
3,6-Dimethyl-	178–180°	28
6,8-Dimethyl-	263–265°	243
6,9-Dimethyl-	76–80° or 103–108°	7, 28, 83
7,8-Dimethyl-	196–197°	274
7,9-Dimethyl-	iod. 225–226°	707
2,8-Diphenyl-	259°	18
7,8-Diphenyl-	178–179°	274
1-Ethyl-	198°	500
7-Ethyl-	108–110°	366, 500
9-Ethyl-	53–56°	60, 110, 500
7-Ethyl-8-methyl-	132–133°	274
7-Ethyl-8-phenyl-	137–138°	274
1-Methyl-	234–235°	180
2-Methyl-	286°	243, 391
3-Methyl-	184–185°	401
6-Methyl-	220° or 235–236°	174, 378, 379
7-Methyl-	181–184°	191, 366, 367, 397
8-Methyl-	265° to 273°	6, 203, 242, 273

TABLE 21 (*continued*)

Purine	M.p.	References
9-Methyl-	158° to 164°	129, 191, 364, 367, 397, 398, 707
2-Methyl-8-phenyl-	256–258°	256
3-Methyl-8-phenyl-	245°	23
6-Methyl-8-phenyl-	319°	66,
7-Methyl-8-phenyl-	213–214°	274
8-Methyl-7-phenyl-	171–172°	274
2-Phenyl-	228°	18
6-Phenyl-	—	803
7-Phenyl-	184–186°	366
8-Phenyl-	261–265°	6, 23, 66
9-Phenyl-	161–163°	399
2,6,8-Trimethyl-	222°	243

TABLE 22. Aminopurines.

Purine	M.p.	References
6-Acetamido-	> 260°	612, 613, 615, 617, 619
6-Acetamido-2-amino-	204°	305
6-Acetamido-3-benzyl-	218°	24, 811
6-Acetamido-9-benzyl-	166–168°	811
3-Allyl-6-amino-	204–205°	629
3-Allyl-6-amino-7-methyl-	iod. 256–258°	629
3-Allyl-6-benzamido-	184°	811
9-Allyl-6-benzamido-	134–135°	811
6-Allylamino-	221–222°	585
2-Amino-	277–278°	6, 178, 505, 642
6-Amino- (adenine)	352–354°	52, 53, 103, 115, 136, 178, 183, 184, 224, 267, 343, 344, 636, 790
8-Amino-	> 300°	6, 32
9-Amino-	> 300°	264
6-Amino-3-allyl-	brom. 197°	210
6-Amino-9-allyl-	143–145°	29
6-Amino-9-(2-aminoethyl)-	219–221°	16
2-Amino-6-anilino-	283–285°	507
2-Amino-6-aziridinyl-9-benzyl-	108°	470

continued

TABLE 22 (*continued*)

Purine	M.p.	References
2-Amino-6-aziridinyl-9-methyl-	181–183°	658
6-Amino-1-benzyl-	HBr 248–250°	635, 647
6-Amino-2-benzyl-	260–261°	52
6-Amino-3-benzyl-	275 to 287°	24, 34, 218, 629, 638, 811, 826
6-Amino-7-benzyl-	234 to 242°	34, 212, 400, 629, 638
6-Amino-8-benzyl-	276°	790
6-Amino-9-benzyl-	230–235°	24, 212, 400, 707, 797, 811, 826
6-Amino-1-benzyl-9-methyl-	iod. 233–235°	33
6-Amino-3-benzyl-7-methyl-	iod. 261–262°	629
6-Amino-9-benzyl-1-methyl-	tos. 279–282°; iod. 268–270°	33, 797
2-Amino-6-benzylamino-	239°	49, 661
6-Amino-9-butyl-	138–139°	265
2-Amino-6-butylamino-	165–166°	61
6-Amino-2-butylamino-	218°	61
2-Amino-6-butylamino-3-methyl-	255–258°	44
6-Amino-9-(cyclohex-2-enyl)-	196°	266
6-Amino-9-cyclohexyl-	199–200°	265, 661, 797
2-Amino-6-cyclohexylamino-	> 300°	661
6-Amino-7-cyclopentyl-	hy. 242°	34, 638
6-Amino-9-cyclopentyl-	156°	265
6-Amino-9-cyclopropyl-	HCl 256–258°	263
6-Amino-3,7-dibenzyl-	brom. 205–207°	629
6-Amino-3,9-dibenzyl-	brom. 145°	24
6-Amino-2,8-diethyl-	242°	53
6-Amino-9-(2-diethylaminoethyl)-	180°	404
2-Amino-6,9-dimethyl-	223–225°	654
6-Amino-1,9-dimethyl-	tos. —	607
6-Amino-2,7-dimethyl-	338°	225
6-Amino-2,8-dimethyl-	315° or 342°	53, 230
6-Amino-2,9-dimethyl-	238°	225
6-Amino-3,7-dimethyl-	iod. > 300°	607
2-Amino-6-dimethylamino-	> 300°	61
6-Amino-2-dimethylamino-	295°	61, 222
6-Amino-9-dimethylamino-8-methyl-	239–240°	19
6-Amino-9-(3-dimethylaminopropyl)-	135–136°	16
6-Amino-2-ethyl-	304–305°	52
6-Amino-3-ethyl-	233°	34
6-Amino-9-ethyl-	194–195°	60, 202, 797
6-Amino-9-ethyl-1-methyl-	tos. 275–278°	797

TABLE 22 (*continued*)

Purine	M.p.	References
6-Amino-3-furfuryl-	245°	34
6-Amino-7-furfuryl-	203°	34
6-Amino-9-furfuryl-	192 to 205°	48, 69, 726
2-Amino-9-furfuryl-6-methyl-	114°	47
2-Amino-6-furfurylamino-	206–208°	49
6-Amino-9-hexyl-	146°	263
2-Amino-6-hydrazino-	> 300°	460
6-Amino-2-hydrazino-	> 300°	460
2-Amino-6-hydroxyamino-	310°	36
1-Amino-6-imino-7-methyl-	199–200°	54
6-Amino-3-isopentyl-	230–231°	55, 629
6-Amino-9-isopropyl-	HCl 235–236°	263
2-Amino-6-methyl-	305–315°	165, 174, 178, 204
2-Amino-7-methyl-	283°	367
2-Amino-9-methyl-	242–247°	95, 115, 367
6-Amino-1-methyl-	302–312°	29, 606, 647
6-Amino-2-methyl-	> 300°	52, 178, 221, 230, 303, 456, 608, 695
6-Amino-3-methyl-	291 to 313°	25, 34, 101, 606
6-Amino-7-methyl-	344–351°	34, 54, 101, 258, 347, 356, 415, 629, 638
6-Amino-8-methyl-	> 300°	230, 242, 273
6-Amino-9-methyl-	296 to 310°	101, 103, 202, 206, 223, 261, 283, 353, 415, 579, 580, 797
8-Amino-7-methyl-	—	455
6-Amino-7-methyl-3-(3-methylbut-2-enyl)-	iod. 236–239°	629
2-Amino-6-methyl-8-phenyl-	HCl > 350°	256
6-Amino-1-methyl-9-propenyl-	iod. 298–300°	29
2-Amino-6-methylamino-	> 300°	61
6-Amino-2-methylamino-	> 300°	61, 101
6-Amino-1-(3-methylbut-2-enyl)-	237–239°	33
6-Amino-3-(3-methylbut-2-enyl)- (triacanthine)	228–231°	34, 55, 628, 629, 638
6-Amino-7-(3-methylbut-2-enyl)-	195–196°	34, 55, 638

continued

TABLE 22 (*continued*)

Purine	M.p.	References
6-Amino-9-(3-methylbut-2-enyl)-	167–168°	55, 811
2-Amino-6-(2-methylhydrazino)-	300°	41
6-Amino-9-pentyl-	132–135°	263
2-Amino-8-phenyl-	268°; HCl 333°	254, 256
6-Amino-2-phenyl-	319–321°	52, 303, 304, 402
6-Amino-8-phenyl-	310–311°	66
6-Amino-9-phenyl-	235–238°	223, 797
6-Amino-9-propenyl-	197°	29
6-Amino-3-propyl-	brom. 252°	218
6-Amino-9-propyl-	168–173°	202, 263
6-Amino-9-vinyl-	196–197°	219
6-(2-Aminoethyl)-	sub. 200°	76
6-Aminomethyl-	183–185°; HCl 272–274°	50, 158
	HCl 263–265°	165
6-(3-Aminopropyl)-	HCl 200°	76
6-Anilino-	278–285°	51, 179, 223, 584, 608, 623
8-Anilino-7-ethyl-	263–265°	274
6-Anilino-9-furfuryl-	151–152°	726
8-Anilino-7-methyl-	229–230°	274
2-Anilino-9-phenyl-	215–216°	820
8-Anilino-7-phenyl-	258–260°	274
6-Aziridinyl-9-benzyl-	137°	470
6-Aziridinyl-9-ethyl-	126°	470
6-Benzamido-	240–243°	618, 619, 620
6-Benzamido-3-benzyl-	195–196°; HCl 258°	24, 811
6-Benzamido-7-benzyl-	238°	24
6-Benzamido-9-benzyl-	159–161°	811
6-Benzamido-3-(3-methylbut-2-enyl)-	167–168°, HBr 181°	38, 811
6-Benzamido-9-(3-methylbut-2-enyl)-	125–127°	38, 811
6-Benzimido-3,7-dibenzyl-	217–218°	24, 811
1-Benzyl-6-benzylamino-	200–202°	212
3-Benzyl-6-benzylamino-	179–180° or 212–214°	24, 212
7-Benzyl-6-benzylamino-	hy. 118–132°	212
9-Benzyl-6-benzylamino-	174–175°	212
3-Benzyl-6-dimethylamino-	142°; pic. 187–188°	288*, 826
7-Benzyl-6-dimethylamino-	134–135°	400
9-Benzyl-6-dimethylamino-	117° or 131–132°	400, 470, 826
7-Benzyl-6-ethylamino-	249–250°	400
9-Benzyl-6-ethylamino-	—	400
7-Benzyl-6-hydrazino-	—	400
9-Benzyl-6-hydrazino-	209–210°	400

TABLE 22 (*continued*)

Purine	M.p.	References
6-Benzylamino-	228–230°	14, 51, 223, 465, 468, 600, 608, 618, 620, 621, 622
2-Benzylamino-6-benzylaminomethyl-9-methyl-	171–174°	83
2-Benzylamino-6,9-dimethyl-	132–134°	654
6-Benzylamino-8-ethyl-	210–211°	608
6-Benzylamino-9-furfuryl-	124–125°	726
6-Benzylamino-2-methyl-	285–286°	608
6-Benzylamino-8-methyl-	294–295°	608
6-Benzylamino-9-methyl-	237–238°	608
6-Benzylamino-3-(3-methylbut-2-enyl)-	150°	55
6-(*N*-Benzyl)methylamino-	212°	14
2,6-bis-Butylamino-	274–275°	61
2,6-bis-Diethylamino-	116–117°	413
2,6-bis-Dimethylamino-	254–255°	284, 413
6,8-bis-Dimethylamino-	291–293°	58
6,9-bis-Dimethylamino-8-methyl-	62°; 120°/0.5 mm	19
6,8-bis-Ethylamino-	215–217°	58
2,6-bis-Methylamino-	> 300°	61
6,8-bis-Methylamino-	HCl 305°	58
7-Butyl-6-butylamino-8-propyl-	82–83°	345
1-Butyl-6-imino-7-methyl-	125–126°	54
7-Butyl-6-methylamino-8-propyl-	129–130°	345
6-Butylamino-	233–234°	179, 584, 585
2-Butylamino-6,9-dimethyl-	77°; HCl 182–185°	461, 654
6-Butylamino-2,9-dimethyl-	97–98°	7
8-Butylamino-2,9-dimethyl-	155–156°	7
2-Butylamino-6-dimethylamino-	177°	61
6-Butylamino-2-dimethylamino-	172°	
6-Butylamino-9-ethyl-	60–61°	60
6-Butylamino-7-methyl-	144–146°	54
2-Butylamino-6-methylamino-	254°	61
6-Butylamino-2-methylamino-	152°	61
6-Butyramido-	215–217°	613, 614
6-Butyramido-9-butyryl-	152–154°	613, 614
9-(Cyclohex-2-enyl)-6-hydrazino-	146°	266
6-Cyclohexylamino-	210–211° or 235–236°	466, 661
6-Cyclohexylamino-7-isobutyl-8-isopropyl-	209–210°	345
2,6-Diacetamido-	295–300°	612

continued

TABLE 22 (*continued*)

Purine	M.p.	References
2,6-Diamino-	300–302°	52, 178, 184, 188, 303, 612
2,8-Diamino-	sul. 300°	32
6,8-Diamino-	—	58, 426
2,6-Diamino-9-benzyl-	180–181° (or > 300°?)	427, 661
2,6-Diamino-9-cyclohexyl-	> 300°	661
2,6-Diamino-3,9-dimethyl-	sul. > 300°	39
2,6-Diamino-3-methyl-	> 300°	44
2,6-Diamino-7-methyl-	> 350°	112, 415, 771
2,6-Diamino-8-phenyl-	HCl 342–343°	256
2,6-Diamino-9-phenyl-	283–285°	507
2,6-Dibenzamido-	320°	612
3,7-Dibenzyl-6-benzylimino-	128–130°	24
3,9-Dibenzyl-6-dimethylamino-	brom. 220°	100
1,7-Dibenzyl-6-imino-	125–127°	212
1,9-Dibenzyl-6-imino-	163 to 172°	212, 826
3,7-Dibenzyl-6-imino-	125–127°	113, 212
6-Dibenzylamino-	184°	14
6-Dibutylamino-2,8-dimethyl-	152°	228, 663
6-Dibutylamino-8-ethyl-2-methyl-	140°	231
6-Dibutylamino-2-methyl-	128°	228, 663
6-Dibutylamino-8-methyl-	109°	229
2-Diethylamino-	228–230°	178
6-Diethylamino-	212 to 223°	189, 413, 585
6-Diethylamino-2,8-dimethyl-	162–166°	228, 238, 663
6-Diethylamino-3,9-dimethyl-	iod. 200–208°	630
6-Diethylamino-9-dimethylamino-8-methyl-	85–86°	19
6-Diethylamino-8-ethyl-2-methyl-	179°	231, 663
9-Diethylamino-9-furfurylamino-8-methyl-	94–95°	19
6-Diethylamino-2-methyl-	212°	228, 663
6-Diethylamino-7-methyl-	HCl 201°	258
6-Diethylamino-8-methyl-	171°	229, 663
6-Diethylamino-9-methyl-	48–50°; HCl 176–179°	206, 630
9-(2-Diethylaminoethyl)-	HCl 182°	404
9-(2-Diethylaminoethyl)-6-hydrazino-	133–134°	404
9-(2-Diethylaminoethyl)-6-methylamino-	134–135°	404
6-Diethylaminomethyl-2-dimethylamino-9-methyl-	HCl 210–214°	654
6-Diethylaminomethyl-9-methyl-	138–143°	83
2,6-Difurfurylamino-	162–163°	105
2,6-Dihydrazino-	> 300°	460
6,8-Dihydrazino-	> 300°	58

TABLE 22 (*continued*)

Purine	M.p.	References
2,6-Dihydrazino-7-methyl-	258–260°	112
2,6-Dihydroxyamino-	260°	599
6-Diisobutylamino-2,8-dimethyl-	161°	228
6-Diisobutylamino-2-methyl-	131°	228, 663
6-Diisobutylamino-8-methyl-	136°	229
6,9-Dimethyl-2-methylamino-	167–169°	118
7,9-Dimethyl-6-methylamino-	iod. 269°	12
2-Dimethylamino-	222–223° or 245–249°	6, 305
6-Dimethylamino-	251 to 263°	6, 179, 187, 584, 585
8-Dimethylamino-	292°	6
2-Dimethylamino-6,9-dimethyl-	106–109°	654
6-Dimethylamino-2,8-dimethyl-	266°	228, 663
2-Dimethylamino-6-dimethylamino-methyl-9-methyl-	HCl 246°	83
6-Dimethylamino-3-ethyl-	135–136°	288*
6-Dimethylamino-9-ethyl-	79–84°	60, 288
6-Dimethylamino-8-ethyl-2-methyl-	227°	231, 663
6-Dimethylamino-9-furfuryl-	117–118°	47
2-Dimethylamino-9-methyl-	199–207°	83
6-Dimethylamino-1-methyl-	199–200°	109
6-Dimethylamino-2-methyl-	282–285°	408
6-Dimethylamino-3-methyl-	161 to 170°	9, 11, 23, 109, 288*, 802
6-Dimethylamino-7-methyl-	111–112°	109
6-Dimethylamino-8-methyl-	235°	229, 663
6-Dimethylamino-9-methyl-	114–120° (or 132°)	9, 206, 288, 802, (305)
6-Dimethylamino-3-methyl-8-phenyl-	pic. 230–232°	23
2-Dimethylamino-6-methylamino-	> 300°	61
6-Dimethylamino-2-methylamino-	234°	61
8-Dimethylamino-6-methylamino-	278°	58
6-Dimethylamino-9-phenyl-	168–169°	462
6-Dipropylamino-2,8-dimethyl-	178°	228, 663
6-Dipropylamino-8-ethyl-2-methyl-	165°	231, 663
6-Dipropylamino-2-methyl-	177°	228, 663
6-Dipropylamino-8-methyl-	161°	229, 663
6-Ethoxyamino-9-ethyl-	189–190°	656, 693
9-Ethyl-6-hydrazino-	160–162°	60
6-Ethylamino-	235–236°	179, 468, 584, 585, 621
6-Ethylamino-9-furfuryl-	125–127°	726
2-Ethylamino-6-imino-1,9-dimethyl-	186–188°	418

continued

TABLE 22 (*continued*)

Purine	M.p.	References
6-Ethylamino-2-methyl-	300°	408
6-Ethylamino-9-methyl-	157–158°	206
6-Ethylimino-2-ethylamino-1,9-dimethyl-	184–186°	418
6-Ethylimino-1,2,9-trimethyl-	146–170°	418
2-Formamido-6-methyl-	> 300°	68
9-Formamido-6-methyl-	199–200°	294
6-Furfurylamino- (kinetin)	266–267°	51, 69, 223, 465, 583, 600, 604, 618, 619, 620, 623
6-Furfurylamino-2,8-dimethyl-	197°	228
9-Furfuryl-6-furfurylamino-	127–129° or 140°	47, 48, 726
6-Furfurylamino-2-hydrazino-	212–214°	105
6-Furfurylamino-2-methyl-	269–270°	228, 408
6-Furfurylamino-7-methyl-	214–215°	347
6-Furfurylamino-8-methyl-	264°	229
6-Furfurylamino-9-methyl-	175–177°	202, 206
6-Hexylamino-	176–178°	466, 585, 618
2-Hexylamino-6,9-dimethyl-	69°	461
2-Hydrazino-	> 300°	460
6-Hydrazino-	230° or 246–247°	179, 460, 584
6-Hydrazino-7-methyl-	243°	347
6-Hydrazino-9-methyl-	210–211°	206
6-Hydroxyamino-	260°	50, 725
8-Hydroxyamino-	315°	41
6-Hydroxyamino-9-methyl-	244°	36
6-Imino-1,2,9-trimethyl-	154–155°	418
6-Isopropylamino-9-methyl-	136–137°	206
6-Methoxyamino-	196°	41, 158, 725
6-Methoxyamino-9-methyl-	239°	656, 693
2-Methyl-6-methylamino-	300°	408
3-Methyl-6-methylamino-	314–315°	25
6-Methyl-2-methylamino-	> 300°	243
7-Methyl-6-methylamino-	300°	347
8-Methyl-6-methylamino-	> 300°	242
9-Methyl-6-methylamino-	190–191°	12, 206
3-Methyl-6-methylamino-8-phenyl-	190–192°	23
7-Methyl-6-propylamino-	178°	347
2-Methylamino-	276°	12
6-Methylamino-	308 to 321°	179, 585, 607
8-Methylamino-	332–334°	6, 32
6-Methylamino-7-methyl-	311°	54
6-Methylamino-9-phenyl-	155–156°	462
6-(3-Methylbut-2-enylamino)-	213–215°	33, 639
6-(3-Methylbut-3-enylamino)-	181–182°	73

TABLE 22 (*continued*)

Purine	M.p.	References
6-(2-Methylhydrazino)-	251°	41
6-(2-Methylhydrazino)-9-methyl-	100–101°	206
6-(N-Methyl)hydroxyamino-	265°	41
6-Pentylamino-	164–165°	585
6-Propionamido-	235–237°	613, 614
9-Propyl-2-propylamino-	85–86°	820
6-Propylamino-	234°	585
6-Propylamino-9-methyl-	130–131°	206
2,6,8-Triamino-7-methyl-	335–340°	112
2,6,8-Tributylamino-	206–207°	105
2,6,8-Trifurfurylamino-	160–161°	105
2,6,8-Trihexylamino-	159–160°	105
2,6,8-Trihydrazino-	209°	105
2,6,8-Trihydrazino-7-methyl-	260–262°	112
1,2,9-Trimethyl-6-methylimino-	150–151°	418
6-Ureido-	330–335°	50

TABLE 23. Carboxypurines.

Purine	M.p.	References
9-Acetyl-	167–168°	64
7(9)-Acetyl-6-thioformyl-	182–184°	378
6-Azidocarbonyl-	expl. 156°	50
6-Carbamoyl-	315–325°	50, 374, 375, 666
6-Carbamoyl-9-methyl-	289–291°	431
6-(2-Carbamoylethyl)-	250°	76
6-Carbamoylmethyl-	242°	171
9-Carbamoylmethyl-	245°	400
6-Carboxy-	198–202°	165, 375, 666, 666, 672, 791
8-Carboxy-	210–212°	373
6-(2-Carboxyethyl)-	227°	76, 492
6-(2-Carboxyvinyl)-	230°	76
6-Cyano-	177–178°	165, 374, 666, 674
6-Cyano-2-methyl-	239–241°	431
6-Cyano-9-methyl-	153–154°	431
6-(NN-Dimethyl)carbamoyl-	210–211°	50
2,6-Dithiocyanato-	> 240°	164
6,8-Dithiocyanato-	240°	809
6-(2-Ethoxycarbonylethyl)-	150°	76

continued

TABLE 23 (*continued*)

Purine	M.p.	References
6-Ethoxycarbonylmethyl-	135°	171
9-Ethoxycarbonylmethyl-	122°	400
6-Ethoxycarbonylmethyl-9-methyl-	HCl 165–166°	654
6-(2-Ethoxycarbonylvinyl)-	208–213°	76
6-Formyl-	199°; HCl 235°	132, 165
6-Hydrazinocarbonyl-	292–294°	50
6-Hydrazinocarbonylmethyl-	246°	171
9-Hydrazinocarbonylmethyl-	177–178°	400
6-(N-Hydroxy)carbamoyl-	219–220°	158
6-Methoxycarbonyl-	226–227°	375
9-Methyl-2,6-dithiocyanato-	240°	809
6-(N-Methyl)carbamoyl-	308–310°	50
6-Thiocarbamoyl-	240–242°	374, 666, 674
6-Thiocyanato-	235° or 225–226°	63, 164
6-Thiocyanatomethyl-	208–210°	42
2,6,8-Trithiocyanato-	240°	809

TABLE 24. Carboxythiopurines.

Purine	M.p.	References
6-Acetonylthio-	184–186°	63
6,8-bis-Carboxymethylthio-	> 230°	240
9-Butyl-6-(4-carboxybutylthio)-	56–57°	684
6-(4-Carbamoylbutylthio)-	184–186°	681
6-Carbamoylmethylthio-	264°	492
6-(3-Carbamoylpropylthio)-	208–209°	680
6-(4-Carboxybutylthio)-	204–205°	680
6-(4-Carboxybutylthio)-9-cyclohexyl-	99–100°	684
6-(4-Carboxybutylthio)-9-cyclopentyl-	110–111°	684
6-(4-Carboxybutylthio)-9-hexyl-	49–51°	684
6-(4-Carboxybutylthio)-2-methyl-	182–183°	681
6-(4-Carboxybutylthio)-8-methyl-	202–203°	681
6-(4-Carboxybutylthio)-9-methyl-	176–177°	684
6-(1-Carboxyethylthio)-	199–203°	678
6-(2-Carboxyethylthio)-	219–220° or 233–234°	678, 680
2-Carboxymethylthio-	200°	6
6-Carboxymethylthio-	230–260°	63, 492, 623, 678, 680
8-Carboxymethylthio-	220°	6
6-Carboxymethylthio-9-methyl-	225–226°	59
6-(5-Carboxypentylthio)-	197–198°	680
6-(3-Carboxypropylthio)-	203–204°	680

TABLE 24 (*continued*)

Purine	M.p.	References
6-Cyanomethylthio-	258°	492
6-(3-Cyanopropylthio)-	147–148°	680
6-(*NN*-Diethylcarbamoyl)methylthio-	hy. 80–81°	114
6-(4-Ethoxycarbonylbutylthio)-	89–90°	680
6-(4-Ethoxycarbonylbutylthio)-2-methyl-	131–132°	681
6-(4-Ethoxycarbonylbutylthio)-8-methyl-	92–93°	681
6-Ethoxycarbonylmethylthio-	127–128°	451
6-(3-Ethoxycarbonylpropylthio)-	98–99°	680
6-Ethoxycarbonylthio-9-methyl-	100–101°	382
6-(4-Hydrazinocarbonylbutylthio)-	200–201°	681
6-Methoxycarbonylmethylthio-	169°	492
6-Methoxycarbonylthio-9-methyl-	136–137°	382

TABLE 25. Halogenopurines.

Purine	M.p.	References
7-Benzyl-6-chloro-	152–154°	113, 400, 454
9-Benzyl-6-chloro-	85–88°	400, 454, 462, 707
9-Benzyl-2-chloro-6-trifluoromethyl-	118–120°	208
9-Benzyl-6-chloro-2-trifluoromethyl-	98–99°	208
7-Benzyl-2,6-dichloro-	148°	200
9-Benzyl-6,8-dichloro-	92°	470
9-Benzyl-2,6-difluoro-	oil	10
9-Benzyl-6-fluoro-	127–131°	397
2-Bromo-	243°	207
6-Bromo-	194°	207, 406, 407
8-Bromo-	—	207
6-Bromomethyl-	> 140°	375
9-Butyl-6-chloro-	142°/0.2 mm	265
7-Butyl-6-chloro-8-propyl-	oil	345
2-Chloro-	231–234°	105, 108, 111, 259, 454
6-Chloro-	—	50, 108, 191, 259, 407, 416
8-Chloro-	> 200°	108
2-Chloro-6,8-bis-trifluoromethyl-	149°	208
6-Chloro-9-(2-chloroethyl)-	106–108°	400, 430
2-Chloro-6-chloromethyl-9-methyl-	128–130°	83
6-Chloro-9-(cyclohex-2-enyl)-	134–136°	266
6-Chloro-9-cyclohexyl-	114°	265
6-Chloro-9-cyclopentyl-	96°	265

continued

TABLE 25 (*continued*)

Purine	M.p.	References
6-Chloro-9-cyclopropyl-	HCl 124°	263
6-Chloro-2,9-diethyl-	140°/0.5 mm	170
2-Chloro-6,8-diiodo-	223–224°	111
2-Chloro-6,9-dimethyl-	155–157°	83, 461
6-Chloro-2,8-dimethyl-	212°	226, 243, 409
6-Chloro-2-ethyl-	235°	170
6-Chloro-7-ethyl-	122–123°	400
6-Chloro-8-ethyl-	170–172°	64
6-Chloro-9-ethyl-	81–84°	60, 400
6-Chloro-8-ethyl-2-methyl-	128°	226, 409
2-Chloro-9-ethyl-6-trifluoromethyl-	78–79°	208
6-Chloro-9-ethyl-2-trifluoromethyl-	63–65°	208
6-Chloro-2-fluoro-	174°	427
6-Chloro-9-furfuryl-	97–99° or 109–112°	69, 726
6-Chloro-9-hexyl-	160°/0.2 mm	263
2-Chloro-6-iodo-	208°	406
2-Chloro-8-iodo-	238–240°	32
6-Chloro-7-isobutyl-8-isopropyl-	134–135°	345
6-Chloro-9-isopropyl-	HCl 153°	263
6-Chloro-8-isopropyl-2-methyl-	127°	226, 409
2-Chloro-6-methyl-	275°	243
2-Chloro-7-methyl-	200–207°	367, 397
2-Chloro-9-methyl-	132–137°	95, 97, 115, 159, 367, 397
6-Chloro-2-methyl-	175–185°	226, 408, 409
6-Chloro-3-methyl-	91–92°	23
6-Chloro-7-methyl-	198–199°	347, 109
6-Chloro-8-methyl-	200 to 227°	64, 229, 242
6-Chloro-9-methyl-	124 to 144°	83, 159, 206
8-Chloro-7-methyl-	*ca* 150°	397
8-Chloro-9-methyl-	106–108°	159, 397
6-Chloro-2-methyl-8-propyl-	159°	226, 409
6-Chloro-3-methyl-8-phenyl-	225–227°	23
2-Chloro-9-methyl-6-trifluoromethyl-	126–127°	208
6-Chloro-9-methyl-2-trifluoromethyl-	73–74°	208
6-Chloro-9-pentyl-	152–156°/0.7 mm	263
2-Chloro-9-phenyl-	162–163°	399
6-Chloro-2-phenyl-	—	402
6-Chloro-9-phenyl-	202–203°	462
6-Chloro-9-propyl-	HCl 193–195°	263
2-Chloro-6-trifluoromethyl-	240°	208
6-Chloro-2-trifluoromethyl-	200–201°	208
8-Chloro-2-trifluoromethyl-	141–143°	329
6-Chloro-9-vinyl-	166–167°	110
6-Chloromethyl-9-methyl-	107–109°	83
7-Chloromethyl-2-methyl-	—	391
6-(3-Chloropropyl)-	239–243°	76

TABLE 25 (*continued*)

Purine	M.p.	References
9-Cyclopentyl-6-fluoro-	55–58°	261
2,6-Dibromo-	207°	207, 406
2,8-Dibromo-	—	207
6,8-Dibromo-	223°	207
6-Dibromomethyl-	165–166°	375
2,6-Dichloro-	179–181° or 188–190°	61, 108, 259, 406, 700
2,8-Dichloro-	145–150°	108, 111
6,8-Dichloro-	175–178°	58, 108, 143
2,6-Dichloro-7-benzyl-	—	470
2,6-Dichloro-9-benzyl-	148°	470
2,6-Dichloro-7-methyl-	195–201°	258, 347, 397, 417
2,6-Dichloro-9-methyl-	152–153°	320, 397, 403
6,8-Dichloro-2-methyl-	205–206°	410
6,8-Dichloro-3-methyl-	188°	72
2,6-Dichloro-9-phenyl-	244–246°	57
2,6-Dichloro-9-vinyl-	126–127°	110
6-Dichloromethyl-	176–177°	375
2,6-Difluoro-7-methyl-	154–161°	397
2,6-Diiodo-	205–208° or 224°	367, 432
2-Fluoro-	216°	427, 433
6-Fluoro-	125 126°	99
2-Fluoro-6-methyl-	194°	427
2-Fluoro-9-methyl-	151°	397
6-Fluoro-9-methyl-	135–136° or 125–127°	261, 397
8-Fluoro-9-methyl-	111–112°	397
6-Iodo-	hy. 167°	406
2-Iodo-7-methyl-	229°	367
2-Iodo-9-methyl-	171–172°	367
6-Iodo-2-methyl-	195–197°	431
6-Iodo-9-methyl-	214–215°	431
2-Iodo-9-phenyl-	165–166°	399
6-Methyl-8-trifluoromethyl-	241–243°	118
2,6,8-Tribromo-	223°	207
6-Tribromomethyl-	194–195° or 183–185°	375, 791
2,6,8-Trichloro-	177 to 189°	84, 102, 375, 411, 412, 432
2,6,8-Trichloro-7-methyl-	155–161°	102, 104, 310, 452
2,6,8-Trichloro-9-methyl-	176°	102, 452
2,6,8-Trichloro-9-phenyl-	210–211°	399
6-Trichloromethyl-	204–206°	375
6-Trichloromethyl-8-trifluoromethyl-	185–186°	118
6-(3,3,3-Trichloropropenyl)-	HCl 160°	76
6-Trifluoromethyl-	254–255°	56
8-Trifluoromethyl-	192°	56

TABLE 26.   Oxopurines, Including Those with *C*-, but not *N*-, Alkyl Groups.

Purine	M.p.	References
6-Acetoxymethyl-	239°	379
6-Benzyloxy-	170–171°	471
6-Butoxy-	163–164°	468, 475
2,6-Diethoxy-	192–194°	708
2,3-Dihydro-6,8-dimethyl-2-oxo-	> 300°	804
2,3-Dihydro-6-methoxy-2-oxo-	—	140
2,3-Dihydro-6-methyl-2-oxo-	> 350°	178, 818
2,3-Dihydro-8-methyl-2-oxo-	> 310°	193, 273
7,8-Dihydro-2-methyl-8-oxo-	> 300°	328, 521, 705
7,8-Dihydro-6-methyl-8-oxo-	> 325°	174
1,6-Dihydro-6-oxo- (hypoxanthine q.v.)		
2,3-Dihydro-2-oxo-	> 300°; pic. 245°	6, 193, 505, 642
7,8-Dihydro-8-oxo-	*ca.* 317°	2, 203, 327, 523, 791, 827
7,8-Dihydro-8-oxo-2-phenyl-	> 300°	18
2,6-Dimethoxy-	> 300°	516
2,8-Dimethoxy-	—	32
2-Ethoxy-	197–198°	97
6-Ethoxy-	223–224°	468, 475
8-Ethoxy-	191–192°	97
2-Ethoxy-6-methyl-	236°	243
6-Ethoxy-9-methyl-	107–108°	97
1,2,3,6,7,8-Hexahydro-2,6,8-trioxo- (uric acid q.v.)		
6-Hydroxymethyl-	> 350°	165
8-Hydroxymethyl-	262°	77, 237
8-Hydroxymethyl-6-methyl-	252°	790
6-(3-Hydroxypropyl)-	173–176°	76
Hypoxanthine	> 350°	53, 103, 136, 178, 179, 217, 566, 598, 791, 706
Hypoxanthine/2,8-diethyl-	282°	53
Hypoxanthine/2,8-dimethyl-	> 300°	53, 226, 243, 409, 806
Hypoxanthine/2,8-diphenyl-	300°	53
Hypoxanthine/2-ethyl-	> 250°	706
Hypoxanthine/8-ethyl-2-methyl-	336°	226, 409
Hypoxanthine/8-ethyl-2-phenyl-	328–330°	227
Hypoxanthine/2-furyl-	350°	227
Hypoxanthine/8-hexyl-	> 280°	342
Hypoxanthine/(1-hydroxy-1-methylethyl)-	—	711
Hypoxanthine/8-hydroxymethyl-	294–296°	239, 790
Hypoxanthine/8-isopropyl-2-methyl-	334°	226, 409
Hypoxanthine/2-methyl-	> 350°	53, 178, 181, 221, 225, 226, 409, 806

TABLE 26 (*continued*)

Purine	M.p.	References
Hypoxanthine/8-methyl-	288–289°	17, 229, 242, 253, 790
Hypoxanthine/2-methyl-8-phenyl-	350°	66
Hypoxanthine/8-methyl-2-phenyl-	328–330°	227
Hypoxanthine/2-methyl-8-propyl-	326°	226, 409
Hypoxanthine/2-phenyl-	> 300°	18, 227, 402
Hypoxanthine/8-phenyl-	> 350°	256
Hypoxanthine/2-phenyl-8-propyl-	325–327°	227
6-Isopropoxy-	192–193°	475
2-Methoxy-	205–206°	6
6-Methoxy-	194–195°	416, 475
8-Methoxy-	153–154°	166
2-Methoxy-6-methyl-	283–284°	243
6-Phenoxy-	—	416
6-Propoxy-	180–181°	475
1,2,3,6-Tetrahydro-2,6-dioxo- (xanthine q.v.)		
2,3,7,8-Tetrahydro-2,8-dioxo-	> 350°	6, 328, 705, 819
1,6,7,8-Tetrahydro-6,8-dioxo-	> 350°	2, 6, 58, 143, 210, 289, 523, 823
1,6,7,8-Tetrahydro-6,8-dioxo-2-phenyl-	> 300°	18
1,6,7,8-Tetrahydro-2-methyl-6,8-dioxo-	> 300°	245, 410
2,3,7,8-Tetrahydro-6-methyl-2,8-dioxo-	> 300°	818
6-Trimethoxymethyl-	179–180°	375
Uric acid	350°	13, 276, 278, 307, 325, 362
Xanthine	> 350°	13, 103, 178, 217, 278, 302, 325, 337, 343, 358, 370, 611, 640, 641, 645, 706, 806
Xanthine/8-benzyl-	> 300°	361
Xanthine/8-ethyl-	> 350°	216, 775
Xanthine/8-hydroxymethyl-	—	75
Xanthine/8-isopropyl-	> 340°	775
Xanthine/8-methyl-	> 350°	17, 63, 242, 244, 249, 250, 270, 325, 561, 775
Xanthine/8-phenyl-	> 360°	91, 256, 299, 359

TABLE 27.   Oxopurines with *N*-Alkyl Groups.

Purine	M.p.	References
8-Acetoxy-1,2,3,6-tetrahydro-1,3-dimethyl-2,6-dioxo-	> 360°	699
8-Acetoxy-1,2,3,6-tetrahydro-1,7-dimethyl-2,6-dioxo-	—	519
8-Acetoxy-1,2,3,6-tetrahydro-7-methyl-2,6-dioxo-	—	519
8-Acetoxy-1,2,3,6-tetrahydro-1,3,7-trimethyl-2,6-dioxo-	—	519
7-Acetoxymethyl-1,2,3,6-tetrahydro-1,3-dimethyl-2,6-dioxo-	165°	609
8-Allyloxy-1,2,3,6-tetrahydro-1,3,7-trimethyl-2,6-dioxo-	124–126°	807
3-Benzyl-2,3-dihydro-2-oxo-	> 240°	37
7-Benzyl-6-methoxy-	126–127°	400
9-Benzyl-6-methoxy-	128°	400
3-Benzyl-1,6,7,8-tetrahydro-6,8-dioxo-	> 300°	37
8-Benzyloxy-1,2,3,6-tetrahydro-1,3,7-trimethyl-2,6-dioxo-	172–173°	807
8-Benzyloxymethyl-1,2,3,6-tetrahydro-1,3,7-trimethyl-2,6-dioxo-	134–135°	501
2-Butoxy-1,6-dihydro-1,7-dimethyl-6-oxo-	83–85°	671
2-Butoxy-3,6-dihydro-3,7-dimethyl-6-oxo-	129–130°	671
8-Butoxy-1,2,3,6-tetrahydro-1,3,7-trimethyl-2,6-dioxo-	89–90°	807
8-*s*-Butoxy-1,2,3,6-tetrahydro-1,3,7-trimethyl-2,6-dioxo-	122–124°	807
8-*t*-Butoxy-1,2,3,6-tetrahydro-1,3,7-trimethyl-2,6-dioxo-	158–161°	807
2,6-Diallyloxy-7-methyl-	111–112°	477
2,6-Diethoxy-7,8-dihydro-7,9-dimethyl-8-oxo-	129°	102
2,6-Diethoxy-7-methyl-	147–149°	714
2,6-Diethoxy-9-methyl-	87–89°	658
2,3-Dihydro-3,7-dimethyl-2-oxo-	256–257°	528, 531
2,3-Dihydro-6,9-dimethyl-2-oxo-	280–284° or > 320°	83, 182
2,3-Dihydro-8,9-dimethyl-2-oxo-	> 320°	182
3,6-Dihydro-3,7-dimethyl-6-oxo-2-propoxy-	127–129°	671
8,9-Dihydro-7-isopropenyl-8-oxo-	196–197°	827
7,8-Dihydro-9-isopropyl-8-oxo-	214°	827
8,9-Dihydro-7-isopropyl-8-oxo-	160–161°	827
3,6-Dihydro-2-methoxy-3,7-dimethyl-6-oxo-	180°	671
1,6-Dihydro-7-methyl-2-methoxy-6-oxo-	> 220°	248
1,2-Dihydro-1-methyl-2-oxo-	> 280°	125, 194
2,3-Dihydro-3-methyl-2-oxo-	297–300°	522, 704
2,3-Dihydro-7-methyl-2-oxo-	hy. 323°	367, 704, 717

TABLE 27 (*continued*)

Purine	M.p.	References
2,3-Dihydro-9-methyl-2-oxo-	305–310°	125
7,8-Dihydro-3-methyl-8-oxo-	> 300°	522
7,8-Dihydro-9-methyl-8-oxo-	233°	166, 407
8,9-Dihydro-7-methyl-8-oxo-	266–267°	104, 166, 274
3,6-Dihydro-3-methyl-6-oxo-8-phenyl-	> 300°	23
8,9-Dihydro-7-methyl-8-oxo-9-phenyl-	151–152°	274
7,8-Dihydro-8-oxo-	305–307°	6
8,9-Dihydro-8-oxo-7-phenyl-	237–238°	274
2,3-Dihydro-6,8,9-trimethyl-2-oxo-	275°	182
2,6-Dimethoxy-7-methyl-	199°	476
2,6-Dipropyloxy-7-methyl-	92°	477
8-Ethoxy-3,7-diethyl-1,2,3,6-tetrahydro-2,6-dioxo-	212°	459
8-Ethoxy-3,7-diethyl-1,2,3,6-tetrahydro-1-methyl-2,6-dioxo-	112°	459
2-Ethoxy-1,6-dihydro-1,7-dimethyl-6-oxo-	158–160°	714
2-Ethoxy-1,6-dihydro-1,9-dimethyl-6-oxo-	192–194°	403
2-Ethoxy-1,6-dihydro-8,9-dimethyl-6-oxo-	227–229°	384
2-Ethoxy-3,6-dihydro-3,7-dimethyl-6-oxo-	—	671
2-Ethoxy-1,6-dihydro-8-hydroxymethyl-1,9-dimethyl-6-oxo-	192–194°	712
2-Ethoxy-1,6-dihydro-8-hydroxymethyl-9-methyl-6-oxo-	220–222°	712
2-Ethoxy-1,6-dihydro-7-methyl-6-oxo-	240–242°	714
2-Ethoxy-1,6-dihydro-1,8,9-trimethyl-6-oxo-	98–99°	384
2-Ethoxy-8-ethoxymethyl-1,6-dihydro-1,9-dimethyl-6-oxo-	114–116°	712
2-Ethoxy-8-ethoxymethyl-1,6-dihydro-9-methyl-6-oxo-	189–191°	712
8-Ethoxy-1-ethyl-1,2,3,6-tetrahydro-3,7-dimethyl-2,6-dioxo-	153–154°	483
6-Ethoxy-9-furfuryl-	118°	69
2-Ethoxy-9-methyl-	111–112°	97
6-Ethoxy-3-methyl-	198–200°	23
8-Ethoxy-9-methyl-	95–96°	97
8-Ethoxy-1,2,3,6-tetrahydro-1,9-dimethyl-2,6-dioxo-	261–262°	437
8-Ethoxy-1,2,3,6-tetrahydro-7-isopropyl-1,3-dimethyl-2,6-dioxo-	270–272°	699
8-Ethoxy-1,2,3,6-tetrahydro-1,3,7-trimethyl-2,6-dioxo-	138–142°	367, 385, 504, 603, 807
8-Ethoxy-1,2,3,6-tetrahydro-1,3,9-trimethyl-2,6-dioxo-	260–261°	437
2-Ethoxy-1,6,7,8-tetrahydro-1,7,9-trimethyl-6,8-dioxo-	149°	518

continued

TABLE 27 (*continued*)

Purine	M.p.	References
2-Ethoxy-1,7,9-triethyl-1,6,7,8-tetrahydro-6,8-dioxo-	114°	253
8-Ethoxymethyl-1,2,3,6-tetrahydro-1,3-dimethyl-2,6-dioxo-	—	75
8-Ethoxymethyl-1,2,3,6-tetrahydro-1,9-dimethyl-2,6-dioxo-	240°	712
8-Ethoxymethyl-1,2,3,6-tetrahydro-1,3,7-trimethyl-2,6-dioxo-	127–128°	501
8-Ethoxymethyl-1,2,3,6-tetrahydro-1,3,9-trimethyl-2,6-dioxo-	153–155°	440, 669
9-Ethyl-2,3-dihydro-6,8-dimethyl-2-oxo-	>230°	241
9-Ethyl-2,3-dihydro-6-methyl-2-oxo-	>250°	241
7-Ethyl-8,9-dihydro-8-oxo-	196–197°	274
9-Ethyl-7,8-dihydro-8-oxo-	250–251°	331
9-Ethyl-2,3,7,8-tetrahydro-2,8-dioxo-	>300°	816
1-Ethyl-1,2,3,6-tetrahydro-8-methoxy-3,7-dimethyl-2,6-dioxo-	164–165°	482
7-Ethyl-1,2,3,6-tetrahydro-8-methoxy-1,3-dimethyl-2,6-dioxo-	122–124° or 135°	518, 699
9-Ethyl-2,3,7,8-tetrahydro-6-methyl-2,8-dioxo-	>320°	241, 814
9-Furfuryl-6-methoxy-	103–104°	726
7-(2-Hydroxyethyl)-	147–148°	366
Hypoxanthine/1-benzyl-	268–270°	30
Hypoxanthine/3-benzyl-	245–247° or 282–283°	26, 124, 169
Hypoxanthine/7-benzyl-	260–270°	200, 400
Hypoxanthine/9-benzyl-	254–260° or 295–297°	26, 100, 223, 400
Hypoxanthine/3-benzyl-1-methyl-	iod. 185–187°	100
Hypoxanthine/7-benzyl-1-methyl-	159–160°	180
Hypoxanthine/9-benzyl-1-methyl-	207–208°	180
Hypoxanthine/7-butyl-	183°	200
Hypoxanthine/9-butyl-	259–260°	265
Hypoxanthine/7-butyl-8-propyl-	202–203°	345
Hypoxanthine/9-(cyclohex-2-enyl)-	261°	266
Hypoxanthine/9-cyclohexyl-	273–275°	265
Hypoxanthine/9-cyclohexyl-8-methyl-	285°	341
Hypoxanthine/9-cyclopentyl-	230°	265
Hypoxanthine/9-cyclopropyl-	291–293°	263
Hypoxanthine/1,3-dibenzyl-	brom. 196°	100
Hypoxanthine/1,7-dibenzyl-	105–110°	100, 124
Hypoxanthine/1,9-dibenzyl-	208°	100, 124
Hypoxanthine/3,7-dibenzyl-	181–183°	113, 124
Hypoxanthine/7,9-dibenzyl-	>260°; brom. 211°	100

TABLE 27 (*continued*)

Purine	M.p.	References
Hypoxanthine/1,7-dimethyl-	245–246°	671
Hypoxanthine/1,9-dimethyl-	255–256°	101
Hypoxanthine/2,9-dimethyl-	> 300°	221, 225
Hypoxanthine/3,7-dimethyl-	242–245°	671
Hypoxanthine/3,8-dimethyl-	*ca.* 300°	273
Hypoxanthine/3,9-dimethyl-	—	11
Hypoxanthine/7,9-dimethyl-	bet. 330–332°; tos. 260°	25, 532, 715
Hypoxanthine/1,3-dimethyl-8-phenyl-	iod. —	715
Hypoxanthine/1-ethyl-	275–276°	500
Hypoxanthine/9-ethyl-	263–265°	60
Hypoxanthine/7-ethyl-8-propyl-	217–218°	345
Hypoxanthine/9-furfuryl-	265°	726
Hypoxanthine/9-hexyl-	254°	263
Hypoxanthine/2-hydroxymethyl-1,9-dimethyl-	210–212°	383
Hypoxanthine/2-hydroxymethyl-3,7-dimethyl-	250–252°	821
Hypoxanthine/7-(3-hydroxypropyl)-	176–178°	464
Hypoxanthine/9-isopropyl-	222°	263
Hypoxanthine/7-isobutyl-8-isopropyl-	181–182°	345
Hypoxanthine/1-methyl-	311–312°	29, 101, 180
Hypoxanthine/3-methyl-	280° or 307–309°	25, 101, 162, 522
Hypoxanthine/7-methyl-	> 350°	347, 415, 417, 458
Hypoxanthine/9-methyl-	> 360°	101, 166, 206, 261, 340, 415
Hypoxanthine/1-methyl-8-phenyl-	> 300°	715
Hypoxanthine/1-methyl-9-propenyl-	220°	29
Hypoxanthine/9-pentyl-	262°	263
Hypoxanthine/9-phenyl-	306–308°	223, 462
Hypoxanthine/9-propenyl-	301–303°	29
Hypoxanthine/9-propyl-	259–260°	263
Hypoxanthine/1,7,9-tribenzyl-	brom. —	100
Hypoxanthine/1,2,7-trimethyl-	—	181
Hypoxanthine/1,2,9-trimethyl-	223–225°	383
Hypoxanthine/1,3,7-trimethyl-	nit. 235°	649
Hypoxanthine/2,3,7-trimethyl-	248–250°	821
7-Isopropyl-1,2,3,6-tetrahydro-8-methoxy-1,3-dimethyl-2,6-dioxo-	180–182°	520
6-Methoxy-3-methyl-	162–163°	23
6-Methoxy-7-methyl-	200°	347
6-Methoxy-9-methyl-	152–153°	206
6-Methoxy-3-methyl-8-phenyl-	174–176°	23

continued

TABLE 27 (*continued*)

Purine	M.p.	References
1,2,7,8-Tetrahydro-1,6-dimethyl-2,8-dioxo-	> 300°	812
1,2,7,8-Tetrahydro-1,9-dimethyl-2,8-dioxo-	—	555
1,2,8,9-Tetrahydro-1,7-dimethyl-2,8-dioxo-	> 300°	555
1,6,7,8-Tetrahydro-1,9-dimethyl-6,8-dioxo-	> 350°	491
2,3,7,8-Tetrahydro-6,9-dimethyl-2,8-dioxo-	> 300°	813
1,2,3,6-Tetrahydro-8-isopropoxy-1,3,7-trimethyl-2,6-dioxo-	157–159°	699
1,2,3,6-Tetrahydro-7-isopropyl-8-methoxy-1,3-dimethyl-2,6-dioxo-	182–184°	699
1,2,3,6-Tetrahydro-8-methoxy-1,3-dimethyl-2,6-dioxo-	264–266°	699
1,2,3,6-Tetrahydro-8-methoxy-3,7-dimethyl-2,6-dioxo-	282°	479
1,6,7,8-Tetrahydro-2-methoxy-1,9-dimethyl-6,8-dioxo-	275°	519
1,6,8,9-Tetrahydro-2-methoxy-1,7-dimethyl-6,8-dioxo-	282°	662
1,2,3,6-Tetrahydro-8-methoxy-1,3-dimethyl-2,6-dioxo-7-propyl-	72–74°	699
1,2,3,6-Tetrahydro-8-methoxy-1,3,7-trimethyl-2,6-dioxo-	175–177°	518, 699, 807
1,2,3,6-Tetrahydro-8-methoxy-1,3,9-trimethyl-2,6-dioxo-	230°	309
1,6,7,8-Tetrahydro-2-methoxy-1,7,9-trimethyl-6,8-dioxo-	186–197°	253, 518, 662
1,2,7,8-Tetrahydro-1-methyl-2,8-dioxo-	> 300°	74, 813
1,6,7,8-Tetrahydro-9-methyl-6,8-dioxo-	> 350°	491
1,6,8,9-Tetrahydro-7-methyl-6,8-dioxo-	> 350°	455
2,3,7,8-Tetrahydro-3-methyl-2,8-dioxo-	> 300°	75, 484, 522
2,3,7,8-Tetrahydro-9-methyl-2,8-dioxo-	> 300°	74, 815
3,6,7,8-Tetrahydro-3-methyl-6,8-dioxo-	hy. > 300°	522
1,2,7,8-Tetrahydro-1,7,9-trimethyl-2,8-dioxo-	237–240°	555
1,6,7,8-Tetrahydro-1,7,9-trimethyl-6,8-dioxo-	229–230°	2, 455
2,3,7,8-Tetrahydro-3,7,9-trimethyl-2,8-dioxo-	254°	455, 484
1,2,3,6-Tetrahydro-1,3,7-trimethyl-2,6-dioxo-8-phenoxy-	140–143°	807
1,2,3,6-Tetrahydro-1,3,7-trimethyl-2,6-dioxo-8-propoxy-	129–130°	807
Uric acid/9-allyl-1,3,7-trimethyl-	145–148°	806
Uric acid/7-benzyl-1,3,9-trimethyl-	177–178°	699
Uric acid/9-benzyl-1,3,7-trimethyl-	187–189°	806
Uric acid/1-butyl-3,7-dimethyl-	270°	549
Uric acid/1,3-diethyl-	> 300°	317, 568

TABLE 27 (*continued*)

Purine	M.p.	References
Uric acid/3,7-diethyl-	350–355° or 371–376°	295, 459, 557
Uric acid/(?),9-diethyl-	314°	331
Uric acid/1,9-diethyl-3,7-dimethyl-	158°	483
Uric acid/3,7-diethyl-1-methyl-	266°	459, 557
Uric acid/1,3-dimethyl-	>350°	1, 13, 307, 311, 316, 520, 568, 699
Uric acid/1,7-dimethyl-	>350°	253, 311, 326, 362
Uric acid/1,9-dimethyl-	>350°	137, 491
Uric acid/3,7-dimethyl-	—	104, 326
Uric acid/3,9-dimethyl-	>350°	151, 565, 570
Uric acid/7,9-dimethyl-	>350°	140, 142, 326, 533
Uric acid/1,7-dimethyl-9-phenyl-	>280°	315
Uric acid/3,7-dimethyl-1-propyl-	293°	549
Uric acid/1,3-diphenyl-	>300°	332
Uric acid/3-ethyl-	>360°	295, 556
Uric acid/7-ethyl-	361–362°	313
Uric acid/9-ethyl-	>350°	331
Uric acid/1-ethyl-3,7-dimethyl-	—	504
Uric acid/7-ethyl-1,3-dimethyl-	283°	314
Uric acid/7-ethyl-9-methyl-	—	297
Uric acid/9-ethyl-1-methyl-	>350°	149
Uric acid/1-ethyl-3,7,9-trimethyl-	176–177°	482
Uric acid/3-ethyl-1,7,9-trimethyl-	240–241°	295, 518
Uric acid/7-ethyl-1,3,9-trimethyl-	215°	518
Uric acid/9-ethyl-1,3,7-trimethyl-	199–202°	480, 481, 806
Uric acid/3-(2-hydroxyethyl)-	325–344°	293
Uric acid/9-(2-hydroxyethyl)-	>350°	292
Uric acid/7-hydroxymethyl-1,3-dimethyl-	—	569
Uric acid/7-hydroxymethyl-3-methyl-		572
Uric acid/1-methyl-	>350°	173,* 253, 311, 326
Uric acid/3-methyl-	—	13, 302, 568
Uric acid/7-methyl-	>350°	104, 307, 312, 326
Uric acid/9-methyl-	>385°	388, 567
Uric acid/1-methyl-3-phenyl-	>340°	660
Uric acid/3-phenyl-	300°	660, 679
Uric acid/9-phenyl-	—	181, 308
Uric acid/1,3,7,9-tetraethyl-	82°	557
Uric acid/1,3,7,9-tetramethyl-	226–228°	248, 253, 281, 302, 307, 321, 479, 480
Uric acid/1,3,7-triethyl-	218–219°	557

continued

TABLE 27 (*continued*)

Purine	M.p.	References
Uric acid/1,7,9-triethyl-	221°	253
Uric acid/3,7,9-triethyl-	204°	557
Uric acid/3,7,9-triethyl-1-methyl-	104°	557
Uric acid/1,3,7-trimethyl-	> 300°	307, 314, 367, 385, 504, 603, 699
Uric acid/1,3,9-trimethyl-	> 300°	1, 309, 479, 519
Uric acid/1,7,9-trimethyl-	339–348°	253, 491, 533
Uric acid/3,7,9-trimethyl-	373–375°	479, 792
Uric acid/1,3,7-trimethyl-9-phenyl-	258–266°	308, 315
Uric acid/1,7,9-trimethyl-3-phenyl-	229°	660
Uric acid/1,3,7-trimethyl-9-propyl-	138–140°	806
Xanthine/7-allyl-1,3-diethyl-8-hydroxy-methyl-	124–125°	808
Xanthine/3-allyl-1,7-dimethyl-	114–116°	349
Xanthine/7-allyl-1,3-dimethyl	101–105°	450, 784
Xanthine/8-allyl-1,3-dimethyl-	284–285°	233
Xanthine/1-allyl-3-ethyl-7-(2-hydroxyethyl)-	94–96°	512
Xanthine/3-allyl-1-ethyl-7-(2-hydroxyethyl)-	113–116°	512
Xanthine/7-allyl-8-hydroxymethyl-1,3-dimethyl-	131–132°	808
Xanthine/3-benzyl-	> 300°	37
Xanthine/7-benzyl-	295°	181
Xanthine/9-benzyl-	316°	92
Xanthine/7-benzyl-1,3-diethyl-8-hydroxymethyl-	141–142°	808
Xanthine/1-benzyl-3,7-dimethyl-	139–141°	157
Xanthine/1-benzyl-3,9-dimethyl-	219–221°	539
Xanthine/3-benzyl-1,7-dimethyl-	166–168°	349
Xanthine/7-benzyl-1,3-dimethyl-	158°	181, 390, 764
Xanthine/8-benzyl-1,3-dimethyl-	287 to 300°	91, 234, 496, 716
Xanthine/8-benzyl-1,7-dimethyl-	252°	361
Xanthine/8-benzyl-3,7-dimethyl-	251–252°	716
Xanthine/9-benzyl-1,3-dimethyl-	167–169°	46
Xanthine/7-benzyl-1,3-dimethyl-8-phenyl-	221°	298
Xanthine/7-benzyl-8-hydroxymethyl-1,3-dimethyl-	191–193°	808
Xanthine/7-benzyl-1-methyl-	250°	181
Xanthine/7-benzyl-3-methyl-	273°	181
Xanthine/7-benzyl-8-methyl-	325–327°	796
Xanthine/9-benzyl-7-methyl-	bet. 280–282°	554
Xanthine/7-benzyl-3-methyl-8-phenyl-	—	298
Xanthine/1-benzyl-8-phenyl-	> 360°	91
Xanthine/3-benzyl-8-phenyl-	> 360°	91
Xanthine/8-benzyl-1-phenyl-	318°	359
Xanthine/7-benzyl-1,3,8-trimethyl-	159–160°	777
Xanthine/8-benzyl-1,3,7-trimethyl-	161–163°	234, 361, 716

TABLE 27 (*continued*)

Purine	M.p.	References
Xanthine/7-butyl-1,3-diethyl-8-hydroxymethyl-	94–95°	808
Xanthine/1-butyl-3,7-dimethyl-	119°	549
Xanthine/3-butyl-1,7-dimethyl-	90–92°	349
Xanthine/3-s-butyl-1,7-dimethyl-	98–99°	349
Xanthine/8-butyl-1,3-dimethyl-	233–235°	233, 389
Xanthine/8-t-butyl-1,3-dimethyl-	277–278°	722
Xanthine/9-butyl-1,3-dimethyl-	170–171°	46
Xanthine/3-butyl-1-ethyl-	166°	232
Xanthine/7-butyl-8-hydroxymethyl-1,3-dimethyl-	132–133°	808
Xanthine/1-butyl-3-methyl-	207–210°	527
Xanthine/7-butyl-8-propyl-	212–213°	345
Xanthine/8-butyl-1,3,7-trimethyl-	123–125°	386
Xanthine/8-crotyl-1,3-dimethyl-	233–234°	233
Xanthine/8-crotyl-1,3,7-trimethyl-	173–174°	233
Xanthine/9-cyclohexyl-1,3-dimethyl-	316–319°	729
Xanthine/1,3-diallyl-7-(2-hydroxyethyl)-	122–124°	512
Xanthine/1,3-diallyl-7-methyl-	277–278°	477
Xanthine/1,3-dibenzyl-	225°	14
Xanthine/1,3-dibenzyl-8-hydroxymethyl-	210°	75
Xanthine/1,7-dibenzyl-3-methyl-	105–107°	539
Xanthine/1,3-dibutyl-	192–193°	91, 232
Xanthine/3,7-dibutyl-	127°	459
Xanthine/1,3-dibutyl-8-ethyl-	124°	232
Xanthine/1,3-dibutyl-7-(2-hydroxyethyl)-	89–91°	512
Xanthine/1,3-dibutyl-8-methyl-	158–159°	232
Xanthine/1,3-dibutyl-8-phenyl-	264–266°	91
Xanthine/1,3-diethyl-	208 to 220°	163, 193, 232
Xanthine/1,7-diethyl-	191–192° or 224–284°	5, 190
Xanthine/3,7-diethyl-	183°	295
Xanthine/1,3-diethyl-7-(4-hydroxybutyl)-	195–198/0.35 mm, hy. 74°	512
Xanthine/1,3-diethyl-8-hydroxymethyl-	227–230°	808
Xanthine/1,3-diethyl-8-hydroxymethyl-7-methyl-	130–132°	808
Xanthine/1,3-diethyl-7-(3-hydroxypropyl)-	86–88°	512
Xanthine/1,3-diethyl-8-isobutyl-	152–153°	232
Xanthine/1,3-diethyl-8-isopropyl-	156–157°	232
Xanthine/1,3-diethyl-8-methyl-	231–232°	232
Xanthine/1,7-diethyl-3-methyl-	127–128°	772
Xanthine/1,7-diethyl-8-methyl-	235–236°	5
Xanthine/1,3-diethyl-8-propyl-	170–171°	232

continued

TABLE 27 (*continued*)

Purine	M.p.	References
Xanthine/1,3-dimethyl- (theophylline)	269–274°	13, 91, 195, 197–199, 216, 217, 270, 302
Xanthine/1,7-dimethyl- (paraxanthine)	294–299°	5, 349, 361, 417, 421, 506
Xanthine/1,8-dimethyl-	—	359, 360
Xanthine/1,9-dimethyl-	351–352°	280, 319, 330, 422, 442, 657
Xanthine/3,7-dimethyl- (theobromine)	351°	13, 199, 253, 270, 348, 485
Xanthine/3,8-dimethyl-	338 to 355°	181, 216, 250, 302, 562, 775
Xanthine/3,9-dimethyl-	> 300°	39, 539, 657
Xanthine/7,8-dimethyl-	—	714
Xanthine/7,9-dimethyl-	bet. 350–370°; tos. 272–273°	25, 39, 262, 532, 553, 682
Xanthine/8,9-dimethyl-	354–355°	262, 384
Xanthine/1,3-dimethyl-8-pentyl-	216–217°	233
Xanthine/1,3-dimethyl-7-phenyl-	193–195°	563
Xanthine/1,3-dimethyl-8-phenyl-	> 360°	91, 234, 298, 299, 301, 520, 822
Xanthine/1,7-dimethyl-3-phenyl-	305–310°	659
Xanthine/1,7-dimethyl-8-phenyl-	326°	361
Xanthine/1,9-dimethyl-8-phenyl-	> 320°	822
Xanthine/1,3-dimethyl-7-(prop-1-enyl)-	103–105°	450
Xanthine/1,3-dimethyl-7-(prop-2-enyl)-	212°	538, 787
Xanthine/3,7-dimethyl-1-(prop-2-enyl)-	208°	787
Xanthine/1,3-dimethyl-7-propyl-	99–100°	764
Xanthine/1,3-dimethyl-8-propyl-	256°	232
Xanthine/1,7-dimethyl-3-propyl-	105–107°	349
Xanthine/3,7-dimethyl-1-propyl-	136°	549, 763
Xanthine/1,3-dimethyl-7-vinyl-	173–174°	463
Xanthine/1,8-diphenyl-	> 360°	91
Xanthine/3-ethyl-	> 275°	190, 295
Xanthine/7-ethyl-	—	190
Xanthine/9-ethyl-	345° or 364–367°	57, 92, 319
Xanthine/1-ethyl-3,7-dimethyl-	164–165°	535, 537, 549, 763
Xanthine/3-ethyl-1,7-dimethyl-	128–130°	295, 349
Xanthine/7-ethyl-1,3-dimethyl-	154°	535–537, 764
Xanthine/8-ethyl-1,3-dimethyl-	277°	91, 232
Xanthine/9-ethyl-1,3-dimethyl-	233–235°	46, 319
Xanthine/9-ethyl-3,7-dimethyl-	bet. 252°; tos. 227°	553, 682

TABLE 27 (*continued*)

Purine	M.p.	References
Xanthine/1-ethyl-7-(2-hydroxyethyl)-3-isopropyl-	76–80°	512
Xanthine/1-ethyl-7-(2-hydroxyethyl)-3-propyl-	120–122°	512
Xanthine/3-ethyl-8-hydroxymethyl-	233–235°	808
Xanthine/7-ethyl-8-hydroxymethyl-1,3-dimethyl-	199–201°	808
Xanthine/1-ethyl-7-methyl-	238–239°	5
Xanthine/7-ethyl-1-methyl-	225–226°	5
Xanthine/7-ethyl-3-methyl-	282–283°	772
Xanthine/7-ethyl-8-methyl-	—	351, 357
Xanthine/9-ethyl-1-methyl-	335°	319
Xanthine/7-ethyl-8-propyl-	265–266°	345
Xanthine/8-ethyl-7-propyl-	224–225°	345
Xanthine/8-ethyl-1,3,7-trimethyl-	186–191°	216, 251, 334, 381, 670, 777
Xanthine/9-ethyl-1,3,7-trimethyl-	perchlor. 181°	553, 682
Xanthine/1-hexyl-3,7-dimethyl-	82–83°	546
Xanthine/3-hexyl-1,7-dimethyl-	101–102°	349
Xanthine/8-hexyl-1,3-dimethyl-	268°	722
Xanthine/7-(4-hydroxybutyl)1,3-dimethyl-	117–118°	211
Xanthine/9-(2-hydroxyethyl)-	308°	92
Xanthine/7-(2-hydroxyethyl)-1,3-diisobutyl-	108–110°	512
Xanthine/1-(2-hydroxyethyl)-3,7-dimethyl-	195°	514
Xanthine/7-(2-hydroxyethyl)-1,3-dimethyl-	157–160°	156, 514, 716
Xanthine/8-(2-hydroxyethyl)-3,7-dimethyl-	275–277°	449
Xanthine/1-hydroxymethyl-3,7-dimethyl-	—	571
Xanthine/7-hydroxymethyl-1,3-dimethyl-	—	571
Xanthine/8-hydroxymethyl-1,3-dimethyl-	243–244°	91, 808
Xanthine/8-hydroxymethyl-1,7-dimethyl-	—	724
Xanthine/8-hydroxymethyl-1,9-dimethyl-	310–312°	712
Xanthine/8-hydroxymethyl-3,7-dimethyl-	297°	75, 446
Xanthine/8-hydroxymethyl-3-methyl-	276–277°	75, 808
Xanthine/8-hydroxymethyl-7-methyl-	—	714
Xanthine/8-hydroxymethyl-9-methyl-	300°	712
Xanthine/8-hydroxymethyl-1,3,7-trimethyl-	223–226°	75, 444, 448, 501, 509, 669, 671, 808
Xanthine/8-hydroxymethyl-1,3,9-trimethyl-	253–255°	448, 699
Xanthine/7-(5-hydroxypentyl)-1,3-dimethyl-	121–122°	211
Xanthine/1-(3-hydroxypropyl)-3,7-dimethyl-	140–142°	517
Xanthine/7-(3-hydroxypropyl)-1,3-dimethyl-	151–152°	540
Xanthine/7-(3-hydroxypropyl)-1,3-dipropyl-	80–82°	512
Xanthine/8-(3-hydroxypropyl)-1,3,7-trimethyl-	88–91°	381, 670

continued

TABLE 27 (*continued*)

Purine	M.p.	References
Xanthine/3-isobutyl-	299–301°	527
Xanthine/9-isobutyl-	>300°	57
Xanthine/1-isobutyl-3,7-dimethyl-	129–130°	763
Xanthine/3-isobutyl-1,7-dimethyl-	97–99°	349
Xanthine/8-isobutyl-1,3-dimethyl-	227–234°	91, 232, 298
Xanthine/3-isobutyl-1-ethyl-	195–197°	527
Xanthine/7-isobutyl-8-isopropyl-	246–247°	345
Xanthine/3-isobutyl-7-methyl-	239–241°	527
Xanthine/8-isobutyl-3-methyl-	—	298
Xanthine/3-isobutyl-1-propyl-	189–192°	527
Xanthine/3-isopentyl-1,7-dimethyl-	110–112°	349
Xanthine/9-isopropyl-	325°	46
Xanthine/3-isopropyl-1,7-dimethyl-	185–186°	349
Xanthine/7-isopropyl-1,3-dimethyl-	140°	764
Xanthine/8-isopropyl-1,3-dimethyl-	271° or 281°	91, 231
Xanthine/9-isopropyl-1,3-dimethyl-	304–306°	729
Xanthine/8-isopropyl-1,3,7-trimethyl-	143–144°	119, 334
Xanthine/1-methyl-	>350°	101, 192, 271, 506, 657
Xanthine/3-methyl-	>350°	13, 44, 101, 217, 298, 302, 419
Xanthine/7-methyl- (heteroxanthine)	>360°	31, 253, 258, 335, 355, 417
Xanthine/9-methyl-	>350°	57, 92, 319, 336, 337, 443
Xanthine/1-methyl-3-phenyl-	>310°	659
Xanthine/1-methyl-8-phenyl-	>360°	91, 360
Xanthine/3-methyl-8-phenyl-	>360°	91
Xanthine/3-methyl-9-phenyl-	—	181
Xanthine/7-methyl-8-phenyl-	340°	359, 361
Xanthine/8-methyl-3-phenyl-	300°	679
Xanthine/1-methyl-7-propyl-	204–205°	5
Xanthine/3-phenyl-	300°	679
Xanthine/8-phenyl-	*ca.* 360°	361
Xanthine/9-phenyl-	>300°	57, 507
Xanthine/8-propyl-1,3-dimethyl-	264°	91
Xanthine/1,3,7,8-tetramethyl-	210–212°	79, 119, 215, 216, 244, 246, 251, 252, 270, 334, 363, 395, 520, 534, 714
Xanthine/1,3,7,9-tetramethyl-	perchlor. 198°: iod. 137°	248, 553, 682
Xanthine/1,3,8,9-tetramethyl-	254 to 263°	79, 216, 244, 262, 350, 561
Xanthine/1,3,7-tributyl-	41–42°	459

TABLE 27 (*continued*)

Purine	M.p.	References
Xanthine/1,3,7-triethyl-	115°	163, 459, 535
Xanthine/1,3,8-triethyl-	193°	232
Xanthine/3,7,8-triethyl-	210–212°	232, 777
Xanthine/1,3,7-triethyl-8-hydroxymethyl-	144–145°	808
Xanthine/1,3,7-triethyl-8-methyl-	132–133°	777
Xanthine/1,3,7-trimethyl- (caffeine)	234–236°	91, 119, 195, 196, 199, 202, 217, 270, 348, 361, 536
Xanthine/1,3,8-trimethyl-	325–330°	91, 181, 216, 232, 244, 300, 301, 334, 520, 699, 775
Xanthine/1,3,9-trimethyl-	285–287°	46, 216, 269, 319, 337, 657
Xanthine/1,7,8-trimethyl-	344°	363
Xanthine/1,7,9-trimethyl-	bet. 330–340°; tos. 235°	262, 553, 682
Xanthine/1,8,9-trimethyl-	336–338°	384
Xanthine/3,7,8-trimethyl-	—	252, 534
Xanthine/3,7,9-trimethyl-	bet. 245 to 270°; tos. 257°	262, 553, 682
Xanthine/1,3,7-trimethyl-8-pentyl-	78–79°	386
Xanthine/1,3,7-trimethyl-8-phenyl-	185–187°	235, 334, 361
Xanthine/1,3,7-trimethyl-8-propyl-	116–117°	381, 386, 670
Xanthine/1,3,8-triphenyl-	> 360°	91

TABLE 28. Sulphonylpurines.

Purine	M.p.	References
6-(N-Allylsulphamoyl)-	222°	114
6-(N-Benzylsulphamoyl)-	200°	114
2,6-bis-Methylsulphonyl-	242 or 258°	284, 591
6-(N-Butylsulphamoyl)-	217°	114
6-Butylsulphonyl-	159°	284
2,6-Difluorosulphonyl-	182°	127
2,9-Dimethyl-6-methylsulphonyl-	202–204°	785
2,9-Dimethyl-8-methylsulphonyl-	139–140°	785
2-(NN-Dimethylsulphamoyl)-	271°	127
6-(NN-Dimethylsulphamoyl)-	179–181°	127
6-(N-Ethylsulphamoyl)-	227°	127

continued

TABLE 28 (*continued*)

Purine	M.p.	References
6-Ethylsulphonyl-	186°	284
2-Fluorosulphonyl-	178–179°	127
6-Fluorosulphonyl-	—	127
6-(*N*-Isobutylsulphamoyl)-	224°	114
9-Methyl-2-methylsulphinyl-	186–187°	785
9-Methyl-6-methylsulphinyl-	168–170°	785
9-Methyl-8-methylsulphinyl-	144–145°	785
9-Methyl-2-methylsulphonyl-	167–168°	785
9-Methyl-6-methylsulphonyl-	210–212°	785
9-Methyl-8-methylsulphonyl-	133–135°	785
6-(*N*-Methylsulphamoyl)-	225°	127
2-Methylsulphonyl-	226°	284
6-Methylsulphonyl-	205–208°	284, 591
8-Methylsulphonyl-	*ca.* 300°	166
6-(*N*-Propylsulphamoyl)-	226°	114
6-Propylsulphonyl-	175°	284
2-Sulphamoyl-	>320°	127
6-Sulphamoyl-	258°	127
2-Sulphamoyl-6-sulpho-	—	127
6-Sulphino-	sod. >175°	376
6-Sulpho-	pot. >300°	376
2,6,8-Tris-methylsulphonyl-	153°	284

TABLE 29. Thiopurines without *N*-Alkyl Groups.

Purine	M.p.	References
6-Acetylthiomethyl-	178–180°	42
6-Allylthio-	176°	93, 492
6-Allylthio-2,8-dimethyl-	189°	228
6-Allylthio-8-ethyl-2-methyl-	125°	231
6-Allylthio-2-methyl-	212°	228
2-Benzylthio-	202°	500
6-Benzylthio-	193°	205, 465, 467, 492
8-Benzylthio-	204°	500
6-Benzylthiomethyl-	94–96°	42
2,6-bis-Benzylthio-	194–196°	40, 532, 574
2,6-bis-Methylthio-	254 to 261°	284, 516, 574
2,8-bis-Methylthio-	>300° or 223°	209, 284
6,8-bis-Methylthio-	254–256°	58
2,8-bis-Methylthio-1,6-dihydro-6-thio-	>300°	45
7-Butyl-1,6-dihydro-8-propyl-6-thio-	262–264°	345
6-Butylthio-	150–151°	63, 93, 492

TABLE 29 (*continued*)

Purine	M.p.	References
6-*s*-Butylthio-	197–198°	93
6-Butylthio-2,8-dimethyl-	172°	228, 663
6-Butylthio-8-ethyl-2-methyl-	102°	231, 663
6-Butylthio-8-methyl-	179–180°	229, 242, 663
6-Cyclopentylthio-	228°	492
7,8-Dihydro-2,6-bis-methylthio-8-thio-	295°	284
1,6-Dihydro-2,8-dimethyl-6-thio-	312°	228, 243, 245, 663
1,6-Dihydro-8-hydroxyethyl-6-thio-	296–298°	240
1,6-Dihydro-8-hydroxymethyl-6-thio-	>270°	240
1,6-Dihydro-8-methyl-2-methyl-thio-6-thio-	>300°	242
1,6-Dihydro-2-methyl-8-phenyl-6-thio-	hy. 328°	256
2,3-Hydro-6-methyl-8-phenyl-2-thio-	246–247°	256
1,6-Dihydro-2-methyl-6-thio-	312–315°	228, 245, 408
1,6-Dihydro-8-methyl-6-thio-	312–315°	58, 63, 229, 242, 273, 663
2,3-Dihydro-6-methyl-2-thio-	>320°	378
7,8-Dihydro-6-methyl-8-thio-	>350°	174, 378
7,8-Dihydro-9-methyl-8-thio-	314°	166
1,6-Dihydro-2-methylthio-6-thio-	—	410
7,8-Dihydro-6-methylthio-8-thio-	>300°	58
1,6-Dihydro-2-phenyl-6-thio-	>300°	18
1,6-Dihydro-8-phenyl-6-thio-	330°	66, 106, 123
1,6-Dihydro-6-thio- (6-mercaptopurine)	313–315°	35, 179, 191, 432, 453, 593–597
2,3-Dihydro-2-thio-	>300°	6, 178
7,8-Dihydro-8-thio-	295–297° or 317°	6, 159, 203, 287, 378
2,8-Dimethyl-6-methylthio-	245°	228, 663
2,8-Dimethyl-6-propylthio-	168°	228, 663
8-Ethyl-1,6-dihydro-2-methyl-6-thio-	270°	231, 663
2-Ethyl-1,6-dihydro-6-thio-	>260°	170
6-Ethyl-7,8-dihydro-8-thio-	>300°	577
8-Ethyl-1,6-dihydro-6-thio-	260°	64
8-Ethyl-6-ethylthio-2-methyl-	131°	231, 663
8-Ethyl-2-methyl-6-methylthio-	235°	231, 663
8-Ethyl-2-methyl-6-propylthio-	134°	231, 663
2-Ethyl-6-propylthio-	202–203°	170
6-Ethylthio-	196 to 204°	63, 93, 492
6-Ethylthio-2,8-dimethyl-	221°	228, 663
6-Ethylthio-2-methyl-	221°	228, 663
6-Ethylthio-8-methyl-	206–212°	229, 242, 663
6-Ethylthio-8-methylthio-	175–177°	58
6-Ethylthiomethyl-	154°	42

continued

TABLE 29 (*continued*)

Purine	M.p.	References
1,2,3,6,7,8-Hexhydro-2,6,8-trithio-	> 360°	207, 284, 432, 458, 774
6-Hexylthio-	95–96°	93
6-Isobutylthio-	202–203°	93
6-Isopentylthio-	124–125° or 155–156°	93, 114
6-Isopropylthio-	243–244°	93, 492
6-Isopropylthio-2,8-dimethyl-	191°	228, 663
6-Mercaptomethyl-	146–147°	42
2-Methyl-6-methylthio-	249–250°	228, 408
6-Methyl-2-methylthio-	290–294°	243
8-Methyl-6-methylthio-	223–224°	229, 242, 663
2-Methyl-8-propylthio-	197°	663
8-Methyl-6-propylthio-	213°	229, 663
6-Methylseleno-	193–194°	493
2-Methylthio-	250–255°	6
6-Methylthio-	hy. 218–220°	179, 432, 584, 600
8-Methylthio-	257–259°	6
6-Methylthio-2-phenyl-	257°	18
6-Methylthio-8-phenyl-	273–274°	23
6-Methylthiomethyl-	175–176°	42
6-Pentylthio-	107–115°	93, 492
6-Phenylthio-	244–245°	63
6-Phenylthiomethyl-	150°	42
6-Propylthio-	184–185°	93, 492
6-Propylthio-2-methyl-	197°	228
6-Propylthio-8-methyl-	214–215°	242
6-Seleno-	280–282°	493
1,2,3,6-Tetrahydro-2,6-dithio-	> 350°	40, 370, 432, 516
1,6,7,8-Tetrahydro-6,8-dithio-	> 300°	58, 63, 86, 207, 240, 521
2,3,7,8-Tetrahydro-2,8-dithio-	—	284
1,2,3,6-Tetrahydro-8-methyl-2,6-dithio-	> 300°	242, 245
1,6,7,8-Tetrahydro-2-methyl-6,8-dithio-	—	245, 410
1,6,7,8-Tetrahydro-2-methylthio-6,8-dithio-	—	410
1,2,3,6-Tetrahydro-8-phenyl-2,6-dithio-	270°	66
1,6,7,8-Tetrahydro-2-phenyl-6,8-dithio-	> 300°	18
2,6,8-Tris-methylthio-	284°	284

TABLE 30. Thiopurines with *N*-Alkyl Groups.

Purine	M.p.	References
3-Benzyl-6-benzylthio-	147–148°	124
7-Benzyl-6-benzylthio-	118–120°	200, 492
9-Benzyl-6-benzylthio-	108°	492
7-Benzyl-1,6-dihydro-1-methyl-6-thio-	171°	180
3-Benzyl-3,6-dihydro-7-methyl-6-thio-	220–225°	37
1-Benzyl-1,6-dihydro-6-thio-	269–271°	124
3-Benzyl-3,6-dihydro-6-thio-	—	124
7-Benzyl-1,6-dihydro-6-thio-	265–266°	400
9-Benzyl-1,6-dihydro-6-thio-	>260°	400
9-Benzyl-7,8-dihydro-8-thio-	234–235°	7
1-Benzyl-6-methylthio-	169–171°	212
3-Benzyl-6-methylthio-	156–157°	113
7-Benzyl-6-methylthio-	121° or 184–185°	113, 200
9-Benzyl-6-methylthio-	117–118°	113
9-Benzyl-8-methylthio-	126–127°	7
2-Benzylthio-1,6-dihydro-1-methyl-6-thio-	257–258°	101
6-Benzylthio-1,6-dihydro-1-methyl-	144–145°	180
8-Benzylthio-9-ethyl-	*ca.* 32°	500
6-Benzylthio-7-methyl-	155–157°	503
6-Benzylthio-9-methyl-	121–123°	503
7-Butyl-1,6-dihydro-1-methyl-6-thio-	124–125°	180
7-Butyl-1,6-dihydro-6-thio-	220°	200
9-Butyl-1,6-dihydro-6-thio-	311–312°	265, 684
6-Butylthio-7-methyl-	118–119°	503
9-(Cyclohex-2-enyl)-1,6-dihydro-6-thio-	291–294°	266
9-Cyclohexyl-1,6-dihydro-6-thio-	320–324°	265, 684
9-Cyclopentyl-1,6-dihydro-6-thio-	310–312°	265, 684
9-Cyclopropyl-1,6-dihydro-6-thio-	345°	263
3,7-Dibenzyl-6-benzylthio-	brom. >110°	113
1,7-Dibenzyl-1,6-dihydro-6-thio-	155°	100, 124
1,9-Dibenzyl-1,6-dihydro-6-thio-	163–165°	27
3,7-Dibenzyl-3,6-dihydro-6-thio-	181–182°	113
3,7-Dibenzyl-6-methylthio-	iod. 80°	113
7,9-Dibenzyl-6-thio-	bet. 205–209°	113
2,9-Diethyl-1,6-dihydro-6-thio-	211–213°	170
2,9-Diethyl-6-propylthio-	140°/0.1 mm	170
1,6-Dihydro-1,9-dimethyl-6-thio-	250°	11
1,6-Dihydro-7,9-dimethyl-2-methylthio-6-thio-	bet. 265°	532
1,6-Dihydro-7,9-dimethyl-6-thio-	bet. 283°	532
1,6-Dihydro-8,9-dimethyl-6-thio-	250°	353
2,3-Dihydro-6,9-dimethyl-2-thio-	285°	7
2,3-Dihydro-8,9-dimethyl-2-thio-	274°	785
3,6-Dihydro-2,3-dimethyl-6-thio-	295–297°	11

continued

TABLE 30 (*continued*)

Purine	M.p.	References
3,6-Dihydro-3,7-dimethyl-6-thio-	282–283°	11
3,6-Dihydro-3,8-dimethyl-6-thio-	300°	106
3,6-Dihydro-3,9-dimethyl-6-thio-	276–278°	11
7,8-Dihydro-2,9-dimethyl-8-thio-	> 300°	7
7,8-Dihydro-7,9-dimethyl-8-thio-	162–163°	274
1,6-Dihydro-7-isobutyl-8-isopropyl-6-thio-	244–245°	345
1,6-Dihydro-9-isopropyl-6-thio-	335–338°	263
1,6-Dihydro-1-(2-mercaptoethyl)-6-thio-	188–190°	573
3,6-Dihydro-3-methyl-8-phenyl-6-thio-	280–285°	23, 106
7,8-Dihydro-7-methyl-9-phenyl-8-thio-	221–222°	274
1,6-Dihydro-1-methyl-6-thio-	288–289°	101, 180
1,6-Dihydro-7-methyl-6-thio-	306–308°	101, 347, 458
1,6-Dihydro-9-methyl-6-thio-	338–342°	101, 206, 353, 658, 684
2,3-Dihydro-7-methyl-2-thio-	295°	367
2,3-Dihydro-9-methyl-2-thio-	220°	785
3,6-Dihydro-3-methyl-6-thio-	322–323°	101, 522
7,8-Dihydro-9-methyl-8-thio-	314°	166
8,9-Dihydro-7-methyl-8-thio-	248–249°	458
1,6-Dihydro-9-pentyl-6-thio-	309–312°	263
1,6-Dihydro-9-phenyl-6-thio-	> 300°	462
8,9-Dihydro-7-phenyl-8-thio-	250–251°	274
1,6-Dihydro-9-propyl-6-thio-	321–323°	263
3,6-Dihydro-2,3,7-trimethyl-6-thio-	298–299°	11
3,6-Dihydro-3,7,8-trimethyl-6-thio-	175°	11
1,9-Dimethyl-6-methylthio-	iod. 217–219°	11
2,3-Dimethyl-6-methylthio-	192–194°	11
2,9-Dimethyl-6-methylthio-	106–107°	7
2,9-Dimethyl-8-methylthio-	136–137°	7
3,7-Dimethyl-6-methylthio-	iod. 194°	11
3,8-Dimethyl-6-methylthio-	195°	106
3,9-Dimethyl-6-methylthio-	iod. 222°	11
6,7-Dimethyl-2-methylthio-	174–175°	7
6,9-Dimethyl-2-methylthio-	128°	7
8,9-Dimethyl-2-methylthio-	144°	785
7-Ethyl-6-benzylthio-	99°	200
7-Ethyl-7,8-dihydro-9-methyl-8-thio-	94–95°	274
7-Ethyl-7,8-dihydro-9-phenyl-8-thio-	172–173°	274
9-Ethyl-1,6-dihydro-6-thio-	333–337°	60
9-Ethyl-7,8-dihydro-8-thio-	264°	287
9-Ethyl-6-methylthio-	116–118°	60
6-Ethylthio-7-methyl-	158°	347
9-Furfuryl-1,6-dihydro-6-thio-	315°	726
1,2,3,6,7,8-Hexahydro-1,3-dimethyl-2,6,8-trithio-	304–310°	88
1,2,3,6,8,9-Hexahydro-7-methyl-2,6,8-trithio-	> 320°	458, 774

TABLE 30 (*continued*)

Purine	M.p.	References
9-Hexyl-1,6-dihydro-6-thio-	284–287° or 304–306°	263, 684
7-Methyl-2,6-bis-methylthio-	176–178°	347
1-Methyl-6-methylthio-	pic. 206–208°	161
3-Methyl-6-methylthio-	163–165°; tos. 270°	25, 522
3-Methyl-8-methylthio-	166–167°	785
7-Methyl-6-methylthio-	212–213°	458
9-Methyl-2-methylthio-	131–132°	785
9-Methyl-6-methylthio-	169–170°	166, 206, 658, 785
9-Methyl-8-methylthio-	151°	166, 785
3-Methyl-6-methylthio-8-phenyl-	196–198°	106
7-Methyl-8-methylthio-6-phenyl-	—	355
7-(3-Methylbut-2-enyl)-6-methylthio-	141–142°	34
6-Methylthio-9-phenyl-	148–149°	462
8-Methylthio-7-phenyl-	201–202°	274
1,2,3,6-Tetrahydro-1,3-dimethyl-2,6-dithio-	267–269°	214
1,2,3,6-Tetrahydro-1,9-dimethyl-2-thio-	289–290°	403
1,2,3,6-Tetrahydro-3,7-dimethyl-2,6-dithio-	290–292°	530
1,2,3,6-Tetrahydro-7,9-dimethyl-2,6-dithio-	bet. 283°	532
1,2,3,6-Tetrahydro-1-methyl-2,6-dithio-	292–294°	101
1,2,3,6-Tetrahydro-3-methyl-2,6-dithio-	>300°	101
1,2,3,6-Tetrahydro-7-methyl-2,6-dithio-	>300°	347, 458, 774
3,6,7,8-Tetrahydro-3-methyl-6,8-dithio-	>300°	72
1,2,3,6-Tetrahydro-1-methyl-2-methyl-thio-6-thio-	310–315°	101
1,2,3,6-Tetrahydro-9-phenyl-2,6-dithio-	>300°	57
1,2,3,6-Tetrahydro-1,3,7-trimethyl-2,6-dithio-	227–228° or 380°	529, 651, 774
1,2,3,6-Tetrahydro-1,7,9-trimethyl-2,6-dithio-	bet. 255°	532
3,7,8-Trimethyl-6-methylthio-	iod. 213–214°	11

TABLE 31. Amino-Carboxypurines.

Purine	M.p.	References
6-Acetamido-9-acetyl-	195°	613, 614
6-Amino-1-benzyl-9-(2-cyanoethyl)-	brom. 245–247°	647
6-Amino-9-(2-carbamoylethyl)-	275–281°	20
6-Amino-9-carbamoylmethyl-	hy. >260°	400

continued

TABLE 31 (*continued*)

Purine	M.p.	References
6-Amino-8-carboxy-	340–342°	790
6-Amino-9-carboxy-	184–185°	431
6-Amino-9-(2-carboxyethyl)-	284–288°	20, 126, 634
6-(2-Amino-2-carboxyethyl)-9-methyl-	235–239°	116
6-Amino-3-carboxymethyl-	330–333°	128
6-Amino-9-carboxymethyl-	350°	128, 798
6-Amino-9-(3-carboxypropyl)-	240–242°	798
2-Amino-6-cyano-	310°	674
6-Amino-9-(2-cyanoethyl)-	247–250° or 258–261°	20, 126
6-Amino-9-(2-cyanoethyl)-1-methyl-	iod. 256–258°	647
6-Amino-9-(3-cyanopropyl)-	196°	798
2-Amino-6-ethoxycarbonyl-	267°	185
6-Amino-9-(2-ethoxycarbonylethyl)-	167–168°	20
6-Amino-3-ethoxycarbonylmethyl-	237–240°	128
6-Amino-9-ethoxycarbonylmethyl-	228–231°	128
6-Amino-9-ethylthiocarbonyl-	> 250°	431
2-Amino-6-formyl-	hydrazone 258–260°	165
6-Amino-9-methoxycarbonylmethyl-	241–242°	798
6-Amino-9-(3-methoxycarbonylpropyl)-	140°	798
2-Amino-6-thiocarbamoyl-	300°	674
2-Amino-6-thiocyanato-	> 240°	164
6-Amino-8-thiocyanato-	> 240°	164
6-Carbamoylmethyl-2-dimethylamino-9-methyl-	198–201°	654
8-Cyano-6-diethylamino-2-methyl-	302°	238
9-(2-Cyanoethyl)-6-diethylamino-	HCl 173–176°	126
2,6-Diacetamido-9-acetyl-	243–245°	617
2-Dimethylamino-6-ethoxycarbonyl-	216°	185
2-Dimethylamino-6-ethoxycarbonyl-8-methyl-	158°	185
6-Dimethylamino-9-ethoxycarbonyl-	91–92°	683
2-Dimethylamino-6-ethoxycarbonyl-methyl-9-methyl-	HCl 197–198°	654
6-Diethylamino-8-formyl-2-methyl-	210–211°	238
6-Hydrazino-9-hydrazinocarbonyl-methyl-	> 260°	400
6-Propionamido-9-propionyl-	180–182° or 145–148°	613, 614

TABLE 32. Amino-Carboxythiopurines.

Purine	M.p.	References
6-Acetonylthio-2-amino-	198–199°	63
8-Acetonylthio-2,6-diamino-	HCl 204–205°	578
2-Amino-6-benzoylmethylthio-	208–209°	62
2-Amino-6-carbamoylmethylthio-	285°	62
2-Amino-6-(4-carboxybutylthio)-	220–221°	681
2-Amino-6-carboxyethylthio-	250°	62
2-Amino-6-carboxymethylthio-	> 300°	62, 63
6-Amino-2-carboxymethylthio-	—	184
2-Amino-6-carboxymethylthio-9-butyl-	199°	8
2-Amino-6-carboxymethylthio-9-isobutyl-	197°	8
2-Amino-6-carboxymethylthio-9-pentyl-	209°	8
2-Amino-6-carboxymethylthio-9-propyl-	200°	8
2-Amino-6-cyanomethylthio-	265°	62
2-Amino-9-ethoxycarbonyl-6-ethoxycarbonylthio-	158–161°	683
2-Amino-6-(4-ethoxycarbonylbutylthio)-	113–115°	681
2,6-Diamino-8-carboxymethylthio-	> 300°	578
2,6-Diamino-8-ethoxycarbonylmethylthio-	HCl 222–224°	578

TABLE 33. Amino-Halogenopurines.

Purine	M.p.	References
2-Acetamido-6-chloro-	> 300°	471
2-Amino-7-benzyl-6-chloro-	> 260°	470
2-Amino-9-benzyl-6-chloro-	212°	470
6-Amino-9-benzyl-2-fluoro-	272°	427
2-Amino-9-benzyl-6-trifluoromethyl-	183–184°	208
6-Amino-9-benzyl-2-trifluoromethyl-	181–182°	208
2-Amino-6,8-bis-trifluoromethyl-	230°	56
2-Amino-6-bromo-	—	207
6-Amino-8-bromo-	—	207
2-Amino-6-bromo-8-phenyl-	—	255
2-Amino-6-chloro-	275°	62, 425
6-Amino-2-chloro-	—	111, 456, 472
6-Amino-8-chloro-	—	58, 210, 426
9-Amino-6-chloro-	173–176°	172
2-Amino-6-chloro-7,9-dimethyl-	chlor. 293°	532
2-Amino-6-chloro-3-methyl-	280°	44
2-Amino-6-chloro-9-methyl-	262–263°	658
6-Amino-2-chloro-7-methyl-	284°	112, 257, 415
6-Amino-2-chloro-9-methyl-	—	473
2-Amino-6-chloro-8-phenyl-	> 350°	236

continued

TABLE 33 (*continued*)

Purine	M.p.	References
6-Amino-9-(2-chloroethyl)-	203–205°; HCl 223°	16, 219
6-Amino-9-(3-chloropropyl)-	HCl 300°	20
6-Amino-2,8-dichloro-	>300°	84, 103
8-Amino-2,6-dichloro-	>300°	452
6-Amino-2,8-dichloro-7-methyl-	>270°	415
6-Amino-2,8-dichloro-9-methyl-	253°	103, 405
8-Amino-2,6-dichloro-7-methyl-	—	455
8-Amino-2,6-dichloro-9-methyl-	260 or 270°	405, 415
2-Amino-9-ethyl-6-trifluoromethyl-	162–164°	208
6-Amino-9-ethyl-2-trifluoromethyl-	191–192°	208
6-Amino-9-methyl-2-trifluoromethyl-	232–235°	208
6-Amino-2-fluoro-	>350°	427
2-Amino-6-iodo-	>245°	86
2-Amino-8-iodo-	sul. >160°	86
6-Amino-8-iodo-	HCl >250°	86
2-Amino-9-methyl-6-trifluoromethyl-	243–244°	208
2-Amino-6-trichloromethyl-	—	118
6-Amino-9-(2,2,2-trifluoroethyl)-	240°	634
2-Amino-6-trifluoromethyl-	>300°	56, 65
2-Amino-8-trifluoromethyl-	300°	209
6-Amino-2-trifluoromethyl-	360–362°	56
6-Amino-8-trifluoromethyl-	>300°	56
9-(2-Aminoethyl)-6-chloro-	270°	798
6-Aminomethyl-2-chloro-9-methyl-	HCl >300°	83
6-Anilino-9-phenyl-2-trifluoromethyl-	289–290°	208
6-Aziridinyl-9-benzyl-2-chloro-	128°	470
6-Aziridinyl-9-benzyl-8-chloro-	307°	470
6-Aziridinyl-2-chloro-7-methyl-	360°	258
6-Aziridinyl-2-chloro-9-methyl-	130–132°	658
9-Benzyl-6-benzylamino-2-trifluoro- methyl-	271–272°	208
9-Benzyl-6-dimethylamino-2-fluoro-	137°	427
2-Benzylamino-6-chloro-9-methyl-	142–143°	658
6-Benzylamino-2,8-dichloro-	220–221°	468
6-Benzylamino-2-fluoro-	222–224°	434
9-Butyl-6-butylamino-2-trifluoromethyl-	255°	208
2-Chloro-6-butylamino-	>300°	61
2-Chloro-6-diethylamino-	224–226°	413
2-Chloro-6-diethylamino-7-methyl-	107°	257
6-Chloro-2-diethylamino-9-methyl-	86–88°	658
8-Chloro-6-diethylamino-2-methyl-	HCl 163–165°	410
6-Chloro-9-(2-diethylaminoethyl)-	liq.	404
2-Chloro-6-diethylaminomethyl-9-methyl-	102–105°	83
2-Chloro-6,8-dihydrazino-7-methyl-	198°	112
2-Chloro-6-dimethylamino-	240–280°	61
8-Chloro-6-dimethylamino-	264–266°	58

TABLE 33 (*continued*)

Purine	M.p.	References
2-Chloro-6-dimethylamino-9-methyl-	140–142°	802
6-Chloro-2-dimethylamino-9-methyl-	150–152°	658
6-Chloro-9-dimethylamino-8-methyl-	133–134°	19
8-Chloro-6-dimethylamino-3-methyl-	223–225°	802
8-Chloro-6-dimethylamino-9-methyl-	89–92°	802
2-Chloro-6-dimethylamino-9-phenyl-	168–169°	57
8-Chloro-6-dipropylamino-2-methyl-	HCl 163–165°	410
8-Chloro-6-ethylamino-	276–278°	58
2-Chloro-6-ethylamino-9-methyl-	182–185°	658
6-Chloro-2-ethylamino-9-methyl-	187–189°	658
2-Chloro-6-furfurylamino-	263–266°	105
2-Chloro-6-hydrazino-	> 300°	460
2-Chloro-6-hydrazino-7-methyl-	> 200°	112, 415
2-Chloro-6-hydroxyamino-	320°	41
2-Chloro-7-methyl-6-methylamino-	269°	415
2-Chloro-6-methylamino-	> 220°	61
2-Chloro-8-methylamino-	—	32
8-Chloro-6-methylamino-	300°	58
2-Chloro-9-phenyl-6-propylamino-	121–122°	57
8-Chloro-6-propylamino-	290 292°	58
6-Chloromethyl-2-dimethylamino-9-methyl-	123–125°	654
2,6-Diamino-8-bromo-	—	207
6,8-Diamino-2-chloro-7-methyl-	311–312°	112
2,6-Diamino-8-trifluoromethyl-	350°	56
6-Dibutylamino-2,8-dichloro-	168–169°	413
2,8-Dichloro-6-dibenzylamino-9-methyl-	152°	14
2,8-Dichloro-6-diethylamino-	225°	413
2,8-Dichloro-6-dimethylamino-	287–288°	105
2,8-Dichloro-6-dimethylamino-3-methyl-	238°	802
2,8-Dichloro-6-dimethylamino-9-methyl-	208–209°	802
2,8-Dichloro-6-ethylamino-	264° or 273–274°	468, 719
2,8-Dichloro-6-furfurylamino	248–249°	105
6-Dimethylamino-2-fluoro-	> 280°	427
9-Formamido-6-chloro-	> 260°	172
9-Methyl-2-methylamino-6-trichloromethyl-	200–201°	118
6-Methylamino-2-trifluoromethyl-	282°	728

TABLE 34.   Amino-Oxopurines without *N*-Alkyl Groups.

Purine	M.p.	References
2-Acetamido-3-benzyl-1,6-dihydro-6-oxo-	285°	43
2-Acetamido-6-benzyloxy-	241–242°	471
2-Acetamido-1,6-dihydro-8-methyl-6-oxo-	350°	786
2-Acetamido-1,6-dihydro-6-oxo-	—	616
2-Amino-6-benzyloxy-	202–204°	471
2-Amino-6-butoxy-	175°	425
2-Amino-7,8-dihydro-6-isopropoxy-8-oxo-	277°	253
2-Amino-7,8-dihydro-6-methoxy-8-oxo-	> 360°	253
9-Amino-1,6-dihydro-8-methyl-6-oxo-	260–262°	17
8-Amino-1,6-dihydro-2-methylamino-6-oxo-	—	210
2-Amino-1,6-dihydro-6-oxo- (guanine, q.v.)		
2-Amino-7,8-dihydro-8-oxo-	> 300°	32, 74, 328, 705
6-Amino-2,3-dihydro-2-oxo- (isoguanine, q.v.)		
6-Amino-7,8-dihydro-8-oxo-	> 300°	2, 34, 58, 289
8-Amino-1,6-dihydro-6-oxo-	> 300°	58
8-Amino-2,3-dihydro-2-oxo-		32
9-Amino-1,6-dihydro-6-oxo-	> 260°	264
2-Amino-6-ethoxy-	293°	425
6-Amino-8-(1-hydroxy-1-methylethyl)-	249–251°	711
6-Amino-8-hydroxymethyl-	320–322°	790
2-Amino-6-isopropoxy-	177°	425
2-Amino-6-methoxy-	245–248°	305, 425
6-Amino-2-methoxy-	275°	478, 646
6-Amino-8-methoxy-	—	58
2-Amino-6-propoxy-	208°	425
2-Amino-1,6,7,8-tetrahydro-6,8-dioxo-	> 350°	140, 210, 289, 307, 414, 644
6-Amino-2,3,7,8-tetrahydro-2,8-dioxo-	> 360°	103, 289, 771
8-Amino-1,2,3,6-tetrahydro-2,6-dioxo-	—	210
9-Amino-1,2,3,6-tetrahydro-8-methyl-2,6-dioxo-	> 350°	17
8-Anilino-1,6-dihydro-2-methyl-6-oxo-	HCl > 310°	279
9-Anilino-1,6-dihydro-8-methyl-6-oxo-	> 360°	17
2-Anilino-1,6-dihydro-6-oxo-	—	526
6-Anilino-2,3-dihydro-2-oxo-	> 300°	586, 661
6-Benzylamino-2,3-dihydro-2-oxo-	> 320°	661
3-Benzylamino-1,2,3,6-tetrahydro-2,6-dioxo-	300–301°	652
6-Butylamino-2,3-dihydro-2-oxo-	> 300°	574
6-Cyclohexylamino-2,3-dihydro-2-oxo-	> 320°	661
2,6-Diamino-7,8-dihydro-8-oxo-	> 300°	2, 289, 771
2,8-Diamino-1,6-dihydro-6-oxo-	—	691
6,8-Diamino-2,3-dihydro-2-oxo-	—	643, 692
2,6-Diamino-8-hydroxymethyl-	—	701, 825
6-Dibutylamino-2,3-dihydro-2-oxo-	279–280°	413
6-Diethylamino-8-hydroxymethyl-2-methyl-	210°	238
1,6-Dihydro-2-dimethylamino-6-oxo-	HCl —	107
1,6-Dihydro-2-hydroxyamino-6-oxo-	260°	41, 428

TABLE 34 (*continued*)

Purine	M.p.	References
2,3-Dihydro-6-hydroxyamino-2-oxo-	355°	36
1,6-Dihydro-2-methylamino-6-oxo-	—	107, 210, 526
1,6-Dihydro-8-methylamino-6-oxo-(?)	>350°	239
2,3-Dihydro-6-methylamino-2-oxo-	>300°	574, 709
2,3-Dihydro-8-methylamino-2-oxo-	>300°	581
7,8-Dihydro-6-methylamino-8-oxo-	>300°	58
9-Dimethylamino-1,6-dihydro-8-methyl-6-oxo-	335–336°	19
2-Dimethylamino-1,6-dihydro-6-oxo-	300°	526, 710
6-Dimethylamino-2,3-dihydro-2-oxo-	>300°	574, 709
2-Dimethylamino-6-methoxy-	125°	305
3-Dimethylamino-1,2,3,6-tetrahydro-2,6-dioxo-	310°	513
6-Dipropylamino-2,3-dihydro-2-oxo-	290–291°	413
8-Ethyl-1,6-dihydro-6-oxo-2-propionamido-	350°	786
2-Ethylamino-1,6-dihydro-6-oxo-	—	526
6-Ethylamino-2,3-dihydro-2-oxo-	>260°	709
6-Furfurylamino-2,3-dihydro-2-oxo	>300°	586
Guanine	>300°	4, 62, 103, 178, 217, 305
Guanine/8-benzyl-	303–305°	727
Guanine/8-ethyl-	—	181
Guanine/8-hydroxymethyl-	330°	788
Guanine/8-methyl-	350°; HCl 320°	62, 181, 242, 786, 796
Guanine/8-phenyl-	>350°	236
2-Hydrazino-1,6-dihydro-6-oxo-	>300°	460
Isoguanine	>300°	6, 52, 103, 184, 222, 303, 343, 456, 611, 637, 645, 709
8-Methoxy-6-methylamino-	—	58
1,2,3,6-Tetrahydro-8-methylamino-2,6-dioxo-	hy. >360°	282

TABLE 35. Amino-Oxopurines with *N*-Alkyl Groups.

Purine	M.p.	References
2-Acetamido-3-benzyl-3,6-dihydro-6-oxo-	282–285°	826
2-Acetamido-7-benzyl-1,6-dihydro-6-oxo-	241°	43, 826
2-Acetamido-9-benzyl-1,6-dihydro-6-oxo-	229°	43, 826
2-Acetamido-1,6-dihydro-1,8-dimethyl-6-oxo-	283–285°	786
2-Acetamido-1,6-dihydro-7,8-dimethyl-6-oxo-	350°	786
6-Amino-9-benzyl-7,8-dihydro-8-oxo-	270°	155

continued

TABLE 35 (*continued*)

Purine	M.p.	References
6-Amino-2-benzyloxy-9-methyl-	200°	222
8-Amino-3-butyl-1,2,3,6-tetrahydro-1,7-dimethyl-2,6-dioxo-	249–250°	436
2-Amino-6-ethoxy-3-methyl-	243–244°	44
6-Amino-2-ethoxy-9-methyl-	214–215° or 252–254°	222, 473
8-Amino-3-ethyl-1,2,3,6-tetrahydro-1,7-dimethyl-2,6-dioxo-	293–294°	436
2-Amino-6-methoxy-3-methyl-	> 300°	44
8-Amino-1,2,3,6-tetrahydro-1,3-dimethyl-2,6-dioxo-	> 300°	90, 210, 690, 691, 781
8-Amino-1,2,3,6-tetrahydro-1,7-dimethyl-2,6-dioxo-	> 350°	782
8-Amino-1,2,3,6-tetrahydro-3,7-dimethyl-2,6-dioxo-	> 350°	112, 688, 783
6-Amino-2,3,8,9-tetrahydro-3,7-dimethyl-2,8-dioxo-	—	485
2-Amino-1,6,8,9-tetrahydro-7-methyl-6,8-dioxo-	—	310
8-Amino-1,2,3,6-tetrahydro-7-methyl-2,6-dioxo-	—	455
8-Amino-1,2,3,6-tetrahydro-9-methyl-2,6-dioxo-	—	210
6-Amino-2,3,7,8-tetrahydro-3-methyl-2,8-dioxo-	—	484
6-Amino-2,3,8,9-tetrahydro-7-methyl-2,8-dioxo-	> 300°	415
8-Amino-1,2,3,6-tetrahydro-1,3,7-trimethyl-2,6-dioxo-	> 360°	486, 504
7-(2-Aminoethyl)-1,2,3,6-tetrahydro-1,3-dimethyl-2,6-dioxo-	144–146°	499
8-Aminoethyl-1,2,3,6-tetrahydro-1,3,7-trimethyl-2,6-dioxo-	143–144°	447
8-Aminoethyl-1,2,3,6-tetrahydro-1,3,9-trimethyl-2,6-dioxo-	177–179°	350
8-Aminomethyl-1,2,3,6-tetrahydro-1,3-dimethyl-2,6-dioxo-	252°	81
8-Aminomethyl-1,2,3,6-tetrahydro-3,7-dimethyl-2,6-dioxo-	HCl 305°	446
8-Aminomethyl-1,2,3,6-tetrahydro-1,3,7-trimethyl-2,6-dioxo-	203–205°	447
8-Aminomethyl-1,2,3,6-tetrahydro-1,3,9-trimethyl-2,6-dioxo-	191–193°	497
6-Anilino-9-(3-hydroxypropyl)-	185–187°	502
8-Anilino-1,2,3,6-tetrahydro-1,3-dimethyl-2,6-dioxo-	> 320°	781

TABLE 35 (*continued*)

Purine	M.p.	References
8-Anilino-1,2,3,6-tetrahydro-1,7-dimethyl-2,6-dioxo-	> 340°	782
8-Anilino-1,2,3,6-tetrahydro-3,7-dimethyl-2,6-dioxo-	> 350°	783
8-Anilino-1,2,3,6-tetrahydro-1,3,7-trimethyl-2,6-dioxo-	> 260°	487
2-Aziridinyl-7-benzyl-1,6-dihydro-6-oxo-	275°	470
2-Aziridinyl-9-benzyl-1,6-dihydro-6-oxo-	92°	470
2-Aziridinyl-1,6-dihydro-1,9-dimethyl-6-oxo-	163°	383
9-Benzyl-6-benzylamino-7,8-dihydro-8-oxo-	227–228°	155
8-Benzyl-1,6-dihydro-1-methyl-2-methyl-amino-6-oxo-	315°	282
9-Benzyl-2-dimethylamino-1,6-dihydro-6-oxo-	282°	470
6-Benzylamino-2-benzyloxy-9-methyl-	170–171°	222
2-Benzylamino-1,6-dihydro-1,9-dimethyl-6-oxo-	206–208°	403
2-Benzylamino-1,6-dihydro-9-methyl-6-oxo-	334–335°	658
6-Benzylamino-2,3-dihydro-9-methyl-2-oxo-	266–267°	222
6-Benzylamino-7,8-dihydro-3-methyl-8-oxo-	> 300°	72
9-Benzylamino-1,6-dihydro-6-oxo-	> 264°	264
6-Benzylamino-9-(3-hydroxypropyl)-	116–118°	502
8-Benzylamino-1,2,3,6-tetrahydro-1,3-dimethyl-2,6-dioxo-	234–235°	90
2-Benzylamino-1,6,8,9-tetrahydro-7-methyl-6,8-dioxo-	—	310
8-Benzylamino-1,2,3,6-tetrahydro-1,3,7-trimethyl-2,6-dioxo-	231°	488
3-Butyl-8-diethylamino-1,2,3,6-tetrahydro-1,7-dimethyl-2,6-dioxo-	82–83°	436
3-Butyl-8-dimethylamino-1,2,3,6-tetrahydro-1,7-dimethyl-2,6-dioxo-	69–70°	436
3-Butyl-8-ethylamino-1,2,3,6-tetrahydro-1,7-dimethyl-2,6-dioxo-	206–207°	436
3-Butyl-1,2,3,6-tetrahydro-1,7-dimethyl-8-methylamino-2,6-dioxo-	230–232°	436
8-Butylamino-1,2,3,6-tetrahydro-1,3-dimethyl-2,6-dioxo-	hy. 222°	90
8-Cyclohexylamino-1,2,3,6-tetrahydro-1,3-dimethyl-2,6-dioxo-	179°	445
8-Diacetamido-1,2,3,6-tetrahydro-1,3,7-trimethyl-2,6-dioxo-	137–142°	486
2-Diethylamino-1,6-dihydro-1,7-dimethyl-6-oxo-	82–85°	671
2-Diethylamino-1,6-dihydro-1,9-dimethyl-6-oxo-	105–107°	403

continued

573

TABLE 35 (*continued*)

Purine	M.p.	References
2-Diethylamino-3,6-dihydro-3,7-dimethyl-6-oxo-	124–126°	671
2-Diethylamino-1,6-dihydro-9-methyl-6-oxo-	260–262°	658
8-Diethylamino-3-ethyl-1,2,3,6-tetrahydro-1,7-dimethyl-2,6-dioxo-	125–127°	436
8-Diethylamino-1,2,3,6-tetrahydro-1,3-dimethyl-2,6-dioxo-	255–256° or 269°	90, 489
8-Diethylamino-1,2,3,6-tetrahydro-1,3,7-trimethyl-2,6-dioxo-	109°	488
8-Diethylaminomethyl-2-ethoxy-1,6-dihydro-1,9-dimethyl-6-oxo-	74–76°	712
8-Diethylaminomethyl-2-ethoxy-1,6-dihydro-9-methyl-6-oxo-	201–203°	712
7-Diethylaminomethyl-1,2,3,6-tetrahydro-1,3-dimethyl-2,6-dioxo-	116–117°	71, 558
8-Diethylaminomethyl-1,2,3,6-tetrahydro-1,3-dimethyl-2,6-dioxo-	176–177°	445
8-Diethylaminomethyl-1,2,3,6-tetrahydro-1,7-dimethyl-2,6-dioxo-	192–194°	724
8-Diethylaminomethyl-1,2,3,6-tetrahydro-3,7-dimethyl-2,6-dioxo-	175°	446
8-Diethylaminomethyl-1,2,3,6-tetrahydro-1,3,7-trimethyl-2,6-dioxo-	111–113°	447
8-Diethylaminomethyl-1,2,3,6-tetrahydro-1,3,9-trimethyl-2,6-dioxo-	142–143°	497
1,6-Dihydro-1,7-dimethyl-2-methylamino-6-oxo-	341–344°	671
3,6-Dihydro-3,7-dimethyl-2-methylamino-6-oxo-	336–338°	671
1,6-Dihydro-1-methyl-2-methylamino-6-oxo-	345–350°	282
1,6-Dihydro-1-methyl-2-methylamino-6-oxo-8-phenyl-	360°	282
2-Dimethylamino-1,6-dihydro-1,9-dimethyl-6-oxo-	149–151°	403
2-Dimethylamino-3,6-dihydro-3,7-dimethyl-6-oxo-	175–177°	671
2-Dimethylamino-1,6-dihydro-1-methyl-6-oxo-	280°	710
2-Dimethylamino-1,6-dihydro-7-methyl-6-oxo-	320°	553, 682
2-Dimethylamino-1,6-dihydro-9-methyl-6-oxo-	331–333°	658
2-Dimethylamino-7,9-dimethyl-6-oxido-	tos. 221–222°	682
8-Dimethylamino-3-ethyl-1,2,3,6-tetrahydro-1,7-dimethyl-2,6-dioxo-	130–132°	436
3-Dimethylamino-1,2,3,6-tetrahydro-1,9-dimethyl-2,6-dioxo-	270°	513, 653
8-Dimethylamino-1,2,3,6-tetrahydro-1,3-dimethyl-2,6-dioxo-	337–338°	82, 781

TABLE 35 (*continued*)

Purine	M.p.	References
8-Dimethylamino-1,2,3,6-tetrahydro-1,7-dimethyl-2,6-dioxo-	225°	782
8-Dimethylamino-1,2,3,6-tetrahydro-3,7-dimethyl-2,6-dioxo-	270°	783
3-Dimethylamino-1,2,3,6-tetrahydro-2,6-dioxo-8-phenyl-	>360°	718
3-Dimethylamino-1,2,3,6-tetrahydro-1-methyl-2,6-dioxo-	256–258°	513, 652
3-Dimethylamino-1,2,3,6-tetrahydro-1-methyl-2,6-dioxo-8-phenyl-	>360°	718
8-(2-Dimethylaminoethyl)-1,2,3,6-tetrahydro-3,7-dimethyl-2,6-dioxo-	hy. 156°	449
2-Dimethylaminomethyl-1,6-dihydro-1,9-dimethyl-6-oxo-	234–237°	383
8-Dimethylaminomethyl-1,2,3,6-tetrahydro-1,3-dimethyl-2,6-dioxo-	180°	81
8-Dimethylaminomethyl-1,2,3,6-tetrahydro-1,3,7-trimethyl-2,6-dioxo-	125–126°	447
8-Dipropylamino-1,2,3,6-tetrahydro-1,3,7-trimethyl-2,6-dioxo-	95°	488
3-Ethyl-8-ethylamino-1,2,3,6-tetrahydro-1,7-dimethyl-2,6-dioxo-	240–242°	436
3-Ethyl-1,2,3,6-tetrahydro-1,7-dimethyl-8-methylamino-2,6-dioxo-	267–269°	436
2-Ethylamino-1,6-dihydro-1,9-dimethyl-6-oxo-	199–200°	403
8-Ethylamino-1,2,3,6-tetrahydro-1,3-dimethyl-2,6-dioxo-	318–319°	90
8-Ethylamino-1,2,3,6-tetrahydro-1,3,7-trimethyl-2,6-dioxo-	226–230°	487
7-(2-Ethylaminoethyl)-1,2,3,6-tetrahydro-1,3-dimethyl-2,6-dioxo-	258°	499
2-Formamido-9-(2-hydroxyethyl)-	172–173°	820*
Guanine/3-benzyl-	*ca.* 300°	43, 826
Guanine/7-benzyl-	*ca.* 300°	43, 635
Guanine/9-benzyl-	303–304°	8, 43, 470, 507, 826
Guanine/7-benzyl-8-methyl-	*ca.* 300°	796
Guanine/9-benzyl-7-methyl-	tosyl. 225–227°	553, 554, 682
Guanine/7-benzyl-8-phenyl-	—	298
Guanine/9-butyl-	347–349°	8
Guanine/9-cyclohexyl-	>350°	8, 507
Guanine/9-cyclohexyl-7-methyl-	tosyl. 295–297°	39
Guanine/9-cyclopentyl-	>350°	8

continued

TABLE 35 *(continued)*

Purine	M.p.	References
Guanine/7,9-diethyl-	bet. 274°; tos. 248°	553, 682
Guanine/7,9-di(2-hydroxyethyl)-	bet. 320°	633
Guanine/1,7-dimethyl-	343–345°	3, 5, 417, 506, 631
Guanine/1,8-dimethyl-	320°	786
Guanine/1,9-dimethyl-	272 to 287°	403, 631
Guanine/3,7-dimethyl-	328–330°	671
Guanine/7,8-dimethyl-	350°	786
Guanine/7,9-dimethyl-	bet. 312°; tos. 283°	25, 39, 632
Guanine/8,9-dimethyl-	350°	786, 796
Guanine/7-ethyl-	>250°	633
Guanine/9-ethyl-	>300°	57
Guanine/7-ethyl-1-methyl-	256–257°	5
Guanine/9-ethyl-7-methyl-	bet. 277°; tos. 274°	553, 682
Guanine/7-(4-hydroxybutyl)-	308–315°	633
Guanine/9-(4-hydroxybutyl)-	220–222°	655
Guanine/7-(2-hydroxyethyl)-	>325°	633
Guanine/9-(2-hydroxyethyl)-	308–309°	655
Guanine/7-(2-hydroxyethyl)-1-methyl-	250–260°	5
Guanine/9-isobutyl-	362–365°	8, 57
Guanine/9-isopentyl-	357–359°	8
Guanine/7-isopropyl-1-methyl-	sul. 249–250°	5
Guanine/1-methyl-	>350°	101, 506, 631
Guanine/3-methyl-	>300°	44, 101, 310, 605
Guanine/7-methyl-	>350°	3, 320, 417, 605, 771
Guanine/9-methyl-	>350°	8, 57, 202, 320, 605, 631, 658
Guanine/1-methyl-7-propyl-	sul. 231–233°	5
Guanine/9-pentyl-	303–305°	8
Guanine/9-phenyl-	>360°	507
Guanine/9-propyl-	373–375°	8
Guanine/1,7,9-trimethyl-	iod. 314–333°	506, 631
8-Hydrazino-1,2,3,6-tetrahydro-1,3-dimethyl-2,6-dioxo-	325°	213, 429
8-Hydrazino-1,2,3,6-tetrahydro-1,7-dimethyl-2,6-dioxo-	265–268°	429
8-Hydrazino-1,2,3,6-tetrahydro-3,7-dimethyl-2,6-dioxo-	306°	429
8-Hydrazino-i,2,3,6-tetrahydro-1,3,7-trimethyl-2,6-dioxo-	288–290° or 320–322°	429, 487, 624, 650

TABLE 35 (*continued*)

Purine	M.p.	References
6-Hydroxyamino-9-(2-hydroxyethyl)-	232°	41
Isoguanine/9-benzyl-	312°	661
Isoguanine/9-butyl-	302°	661
Isoguanine/9-cyclohexyl-	>320°	661
Isoguanine/3,7-dimethyl-	350°	485
Isoguanine/3,9-dimethyl-	sul. 300°	39
Isoguanine/9-ethyl-	263°	661
Isoguanine/7-methyl-	—	31
Isoguanine/9-methyl-	>250°	222, 473
Isoguanine/9-phenyl-	>320°	661
Isoguanine/9-propyl-	305°	661
1,2,3,6-Tetrahydro-1,3-dimethyl-8-methylamino-2,6-dioxo-	364–366°	90
1,2,3,6-Tetrahydro-1,7-dimethyl-8-methylamino-2,6-dioxo-	>350°	782
1,2,3,6-Tetrahydro-1,3,7-trimethyl-8-methylamino-2,6-dioxo-	>310°	487

TABLE 36. Amino-Sulphonylpurines.

Purine	M.p.	References
6-Amino-2-fluorosulphonyl-	—	127
6-Amino-9-methyl-2-methylsulphinyl-	295°	222
6-Amino-9-methyl-2-methylsulphonyl-	313–314°	222
6-Amino-2-methylsulphinyl-	>300°	354
2-Amino-8-methylsulphonyl-	>300°	209
6-Amino-2-methylsulphonyl-	>300°	222, 354, 591
6-Amino-2-sulphamoyl-	—	127
2-Amino-6-sulphino-	sod. *ca.* 210°	376
2-Amino-6-sulpho-	pot. >300°	376
6-Benzylamino-9-methylsulphonyloxy-	163–164°	608
6-Dimethylamino-2,8-bis-methylsulphonyl-	253°	284
6-Dimethylamino-2-(*NN*-dimethylsulphamoyl)-	305°	127
6-Dimethylamino-2-methylsulphonyl-	>300°	284

TABLE 37.   Amino-Thiopurines.

Purine	M.p.	References
2-Acetamido-6-benzylthio-	257–259°	471
9-Acetamido-1,6-dihydro-6-thio-	312°	264
2-Amino-6-allylthio-	198–200°	62
2-Amino-9-benzyl-6-benzylthio-	157°	8
6-Amino-9-benzyl-2,8-bis-methylthio-	206°	14
2-Amino-9-benzyl-1,6-dihydro-6-thio-	303–304°	507
2-Amino-9-benzyl-6-methylthio-	210°	8
2-Amino-6-benzylthio-	205 to 214°	49, 62, 63, 574
6-Amino-2-benzylthio-	238–240°	267
2-Amino-6-benzylthio-9-butyl-	163°	8
2-Amino-6-benzylthio-9-cyclohexyl-	213°	8
2-Amino-6-benzylthio-9-cyclopentyl-	215°	8
2-Amino-6-benzylthio-9-isobutyl-	185°	8
2-Amino-6-benzylthio-3-methyl-	274°	44
2-Amino-6-benzylthio-7-methyl-	192–193°	471
2-Amino-6-benzylthio-8-methyl-	185–186°	62
2-Amino-6-benzylthio-9-methyl-	131–133°	471
6-Amino-2-benzylthio-9-methyl-8-methylthio-	198°	283
6-Amino-8-benzylthio-9-methyl-2-methylthio-	199–200°	283
6-Amino-2-benzylthio-8-methylthio-9-phenyl-	174°	283
2-Amino-6-benzylthio-9-pentyl-	149°	8
2-Amino-6-benzylthio-9-propyl-	154°	8
2-Amino-6,8-bis-methylthio-	283–284°	62
6-Amino-2,8-bis-methylthio-	254°	14
6-Amino 2,8-bis-methylthio-9-phenyl-	228–229°	283
2-Amino-9-butyl-1,6-dihydro-6-thio-	290–291°	8
2-Amino-9-butyl-6-isopropylthio-	112°	8
2-Amino-6-butylthio-	204–206°	62, 63
2-Amino-6-butylthio-3-methyl-	258–259°	44
2-Amino-6-butylthio-9-methyl-	109°	8
2-Amino-9-cyclohexyl-1,6-dihydro-6-thio-	357–359°	8
2-Amino-9-cyclopentyl-1,6-dihydro-6-thio-	340–342°	8
2-Amino-1,6-dihydro-7,9-dimethyl-6-thio-	bet. 295°	532
2-Amino-1,6-dihydro-9-isobutyl-6-thio-	330–332°	8
2-Amino-1,6-dihydro-9-isopentyl-6-thio-	317–319°	8
6-Amino-2,3-dihydro-9-methyl-8-methylthio-2-thio-	270°; pic. 246°	283
6-Amino-7,8-dihydro-9-methyl-2-methylthio-8-thio-	280–282°	283
2-Amino-1,6-dihydro-1-methyl-6-thio-	340–342°	122
2-Amino-1,6-dihydro-7-methyl-6-thio-	> 300°	347
2-Amino-1,6-dihydro-8-methyl-6-thio-	> 300°	62
2-Amino-1,6-dihydro-9-methyl-6-thio-	> 300°	57, 658

TABLE 37 (*continued*)

Purine	M.p.	References
2-Amino-3,6-dihydro-3-methyl-6-thio-	> 300°	44
6-Amino-2,3-dihydro-3-methyl-2-thio-	> 300°	101
6-Amino-2,3-dihydro-8-methylthio-9-phenyl-2-thio-	230–235°	283
2-Amino-7,8-dihydro-6-methylthio-8-thio-	> 300°	62
6-Amino-7,8-dihydro-2-methylthio-8-thio-	> 305°	14, 15
2-Amino-1,6-dihydro-9-pentyl-6-thio-	302–304°	8
2-Amino-1,6-dihydro-9-phenyl-6-thio-	304–305°	263, 507
2-Amino-1,6-dihydro-9-propyl-6-thio-	313–315°	8
2-Amino-1,6-dihydro-6-seleno-	—	494
2-Amino-1,6-dihydro-6-thio- (thioguanine)	> 360°	62, 205, 525
2-Amino-7,8-dihydro-8-thio-	350°	32, 378
6-Amino-2,3-dihydro-2-thio-	—	136, 178, 184
6-Amino-7,8-dihydro-8-thio-	> 300°	58
8-Amino-1,6-dihydro-6-thio-	> 300°	58, 170
2-Amino-7,9-dimethyl-6-methylthio-	tos. 293–294°	25
2-Amino-8-ethyl-1,6-dihydro-6-thio-	> 300°	245
2-Amino-9-ethyl-1,6-dihydro-6-thio-	299–302°	8
8-Amino-2-ethyl-1,6-dihydro-6-thio-	—	170
2-Amino-6-ethylthio-	206–208°	62, 63
2-Amino-6-ethylthio-3-methyl-	hy. 261–262°	44
2-Amino-6-ethylthio-9-methyl-	165°	8
2-Amino-6-hexylthio-	180–182°	62
2-Amino-9-isobutyl-6-isobutylthio-	113°	8
2-Amino-9-isobutyl-6-isopropylthio-	149°	8
2-Amino-9-isobutyl-6-propylthio-	103°	8
2-Amino-6-isobutylthio-	188–191°	62
2-Amino-6-isopentylthio-	201–203°	62
2-Amino-6-isopropylthio-	hy. 164–165°	62
2-Amino-6-isopropylthio-9-propyl-	98°	8
6-Amino-7-methyl-2,8-bis-methylthio-	263°	14
6-Amino-9-methyl-2,8-bis-methylthio-	235°	14, 283
2-Amino-3-methyl-6-methylthio-	hy. 289–292°	44
2-Amino-8-methyl-6-methylthio-	292–293°	62
2-Amino-9-methyl-6-methylthio-	190°	8, 658
6-Amino-2-methyl-6-methylthio-	261–262°	221
6-Amino-7-methyl-2-methylthio-	283°	14
2-Amino-6-methylthio-	238–241°	49, 62, 63, 460
2-Amino-8-methylthio-	235–236°	209
6-Amino-2-methylthio-	295–296°	52, 178, 221, 305
6-Amino-8-methylthio-	288–290°	58
2-Amino-6-pentylthio-	202°	62
2-Amino-6-propylthio-	191–193°	62, 63
2-Amino-1,6,7,8-tetrahydro-6,8-dithio-	> 300°	63, 86

continued

TABLE 37 (*continued*)

Purine	M.p.	References
6-Amino-2,3,7,8-tetrahydro-2,8-dithio-	320°	14
9-(2-Aminoethyl)-1,6-dihydro-6-thio-	HCl >270°	622
2-Anilino-1,6-dihydro-6-thio-	—	526
2-Benzylamino-1,6-dihydro-9-methyl-6-thio-	292–294°	658
2-Benzylamino-9-methyl-6-methylthio-	131–133°	658
2-Benzylthio-6-butylamino-	247–248°	574
2-Benzylthio-6-dimethylamino-	270°	574
8-Benzylthio-6-dimethylamino-2-methylthio-	230–232°	187
2-Benzylthio-6-methylamino-	283–284°	574
6-Butylamino-2,3-dihydro-2-thio-	272–277°	574
6-Butylamino-2-methylthio-	254°	574
2,6-Diamino-7,8-dihydro-8-thio-	—	578
2,8-Diamino-1,6-dihydro-6-thio-	—	210
2,8-Diamino-6-methylthio-	hy. 306–308°	210
6-Dibenzylamino-2-methylthio-	206°	14
2-Diethylamino-1,6-dihydro-9-methyl-6-thio-	225–228°	658
6-Diethylamino-7,8-dihydro-2-methylthio-8-thio-	264–265°	189, 286
2-Diethylamino-9-methyl-6-methylthio-	105–106°	658
6-Diethylamino-8-methyl-2-methylthio-	216–218°	242
6-Diethylamino-2-methylthio-	198–200°	189
9-(2-Diethylaminoethyl)-1,6-dihydro-6-thio-	247–251°	404
1,6-Dihydro-2-hydroxyamino-6-methylthio-	250°	41
1,6-Dihydro-2-methylamino-6-thio-	—	526
1,6-Dihydro-8-methylamino-6-thio-	—	240
2,3-Dihydro-6-methylamino-2-thio-	>300°	574
7,8-Dihydro-6-methylamino-8-thio-	>300°	58
6-Dimethylamino-2,8-bis-methylthio-	257–259° or 272°	187, 284
2-Dimethylamino-1,6-dihydro-9-methyl-6-thio-	273–275°	658
9-Diethylamino-1,6-dihydro-8-methyl-6-thio-	277–278°	
6-Dimethylamino-7,8-dihydro-2-methylthio-8-thio-	>350°	187, 286
2-Dimethylamino-1,6-dihydro-6-thio-	—	526
6-Dimethylamino-2,3-dihydro-2-thio-	>300°	574
6-Dimethylamino-3-ethyl-2,8-bis-methylthio-	161–163°	288*
6-Dimethylamino-9-ethyl-2,8-bis-methylthio-	126–128°	288
6-Dimethylamino-9-ethyl-7,8-dihydro-2-methylthio-8-thio-	221–223°	288
6-Dimethylamino-9-ethyl-2-methylthio-	75–77°	627

TABLE 37 (*continued*)

Purine	M.p.	References
6-Dimethylamino-3-methyl-2,8-bis-methylthio-	165–166°	288*
6-Dimethylamino-9-methyl-2,8-bis-methylthio-	124–125°	288
2-Dimethylamino-9-methyl-6-methylthio-	152–153°	658
6-Dimethylamino-8-methyl-2-methylthio-	—	242
6-Dimethylamino-2-methylthio-	285–286°; HCl 299°	187, 260, 410, 574
6-Dimethylamino-8-methylthio-	260°	58
6-(2-Dimethylhydrazino)-8-methyl-2-methylthio-	289–291°	242
2-Ethylamino-1,6-dihydro-9-methyl-6-thio-	> 300°	658
2-Ethylamino-1,6-dihydro-6-thio-	—	526
2-Ethylamino-9-methyl-6-methylthio-	143–145°	658
6-Ethylamino-8-methylthio-	235–236°	58
2-Ethylthio-6-methylamino-	293–294°	728
8-Methyl-6-methylamino-2-methylthio-	209°	242
9-Methyl-6-methylamino-2-methylthio-	239°	12
6-Methylamino-2-methylthio-	> 300°	574

TABLE 38. Carboxy-Halogenopurines.

Purine	M.p.	References
9(7)-Acetyl-2-chloro-	192–195°	259
9(7)-Acetyl-6-chloro-	140–142°	259
9(7)-Acetyl-2,6-dichloro-	154–158°	259
9-Benzyloxycarbonyl-6-chloro-	102°	683
9-Butyl-6-chloro-	132°	683
9-Carbamoylmethyl-6-chloro-	229–231°	400
6-Carbamoylmethyl-2-chloro-9-methyl	218–220°	654
6-Carboxy-2-chloro-9-(2-chloroethyl)-	206°	272
6-Carboxymethyl-2-chloro-9-methyl-	152–154°	654
6-Chloro-9-(2-cyanoethyl)-	145–146°	126
6-Chloro-9-cyanomethyl-	134–135°	400
6-Chloro-9-(3-cyanopropyl)-	86°	798
2-Chloro-6-ethoxycarbonyl-	254°	68
6-Chloro-9-ethoxycarbonyl-	125–126°	683
2-Chloro-6-ethoxycarbonyl-9-ethoxycarbonylmethyl-	135°	68
2-Chloro-6-ethoxycarbonyl-9-methyl-	117–120°	654
6-Chloro-9-ethoxycarbonylmethyl-	93–95°	400
6-Chloro-9-propionyl-	139°	683
9-(2-Methoxycarbonylethyl)-6-chloro-	74–76°	400

TABLE 39. Carboxy-Oxopurines.

Purine	M.p.	References
7-Acetyl-1,6,7,8-tetrahydro-2-methoxy-1,9-dimethyl-6,8-dioxo-	150°	519
9-Acetyl-1,6,8,9-tetrahydro-2-methoxy-1,7-dimethyl-6,8-dioxo-	162°	662
6-Carboxy-7,8-dihydro-8-oxo-	265°	791
8-Carboxymethyl-2-ethoxy-1,6-dihydro-1,9-dimethyl-6-oxo-	98–99°	384
8-Carboxymethyl-2-ethoxy-1,6-dihydro-9-methyl-6-oxo-	227–229°	384
6-(2-Carboxy-1-hydroxyethyl)-	165°	76
9-Carboxymethyl-6-methoxy-	246–248°	400
2-Ethoxy-6-ethoxycarbonyl-	162°	185
6-Ethoxycarbonyl-2,3-dihydro-2-oxo-	243°	185
Hypoxanthine/1-acetyl-	238–240°	559
Hypoxanthine/2-carbamoyl-1,9-dimethyl-	268–270°	383
Hypoxanthine/8-carboxy-	> 350°	239
Hypoxanthine/9-(2-carboxyethyl)-	284–287°	126
Hypoxanthine/9-carboxymethyl-	264°	400
Hypoxanthine/2-carboxymethyl-1,9-dimethyl-	221–223°	383
Hypoxanthine/2,8-dithiocyanato-	240°	809
Hypoxanthine/8-ethoxycarbonyl-	310°	239
Hypoxanthine/2-methoxycarbonylmethyl-1,9-dimethyl-	180°	383
Hypoxanthine/2-thiocyanato-	> 240°	164
Uric acid/7-acetyl-	—	519
Uric acid/7-acetyl-1,3-dimethyl-	304°	519
Uric acid/7-acetyl-1,9-dimethyl-	282°	519
Uric acid/7-acetyl-9-methyl-	> 300°	519
Uric acid/7-acetyl-1,3,9-trimethyl-	235°	519
Xanthine/7-acetonyl-1,3-dimethyl-	160–162°	490, 538, 542, 545, 547
Xanthine/7-acetonyl-3,7-dimethyl-	167°	490
Xanthine/8-acetoxy-7-acetyl-	> 300°	519
Xanthine/8-acetoxy-7-acetyl-1,3-dimethyl-	125°	519
Xanthine/8-acetoxy-7-acetyl-1-methyl-	> 260°	519
Xanthine/1-acetyl-3,7-dimethyl-	165°	157
Xanthine/7-acetyl-1,3-dimethyl-	156–157°	120, 560
Xanthine/8-acetyl-1,3-dimethyl-	288°	116
Xanthine/8-acetyl-1,3,7-trimethyl-	200°	116
Xanthine/7-(2-acetylethyl)-1,3-dimethyl-	140–141°	610
Xanthine/7-allyl-1,3-diethyl-8-isothiocyanato-methyl-	55–57°	808
Xanthine/7-allyl-8-isothiocyanatomethyl-1,3-dimethyl-	126–128°	808
Xanthine/7-benzoyl-1,3-dimethyl-	205°	157
Xanthine/1-benzoyl-3,7-dimethyl-	206°	157

TABLE 39 (*continued*)

Purine	M.p.	References
Xanthine/7-(2-benzoylethyl)-1,3-dimethyl-	190–192°	610
Xanthine/1-(2-benzoylethyl)-3,7-dimethyl-	170–172°	610
Xanthine/1-benzyl-8-carbamoylmethyl-3,7-dimethyl-	275–277°	714
Xanthine/1-benzyl-8-carboxymethyl-3,7-dimethyl-	145–147°	714
Xanthine/8-benzyl-7-carboxymethyl-1,3-dimethyl-	222–223°	234
Xanthine/7-benzyl-1,3-diethyl-8-isocyanatomethyl-	—	808
Xanthine/1-benzyl-8-ethoxycarbonylmethyl-3,7-dimethyl-	135°	714
Xanthine/7-benzyl-8-isothiocyanatomethyl-1,3-dimethyl-	151–152°	808
Xanthine/8-(*N*-benzylcarbamoylmethyl)-1,3,9-trimethyl-	248–250°	668
Xanthine/8-benzylthiocarbonyl-1,3,7-trimethyl-	175–176°	509
Xanthine/8-benzylthiocarbonylethyl-1,3,7-trimethyl-	126–127°	381, 670
Xanthine/7-butoxycarbonyl-1,3-dimethyl-	71–75°	589
Xanthine/7-butyl-1,3-diethyl-8-isothiocyanato-methyl-	52–54°	808
Xanthine/7-butyl-8-isothiocyanatomethyl-1,3-dimethyl-	132–133°	808
Xanthine/7-butyryl-1,3-dimethyl-	88–90°	120
Xanthine/1-carbamoyl-3,7-dimethyl-	301–302°	545
Xanthine/7-carbamoyl-1,3-dimethyl-	260°	540, 545
Xanthine/8-carbamoyl-1,3-dimethyl-	> 360°	673
Xanthine/8-carbamoyl-1,3,7-trimethyl-	> 360°	667
Xanthine/7-(2-carbamoylethyl)-1,3-dimethyl-	269°	540
Xanthine/8-(2-carbamoylethyl)-3-methyl-	—	181
Xanthine/8-(2-carbamoylethyl)-1,3,7-trimethyl-	233–235°	648
Xanthine/8-(2-carbamoylethyl)-1,3,9-trimethyl-	252–254°	497
Xanthine/7-carbamoylmethyl-1,3-dimethyl-	280°	589
Xanthine/8-carbamoylmethyl-1,9-dimethyl-	308–310°	384
Xanthine/8-carbamoylmethyl-9-methyl-	325–327°	384
Xanthine/8-carbamoylmethyl-1,3,7-trimethyl-	273°	385
Xanthine/8-carbamoylmethyl-1,3,9-trimethyl-	271–272°	668
Xanthine/8-carboxy-1,3-dimethyl-	273°	673
Xanthine/8-carboxy-3,7-dimethyl-	345°	117
Xanthine/8-carboxy-3-methyl-	160°	117, 135, 181
Xanthine/8-carboxy-1,3,7-trimethyl-	227–232°	501, 667
Xanthine/8-carboxy-1,3,9-trimethyl-	250–260°	448, 669
Xanthine/1-(2-carboxyethyl)-3,7-dimethyl-	203–205°	610
Xanthine/7-(1-carboxyethyl)-1,3-dimethyl-	200–205°	677
Xanthine/7-(2-carboxyethyl)-1,3-dimethyl-	204–205°	540, 551, 610

continued

TABLE 39 (*continued*)

Purine	M.p.	References
Xanthine/7-(2-carboxyethyl)-1-methyl-	309°	664
Xanthine/8-(2-carboxyethyl)-3-methyl-	—	135, 181
Xanthine/8-(2-carboxyethyl)-1,3,7-trimethyl-	232–233°	648
Xanthine/8-(2-carboxyethyl)-1,3,9-trimethyl-	243–245°	497
Xanthine/1-carboxymethyl-3,7-dimethyl-	260°	664
Xanthine/7-carboxymethyl-1,3-dimethyl-	271°	543, 664
Xanthine/8-carboxymethyl-1,3-dimethyl-	260°	181, 789
Xanthine/8-carboxymethyl-1,9-dimethyl-	335–337°	384
Xanthine/7-carboxymethyl-1-methyl-	306°	664
Xanthine/8-carboxymethyl-3-methyl-	—	135, 181
Xanthine/8-carboxymethyl-7-methyl-	—	714
Xanthine/8-carboxymethyl-9-methyl-	353–355°	384
Xanthine/8-carboxymethyl-1,3,9-trimethyl-	255–257°	668
Xanthine/8-(2-carboxyvinyl)-1,3-dimethyl-	320°	673
Xanthine/8-(2-carboxyvinyl)-1,3,7-trimethyl-	269–271°	648
Xanthine/1-chlorocarbonyl-3,7-dimethyl-	137°	134, 550
Xanthine/7-chlorocarbonyl-1,3-dimethyl-	150–155°	540, 545
Xanthine/8-chlorocarbonyl-1,3-dimethyl-	—	673
Xanthine/7-chlorocarbonylmethyl-1,3-dimethyl-	152–155°	543
Xanthine/8-chlorocarbonylmethyl-1,3-dimethyl-	270°	789
Xanthine/8-cyano-1,3-dimethyl-	300°	673
Xanthine/8-cyano-3,7-dimethyl-	287–288°	673
Xanthine/8-cyano-1,3,7-trimethyl-	151°	116, 667
Xanthine/1-(2-cyanoethyl)-3,7-dimethyl-	184°	551, 610
Xanthine/7-(2-cyanoethyl)-1,3-dimethyl-	160°	551, 552, 610
Xanthine/8-cyanomethyl-1,3-dimethyl-	250°	789
Xanthine/1,3-diethyl-8-isothiocyanatomethyl-7-methyl-	112–114°	808
Xanthine/1,3-diethyl-8-isothiocyanatomethyl-7-propyl-	64–66°	808
Xanthine/1-(*NN*-diethylcarbamoyl)-3,7-dimethyl-	154–155° or 210–211°	545, 550
Xanthine/7-(*NN*-diethylcarbamoyl)-1,3-dimethyl-	186–187°	545
Xanthine/7-(*NN*-diethylcarbamoylmethyl)-1,3-dimethyl-	184–186°	543, 548
Xanthine/1,3-dimethyl-7-phenacyl-	184–186°	542, 544
Xanthine/1,3-dimethyl-7-propionyl-	129–130°	120
Xanthine/1,3-dimethyl-7-(2-thiocyanatoethyl)-	151–152°	587
Xanthine/3,7-dimethyl-1-(2-thiocyanatoethyl)-	173–174°	588
Xanthine/1-(*NN*-dimethylcarbamoyl)-3,7-dimethyl-	151–152°	550
Xanthine/1-(*NN*-dimethylcarbamoylmethyl)-3,7-dimethyl-	226–227°	550
Xanthine/7-(*NN*-dimethylcarbamoylmethyl)-1,3-dimethyl-	184–185°	540, 543
Xanthine/1-ethoxycarbonyl-3,7-dimethyl-	138°	134, 157

TABLE 39 (*continued*)

Purine	M.p.	References
Xanthine/7-ethoxycarbonyl-1,3-dimethyl-	136 to 141°	157, 540, 545, 589
Xanthine/8-ethoxycarbonyl-1,3-dimethyl-	277–279°	91
Xanthine/8-ethoxycarbonyl-3,7-dimethyl-	300°	117
Xanthine/8-ethoxycarbonyl-3-methyl-	304–305°	117
Xanthine/8-ethoxycarbonyl-1,3,7-trimethyl-	207–208°	667
Xanthine/8-ethoxycarbonyl-1,3,9-trimethyl-	261–262°	393
Xanthine/7-(2-ethoxycarbonylethyl)-1,3-dimethyl-	105–106°	540, 551, 610
Xanthine/8-(2-ethoxycarbonylethyl)-1,3,7-trimethyl-	136°	648
Xanthine/8-(2-ethoxycarbonylethyl)-1,3,9-trimethyl-	152°	497
Xanthine/7-ethoxycarbonylmethyl-1,3-dimethyl-	146°	543
Xanthine/8-ethoxycarbonylmethyl-1,3-dimethyl-	215 or 235°	135, 789
Xanthine/8-ethoxycarbonylmethyl-3,7-dimethyl-	254°	714
Xanthine/8-ethoxycarbonylmethyl-1,3,9-trimethyl-	213–214°	668
Xanthine/8-(2-ethoxycarbonylvinyl)-1,3-dimethyl-	255–256°	673
Xanthine/8-(2-ethoxycarbonylvinyl)-1,3,7-trimethyl-	208°	648
Xanthine/7-ethyl-8-isothiocyanatomethyl-1,3-dimethyl-	131–132°	808
Xanthine/1-(N-ethylcarbamoylmethyl)-3,7-dimethyl-	307°	550
Xanthine/7-(N-ethylcarbamoylmethyl)-1,3-dimethyl-	265–266°	540
Xanthine/8-formyl-3,7-dimethyl-	288–290°	393, 713
Xanthine/8-formyl-1,3,7-trimethyl-	165–167°	501, 676
Xanthine/8-formyl-1,3,9-trimethyl-	203–205°	396, 675
Xanthine/1-hydrazinocarbonylmethyl-3,7-dimethyl-	300°	550
Xanthine/7-hydrazinocarbonylmethyl-1,3-dimethyl-	276–277°	540, 543
Xanthine/7-isopropoxycarbonyl-1,3-dimethyl-	153–154°	589
Xanthine/8-isothiocyanatomethyl-1,3-dimethyl-7-propyl-	151–152°	808
Xanthine/8-isothiocyanatomethyl-1,3,7-trimethyl-	175–177°	808
Xanthine/1-methoxycarbonyl-3,7-dimethyl-	165–168°	157, 545
Xanthine/7-methoxycarbonyl-1,3-dimethyl-	255–260°	157, 589
Xanthine/8-methoxycarbonyl-1,3-dimethyl-	258°	673
Xanthine/8-methoxycarbonyl-3,7-dimethyl-	270°	117
Xanthine/8-methoxycarbonyl-3-methyl-	290–291°	117
Xanthine/8-methoxycarbonyl-1,3,7-trimethyl-	174 or 201°	385, 667
Xanthine/7-methoxycarbonylmethyl-1,3-dimethyl-	148°	543
Xanthine/8-methoxycarbonyl-1,9-dimethyl-	280°	384

continued

TABLE 39 (*continued*)

Purine	M.p.	References
Xanthine/8-methoxycarbonylmethyl-3-methyl-	—	135
Xanthine/8-methoxycarbonylmethyl-9-methyl-	246°	384
Xanthine/8-methoxycarbonylmethyl-1,3,9-trimethyl-	205–206°	668
Xanthine/7-(*N*-methylcarbamoyl)-1,3-dimethyl-	258°	540
Xanthine/1-(*N*-methylcarbamoylmethyl)-3,7-dimethyl-	304–305°	550
Xanthine/7-(*N*-methylcarbamoylmethyl)-1,3-dimethyl-	267°	543
Xanthine/7-propoxycarbonyl-1,3-dimethyl-	65°	589
Xanthine/1,3,7-triethyl-8-isothiocyanatomethyl-	110–112°	808
Xanthine/1,3,7-trimethyl-8-propionyl-	142–143°	116

TABLE 40.   Carboxy-Thiopurines.

Purine	M.p.	References
9-Acetyl-1,6-dihydro-6-thio-	275°	683
9(7)-Acetyl-7,8-dihydro-8-thio-	240–243°	130
9-Acetyl-6-methylthio-	151°	683
9-(*N*-Butylcarbamoyl)-6-methylthio-	95–96°	382
9-Butyryl-1,6-dihydro-6-thio-	251–252°	683
9-Carbamoylmethyl-1,6-dihydro-6-thio-	>260°	400
9-Carboxy-1,6-dihydro-6-thio-	>260°	400
9-(2-Carboxyethyl)-1,6-dihydro-6-thio-	262–264°	126
9-(2-Cyanoethyl)-1,6-dihydro-6-thio-	283–286°	126
1,6-Dihydro-9-propionyl-6-thio-	285–288°	683
7,8-Dihydro-9(7)-propionyl-8-thio-	248–251°	130
9-Ethoxycarbonyl-1,6-dihydro-6-seleno-	189–191°	683
8-Ethoxycarbonyl-1,6-dihydro-6-thio-	>280°	240
9-Ethoxycarbonyl-1,6-dihydro-6-thio-	201–202°	382, 683
9-Ethoxycarbonyl-7,8-dihydro-8-thio-	186–187°	382
9-Ethoxycarbonyl-6-methylthio-	119 or 142°	382, 400
9-Ethoxycarbonylmethyl-1,6-dihydro-6-thio-	>260°	400
9-Hydrazinocarbonyl-1,6-dihydro-6-thio-	>260°	400
9-Methoxycarbonyl-6-methylthio-	147°	382
9-Methoxycarbonyl-8-methylthio-	126–127°	382
6-Methylthio-9-(*N*-phenylcarbamoyl)-	151–152°	382

TABLE 41.   Carboxythio-Oxopurines.

Purine	M.p.	References
6-(4-Carboxybutylthio)-2,3-dihydro-2-oxo-	236–240°	681
8-Carboxyethylthio-1,6-dihydro-2-methyl-6-oxo-	> 300°	94
8-Carboxymethylthio-1,6-dihydro-2-methyl-6-oxo-	> 300°	94
8-Carboxymethylthio-9-ethyl-2,3-dihydro-6-methyl-2-oxo-	> 270°	241
2-Carboxymethylthio-1,6,7,8-tetrahydro-6,8-dioxo-	*ca.* 225°	278
1,6-Dihydro-8-isopropoxycarbonylmethylthio-2-methyl-6-oxo-	223–226°	94
1,6-Dihydro-2-methyl-8-methoxycarbonyl-methylthio-6-oxo-	250–251°	94
1,6-Dihydro-2-methyl-6-oxo-8-propoxycarbonyl-methylthio-	230–231°	94
6-(4-Ethoxycarbonylbutylthio)-2,3-dihydro-2-oxo-	248–250°	681
8-Ethoxycarbonylmethylthio-1,6-dihydro-2-methyl-6-oxo-	233–235°	94
Hypoxanthine/2-(4-carboxybutylthio)-	266–268°	681
Hypoxanthine/2-carboxymethylthio-	*ca.* 240°	278
Hypoxanthine/8-carboxymethylthio-	hy. 265°	239
Hypoxanthine/2,8-dicarboxymethylthio-	240°	278
Hypoxanthine/2-(4-ethoxycarbonylbutylthio)-	195–196°	681
Xanthine/8-acetonylthio-1,3-dimethyl-	204–205°	575
Xanthine/7-carbamoylmethyl-8-carbonyl-methylthio-1,3-dimethyl-	263–264°	582
Xanthine/8-carboxymethylthio-	343°	388
Xanthine/8-carboxymethylthio-1,3-dimethyl-	268 or 275°	46, 388
Xanthine/8-carboxymethylthio-3,7-dimethyl-	302°	388
Xanthine/8-carboxymethylthio-9-ethyl-	289°	388
Xanthine/8-carboxymethylthio-9-methyl-	329°	388
Xanthine/8-carboxymethylthio-3-phenyl-	273–274°	679
Xanthine/8-carboxymethylthio-9-propyl-	354°	388
Xanthine/8-carboxymethylthio-1,3,7-trimethyl-	234°	388
Xanthine/7-ethoxycarbonylmethyl-8-ethoxy-carbonylmethylthio-1,3-dimethyl-	145°	582
Xanthine/8-methoxycarbonylethylthio-1,3-dimethyl-	159–161°	46

TABLE 42. Halogeno-Oxopurines without *N*-Alkyl Groups.

Purine	M.p.	References
9-Bromo-2,6-diethoxy-	175–177°	708
2-Bromo-1,6-dihydro-6-oxo-	—	207
2-Bromo-7,8-dihydro-8-oxo-	—	207
8-Bromo-1,6-dihydro-6-oxo-	—	508
2-Bromo-1,6,7,8-tetrahydro-6,8-dioxo-	—	207
8-Bromo-1,2,3,6-tetrahydro-2,6-dioxo-	—	207
6-Butoxy-2,8-dichloro-	134–135°	468
2-Chloro-6,8-diethoxy-	202–204°	452
8-Chloro-2,6-diethoxy-	209°	103
6-Chloro-7,8-dihydro-2-methyl-8-oxo-	> 300°	410
8-Chloro-1,6-dihydro-2-methyl-6-oxo-	> 300°	410
2-Chloro-1,6-dihydro-6-oxo-	> 300°	111, 460
2-Chloro-7,8-dihydro-8-oxo-	> 300°	32, 108, 277
6-Chloro-7,8-dihydro-8-oxo-	322°	58, 290, 291
8-Chloro-1,6-dihydro-6-oxo-	—	58, 424
8-Chloro-2,3-dihydro-2-oxo-	—	32
6-Chloro-8-ethoxy-	187–190°	333
8-Chloro-6-ethoxy-	197–199°	58
2-Chloro-6-methoxy-	254°	369
8-Chloro-6-methoxy-	203–204°	58
2-Chloro-1,6,7,8-tetrahydro-6,8-dioxo-	—	277, 424
6-Chloro-2,3,7,8-tetrahydro-2,8-dioxo-	> 200°	424
8-Chloro-1,2,3,6-tetrahydro-2,6-dioxo-	> 300°	413, 457
2,6-Dichloro-7,8-dihydro-8-oxo-	> 350°	2, 84, 412
2,8-Dichloro-1,6-dihydro-6-oxo-	> 350°	103, 424
2,6-Dichloro-8-ethoxy-	194–196°	452
2,8-Dichloro-6-ethoxy-	203–204°	103
1,6-Dihydro-8-iodo-6-oxo-	> 245°	86
2,3-Dihydro-8-iodo-2-oxo-	> 250°	86
1,6-Dihydro-6-oxo-2-trifluoromethyl-	324–326°	56
2,3-Dihydro-2-oxo-8-trifluoromethyl-	220°	209
7,8-Dihydro-6-oxo-8-trifluoromethyl-	322–324°	56
7,8-Dihydro-8-oxo-2-trifluoromethyl-	310–311°	329
2-Fluoro-1,6-dihydro-6-oxo-	> 260°	427, 428
1,2,3,6-Tetrahydro-8-iodo-2,6-dioxo-	> 200°	442

TABLE 43. Halogeno-Oxopurines with *N*-Alkyl Groups.

Purine	M.p.	References
7-Benzyl-8-bromo-1,2,3,6-tetrahydro-1,3-dimethyl-2,6-dioxo-	155°	390
7-Benzyl-2-chloro-1,6-dihydro-6-oxo-	285°	470
9-Benzyl-2-chloro-1,6-dihydro-6-oxo-	245°	470

TABLE 43 (*continued*)

Purine	M.p.	References
9-Benzyl-6-chloro-7,8-dihydro-6-oxo-	187°	155
1-Benzyl-8-chloro-1,2,3,6-tetrahydro-3,7-dimethyl-2,6-dioxo-	150–151°	590, 714
7-Benzyl-8-chloro-1,2,3,6-tetrahydro-1,3-dimethyl-2,6-dioxo-	153°	489, 563
8-Bromo-1-(3-bromopropyl)-1,2,3,6-tetrahydro-3,7-dimethyl-2,6-dioxo-	143–144°	87
8-Bromo-1-ethyl-1,2,3,6-tetrahydro-3,7-dimethyl-2,6-dioxo-	153°	504
8-Bromo-7-ethyl-1,2,3,6-tetrahydro-1,3-dimethyl-2,6-dioxo-	173°	536
8-Bromo-1,2,3,6-tetrahydro-1,3-dimethyl-2,6-dioxo-	307 to 320°	390, 420, 439, 441
8-Bromo-1,2,3,6-tetrahydro-1,9-dimethyl-2,6-dioxo-	265–266°	280
8-Bromo-1,2,3,6-tetrahydro-3,7-dimethyl-2,6-dioxo-	296–298° or 322°	257, 441, 444, 703
8-Bromo-1,2,3,6-tetrahydro-1-(3-hydroxypropyl)-3,7-dimethyl-2,6-dioxo-	23–24°	87
8-Bromo-1,2,3,6-tetrahydro-9-methyl-2,6-dioxo-	>290°	319
8-Bromo-1,2,3,6-tetrahydro-1,3,7-trimethyl-2,6-dioxo-	205–207°	367, 441, 504, 536, 650, 702
8-Bromo-1,2,3,6-tetrahydro-1,3,9-trimethyl-2,6-dioxo-	256°	309
7-(2-Bromoethyl)-8-chloro-1,2,3,6-tetrahydro-1,3-dimethyl-2,6-dioxo-	162°	489
1-(2-Bromoethyl)-1,2,3,6-tetrahydro-3,7-dimethyl-2,6-dioxo-	148°	463
7-(2-Bromoethyl)-1,2,3,6-tetrahydro-1,3-dimethyl-2,6-dioxo-	144°	463
8-Bromomethyl-1,2,3,6-tetrahydro-1,3,7-trimethyl-2,6-dioxo-	215–216°	395, 501, 669
3-Butyl-8-chloro-1,2,3,6-tetrahydro-1,7-dimethyl-2,6-dioxo-	69–70°	436
8-Chloro-3-chloromethyl-1,2,3,6-tetrahydro-1,9-dimethyl-2,6-dioxo-	203–205°	422
8-Chloro-3-chloromethyl-1,2,3,6-tetrahydro-3,7-dimethyl-2,6-dioxo-	145–146°	423
8-Chloro-7-chloromethyl-1,2,3,6-tetrahydro-1,3-dimethyl-2,6-dioxo-	150–152°	423, 702
2-Chloro-6,8-diethoxy-7-methyl-	194–195°	455
2-Chloro-6,8-diethoxy-9-methyl-	149–150°	455
8-Chloro-2,6-diethoxy-7-methyl-	140–141°	714

continued

TABLE 43 (*continued*)

Purine	M.p.	References
8-Chloro-2,6-diethoxy-9-methyl-	97–98°	384
8-Chloro-3,7-diethyl-1,2,3,6-tetrahydro-2,6-dioxo-	201–202°	295, 459
8-Chloro-1,7-diethyl-1,2,3,6-tetrahydro-3-methyl-2,6-dioxo-	136°	772
8-Chloro-3,7-diethyl-1,2,3,6-tetrahydro-1-methyl-2,6-dioxo-	113°	459
2-Chloro-1,6-dihydro-1,7-dimethyl-6-oxo-	270°	417
2-Chloro-1,6-dihydro-1,9-dimethyl-6-oxo-	180°	383
2-Chloro-3,6-dihydro-3,7-dimethyl-6-oxo-	255–257°	671
6-Chloro-7,8-dihydro-7,9-dimethyl-8-oxo-	175–178°	397
2-Chloro-1,6-dihydro-7-methyl-6-oxo-	>310°	417
2-Chloro-1,6-dihydro-9-methyl-6-oxo-	>260°	658
2-Chloro-1,6-dihydro-6-oxo-9-phenyl-	280–281°	57
8-Chloro-2-ethoxy-1,6-dihydro-1,9-dimethyl-6-oxo-	148–149°	384
2-Chloro-6-ethoxy-7-methyl-	240°	417
2-Chloro-6-ethoxy-9-methyl-	121–122°	658
2-Chloro-6-ethoxymethyl-9-methyl-	149–151°	83
8-Chloro-3-ethyl-1,2,3,6-tetrahydro-1,7-dimethyl-2,6-dioxo-	111–112°	295, 436
8-Chloro-7-ethyl-1,2,3,6-tetrahydro-1,3-dimethyl-2,6-dioxo-	142°	489
8-Chloro-3-ethyl-1,2,3,6-tetrahydro-2,6-dioxo-	295°	295
8-Chloro-7-ethyl-1,2,3,6-tetrahydro-3-methyl-2,6-dioxo-	225°	772
6-Chloro-9-(2-hydroxyethyl)-	148–149° or 157°	121, 400
6-Chloro-7-(3-hydroxypropyl)-	136°	502
6-Chloro-9-(3-hydroxypropyl)-	120°	502
2-Chloro-6-methoxy-7-methyl-	219°	476
8-Chloro-6-methoxy-3-methyl-	205°	72
8-Chloro-1,3,7-triethyl-1,2,3,6-tetrahydro-2,6-dioxo-	79–80°	459
8-Chloro-1,2,3,6-tetrahydro-7-hexyl-1,3-dimethyl-2,6-dioxo-	75–80°	563
2-Chloro-1,6,7,8-tetrahydro-1,9-dimethyl-6,8-dioxo-	291°	491
2-Chloro-1,6,7,8-tetrahydro-7,9-dimethyl-6,8-dioxo-	312°	491
8-Chloro-1,2,3,6-tetrahydro-1,3-dimethyl-2,6-dioxo-	>300°	89, 420
8-Chloro-1,2,3,6-tetrahydro-1,7-dimethyl-2,6-dioxo-	295°	431, 423, 773
8-Chloro-1,2,3,6-tetrahydro-1,9-dimethyl-2,6-dioxo-	318°	422, 437
8-Chloro-1,2,3,6-tetrahydro-3,7-dimethyl-2,6-dioxo-	228 to 304°	75, 419, 445, 564, 772

TABLE 43 (*continued*)

Purine	M.p.	References
8-Chloro-1,2,3,6-tetrahydro-1,3-dimethyl-2,6- dioxo-7-phenyl-	259–260°	563
8-Chloro-1,2,3,6-tetrahydro-7-(2-hydroxyethyl)- 1,3-dimethyl-2,6-dioxo-	158°	489
2-Chloro-1,6,7,8-tetrahydro-9-methyl-6,8-dioxo-	320°	491
2-Chloro-1,6,8,9-tetrahydro-7-methyl-6,8-dioxo-	—	310
6-Chloro-2,3,7,8-tetrahydro-3-methyl-2,8-dioxo-	> 300°	484
8-Chloro-1,2,3,6-tetrahydro-3-methyl-2,6-dioxo-	> 340°	419, 484, 772
8-Chloro-1,2,3,6-tetrahydro-7-methyl-2,6-dioxo-	> 300°	772
2-Chloro-1,6,7,8-tetrahydro-1,7,9-trimethyl-6,8- dioxo-	251–252°	491
8-Chloro-1,2,3,6-tetrahydro-1,3,7-trimethyl-2,6- dioxo-	189–191°	103, 420, 650, 772
8-Chloro-1,2,3,6-tetrahydro-1,3,9-trimethyl-2,6- dioxo-	255–265°	422, 713
8-Chloro-1,3,7-tris-chloromethyl-1,2,3,6- tetrahydro-2,6-dioxo-	129–130°	423
7-(4-Chlorobutyl)-1,2,3,6-tetrahydro-1,3-dimethyl- 2,6-dioxo-	92°	211
7-(2-Chloroethyl)-1,2,3,6-tetrahydro-1,3- dimethyl-2,6-dioxo-	121–122°	156, 463
8-(2-Chloroethyl)-1,2,3,6-tetrahydro-3,7-dimethyl- 2,6-dioxo-	*ca.* 280°	449
8-Chloromethyl-2-ethoxy-1,6-dihydro-1,9- dimethyl-6-oxo-	127–129°	712
8-Chloromethyl-2-ethoxy-1,6-dihydro-9-methyl-6- oxo-	—	712
8-Chloromethyl-1,2,3,6-tetrahydro-1,3- dimethyl-2,6-dioxo-	249°	75, 445, 498
8-Chloromethyl-1,2,3,6-tetrahydro-1,7- dimethyl-2,6-dioxo-	250–251°	724
8-Chloromethyl-1,2,3,6-tetrahydro-1,9- dimethyl-2,6-dioxo-	—	712
3-Chloromethyl-1,2,3,6-tetrahydro-1,9-dimethyl- 2,6-dioxo-8-trichloromethyl-	177–178°	280
7-Chloromethyl-1,2,3,6-tetrahydro-1,3-dimethyl- 2,6-dioxo-8-trichloromethyl-	162–174°	394
8-Chloromethyl-1,2,3,6-tetrahydro-3-methyl- 2,6-dioxo-	—	75
8-Chloromethyl-1,2,3,6-tetrahydro-1,3,7- trimethyl-2,6-dioxo-	214–215°	447, 509, 713, 780
8-Chloromethyl-1,2,3,6-tetrahydro-1,3,9- trimethyl-2,6-dioxo-	185–186°	448, 669

continued

TABLE 43 (*continued*)

Purine	M.p.	References
7-(5-Chloropentyl)-1,2,3,6-tetrahydro-1,3-dimethyl-2,6-dioxo-	79–80°	211
8-(3-Chloropropyl)-1,2,3,6-tetrahydro-1,3,7-trimethyl-2,6-dioxo-	117–118°	381
3,7-Dibutyl-8-chloro-1,2,3,6-tetrahydro-2,6-dioxo-	145°	459
2,6-Dichloro-7,8-dihydro-7,9-dimethyl-8-oxo-	185–186°	2, 405, 455, 533
2,6-Dichloro-7,8-dihydro-9-methyl-8-oxo-	284°	102, 405
2,6-Dichloro-8,9-dihydro-7-methyl-8-oxo-	265–270°	104, 310, 455
2,6-Dichloro-8-ethoxy-7-methyl-	185–186°	455
2,6-Dichloro-8-ethoxy-9-methyl-	154°	455
2,6-Dichloro-9-ethyl-7,8-dihydro-8-oxo-	263–266°	331
8-Dichloromethyl-1,2,3,6-tetrahydro-1,3,7-trimethyl-2,6-dioxo-	230–232°	780
8-Dichloromethyl-1,2,3,6-tetrahydro-1,3,9-trimethyl-2,6-dioxo-	193–194°	396, 675
1,7-Diethyl-1,2,3,6-tetrahydro-2,6-dioxo-8-trichloromethyl-	149–150°	5
1,6-Dihydro-6-oxo-9-(2,2,2-trifluoroethyl)-	270°	634
9-Ethyl-7,8-dihydro-2-iodo-8-oxo-	247–248°	331
1,2,3,6-Tetrahydro-8-dichloromethyl-3,7-dimethyl-2,6-dioxo-	234–237°	393
1,2,3,6-Tetrahydro-3,7-dimethyl-2,6-dioxo-8-trichloromethyl-	202–212°	392, 780
1,2,3,6-Tetrahydro-8-iodo-1,3-dimethyl-2,6-dioxo-	>250°	442, 536
1,2,3,6-Tetrahydro-8-iodo-1,9-dimethyl-2,6-dioxo-	>280°	442
1,2,3,6-Tetrahydro-8-iodo-3,7-dimethyl-2,6-dioxo-	>280°	442
1,2,3,6-Tetrahydro-8-iodo-9-methyl-2,6-dioxo-	300°	443
1,2,3,6-Tetrahydro-8-iodo-1,3,7-trimethyl-2,6-dioxo-	219 or 230°	442, 702
1,2,3,6-Tetrahydro-8-iodo-1,3,9-trimethyl-2,6-dioxo-	>225°	442
1,2,3,6-Tetrahydro-7-(4-iodobutyl)-1,3-dimethyl-2,6-dioxo-	105°	211
1,2,3,6-Tetrahydro-7-(2-iodoethyl)-1,3-dimethyl-2,6-dioxo-	150–151°	463
1,2,3,6-Tetrahydro-7-(5-iodopentyl)-1,3-dimethyl-2,6-dioxo-	81–82°	211
1,2,3,6-Tetrahydro-1,3,7-trimethyl-2,6-dioxo-8-trichloromethyl-	182–184°	394, 780
1,2,3,6-Tetrahydro-1,3,9-trimethyl-2,6-dioxo-8-trichloromethyl-	214–215°	280, 393
1,3,7-Tributyl-8-chloro-1,2,3,6-tetrahydro-2,6-dioxo-	232–240°/10 mm	459

TABLE 44. Halogeno-Sulphonylpurines.

Purine	M.p.	References
6-Chloro-2,8-bis-methylsulphonyl-	230°	284
8-Chloro-2,6-bis-methylsulphonyl-	240°	284
6-Chloro-2-(NN-dimethylsulphamoyl)-	270°	127
6-Chloro-2-fluorosulphonyl-	205–207°	127
6-Chloro-2-methylsulphonyl-	260°	284
6-Chloro-8-methylsulphonyl-	>180°	284
6-Chloro-2-sulphamoyl-	—	127
6,8-Dichloro-2-methylsulphonyl-	210°	284

TABLE 45. Halogeno-Thiopurines.

Purine	M.p.	References
6-Benzylthio-9-(2-chloroethyl)-	93–94°	430
9-Benzyl-2-ethylthio-6-trifluoromethyl-	103–105°	208
6-Chloro-2,8-bis-methylthio-	258°	284
8-Chloro-2,6-bis-methylthio-	244°	284
2-Chloro-1,6-dihydro-7-methyl-6-thio-	>250°	458, 775
2-Chloro-1,6-dihydro-9-methyl-6-thio-	>300°	658
8-Chloro-1,6-dihydro-2-methyl-6-thio-	—	410
8-Chloro-3,6-dihydro-3-methyl-6-thio-	260°	72
8-Chloro-1,6-dihydro-2-methylthio-6-thio-	—	410
2-Chloro-7,8-dihydro-8-thio-	—	32
8-Chloro-1,6-dihydro-6-thio-	—	58
6-Chloro-2-ethylthio-	185–187°	728
8-Chloro-6-ethylthio-	158–159°	58
2-Chloro-6-mercaptomethyl-9-methyl-	219–225°	83
2-Chloro-9-methyl-6-methylthio-	161–162°	658
6-Chloro-2-methyl-8-methylthio-	231°	279
6-Chloro-8-methyl-2-methylthio-	268–270°	242
8-Chloro-3-methyl-6-methylthio-	250°	72
2-Chloro-8-methylthio-	—	32
6-Chloro-2-methylthio-	274°	410
6-Chloro-8-methylthio-	220–222°	58
8-Chloro-2-methylthio-	208°	410
8-Chloro-6-methylthio-	192–195°	58
9-(2-Chloroethyl)-6-methylthio-	152–153°	430
6,8-Dichloro-2-methylthio-	224–227°	410
1,6-Dihydro-6-thio-2-trifluoromethyl-	274–275°	208
9-Ethyl-2-ethylthio-6-trifluoromethyl-	285°	208
2-Ethylthio-9-methyl-6-trifluoromethyl-	107–109°	208
8-Ethylthio-9-methyl-6-trifluoromethyl-	79–80°	208
2-Fluoro-1,6-dihydro-6-thio-	>360°	427
2-Fluoro-6-methylthio-	247–248° or 260°	427, 434

TABLE 46.   Oxo-Sulphonylpurines.

Purine	M.p.	References
7,8-Dihydro-2,6-bis-methylsulphonyl-8-oxo-	> 300°	284
7,8-Dihydro-3-methyl-6-methylsulphonyl-8-oxo-	> 300°	72
7,8-Dihydro-2-methylsulphonyl-8-oxo-	> 240°	284
7,8-Dihydro-6-methylsulphonyl-8-oxo-	> 300°	284
7,8-Dihydro-8-oxo-2,6-disulpho-	—	424
Hypoxanthine/2,8-bis-methylsulphonyl-	288–290°	284
Hypoxanthine/2-(NN-dimethylsulphamoyl)-	—	127
Hypoxanthine/2-fluorosulphonyl-	—	127
Hypoxanthine/2-methylsulphonyl-	> 300°	284, 591
Hypoxanthine/2-sulphamoyl-	—	127
Xanthine/7-benzyl-1,3-dimethyl-8-(N-methyl-sulphamoyl)-	186–187°	590
Xanthine/7-benzyl-1,3-dimethyl-8-methyl-sulphonyl-	139–141°	590
Xanthine/1-benzyl-3,7-dimethyl-8-sulphamoyl-	297–299°	590
Xanthine/7-benzyl-1,3-dimethyl-8-sulphamoyl-	199–201°	563
Xanthine/9-benzyl-1,3-dimethyl-8-sulphamoyl-	220°	590
Xanthine/7-benzyl-1,3-dimethyl-8-sulpho-	hy. 292–293°	590
Xanthine/8-benzenesulphonyl-1,3,7-trimethyl-	209°	435
Xanthine/8-butylsulphonyl-1,3,7-trimethyl-	127°	435
Xanthine/1,3-dimethyl-7-phenyl-8-sulphamoyl-	281°	563
Xanthine/1,3-dimethyl-8-sulphamoyl-	299°	563
Xanthine/8-ethylsulphonyl-1,3,7-trimethyl-	157°	435
Xanthine/1,3,7-trimethyl-8-methylsulphonyl-	219°	435
Xanthine/1,3,7-trimethyl-8-phenylsulphonyl-	230°	435
Xanthine/1,3,7-trimethyl-8-propylsulphonyl-	133°	435
Xanthine/1,3,9-trimethyl-8-sulphamoyl-	245 or 275°(?)	563
Xanthine/1,3,7-trimethyl-8-sulpho-	—	601, 602

TABLE 47.   Oxo-Thiopurines without N-Alkyl Groups.

Purine	M.p.	References
2-Benzylthio-1,6-dihydro-6-oxo-	262–263°	532
6-Benzylthio-2,3-dihydro-2-oxo-	279–280° or 293–298°	574, 586
6-Benzylthio-7,8-dihydro-8-oxo-	295°	523
1,6-Dihydro-2,8-bis-methylthio-6-oxo-	301°	284
7,8-Dihydro-2,6-bis-methylthio-8-oxo-	285–288°	284, 495
1,6-Dihydro-8-hydroxyethyl-6-thio-	296–298°	240
1,6-Dihydro-8-hydroxymethyl-6-thio-	> 270°	240
1,6-Dihydro-2-methyl-8-methylthio-6-oxo-	> 300°	94

TABLE 47 (*continued*)

Purine	M.p.	References
1,6-Dihydro-2-methylthio-6-oxo-	>310°	302, 526
1,6-Dihydro-8-methylthio-6-oxo-	>300°	58
2,3-Dihydro-6-methylthio-2-oxo-	274°	586, 709
2,3-Dihydro-8-methylthio-2-oxo-	>260°	209, 581
7,8-Dihydro-2-methylthio-8-oxo-	—	410
7,8-Dihydro-6-methylthio-8-oxo-	hy. >300°	58, 523
2-Ethylthio-1,6-dihydro-6-oxo-	254–255°	728
6-Ethylthio-7,8-dihydro-8-oxo-	263°	72
8-Ethylthio-1,2,3,6-tetrahydro-2,6-dioxo-	>290°	536
1,2,3,6,7,8-Hexahydro-2,6-dioxo-8-thio-	hy. >300°	210, 323, 458, 524, 577, 774, 778, 779
1,2,3,6,7,8-Hexahydro-2,8-dioxo-6-thio-	>300°	45, 424, 524
1,2,3,6,7,8-Hexahydro-6,8-dioxo-2-thio-	>300°	278, 410, 524
1,2,3,6,7,8-Hexahydro-2-oxo-6,8-dithio-	>300°	521
1,2,3,6,7,8-Hexahydro-6-oxo-2,8-dithio-	300°	278, 284
1,2,3,6,7,8-Hexahydro-8-oxo-2,6-dithio-	>350°	284, 432, 495
9-Hydroxymethyl-6-methylthio-	160–161°	510
9-Hydroxymethyl-6-pentylthio-	139–140°	510
9-Hydroxymethyl-6-propylthio-	123–125°	510
6-Methoxy-8-methylthio-	205–206°	58
1,6,7,8-Tetrahydro-6,8-dioxo-2-phenyl-	>300°	18
1,2,3,6-Tetrahydro-8-methyl-6-oxo-2-thio-	>300°	242, 245
1,6,7,8-Tetrahydro-2-methyl-6-oxo-8-thio-	>300°	94, 245, 279, 521
1,6,7,8-Tetrahydro-2-methyl-8-oxo-6-thio-	hy. >300°	245, 521
2,3,7,8-Tetrahydro-6-methyl-2-oxo-8-thio-	>300°	818
1,6,7,8-Tetrahydro-2-methylthio-6,8-dioxo-	>300°	275, 410
1,6,7,8-Tetrahydro-2-methylthio-6-oxo-8-thio-	*ca.* 275°	241
1,2,3,6-Tetrahydro-2-oxo-8-phenyl-6-thio-	330°	123
1,2,3,6-Tetrahydro-6-oxo-8-phenyl-2-thio-	225° or >360°	256, 359, 361
1,6,7,8-Tetrahydro-6-oxo-2-phenyl-8-thio-	>300°	18
1,6,7,8-Tetrahydro-8-oxo-2-phenyl-6-thio-	>300°	18
1,2,3,6-Tetrahydro-2-oxo-6-thio-	>330°	370, 586, 709
1,2,3,6-Tetrahydro-6-oxo-2-thio-	325–340°	136, 178, 302, 352, 370, 526
1,6,7,8-Tetrahydro-6-oxo-8-thio-	235° or >300°	58, 63, 143, 239, 287
1,6,7,8-Tetrahydro-8-oxo-6-thio-	>300°	58, 63, 523
2,3,7,8-Tetrahydro-2-oxo-8-thio-	>300°	581
2,3,7,8-Tetrahydro-8-oxo-2-thio-	>300°	410

TABLE 48. Oxo-Thiopurines with *N*-Alkyl Groups.

Purine	M.p.	References
1-Allyl-8-allylthio-1,2,3,6-tetrahydro-3,7-dimethyl-2,6-dioxo-	118°	536
8-Allyl-1,2,3,6-tetrahydro-1,3-dimethyl-2,6-dioxo-	234–235°	582
8-Allylthio-1,2,3,6-tetrahydro-3,7-dimethyl-2,6-dioxo-	212°	536
8-Allylthio-1,2,3,6-tetrahydro-1,3,7-trimethyl-2,6-dioxo-	103°	536
8-Allylthio-1,2,3,6-tetrahydro-1,3-9-trimethyl-2,6-dioxo-	199°	46
8-Benzyl-1,6-dihydro-1-methyl-2-methylthio-6-oxo-	340°	282
7-Benzyl-1,2,3,6,8,9-hexahydro-1,3-dimethyl-2,6-dioxo-8-thio-	288–290°	563
9-Benzyl-1,2,3,6,7,8-hexahydro-1,3-dimethyl-2,6-dioxo-8-thio-	275°	46
8-Benzyl-1,2,3,6-tetrahydro-1,7-dimethyl-6-oxo-2-thio-	>295°	359, 361
8-Benzyl-1,2,3,6-tetrahydro-1-methyl-6-oxo-2-thio-	—	360
8-Benzyl-1,2,3,6-tetrahydro-6-oxo-1-phenyl-2-thio-	>360°	359
3-Benzyl-1,2,3,6-tetrahydro-2-oxo-6-thio-	>250°	37
2-Benzylthio-1,6-dihydro-1-methyl-6-oxo-	244–246°	101
2-Benzylthio-3,6-dihydro-3-methyl-6-oxo-	218°	532
6-Benzylthio-9-(2-hydroxyethyl)-	139–140°	430
8-Benzylthio-1,2,3,6-tetrahydro-1,3,7-trimethyl-2,6-dioxo-	149°	435
8-Benzylthio-1,2,3,6-tetrahydro-1,3,9-trimethyl-2,6-dioxo-	200–201°	46
2,6-bis-Ethylthio-7,8-dihydro-7,9-dimethyl-8-oxo-	104°	495
9-Butyl-1,2,3,6,7,8-hexahydro-1,3-dimethyl-2,6-dioxo-8-thio-	228–230°	46
8-Butyl-1,2,3,6-tetrahydro-1,3-dimethyl-2-oxo-6-thio-	204°	285
6-Butylthio-9-hydroxymethyl-	83–84°	510
8-Butylthio-1,2,3,6-tetrahydro-1,3-dimethyl-6-oxo-2-thio-	214–215°	285
8-Butylthio-1,2,3,6-tetrahydro-1,3,7-trimethyl-2,6-dioxo-	70–73°	435
8-Butylthio-1,2,3,6-tetrahydro-1,3,9-trimethyl-2,6-dioxo-	171–172°	46
9-Cyclohexyl-1,2,3,6,7,8-hexahydro-1,3-dimethyl-2,6-dioxo-8-thio-	>254°	729
9-Cyclohexyl-1,2,3,6-tetrahydro-1,3-dimethyl-8-methylthio-2,6-dioxo-	213–215°	729
8-Cyclohexylthio-1,2,3,6-tetrahydro-1,3,9-trimethyl-2,6-dioxo-	222–223°	46
7,9-Dibenzyl-7,8-dihydro-6-oxido-8-thio-	bet. 218°	576

TABLE 48 (*continued*)

Purine	M.p.	References
2,6-Dibenzylthio-7,8-dihydro-7,9-dimethyl-8-oxo-	158°	495
7,9-Diethyl-1,2,3,6,7,8-hexahydro-2,6-dioxo-8-thio-	340°	318
7,8-Dihydro-7,9-dimethyl-2,6-bis-methylthio-8-oxo-	172–173°	495
1,6-Dihydro-1,7-dimethyl-2-methylthio-6-oxo-	246–248°	302
1,6-Dihydro-3,8-dimethyl-2-methylthio-6-oxo-	312–315°	273
2,3-Dihydro-1,3-dimethyl-6-methylthio-2-oxo-	189–191°	106
2,3-Dihydro-3,7-dimethyl-6-methylthio-2-oxo-	260–262° or 299–302°	527, 528
1,6-Dihydro-1,7-dimethyl-2-methylthio-6-oxo-8-phenyl-	235°	282
7,8-Dihydro-7,9-dimethyl-6-oxido-8-thio-	bet. > 300°	576
1,6-Dihydro-9-(2-hydroxyethyl)-6-thio-	283–284°	430
1,6-Dihydro-1-methyl-2-methylthio-6-oxo-	273–276°	101, 302
1,6-Dihydro-2-methyl-8-methylthio-8-oxo-	> 310°	279
1,6-Dihydro-9-methyl-2-methylthio-6-oxo-	332°	221
1,6-Dihydro-9-methyl-8-methylthio-6-oxo-	280–282°	340
3,6-Dihydro-3-methyl-2-methylthio-6-oxo-	321°	44, 101
7,8-Dihydro-3-methyl-6-methylthio-8-oxo-	> 250°	72
1,6-Dihydro-8-methylthio-6-oxo-9-phenyl-	304–306°	283
3,6-Dihydro-2-methylthio-6-oxo-3-phenyl-	300°	679
7,9-Dimethyl-2-methylthio-6-oxo-	330–332°; tos. 257°	25
2-Ethoxy-1,6-dihydro-7-methyl-6-thio-	234°	458
3-Ethyl-6-ethylthio-7,8-dihydro-8-oxo-	250°	72
1-Ethyl-8-ethylthio-1,2,3,6-tetrahydro-3,7-dimethyl-2,6-dioxo-	136°	536
7-Ethyl-8-ethylthio-1,2,3,6-tetrahydro-1,3-dimethyl-2,6-dioxo-	115°	536
3-Ethyl-1,2,3,6,7,8,9-hexahydro-1,7-dimethyl-2,6-dioxo-8-thio-	272°	295
7-Ethyl-1,2,3,6,8,9-hexahydro-1,3-dimethyl-2,6-dioxo-8-thio-	264°	536
9-Ethyl-1,2,3,6,7,8-hexahydro-1,3-dimethyl-2,6-dioxo-8-thio-	275°	46, 319
3-Ethyl-1,2,3,6,7,8-hexahydro-2,6-dioxo-8-thio-	> 365°	295
9-Ethyl-1,2,3,6,7,8-hexahydro-2,6-dioxo-8-thio-	> 350°	319
9-Ethyl-1,2,3,6,7,8-hexahydro-1-methyl-2,6-dioxo-8-thio-	> 350°	319
7-Ethyl-1,2,3,6,7,8-hexahydro-9-methyl-2,6-dioxo-8-thio-	> 330°	297
9-Ethyl-1,2,3,6,7,8-hexahydro-7-methyl-2,6-dioxo-8-thio-	350°	168
7-Ethyl-1,2,3,6-tetrahydro-1,3-dimethyl-8-methylthio-2,6-dioxo-	128°	536

continued

TABLE 48 (*continued*)

Purine	M.p.	References
9-Ethyl-2,3,7,8-tetrahydro-6-methyl-2-oxo-8-thio-	295°	241
6-Ethylthio-9-hydroxymethyl-	122–123°	510
8-Ethylthio-1,2,3,6-tetrahydro-1,3-dimethyl-2,6-dioxo-	250–251°	536, 575
8-Ethylthio-1,2,3,6-tetrahydro-3,7-dimethyl-2,6-dioxo-	217°	536
8-Ethylthio-1,2,3,6-tetrahydro-1,3-dimethyl-2-oxo-6-thio-	223°	285
8-Ethylthio-1,2,3,6-tetrahydro-1,3-dimethyl-6-oxo-2-thio-	290°	285
8-Ethylthio-1,2,3,6-tetrahydro-1,3,7-trimethyl-2,6-dioxo-	138–139°	435, 536
8-Ethylthio-1,2,3,6-tetrahydro-1,3,7-trimethyl-6-oxo-2-thio-	156°	348
8-Ethylthio-1,2,3,6-tetrahydro-1,3,9-trimethyl-2,6-dioxo-	220–222°	46
1,2,3,6,7,8-Hexahydro-1,3-dimethyl-2,6-dioxo-9-phenyl-8-thio-	*ca.* 300°	319
1,2,3,6,8,9-Hexahydro-1,3-dimethyl-2,6-dioxo-7-phenyl-8-thio-	248–252°	563
1,2,3,6,7,8-hexahydro-1,3-dimethyl-2,6-dioxo-9-propyl-8-thio-	243–244°	46
1,2,3,6,7,8-Hexahydro-1,3-dimethyl-2,6-dioxo-8-thio-	322–324°	80, 88, 160, 285, 575, 776
1,2,3,6,7,8-Hexahydro-1,9-dimethyl-2,6-dioxo-8-thio-	> 350°	319, 330, 729
1,2,3,6,7,8-Hexahydro-7,9-dimethyl-2,6-dioxo-8-thio-	362°	248, 296
1,2,3,6,8,9-Hexahydro-3,7-dimethyl-2,6-dioxo-8-thio-	260°	85
1,2,3,6,7,8-Hexahydro-1,3-dimethyl-2-oxo-6,8-dithio-	327–329°	88, 285
1,2,3,6,7,8-Hexahydro-1,3-dimethyl-6-oxo-2,8-dithio-	315–322°	88, 285
1,2,3,6,7,8-Hexahydro-7,9-dimethyl-8-oxo-2,6-dithio-	> 300°	495
1,2,3,6,7,8-Hexahydro-2,6-dioxo-3-phenyl-8-thio-	309–310°	679
1,2,3,6,7,8-Hexahydro-6,8-dioxo-3-phenyl-2-thio-	300°	679
1,2,3,6,8,9-Hexahydro-7-hexyl-1,3-dimethyl-2,6-dioxo-8-thio-	210–214°	563
1,2,3,6,7,8-Hexahydro-9-isobutyl-2,6-dioxo-8-thio-	> 300°	57
1,2,3,6,7,8-Hexahydro-9-isopropyl-2,6-dioxo-8-thio-	> 300°	46
1,2,3,6,7,8-Hexahydro-3-methyl-2,6-dioxo-9-phenyl-8-thio-	—	181

TABLE 48 (*continued*)

Purine	M.p.	References
1,2,3,6,7,8-Hexahydro-3-methyl-2,6-dioxo-8-thio-	> 340°	80, 160, 776
1,2,3,6,7,8-Hexahydro-3-methyl-6,8-dioxo-2-thio-	> 300°	522
1,2,3,6,7,8-Hexahydro-9-methyl-2,6-dioxo-8-thio-	> 350°	319
1,2,3,6,7,8-Hexahydro-9-methyl-6-oxo-2,8-dithio-	> 300°	340
1,2,3,6,7,8-Hexahydro-6-oxo-3-phenyl-2,8-dithio-	300°	679
1,2,3,6,7,8-Hexahydro-1,3,7,9-tetramethyl-2,6-dioxo-8-thio-	255–260°	248, 321
1,2,3,6,7,8-Hexahydro-1,3,7,9-tetramethyl-6,8-dioxo-2-thio-	297–298°	348
1,2,3,6,8,9-Hexahydro-1,3,7-trimethyl-2,6-dioxo-8-thio-	—	85
1,2,3,6,8,9-Hexahydro-1,3,7-trimethyl-6,8-dioxo-2-thio-	343°	348
1,2,3,6,7,8-Hexahydro-1,3,9-trimethyl-2,6-dioxo-8-thio-	331–335°	46, 88, 319
1,2,3,6,7,8-Hexahydro-1,7,9-trimethyl-2,6-dioxo-8-thio-	317°	322
1,2,3,6,7,8-Hexahydro-3,7,9-trimethyl-2,6-dioxo-8-thio-	295–305°	248
1,2,3,6,8,9-Hexahydro-1,3,7-trimethyl-6-oxo-2,8-dithio-	285°	348
9-(2-Hydroxyethyl)-6-methylthio-	197–198°	430
3-Isobutyl-1,2,3,6-tetrahydro-1-methyl-2-oxo-6-thio-	169–172°	527
2,3,7,8-Tetrahydro-9-butyl-2-oxo-8-thio-	330°	74
1,2,3,6-Tetrahydro-1,3-dimethyl-7-(4-mercaptobutyl)-2,6-dioxo-	167–168°	211
1,2,3,6-Tetrahydro-1,3-dimethyl-7-(5-mercaptopentyl)-2,6-dioxo-	129–130°	211
1,2,3,6-Tetrahydro-1,3-dimethyl-8-methylthio-2,6-dioxo-	307–310°	88, 536
1,2,3,6-Tetrahydro-3,7-dimethyl-8-methylthio-2,6-dioxo-	263°	536
1,2,3,6-Tetrahydro-1,3-dimethyl-8-methylthio-2-oxo-6-thio-	253°	285
1,2,3,6-Tetrahydro-1,3-dimethyl-8-methylthio-6-oxo-2-thio-	338–340°	88, 285
1,2,3,6-Tetrahydro-1,7-dimethyl-6-oxo-8-phenyl-2-thio-	342°	359, 361
1,2,3,6-Tetrahydro-1,3-dimethyl-2-oxo-8-propylthio-6-thio-	231°	285
1,2,3,6-Tetrahydro-1,3-dimethyl-6-oxo-8-propylthio-2-thio-	214–215°	285
1,2,3,6-Tetrahydro-1,3-dimethyl-2-oxo-6-thio-	311 to 325°	214, 515, 527

continued

TABLE 48 (*continued*)

Purine	M.p.	References
1,2,3,6-Tetrahydro-1,3-dimethyl-6-oxo-2-thio-	344–348°	214, 515, 527
1,2,3,6-Tetrahydro-1,7-dimethyl-6-oxo-2-thio-	346–348°	361, 671
1,2,3,6-Tetrahydro-1,8-dimethyl-6-oxo-2-thio-	—	359, 360
1,6,7,8-Tetrahydro-1,9-dimethyl-6-oxo-8-thio-	> 300°	576
1,2,3,6-Tetrahydro-3,7-dimethyl-2-oxo-6-thio-	270–275°	528
1,2,3,6-Tetrahydro-3,7-dimethyl-6-oxo-2-thio-	306–308°	532, 671
1,2,3,6-Tetrahydro-7,9-dimethyl-2-oxo-6-thio-	bet. 285°	532
1,2,3,6-Tetrahydro-7,9-dimethyl-5-oxo-2-thio-	bet. 297°	532
2,3,7,8-Tetrahydro-6,9-dimethyl-2-oxo-8-thio-	—	581
1,2,3,6-Tetrahydro-7-(2-hydroxyethyl)-1,3-dimethyl-2-oxo-6-thio-	137° or 236–240°	88, 515
1,2,3,6-Tetrahydro-8-(2-hydroxyethylthio)-1,3,9-trimethyl-2,6-dioxo-	250°	46
1,2,3,6-Tetrahydro-8-hydroxymethyl-1,3-dimethyl-2,6-dioxo-	—	380
1,2,3,6-Tetrahydro-8-isopropylthio-1,3-dimethyl-2-oxo-6-thio-	255–256°	285
1,2,3,6-Tetrahydro-8-isopropylthio-1,3,7-trimethyl-2,6-dioxo-	126–128°	435
1,2,3,6-Tetrahydro-8-isopropylthio-1,3,9-trimethyl-2,6-dioxo-	182°	46
1,2,3,6-Tetrahydro-8-mercaptomethyl-3,7-dimethyl-2,6-dioxo-	> 350°	446
1,2,3,6-Tetrahydro-8-mercaptomethyl-1,3,9-trimethyl-2,6-dioxo-	206–207°	497
1,2,3,6-Tetrahydro-8-methoxy-1,3,7-trimethyl-6-oxo-2-thio-	174°	348
1,2,3,6-Tetrahydro-9-methyl-8-methylthio-2,6-dioxo-	320°	340, 443
1,6,7,8-Tetrahydro-1-methyl-2-methylthio-6,8-dioxo-	> 300°	817
1,2,3,6-Tetrahydro-9-methyl-8-methylthio-6-oxo-2-thio-	> 300°	283
1,2,3,6-Tetrahydro-1-methyl-6-oxo-8-phenyl-2-thio-	> 360°	360
1,2,3,6-Tetrahydro-8-methyl-6-oxo-3-phenyl-2-thio-	300°	679
1,2,3,6-Tetrahydro-1-methyl-2-oxo-6-thio-	323–325°	101
1,2,3,6-Tetrahydro-1-methyl-6-oxo-2-thio-	> 300°	101, 359, 360
1,2,3,6-Tetrahydro-3-methyl-2-oxo-6-thio-	320–322°	101, 522
1,2,3,6-Tetrahydro-3-methyl-6-oxo-2-thio-	> 300°	101, 162, 522
1,2,3,6-Tetrahydro-7-methyl-2-oxo-6-thio-	343–344°	347
1,2,3,6-Tetrahydro-7-methyl-6-oxo-2-thio-	> 360°	347
1,2,3,6-Tetrahydro-9-methyl-6-oxo-2-thio-	315°	283, 340
1,2,7,8-Tetrahydro-1-methyl-2-oxo-8-thio-	> 360°	74
1,6,8,9-Tetrahydro-7-methyl-2-oxo-6-thio-	343°	458

TABLE 48 (*continued*)

Purine	M.p.	References
2,3,7,8-Tetrahydro-9-methyl-2-oxo-8-thio-	> 360°	74
3,6,7,8-Tetrahydro-3-methyl-8-oxo-6-thio-	> 300°	72
1,2,3,6-Tetrahydro-8-methylthio-2,6-dioxo-3-phenyl-	250°	679
3,6,7,8-Tetrahydro-2-methylthio-6,8-dioxo-3-phenyl-	300°	679
1,2,3,6-Tetrahydro-2-oxo-9-phenyl-6-thio-	> 300°	57
1,2,3,6-Tetrahydro-6-oxo-3-phenyl-2-thio-	hy. 323°	679
1,2,3,6-Tetrahydro-1,3,7-trimethyl-2,6-dioxo-8-phenylthio-	147°	435
1,2,3,6-Tetrahydro-1,3,7-trimethyl-2,6-dioxo-8-propylthio-	131°	435
1,2,3,6-Tetrahydro-1,3,9-trimethyl-2,6-dioxo-8-propylthio-	203°	46
1,2,3,6-Tetrahydro-1,3,7-trimethyl-8-methylthio-2,6-dioxo-	183–185°	321, 435, 536
1,2,3,6-Tetrahydro-1,3,9-trimethyl-8-methylthio-2,6-dioxo-	240–242°	46
1,2,3,6-Tetrahydro-1,3,7-trimethyl-8-methylthio-6-oxo-2-thio-	183°	348
1,2,3,6-Tetrahydro-1,3,7-trimcthyl-2-oxo-6-thio-	246–247°	515, 527, 592, 651
1,2,3,6-Tetrahydro-1,3,7-trimethyl-6-oxo-2-thio-	231–232°	348, 649, 671
1,2,3,6-Tetrahydro-1,7,8-trimethyl-6-oxo-2-thio-	365°	363
1,2,3,6-Tetrahydro-1,7,9-trimethyl-2-oxo-6-thio-	bet. 355°	532
1,2,3,6-Tetrahydro-1,7,9-trimethyl-6-oxo-2-thio-	bet. 255°	532

TABLE 49.   Sulphonyl-Thiopurines.

Purine	M.p.	References
1,6-Dihydro-2-sulphamoyl-6-thio-	—	127
6-(4-Sulphobutylthio)-	259–261°	681

TABLE 50.   Amino-Oxopurines with a Functional Group.

Purine	M.p.	References
2-Acetamido-7-acetyl-1,6-dihydro-8-methyl-6-oxo-	320°	786
9-Acetamido-6-bromo-7,8-dihydro-8-oxo-	153°	291
1-Acetonyl-8-diethylamino-1,2,3,6-tetrahydro-3,7-dimethyl-2,6-dioxo-	110–112°	511
1-Acetonyl-8-dimethylamino-1,2,3,6-tetrahydro-3,7-dimethyl-2,6-dioxo-	164–165°	511
2-Amino-8-bromo-1,6-dihydro-6-oxo-	—	439, 685
6-(2-Amino-2-carboxyethyl)-2,3-dihydro-9-methyl-2-oxo-	239–243°	116
6-Amino-8-chloro-2,3-dihydro-3,7-dimethyl-2-oxo-	—	485
6-Amino-2-chloro-7,8-dihydro-9-methyl-8-oxo-	>360°	415
6-Amino-2-chloro-8,9-dihydro-7-methyl-8-oxo-	335°	112, 415
2-Amino-6-chloro-7,8-dihydro-8-oxo-	>300°	63, 414
6-Amino-2-chloro-7,8-dihydro-8-oxo-	—	2, 771
9-Amino-6-chloro-7,8-dihydro-8-oxo-	191–192°	291
6-Amino-8-chloro-2-ethoxy-	275–280°	103
6-Amino-2-chloro-8-ethoxy-7-methyl-	242–243°	112
2-Amino-1,6-dihydro-1-methyl-6-oxo-8-trifluoro-methyl-	350°	786
2-Amino-1,6-dihydro-7-methyl-6-oxo-8-trifluoro-methyl-	350°	786
2-Amino-1,6-dihydro-6-oxo-8-trifluoromethyl-	350°	786
2-Amino-8-ethylthio-1,6-dihydro-6-oxo-	>300°	245
6-Amino-7-formyl-8,9-dihydro-8-oxo-	>300°	34
6-Azido-2-chloro-8-ethoxy-7-methyl-	160–163°	112
2-Amino-1,6-dihydro-8-iodo-6-oxo-	—	86
2-Amino-7,8-dihydro-6-iodo-8-oxo-	—	414
6-Amino-2,3-dihydro-8-iodo-2-oxo-	>235°	86
2-Amino-1,6-dihydro-9-methyl-6-oxo-8-trifluoro-methyl-	350°	786
6-Carbamoyl-2-dimethylamino-7,8-dihydro-8-oxo-	311–316°	185
2-Chloro-6-diethylamino-7,8-dihydro-8-oxo-	>225°	257
2-Chloro-8-ethoxy-6-hydrazino-7-methyl-	207°	112
6-Dibutylamino-8-chloro-2-ethoxy-	164–165°	413
6-Dimethylamino-7,8-dihydro-2-methylsulphonyl-8-oxo-	—	284
2-Dimethylamino-6-ethoxycarbonyl-7,8-dihydro-8-oxo-	303–305°	185
Guanine/8-(2-carboxyethyl)-	—	135, 181
Guanine/8-(2-ethoxycarbonylethyl)-	—	135

TABLE 51.   Amino-Thiopurines with a Functional Group.

Purine	M.p.	References
6-Amino-9-benzyloxycarbonyl-2-methylthio-	172°	14
2-Amino-6-benzylthio-9-(2-chloroethyl)-	123–124°	430
6-Amino-8-carboxymethyl-9-methyl-2-methylthio-	244°	283
6-Amino-2-chloro-9-methyl-8-methylthio-	283°	14
2-Amino-9-(2-chloroethyl)-6-methylthio-	141–142°	430
2-Amino-6-ethoxycarbonyl-7,8-dihydro-8-thio-	275°	185
8-Chloro-6-dimethylamino-2-methylthio-	291°	410
2-Dimethylamino-6-ethoxycarbonyl-7,8-dihydro-8-thio-	246–248°	185

TABLE 52.   Amino-Oxo-Thiopurines.

Purine	M.p.	References
2-Amino-6-benzylthio-9-(2-hydroxyethyl)-	181–182°	430
2-Amino-1,6-dihydro-8-hydroxymethyl-6-thio-	320°	788
2-Amino-1,6-dihydro-8-methylthio-6-oxo-	> 300°	63
8-Amino-1,6-dihydro-2-methylthio-6-oxo-	> 320°	275
2-Amino-9-(2-hydroxyethyl)-6-methylthio-	164–166°	430
2-Amino-1,6,7,8-tetrahydro-6-oxo-8-thio-	> 300°	58, 63, 245
2-Amino-1,6,7,8-tetrahydro-8-oxo-6-thio-	> 300°	63
1,6-Dihydro-8-methylamino-2-methylthio-6-oxo-	hy. > 326°	282

TABLE 53.   Aminopurines with Two Minor Groups.

Purine	M.p.	References
6-(2-Amino-2-carboxyethyl)-2-chloro-9-methyl-	*ca.* 260°	116
8-Chloro-6-dimethylamino-2-methylsulphonyl-	254°	284

TABLE 54.   Oxo-Thiopurines with a Functional Group.

Purine	M.p.	References
1-Acetonyl-8-carboxymethylthio-1,2,3,6-tetrahydro-3,7-dimethyl-2,6-dioxo-	255–256°	511
1-Acetonyl-8-ethylthio-1,2,3,6-tetrahydro-3,7-dimethyl-2,6-dioxo-	174–175°	511
1-Acetonyl-1,2,3,6,8,9-hexahydro-3,7-dimethyl-2,6-dioxo-8-thio-	300°	511
1-Acetonyl-1,2,3,6-tetrahydro-3,7-dimethyl-8-methylthio-2,6-dioxo-	193–194°	511
6-Chloro-7,8-dihydro-2-methylthio-8-oxo-	>300°	410
8-Chloro-1,6-dihydro-2-methylthio-6-oxo-	>300°	410
2-Chloro-1,6,7,8-tetrahydro-7,9-dimethyl-8-oxo-6-thio-	—	495
2-Chloro-1,6,7,8-tetrahydro-8-oxo-6-thio-	>300°	45
8-Chloro-1,2,3,6-tetrahydro-1,3,7-trimethyl-6-oxo-2-thio-	186–187°	348
2-Ethoxy-6-ethoxycarbonyl-7,8-dihydro-8-thio-	244–246°	185
1,6,7,8-Tetrahydro-8-oxo-2-sulpho-6-thio-	—	424

TABLE 55.   Oxopurines with Two Minor Groups.

Purine	M.p.	References
7-Acetonyl-8-bromo-1,2,3,6-tetrahydro-1,3-dimethyl-2,6-dioxo-	203°	490
1-Acetonyl-8-bromo-1,2,3,6-tetrahydro-3,7-dimethyl-2,6-dioxo-	192–193°	511
8-Bromo-7-carbamoylmethyl-1,2,3,6-tetrahydro-1,3-dimethyl-2,6-dioxo-	280–281°	440
8-Bromo-7-carboxymethyl-1,2,3,6-tetrahydro-1,3-dimethyl-2,6-dioxo-	214°	440
8-Bromo-7-ethoxycarbonyl-1,2,3,6-tetrahydro-1,3-dimethyl-2,6-dioxo-	153°	440
6-Chloro-7,8-dihydro-2-methylsulphonyl-8-oxo-	308°	284
6-Chloro-7,8-dihydro-8-oxo-2-sulpho-	—	424

TABLE 56. Purines with Carboxyamino Groups.

Purine	M.p.	References
9-Benzyloxycarbonylamino-6-chloro-7,8-dihydro-8-oxo-	220–221°	291
6-Butoxycarbonylmethylamino-	203–204°	133
6-(4-Carboxybutylamino)-	240°	468
6-(1-Carboxyethylamino)-	hy. 235–237°	721
6-(2-Carboxyethylamino)-	237–238°	468, 469, 721
6-(1-Carboxyethylamino)-2,8-dichloro-	—	719
6-(2-Carboxyethylamino)-2,8-dichloro-	—	719
6-Carboxymethylamino-	> 300°	465, 469, 720, 721
6-(5-Carboxypentylamino)-	234–235°	468
6-(3-Carboxypropylamino)-	222–223°	468
6-Chloro-8-ethoxycarbonylamino-	137–138°	96
6-Chloro-9-ethoxycarbonylamino-	137–138°	96
2-Chloro-6-ethoxycarbonylmethylamino-7-methyl-	210°	258
6-Ethoxycarbonylmethylamino-	254–255°	133
6-Ethoxycarbonylmethylamino-7-methyl-	HCl 218°	258
6-Formylmethylamino-	HCl 283°	824
Hypoxanthine/2-(1-carboxyethylamino)-	—	107
6-Methoxycarbonylmethylamino-	241–242°	133, 465
6-Propoxycarbonylmethylamino-	225–227°	133

TABLE 57. Nitropurines.

Purine	M.p.	References
2-Amino-6-azido-	> 260°	201
6-Amino-2-azido-	> 260°	201
6-Amino-8-diazo-	—	210
2-Amino-8-diazo-1,6-dihydro-6-oxo-	—	210
2-Amino-1,6-dihydro-8-nitro-6-oxo-	—	210
2-Azido-	240–250°	626
6-Azido-7-benzyl-	145–146°	201
6-Azido-9-benzyl-	160–161°	201
6-Azido-2-chloro-	160°	201
6-Azido-2-chloro-7-methyl-	190–191°	112
2-Azido-1,6-dihydro-6-oxo-	> 260°	201
8-Azido-1,2,3,6-tetrahydro-3,7-dimethyl-2,6-dioxo-	170–180°	112
8-Azido-1,2,3,6-tetrahydro-1,3,7-trimethyl-2,6-dioxo-	—	487
7-Benzyl-1,2,3,6-tetrahydro-1,3-dimethyl-8-nitro-2,6-dioxo-	hy. 145–147°	390

continued

TABLE 57 (*continued*)

Purine	M.p.	References
8-Diazo-1,6-dihydro-2-methylamino-6-oxo-	—	210
2,6-Diazido-	207–210°	201, 474
6,8-Diazido-2-chloro-7-methyl-	190–195°	112
2,6-Diazido-7-methyl-	175–180°	112
8-Diazo-1,6-dihydro-6-oxo-	bet. expl. > 100°	210
8-Diazo-1,2,3,6-tetrahydro-2,6-dioxo-	—	210
1,6-Dihydro-8-nitro-6-oxo-	hy. —	210
1,6-Dihydro-2-(*N*-nitrosoethylamino)-6-oxo-	> 210°	723
6-Nitroso-	expl. 220°	158, 625
6-(*N*-Nitrosobenzylamino)-	226°	608
6-(*N*-Nitrosohydroxyamino)-	expl. 118–120°	158, 625
6-(*N*-Nitrosomethylamino)-	> 235°	723
1,2,3,6-Tetrahydro-8-diazo-1,3-dimethyl-2,6-dioxo-	—	210
1,2,3,6-Tetrahydro-1,3-dimethyl-8-nitro-2,6-dioxo-	280°	390, 689
1,2,3,6-Tetrahydro-3,7-dimethyl-8-nitro-2,6-dioxo-	270° or 282–283°	89, 688, 689
1,2,3,6-Tetrahydro-9-methyl-8-nitro-2,6-dioxo-	—	210
1,2,3,6-Tetrahydro-8-nitro-2,6-dioxo-	—	210
1,2,3,6-Tetrahydro-1,3,7-trimethyl-8-nitro-2,6-dioxo-	166–168°	89, 686
2,6,8-Triazido-	180–190°	474
2,6,8-Triazido-7-methyl-	155°	112

TABLE 58. Purine-*N*-Oxides.

Purine	M.p.	References
(*a*) 1-*Oxides:*		
1-Acetoxy-6-amino-	220–224°	695
1-Acetoxy-6-amino-2-methyl-	—	695
6-Amino-9-benzyl-1-benzyloxy-	brom. 218°	147, 148
6-Amino-9-benzyl-1-ethoxy-	brom. 130°; iod. 168°	146
6-Amino-9-benzyl-1-methoxy-	iod. 213–215°	146
6-Amino-9-benzyl-1-oxido-	280–285°	147, 148
6-Amino-1-benzyloxy-	164–166°; brom. 214–219°	147
6-Amino-1-benzyloxy-9-ethyl-	brom. 203–204°	148
6-Amino-1-benzyloxy-9-methyl-	brom. 207°	148
6-Amino-2-chloro-1-oxido-	236°	354
6-Amino-1,2-dihydro-1-hydroxy-2-oxo-	> 300°	154, 354

TABLE 58 (*continued*)

Purine	M.p.	References
6-Amino-1,2-dihydro-1-hydroxy-2-thio-	> 300°	154, 354
6-Amino-7,8-dihydro-1-oxido-8-oxo-	325°	694
6-Amino-1-ethoxy-	219°; iod. 208°	147
6-Amino-1-ethoxy-9-ethyl-	iod. 186–188°	147, 148
6-Amino-1-ethoxy-9-methyl-	iod. 204°	146
6-Amino-9-ethyl-1-methoxy-	iod. 184°	148
6-Amino-9-ethyl-1-oxido-	280–283°	148
6-Amino-8-hydroxymethyl-1-oxido-	350°	790
6-Amino-1-methoxy-	255–257°; iod. 222°	147
6-Amino-1-methoxy-9-methyl-	iod. 210–215°	147, 148
6-Amino-2-methyl-1-oxido-	306°	695
6-Amino-8-methyl-1-oxido-	295–297°	790
6-Amino-9-methyl-1-oxido-	292–293°	148
6-Amino-2-methylsulphinyl-1-oxido-	275°	354
6-Amino-2-methylthio-1-oxido-	279–280°	354
6-Amino-1-oxido-	297–301°	152, 153, 354, 697
6-Amino-1-oxido-2-sulpho-	amm. 296–300°	154
7-Benzyl-1,2,3,6-tetrahydro-1-hydroxy-8-methylthio-2,6-dioxo-	269–270°	145, 339
6-Carboxy-1-oxido-	—	829
2,6-Diamino-1-oxido-	—	153
7,8-Dihydro-6-methyl-1-oxido-8-oxo-	288–298°	379
7,8-Dihydro-1-oxido-8-oxo-	293°	379
6-Formyl-1-oxido-	—	829
Hypoxanthine/1-hydroxy-	> 360°	145, 346
6-Methyl-1-oxido-	265°	379, 829
1-Oxido-	—	829
1,2,3,6-Tetrahydro-1-hydroxy-7-methyl-8-methylthio-2,6-dioxo-	295°	145, 339
Xanthine/7-benzyl-1-hydroxy-	273–274°	145, 339
Xanthine/1-hydroxy-	350°	154, 338
Xanthine/1-hydroxy-7-methyl-	339°	145, 339
(*b*) 3-*Oxides*:		
6-Amino-3-benzyloxymethyl-	—	167
2-Amino-7,8-dihydro-3-hydroxy-6,8-dioxo-	—	828
6-Amino-7-methyl-3-oxido-	278°	356
6-Amino-3-oxido-	350°	795
6-Bromo-3-oxido-	expl. 178°	144
6-Carboxy-3-oxido-	285–287°	829
6-Chloro-3-oxido-	160°	144, 177
6-Cyano-3-oxido-	316–318°	829
3,6-Dihydro-3-hydroxy-6-thio-	hy. 230°	139, 144, 371

continued

TABLE 58 (*continued*)

Purine	M.p.	References
6-Ethoxy-3-oxido-	213°	143, 177
Guanine/3-hydroxy-	—	175*, 696
6-Hydroxyamino-3-oxido-	215°	176
Hypoxanthine/3-hydroxy-	> 300°	143, 144, 177, 795
6-Iodo-3-oxido-	expl. 175°	144
6-Methoxy-3-oxido-	216–218° or 232°	143, 144, 177, 795
6-Methyl-3-oxido-	240°	829
6-Methylsulphonyl-3-oxido-	192–198°	177, 795
6-Methylthio-3-oxido-	246–249°	371
3-Oxido-	288–289°	829
3-Oxido-6-sulphino-	—	795, 829
3-Oxido-6-sulpho-	400°	144, 176, 795
Xanthine/3-hydroxy-	> 350°	141, 175*, 186, 696
Xanthine/3-hydroxy-7,9-dimethyl-	hy. —	141
(*c*) 7-*Oxides:*		
Xanthine/8-benzyl-7-hydroxy-1,3-dimethyl-	215–218°	91, 306
Xanthine/8-benzyl-7-methoxy-1,3-dimethyl-	152–154°	699
Xanthine/7-butoxy-1,3,8-trimethyl-	68–69°	699
Xanthine/7-ethoxy-8-isopropyl-1,3-dimethyl-	106–109°	699
Xanthine/7-ethoxy-1,3,8-trimethyl-	144–146°	699
Xanthine/8-ethyl-7-hydroxy-1,3-dimethyl-	171–175°	91, 306
Xanthine/8-ethyl-7-methoxy-1,3-dimethyl-	164–166°	699
Xanthine/7-hydroxy-1,3-dimethyl-	210–218°	150, 520, 699
Xanthine/7-hydroxy-1,3-dimethyl-8-phenyl-	342–344°	91, 520, 698
Xanthine/7-hydroxy-1,3-dimethyl-8-propyl-	154–155°	91, 306
Xanthine/7-hydroxy-8-isopropyl-1,3-dimethyl-	221–223°	91
Xanthine/7-hydroxy-1,3,8-trimethyl-	180–184°	91, 306
Xanthine/8-isopropyl-7-methoxy-1,3-dimethyl-	124–126°	91, 520, 699
Xanthine/7-methoxy-1,3-dimethyl-	182–184°	699
Xanthine/7-methoxy-1,3-dimethyl-8-phenyl-	168–169°	91, 698, 699
Xanthine/7-methoxy-1,3-dimethyl-8-propyl-	120–121°	699
Xanthine/7-methoxy-1,3,8-trimethyl-	187°	150, 520, 699

TABLE 59.   Isotopically Labelled Purines.

Purine	Atoms Labelled	References
Unsubstituted	$^2H_{(2)}$	369, 731, 762
	$^2H_{(6)}$	369, 730, 731, 734, 762
	$^2H_{(8)}$	730, 734, 762, 765
	$^2H_{(9)}$	762
	$^2H_{(2)} \, ^2H_{(6)}$	369
	$^2H_{(6)} \, ^2H_{(8)}$	765
	$^{14}C_{(8)}$	732, 733
2-Methyl-	$^2H_{(6)}$	731
6-Methyl-	$^2H_{(2)}$	731, 766
8-Methyl-	$^2H_{(2)}$	731
	$^2H_{(6)}$	731
6-Amino- (adenine)	$^{14}C_{(2)}$	735, 736
	$^{14}C_{(8)}$	115, 739
	$^{14}C_{(4)} \, ^{14}C_{(6)}$	736, 738, 742
	$^{13}C_{(8)}$	740, 741
	$^{13}C_{(4)} \, ^{13}C_{(6)}$	637, 742
	$^3H_{(2)}$	793
	$^3H_{(8)}$	793
	$^3H_{(?)}$	769, 794
	$^{15}N_{(1)} \, ^{15}N_{(3)}$	637, 736, 742, 743
2,6-Diamino-	$^{14}C_{(2)}$	745
6-Amino-2,3-dihydro-2-oxo-		
(isoguanine)	$^{15}N_{(1)} \, ^{15}N_{(3)}$	184
2-Amino-1,6-dihydro-6-thio-	^{35}S	767, 768
6-Amino-7,8-dihydro-8-thio-	$^{14}C_{(8)}$	737
6-Amino-1-oxido-	$^{14}C_{(8)}$	744
6-Benzylamino-	$^{14}C_{(8)}$	746
6-Chloro-	$^{14}C_{(8)}$	750
6-Furfurylamino- (kinetin)	$^{14}C_{(8)}$	750
Guanine	$^{14}C_{(2)}$	745
	$^{14}C_{(4)}$	736, 738
	$^{14}C_{(5)}$	749
	$^{14}C_{(8)}$	736, 747, 748
	$^{13}C_{(4)}$	637
	$^3H_{(8)}$	794
	$^{15}N_{(7)}$	605
	$^{15}N_{(1)} \, ^{15}N_{(3)}$	637
	$^{15}N_{(1)} \, ^{15}N_{(2)} \, (^{15}NH_2)$	637, 641, 736
Guanine/7-methyl-	$^{15}N_{(7)}$	605
Guanine/3-hydroxy-	$^{14}C_{(8)}$	828
1,6-Dihydro-2-methyl-6-thio-	^{35}S	767
1,6-Dihydro-6-thio-	$^{14}C_{(8)}$	761
(6-mercaptopurine)	^{35}S	761, 767

continued

TABLE 59 (*continued*)

Purine	Atoms Labelled	References
Hypoxanthine	$^{14}C_{(2)}$	736
	$^{14}C_{(4)}$	754
	$^{14}C_{(8)}$	736, 745, 753
	$^{15}N_{(1)}$ $^{15}N_{(3)}$	755
6-Methoxy-	$^{2}H_{(2)}$	766
1,2,3,6-Tetrahydro-2-oxo-6-thio-	^{35}S	767, 768
Uric acid	$^{14}C_{(2)}$	756, 757
	$^{14}C_{(4)}$	754, 757, 758, 759
	$^{14}C_{(5)}$	754, 756, 759
	$^{14}C_{(6)}$	757, 758, 759
	$^{14}C_{(8)}$	736, 756, 757
	$^{15}N_{(1)}$ $^{15}N_{(3)}$	760
	$^{15}N_{(7)}$	757
	$^{15}N_{(9)}$	736, 760
Xanthine	$^{14}C_{(2)}$	736
	$^{14}C_{(5)}$	749
	$^{14}C_{(8)}$	736, 753
	$^{15}N_{(1)}$ $^{15}N_{(3)}$	637, 641, 736 755,

General References

1. Fischer and Ach, *Ber.*, **28**, 2473 (1895).
2. Fischer and Ach, *Ber.*, **30**, 2208 (1897).
3. Fischer, *Ber.*, **31**, 542 (1898).
4. Traube, *Ber.*, **33**, 1371 (1900).
5. Mann and Porter, *J. Chem. Soc.*, **1945**, 751.
6. Albert and Brown, *J. Chem. Soc.*, **1954**, 2060.
7. Brown, Ford, and Tratt, *J. Chem. Soc. (C)*, **1967**, 1445.
8. Noell and Robins, *J. Medicin. Chem.*, **5**, 558 (1962).
9. Pal and Horton, *J. Chem. Soc.*, **1964**, 400.
10. Montgomery and Hewson, *J. Org. Chem.*, **34**, 1396 (1969).
11. Neiman and Bergmann, *Israel J. Chem.*, **3**, 161 (1965).
12. Brown and Jacobsen, *J. Chem. Soc.*, **1965**, 3770.
13. Traube, *Ber.*, **33**, 3035 (1900).
14. Blackburn and Johnson, *J. Chem. Soc.*, **1960**, 4347.
15. Blackburn and Johnson, *J. Chem. Soc.*, **1960**, 4358.
16. Lister and Timmis, *J. Chem. Soc.*, **1960**, 327.
17. Naylor, Shaw, Wilson, and Butler, *J. Chem. Soc.*, **1961**, 4845.
18. Bergmann, Kalmus, Ungar-Waron, and Kwietny-Govrin, *J. Chem. Soc.*, **1963**, 3729.
19. Leese and Timmis, *J. Chem. Soc.*, **1961**, 3818.
20. Lira and Huffman, *J. Org. Chem.*, **31**, 2188 (1966).
21. Johnston and Gallagher, *J. Org. Chem.*, **28**, 1305 (1963).
22. Johnston, Fikes, and Montgomery, *J. Org. Chem.*, **27**, 973 (1962).
23. Bergmann, Neiman, and Kleiner, *J. Chem. Soc. (C)*, **1966**, 10.
24. Montgomery and Thomas, *J. Heterocyclic Chem.*, **1**, 115 (1964).
25. Jones and Robins, *J. Amer. Chem. Soc.*, **84**, 1914 (1962).
26. Shaw, *J. Org. Chem.*, **30**, 3371 (1965).
27. Montgomery and Thomas, *J. Org. Chem.*, **31**, 1411 (1966).
28. Vincze and Cohen, *Israel J. Chem.*, **4**, 23 (1966).
29. Montgomery and Thomas, *J. Org. Chem.*, **30**, 3235 (1965).
30. Shaw, *J. Amer. Chem. Soc.*, **80**, 3899 (1958).
31. Shaw, *J. Org. Chem.*, **27**, 883 (1962).
32. Lewis, Beaman, and Robins, *Canadian J. Chem.*, **41**, 1807 (1963).
33. Leonard and Fujii, *Proc. Nat. Acad. Sci. U.S.A.*, **51**, 73 (1964).
34. Denayer, *Bull. Soc. chim. (France)*, **1962**, 1358.
35. Goodman, Salce, and Hitchings, *J. Medicin. Chem.*, **11**, 516 (1968).
36. Giner-Sorolla, O'Bryant, Burchenal, and Bendich, *Biochemistry*, **5**, 3057 (1966).
37. Neiman and Bergmann, *Israel J. Chem.*, **6**, 9 (1968).
38. Miyaki, Iwasi, and Shimizu, *Chem. and Pharm. Bull. (Japan)*, **14**, 87 (1966).
39. Okano, Goya, and Kaizu, *J. Pharm. Soc. Japan*, **87**, 469 (1967).
40. Lewis, Noell, Beaman, and Robins, *J. Medicin. Chem.*, **5**, 607 (1962).
41. Giner-Sorolla, Nanos, Burchenal, Dollinger, and Bendich, *J. Medicin. Chem.*, **11**, 521 (1968).
42. Giner-Sorolla and Bendich, *J. Medicin. Chem.*, **8**, 667 (1965).

43. Shimizu and Miyaki, *Chem. and Pharm. Bull.* (*Japan*), **15**, 1066 (1967).
44. Townsend and Robins, *J. Amer. Chem. Soc.*, **84**, 3008 (1962).
45. Elion, Mueller, and Hitchings, *J. Amer. Chem. Soc.*, **81**, 3042 (1959).
46. Blicke and Schaaf, *J. Amer. Chem. Soc.*, **78**, 5857 (1956).
47. Hull, *J. Chem. Soc.*, **1959**, 481.
48. Leese and Timmis, *J. Chem. Soc.*, **1958**, 4107.
49. Leonard, Skinner, Lansford, and Shive, *J. Amer. Chem. Soc.*, **81**, 907 (1959).
50. Giner-Sorolla and Bendich, *J. Amer. Chem. Soc.*, **80**, 3932 (1958).
51. Whitehead and Traverso, *J. Amer. Chem. Soc.*, **82**, 3971 (1960).
52. Taylor, Vogel, and Cheng, *J. Amer. Chem. Soc.*, **81**, 2442 (1959).
53. Richter, Loeffler, and Taylor, *J. Amer. Chem. Soc.*, **82**, 3144 (1960).
54. Taylor and Loeffler, *J. Amer. Chem. Soc.*, **82**, 3147 (1960).
55. Leonard and Deyrup, *J. Amer. Chem. Soc.*, **84**, 2148 (1962).
56. Giner-Sorolla and Bendich, *J. Amer. Chem. Soc.*, **80**, 5744 (1958).
57. Koppel and Robins, *J. Amer. Chem. Soc.*, **80**, 2751 (1958).
58. Robins, *J. Amer. Chem. Soc.*, **80**, 6671 (1958).
59. Taylor, Knopf, Cogliano, Barton, and Pfleiderer, *J. Amer. Chem. Soc.*, **82**, 6058 (1960).
60. Montgomery and Temple, *J. Amer. Chem. Soc.*, **79**, 5238 (1957).
61. Montgomery and Holum, *J. Amer. Chem. Soc.*, **80**, 404 (1958).
62. Davies, Noell, Robins, Koppel, and Beaman, *J. Amer. Chem. Soc.*, **82**, 2633 (1960).
63. Elion, Goodman, Lange, and Hitchings, *J. Amer. Chem. Soc.*, **81**, 1898 (1959).
64. Montgomery and Temple, *J. Org. Chem.*, **25**, 395 (1960).
65. Kaiser and Burger, *J. Org. Chem.*, **24**, 113 (1959).
66. Fu, Chinoporos, and Terzian, *J. Org. Chem.*, **30**, 1916 (1965).
67. Temple, Kussner, and Montgomery, *J. Org. Chem.*, **30**, 3601 (1965).
68. Clark and Lister, *J. Chem. Soc.*, **1961**, 5048.
69. Hull, *J. Chem. Soc.*, **1958**, 2746.
70. Baddiley, Buchanan, Hawker, and Stephenson, *J. Chem. Soc.*, **1956**, 4659.
71. Burkhalter and Dill, *J. Org. Chem.*, **24**, 562 (1959).
72. Dille, Neiman, and Bergmann, *J. Chem. Soc.*, **1968**, 878.
73. Leonard and Hecht, *Chem. Comm.*, **1967**, 973.
74. Brown, *J. Appl. Chem.*, **9**, 203 (1959).
75. Bredereck, Siegel, and Föhlisch, *Chem. Ber.*, **95**, 403 (1962).
76. Lettré and Woenckhaus, *Annalen*, **649**, 131 (1961).
77. Albert, *Chem. and Ind.*, **1955**, 202.
78. Chu and Mautner, *J. Org. Chem.*, **26**, 4498 (1961).
79. Golovchinskaya, *Zhur. obshchei Khim.*, **24**, 136 (1954).
80. Boehringer, Ger. Pat., 128,117 (1902); through *Chem. Zentr.*, **I**, 548 (1902).
81. Bayer, Ger. Pat., 209,728 (1908).
82. Taylor and Sowisky, *J, Amer. Chem. Soc.*, **90**, 1374 (1968).
83. Chaman and Golovchinskaya, *Zhur. obshchei Khim.*, **33**, 3342 (1963).
84. Boldyrev and Makita, *J. Appl. Chem. U.S.S.R.* (Eng. Edn.), **28**, 399 (1955).
85. Berdichevskii, Rachinskii, and Novoselova, *J. Appl. Chem. U.S.S.R.* (Eng. Edn.), **31**, 670 (1958).
86. Koda, Biles, and Wolf, *J. Pharm. Sci.*, **57**, 2056 (1968).
87. Merz and Stähle, *Arch. Pharm.*, **293**, 801 (1965).
88. Mertz and Stahl, *Arzneim.-Forsch.*, **15**, 10 (1965).
89. Cacace and Masironi, *Ann. Chim.* (*Italy*), **47**, 366 (1957).
90. Cacace and Masironi, *Ann. Chim.* (*Italy*), **47**, 362 (1957).

91. Goldner, Dietz, and Carstens, *Annalen,* **691,** 142 (1966).
92. Pfleiderer and Nübel, *Annalen,* **631,** 168 (1960).
93. Koppel, O'Brien, and Robins, *J. Org. Chem.,* **24,** 259 (1959).
94. Elderfield and Prasad, *J. Org. Chem.,* **24,** 1410 (1959).
95. Beaman, Tautz, Duschinsky, and Grundberg, *J. Medicin. Chem.,* **9,** 373 (1966).
96. Temple, McKee, and Montgomery, *J. Org. Chem.,* **28,** 2257 (1963).
97. Barlin and Chapman, *J. Chem. Soc.* (*B*), **1967,** 954.
98. Brown, *J. Appl. Chem.,* **5,** 358 (1955).
99. Kiburis and Lister, *Chem. Comm.,* **1969,** 381.
100. Montgomery, Hewson, Clayton, and Thomas, *J. Org. Chem.,* **31,** 2202 (1966).
101. Elion, *J. Org. Chem.,* **27,** 2478 (1962).
102. Fischer, *Ber.,* **30,** 2220 (1897).
103. Fischer, *Ber.,* **30,** 2226 (1897).
104. Fischer, *Ber.,* **28,** 2480 (1895).
105. Breshears, Wang, Bechtolt, and Christensen, *J. Amer. Chem. Soc.,* **81,** 3789 (1959).
106. Neiman and Bergmann, *Israel J. Chem.,* **3,** 85 (1961).
107. Gerster and Robins, *J. Amer. Chem. Soc.,* **87,** 3752 (1965).
108. Beaman and Robins, *J. Appl. Chem.,* **12,** 432 (1962).
109. Townsend, Robins, Loeppky, and Leonard, *J. Amer. Chem. Soc.,* **86,** 5320 (1964).
110. Pitha and Ts'o, *J. Org. Chem.,* **33,** 1341 (1968).
111. Ballweg, *Annalen,* **649,** 114 (1961).
112. Itai and Ito, *Chem. and Pharm. Bull.* (*Japan*), **10,** 1141 (1962).
113. Neiman and Bergmann, *Israel J. Chem.,* **5,** 243 (1967).
114. Lewis, Noell, Beaman, and Robins, *J. Medicin. Chem.,* **5,** 607 (1962).
115. Clark and Kalcker, *J. Chem. Soc.,* **1950,** 1029.
116. Ehrhart and Hennig, *Arch. Pharm.,* **289,** 453 (1956).
117. Boehringer, Ger. Pat., 153,121 (1902).
118. Cohen and Vincze, *Israel J. Chem.,* **2,** 1 (1964).
119. Goldner, Dietz, and Carstens, *Tetrahedron Letters,* **1965,** 2701.
120. Ishido, Hosono, Isome, Maruyama, and Sato, *Bull. Chem. Soc. Japan,* **37,** 1389 (1964).
121. Ikehara and Ohtsuka, *Chem. and Pharm. Bull.* (*Japan*), **9,** 27 (1961).
122. Noell, Smith, and Robins, *J. Medicin. Chem.,* **5,** 996 (1962).
123. Bergmann, Rashi, Kleiner, and Knafo, *J. Chem. Soc.* (*C*), **1967,** 1254.
124. Montgomery and Thomas, *J. Org. Chem.,* **28,** 2304 (1963).
125. Fox and Van Praag, *J. Org. Chem.,* **26,** 526 (1961).
126. Baker and Tanna, *J. Org. Chem.,* **30,** 2857 (1965).
127. Beaman and Robins, *J. Amer. Chem. Soc.,* **83,** 4038 (1961).
128. Lira, *J. Heterocyclic Chem.,* **6,** 955 (1969).
129. Brown, *J. Appl. Chem.,* **7,** 109 (1957).
130. Dyer and Minnier, *J. Heterocyclic Chem.,* **6,** 23 (1969).
131. Dikstein, Bergmann, and Henis, *J. Biol. Chem.,* **224,** 67 (1959).
132. Giner-Sorolla, Zimmerman, and Bendich, *J. Amer. Chem. Soc.,* **81,** 2515 (1959).
133. Cherkasov, Tret'yakova, Kapran, and Nedel'kina, *Khim. geterotsikl. Soedinenii* **1967,** 170.
134. Merck, Ger. Pat., 290,910 (1914).
135. Bayer, Ger. Pat., 213,711 (1908); *Frdl.,* **9,** 1010 (1908–1910).
136. Traube, *Annalen,* **331,** 64 (1904).
137. Biltz and Strufe, *Annalen,* **423,** 227 (1921).
138. Okumura, Kotarn, Ariga, Masumura, and Kuraisui, *Bull. Chem. Soc. Japan,* **32,** 883 (1959).

139. Levin and Brown, *J. Medicin. Chem.*, **6**, 825 (1963).
140. Wölke, Pfleiderer, Delia, and Brown, *J. Org. Chem.*, **34**, 981 (1969).
141. Wölke and Brown, *J. Org. Chem.*, **34**, 978 (1969).
142. Brown, Pfleiderer, and Delia, *J. Org. Chem.*, in press.
143. Ajinomoto Co. Inc., Fr. Pat., 1,500,662 (1967).
144. Giner-Sorolla, Gryte, Bendich, and Brown, *J. Org. Chem.*, **34**, 2153 (1969).
145. Parham, Fissekis, and Brown, *J. Org. Chem.*, **31**, 966 (1966).
146. Fujii, Itaya, and Yamada, *Chem. and Pharm. Bull.* (*Japan*), **14**, 1452 (1966).
147. Fujii, Itaya, and Yamada, *Chem. and Pharm. Bull.* (*Japan*), **13**, 1017 (1965).
148. Fujii, Wu, Itaya, and Yamada, *Chem. and Ind.*, **1966**, 1598.
149. Biltz and Strufe, *Annalen*, **423**, 237 (1921).
150. Goldner, Dietz, and Carstens, *Z. Chem.*, **12**, 454 (1964).
151. Biltz and Krzikalla, *Annalen*, **423**, 255 (1921).
152. Stevens and Brown, *J. Amer. Chem. Soc.*, **80**, 2759 (1958).
153. Stevens, Magrath, Smith, and Brown, *J. Amer. Chem. Soc.*, **80**, 2755 (1958).
154. Parham, Fissekis, and Brown, *J. Org. Chem.*, **32**, 1151 (1967).
155. Altman and Ben-Ishai, *J. Heterocyclic Chem.*, **5**, 679 (1968).
156. DiPaco and Tauro, *Ann. Chim.* (*Italy*), **47**, 698 (1957).
157. Vieth and Leube, *Biochem. Z.*, **163**, 13 (1925).
158. Giner-Sorolla, *Galenica Acta*, **19**, 97 (1966).
159. Barlin and Chapman, *J. Chem. Soc.*, **1965**, 3017.
160. Boehringer, Ger. Pat., 133,300 (1902); through *Chem. Zentr.*, **II**, 314 (1902).
161. Bergmann, Kleiner, Neiman, and Rashi, *Israel J. Chem.*, **2**, 185 (1964).
162. Traube and Winter, *Arch. Pharm.*, **244**, 11 (1906).
163. Scarlat, *Bull. Soc. Sci. Bucarest*, **13**, 155 (1904).
164. Saneyoshi and Chihari, *Chem. and Pharm. Bull.* (*Japan*), **15**, 909 (1967).
165. Giner-Sorolla, *Chem. Ber.*, **101**, 611 (1968).
166. Brown and Mason, *J. Chem. Soc.*, **1957**, 682.
167. Montgomery and Thomas, *J. Amer. Chem. Soc.*, **87**, 5442 (1965).
168. Biltz and Bülow, *Annalen*, **426**, 265 (1921).
169. Thomas and Montgomery, *J. Org. Chem.*, **31**, 1413 (1960).
170. Temple and Montgomery, *J. Org. Chem.*, **31**, 1417 (1966).
171. Montgomery and Hewson, *J. Org. Chem.*, **30**, 1528 (1965).
172. Temple, McKee, and Montgomery, *J. Org. Chem.*, **28**, 925 (1963).
173. McNaught and Brown, *J. Org. Chem.*, **32**, 3689 (1967).
174. Gabriel and Colman, *Ber.*, **34**, 1234 (1901).
175. Delia and Brown, *J. Org. Chem.*, **31**, 178 (1966).
176. Giner-Sorolla, *J. Medicin. Chem.*, **12**, 717 (1969).
177. Kawashima and Kumashiro, *Bull. Chem. Soc. Japan*, **42**, 750 (1969).
178. Robins, Dille, Willits, and Christensen, *J. Amer. Chem. Soc.*, **75**, 263 (1953).
179. Elion, Burgi, and Hitchings, *J. Amer. Chem. Soc.*, **74**, 411 (1952).
180. Townsend and Robins, *J. Org. Chem.*, **27**, 990 (1962).
181. Traube and co-workers, *Annalen*, **432**, 266 (1923).
182. Johns, *J. Biol. Chem.*, **12**, 91 (1912).
183. Hoffer, jubilee volume dedicated to Emil Christoph Barrel, 1946, p. 428; through *Chem. Abstracts*, **41**, 4108 (1947).
184. Bendich, Tinker, and Brown, *J. Amer. Chem. Soc.*, **70**, 3109 (1948).
185. Clark, Kernick, and Layton, *J. Chem. Soc.*, **1964**, 3221.
186. Cresswell, Maurer, Strauss, and Brown, *J. Org. Chem.*, **30**, 408 (1965).
187. Baker, Joseph, and Schaub, *J. Org. Chem.*, **19**, 631 (1954).
188. Traube, *Ber.*, **37**, 4544 (1904).

189. Baker, Joseph, and Williams, *J. Org. Chem.*, **19**, 1793 (1954).
190. Ruttink, *Rec. Trav. chim.*, **65**, 751 (1946).
191. Bendich, Russell, and Fox, *J. Amer. Chem. Soc.*, **76**, 6073 (1954).
192. Engelmann, *Ber.*, **42**, 177 (1909).
193. Johns, *J. Biol. Chem.*, **11**, 67 (1912).
194. Johns, *J. Biol. Chem.*, **11**, 73 (1912).
195. Bobransky and Synowiedski, *J. Amer. Pharm. Assoc. Sci. Edn.*, **37**, 62 (1948).
196. Comte, U.S. Pat., 2,542,396 (1951); through *Chem. Abstracts*, **45**, 6657 (1951).
197. Ballantyne, U.S. Pat., 2,564,351 (1951); through *Chem. Abstracts*, **46**, 2574 (1952).
198. Homeyer, U.S. Pat., 2,646,432 (1953); through *Chem. Abstracts*, **48**, 8819 (1954).
199. Gepner and Krebs, *Zhur. obshchei Khim.*, **16**, 179 (1946).
200. Montgomery and Hewson, *J. Org. Chem.*, **26**, 4469 (1961).
201. Temple, Kussner, and Montgomery, *J. Org. Chem.*, **31**, 2210 (1966).
202. Myers and Zeleznick, *J. Org. Chem.*, **28**, 2087 (1963).
203. Isay, *Ber.*, **39**, 250 (1906).
204. Rose, *J. Chem. Soc.*, **1952**, 3448.
205. Elion, Lange, and Hitchings, *J. Amer. Chem. Soc.*, **78**, 2858 (1956).
206. Robins and Lin, *J. Amer. Chem. Soc.*, **79**, 490 (1957).
207. Beaman, Gerster, and Robins, *J. Org. Chem.*, **27**, 986 (1962).
208. Nagano, Inoue, Saggiomo, and Nodiff, *J. Medicin. Chem.*, **7**, 215 (1968).
209. Albert, *J. Chem. Soc. (B)*, **1966**, 438.
210. Jones and Robins, *J. Amer. Chem. Soc.*, **82**, 3773 (1960).
211. Parikh and Burger, *J. Amer. Chem. Soc.*, **77**, 2386 (1955).
212. Leonard, Carraway, and Helgeson, *J. Heterocyclic Chem.*, **2**, 291 (1965).
213. Libermann and Rouaix, *Bull. Soc. chim. France*, **1959**, 1793.
214. Merz and Stähle, *Beitrage zur Biochemie und Physiologie von Naturstoffen*, Fischer-Verlag, Jena, 1965, p. 285.
215. Bredereck, Hennig, and Pfleiderer, *Chem. Ber.*, **86**, 321 (1953).
216. Bredereck, Hennig, Pfleiderer, and Weber, *Chem. Ber.*, **86**, 333 (1953).
217. Liau, Yamashita, and Matsui, *Agric. and Biol. Chem. (Japan)*, **26**, 624 (1962).
218. Abshire and Berlinquet, *Canad. J. Chem.*, **42**, 1599 (1964).
219. Ueda, Kondo, Kono, Takemoto, and Imoto, *Makromol. Chem.*, **120**, 3 (1968).
220. Lettré and Ballweg, *Annalen*, **649**, 124 (1961).
221. Baddiley, Lythgoe, McNeil, and Todd, *J. Chem. Soc.*, **1943**, 383.
222. Andrews, Anand, Todd, and Topham, *J. Chem. Soc.*, **1949**, 2490.
223. Daly and Christensen, *J. Org. Chem.*, **21**, 177 (1956).
224. Baddiley, Lythgoe, and Todd, *J. Chem. Soc.*, **1943**, 386.
225. Baddiley, Lythgoe, and Todd, *J. Chem. Soc.*, **1944**, 318.
226. Craveri and Zoni, *Chimica*, **33**, 473 (1957).
227. Pappelardo and Conderelli, *Ann. Chim. (Italy)*, **43**, 727 (1953).
228. Craveri and Zoni, *Chimica*, **34**, 185 (1958).
229. Craveri and Zoni, *Chimica*, **34**, 267 (1958).
230. Craveri and Zoni, *Chimica*, **34**, 407 (1958).
231. Craveri and Zoni, *Chimica*, **34**, 239 (1958).
232. Speer and Raymond, *J. Amer. Chem. Soc.*, **75**, 114 (1953).
233. Fürst and Ebert, *Chem. Ber.*, **93**, 99 (1960).
234. Hager, Krantz, Harmond, and Burgison, *J. Amer. Pharm. Assoc. Sci. Edn.*, **43**, 152 (1954).
235. Hager and Kaiser, *J. Amer. Pharm. Assoc. Sci. Edn.*, **43**, 148 (1954).
236. Fu, Hargis, Chinoporos, and Malkiel, *J. Medicin. Chem.*, **10**, 109 (1967).
237. Albert, *J. Chem. Soc.*, **1955**, 2690.

238. Hull, *J. Chem. Soc.*, **1958**, 4069.
239. Ishidate and Yuki, *Chem. and Pharm. Bull.* (*Japan*), **5**, 240 (1957).
240. Ishidate and Yuki, *Chem. and Pharm. Bull.* (*Japan*), **5**, 244 (1957).
241. Johns and Baumann, *J. Biol. Chem.*, **15**, 515 (1913).
242. Koppel and Robins, *J. Org. Chem.*, **23**, 1457 (1958).
243. Prasad, Noell, and Robins, *J. Amer. Chem. Soc.*, **81**, 193 (1959).
244. Golovchinskaya, *Zhur. obschei. Khim.*, **24**, 146 (1954); through *Chem. Abstracts*, **49**, 3205 (1955).
245. Ueda, Tsuji, and Momona, *Chem. and Pharm. Bull.* (*Japan*), **11**, 912 (1963).
246. Bredereck, Hennig, Pfleiderer, and Deschler, *Chem. Ber.*, **86**, 845 (1953).
247. Bredereck, Herlinger, and Resemann, *Chem. Ber.*, **93**, 236 (1960).
248. Bredereck, Küpsch, and Wieland, *Chem. Ber.*, **92**, 566 (1959).
249. Khemelevski, *Zhur. obshchei. Khim.*, **31**, 3123 (1961); through *Chem. Abstracts*, **56**, 15505 (1962).
250. Biltz and Schmidt, *Annalen*, **431**, 70 (1923).
251. Huston and Allen, *J. Amer. Chem. Soc.*, **56**, 1793 (1934).
252. Golovchinskaya, *Zhur. obshchei Khim.*, **19**, 1173 (1946); through *Chem. Abstracts*, **42**, 2580 (1948).
253. Birkofer, Ritter, and Kühlthau, *Chem. Ber.*, **97**, 934 (1964).
254. Falco, Elion, Burgi, and Hitchings, *J. Amer. Chem. Soc.*, **74**, 4897 (1952).
255. Elion, Burgi, and Hitchings, *J. Amer. Chem. Soc.*, **73**, 5235 (1951).
256. Fu and Chinoporos, *J. Heterocyclic Chem.*, **3**, 476 (1966).
257. Adams and Whitmore, *J. Amer. Chem. Soc.*, **67**, 1271 (1945).
258. Uretskaya, Rybkina, and Menshikov, *Zhur. obshchei Khim.*, **30**, 327 (1960).
259. Montgomery, *J. Amer. Chem. Soc.*, **78**, 1928 (1956).
260. Goldman, Marsico, and Gazzola, *J. Org. Chem.*, **21**, 599 (1956).
261. Beaman and Robins, *J. Medicin. Chem.*, **5**, 1067 (1962).
262. Pfleiderer, *Annalen*, **647**, 161 (1961).
263. Temple, Kussner, and Montgomery, *J. Medicin. Chem.*, **5**, 866 (1962).
264. Montgomery and Temple, *J. Amer. Chem. Soc.*, **82**, 4592 (1960).
265. Montgomery and Temple, *J. Amer. Chem. Soc.*, **80**, 409 (1958).
266. Schaeffer and Weimar, *J. Amer. Chem. Soc.*, **81**, 197 (1959).
267. Berezovski and Yurkevich, *Zhur. obshchei Khim.*, **32**, 1655 (1962).
268. Lister, *J. Chem. Soc.*, **1960**, 899.
269. Pfleiderer and Schundehütte, *Annalen*, **612**, 158 (1958).
270. Bredereck, Von Schuh, and Martini, *Chem. Ber.*, **83**, 201 (1950).
271. Dikstein, Bergmann, and Chaimovitz, *J. Biol. Chem.*, **221**, 239 (1956).
272. Clark and Ramage, *J. Chem. Soc.*, **1958**, 2821.
273. Bergmann and Tamari, *J. Chem. Soc.*, **1961**, 4468.
274. Bredereck, Effenberger, and Österlin, *Chem. Ber.*, **100**, 2280 (1967).
275. Johns and Bauman, *J. Biol. Chem.*, **14**, 381 (1913).
276. Clusius and Vecchi, *Helv. Chim. Acta*, **36**, 1324 (1953).
277. Bergmann, Ungar, and Kalmus, *Biochim. Biophys. Acta*, **45**, 49 (1960).
278. Johns and Hagan, *J. Biol. Chem.*, **14**, 299 (1913).
279. King and King, *J. Chem. Soc.*, **1947**, 943.
280. Golovchinskaya, Ovcharova, and Cherkasova, *Zhur. obshchei Khim.*, **30**, 3332 (1960).
281. Bredereck, Küpsch, and Wieland, *Chem. Ber.*, **92**, 583 (1959).
282. Cook and Thomas, *J. Chem. Soc.*, **1950**, 1888.
283. Cook and Smith, *J. Chem. Soc.*, **1949**, 3001.
284. Noell and Robins, *J. Amer. Chem. Soc.*, **81**, 5997 (1959).

285. Dietz and Burgison, *J. Medicin. Chem.*, **9**, 160 (1966).
286. Baker and Schaub, U.S. Pat., 2,705,715 (1955); through *Chem. Abstracts*, **50**, 5041 (1956).
287. Balsiger, Fikes, Johnson, and Montgomery, *J. Org. Chem.*, **26**, 3386 (1961).
288. Baker, Schaub, and Joseph, *J. Org. Chem.*, **19**, 638 (1954).
289. Cavalieri and Bendich, *J. Amer. Chem. Soc.*, **72**, 2587 (1950).
290. Krackov and Christensen, *J. Org. Chem.*, **28**, 2677 (1963).
291. Temple, Smith, and Montgomery, *J. Org. Chem.*, **33**, 530 (1968).
292. Forrest, Hatfield, and Lagowski, *J. Chem. Soc.*, **1961**, 963.
293. Lohrmann, Lagowski, and Forrest, *J. Chem. Soc.*, **1964**, 451.
294. Taylor, Barton, and Paudler, *J. Org. Chem.*, **26**, 4961 (1961).
295. Biltz and Peukert, *Ber.*, **58**, 2190 (1925).
296. Biltz and Bülow, *Annalen*, **426**, 246 (1922).
297. Biltz and Heidrich, *Annalen*, **426**, 269 (1922).
298. Traube and Nithack, *Ber.*, **39**, 227 (1906).
299. Jerchel, Kracht, and Krucker, *Annalen*, **590**, 232 (1954).
300. Ridley, Spickett, and Timmis, *J. Heterocyclic Chem.*, **2**, 453 (1965).
301. Ried and Torinus, *Chem. Ber.*, **92**, 2902 (1959).
302. Bredereck and Edenhofer, *Chem. Ber.*, **88**, 1306 (1955).
303. Vogel and Taylor, *J. Amer. Chem. Soc.*, **79**, 1518 (1957).
304. Taylor and Morrison, *J. Amer. Chem. Soc.*, **87**, 1976 (1965).
305. Kempter, Rokos, and Pfleiderer, *Angew. Chem. Internat. Edn.*, **6**, 258 (1967).
306. Goldner, Dietz, and Carstens, *Annalen*, **699**, 145 (1966).
307. Fischer, *Ber.*, **30**, 559 (1897).
308. Fischer, *Ber.*, **33**, 1701 (1900).
309. Biltz and Strufe, *Annalen*, **423**, 242 (1921).
310. Borowitz, Bloom, Rothschild, and Sprinson, *Biochemistry*, **4**, 650 (1965).
311. Fischer and Clemm, *Ber.*, **30**, 3089 (1897).
312. Biltz, Marwitzky, and Heyn, *Annalen*, **423**, 122 (1921).
313. Biltz, Marwitzky, and Heyn, *Annalen*, **423**, 147 (1921).
314. Biltz and Zellner, *Annalen*, **423**, 192 (1923).
315. Gatewood, *J. Amer. Chem. Soc.*, **45**, 3056 (1921).
316. Biltz and Heyn, *Annalen*, **423**, 185 (1921).
317. Sembritsky, *Ber.*, **30**, 1814 (1897).
318. Biltz and Bülow, *Annalen*, **426**, 299 (1922).
319. Biltz and Strufe, *Annalen*, **423**, 200 (1921).
320. Gulland and Story, *J. Chem. Soc.*, **1938**, 692.
321. Biltz and Heidrich, *Annalen*, **426**, 290 (1922).
322. Biltz and Bülow, *Annalen*, **426**, 283 (1922).
323. Fischer and Tullner, *Ber.*, **35**, 2563 (1902).
324. Viout and Rumpf, *Bull. Soc. chim. France*, **1962**, 1250.
325. Fischer, Neumann, and Roch, *Chem. Ber.*, **85**, 752 (1952).
326. Prusse, *Annalen*, **441**, 203 (1925).
327. Bredereck, Effenberger, and Schweizer, *Chem. Ber.*, **95**, 956 (1962).
328. Dornow and Hinz, *Chem. Ber.*, **91**, 1834 (1958).
329. Barone, *J. Medicin. Chem.*, **6**, 39 (1963).
330. Cook, Downer, and Heilbron, *J. Chem. Soc.*, **1949**, 1069.
331. Armstrong, *Ber.*, **33**, 2308 (1900).
332. Whitely, *J. Chem. Soc.*, **91**, 1330 (1907).
333. Temple, McKee, and Montgomery, *J. Amer. Chem. Soc.*, **27**, 1671 (1962).
334. Goldner, Dietz, and Carstens, *Annalen*, **692**, 134 (1966).

335. Sarasin and Wegmann, *Helv. Chim. Acta*, **7**, 713 (1924).
336. Baxter, Gowenlock, Newbold, Woods, and Spring, *Chem. and Ind.*, **23**, 77 (1945).
337. Baxter and Spring, *J. Chem. Soc.*, **1945**, 232.
338. Bauer and Dhawan, *J. Heterocyclic Chem.*, **2**, 220 (1965).
339. Bauer, Nambury, and Dhawan, *J. Heterocyclic Chem.*, **1**, 275 (1964).
340. Cook and Smith, *J. Chem. Soc.*, **1949**, 2329.
341. Shaw, Warrener, Butler, and Ralph, *J. Chem. Soc.*, **1959**, 1648.
342. Nakata, *Meiji Yakka Daigaku Kenkyu Kiyo*, **2**, 66 (1963); through *Chem. Abstracts*, **61**, 1864 (1964).
343. Shaw, *J. Biol. Chem.*, **185**, 439 (1950).
344. Ichikawa, Kato, and Takenishi, *J. Heterocyclic Chem.*, **2**, 253 (1965).
345. Trout and Levy, *Rec. Trav. chim.*, **85**, 1254 (1966).
346. Taylor, Cheng, and Vogel, *J. Org. Chem.*, **24**, 2019 (1959).
347. Prasad and Robins, *J. Amer. Chem. Soc.*, **79**, 6401 (1957).
348. Biltz and Rakett, *Ber.*, **61**, 1409 (1928).
349. Blicke and Godt, *J. Amer. Chem. Soc.*, **76**, 3653 (1954).
350. Golovchinskaya, Kolganova, Nikolaeva, and Chaman, *Zhur. obshchei Khim.*, **33**, 1650 (1963).
351. Montequi, *Anales real Soc. espan. Fis. Quim.*, **24**, 731 (1926).
352. Yamazaki, Kumashiro, and Takenishi, *J. Org. Chem.*, **32**, 3032 (1967).
353. Shaw and Butler, *J. Chem. Soc.*, **1959**, 4040.
354. Cresswell and Brown, *J. Org. Chem.*, **28**, 2560 (1963).
355. Gompper, Gäng, and Saygin, *Tetrahedron Letters*, **1966**, 1885.
356. Taylor and Loeffler, *J. Org. Chem.*, **24**, 2035 (1959).
357. Montequi, *Anales real Soc. espan. Fis. Quim.*, **25**, 182 (1927).
358. Allsebrook, Gulland, and Story, *J. Chem. Soc.*, **1942**, 232.
359. Heilbron and Cook, Brit. Pat. 683,523 (1952); through *Chem. Abstracts*, **48**, 2093 (1954).
360. Cook, Davis, Heilbron, and Thomas, *J. Chem. Soc.*, **1949**, 1071.
361. Cook and Thomas, *J. Chem. Soc.*, **1950**, 1884.
362. Bills, Gebura, Meek, and Sweeting, *J. Org. Chem.*, **27**, 4633 (1962).
363. Bader and Downer, *J. Chem. Soc.*, **1953**, 1641.
364. Bredereck, Ulmer, and Waldmann, *Chem. Ber.*, **89**, 12 (1956).
365. Bredereck, Effenberger, Rainer, and Schosser, *Annalen*, **659**, 133 (1962).
366. Bredereck, Effenberger, and Rainer, *Annalen*, **673**, 82 (1964).
367. Fischer, *Ber.*, **31**, 2550 (1898).
368. Bredereck, Herlinger, and Graudums, *Chem. Ber.*, **95**, 54 (1962).
369. Coburn, Thorpe, Montgomery, and Hewson, *J. Org. Chem.*, **30**, 1110 (1965).
370. Beaman, *J. Amer. Chem. Soc.*, **76**, 5633 (1954).
371. Brown, Levin, Murphy, Sele, Reilly, Tarnowski, Schmid, Teller, and Stock, *J. Medicin. Chem.*, **8**, 190 (1965).
372. Kossel, *Z. physiol. Chem.*, **10**, 248 (1886).
373. Albert, *J. Chem. Soc.*, **1960**, 4705.
374. Mackay and Hitchings, *J. Amer. Chem. Soc.*, **78**, 3511 (1956).
375. Cohen, Thom, and Bendich, *J. Org. Chem.*, **27**, 3545 (1962).
376. Doerr, Wempen, Clarke, and Fox, *J. Org. Chem.*, **26**, 3401 (1961).
377. Chan, Schweizer, Ts'o, and Hemlkamp, *J. Amer. Chem. Soc.*, **86**, 4182 (1964).
378. Giner-Sorolla, Thom, and Bendich, *J. Org. Chem.*, **29**, 3209 (1964).
379. Stevens, Giner-Sorolla, Smith, and Brown, *J. Org. Chem.*, **27**, 567 (1962).
380. Goldner, Dietz, and Carstens, *Naturwiss.*, **51**, 137 (1964).
381. Golovchinskaya and Chaman, *Zhur. obshchei Khim.*, **22**, 2220 (1952).

382. Dyer and Bender, *J. Medicin. Chem.*, **7**, 10 (1964).
383. Ovcharova and Golovchinskaya, *Zhur. obshchei Khim.*, **34**, 3254 (1964).
384. Nikolaeva and Golovchinskaya, *Zhur. obshchei Khim.*, **34**, 1137 (1964).
385. Golovchinskaya, *Sbornik Statei obshchei Khim. Akad. Nauk. S.S.S.R.*, **1**, 692 (1953).
386. Golovchinskaya, *Sbornik Statei obshchei Khim. Akad. Nauk. S.S.S.R.*, **1**, 702 (1953).
387. Shioi, Jap. Pat., 177,356 (1949); through *Chem. Abstracts*, **45**, 7590 (1951).
388. Biltz and Sauer, *Ber.*, **64**, 752 (1931).
389. Donat and Carstens, *Chem. Ber.*, **92**, 1500 (1959).
390. Serchi, Sancio, and Bichi, *Farmaco. Ed. Sci.*, **10**, 733 (1955).
391. Shioi, Jap. Pat., 177,355 (1949); through *Chem. Abstracts*, **45**, 7590 (1951).
392. Golovchinskaya, Fedosova, and Cherkasova, *Zhur. priklad. Khim.*, **31**, 1241 (1958).
393. Chaman, Cherkasova, and Golovchinskaya, *Zhur. obshchei Khim.*, **30**, 1878 (1960).
394. Golovchinskaya, *Zhur. priklad. Khim.*, **31**, 918 (1958).
395. Zimmer and Atchley, *Arzneim.-Forsch.*, **16**, 541 (1966).
396. Chaman and Golovchinskaya, *Zhur. obshchei Khim.*, **32**, 2015 (1962).
397. Beaman and Robins, *J. Org. Chem.*, **28**, 2310 (1963).
398. Brown, *J. Appl. Chem.*, **7**, 109 (1957).
399. Fischer and Loeben, *Ber.*, **33**, 2278 (1900).
400. Montgomery and Temple, *J. Amer. Chem. Soc.*, **83**, 630 (1961).
401. Townsend and Robins, *J. Heterocyclic Chem.*, **3**, 241 (1966).
402. Traube and Herrman, *Ber.*, **37**, 2267 (1904).
403. Ovcharova, Nikolaeva, Chaman, and Golovchinskaya, *Zhur. obshchei Khim.*, **32**, 2010 (1962).
404. Lin and Price, *J. Org. Chem.*, **26**, 108 (1961).
405. Fischer, *Ber.*, **32**, 267 (1899).
406. Elion and Hitchings, *J. Amer. Chem. Soc.*, **78**, 3508 (1956).
407. Fujimoto, Jap. Pat., 6918 (1967); through *Chem. Abstracts*, **67**, 82224 (1967).
408. Robins, Jones, and Lin, *J. Org. Chem.*, **21**, 695 (1956).
409. Craveri and Zoni, *Boll. chim. farm.*, **97**, 393 (1958).
410. Noell and Robins, *J. Org. Chem.*, **24**, 320 (1959).
411. Davoll and Lowy, *J. Amer. Chem. Soc.*, **73**, 2936 (1951).
412. Davoll, Lythgoe, and Todd, *J. Chem. Soc.*, **1946**, 833.
413. Robins and Christensen, *J. Amer. Chem. Soc.*, **74**, 3624 (1952).
414. Fischer, *Ber.*, **31**, 2619 (1898).
415. Fischer, *Ber.*, **31**, 104 (1898).
416. Hitchings and Elion, U.S. Pat., 2,746,961 (1956); through *Chem. Abstracts*, **51**, 1258 (1957).
417. Fischer, *Ber.*, **30**, 2400 (1897).
418. Ovcharova and Golovchinskaya, *Zhur. obshchei Khim.*, **34**, 2472 (1964).
419. Fischer and Ach, *Ber.*, **31**, 1980 (1898).
420. Fischer and Ach, *Ber.*, **28**, 3135 (1895).
421. Fischer and Clemm, *Ber.*, **31**, 2622 (1898).
422. Golovchinskaya and Chaman, *Zhur. obshchei Khim.*, **30**, 1873 (1960).
423. Fischer and Ach, *Ber.*, **39**, 423 (1906).
424. Robins, *J. Org. Chem.*, **26**, 447 (1961).
425. Balsiger and Montgomery, *J. Org. Chem.*, **25**, 1573 (1960).
426. Usbeck, Jones, and Robins, *J. Amer. Chem. Soc.*, **83**, 1113 (1961).
427. Montgomery and Hewson, *J. Amer. Chem. Soc.*, **82**, 463 (1960).
428. Gerster and Robins, *J. Org. Chem.*, **31**, 3258 (1966).

429. Priewe and Poljack, *Chem. Ber.*, **88**, 1932 (1955).
430. O'Brien, Westover, Robins, and Cheng, *J. Medicin. Chem.*, **8**, 182 (1965).
431. Dyer, Reitz, and Farris, *J. Medicin. Chem.*, **6**, 289 (1963).
432. Garkusa, *Zhur. obshchei Khim.*, **27**, 1712 (1957).
433. Montgomery and Hewson, *J. Amer. Chem. Soc.*, **79**, 4559 (1957).
434. Leonard, Skinner, and Shive, *Arch. Biochem. Biophys.*, **92**, 33 (1961).
435. Long, *J. Amer. Chem. Soc.*, **69**, 2939 (1947).
436. Blicke and Godt, *J. Amer. Chem. Soc.*, **76**, 3655 (1954).
437. Ovcharova and Golovchinskaya, *Zhur. obshchei Khim.*, **30**, 3339 (1960).
438. Ovcharova and Golovchinskaya, U.S.S.R. Pat., 135,084 (1961); through *Chem. Abstracts*, **55**, 16596 (1961).
439. Duval and Ebel, *Bull. Soc. Chim. biol.*, **46**, 1059 (1964).
440. Cacace, Criserà, and Zifferero, *Ann. Chim. (Italy)*, **46**, 99 (1956).
441. Lespagnol and Gaumeton, *Bull. Soc. chim. France*, **1961**, 253.
442. Biltz and Beck, *J. prakt. Chem.*, **118**, 149 (1928).
443. Biltz and Bülow, *Annalen*, **426**, 306 (1922).
444. Koppel, Springer, Robins, Schneider, and Cheng, *J. Org. Chem.*, **27**, 2173 (1962).
445. Kallischnigg, U.S. Pat., 2,879,271 (1959); through *Chem. Abstracts*, **54**, 591 (1960).
446. Golovchinskaya, Ebed, and Chaman, *Zhur. obshchei Khim.*, **32**, 4097 (1962).
447. Golovchinskaya and Chaman, *Zhur. obshchei Khim.*, **22**, 535 (1952).
448. Chaman and Golovchinskaya, *Zhur. obshchei Khim.*, **31**, 2645 (1961).
449. Ebed, Chaman, and Golovchinskaya, *Zhur. obshchei Khim.*, **36**, 816 (1966).
450. Zelnik, *Bull. Soc. chim. France*, **1960**, 1917.
451. Huber, *Angew Chem.*, **68**, 706 (1956).
452. Sutcliffe and Robins, *J. Org. Chem.*, **28**, 1662 (1963).
453. Garkusa, U.S.S.R. Pat., 104,281 (1956); through *Chem. Abstracts*, **51**, 5847 (1957).
454. Panagapoulos and co-workers, *Arzneim.-Forsch.*, **15**, 204 (1965).
455. Fischer, *Ber.*, **30**, 1846 (1897).
456. Davoll and Lowy, *J. Amer. Chem. Soc.*, **74**, 1563 (1952).
457. Lloyd, *Chem. and Ind.*, **1963**, 953.
458. Fischer, *Ber.*, **31**, 431 (1898).
459. MacCorquodale, *J. Amer. Chem. Soc.*, **51**, 2245 (1929).
460. Montgomery and Holum, *J. Amer. Chem. Soc.*, **79**, 2185 (1957).
461. Brown, England, and Lyall, *J. Chem. Soc. (C)*, **1966**, 266.
462. Greenberg, Ross, and Robins, *J. Org. Chem.*, **24**, 1314 (1959).
463. Cacace, Fabrizi, and Zifferero, *Ann. Chim. (Italy)*, **46**, 91 (1956).
464. Schaeffer and Vince, *J. Medicin. Chem.*, **8**, 710 (1965).
465. Bullock, Hand, and Stokstad, *J. Amer. Chem. Soc.*, **78**, 3693 (1956).
466. Sutherland and Christensen, *J. Amer. Chem. Soc.*, **79**, 2251 (1957).
467. Skinner, Shive, Ham, Fitzgerald, and Eakin, *J. Amer. Chem. Soc.*, **78**, 5097 (1956).
468. Lettré and Ballweg, *Annalen*, **633**, 171 (1960).
469. Ballio and DiVittorio, *Gazzetta*, **90**, 501 (1960).
470. Montgomery, Hewson, and Temple, *J. Medicin. Chem.*, **5**, 15 (1962).
471. Bowles, Schneider, Lewis, and Robins, *J. Medicin. Chem.*, **6**, 471 (1963).
472. Brown and Weliky, *J. Org. Chem.*, **23**, 125 (1958).
473. Falconer, Gulland, and Story, *J. Chem. Soc.*, **1939**, 1784.
474. Smirnova and Postovski, *Zhur. Vsesoyuz Khim. obshch. im D.I. Mendeleeva*, **9**, 711 (1964).
475. Huber, *Chem. Ber.*, **90**, 698 (1957).
476. Bergmann and Heimhold, *J. Chem. Soc.*, **1935**, 955.
477. Bergmann and Heimhold, *J. Chem. Soc.*, **1935**, 1365.

478. Bergmann and Burke, *J. Org. Chem.*, **21**, 226 (1956).
479. Biltz and Pardon, *Ber.*, **63**, 2876 (1930).
480. Wislicenus and Körber, *Ber.*, **35**, 1991 (1902).
481. Biltz and Bergius, *Annalen*, **414**, 54 (1917).
482. Biltz and Max, *Annalen*, **414**, 68 (1917).
483. Biltz and Max, *Annalen*, **414**, 79 (1917).
484. Fischer and Ach, *Ber.*, **32**, 2721 (1899).
485. Fischer, *Ber.*, **30**, 1839 (1897).
486. Zimmer and Mettalia, *J. Org. Chem.*, **24**, 1813 (1959).
487. Cramer, *Ber.*, **27**, 3089 (1894).
488. Einhorn and Baumeister, *Ber.*, **31**, 1138 (1898).
489. Damiens and Delaby, *Bull. Soc. chim. France*, **1955**, 888.
490. Polonovski, Pesson, and Zelnik, *Compt. rend.*, **236**, 2519 (1953).
491. Fischer, *Ber.*, **32**, 250 (1899).
492. Johnston, Holum, and Montgomery, *J. Amer. Chem. Soc.*, **80**, 6265 (1958).
493. Mautner, *J. Amer. Chem. Soc.*, **78**, 5292 (1956).
494. Mautner, Chu, Jaffe, and Sartorelli, *J. Medicin. Chem.*, **6**, 36 (1963).
495. Ray, Chakravarti, and Bose, *J. Chem. Soc.*, **1923**, 1957.
496. von Schuh, Ger. Pat., 1,091,570 (1960); through *Chem. Abstracts*, **56**, 14305 (1962).
497. Golovchinskaya and Chaman, *Zhur. obshchei Khim.*, **32**, 3245 (1962).
498. Kallischnigg, Ger. Pat., 1,001,273 (1957); through *Chem. Abstracts*, **54**, 1569 (1960).
499. Klinger and Kohlstädt, Ger. Pat., 1,011,424 (1957); through *Chem. Abstracts*, **53**, 18071 (1959).
500. Balsiger, Fikes, Johnston, and Montgomery, *J. Org. Chem.*, **26**, 3446 (1961).
501. Golovchinskaya, *Zhur. obshchei Khim.*, **18**, 2129 (1948).
502. Schaeffer and Vince, *J. Medicin. Chem.*, **8**, 33 (1965).
503. Robins, Godefroi, Taylor, Lewis, and Jackson, *J. Amer. Chem. Soc.*, **83**, 2574 (1961).
504. Fischer, *Annalen*, **215**, 253 (1882).
505. Tafel and Ach, *Ber.*, **34**, 1170 (1901).
506. Traube and Dudley, *Ber.*, **46**, 3839 (1913).
507. Koppel, O'Brien, and Robins, *J. Amer. Chem. Soc.*, **81**, 3046 (1959).
508. Kruger, *Ber.*, **26**, 1914 (1893).
509. Golovchinskaya and Chaman, *Zhur. obshchei Khim.*, **22**, 2225 (1952).
510. Bryant and Harmon, *J. Medicin. Chem.*, **10**, 104 (1967).
511. Gräfe, *Arch. Pharm.*, **300**, 111 (1967).
512. Stoll and Schmidt, U.S. Pat., 2,756,229 (1956).
513. Kazmirowsky, Dietz, and Carstens, *J. prakt. Chem.*, **19**, 162 (1963).
514. Roth, *Arch. Pharm.*, **292**, 234 (1959).
515. Seyden-Penne, Le Thi Minh, and Chabrier, *Bull. Soc. chim. France*, **1966**, 3934.
516. Dille and Christensen, *J. Amer. Chem. Soc.*, **76**, 5087 (1954).
517. Chemiewerk Homberg, A. G. Brit. Pat., 816,299 (1959); through *Chem. Abstracts*, **54**, 1571 (1960).
518. Biltz and Max, *Ber.*, **53**, 2327 (1920).
519. Biltz and Pardon, *J. prakt. Chem.*, **134**, 310 (1932).
520. Goldner, Dietz, and Carstens, *Z. Chem.*, **4**, 454 (1964).
521. Bergmann and Kalmus, *J. Chem. Soc.*, **1962**, 860.
522. Bergmann, Levin, Kalmus, and Kwietny-Govrin, *J. Org. Chem.*, **26**, 1504 (1961).
523. Bergmann and Kalmus, *J. Org. Chem.*, **26**, 1660 (1961).
524. Loo, Michael, Garceau, and Reid, *J. Amer. Chem. Soc.*, **81**, 3039 (1959).

525. Elion and Hitchings, *J. Amer. Chem. Soc.*, **77**, 1676 (1955).
526. Elion, Lange, and Hitchings, *J. Amer. Chem. Soc.*, **78**, 217 (1956).
527. Wooldridge and Slack, *J. Chem. Soc.*, **1962**, 1862.
528. Kalmus and Bergmann, *J. Chem. Soc.*, **1960**, 3679.
529. Khaletski and Eshmann, *Zhur. obshchei Khim.*, **18**, 2116 (1948).
530. Khaletski and Eshmann, *Zhur. obshchei Khim.*, **20**, 1246 (1950).
531. Tafel, *Ber.*, **32**, 3194 (1899).
532. Bredereck, Schellenberg, Nast, Heise, and Christmann, *Chem. Ber.*, **99**, 944 (1966).
533. Biltz and Bülow, *Annalen*, **423**, 159 (1921).
534. Golovchinskaya, *Zhur. priklad. Khim.*, **30**, 1374 (1957).
535. Frydman and Troparesky, *Anales. Asoc. quim. argentina*, **45**, 79 (1957).
536. Biltz and Beck, *J. prakt. Chem.*, **118**, 198 (1928).
537. Alexander and Marienthal, *J. Pharm. Sci.*, **53**, 962 (1964).
538. Roche Products Ltd., Brit. Pat., 750,588 (1956); through *Chem. Abstracts*, **51**, 2888 (1957).
539. Vel'kina, Chaman, and Ebed, *Zhur. obshchei Khim.*, **37**, 508 (1967).
540. Cacace, Fabrizi, and Zifferero, *Ann. Chim. (Italy)*, **45**, 983 (1955).
541. Stieglitz and Stamm, Ger. Pat., 1,122,534 (1962); through *Chem. Abstracts*, **56**, 14306 (1962).
542. Klosa, *Arch. Pharm.*, **288**, 301 (1955).
543. Klosa, *Arch. Pharm.*, **288**, 114 (1955).
544. Zelnik, Pesson, and Polonovski, *Bull. Soc. chim. France*, **1956**, 888.
545. McMillan and Wuest, *J. Amer. Chem. Soc.*, **75**, 1998 (1953).
546. Eidenbenz and Von Schuh, Ger. Pat., 860,217 (1952); through *Chem. Abstracts*, **47**, 11238 (1953).
547. Serchi and Bichi, *Farmaco, Ed. Sci.*, **11**, 501 (1956).
548. Weissenburger, *Arch. Pharm.*, **288**, 532 (1955).
549. Biltz and Max, *Annalen*, **423**, 318 (1921).
550. Cacace and Zifferero, *Ann. Chim.*, **45**, 1026 (1955).
551. Polonovski, Pesson, and Zelnik, *Compt. rend.*, **241**, 215 (1955).
552. Doebel and Spiegelberger, U.S. Pat., 2,761,862 (1956); through *Chem. Abstracts*, **51**, 3676 (1957).
553. Bredereck, Christmann, Koser, Schellenberg, and Nast, *Chem. Ber.*, **95**, 1812 (1962).
554. Pfleiderer and Sagi, *Annalen*, **673**, 78 (1964).
555. Johns, *J. Biol. Chem.*, **17**, 1 (1914).
556. Biilman and Bjerrum, *Ber.*, **50**, 837 (1917).
557. Biltz and Sedlatschek, *Ber.*, **57**, 175 (1924).
558. Okuda, *J. Pharm. Soc. Japan*, **80**, 205 (1960).
559. Hashizume and Iwamura, *Tetrahedron Letters*, **1965**, 3095.
560. Biltz and Strufe, *Annalen*, **404**, 170 (1914).
561. Golovchinskaya, *Zhur. obshchei Khim.*, **29**, 1213 (1959).
562. Golovchinskaya and Kolodkin, *Zhur. obshchei Khim.*, **29**, 1650 (1959).
563. Dolman, Van der Goot, Mos, and Moed, *Rec. Trav. chim.*, **83**, 1215 (1964).
564. Biltz and Topp, *Ber.*, **44**, 1524 (1911).
565. Biltz and Krzikalla, *Annalen*, **457**, 131 (1927).
566. Taylor and Cheng, *J. Org. Chem.*, **25**, 148 (1960).
567. Falconer and Gulland, *J. Chem. Soc.*, **1939**, 1369.
568. Bergmann and Dikstein, *J. Amer. Chem. Soc.*, **77**, 691 (1955).
569. Boehringer, Ger. Pat., 106,493 (1893), *Frdl.*, **5**, 829 (1897–1900).
570. Mabery and Hill, *Ber.*, **11**, 1329 (1878).

571. Bayer, Ger. Pat., 254,488 (1911); through *Chem. Zentr.*, **I**, 197 (1913).

572. Boehringer, Ger. Pat., 102,158 (1897); *Frdl.*, **5**, 827 (1897–1900).

573. Montgomery, Balsiger, Fikes, and Johnston, *J. Org. Chem.*, **27**, 195 (1962).

574. Montgomery, Holum, and Johnston, *J. Amer. Chem. Soc.*, **81**, 3963 (1959).

575. Ochiai, *Ber.*, **69**, 1650 (1936).

576. Neiman, *Chem. Comm.*, **1968**, 200.

577. Todd and Bergel, *J. Chem. Soc.*, **1936**, 1559.

578. Gordon, *J. Amer. Chem. Soc.*, **73**, 984 (1951).

579. Howard, Lythgoe, and Todd, *J. Chem. Soc.*, **1945**, 556.

580. Krüger, *Z. physiol. Chem.*, **18**, 434 (1894).

581. Johns, *J. Biol. Chem.*, **21**, 319 (1915).

582. Cacace and Masironi, *Ann. Chim. (Italy)*, **46**, 806 (1956).

583. Miller, Skoog, Okumura, Von Saltza, and Strong, *J. Amer. Chem. Soc.*, **77**, 2662 (1955).

584. Wellcome Foundation Ltd., Brit. Pat., 713,259 (1954); through *Chem. Abstracts*, **49**, 13301 (1955).

585. Okumura, Enishi, Itoh, Masumura, and Kuraishi, *Chem. and Pharm. Bull. (Japan)*, **32**, 886 (1959).

586. Leonard, Orme-Johnson, McMurtray, Skinner, and Shive, *Arch. Biochem. Biophys.*, **99**, 16 (1962).

587. Eckstein, Gorczyca, Kocwa, and Zejc, *Dissertationes Pharm. (Poland)*, **10**, 239 (1958); through *Chem. Abstracts*, **53**, 18046 (1959).

588. Eckstein and Sulko, *Dissertationes Pharm. (Poland)*, **13**, 97 (1961); through *Chem. Abstracts*, **55**, 23548 (1961).

589. Giani and Molteni, *Farmaco. Ed. Sci.*, **12**, 1016 (1957); through *Chem. Abstracts*, **52**, 12874 (1958).

590. Dolman, Van der Goot, and Moed, *Rec. Trav. chim.*, **84**, 193 (1965).

591. Ikehara, Yamazaki, and Fujieda, *Chem. and Pharm. Bull. (Japan)*, **10**, 1075 (1962).

592. Walter and Voss, *Annalen*, **698**, 113 (1966).

593. Wellcome Foundation Ltd., Brit. Pat., 713,286 (1954); through *Chem. Abstracts*, **49**, 12546 (1955).

594. Hitchings and Elion, U.S. Pat., 2,691,654 (1955); through *Chem. Abstracts*, **50**, 1933 (1956).

595. Elion and Hitchings, U.S. Pat., 2,724,711 (1955); through *Chem. Abstracts*, **50**, 8748 (1956).

596. Hitchings and Elion, U.S. Pat., 2,756,228 (1956); through *Chem. Abstracts*, **51**, 2887 (1956).

597. Elion and Hitchings, U.S. Pat., 2,721,866 (1955); through *Chem. Abstracts*, **50**, 8748 (1956).

598. Ikehara, Nakazawa, and Nakayama, *Chem. and Pharm. Bull. (Japan)*, **10**, 660 (1962).

599. Giner-Sorolla, Nanos, Burchenal, Dollinger, and Bendich, *J. Medicin. Chem.*, **11**, 52 (1968).

600. Skinner and Shive, *J. Amer. Chem. Soc.*, **77**, 6692 (1955).

601. Fritz, *Pharm. Post.*, **28**, 130 (1895).

602. Meister, Lucius, Brüning, and Hoechst, Ger. Pat., 74045 (1893); *Frdl.*, **3**, 979 (1890–1894).

603. Balaban, *J. Chem. Soc.*, **1926**, 569.

604. Miller, Skoog, Okumura, Von Saltza, and Strong, *J. Amer. Chem. Soc.*, **78**, 1375 (1956).

605. Litwack and Weissman, *Biochemistry*, **5**, 3007 (1966).

606. Brookes and Lawley, *J. Chem. Soc.*, **1960**, 539.
607. Broom, Townsend, Jones, and Robins, *Biochemistry*, **3**, 494 (1964).
608. Overbeek, U.S. Pat., 3,013,885 (1960); through *Chem. Abstracts*, **56**, 14306 (1962).
609. Roth and Brandes, *Arch. Pharm.*, **298**, 765 (1965).
610. Zelnik and Pesson, *Bull. Soc. chim. France*, **1959**, 1667.
611. Spies, *J. Amer. Chem. Soc.*, **61**, 350 (1939).
612. Davoll and Lowy, *J. Amer. Chem. Soc.*, **73**, 1650 (1951).
613. Birkofer, *Ber.*, **76**, 769 (1943).
614. Schein, *J. Medicin. Chem.*, **5**, 302 (1962).
615. Mehrotra, Jain, and Anand, *Indian J. Chem.*, **4**, 146 (1960).
616. Wulff, *Annalen*, **17**, 468 (1893).
617. Baker and Hewson, *J. Org. Chem.*, **22**, 959 (1957).
618. Okumura, Jap. Pat., 8526 (1959); through *Chem. Abstracts*, **54**, 6769 (1960).
619. Baizer, Clark, Dub, and Loter, *J. Org. Chem.*, **21**, 1276 (1956).
620. Bullock, Hand, and Stokstad, *J. Org. Chem.*, **22**, 568 (1957).
621. Lettré and Ballweg, *Chem. Ber.*, **91**, 345 (1958).
622. Johnston, McCaleb, and Montgomery, *J. Medicin. Chem.*, **6**, 669 (1963).
623. Huber, Ger. Pat., 1,085,532 (1956); through *Chem. Abstracts*, **56**, 3494 (1962).
624. Klosa, *J. prakt. Chem.*, **23**, 34 (1964).
625. Giner-Sorolla, *J. Heterocyclic Chem.*, **7**, 75 (1970).
626. Johnson, Thomas, and Schaeffer, *J. Amer. Chem. Soc.*, **80**, 699 (1958).
627. Baker, Joseph, Schaub, and Williams, *J. Org. Chem.*, **19**, 1780 (1954).
628. Leonard and Deyrup, *J. Amer. Chem. Soc.*, **82**, 6202 (1960).
629. Leonard and Fujii, *J. Amer. Chem. Soc.*, **85**, 3719 (1963).
630. Marsico and Goldman, *J. Org. Chem.*, **30**, 3597 (1965).
631. Pfleiderer, *Annalen*, **647**, 167 (1961).
632. Bredereck, Christmann, and Koser, *Chem. Ber.*, **93**, 1206 (1960).
633. Brookes and Lawley, *J. Chem. Soc.*, **1961**, 3923.
634. Mian and Walker, *J. Chem. Soc.* (*C*), **1968**, 2577.
635. Brookes, Dipple, and Lawley, *J. Chem. Soc.* (*C*), **1968**, 2026.
636. Fujimoto and Ono, *J. Pharm. Soc. Japan*, **86**, 364 (1966).
637. Cavalieri, Tinker, and Brown, *J. Amer. Chem. Soc.*, **71**, 3973 (1949).
638. Denayer, Cave, and Goutarel, *Compt. rend.*, **253**, 2994 (1961).
639. Hall, Robins, Stasink, and Thedford, *J. Amer. Chem. Soc.*, **88**, 2614 (1966).
640. Fischer, *Ber.*, **43**, 805 (1910).
641. Plentl and Schoenheimer, *J. Biol. Chem.*, **153**, 203 (1944).
642. Tafel and Ach, *Ber.*, **34**, 1165 (1901).
643. Spies and Harris, *J. Amer. Chem. Soc.*, **61**, 351 (1939).
644. Karrer, Manunta, and Schwyzer, *Helv. Chim. Acta*, **31**, 1214 (1948).
645. Trattner, Elion, Hitchings, and Sharefkin, *J. Org. Chem.*, **29**, 2674 (1964).
646. Bergmann and Stempien, *J. Org. Chem.*, **22**, 1575 (1957).
647. Lira, *J. Heterocyclic Chem.*, **5**, 863 (1968).
648. Golovchinskaya and Chaman, *Zhur. obshchei Khim.*, **22**, 528 (1952).
649. Walentowski and Wanzlick, *Chem. Ber.*, **102**, 3000 (1969).
650. Klosa, *Arch. Pharm.*, **289**, 211 (1956).
651. Khaletskii and Eshman, *Zhur. obshchei Khim.*, **18**, 2129 (1948).
652. Kazmirowski, Dietz. and Carstens, Ger. Pat. (east), 25,960 (1963); through *Chem. Abstracts*, **61**, 12016 (1964).
653. Kazmirowski, Dietz, and Carstens, Ger. Pat. (east), 26,413 (1964); through *Chem. Abstracts*, **61**. 12016 (1964).
654. Chaman and Golovchinskaya, *Zhur. obshchei Khim.*, **36**, 1608 (1966).

655. Yamazaki, *Chem. and Pharm. Bull.* (*Japan*), **17**, 1268 (1969).
656. Fujii, Itaya, Wu, and Yamada, *Chem. and Ind.*, **1966**, 1967.
657. Pfleiderer and Nubel, *Annalen*, **647**, 161 (1961).
658. Ovcharova and Golovchinskaya, *Zhur. obshchei Khim.*, **34**, 3247 (1964).
659. Hepner and Frankenberg, *Helv. Chim. Acta*, **15**, 350 (1932).
660. Hepner and Frankenberg, *Helv. Chim. Acta*, **15**, 533 (1932).
661. Okano, Goya, Takadate, and Ito, *J. Pharm. Soc. Japan*, **86**, 649 (1966).
662. Biltz and Pardon, *J. prakt. Chem.*, **140**, 209 (1934).
663. Craveri and Zoni, *Boll. sci. fac. chim. ind. Bologna*, **16**, 89 (1958).
664. Merck, Wolfes, and Kornick, Ger. Pat., 352,980; through *Chem. Abstracts*, **17**, 1306 (1923).
665. Polonovski, Pesson, and Zelnick, *Compt. rend.*, **241**, 215 (1955).
666. Hitchings, Elion, and Mackay, U.S. Pat., 3,098,074 (1963); through *Chem. Abstracts*, **60**, 1771 (1964).
667. Gomberg, *Amer. J. Chem.*, **17**, 403 (1895); **14**, 611 (1892).
668. Golovchinskaya and Chaman, *Zhur. obshchei Khim.*, **30**, 3628 (1960).
669. Chaman and Golovchinskaya, *Zhur. obshchei Khim.*, **31**, 2645 (1961).
670. Golovchinskaya and Chaman, *Zhur. obshchei Khim.*, **22**, 2220 (1952).
671. Ovcharova, Nikolaeva, and Golovchinskaya, *Khim.-Farm. Zhur.*, **2**, 18 (1968).
672. Cohen, Thom, and Bendich, *J. Org. Chem.*, **28**, 1379 (1963).
673. Bredereck and Föhlisch, *Chem. Ber.*, **95**, 414 (1962).
674. Hitchings, Elion, and Mackay, U.S. Pat., 3,128,274 (1964); through *Chem. Abstracts*, **60**, 14523 (1964).
675. Chaman and Golovchinskaya, *Zhur. obshchei Khim.*, **32**, 2015 (1962).
676. Lugovkin, *Zhur. obshchei Khim.*, **30**, 2427 (1960).
677. Aliprandi, Cacace, and Montifinale, *Farmaco, Ed. Sci.*, **12**, 751 (1957).
678. Skinner, Claybrook, Ross, and Shive, *J. Org. Chem.*, **23**, 1223 (1958).
679. Kishikawa and Yuki, *Chem. and Pharm. Bull.* (*Japan*), **14**, 1365 (1966).
680. Sermonský, Černý, and Jelinek, *Coll. Czech. Chem. Comm.*, **25**, 1091 (1960).
681. Černý, Sermonsky, and Jelinek, *Coll. Czech. Chem. Comm.*, **27**, 57 (1962).
682. Bredereck, Christmann, Koser, Schellenberg, and Nast, *Chem. Ber.*, **95**, 1812 (1962).
683. Dyer, Farris, Minnier, and Tokizawa, *J. Org. Chem.*, **34**, 973 (1969).
684. Kotva, Černý, Semonský, Vachek, and Jelínek, *Coll. Czech. Chem. Comm.*, **34**, 2114 (1969).
685. Fischer and Reese, *Annalen*, **221**, 336 (1883).
686. Schultzen, *Z. Chem.*, **10**, 614 (1867).
687. Knoll and Co., Ger. Pat., 399,903 (1922); *Frdl.*, **14**, 1322 (1921–1925).
688. Brunner and Leins, *Ber.*, **30**, 2584 (1897).
689. Marquardt and Müller-Ebeling, Ger. Pat., 859,470 (1952); through *Chem. Abstracts*, **47**, 11237 (1953).
690. Kalle and Co., Ger. Pat., 230,401 (1909); through *Chem. Abstracts*, **5**, 2733 (1911).
691. Fischer, *Z. physiol. Chem.*, **60**, 69 (1909).
692. Cherbuliez and Bernhard, *Helv. Chim. Acta*, **15**, 464 (1932).
693. Fujii, Itaya, Wu, and Yamada, *Chem. and Ind.*, **1966**, 1967.
694. Brown, Stevens, and Smith, *J. Biol. Chem.*, **233**, 1513 (1958).
695. Stevens, Smith, and Brown, *J. Amer. Chem. Soc.*, **82**, 1148 (1960).
696. Brown, Suguira, and Cresswell, *Cancer Res.*, **25**, 986 (1965).
697. Von Euler and Hasselquist, *Arkiv. Kemi.*, **13**, 185 (1958).
698. Taylor and Garcia, *J. Amer. Chem. Soc.*, **86**, 4721 (1964).
699. Goldner, Dietz, and Carstens, *Annalen*, **693**, 233 (1966).

700. Kawashima and Kumashiro, *Bull. Chem. Soc. Japan*, **40**, 639 (1967).
701. Baker and Santi, *J. Heterocyclic Chem.*, **4**, 216 (1967).
702. Yoshitomi, *J. Pharm. Soc. Japan*, **512**, 839 (1924).
703. Yoshitomi, *J. Pharm. Soc. Japan*, **524**, 884 (1925).
704. Tafel and Weinshenk, *Ber.*, **33**, 3369 (1900).
705. Dornow, Ger. Pat., 1,064,950 (1959); through *Chem. Abstracts*, **55**, 10483 (1961).
706. Yamazaki, Kumashiro, and Takenishi, *J. Org. Chem.*, **32**, 3258 (1967).
707. Taylor, Maki, and McKillop, *J. Org. Chem.*, **34**, 1170 (1969).
708. Cassidy, Olsen, and Robins, *J. Heterocyclic Chem.*, **5**, 461 (1968).
709. Yamazaki, Kumashiro, Takenishi, and Ikehara, *Chem. and Pharm. Bull. (Japan)*, **16**, 2172 (1968).
710. Tamari and Awruch, *Tetrahedron*, **24**, 2611 (1968).
711. Steinmaus, Rosenthal, and Elad, *J. Amer. Chem. Soc.*, **91**, 4921 (1969).
712. Nikolaeva and Golovchinskaya, *Probl. Org. Sinteza. Akad. Nauk., S.S.S.R., Otd. Obshchei i Tekhn. Khim.*, **1965**, 192; through *Chem. Abstracts*, **64**, 9725 (1966).
713. Ebed, Chaman, and Golovchinskaya, *Probl. Org. Sinteza Akad. Nauk. S.S.S.R., Otd. obshchei i Tekhn. Khim.*, **1965**, 198; through *Chem. Abstracts*, **64**, 9725 (1966).
714. Nikolaeva, Ebed, and Golovchinskaya, *Sintez Prirodn. Soedin. ikh Analogov. i Fragmentov, Akad. Nauk. S.S.S.R., Otd. obshchei i Tekhn. Khim.*, **1965**, 245; through *Chem. Abstracts*, **65**, 7178 (1966).
715. Neiman, *J. Chem. Soc. (C)*, **1970**, 91.
716. Kostolansky, Mokry, and Tamchyna, *Chem. Zvesti*, **10**, 96 (1956).
717. Shapiro, *J. Amer. Chem. Soc.*, **73**, 3526 (1951).
718. Kazmirowski, Goldner, and Carstens, *J. prakt. Chem.*, **32**, 43 (1966).
719. Baddiley, Buchanan, and Stephenson, *Arch. Biochem. Biophys.*, **83**, 54 (1959).
720. Chheda and Hall, *Biochemistry*, **5**, 2082 (1966).
721. Ward, Wade, Walborg, and Osdene, *J. Org. Chem.*, **26**, 5000 (1961).
722. Pfleiderer and Kempter, *Angew. Chem. Internat. Edn.*, **6**, 259 (1967).
723. Shapiro and Shiuey, *Biochim. Biophys. Acta*, **174**, 403 (1969).
724. Ovcharova, Chaman, and Golovchinskaya, *Khim. geterotsikl. Soedinenii*, **1967**, 1129.
725. Budowsky, Sverdlov, and Monastyrskaya, *J. Mol. Biol.*, **44**, 205 (1969).
726. Hamann, Spaziano, Chou, Price, and Lin, *Canad. J. Chem.*, **46**, 419 (1968).
727. Tirzit, Pengerote, Ziderman, and Dubur, *Khim. geterotsikl. Soedinenii*, **1967**, 1132.
728. Gough and Maguire, *J. Medicin. Chem.*, **10**, 475 (1967).
729. Bühler and Pfleiderer, *Chem. Ber.*, **100**, 492 (1967).
730. Schweizer, Chan, Helmkamp, and Ts'o, *J. Amer. Chem. Soc.*, **86**, 696 (1964).
731. Matsuura and Goto, *J. Chem. Soc.*, **1965**, 623.
732. Gordon and Brown, *J. Biol. Chem.*, **220**, 927 (1956).
733. Gordon, Weliky, and Brown, *J. Amer. Chem. Soc.*, **79**, 3245 (1957).
734. Bullock and Jardetzky, *J. Org. Chem.*, **29**, 1988 (1964).
735. Paterson and Zbarsky, *J. Amer. Chem. Soc.*, **75**, 5753 (1953).
736. Korn, *Adv. Enzymol.*, **4**, 615 (1957).
737. Fel'dman and Zlobina, *Mechenye Biol. Aktivn. Veshchestva Sbornik. Statei*, **1962**, 52; through *Chem. Abstracts*, **59**, 7527 (1963).
738. Bennett, *J. Amer. Chem. Soc.*, **74**, 2420 (1952).
739. Abrams and Clark, *J. Amer. Chem. Soc.*, **73**, 4609 (1951).
740. Gordon, *J. Chem. Soc.*, **1954**, 757.
741. Cavalieri and Brown, *J. Amer. Chem. Soc.*, **71**, 2246 (1949).
742. Cavalieri, Tinker, and Bendich, *J. Amer. Chem. Soc.*, **71**, 533 (1949).
743. Brown, Roll, Plentl, and Cavalieri, *J. Biol. Chem.*, **172**, 469 (1948).

744. Dunn, Maguire, and Brown, *J. Biol. Chem.*, **234**, 620 (1959).
745. Bennett and Skipper, *Arch. Biochem. Biophys.*, **54**, 566 (1955).
746. Fillippi and Guern, *Bull. Soc. chim. France*, **1966**, 2617.
747. Weygand and Grossinsky, *Chem. Ber.*, **84**, 839 (1951).
748. Balis, Brown, Elion, Hitchings, and Vanderwerff, *J. Biol. Chem.*, **188**, 217 (1951).
749. Korte and Barkmeyer, *Chem. Ber.*, **89**, 2400 (1956).
750. Schütte, Schaaf, Liebisch, Benes, Kozel, and Veres, *Z. Chem.*, **4**, 430 (1964).
751. Duggan and Titus, *J. Pharmacol.*, **130**, 375 (1960).
752. Weygand and Grosskinsky, *Chem. Ber.*, **84**, 839 (1951).
753. Weygand, Klebe, Trebst, and Simon, *Z. Naturforsch*, **9**, 450 (1954).
754. Hartman and Fellig, *J. Amer. Chem. Soc.*, **77**, 1051 (1955).
755. Getler, Roll, Tinker, and Brown, *J. Biol. Chem.*, **178**, 259 (1949).
756. Canellakis and Cohen, *J. Biol. Chem.*, **213**, 379 (1955).
757. Brandenberger, *Biochim. Biophys. Acta*, **15**, 108 (1954).
758. Brandenberger, *Helv. Chim. Acta*, **37**, 641 (1954).
759. Dalgleish and Neuberger, *J. Chem. Soc.*, **1954**, 3407.
760. Cavalieri, Blair, and Brown, *J. Amer. Chem. Soc.*, **70**, 1240 (1948).
761. Elion and Hitchings, *J. Amer. Chem. Soc.*, **76**, 4027 (1954).
762. Goto and Matsuura, *J. Chem. Soc. Japan*, **87**, 71 (1966).
763. Van der Slooten, *Arch. Pharm.*, **235**, 469 (1897).
764. Schmidt and Schwabe, *Arch. Pharm.*, **245**, 312 (1907).
765. Pugmire, Grant, Robins, and Rhodes, *J. Amer. Chem. Soc.*, **87**, 2225 (1965).
766. Coburn, Thorpe, Montgomery, and Hewson, *J. Org. Chem.*, **30**, 1114 (1965).
767. Morávek and Nejedlý, *Chem. and Ind.*, **1960**, 530.
768. Chiotan and Zamfir, *J. Labelled Compounds*, **4**, 356 (1968).
769. Măntescu, Genunche, and Balaban, *J. Labelled Compounds*, **2**, 261 (1966).
770. Kruger and Saloman, *Z. physiol. Chem.*, **24**, 364 (1898); **26**, 367 (1899).
771. Boehringer, Ger. Pat., 96,926 (1897); *Frdl.*, **5**, 834 (1897–1900).
772. Boehringer, Ger. Pat., 99,123 (1897); *Frdl.*, **5**, 854 (1897–1900).
773. Boehringer, Ger. Pat., 107,507 (1898); *Frdl.*, **5**, 859 (1897–1900).
774. Boehringer, Ger. Pat., 100,875 (1898); *Frdl.*, **5**, 860 (1897–1900).
775. Boehringer, Ger. Pat., 121,224 (1899); *Frdl.*, **6**, 1182 (1900–1902).
776. Boehringer, Ger. Pat., 133,300 (1901); *Frdl.*, **6**, 1189 (1900–1902).
777. Boehringer, Ger. Pat., 128,212 (1901); *Frdl.*, **6**, 1187 (1900–1902).
778. Boehringer, Ger. Pat., 142,468 (1902); *Frdl.*, **7**, 668 (1902–1904).
779. Boehringer, Ger. Pat., 141,974 (1902); *Frdl.*, **7**, 668 (1902–1904).
780. Boehringer, Ger. Pat., 146,714 (1902); *Frdl.* **7**, 670 (1902–1904).
781. Boehringer, Ger. Pat., 156,900 (1903); *Frdl.*, **7**, 677 (1902–1904).
782. Boehringer, Ger. Pat., 156,901 (1903); *Frdl.*, **7**, 678 (1902–1904).
783. Boehringer, Ger. Pat., 164,425 (1903); *Frdl.*, **8**, 1146 (1905–1907).
784. Boehringer, Ger. Pat., 1,000,942 (1921); *Frdl.*, **14**, 1325 (1921–1925).
785. Brown and Ford, *J. Chem. Soc. (C)*, **1969**, 2620.
786. Pfleiderer and Shanshal, *Annalen*, **726**, 201 (1969).
787. Reisch, *Arzneim.-Forsch.*, **18**, 1485 (1968).
788. Studentov and Nemets, *Khim. geterotsikl Soedinenii*, **1968**, 732.
789. Studentov and Nemets, *Khim. geterotsikl Soedinenii*, **1968** 933.
790. Giner-Sorolla and Brown, *J. Chem. Soc. (C)*, **1971**, 126
791. Brown and Giner-Sorolla, *J. Chem. Soc. (C)*, **1971**, 128
792. Fischer, *Ber.*, **17**, 1776 (1884).
793. Evans, Shepherd, and Turner, *J. Labelled Compounds*, **6**, 76 (1970).
794. Eidinoff and Knoll, *J. Amer. Chem. Soc.*, **75**, 1992 (1953).

795. Scheinfeld, Parham, Murphy, and Brown, *J. Org. Chem.*, **34**, 2153 (1969).
796. Barlin and Pfleiderer, *Chem. Ber.*, **102**, 4032 (1969).
797. Takahashi, *J. Pharm. Soc. Japan*, **89**, 591 (1969).
798. Chakraborti, *Indian J. Chem.*, **7**, 426 (1969).
799. Yamazaki, Kumashiro, and Takenishi, *Chem. and Pharm. Bull.* (*Japan*), **16**, 1561 (1968).
800. Carroll and Philip, *J. Org. Chem.*, **33**, 3776 (1968).
801. Rousseau, Robins, and Townsend, *J. Amer. Chem. Soc.*, **90**, 2661 (1968).
802. Kiburis and Lister, unpublished results.
803. Lettré, Ballweg, Maurer, and Rehberger, *Naturwiss.*, **50**, 224 (1963).
804. Johns, *J. Biol. Chem.*, **14**, 1 (1913).
805. Acker and Castle, *J. Org. Chem.*, **23**, 2010 (1958).
806. Huston and Allen, *J. Amer. Chem. Soc.*, **56**, 1358 (1934).
807. Huston and Allen, *J. Amer. Chem. Soc.*, **56**, 1356 (1934).
808. Rybár and Antos, *Coll. Czech. Chem. Comm.*, **35**, 1415 (1970).
809. Saneyoshi, Jap. Pat., 7,002,179 (1970); through *Chem. Abstracts*, **72**, 90522 (1970).
810. Lawley and Brookes, *Biochem. J.*, **92**, 19c (1964).
811. Shimizu and Miyaki, *Chem. and Pharm. Bull.* (*Japan*), **18**, 570 (1970).
812. Johns and Baumann, *J. Biol. Chem.*, **16**, 135 (1913).
813. Johns, *J. Biol. Chem.*, **11**, 393 (1912).
814. Johns and Baumann, *J. Biol. Chem.*, **15**, 119 (1913).
815. Johns, *J. Biol. Chem.*, **9**, 161 (1911).
816. Johns and Hendrix, *J. Biol. Chem.*, **19**, 25 (1914).
817. Johns and Hendrix, *J. Biol. Chem.*, **20**, 153 (1915).
818. Johns, *Amer. Chem. J.*, **41**, 58 (1909).
819. Johns, *Amer. Chem. J.*, **45**, 79 (1911).
820. Dille, Sutherland, and Christensen, *J. Org. Chem.*, **20**, 171 (1955).
821. Ovcharova, Babenko, and Golovchinskaya, *Khim-Farm. Zhur.*, **7**, 26 (1970).
822. Yoneda, Ogiwara, Kanahori, and Nishigaki, *Chem. Comm.*, **1970**, 1068.
823. Ohtsuka and Sugimoto, *Bull. Chem. Soc. Japan*, **43**, 2281 (1970).
824. Shaw and Smallwood, *J. Chem. Soc.* (*C*), **1970**, 2206.
825. Weinstock, Grabowski, and Cheng, *J. Medicin. Chem.*, **13**, 995 (1970).
826. Miyaki and Shimizu, *Chem. and Pharm. Bull.* (*Japan*), **18**, 1446 (1970).
827. Israel, Tinter, Trites, and Modest, *J. Heterocyclic Chem.*, **7**, 1029 (1970).
828. Stöhrer and Brown, *J. Biol. Chem.*, **244**, 2498 (1969).
829. Giner-Sorolla, Gryte, Cox, and Parham, *J. Org. Chem.*, (in press).

Index

The Index covers in detail all Chapters and the Tables contained in them. Not included, however, is material from the Appendix Tables. In the compilation no attempt has been made to include all compounds mentioned in the text but the parent members of all main classes of purines have been given. By reference, therefore, to the particular type of purine the properties or reactions of a derivative of it should be readily located. Because of the limitations governing the inclusion of compounds in the Appendix Tables (see pp. 529, 530) the opportunity has been taken to give more complete coverage by including in the Index the more complex purines and related heterocycles which are outside the above Tables. Only the names of workers specifically mentioned in the text of Chapters I–XII have been listed.

A number in parentheses indicates that the item is either obliquely referred to or is not named in full. As an example isocaffeine (1,3,9-trimethylxanthine) on p. 226 is denoted by the statement "Methylation at $N_{(3)}$ occurs with 1,9-dimethylxanthine ...." If the letter f follows a number this signifies that discussion of the subject is continued on the following page. Where a reaction listed is followed by 'attempted' this is an indication either that no product, or not that expected, has been obtained.

629